FOURTH EDITION

Quality Management

Introduction to Total Quality Management for Production, Processing, and Services

David L. Goetsch

Stanley B. Davis

Prentice Hall

Upper Saddle River, New Jersey
Columbus, Ohio

Library of Congress Cataloging-in-Publication Data

Goetsch, David L.
   Quality management : introduction to total quality management for production,
 processing, and services / David L. Goetsch, Stanley B. Davis.--4th ed.
      p. cm.
   Includes bibliographical references and index.
   ISBN 0-13-093387-2
   I. Total quality management. I. Davis, Stanley, 1931-II. Title.

HD62.15 .G64 2003
658.4'013--dc21

2001059311

Editor in Chief: Stephen Helba
Executive Editor: Debbie Yarnell
Media Development Editor: Michelle Churma
Production Editor: Louise N. Sette
Production Supervision: Clarinda Publication Services
Design Coordinator: Diane Ernsberger
Cover Designer: Ceri Fitzgerald
Cover Art: Ceri Fitzgerald
Production Manager: Brian Fox
Marketing Manager: Jimmy Stephens

This book was set in Goudy Sans by The Clarinda Company. It was printed and bound by R.R.
Donnelley & Sons Company. The cover was printed by Phoenix Color Corp.

Earlier editions, entitled *Introduction to Total Quality: Quality Management for Production, Processing, and Services*, © 1997 by Prentice-Hall, Inc. and © 1994 by Macmillan College Publishing Company.

Pearson Education Ltd., *London*
Pearson Education Australia Pty. Limited, *Sydney*
Pearson Education Singapore Pte. Ltd.
Pearson Education North Asia Ltd., *Hong Kong*
Pearson Education of Canada, Ltd., *Toronto*
Pearson Educación de Mexico, S. A., de C. V.
Pearson Education—Japan, *Tokyo*
Pearson Education Malaysia Pte. Ltd.
Pearson Education, *Upper Saddle River, New Jersey*

Prentice
Hall

10 9 8 7 6 5 4 3 2 1

ISBN: 0-13-093387-2

To Savannah Marie and Deborah Marie Goetsch

DLG

To my wife, Sammie, who shared with me the long months of effort that went into this book

SBD

PREFACE

BACKGROUND

At one time in history, Great Britain was the world's leader in commerce and industry. Eventually, the United States emerged as a major friendly competitor. Then, following World War II, the United States took over as the undisputed world leader of commerce and industry. During these postwar years, while the United States was enjoying unparalleled prosperity, Japan and Germany were rebuilding from the ashes of the war. With a great deal of help from the United States, Japan was able to rebound and during the 1970s began to challenge the United States in such key manufacturing sectors as automobiles, computers, and consumer electronics. By 1980 Japan had emerged as a world-class competitor and a global leader in selected areas of commerce and industry. German industry had also reemerged by this time.

As a result, the United States found itself losing market share in economic sectors it had dominated (and taken for granted) for decades. At first, industrialists in the United States turned their backs on the lesson their counterparts in Japan, Germany, and other leading industrialized countries had learned. This lesson was that the key to competing in the international marketplace was to simultaneously improve both quality and productivity on a continual basis. However, as more and more market share slipped away, the message started to sink in to the United States. This belated awareness gave rise to a quality movement that began to take hold. Its progress was slow at first. However, an approach to doing business known as the *total quality approach* to quality management has caught on and is now widely practiced.

The total quality philosophy is an approach to doing business that focuses all of the resources of an organization on the continual and simultaneous improvement of both quality and productivity. The purpose of the total quality approach is to continually improve the organization's performance and, in turn, competitiveness.

WHY WAS THIS BOOK WRITTEN, AND FOR WHOM?

This book was written in response to the need for a practical teaching resource that encompasses all of the various elements of the total quality approach and pulls them together in a coherent format that allows the reader to understand both the big picture and the specific details of total quality. It is intended for use in universities, colleges, community colleges, corporate environments, and any other settings in which people want to learn to be effective agents of the total quality approach or are attempting to implement total quality. Students enrolled in technology, engineering, or management programs will find this book both valuable and easy to use. Practitioners in corporate settings will find it a valuable guide in helping them understand and implement total quality.

The direct, straightforward presentation of material focuses on making the theories and principles of total quality practical and useful in a real-world setting. Up-to-date research has been integrated throughout in a down-to-earth manner.

ORGANIZATION OF THIS BOOK

The text consists of 22 chapters. A standard format is used throughout the book. Each chapter begins with a list of its major topics and ends with a comprehensive summary. Following the summary, each chapter contains end material consisting of key terms and concepts, factual review questions, a critical thinking activity, discussion assignments, and endnotes. The endnotes provide readers with comprehensive lists of additional reading and research material that can be pursued at the discretion of the student and/or the instructor. The other end material is provided to encourage review, stimulate additional thought, promote discussion, and facilitate additional research.

HOW THIS BOOK DIFFERS FROM OTHERS

Most books on the market deal with one of the several elements of total quality, such as teamwork, just-in-time manufacturing, scientific measurement (SPC or quality tools), continual improvement, employee involvement, and so on. Many of the books available were developed with the advanced-level practitioner in mind rather than the beginner. Few of the books on the market were formatted for use in a classroom setting. This book was written to be both comprehensive and in depth. All of the elements of total quality are covered, including several that receive little or no attention in other total quality books (i.e., partnering, manufacturing networks, and how to implement total quality). Each of these subjects is covered in sufficient depth to allow a beginner to learn everything necessary to understand and implement total quality without having to look to any other source of information.

New in the Fourth Edition

The fourth edition contains major improvements that reflect both changes in the global marketplace and feedback from customers. These include the following:

1. Chapters 1 and 2 have been reversed so that the old Chapter 1 is now Chapter 2 and the old Chapter 2 is now Chapter 1.
2. "Quality Cases" were added, where applicable, to most chapters. These cases are real-world examples that illustrate concepts presented in their corresponding chapters. Each case highlights a well-known and widely recognized business that competes on a global basis.
3. The following new material was added to the chapters noted:
   - *Chapter 1:* (was Chapter 2): New sections were added: Two Views of Quality and Six Sigma Concept.
   - *Chapter 2:* (was Chapter 1): New sections were added: Costs of Poor Quality; Characteristics of World-class Organizations; E-Commerce, Information Quality, and Competitiveness; Key Global Trends; and Strengths and Weaknesses of U.S. Companies in the Global Marketplace.
   - *Chapter 3:* New sections were added: Creative Thinking in Strategic Planning and Strategic Planning in Action (examples from the real world of business).
   - *Chapter 4:* New section was added: Ethical Dilemmas, about concerns often faced in the workplace.

- *Chapter 5:* New section was added: Innovative Alliances and Partnerships.
- *Chapter 6:* New sections were added: Quality Culture versus Traditional Culture and Maintaining a Quality Culture.
- *Chapter 8:* New sections were added: How to Recognize Empowered Employees and Avoiding Empowerment Traps.
- *Chapter 9:* New sections were added: Lessons from Distinguished Leaders and Servant Leadership and Stewardship.
- *Chapter 11:* New section was added: Personality and Communication.
- *Chapters 12 through* 22. No new sections were added, but copious amounts of new material were added within existing sections.

ANCILLARY MATERIAL

Also available from the publisher is an instructor's resource manual that contains a test bank of questions, suggested answers to review questions, and a set of transparency masters.

ABOUT THE AUTHORS

David L. Goetsch is provost of the joint campus of the University of West Florida and Okaloosa-Walton Community College in Fort Walton Beach, Florida, where he also serves as professor of quality and safety management. He also administers Florida's Center for Manufacturing Competitiveness that is located on this campus and is president of The Management Institute, a private company dedicated to the continual improvement of quality, productivity, and competitiveness. Dr. Goetsch is cofounder of The Quality Institute, a partnership of the University of West Florida, Okaloosa-Walton Community College, and the Okaloosa Economic Development Council. He currently serves on the executive board of the Institute.

Stanley B. Davis was a manufacturing executive with Harris Corporation until his retirement in 1991. He was founding managing director of The Quality Institute and is a well-known expert in the areas of implementing total quality, statistical process control, just-in-time manufacturing, benchmarking, ISO 9000, and ISO 14000. He currently serves as professor of quality at the Institute and heads his own firm, Stan Davis Consulting.

ACKNOWLEDGMENTS

The authors wish to gratefully acknowledge the contributions of numerous individuals whose assistance was invaluable to the development of this book. Our appreciation goes to Kay Rasmusser, Wannda Edwards, and Faye Crawford for their assistance in word processing of the manuscript, and to Savannah M. Goetsch and Caleb Sutton for proofreading and copy editing. Thanks also to the following reviewers for their invaluable input: Christopher Gazaway, University of Phoenix–Tulsa; Michael Godt, Ranken Technical College; Peter Gutierrez, principal of Management Advisory Services and core adjunct professor, National University; Nicholas Louis, Kansas State University; Joseph A. Phillips, DeVry Institute of Technology (OH); Elaine M. Simmons, Guildford Technical Community College; and Jerry D. Westbrook, professor, Department of Industrial Systems, Engineering, and Engineering Management, University of Alabama in Huntsville. The author also thanks Professor Sean Ahard for contributing material on Service Leadership and Stewardship.

CONTENTS

CHAPTER 5
Partnering and Strategic Alliances 160

CHAPTER 6
Quality Culture 188

CHAPTER 7
Customer Satisfaction and Retention   214

CHAPTER 8
Employee Empowerment   241

CHAPTER 9
Leadership and Change 286

CHAPTER 12
Education and Training   403

CHAPTER 13
Overcoming Politics, Negativity, and Conflict in the Workplace   461

CHAPTER 14
ISO 9000 and Total Quality: The Relationship   506

PART TWO
Tools and Techniques   521

CHAPTER 15
Overview of Total Quality Tools   522

CHAPTER 16
Problem Solving and Decision Making   574

CHAPTER 17
Quality Function Deployment (QFD)   607

CHAPTER 18
Optimizing and Controlling Processes through Statistical Process Control (SPC)   629

CHAPTER 19
Continual Improvement   678

CHAPTER 20
Benchmarking   711

CHAPTER 21
Just-in-Time Manufacturing (JIT)   738

CHAPTER 22
Implementing Total Quality Management   791

APPENDIX
Manufacturing Networks in the United States   838

INDEX   843

PART ONE

Philosophy and Concepts

CHAPTER ONE

The Total Quality Approach to Quality Management

"There are really only three types of people: those who make things happen, those who watch things happen, and those who say, 'What happened?'"
   Ann Landers

MAJOR TOPICS

- What Is Quality?
- The Total Quality Approach Defined
- Two Views of Quality
- Key Elements of Total Quality
- Total Quality Pioneers
- Why Total Quality Efforts Succeed
- Six Sigma Concept
- The Future of Quality Management

The total quality concept as an approach to doing business began to gain wide acceptance in the United States in the late 1980s and early 1990s. However, individual elements of the concept—such as the use of statistical data, teamwork, continual improvement, customer satisfaction, and employee involvement—have been used by visionary organizations for years. It is the pulling together and coordinated use of these and other previously disparate elements that gave birth to the comprehensive concept known as *total quality*. This chapter provides an overview of that concept, laying the foundation for the study of all remaining chapters.

WHAT IS QUALITY?

To understand total quality, one must first understand *quality*. Customers that are businesses will define quality very clearly using specifications, standards, and other measures. This makes the point that quality can be defined and measured. Although few

3

consumers could define *quality* if asked, all know it when they see it. This makes the critical point that quality is in the eye of the beholder. With the total quality approach, customers ultimately define quality.

People deal with the issue of quality continually in their daily lives. We concern ourselves with quality when grocery shopping, eating in a restaurant, and making a major purchase such as an automobile, a home, a television, or a personal computer. Perceived quality is a major factor by which people make distinctions in the marketplace. Whether we articulate them openly or keep them in the back of our minds, we all apply a number of criteria when making a purchase. The extent to which a purchase meets these criteria determines its quality in our eyes.

One way to understand quality as a consumer-driven concept is to consider the example of eating at a restaurant. How will you judge the quality of the restaurant? Most people apply such criteria as the following:

- Service
- Response time
- Food preparation
- Environment/atmosphere
- Price
- Selection

This example gets at one aspect of quality—the *results* aspect. Does the product or service meet or exceed customer expectations? This is a critical aspect of quality, but it is not the only one. *Total quality* is a much broader concept that encompasses not just the results aspect but also the quality of people and the quality of processes.

Quality has been defined in a number of different ways by a number of different people and organizations. Consider the following definitions:

- Fred Smith, CEO of Federal Express, defines quality as "performance to the standard expected by the customer."[1]
- The General Services Administration (GSA) defines quality as "meeting the customer's needs the first time and every time."[2]
- Boeing defines quality as "providing our customers with products and services that consistently meet their needs and expectations."[3]
- The U.S. Department of Defense (DOD) defines quality as "doing the right thing right the first time, always striving for improvement, and always satisfying the customer."[4]

In his landmark book *Out of the Crisis*, W. Edwards Deming has this to say about quality:

Quality can be defined only in terms of the agent. Who is the judge of quality? In the mind of the production worker, he produces quality if he can take pride in his work. Poor quality, to him, means loss of business, and perhaps of his job. Good quality, he thinks, will keep the company in business. Quality to the plant manager means to get the numbers out and to meet specifications. His job is also, whether he knows it or not, continual improvement of leadership.[5]

Deming goes on to make the point that quality has many different criteria and that these criteria change continually.[6] To complicate matters even further, different people value the various criteria differently. For this reason, it is important to measure consumer preferences and to remeasure them frequently. Deming gives an example of the criteria that are important to him in selecting paper:[7]

- It is not slick and, therefore, takes pencil or ink well.
- Writing on the back does not show through.
- It fits into a three-ring notebook.
- It is available at most stationery stores and is, therefore, easily replenished.
- It is reasonably priced.

Each of these preferences represents a variable the manufacturer can measure and use to continually improve decision making. Deming is well known for his belief that 94% of workplace problems are caused by management and especially for his role in helping Japan rise up out of the ashes of World War II to become a major industrial power. Deming's contributions to the quality movement are explained in greater depth later in this chapter.

Although there is no universally accepted definition of quality, enough similarity does exist among the definitions that common elements can be extracted:

- Quality involves meeting or exceeding customer expectations.
- Quality applies to products, services, people, processes, and environments.
- Quality is an ever-changing state (i.e., what is considered quality today may not be good enough to be considered quality tomorrow).

With these common elements extracted, the following definition of *quality* can be set forth:

> *Quality* is a dynamic state associated with products, services, people, processes, and environments that meets or exceeds expectations.

Consider the individual elements of this definition: The *dynamic state* element speaks to the fact that what is considered quality can and often does change as time

QUALITY TIP

Consumer Preference Studies

"The purpose of studies in consumer preference is to adjust the product to the public, rather than, as in advertising, to adjust the public to the product."

Source: Irwin Bross, *Design for Decision* (New York: Macmillan, 1953), 95; quoted in W. Edwards Deming, *Out of the Crisis* (Cambridge: Massachusetts Institute of Technology Center for Advanced Engineering Study, 1986), 168.

passes and circumstances are altered. For example, gas mileage is an important criterion in judging the quality of modern automobiles. However, in the days of 20¢–per-gallon gasoline, consumers were more likely to concern themselves with horsepower, cubic inches, and acceleration rates than with gas mileage.

The *products, services, people, processes, and environments* element is critical. It makes the point that quality applies not just to the products and services provided but to the people and processes that provide them and the environments in which they are provided. In the short term, two competitors who focus on continual improvement might produce a product of comparable quality. But the competitor who looks beyond just the quality of the finished product and also focuses on the continual improvement of the people who produce the product, the processes they use, and the environment in which they work will win in the long run and, most frequently, in the short run. This is because quality products are produced most consistently by quality organizations. Discussion Assignment 1–1 speaks to this phenomenon.

THE TOTAL QUALITY APPROACH DEFINED

Just as there are different definitions of *quality*, there are different definitions of *total quality*. For example, the DOD defines the total quality approach as follows:

> TQ consists of continuous improvement activities involving everyone in the organization—managers and workers—in a totally integrated effort toward improving performance at every level. This improved performance is directed toward satisfying such cross-functional goals as quality, cost, schedule, mission need, and suitability. TQ integrates fundamental management techniques, existing improvement efforts, and technical tools under a disciplined approach focused on continued process improvement. The activities are ultimately focused on increased customer/user satisfaction.[8]

This definition by the DOD is confusing because it attempts to combine both a definition of total quality and how it is achieved. An easier way to grasp the concept of total quality is to consider the analogy of a three-legged stool, as shown in Figure 1–1. The seat of the stool is customer focus. This means with total quality the customer is in the "driver's seat" as the primary arbiter of what is acceptable in terms of quality. Each of the three legs is a broad element of the total quality philosophy (i.e., measures, people, and processes). The "measures" leg of the stool makes the point that quality can and must be measured. The "people" leg of the stool makes the point that quality cannot be inspected into a product or service. Rather, it must be built in by people who are empowered to do their jobs the right way. The "processes" leg of the stool makes the point that processes must be improved, continually and forever. What is considered excellent today may be just mediocre tomorrow. Consequently, "good enough" is never good enough.

Another way to understand total quality as a concept is shown in Figure 1–2. Notice that the first part of the definition in Figure 1–2 explains the *what* of total quality; the second part explains the *how*. In the case of total quality, the *how* is important because it is what separates this approach to doing business from all of the others.

The *total* in total quality indicates a concern for quality in the broadest sense—what has come to be known as the "Big Q." Big Q refers to quality of products, services, people, processes, and environments. Correspondingly, "Little Q" refers to a nar-

Figure 1–1
Three-Legged Stool of Total
Quality

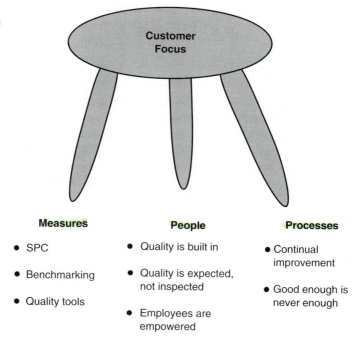

Measures

- SPC

- Benchmarking

- Quality tools

People

- Quality is built in

- Quality is expected,
  not inspected

- Employees are
  empowered

Processes

- Continual
  improvement

- Good enough is
  never enough

What It Is

Total quality is an approach to doing business that attempts to maximize the
competitiveness of an organization through the continual improvement of the quality
of its products, services, people, processes, and environments.

How It Is Achieved

The total quality approach has the following characteristics:
- Strategically based
- Customer focus (internal and external)
- Obsession with quality
- Scientific approach to decision making and problem solving
- Long-term commitment
- Teamwork
- Continual process improvement
- Education and training
- Freedom through control
- Unity of purpose
- Employee involvement and empowerment

Figure 1–2
Total Quality Defined

rower concern that focuses on the quality of one of these elements or individual quality criteria within an individual element.

How Is Total Quality Different?

What distinguishes the total quality approach from traditional ways of doing business can be found in how it is achieved. The distinctive characteristics of total quality are these: customer focus (internal and external), obsession with quality, use of the scientific approach in decision making and problem solving, long-term commitment, teamwork, employee involvement and empowerment, continual process improvement, bottom-up education and training, freedom through control, and unity of purpose, all deliberately aimed at supporting the organizational strategy. Each of these characteristics is explained later in this chapter.

The Historic Development of Total Quality

The total quality movement had its roots in the time and motion studies conducted by Frederick Taylor in the 1920s. Table 1–1 is a timeline that shows some of the major events in the evolution of the total quality movement since the days of Taylor. Taylor is now known as "the father of scientific management."

The most fundamental aspect of scientific management was the separation of planning and execution. Although the division of labor spawned tremendous leaps forward in productivity, it virtually eliminated the old concept of craftsmanship in which one highly skilled individual performed all the tasks required to produce a quality product. In a sense, a craftsman was CEO, production worker, and quality controller all rolled into one person. Taylor's scientific management did away with this by making planning the job of management and production the job of labor. To keep quality from falling through the cracks, it was necessary to create a separate quality department. Such departments had shaky beginnings, and just who was responsible for quality became a clouded issue (see Discussion Assignment 1-3).

As the volume and complexity of manufacturing grew, quality became an increasingly difficult issue. Volume and complexity together gave birth to quality engineering in the 1920s and reliability engineering in the 1950s. Quality engineering, in turn, resulted in the use of statistical methods in the control of quality, which eventually led to the concepts of *control charts* and *statistical process control*, which are now fundamental aspects of the total quality approach.

Joseph M. Juran, writing on the subject of quality engineering, says:

This specialty traces its origin to the application of statistical methods for control of quality in manufacture. Much of the pioneering theoretical work was done in the 1920s by the quality assurance department of the Bell Telephone laboratories. The staff members included Shewhart, Dodge, and Edwards. Much of the pioneering application took place (also in the 1920s) within the Hawthorne Works of the Western Electric Company.[9]

Reliability engineering emerged in the 1950s. It began a trend toward moving quality control away from the traditional after-the-fact approach and toward inserting it throughout design and production. However, for the most part, quality control in the

Table 1–1

Selected Historic Milestones in the Quality Movement in the United States

Year	Milestone
1911	Frederick W. Taylor publishes *The Principles of Scientific Management,* giving birth to such techniques as time and motion studies.
1931	Walter A. Shewhart of Bell Laboratories introduces statistical quality control in his book *Economic Control of Quality of Manufactured Products.*
1940	W. Edwards Deming assists the U.S. Bureau of the Census in applying statistical sampling techniques.
1941	W. Edwards Deming joins the U.S. War Department to teach quality control techniques.
1950	W. Edwards Deming addresses Japanese scientists, engineers, and corporate executives on the subject of quality.
1951	Joseph M. Juran publishes the *Quality Control Handbook.*
1961	Martin Company (later Martin-Marietta) builds a Pershing missile that has zero defects.
1970	Philip Crosby introduces the concept of *zero defects.*
1979	Philip Crosby publishes *Quality Is Free.*
1980	Television documentary *If Japan Can . . . Why Can't We?* airs, giving W. Edwards Deming renewed recognition in the United States.
1981	Ford Motor Company invites W. Edwards Deming to speak to its top executives, which begins a rocky but productive relationship between the automaker and the quality expert.
1982	W. Edwards Deming publishes *Quality, Productivity, and Competitive Position.*
1984	Philip Crosby publishes *Quality without Tears: The Art of Hassle-Free Management.*
1987	• U.S. Congress creates the Malcolm Baldrige National Quality Award. • Motorola introduces "Six Sigma" Method.
1988	• Secretary of Defense Frank Carlucci directs the U.S. Department of Defense to adopt total quality. • Tom Peters writes *In Search of Excellence.*
1989	Florida Power and Light wins Japan's coveted Deming Prize, the first non-Japanese company to do so.
1993	The total quality approach is widely taught in U.S. colleges and universities.
2000	The ISO 9000 standard is rewritten to incorporate total quality concepts.
2001	E-Commerce (information quality) and mass customization are important considerations.

1950s and 1960s involved inspections that resulted in nothing more than cutting out bad parts (see Discussion Assignment 1-4).

World War II had an impact on quality that is still being felt. In general, the effect was negative for the United States and positive for Japan. Because of the urgency to meet production schedules during the war, U.S. companies focused more on meeting delivery dates than on quality. This approach became a habit that carried over even after the war.

Japanese companies, on the other hand, were forced to learn to compete with the rest of the world in the production of nonmilitary goods. At first their attempts were unsuccessful, and "Made in Japan" remained synonymous with poor quality, as it had been before World War II. Around 1950, however, Japan decided to get serious about quality and establishing ways to produce quality products. Here is how Juran describes the start of the Japanese turnaround:

> To solve their quality problems the Japanese undertook to learn how other countries managed for quality. To this end the Japanese sent teams abroad to visit foreign companies and study their approach, and they translated selected foreign literature into Japanese. They also invited foreign lecturers to come to Japan and conduct training courses for managers.
>
> From these and other inputs the Japanese devised some unprecedented strategies for creating a revolution in quality. Several of those strategies were decisive:
>
> 1. The upper managers personally took charge of leading the revolution.
> 2. All levels and functions underwent training in managing for quality.
> 3. Quality improvement was undertaken at a continuing, revolutionary pace.
> 4. The workforce was enlisted in quality improvement through the QC-concept.[10]

Discussion Assignment 1-5 explains how Japanese manufacturers overcame a reputation for producing cheap, shabby products and developed a reputation as world leaders in the production of quality products. More than any other single factor, it was the Japanese miracle—which was not a miracle at all but the result of a concerted effort that took 20 years to really bear fruit—that got the rest of the world to focus on quality. When Western companies finally realized that quality was the key factor in global competition, they responded. Unfortunately, their first responses were the opposite of what was needed.

Juran describes those initial responses as follows:

> The responses to the Japanese quality revolution took many directions. Some of these directions consisted of strategies that had no relation to improving American competitiveness in quality. Rather, these were efforts to block imports through restrictive legislation and quotas, criminal prosecutions, civil lawsuits, and appeals to buy American.[11]

In spite of these early negative reactions, Western companies began to realize that the key to competing in the global marketplace was to improve quality. With this realization, the total quality movement finally began to gain momentum.

TWO VIEWS OF QUALITY

The total quality philosophy introduced a whole new way of looking at quality. The traditional view of quality measured process performance in defective parts per hundred produced. With total quality the same measurement is thought of in parts per million. The traditional view focused on after-the-fact inspections of products. With total quality the emphasis is on continuous improvement of products, processes, and people in order to prevent problems before they occur. The traditional view of quality saw employees as passive workers who followed orders given by supervisors and managers. It was their labor, not their brains that was wanted. With total quality, employees are empowered to think and make recommendations for continual improvement. They are also shown the control boundaries within which they must work and are given freedom to make decisions within those boundaries.

The traditional view of quality expected one improvement per year per employee. Total quality organizations expect to make at least 10 or more improvements per employee per year. Organizations that think traditionally focus on short-term profits. The total quality approach focuses on long-term profits and continual improvement.

The following statements summarize some of the major differences between the traditional view of quality and the total quality perspective:

- *Productivity versus quality.* The traditional view is that productivity and quality are always in conflict. You cannot have both. The total quality view is that lasting productivity gains are made only as a result of quality improvements.

- *How quality is defined.* The traditional view is that quality is defined solely as meeting customer specifications. The total quality view is that quality means satisfying customer needs and exceeding customer expectations.

- *How quality is measured.* The traditional view is that quality is measured by establishing an acceptable level of nonconformance and measuring against that benchmark. The total quality view is that quality is measured by establishing high-performance benchmarks for customer satisfaction and then continually improving performance.

- *How quality is achieved.* The traditional view is that quality is inspected into the product. The total quality view is that quality is determined by product design and achieved by effective control techniques.

- *Attitude toward defects.* The traditional view is that defects are an expected part of producing a product. Measuring defects per hundred is an acceptable standard. The total quality view is that defects are to be prevented using effective control systems and should be measured in defects per million (Six Sigma).

- *Quality as a function.* The traditional view is that quality is a separate function. The total quality view is that quality should be fully integrated throughout the organization—it should be everybody's responsibility.

- *Responsibility for quality.* The traditional view is that employees are blamed for quality. The total quality view is that 80 percent of quality problems are management's fault.

- *Supplier relationships.* The traditional view is that supplier relationships are short term and cost driven. The total quality view is that supplier relationships are long term and quality oriented.

KEY ELEMENTS OF TOTAL QUALITY

The total quality approach was defined in Figure 1–2. This definition has two components: the *what* and the *how* of total quality. What distinguishes total quality from other approaches to doing business is the *how* component of the definition. This component has eleven critical elements, each of which is explained in the remainder of this section and all of which relate to one of the components of the three-legged stool in Figure 1–1.

Strategically Based

Total quality organizations have a comprehensive strategic plan that contains at least the following elements: vision, mission, broad objectives, and activities that must be completed to accomplish the broad objectives. The strategic plan of a total quality organization is designed to give it a *sustainable competitive advantage* in the marketplace. The competitive advantages of a total quality organization are geared toward achieving world-leading quality and improving on it, continually and forever.

Customer Focus

In a total quality setting, the customer is the driver. This point applies to both internal and external customers. External customers define the quality of the product or service delivered. Internal customers help define the quality of the people, processes, and environments associated with the products or services.

Quality and teamwork expert Peter R. Scholtes explains the concept of *customer focus* as follows:

> Whereas Management by Results begins with profit and loss and return on investment, Quality Leadership starts with the customer. Under Quality Leadership, an organization's goal is to meet and exceed customer needs, to give lasting value to the customer. The return will follow as customers boast of the company's quality and service. Members of a quality organization recognize both external customers, those who purchase or use the products or services, and internal customers, fellow employees whose work depends on the work that precedes them.[12]

Obsession with Quality

In a total quality organization, internal and external customers define quality. With quality defined, the organization must then become obsessed with meeting or exceed-

QUALITY TIP

Quality and Competitiveness

"Quality is about doing things right the first time and about satisfying customers. But quality is also about costs, revenues, and profits. Quality plays a key role in keeping costs low, revenues high, and profits robust."

Source: Perry L. Johnson, *Total Quality Management* (Southfield, MI: Perry Johnson, Inc., 1991), 9–1.

QUALITY TIP

Quality and Customers

"In fact, customer satisfaction is regarded as the only relevant objective for ensuring stable and continuously increasing business."

Source: Giorgio Merli, *Total Manufacturing Management* (Cambridge, MA: Productivity Press, 1990), 84.

ing this definition. This means all personnel at all levels approach all aspects of the job from the perspective of "How can we do this better?" When an organization is obsessed with quality, "good enough" is never good enough.

Scientific Approach

Total quality detractors sometimes view total quality as nothing more than "mushy people stuff."[13] Although it is true that people skills, involvement, and empowerment are important in a total quality setting, they represent only a part of the equation. Another important part is the use of the scientific approach in structuring work and in decision making and problem solving that relates to the work. This means that hard data are used in establishing benchmarks, monitoring performance, and making improvements.

Long-Term Commitment

Organizations that implement management innovations after attending short-term seminars often fail in their initial attempt to adopt the total quality approach. This is because they look at total quality as just another management innovation rather than as a whole new way of doing business that requires an entirely new corporate culture.

Too few organizations begin the implementation of total quality with the long-term commitment to change that is necessary for success. Quality consultant Jim Clemmer of Toronto-based Achieve International describes mistakes that organizations frequently make when starting quality initiatives, the first of which is as follows:

Senior managers decide they want all of the benefits of total quality, so they hire an expert or throw some money at a particular department. Why that approach doesn't work has been widely discussed; I won't belabor the point.[14]

Teamwork

In traditionally managed organizations, the best competitive efforts are often among departments within the organization. Internal competition tends to use energy that should be focused on improving quality and, in turn, external competitiveness. Scholtes describes the need for teamwork as follows:

Where once there may have been barriers, rivalries, and distrust, the quality company fosters teamwork and partnerships with the workforce and their representatives. This partnership is not a pretense, a new look to an old battle. It is a common struggle for the customers, not separate struggles for power. The nature of a common struggle for quality also applies to relationships with suppliers, regulating agencies, and local communities.[15]

Continual Process Improvement

Products are developed and services delivered by people using processes within environments (systems). To continually improve the quality of products or services—which is a fundamental goal in a total quality setting—it is necessary to continually improve systems.

Education and Training

Education and training are fundamental to total quality because they represent the best way to improve people on a continual basis. According to Scholtes:

> In a quality organization everyone is constantly learning. Management encourages employees to constantly elevate their level of technical skill and professional expertise. People gain an ever-greater mastery of their jobs and learn to broaden their capabilities.[16]

It is through education and training that people who know how to work hard learn how to also work smart.

Freedom through Control

Involving and empowering employees is fundamental to total quality as a way to simultaneously bring more minds to bear on the decision-making process and increase the ownership employees feel about decisions that are made. Total quality detractors sometimes mistakenly see employee involvement as a loss of management control, when in fact control is fundamental to total quality. The freedoms enjoyed in a total quality setting are actually the result of well-planned and carried-out controls. Controls such as scientific methodologies lead to freedom by empowering employees to solve problems within their scope of control.

QUALITY TIP

Continually Improving Processes and Systems

"Quality leadership recognizes—as Dr. Joseph M. Juran and Dr. W. Edwards Deming have maintained since the early 1950s—that at least 85% of an organization's failures are the fault of management-controlled systems. Workers can control fewer than 15% of the problems. In quality leadership, the focus is on constant and rigorous improvement of every system, not on blaming individuals for problems."

Source: Peter R. Scholtes, *Total Quality Management* (Southfield, MI: Peter Scholtes, Inc., 1991), 1–12.

QUALITY CASE

Federal Express Corporation, Internal Customers, and Quality

The name "Federal Express" is synonymous with quality. The Federal Express Company is widely recognized as a world leader in the areas of both transportation and information. Not only can Federal Express deliver packages overnight to almost anywhere in the world, it can also track those packages and tell customers where they are at any given time. Federal Express Corporation is a company worth $16 billion that delivers more than three million packages a day to more than 200 countries. Not bad for an enterprise that started as the result of a "C" term paper written by an undergraduate student at Yale.

While attending Yale, Fred Smith—founder, chairman, president, and CEO of Federal Express Corporation—wrote a paper explaining how a company could be started that could deliver packages anywhere in the country. Smith's professor wasn't impressed and gave the paper a grade of "C." Smith might not have set Yale's College of Business on fire, but he did set the world on fire with one of the most innovative ideas in the history of business. But it took more than vision, strategic thinking, innovation, and analytical ability to build one of the most successful companies in the world.

Fred Smith, who served in the United States Marine Corps from 1966 to 1970, understood the value of "taking care of the troops." While building his company, Smith never lost sight of the fact that its most important asset was its employees. He cared about them, empowered them, and treated them well. To Smith, Federal Express employees were and are just as important as customers. In fact, they are customers—internal customers. Smith gained a well-deserved reputation for taking care of his employees. As a result, when he needed them most, they took care of him.

One year during the Christmas season, the company's busiest and most profitable time of year, Smith and Federal Express faced a debilitating pilot's strike. Either Smith would meet what appeared to be unreasonable demands on the part of his pilots, or the pilots would cripple the company by striking. Smith knew a pilot's strike might cause his company to lose a significant percentage of its business to archrival United Parcel Service. Even so, he held his ground with the pilots. Why Smith was able to refuse to back down shows how critical the concept of treating employees as if they are important internal customers can be. Thousands of Federal Express employees—drivers, sorters, and office personnel—staged a spontaneous rally in support of Smith and the company. The employees let the pilots and the world know that Fred Smith had been a model of fairness over the years. They also showed the pilots that they had no support outside of their cockpits. The pilots backed away from their threat to strike during the holidays, and Federal Express had an excellent Christmas season.

Source: Thomas J. Neff and James M. Citrin, *LESSONS FROM THE TOP.* (New York: Doubleday, 2001), 277–283.

Scholtes explains this paradox as follows:

In quality leadership there is control, yet there is freedom. There is control over the best-known method for any given process. Employees standardize processes and find ways to ensure everyone follows the standard procedures. They reduce variation in output by reducing variation in the way work is done. As these changes take hold, they are freer to spend time eliminating problems, to discover new markets, and to gain greater mastery over processes.[17]

Unity of Purpose

Historically, management and labor have had an adversarial relationship in U.S. industry. One could debate the reasons behind management–labor discord *ad infinitum* without achieving consensus. From the perspective of total quality, who or what is to blame for adversarial management–labor relations is irrelevant. What is important is this: To apply the total quality approach, organizations must have unity of purpose. This means that internal politics have no place in a total quality organization. Rather, collaboration should be the norm.

A question frequently asked concerning this element of total quality is "Does unity of purpose mean that unions will no longer be needed?" The answer is that unity of purpose has nothing to do with whether unions are needed. Collective bargaining is about wages, benefits, and working conditions, not about corporate purpose and vision. Employees should feel more involved and empowered in a total quality setting than in a traditionally managed situation, but the goal of total quality is to enhance competitiveness, not to eliminate unions. For example, in Japan, where companies are known for achieving unity of purpose, unions are still very much in evidence. Unity of purpose does not necessarily mean that labor and management will always agree on wages, benefits, and working conditions.

Employee Involvement and Empowerment

Employee involvement and empowerment is one of the most misunderstood elements of the total quality approach and one of the most misrepresented by its detractors. The basis for involving employees is twofold. First, it increases the likelihood of a good deci-

 ## QUALITY TIP

Total Quality Requires Unity of Purpose

"There is unity of purpose throughout the company in accord with a clear and widely understood vision. This environment nurtures total commitment from all employees. Rewards go beyond benefits and salaries to the belief 'we are family' and 'we do excellent work.'"

Source: Peter R. Scholtes, *Total Quality Management* (Southfield, MI: Peter Scholtes, Inc., 1991), 1–12.

sion, a better plan, or a more effective improvement by bringing more minds to bear on the situation—not just any minds, but the minds of the people who are closest to the work in question. Second, it promotes ownership of decisions by involving the people who will have to implement them.

Empowerment means not just involving people but involving them in ways that give them a real voice. One of the ways this can be done is by structuring work that allows employees to make decisions concerning the improvement of work processes within well-specified parameters. Should a machinist be allowed to unilaterally drop a vendor if the vendor delivers substandard material? No. However, the machinist should have an avenue for offering his or her input into the matter.

Should the same machinist be allowed to change the way she sets up her machine? If by so doing she can improve her part of the process without adversely affecting someone else's, yes. Having done so, her next step should be to show other machinists her innovation so that they might try it.

Total Quality Pioneers

Total quality is not just one individual concept. It is a number of related concepts pulled together to create a comprehensive approach to doing business. Many people contributed in meaningful ways to the development of the various concepts that are known collectively as *total quality*. The three major contributors are W. Edwards Deming, Joseph M. Juran, and Philip B. Crosby. To these three, many would add Armand V. Feigenbaum and a number of Japanese experts, such as Shigeo Shingo.

Deming's Contributions

Of the various quality pioneers in the United States, the best known is W. Edwards Deming. According to Deming biographer Andrea Gabor:

> Deming also has become by far the most influential proponent of quality management in the United States. While both Joseph Juran and Armand V. Feigenbaum have strong reputations and advocate approaches to quality that in many cases overlap with Deming's ideas, neither has achieved the stature of Deming. One reason is that while these experts have often taken very nuts-and-bolts, practical approaches to quality improvement, Deming has played the role of visionary, distilling disparate management ideas into a compelling new philosophy.[18]

Deming came a long way to achieve the status of internationally acclaimed quality expert. During his formative years, Deming's family bounced from small town to small town in Iowa and Wyoming, trying in vain to rise out of poverty. These early circumstances gave Deming a lifelong appreciation for economy and thrift. In later years, even after he was generating a substantial income, Deming maintained only a simple office in the basement of his modest home out of which he conducted his international consulting business.

Working as a janitor and at other odd jobs, Deming worked his way through the University of Wyoming, where he earned a bachelor's degree in engineering. He went on to receive a master's degree in mathematics and physics from the University of Colorado, and a doctorate in physics from Yale.

His only full-time employment for a corporation was with Western Electric. Many feel that what he witnessed during his employment there had a major impact on the direction the rest of his life would take. Deming was disturbed by the amount of waste he saw at Western Electric's Hawthorne plant. It was there that he pioneered the use of statistics in quality.

Although Deming was asked in 1940 to help the U.S. Bureau of the Census adopt statistical sampling techniques, his reception in the United States during these early years was not positive. With little real competition in the international marketplace, major U.S. corporations felt little need for his help, nor did corporations from other countries. However, World War II changed all this and put Deming on the road to becoming, in Andrea Gabor's words, "the man who discovered quality."[19]

During World War II, almost all of Japan's industry went into the business of producing war materials. After the war, those firms had to convert to the production of consumer goods, and the conversion was not very successful. To have a market for their products, Japanese firms had to enter the international marketplace. This move put them in direct competition with companies from the other industrialized countries of the world, and the Japanese firms did not fare well.

By the late 1940s, key industrial leaders in Japan had finally come to the realization that the key to competing in the international marketplace is quality. At this time, Shigeiti Mariguti of Tokyo University, Sizaturo Mishibori of Toshiba, and several other Japanese leaders invited Deming to visit Japan and share his views on quality. Unlike their counterparts in the United States, the Japanese industrialists accepted Deming's views, learned his techniques, and adopted his philosophy. So powerful was Deming's impact on industry in Japan that the most coveted award a company there can win is the Deming Prize. In fact, the standards that must be met to win this prize are so difficult and so strenuously applied that it is now being questioned by some Japanese companies (see Discussion Assignment 1-6).

By the 1980s, leading industrialists in the United States were where their Japanese counterparts had been in the late 1940s. At last, Deming's services began to be requested in his own country. By this time, Deming was over 80 years old. He has not been received as openly and warmly in the United States as he was in Japan. Deming's attitude toward corporate executives in the United States can be described as cantankerous at best.

Gabor gives the following example of Deming's dealings with the U.S. executives from Ford Motor Company:

> The initial contacts were unsettling for Ford. Instead of delivering a slick presentation on how the automaker could solve its quality problems—the sort of thing that became the stock in trade of U.S. quality experts during the 1980s, Deming questioned, rambled, and seemed to take pleasure in making a laughingstock of his listeners. During the first meeting, wearing one of his signature timeworn three-piece suits, Deming glowered at the car executives with steely blue eyes.[20]

Deming's contributions to the quality movement would be difficult to overstate. Many consider him the father of the movement. The things for which he is most widely known are the Deming Cycle, his Fourteen Points, and the Seven Deadly Diseases.

The Deming Cycle

Summarized in Figure 1-3, the Deming Cycle was developed to link the production of a product with consumer needs and focus the resources of all departments (research, design, production, marketing) in a cooperative effort to meet those needs. The Deming Cycle proceeds as follows:

1. Conduct consumer research and use it in planning the product (plan).
2. Produce the product (do).
3. Check the product to make sure it was produced in accordance with the plan (check).
4. Market the product (act).
5. Analyze how the product is received in the marketplace in terms of quality, cost, and other criteria (analyze).

Deming's Fourteen Points

Deming's philosophy is both summarized and operationalized by his Fourteen Points. Scholtes describes them as follows:

> Over the years, Dr. Deming has developed 14 points that describe what is necessary for a business to survive and be competitive today. At first encounter, their meaning may not be clear. But they are the very heart of Dr. Deming's philosophy. They contain the essence of all his teachings. Read them, think about them, talk about them with your co-workers or with experts who deeply understand the concepts. And then come back to think about them again. Soon you will start to understand how they work together and their significance in the true quality organization. Understanding the 14 points can shape a new attitude toward work and the work environment that will foster continuous improvement.[21]

Figure 1–3
The Deming Cycle

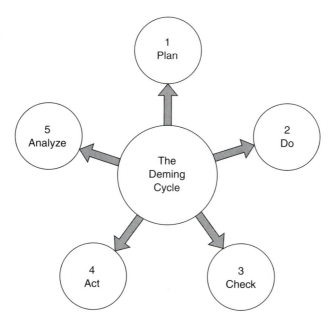

1. Create constancy of purpose toward the improvement of products and services in order to become competitive, stay in business, and provide jobs.
2. Adopt the new philosophy. Management must learn that it is a new economic age and awaken to the challenge, learn their responsibilities, and take on leadership for change.
3. Stop depending on inspection to achieve quality. Build in quality from the start.
4. Stop awarding contracts on the basis of low bids.
5. Improve continuously and forever the system of production and service, to improve quality and productivity, and thus constantly reduce costs.
6. Institute training on the job.
7. Institute leadership. The purpose of leadership should be to help people and technology work better.
8. Drive out fear so that everyone may work effectively.
9. Break down barriers between departments so that people can work as a team.
10. Eliminate slogans, exhortations, and targets for the workforce. They create adversarial relationships.
11. Eliminate quotas and management by objectives. Substitute leadership.
12. Remove barriers that rob employees of their pride of workmanship.
13. Institute a vigorous program of education and self-improvement.
14. Make the transformation everyone's job and put everyone to work on it.

Figure 1–4
Deming's Fourteen Points

Deming's Fourteen Points are contained in Figure 1–4. Deming has modified the specific wording of various points over the years, which accounts for the minor differences among the Fourteen Points as described in various publications. Deming has stated repeatedly in his later years that if he had it all to do over again, he would leave off the numbers. Many people, in Deming's opinion, interpret numbers as an order of priority or progression when this, in fact, is not the point: the numbers represent neither an order of progression nor relative priorities.[22]

Deming's Seven Deadly Diseases

The Fourteen Points summarize Deming's views on what a company must do to effect a positive transition from business as usual to world-class quality. The Seven Deadly Diseases summarize the factors that he believes can inhibit such a transformation (see Figure 1–5).

The description of these factors rings particularly true when viewed from the perspective of U.S. firms trying to compete in the global marketplace. Some of these factors can be eliminated by adopting the total quality approach, but three cannot. This does not bode well for U.S. firms trying to regain market share. Total quality can eliminate or reduce the impact of a lack of consistency, personal review systems, job hopping, and using only visible data. However, total quality will not free corporate execu-

---

1. Lack of constancy of purpose to plan products and services that have a market sufficient to keep the company in business and provide jobs.
2. Emphasis on short-term profits; short-term thinking that is driven by a fear of unfriendly takeover attempts and pressure from bankers and shareholders to produce dividends.
3. Personal review systems for managers and management by objectives without providing methods or resources to accomplish objectives. Performance evaluations, merit ratings, and annual appraisals are all part of this disease.
4. Job hopping by managers.
5. Using only visible data and information in decision making with little or no consideration given to what is not known or cannot be known.
6. Excessive medical costs.
7. Excessive costs of liability driven up by lawyers that work on contingency fees.

Figure 1–5
Deming's Seven Deadly Diseases

tives from pressure to produce short-term profits, excessive medical costs, or excessive liability costs. These are diseases of the nation's financial, health care, and legal systems, respectively.

By finding ways for business and government to cooperate appropriately without collaborating inappropriately, other industrialized countries have been able to focus their industry on long-term rather than short-term profits, hold down health care costs, and prevent the proliferation of costly litigation that has occurred in the United States. Excessive health care and legal costs represent non-value-added costs that must be added to the cost of products produced and services delivered in the United States.

Juran's Contributions

Joseph M. Juran ranks near Deming in the contributions he has made to quality and the recognition he has received as a result. His Juran Institute, Inc., in Wilton, Connecticut, is an international leader in conducting training, research, and consulting activities in the area of quality management (see Figure 1–6). Quality materials produced by Juran have been translated into 14 different languages.

Juran holds degrees in both engineering and law. The emperor of Japan awarded him the Order of the Sacred Treasure medal, in recognition of his efforts to develop quality in Japan and to promote friendship between Japan and the United States. Juran is best known for the following contributions to the quality philosophy:

- Juran's Three Basic Steps to Progress
- Juran's Ten Steps to Quality Improvement
- The Pareto Principle
- The Juran Trilogy

Research and Development
We dedicate ourselves to identifying proven roads that lead to enviable results, and to creating products and services that reflect the experience of successful quality efforts.

Consulting Services
Our quality professionals counsel and assist managers in assessing needs, developing strategies, and applying specific methods for achieving quality goals.

Seminars and Workshops
Juran courses are comprehensive and practical, addressing a full spectrum of needs, industries, and functions. Courses are presented worldwide at client organizations and as public offerings.

Training and Support Materials
Juran videos, workbooks, texts, and reference books are more widely used and translated than those of any other source.

Figure 1–6
Juran Institute, Inc., Quality-Related Services Provided Worldwide
Juran Institute, Inc., promotional literature, 2001.

Juran's Three Basic Steps to Progress

Juran's Three Basic Steps to Progress (listed in Figure 1–7) are broad steps that, in Juran's opinion, companies must take if they are to achieve world-class quality. He also believes there is a point of diminishing return that applies to quality and competitiveness. An example illustrates his observation:

> Say that an automobile maker's research on its midrange line of cars reveals that buyers drive them an average of 50,000 miles before trading them in. Applying Juran's theory, this automaker should invest the resources necessary to make this line of cars run trouble free for perhaps 60,000 miles. According to Juran, resources devoted to improving quality beyond this point will run the cost up higher than the typical buyer is willing to pay.

---

1. Achieve structured improvements on a continual basis combined with dedication and a sense of urgency.
2. Establish an extensive training program.
3. Establish commitment and leadership on the part of higher management.

Figure 1–7
Juran's Three Basic Steps to Progress
Stephen Uselac, *Zen Leadership: The Human Side of Total Quality Team Management* (Loudonville, OH: Mohican, 1993), 37.

QUALITY TIP

Quality and the Law of Diminishing Returns

"Juran favors the concept of quality circles because they improve communication between management and labor. Furthermore, he recommends the use of statistical process control, but does believe that quality is not free. He explains that within the law of diminishing returns, quality will optimize, and beyond that point conformance is more costly than the value of the quality obtained."

Source: Stephen Uselac, *Zen Leadership: The Human Side of Total Quality Team Management* (Loudonville, OH: Mohican, 1993), 37.

Juran's Ten Steps to Quality Improvement

Examining Juran's Ten Steps to Quality Improvement (in Figure 1–8), you will see some overlap between them and Deming's Fourteen Points. They also mesh well with the philosophy of quality experts whose contributions are explained later in this chapter.

The Pareto Principle

The Pareto Principle espoused by Juran shows up in the views of most quality experts, although it often goes by other names. According to this principle, organizations should concentrate their energy on eliminating the vital few sources that cause the majority of problems. Further, both Juran and Deming believe that systems that are controlled by management are the systems in which the majority of problems occur.

1. Build awareness of both the need for improvement and opportunities for improvement.
2. Set goals for improvement.
3. Organize to meet the goals that have been set.
4. Provide training.
5. Implement projects aimed at solving problems.
6. Report progress.
7. Give recognition.
8. Communicate results.
9. Keep score.
10. Maintain momentum by building improvement into the company's regular systems.

Figure 1–8
Juran's Ten Steps to Quality Improvement
Stephen Uselac, *Zen Leadership: The Human Side of Total Quality Team Management* (Loudonville, OH: Mohican, 1993), 37.

QUALITY TIP

The Pareto Principle

"This principle is sometimes called the 80/20 rule: 80% of the trouble comes from 20% of the problems. Though named for turn-of-the-century economist Vilfredo Pareto, it was Dr. Juran who applied the idea to management. Dr. Juran advises us to concentrate on the 'vital few' sources of problems and not be distracted by those of lesser importance."

Source: Peter R. Scholtes, *The Team Handbook* (Madison, WI: Joiner Associates, 1992), 2–9.

The Juran Trilogy

The Juran Trilogy (Figure 1–9) summarizes the three primary managerial functions. Juran's views on these functions are explained in the following sections.[23]

Quality Planning   Quality planning involves developing the products, systems, and processes needed to meet or exceed customer expectations. The following steps are required:

1. Determine who the customers are.
2. Identify customers' needs.
3. Develop products with features that respond to customer needs.
4. Develop systems and processes that allow the organization to produce these features.
5. Deploy the plans to operational levels.

Figure 1–9
The Juran Trilogy
The Juran Trilogy® is a registered
trademark of Juran Institute, Inc.

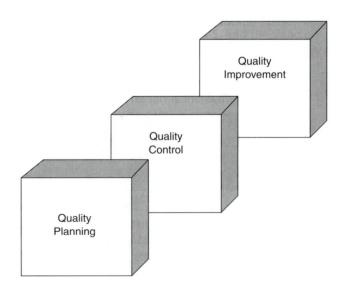

Quality Control   The control of quality involves the following processes:

1. Assess actual quality performance.
2. Compare performance with goals.
3. Act on differences between performance and goals.

Quality Improvement   The improvement of quality should be ongoing and continual:

1. Develop the infrastructure necessary to make annual quality improvements.
2. Identify specific areas in need of improvement, and implement improvement projects.
3. Establish a project team with responsibility for completing each improvement project.
4. Provide teams with what they need to be able to diagnose problems to determine root causes, develop solutions, and establish controls that will maintain gains made.

Crosby's Contributions

Philip B. Crosby started his career in quality later than Deming and Juran. His corporate background includes 14 years as director of quality at ITT (1965–1979). He left ITT in 1979 to form Philip Crosby Associates, an international consulting firm on quality improvement, which he ran until 1992, when he retired as CEO to devote his time to lecturing on quality-related issues.

Crosby, who defines quality simply as conformance, is best known for his advocacy of zero defects management and prevention as opposed to statistically acceptable levels of quality. He is also known for his Quality Vaccine and Crosby's Fourteen Steps to Quality Improvement.

Crosby's Quality Vaccine consists of three ingredients:[24]

1. Determination
2. Education
3. Implementation

His Fourteen Steps to Quality Improvement are listed in Figure 1–10.

WHY TOTAL QUALITY EFFORTS SUCCEED

Organizations that succeed never approach total quality as just another management innovation or, even worse, as a quick fix. Rather, they approach total quality as a new way of doing business. What follows are common errors organizations make when implementing total quality. The successful organizations avoid these errors.

■ *Senior management delegation and poor leadership.*   Some organizations attempt to start a quality initiative by delegating responsibility to a hired expert rather than applying the leadership necessary to get everyone involved.

■ *Team mania.*   Ultimately teams should be established, and all employees should be involved with them. However, working in teams is an approach that must be learned. Supervisors must learn how to be effective coaches and employees must learn how to be team players. The organization must undergo a cultural change before teamwork can succeed. Rushing in and putting everyone in teams before

1. Make it clear that management is committed to quality for the long term.
2. Form cross-departmental quality teams.
3. Identify where current and potential problems exist.
4. Assess the cost of quality and explain how it is used as a management tool.
5. Increase the quality awareness and personal commitment of all employees.
6. Take immediate action to correct problems identified.
7. Establish a zero defects program.
8. Train supervisors to carry out their responsibilities in the quality program.
9. Hold a Zero Defects Day to ensure all employees are aware there is a new direction.
10. Encourage individuals and teams to establish both personal and team improvement goals.
11. Encourage employees to tell management about obstacles they face in trying to meet quality goals.
12. Recognize employees who participate.
13. Implement quality councils to promote continual communication.
14. Repeat everything to illustrate that quality improvement is a never-ending process.

Figure 1–10
Crosby's Fourteen Steps to Quality Improvement
Stephen Uselac, *Zen Leadership: The Human Side of Total Quality Team Management* (Loudonville, OH: Mohican, 1993), 39.

learning has occurred and the corporate culture has changed will create problems rather than solve them.

■ *Deployment process.*   Some organizations develop quality initiatives without concurrently developing plans for integrating them into all elements of the organization (i.e., operations, budgeting, marketing, etc.). According to Jim Clemmer, "More time must be spent preparing plans and getting key stakeholders on board, including managers, unions, suppliers, and other production people. It takes time to pull them in. It involves thinking about structure, recognition, skill development, education, and awareness."[25]

■ *Taking a narrow, dogmatic approach.*   Some organizations are determined to take the Deming approach, Juran approach, or Crosby approach and use only the principles prescribed in them. None of the approaches advocated by these and other leading quality experts is truly a one-size-fits-all proposition. Even the experts encourage organizations to tailor quality programs to their individual needs.

■ *Confusion about the differences among education, awareness, inspiration, and skill building.*   According to Clemmer, "You can send people to five days of training in group dynamics, inspire them, teach them managerial styles, and show them all sorts of grids and analysis, but that doesn't mean you've built any skills. There is a time to educate and inspire and make people aware, and there is a time to give them practical tools they can use to do something specific and different than they did last week."[26]

SIX SIGMA CONCEPT

One of the most innovative developments to emerge out of the total quality movement is the Six Sigma concept introduced by Motorola in the mid–1980s. The purpose of Six Sigma is to improve the performance of processes to the point where the defect rate is 3.4 per million or less. It was designed for use in high-volume production settings. Consequently, the concept is more appropriately used in a manufacturing rather than a service organization. Modern manufacturing systems have many built-in opportunities for defects. This can lead to a high defect rate, one of the principal costs of poor quality. Motorola won the Malcolm Baldrige National Quality Award in 1988 for its pioneering efforts in the development of the Six Sigma concept. The central core of the Six Sigma concept is a six-step protocol for process improvement. The six steps are as follows:

1. Identify the product characteristics wanted by customers.
2. Classify the characteristics in terms of their criticality.
3. Determine if the classified characteristics are controlled by part and/or process.
4. Determine the maximum allowable tolerance for each classified characteristic.
5. Determine the process variation for each classified characteristic.
6. Change the design of the product, process, or both to achieve a Six Sigma processes performance.

It is important to note that the Six Sigma concept is a subset of the broader concept of total quality. Six Sigma is a strategy within the context of total quality that moves the target to a much higher level of quality than many organizations have achieved in the past. It is not a concept that supplants or replaces total quality. Rather, it is an innovative way to pursue a high level of quality under the broad umbrella of total quality.

Six Sigma: The Name

The name Six Sigma comes from the concept of standard deviation, a statistically derived value represented by the lower case Greek letter sigma (σ). The variation of processes and their output products is typically measured in the number of standard deviations from the mean (usually the ideal point). (See the section on histograms in Chapter 15.)

The processes of most good companies presently operate between three and four sigma. This means 99.73% of the output of a process will fall between plus and minus three standard deviations at 3 sigma, or 99.9937 percent at 4 sigma. If the specification (such as a required dimension with a tolerance) for parts produced by the process should correspond to the ±3 sigma values, then a 3-sigma process will yield 2,700 defective parts for every 1,000,000 produced. (See Figure 1–11.)

Let's assume that the specifications describing acceptable product remain constant, and through some improvement we are able to decrease process variation to the point that its new 6-sigma deviation corresponds to the positions of the old 3-sigma values. (See Figure 1–12.) Now if everything else remains constant, the process will yield 99.9999998 percent acceptable product, or a mere 0.002 defective parts per million. In this case, the process is performing at a 6-sigma level. Note that this corresponds to one unacceptable part in 500 million produced.

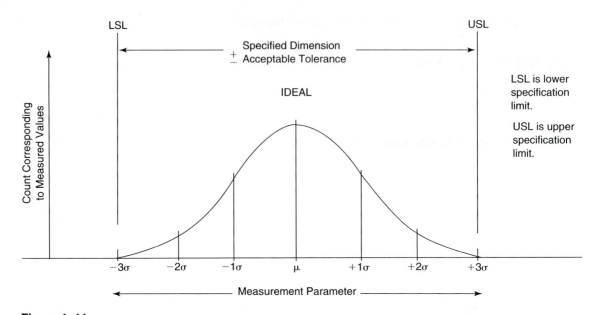

Figure 1–11

Histogram of a 3-Sigma Process

(Requirements match ± 3 sigma values. 99.73% of product produced will fall within the specified limits.)

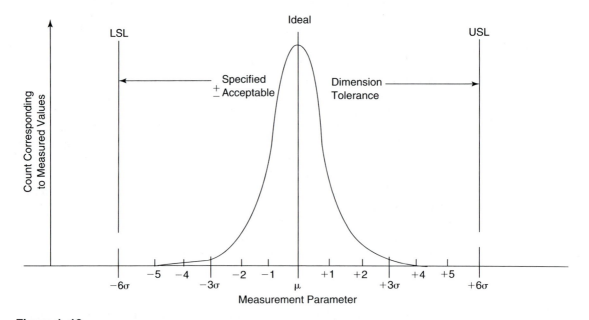

Figure 1–12

Histogram of a 6-Sigma Process

Requirements match ± 6 sigma values. Note that requirements are the same as in Figure 1–11, but the process is improved. 99.9999998% of product will fall within the specified limits.)

When Motorola embarked on its journey to break out of the normal quality level associated with 3-or 4-sigma processes, it targeted 6 sigma; hence the name Six Sigma. Eventually other companies began to adopt the Motorola program and the name has become part of the quality lexicon.

What Is Six Sigma?

Six Sigma is an extension of total quality management which has the aim of taking process and product quality to levels where all customer requirements are met. Depending on which Six Sigma proponent you are listening to, the emphasis may be on the bottom line, or on meeting customer requirements. Of course, the latter is the best way to improve the former.

HOW IS SIX SIGMA ACHIEVED?

Six Sigma can be achieved by improving process performance, but improving processes to this degree can be difficult, and in many cases nearly impossible. On the other hand, Six Sigma can be achieved without improving the process at all if the specification describing acceptable product can be loosened enough to correspond to the original process's ± 6 sigma points, (see Figure 1–13). Note that in this case the process, which is identical to that of Figure 1–11, was not changed in any way. Instead, the definition of what is acceptable in terms of process input was changed. The specification range has been increased from the values corresponding to the ±3 sigma points to the values corresponding to the ±6 sigma points. When we do this, unacceptable product will occur only once in 500 million products.

In order to use this strategy, we must be able to use output that has greater deviation from the ideal value—without sacrificing performance, reliability, or other quality parameters. This is done through engineering processes called *robust design* and *design for manufacture*. Robust design seeks to design products that maintain their reliability, performance, and other quality characteristics even when the component parts of the whole product have great variability of important characteristics. The objective of design for manufacture is to eliminate the possibility of manufacturing errors by simplifying and "error-proofing" the steps required for manufacturing the product. These techniques—improving processes (both incrementally and in giant leaps), and rendering the processes more efficient and foolproof through the use of robust design and design for manufacture—have been around for a long time, and are considered by most to be essential elements of total quality management. The difference here is that Motorola did something no one else had done: they set the target at Six Sigma.

Is Six Sigma Really Six Sigma?

The reader may have noticed that the discussion so far has used numbers relating to acceptable product that are quite different from those found in the Six Sigma literature. For example, at 3 sigma, out of one million chances for success, there will be 2,700 failures. That is a success rate of 99.73 percent. At 6 sigma, the prediction is that 0.002

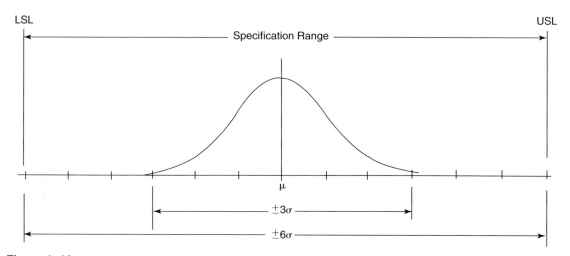

Figure 1–13
Histogram of a 6-sigma process achieved by broadening the specification range for product acceptability

failures will occur out of one million chances. That is a success rate of 99.9999998 percent. However, the most often seen number for failures out of one million chances in 6 sigma is 3.4. This is a significant difference. Statistics predict one failure in 500 million chances at 6 sigma. Motorola uses 3.4 out of one million, or 1,700 times more failures than statistics project. Figure 1–14 illustrates the difference between straight statistics and the Motorola version (*npmo* stands for nonconformances per million opportunities for nonconformance).

Clearly, the numbers used by Motorola in its Six Sigma program are significantly different from those derived from straight statistics. The question is, Which are correct? The answer is that the straight statistics numbers are correct, but the Motorola numbers

Figure 1-14
Sigma Quality Levels, *npmo*

Sigmas	Nonconformances (Statistical)	% Acceptable (Statistical)	Nonconformances (Motorola)	% Acceptable (Motorola)
1	317,400	68.26	697,700	30.23
2	45,400	95.46	308,733	69.1267
3	2,700	99.73	66,803	93.3197
4	63	99.9937	6,200	99.38
5	0.57	99.999943	233	99.9767
6	0.002	99.9999998	3.4	99.99966
7	0.000003	≈ 100	0.019	99.9999981

Nonconformances per Million Opportunities in a Six Sigma Setting are better in a practical sense. What Motorola has done is applied a factor to account for slight changes in environmental conditions, different operators, and so on. The reason Motorola considers this valid is that the sigma values of any process are derived from a statistically valid sample of process operation, necessarily taken over a relatively short period of time. The typical histogram is a snapshot of what is going on at the time the data were collected. Should that period be changed from a few weeks to a few years, one would no doubt encounter new variations resulting from the environment, new operators, and other factors. Motorola believes that the short-term view provides an overly optimistic picture of process variation, and consequently, of the process results at a given sigma level. The company concluded, therefore, that the reality of the long term under which the processes operate should be compensated for. Motorola chose to do this by shifting the process average (μ) from the target, or ideal point, by 1.5 sigma to the right while leaving the normal Six Sigma field and the upper and lower specification limits as they were. (See Figure 1–15.)

Such a shift could take place due to environmental changes or other factors. Drifts and shifts of varying magnitude are the norm in the long term, and Motorola is covering that eventuality by accounting for it before the fact. The values for *npmo* are then taken from the right tail of the histogram. Notice in Figure 1–15 that the +3 sigma point intercepts the shifted curve at the equivalent of its +1½ sigma. At 1½ sigma there should be 66,803 *npmo*, which corresponds exactly with Motorola's expectation for 3 sigma. Similarly, the +6 sigma line intercepts the shifted curve at its 4½ sigma point. 4½ sigma is predicted by statistics to yield 3.4 *npmo*, the value used in the Motorola Six Sigma program. Whether in real life the process should shift to the left or the right makes no difference, since the same results occur in either direction. Practically speaking, and pragmatically anticipated by Motorola, the process may shift back and forth by

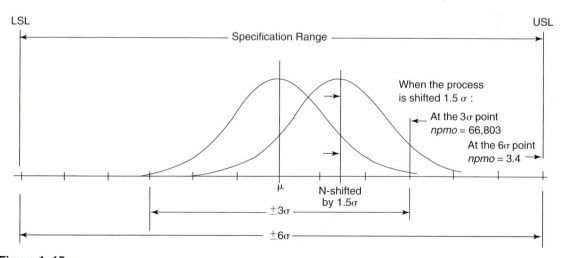

Figure 1–15
The 1½ Sigma Shift
Note that the process histogram is shifted 1½ σ from its ideal position to account for long-term variation.

varying magnitudes over time. By introducing the 1½ sigma shift, Motorola has allowed some long-term variation to enter the picture without causing panic.[27]

We hasten to state that we do not disagree with the route Motorola took with the 1½ sigma shift. Smart people determined that this was the appropriate thing to do. Still, we are not completely comfortable with the Six Sigma name if it does not truly represent Six Sigma. But we will not argue with 3.4 nonconformances out of a million opportunities. This is a quality level that is so far beyond the average that we will gladly accept it, name and all.

History of Six Sigma

In 1981, Motorola President Robert Galvin issued a challenge to his company: *Improve performance tenfold over the next five years.* Motorola responded and achieved the goal. That led the company to its next challenge. In 1987, Galvin, by now versed in some work being done in Motorola's Communications Sector with the goal of Six Sigma quality, called for ten times improvement in each of the next two-year periods, achieving Six Sigma quality throughout the corporation by 1992. That required a staggering 100 times improvement in all operations of the company. It was not achieved by 1992, but by then it was clear that this was the right thing to do. By 1993, many of Motorola's manufacturing operations were operating at or near Six Sigma.[28]

In the ten years following Galvin's Six Sigma challenge, and as a direct result of the Six Sigma efforts, Motorola claims to have saved $414 billion. Sales were up by a factor of five, and profits increased by nearly 20 percent each year.[29]

As the word of Six Sigma spread, other organizations took up the challenge. In 1995, CEO Jack Welch committed General Electric to Six Sigma. Allied Signal was another very large company that took on Six Sigma. By this time, although rooted in total quality management, Six Sigma was developing a life of its own.

THE FUTURE OF QUALITY MANAGEMENT

In an article for *Quality Digest*,[30] quality pioneer Armand Feigenbaum explains several trends that will shape the future of quality management. Those trends are as follows:

■ *Demanding global customers.*   The provision of quality begets an ever-increasing demand for quality. Today's customers share two common characteristics: (a) they are part of regional trade alliances such as the Americas, Europe, and Asia; and (b) they expect both high quality and added value.

■ *Shifting customer expectations.*   Increasingly, today's global customer is interested not just in the quality of a product provided but also the quality of the organization that backs it up. Customers want an excellent product or service from an organization that also provides accurate billing, reliable delivery, and after-purchase support.

■ *Opposing economic pressures.*   The global marketplace exerts enormous, unrelenting pressure on organizations to continually improve quality while simultaneously reducing the prices they charge for goods and services. The key to achieving higher quality and lower prices for customers is the reduction of the expenses associated with satisfying unhappy customers—expenses that amount to as much as 25% of the cost of sales in many companies.

■ *New approaches to management.* Companies that succeed in the global market-place have learned that *you manage budgets, but lead people.* The old approach of providing an occasional seminar or motivational speech for employees without making any fundamental changes in the way the organization operates will no longer work.

Quality Management Characteristics for the Future

To succeed in the global marketplace for now and in the future, organizations will have to operate according to the principles of quality management. Such companies will have the following characteristics:

■ A total commitment to continually increasing value for customers, investors, and employees

■ A firm understanding that *market driven* means that quality is defined by customers, not the company

■ A commitment to *leading* people with a bias for continuous improvement and communication

■ A recognition that sustained growth requires the simultaneous achievement of four objectives all the time, forever: (a) customer satisfaction, (b) cost leadership, (c) effective human resources, and (d) integration with the supplier base.

■ A commitment to fundamental improvement through knowledge, skills, problem solving, and teamwork.

Companies that develop these characteristics will be those that fully institutionalize the principles of quality management. Consequently, quality management as both a practice and a profession has a bright future. In fact, in terms of succeeding in the global marketplace, quality management is the future. Consequently, more and more companies are making quality management the way they do business, and more and more institutions of higher education are offering quality management courses and programs.

 SUMMARY

1. *Quality* has been defined in a number of different ways. When viewed from a consumer's perspective, it means meeting or exceeding customer expectations.
2. Total quality is an approach to doing business that attempts to maximize an organization's competitiveness through the continual improvement of the quality of its products, services, people, processes, and environments.

3. Key characteristics of the total quality approach are as follows: strategically based, customer focus, obsession with quality, scientific approach, long-term commitment, teamwork, employee involvement and empowerment, continual process improvement, bottom-up education and training, freedom through control, and unity of purpose.

4. The rationale for total quality can be found in the need to compete in the global marketplace. Countries that are competing successfully in the global marketplace are seeing their quality of living improve. Those that cannot are seeing theirs decline.

5. W. Edwards Deming is best known for his Fourteen Points, the Deming Cycle, and the Seven Deadly Diseases.

6. Joseph M. Juran is best known for Juran's Three Basic Steps to Progress, Juran's Ten Steps to Quality Improvement, the Pareto Principle, and the Juran Trilogy.

7. Common errors made when starting quality initiatives include senior management delegation and poor leadership; team mania; the deployment process; a narrow, dogmatic approach; and confusion about the differences among education, awareness, inspiration, and skill building.

8. Six Sigma is a concept introduced by Motorola in the mid-1980's that seeks to improve processes to the point that the defect rate is just 3.4 per million or less.

9. Trends affecting the future of quality management include demanding global customers, shifting customer expectations, opposing economic pressures, and new approaches to management.

 KEY TERMS AND CONCEPTS

Bottom-up education and training

Continual process improvement

Customer focus

Deming's Fourteen Points

Deming's Seven Deadly Diseases

Employee involvement and empowerment

Freedom through control

Global customer

The Juran Trilogy

Long-term commitment

Obsession with quality

The Pareto Principle

Quality

Quality control

Quality improvement

Quality planning

Quality vaccine

Scientific approach

Six Sigma

Teamwork

Total quality

TQC

TQL

TQM

Unity of purpose

FACTUAL REVIEW QUESTIONS

1. Define the term *quality*.
2. What is total quality?
3. List and explain the key elements of total quality.
4. Explain the rationale for the total quality approach to doing business.
5. Describe the following concepts:
    - Deming's Fourteen Points
    - The Deming Cycle
    - The Seven Deadly Diseases
6. List and explain Juran's main contributions to the quality movement.
7. Why do some quality initiatives fail?
8. For what contributions to the quality movement is Philip B. Crosby known?
9. Summarize the most common errors made when starting quality initiatives.
10. What is Six Sigma and how does it relate to total quality?
11. Explain the trends that are affecting the future of quality management.

 CRITICAL THINKING ACTIVITY

Have We Spoiled Customers?

"If you want to understand how the worldwide quality movement has benefited consumers, just look at automobiles. What used to be considered a luxury option is now just standard," said one quality manager. "That is precisely the problem," said another quality manager. "We have spoiled the consumer. Now customers will never be happy no matter what we do." Join this debate. What is your opinion concerning the following questions?

1. What features in the modern automobile are customer driven?
2. Henry Ford once said something to the effect that the customer can have any color Model T he wants, as long as it's black. How did the world evolve from Henry Ford's attitude toward customers to the modern attitude of customer-driven quality?
3. Are global consumers spoiled and unrealistic in their expectations, or are they finally demanding their rights in the marketplace?
4. How has the worldwide demand for quality driven the concept of innovation? How has innovation changed your life?

DISCUSSION ASSIGNMENT 1–1

Winning and Longevity

A professional baseball team set its sights on winning the World Series. The team owner wanted to win big and win fast. Consequently, the team sank all of its resources into trading for the best players in the league. It was able to obtain enough of them that within two seasons the team was the World Series champion.

However, the team had committed such a high percentage of its financial resources to players' salaries that other important elements of the team began to suffer. Its stadium quickly fell into such a state of disrepair that fans began to stay home. Training facilities also began to suffer, which caused discontent among the players. The money left over to pay the salaries of coaches wasn't enough to hold onto the good ones, most of whom accepted better offers from other teams. In short, by focusing so intently on the desired end result, this organization neglected other important aspects of building a competitive team. As a result, the team's World Series championship was a short-lived once-in-a-lifetime victory. The very next season the team's crumbling infrastructure sent it tumbling to the bottom of its division. Without the people, processes, and environment to turn the situation around, the team was eventually sold at a loss and moved to another city.

Discussion Assignment

Discuss the following questions in class or outside of class with your fellow students:

1. Why would a company that is turning out a satisfactory product want to continually examine its processes and the work environment? What happened to the old adage, "If it's not broke, don't fix it"?

2. Create a manufacturing, processing, or service sector parallel for this activity. Discuss how this assignment would apply to a company.

DISCUSSION ASSIGNMENT 1–2

Commitment and Perseverance at Hub, Inc.

Ben Camp, CEO of Hub, Inc., a metal products manufacturer near Atlanta, found that total quality takes commitment and perseverance. In his words:

"The company studied various gurus and fashioned its own program from them. That program requires, among other things, meticulous attention to detail. It demands perseverance. It involves dismounting from managerial high horses and realizing that employees recognize lip service as soon as the mouths begin to move. Thus, getting down to action is where the easier-said-than-done part comes in.

"Everyone agrees that profitable fleet companies with devoted customers and dynamic employees are good things. But not everyone wants to endure the pain of change, much less admit that, after years of effort, there still may be as much to be done as has been done."

After 4 years of total quality, attention to detail and perseverance had allowed Hub, Inc., to identify and categorize most of its errors, a major step in correcting future performance. The categories of errors identified were defects, carrier/vendor pricing, order entry, wrong material, wrong quality, customer error, surplus, and duplicates. Errors in these categories cost the company over $16,000 in one sample month. Unidentified errors cost Hub, Inc., another $5,161 in the same month. The errors identified can be used for improving the processes that resulted in them. Unidentified errors cannot be corrected. Hub, Inc., is committed to identifying and correcting 100% of its process errors. The company intends to persevere no matter how long this takes. If Hub, Inc., is going to compete, it has no other option.

Discussion Assignment

Discuss the following question in class or outside of class with your fellow students:

1. What types of pressures do you think the CEO had to overcome to persevere for 4 years?

Source: "Total Quality: Management Method Is More Than Words; Time and Commitment Are Required," *Atlanta Constitution*, 11 October 1992, R-1 and R-6.

DISCUSSION ASSIGNMENT 1–3

Early Impact of Scientific Management

"To restore the balance, the factory managers adopted a new strategy: a central inspection department headed by a chief inspector. The various departmental inspectors were then transferred to the new inspection department over the bitter opposition of the production supervisors.

"Note that during this progression of events the priority given to quality declined significantly. In addition, the responsibility for leading the quality function became vague and confused. In the days of the craft shops, the master (then also the chief executive) participated personally in the process of managing for quality. What emerged was a concept in which upper management became detached from the process of managing for quality."

Discussion Assignment

Discuss the following question in class or outside of class with your fellow students:

1. In your opinion, why can no quality effort succeed without the support of upper management?

Source: J. M. Juran, *Juran on Leadership for Quality* (New York: Free Press, 1989), 4.

DISCUSSION ASSIGNMENT 1–4

Early Inspection-Oriented Quality Program

"The central activity of these quality-oriented departments remained that of inspection and test, that is, separating good products from bad. The prime benefit of this activity was to reduce the risk that defective products would be shipped to customers. However, there were serious detriments: This centralized activity of the quality department helped to foster a widespread belief that achievement of quality was solely the responsibility of the quality department. In turn, this belief hampered efforts at eliminating the causes of defective products; the responsibilities were confused. As a result, failure-prone products and incapable processes remained in force and continued to generate high costs of poor quality."

Discussion Assignment
Discuss the following questions in class or outside of class with your fellow students:

1. What does this assignment say about the adage, "Quality is everybody's business"?
2. Why is the total quality approach superior to an after-the-fact inspection approach?

Source: J. M. Juran, *Juran on Leadership for Quality* (New York: Free Press, 1989), 6.

DISCUSSION ASSIGNMENT 1–5

How Japan Caught Up with the United States and How the United States Caught Back Up with Japan

Immediately following World War II, the quality of products produced by Japanese companies was not good enough to compete in the international marketplace. The only advantage Japanese companies had was price. Japanese goods, as a rule, were cheap. For this reason, Western manufacturers, particularly those in the United States, saw the Japanese threat as being rooted in cost rather than quality.

Reading the future more accurately, albeit belatedly, Japanese companies saw quality as the key to success and, in 1950, began doing something about it. While Japanese companies were slowly but patiently and persistently creating a quality-based infrastructure (people, processes, and facilities), American companies were still focusing on cost, shifting the manufacture of labor-intensive products offshore, and, at the same time, neglecting infrastructure improvements.

By the mid-1970s the quality of Japanese manufactured goods in such key areas as automobiles and consumer electronics products was better than that of competing American firms. As a result, Japanese exports increased exponentially while those of Western countries experienced corresponding decreases.[31]

This explains how Japan rose up out of the ashes of World War II to become a world-leading industrial nation. But the story does not end there. After losing market share to the Japanese for more than two decades, companies in the United States began to embrace the principles of quality management. As a result, by the mid-1990s companies in the United States had reasserted themselves in the global marketplace.

Now the two countries are like well-matched heavyweight boxers who slug it out every day in the world of global business. On any given day either can win the global business battle. There are no longer any automatic winners. Regardless of whether they are Japanese or American, those companies that adhere to the principles of quality management and continually improve are the ones that will win in today's marketplace.

Discussion Assignment

Discuss the following question in class or outside of class with your fellow students:

1. Why do you think that companies in the United States were slow to adopt the quality management principles Japanese companies had used to gain market share worldwide?

DISCUSSION ASSIGNMENT 1–6

Japanese Firms Question the Deming Prize

"Some executives contend that even competing for the Deming Prize can sometimes be counterproductive. For example, Shimizu Construction, which won the Deming Prize in 1983, spent so much time and effort chasing the prize that same year its financial results showed a clear drop from the year before. The human toll can also be high. Some companies go so far as to gather their middle management at the foot of Mt. Fuji for a kind of hazing session, subjecting the group to seven days of marathon discussions that last from 5:00 A.M. to 10:00 P.M., during which each participant admits he has been neglecting his work. Sachiaki Nagae of TIT says that for the last year before going for the Deming Prize he and his managers went without vacations and routinely worked seven days a week."

Source: Andrea Gabor, *The Man Who Discovered Quality* (New York: Times Books, Random House, 1990), 97.

ENDNOTES

1. Air Force Development Test Center, *Total Quality Management (TQM) Training Package,* 2nd edition, 2001, 9.
2. Ibid.
3. Ibid.
4. Ibid.
5. W. Edwards Deming, *Out of the Crisis* (Cambridge: Massachusetts Institute of Technology Center for Advanced Engineering Study, 1986), 168.
6. Deming, 169.
7. Deming, 169.
8. Air Force Development Test Center, Elgin Air Force Base, Total Quality Management training package, 2nd edition, 2001, 13.
9. J. M. Juran, *Juran on Leadership for Quality* (New York: Free Press, 1989), 5.
10. Juran, 7–8.
11. Juran, 10.
12. Peter R. Scholtes, *The Team Handbook* (Madison, WI: Joiner Associates, 1992), 1–11.
13. Comment by a participant in a workshop on total quality presented by David Goetsch in Fort Walton Beach, Florida, in January 1993.
14. Jim Clemmer, "Eye on Quality," *Total Quality Newsletter* (3 April 1992): 7.
15. Scholtes, 1–13.
16. Scholtes, 1–13.
17. Scholtes, 1–12.
18. Andrea Gabor, *The Man Who Discovered Quality* (New York: Times Books, Random House, 1990), 16.
19. Gabor, 1.
20. Gabor, 136.
21. Scholtes, 2–4.
22. W. Edwards Deming, comments made during a teleconference on total quality broadcast by George Washington University in Spring 1992.
23. Juran, 20.
24. Uselac, 38.
25. Jim Clemmer, "5 Common Errors Companies Make Starting Quality Initiatives," *Total Quality* 3 (April 1992): 7.
26. Clemmer, 7.
27. Mikel J. Harry, from an undated paper, "The Nature of Six Sigma Quality," 10.
28. Mikel J. Harry and Richard Schroeder, *Six Sigma: The Breakthrough Management Strategy Revolutionizing the World's Top Organizations.* (New York: Doubleday, 2000), 11.
29. Peter S. Pande, Robert P. Neuman, and Roland R. Cavenagh, *The Six Sigma Way,* (New York: McGraw-Hill, 2000), 7.
30. Armand V. Feigenbaum, "The Future of Quality Management," *Quality Digest,* 18(4) (May 1998): 33–38.
31. J. M. Juran, *Juran on Leadership for Quality* (New York: Free Press, 1989), 10.

CHAPTER TWO

Quality and Global Competitiveness

"I'm surprised how many people think you can throw a hand grenade at a competitor and expect he'll stand there and enjoy it."

Frank Lorenzo

MAJOR TOPICS

- The Relationship between Quality and Competitiveness
- Cost of Poor Quality
- Competitiveness and the U.S. Economy
- Factors Inhibiting Competitiveness
- Comparisons of International Competitors
- Industrial Policy and Competitiveness
- Technology and Competitiveness
- Human Resources and Competitiveness
- Characteristics of World-Class Organizations
- E-Commerce, Information Quality, and Competitiveness
- Management by Accounting: Antithesis of Total Quality
- Key Global Trends
- U.S. Companies: Global Strengths and Weaknesses

One of the results of World War II combined with subsequent technological advances was the creation of a global marketplace. Following the war, industrialized countries began looking for markets outside their own borders. Although the war gave the world a boost in this regard, it was advances in technology that really made the global marketplace possible. Advances in communications technology have made people from all over the world electronic neighbors and electronic customers.

Advances in transportation technology allow raw materials produced in one country to be used in the manufacture of products in a second country that are, in turn, sold to end users in a third country. For example, leather produced in Australia might be shipped as raw material to Italy, where it is used in the manufacture of shoes and purses that are sold in the United States, France, and Japan. At the same time, leather produced in South America is sent to shoe manufacturers in Indonesia. These manufacturers, like their Italian counterparts, sell their shoes in the United States, France, and Japan. This means the manufacturers in Italy compete with manufacturers in Indonesia. This simple example demonstrates the kind of competition that takes place on a global scale every day. Such competition has become the norm, and it can be intense.

It used to be only large corporations and multinational corporations that faced global competition; now even small companies are affected. Today no company is immune to the effects of global competition.

THE RELATIONSHIP BETWEEN QUALITY AND COMPETITIVENESS

The relationship between quality and competitiveness is best illustrated by an example from the world of athletics. Consider track star Juan Arballo. In high school, he was his track team's best sprinter. Competing at the district level, Juan easily bested the competition in such events as the 100-, 200-, and 400-meter runs and several relays in which he was the anchor. He did well enough in high school to win a college scholarship. However, at the college level the competition was of a higher quality, and Juan found he had to train harder and run smarter to win. This he did, and although he no longer won every race, Juan did well enough to pursue a spot on the U.S. Olympic team. In the Olympic Trials, the quality of the competition was yet again better than that to which Juan was accustomed. He made the Olympic team but only in two events: the 200-meter dash and the 4 x 100 relay.

In the preliminary events at the Olympics, Juan Arballo found the quality of his competitors to be even better than he had imagined it would be. Some competitors had preliminary times better than the best times he had ever run in meets. Clearly, Juan faced the competitive challenge of his career. When his event was finally run, Juan, for the first time in his life, did not place high enough to win a medal. The quality of the global competition was simply beyond his reach.

In this example, at each successive level of competition the quality of the competitors increased. A similar phenomenon happens to businesses in the marketplace. Companies that used to compete only on a local, regional, or national level now find themselves competing against companies from throughout the world. Like Juan Arballo, some of these companies find the competition to be more intense than any they have ever encountered. Only those who are able to produce world-class quality can compete at this level. In practical terms, it is extremely important for a country's businesses to be able to compete globally. When they can't, jobs are lost and the quality of life in that country declines correspondingly.

COST OF POOR QUALITY[1]

There is a prevailing and persistent attitude in the business world that quality costs money; some claim too much. Many business executives adopt the attitude that ensuring quality is good thing to do until hard times set in and cost cutting is necessary. During tough times, quality initiatives are often the first functions to go. Companies that take this approach are those that have never integrated continuous quality improvement as a normal part of doing business. Rather, they see it as a stand-alone, separate issue. What executives in such companies fail to calculate or to even understand is the costs associated with poor quality. This ironic dilemma is best illustrated with an example of two companies.

A Tale of Two Companies

Two companies, ABC, Inc., and XYZ, Inc., both need to compete in the global marketplace in order to survive. As might be expected, over the years competition has become increasingly intense. In order to be more competitive, ABC's executives undertook a major company-wide cost-cutting initiative. They eliminated quality audits; changed from trusted, proven suppliers to low-bid suppliers; purchased new computer systems; cut back on research and development; and reduced customer service staff.

These cost-cutting strategies did have the desired effect of decreasing the company's overhead, but they also had the unplanned consequences of disrupting the company's ability to satisfy customers and reducing the company's potential to develop new business in the future. The net outcome of all this was unhappy customers, disenchanted employees, and a decline in business. To make matters even worse, the company was still struggling with the poor performance record that caused its executives to want to cut costs in the first place.

The executives of XYZ, Inc., also needed to make some changes in order to stay competitive, but they decided to take a different approach. XYZ's management team set out to identify all of the costs that would disappear if their company improved its performance in key areas. The costs identified included those associated with the following: late deliveries to customers, billing errors, scrap and rework, and accounts payable errors. In other words, XYZ's executives decided to identify the costs associated with poor quality. Having done so, they were able to begin improvement projects in the areas identified without making cuts in functions essential to competitiveness (e.g., product quality, research and development, customer service, etc.).

Cost of Poor Quality and Competitiveness

Few things affect an organization's ability to compete in the global marketplace more than the costs associated with poor quality. According to Joseph A. DeFeo, president and CEO of the Juran Institute, the costs of poor quality account for 15 to 30 percent of a company's overall costs.[2] When an organization does what is necessary to improve its performance by reducing deficiencies in key areas (cycle time, warranty costs, scrap and rework, on-time delivery, billing, etc.), it can reduce overall costs without eliminating essential services, functions, product features, and personnel. Reducing the costs associated with poor quality is mandatory for companies that hope to compete in the global marketplace.

DeFeo explains this concept as follows: "The need to improve an organization's financial condition correlates directly with the process of making and measuring quality improvements.... In fact, the cost of poor quality should be the driver of the project selection process for Six Sigma."[3] According to DeFeo, lower deficiencies in such areas as cycle time, warranty costs, scrap and rework, on-time delivery, and so on will lead to lower total costs. Correspondingly, improvements in product or service features can lead to higher market share at a better price, which, of course, means higher revenue. Feature improvements coupled with deficiency improvements will enhance customer loyalty and add to net profits.[4] Figure 2–1 summarizes both the traditional and the hidden costs of poor quality. The key principle to understand when examining the hidden costs shown in Figure 2–1 is that if every activity in an organization is performed properly every time, these costs simply disappear.

Interpreting the Costs of Poor Quality

Once activities have been identified that exist only or primarily because of poor quality, improvement projects can be undertaken to correct the situation. It is important at this stage to select those projects that have the greatest potential to yield the highest return. The following steps can be used to measure the costs of poor quality so that selected improvement projects have the highest priority:

1. Identify all activities that exist only or primarily because of poor quality.
2. Decide how to estimate the costs of these activities.
3. Collect data on these activities and make the cost estimates.
4. Analyze the results and take necessary corrective actions in the proper order of priority.

According to DeFeo, "The most important step in developing useful cost-of-poor-quality data is simply to identify activities and other factors that affect costs. Any consistent and unbiased method for estimating costs will yield adequate information that will identify key targets for quality improvement."[5]

Figure 2–1
Factors to consider when quantifying the costs of poor quality

Traditional Costs	
_✓_ Waste	_✓_ Customer returns
_✓_ Rejects	_✓_ Inspection
_✓_ Testing	_✓_ Recalls
_✓_ Rework	

Hidden Costs	
_✓_ Excessive overtime	_✓_ Handling complaints
_✓_ Pricing errors	_✓_ Expediting
_✓_ Billing errors	_✓_ System costs
_✓_ Excessive turnover	_✓_ Planning delays
_✓_ Premium freight costs	_✓_ Late paperwork
_✓_ Development cost of the failed product	_✓_ Lack of follow-up
	_✓_ Excess inventory
_✓_ Field service costs	_✓_ Customer allowances
_✓_ Overdue receivables	_✓_ Unused capacity

COMPETITIVENESS AND THE U.S. ECONOMY

The United States came out of World War II as the only major industrialized nation with its manufacturing sector completely intact. A well-oiled manufacturing sector and the availability of abundant raw materials helped the United States become the world leader in the production and export of durable goods. This resulted in a period of unparalleled prosperity and one of the highest standards of living ever experienced by any country.

While the United States was enjoying its position as the world's preeminent economic superpower, the other industrialized nations of the world, particularly Japan and Germany, were busy rebuilding their manufacturing sectors. As Japanese and German manufacturers rebuilt, two things became apparent to them:

1. To succeed, they would have to compete globally.
2. To compete globally, they would have to produce goods of world-class quality, which meant producing better goods but at reasonable, competitive prices.

Basking in their prosperity, U.S. manufacturers were slow to catch on that the game had changed from mass production with acceptable levels of waste to quality production with things done right the first time every time. The old game was best cost. The new game had become best cost *and* best quality. When foreign companies, through a combination of better training, better technology, and better management, began to eat away at markets, U.S. companies, mistakenly seeing cost rather than quality as the issue, began sending work offshore to hold down labor costs. By the time U.S. companies learned that quality was the key to success in the global marketplace, Japan, Germany, Taiwan, and Korea had made major inroads into global markets previously dominated by U.S. manufacturers (i.e., steel, automobiles, computers, and consumer electronics).

In a relatively short period of time, the United States went from the world's leading lender and exporter to the world's biggest debtor, with a huge balance of trade deficit. By 1980, the United States was consuming more than it produced. According to Ray Marshall and Marc Tucker:

> Like any family that consumes more than it earns, the country is making up the difference by borrowing and depreciating our capital. Our economy is propped up by unparalleled levels of public and private debt. Borrowing is not, of course, necessarily bad. When borrowed money is invested in something productive, it can provide handsome returns. But the money we borrowed during the 1980s was used to sustain our consumption. To borrow it, we had to offer high interest rates, which raised the cost of capital here at home, discouraging business investment. As individuals saved less in order to consume more, the sources of investment funds dried up even further. The United States saved and invested 2.9 percent of its gross national product, compared to an average of 7.7 percent for the seven economic-summit nations (Canada, France, Germany, Italy, Japan, the United Kingdom, and the United States).[6]

Impact of Competitiveness on Quality of Life

A nation's ability to compete in the global marketplace has a direct bearing on the quality of life of its citizens. Because the ability to compete translates into the ability to do a better job of producing quality goods, it is critical that nations and individual

QUALITY TIP

The United States and the Global Marketplace

"The future now belongs to societies that organize themselves for learning. What we know and can do holds the key to economic progress, just as command of natural resources once did. Everything depends on what firms can learn from and teach to their customers and suppliers, on what countries can learn from one another, on what workers can learn from each other and the work they do, on the learning environment that families provide, and, of course, on what we learn in school. More than ever before, nations want high incomes and full employment skills by everyone, not just a select few. The prize will go to those countries that are organized as national learning systems, and where all institutions are organized to learn and to act on what they learn.

"Our most formidable competitors know this. Many newly industrialized countries know it and are vaulting forward as a result. But the United States does not. Because this country continues to operate on the premise that, for the country to be successful, only a few need to know or be able to do very much, we are poised on the precipice of a steep decline in national income, with all that this implies for our material well-being and the stability of our society."

Source: Ray Marshall and Marc Tucker, *Thinking for a Living* (New York: Basic Books, 1992), xiii.

organizations within them focus their policies, systems, and resources in a coordinated way on continually improving both quality and competitiveness.

The United States will begin the first decade of the new century poised on the precipice of a decline in its quality of life and a growing gap between the haves and the have-nots. While Canada, France, Germany, Italy, Japan, Sweden, and Great Britain have taken steps to link economics, education, and labor market policy in ways that promote competitiveness, the United States is still debating the need for an industrial policy and struggling to reverse the decline of its public schools (see Discussion Assignment 2–1 at the end of this chapter).

During the 1980s, the United States improved productivity by putting more people to work. Other countries improved their productivity by making the individual worker more efficient. Most new entrants into the workforce during the 1970s and 1980s were people who had not worked previously, primarily women. This influx of new workers helped the United States maintain its traditionally high level of productivity. However, by the 1990s, the gains that could be made by increasing the number of people in the workforce had been made.

From 2000 to 2010, the number of people in the prime working-years age groups in the United States will be on the decline.[7] As the size of the workforce continues the downward trend it began in the early 1990s, the only way to improve productivity will be to do what other industrialized countries have done—concentrate on improving the efficiency of individual workers. In other words, businesses in the United States will need to get more work out of fewer workers. As some businesses have already learned, the best way to do this is to adopt the total quality philosophy.

Figure 2–2 contains several vignettes relating to the quality of life in the United States. This figure presents either a bleak picture of bad times to come or an unprecedented national challenge. According to Marshall and Tucker, "The only way to avoid this chilling scenario is to find a way to improve our productivity dramatically. A major improvement in productivity would enable us to reduce debt, lower interest rates, invest in the future, and provide rising incomes for everyone."[8]

FACTORS INHIBITING COMPETITIVENESS

Improving competitiveness on a national scale is no simple matter. Much can be done at the level of the individual company where the total quality approach can be applied to great advantage, but competitiveness on a national scale requires more than just total quality. Students of quality management must understand this point. Failure to understand the limits of total quality has caused some business leaders to expect too much too soon. This, as a result, has turned them into detractors.

This section describes factors that can inhibit competitiveness but are beyond the scope of total quality. They are socioeconomic and sociopolitical in nature and are indigenous to the United States. In the age of global competition, managers should apply the principles of total quality to help make their individual organizations more competitive. Simultaneously, they should work through the political and social systems as private citizens and community leaders to help level the playing field among nations by correcting the inhibitors explained in this section. These inhibitors fall into the following categories: business/government–related factors, education-related factors, and family-related factors.

- Productivity in the U.S. is stagnant. Consequently, real income is declining.
- The number of people having to work at more than one job to maintain their quality of life has increased continually since the 1960s.
- The most financially rewarding work years have historically been those between ages 40 and 50. In the 1950s, people in this age bracket experienced a 36% increase in real income. By 2001, their counterparts had experienced a decline in real income.
- The average person in the United States is getting poorer; real income has decreased by 13% since 1969.

Figure 2–2
Quality of Life Issues in the United States

Business/Government–Related Factors

U.S. companies trying to compete in the global marketplace are rowing upstream while dragging an anchor. Actually, they drag three anchors. This was pointed out many years ago by W. Edwards Deming when he first set forth his Seven Deadly Diseases. His second, sixth, and seventh deadly diseases are as follows:[9]

- Emphasis on short-term profits fed by fear of unfriendly takeover attempts and pressure from lenders or shareholders
- Excessive medical costs
- Excessive costs of liability inflated by lawyers working on contingency fees

Each of these diseases adds cost to a company's products without adding value. Nothing could be worse when viewed from the perspective of competitiveness. A company might equal all competitors point for point on all quality and productivity criteria and still lose in the marketplace because it is a victim of deadly diseases that drive up the cost of its product. Discussion Assignment 2–2 illustrates how the emphasis on short-term profits has hampered competitiveness in the United States

Excessive medical costs and litigation, primarily related to workers' compensation, have also served to slant the playing field in favor of foreign competitors. The annual cost of workers' compensation to U.S. businesses is almost $30 billion. This is a non-value-added cost that increases the price these businesses must charge for their products.

Overcoming these business-related inhibitors will require business and government to work together in a positive, constructive partnership to enact policies that will reduce these non-value-added costs to a minimum. To accomplish this goal, the United States will have to undertake major restructuring of its financial, legal, and medical systems.

Family-Related Factors

Human resources are a critical part of the competitiveness equation. Just as one of the most important factors in fielding a competitive athletic team is having the best possible players, one of the most important factors in fielding a competitive company is having the best possible employees. Consequently, the quality of the labor pool is important. The more knowledgeable, skilled, motivated, and able to learn members of the labor pool are, the better.

Well-educated, well-trained, motivated members of the labor pool quickly become productive employees when given jobs. Although providing ongoing training for employees is important in the age of global competitiveness, the type of training provided is important. Organizations that can offer training that has immediate and direct application spend less than those that have to begin by providing basic education for functionally illiterate employees. Since the 1970s, U.S. businesses have had to devote increasing amounts of money to basic education efforts, whereas foreign competitors have been able to provide advanced training that very quickly translates into better quality and productivity.

Many factors account for this difference. Some of these can be traced directly to the family. According to Marshall and Tucker, "families, the most important human resource development agencies in any country, are in considerable trouble in the United States."[10] (Discussion Assignment 2–3 describes the state of the family unit in the

United States.) If the family unit, regardless of how it is constituted, is the nation's most important human resource development agency, the labor pool from which U.S. companies must draw their employees cannot match that in competing countries. Although factors such as those set forth in Discussion Assignment 2–3 do not necessarily preclude success, they certainly mitigate against it.

Single parents who must work full-time have little or no time to help their children excel in school. Children with parents who do not value education are not likely to value it themselves. Marshall and Tucker summarize this situation as follows:

> The children of poverty are far more likely than other children to be denied the caring and support from the adult world that is the indispensable condition of the self-confidence on which human development depends, and we are paying a heavy and growing price for that failure in transfer payments, crime, and lost productivity.[11]

If the family has a strong influence—positive or negative, by design or by default—on the attitudes of children toward learning and work, the United States faces deep-seated problems that must be solved if its companies are going to compete in the global marketplace.

Education-Related Factors

The quality of a country's education system is a major determinant of the quality of its labor pool. The higher the quality of the labor pool, the higher the quality of entry-level employees. The higher the quality of entry-level employees, the faster they can become productive employees and contribute to their employers' competitiveness. Consequently, a high-quality education system is an important component of the competitiveness equation.

Unfortunately, public schools in the United States are not playing as positive a role as they might in creating a high-quality labor pool. Although there are many individual pockets of excellence to be found, as a rule public education in the United States faces a difficult quality dilemma. The problems being experienced by the public schools in the United States are not the result of insufficient funding, as is sometimes assumed.

Figure 2–3 compares annual expenditures per pupil for industrialized countries. Of the top seven, the United States spends the most, whereas Japan, its primary competitor

Figure 2–3
Comparison of per-Pupil Funding in Selected Industrialized Countries
The Management Institute. Global Update (January 2001), 11.

Country	Annual Funding ($)
United States	**3,901**
Canada	3,516
Italy	2,687
West Germany	2,510
France	2,502
United Kingdom	2,498
Japan	1,992

Figure 2–4
Comparison of School Days per
Year in Selected Industrialized
Countries
The Management Institute. Global
Update (January 2001), 12.

Country	School Days
Japan	240
Korea	222
Taiwan	222
Israel	215
Scotland	191
Canada	188
United States	**178**

in the global marketplace, spends the least. Figure 2–4 compares the number of school days required of students annually in the leading industrialized countries. With this criteria, the order is reversed when comparing the United States and Japan.

Figures 2–5, 2–6, 2–7, and 2–8 show the relative ranking for student performance of high school students in selected countries. Figure 2–5 shows that high school students in the United States perform 19th of the 20 countries studied in the area of basic mathematics. In basic science, students in the United States ranked 16th (Figure 2–6). These two figures show the rankings for average students. But what about high-achieving students in the United States compared with their global contemporaries?

Figure 2–5
Mean Achievement Scores by
Country in Mathematics Literacy

Country	Mean Achievement
Netherlands	560
Sweden	552
Denmark	547
Switzerland	540
Iceland	534
Norway	528
France	523
New Zealand	522
Australia	522
Canada	519
Austria	518
Slovenia	512
Germany	495
Hungary	483
Italy	476
Russian Federation	471
Lithuania	469
Czech Republic	466
United States	**461**
Cyprus	446

Figure 2–6
Mean Achievement Scores by
Country in Science Literacy

Country	Mean Achievement
Sweden	559
Netherlands	558
Iceland	549
Norway	544
Canada	532
New Zealand	529
Australia	527
Switzerland	523
Austria	520
Slovenia	517
Denmark	509
Germany	497
Czech Republic	487
France	487
Russian Federation	481
United States	**480**
Italy	475
Hungary	471
Lithuania	461
Cyprus	448

Figures 2–7 and 2–8 show that the mean achievement of gifted students in the United States is low when compared with their contemporaries in the areas of advanced mathematics and physics. What all this means is that businesses in the United States are forced to spend money helping their non-college-trained employees learn the basic skills of reading, writing, and mathematics while their competitors are able to devote their training dollars to developing advanced work-related skills. If global competition can be viewed as a footrace, this is the equivalent of starting U.S. firms 100 yards behind in a 200-yard race.

U.S. Manufacturers and Global Competition

The most important sectors in determining the quality of life in a country are manufacturing and agriculture. The United States has led the world in agricultural production for many years and still does. The United States also led the world in manufacturing productivity for many years. Beginning with the 1960s, however, this lead began to slip. The decline continued and accelerated through the 1980s to the point that the U.S. manufacturing sector entered the 1990s struggling uphill to regain ground. In the mid-1990s, however, the United States began to reemerge as a world-class competitor. No longer is the United States, or any other country, the clear-cut leader in terms of manufacturing productivity. With the dawning of the new millennium, Japan, the United States, Germany, Canada, Italy, France, and Great Britain became closely competitive.

Figure 2–7
Mean Achievement Scores by
Country in Advanced Mathematics

Country	Mean Achievement
France	557
Russian Federation	542
Switzerland	533
Australia	525
Denmark	522
Cyprus	518
Lithuania	516
Greece	513
Sweden	512
Canada	509
Slovenia	475
Italy	474
Czech Republic	469
Germany	465
United States	**442**

Figure 2–9 compares the productivity of automobile manufacturers in Japan, the United States, and Europe. In this chapter the term *productivity* is used several times. In this context the term should be viewed as "total factor productivity" (ratio of outputs to inputs from labor, capital, materials, and energy). The graph compares the average hours required to produce one automobile. Japanese plants located in Japan are

Figure 2–8
Mean Achievement Scores by
Country in Advanced Physics

Country	Mean Achievement
Norway	581
Sweden	573
Russian Federation	545
Denmark	534
Slovenia	523
Germany	522
Australia	518
Cyprus	494
Latvia	488
Switzerland	488
Greece	486
Canada	485
France	466
Czech Republic	451
Austria	435
United States	**423**

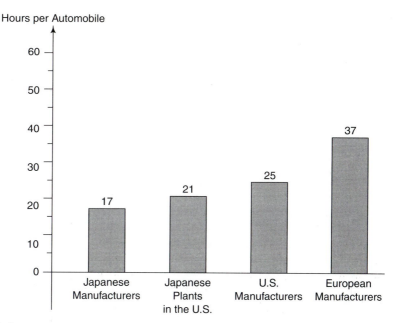

Figure 2–9
Comparative Productivity of Automobile Manufacturers
Source: Congress of the United States, Office of Technology Assessment.

able to produce an automobile in an average of 17 hours. European manufacturers require more than twice that much time. U.S. manufacturers such as General Motors, Ford, and Chrysler require an average of 25 hours per automobile. Japanese manufacturers with assembly plants in the United States using U.S. workers, such as Mazda, average 21 hours per automobile. Because hourly wages in Europe tend to be higher than those in Japan and the U.S., European firms operate at a double competitive disadvantage. U.S. and European firms are nibbling away at these productivity differences to the point that the gap between the best and worst producers is slowly but steadily closing.

Another area in which Japanese firms have gained a competitive advantage is product development. The *product development cycle*—the time it takes to turn an idea into a finished product—is typically shorter in Japan than in the United States and Europe. This allows Japanese firms to get new products to the market faster. Japanese automobile manufacturers take an average of 3 years to complete the product development cycle compared with more than 5 years for their competitors in the United States and Europe.

Another basis for comparison among automobile manufacturers is quality. Productivity gained at the expense of quality yields no competitive advantage. Figure 2–10 compares the major automobile-producing nations in terms of the average number of defects per 100 vehicles manufactured. The quality comparisons follow the same trends found in the earlier productivity comparisons. Japanese manufacturers average the

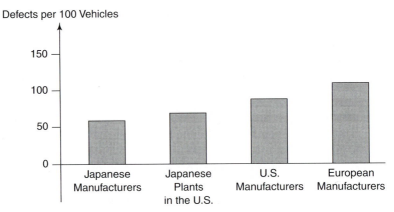

Figure 2–10
Comparative Defect Frequency among Automobile Manufacturers
Source: Congress of the United States, Office of Technology Assessment.

fewest defects; European manufacturers average the most. U.S. manufacturers find it difficult to compete in the global marketplace when their productivity and quality are not up to international standards—a situation that must be reversed if the United States is to regain the preeminent position it has historically enjoyed in the world community.

Even a cursory examination of key economic indicators raises concerns. The ability of a country to compete in the manufacturing arena is a direct determinant of its quality of life. Manufacturing created the great American middle class. If the manufacturing sector dwindles because it cannot compete globally, the middle class dwindles correspondingly. Figure 2–11 contains a number of facts that indicate what has happened to the U.S. economy during the years since World War II. These are the years in which U.S. manufacturers have steadily lost ground to foreign competition.

- Today 25% of children born in the United States will live in poverty. Thirty years ago it was approximately 12 percent.
- The top 1% of income earners in the United States accounted for 60% of the total income increase realized in the 1990s.
- In 1998, the top 1% of the wealthiest U.S. citizens had as many assets as 92% of the poorest U.S. citizens.
- The real hourly wage of a worker in the U.S. today is 14% less than it was in 1979.
- Today the United States has the most unequal distribution of wealth of any industrialized country in the world.

Figure 2–11
Selected Economic Indicators
Source: The Management Institute, *Global Update* (May 2001), 7–9.

Do these comparisons mean that U.S. manufacturers cannot compete? The answer is no. U.S. manufacturers were slow to respond to the international quality revolution. However, in the 1980s and into the new millennium, the realization that quality coupled with productivity was the key to winning global competition caused many U.S. firms to begin adopting the approach set forth in this book while simultaneously pushing for change in areas beyond their control (i.e., cost of capital, industrial policy, etc.). As the total quality approach continues to gain acceptance, companies in the United States are closing the competitiveness gap.

COMPARISONS OF INTERNATIONAL COMPETITORS

According to a report published by the World Economic Forum, the United States dropped from being the second most competitive country in the world community in 1950 to the fifth in 1990, with Japan first, Germany second, Switzerland third, and Denmark fourth. As of 2002 these numbers were unchanged. "The assessment is based on a nation's domestic economic strength, internationalization, government infrastructure, finance management, science and technology, and workforce."[12] Other competitive nations are Canada, France, Great Britain, and Italy. By taking the top two competitive countries separately and the others as a group, one can make useful comparisons with the United States. Four critical indicators of a country's competitive status, shown in Figure 2–12, are used in the comparisons made here: standard of living, manufacturing productivity, investment, and trade.

Standard of Living

A country's *standard of living index* is its gross national product per capita. Although the standard of living index in the United States has improved since 1972, it has not kept pace with those of Japan, Germany, and the other competitive nations of the world (see Figure 2–13). This trend is expected to continue unless the United States is able to

QUALITY TIP

Reason for Shorter Product Development Cycle in Japan

"A key difference is the Japanese emphasis on simultaneous rather than sequential engineering. The people doing research, development, and design of the new model are in constant communication with the people responsible for manufacture. Other factors are involved too, such as the reliance of the major manufacturers on a trusted group of suppliers to do part of the product development work."

Source: U.S. Congress, Office of Technology Assessment, *Making Things Better: Competing in Manufacturing* (Washington, DC: Office of Technology Assessment, 1990), 4.

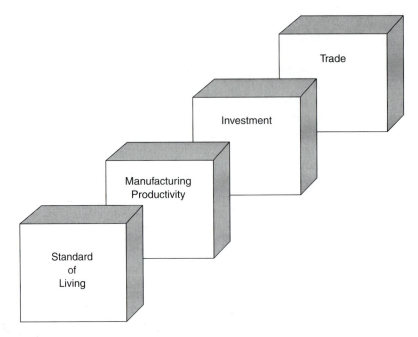

Figure 2–12
Critical Indicators of National Competitive Status

increase its productivity and quality and solve its problems relating to the general cost of doing business (i.e., workers' compensation, litigation, capital, etc.).

Trade: Export Growth

The United States has shown significant export growth since 1985, and the trend continues, but it has not yet erased the huge trade deficit experienced by the United States beginning in the early 1980s.

Even the trend toward a trade surplus is not expected to return the United States to the trade surplus levels enjoyed prior to 1975. The reason is increasingly intense global competition. In this environment, a triple-digit trade surplus measured in billions may no longer be realistic, but a positive trade balance certainly is.

Investment

Investment, from the perspective of global competitiveness, is the percentage of gross national product spent on education, equipment, facilities, and research and development. In the period of 1980–1998, Japan led the list of competitive countries, with an investment level of approximately 29%. The United States invested just over 22% during this period. The other countries, taken as a group, invested slightly more than the United States.

QUALITY TIP

The Importance of Excellence in Manufacturing

"One way or another, however, the United States must regain excellence in the manufacturing process. That is key to raising income for the nation. No longer can U.S. industries count on profiting from new inventions for years before competitors begin to produce them. Many technical inventions cannot be protected from skillful imitators—and the world is now full of manufacturers who can quickly and ably produce things that were invented elsewhere (just as U.S. manufacturers). Over the long run, a country profits from what they can produce competently."

Source: U.S. Congress, Office of Technology Assessment, *Making Things Better: Competing in Manufacturing* (Washington, DC: Office of Technology Assessment, 1990), 6.

When education is removed from the list of investments, the United States invested as much as or more than all other countries except Canada. However, evidence indicates that American dollars invested in education are not used as effectively as those invested by other competing nations. It should be noted also that Japanese investment in equipment and facilities began to decline somewhat in 1992, leveling off in the late 1990s, and is still level in the new millennium.

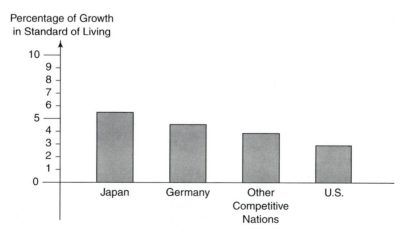

Figure 2–13
Standard of Living Improvement Comparisons
Source: National Council for Occupational Education and the American Association of Community Colleges.

Manufacturing Productivity

During the first three-quarters of the 20th century, the United States enjoyed the highest productivity levels in the world. For example, in 1972 U.S. manufacturing productivity was 56% higher than that of Japan. By 1987, this lead had dwindled to just 6%.[13] By 1993, Japan equaled the United States in productivity. In 2002, the United States and Japan were still approximately even in productivity figures.

Unlike the countries against which it competes, the United States has historically done economic battle in the global marketplace without an industrial policy. The lack of an industrial policy in the United States resulted from a mistrust of government on the part of private sector leaders who feared decisions based on politics rather than sound business judgment and inhibited by a cumbersome government bureaucracy.

However, faced with declining productivity, a widening nondefense research-and-development gap between the United States and foreign competitors and too little investment in equipment, education, and facilities, some business leaders have begun to change their minds. A tentative consensus developed that government can play a role in the United States, in making it easier for private companies to become world-class competitors, as it had in other countries. The controversy surrounding the issue has to do with what role government should play and how much of a role.

Historical Perspective

In spite of opposition to the adoption of an industrial policy, the United States actually has a history of using government incentives to promote commerce, economic development, and business. Land incentives provided by the federal government were key to

QUALITY TIP

Industrial Policy in the United States?

"Should the U.S. have an industrial policy to nurture and promote technology and industry? The answer is 'Yes.' If the words 'industrial policy' set off alarm bells, call it a technology policy, a competitiveness program, or a growth agenda. Whatever the label, the U.S. needs an economic vision geared toward the global economy. . . . Competitive advantage no longer belongs to the biggest or those blessed with abundant natural resources or the most capital. In the global economy, Knowledge is King. And those nations that excel at creating new knowledge and transforming it into new technologies and products will prosper in years to come."

Source: "Industrial Policy," *Business Week*, 6 April 1992, 70.

the development of the transcontinental railroad. The transcontinental railroad, coupled with government land grants, contributed in a major way to the concept of Manifest Destiny. The federal government financed the establishment of the nation's land grant universities and is still a major funding provider for private sector research. Many business leaders have come to the conclusion that the federal government can play a positive role in promoting business without becoming a burden. Again, however, there are stark differences of opinion when it comes to defining this role.

What Can the Government Do?

Whether an industrial policy helps or hinders depends on how it is written and administered. A good industrial policy accomplishes two complementary goals:

- ■ It provides incentives that encourage business to behave in ways that promote competitiveness.
- ■ It removes barriers that mitigate against competitiveness.

A good industrial policy for the United States will help eliminate those inhibitors of competitiveness that even the total quality approach cannot overcome. These are factors that increase the cost of doing business regardless of how well managed a company is. They include inordinately high costs of litigation, capital, health care, and other non-value-added costs. Figure 2–14 shows the minimum components that should be included in an industrial policy for the United States. These components are explained in the following sections.

Investment in Research and Development

Invest in research and development in key frontier technologies such as microelectronics and biotechnology.

QUALITY TIP

Industrial Policy Can Promote Individual Achievement

"A knowledge-based growth policy doesn't call on government to pick winning and losing industries. Nor does it create ponderous bureaucracies or shelter stumbling companies from tough foreign rivals. And while the government may sow a lot of seed corn, it will be up to the private sector to risk its own money to develop commercially valuable ideas. It's the market that picks the winners and losers, not government. This kind of policy builds on the cornerstone of American economic growth: individual achievement."

Source: "Industrial Policy," *Business Week*, 6 April 1992, 70.

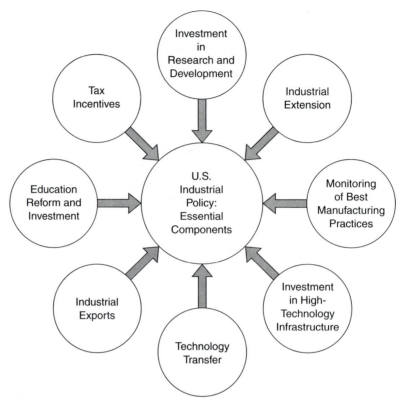

Figure 2–14
U.S. Industrial Technology: Essential Components

Industrial Extension

Invest in industrial extension efforts that provide technical assistance to small and midsized manufacturers. Provide commercial manufacturers with the assistance they need to become world-class companies that can compete in the global marketplace. Companies that have historically relied on the Department of Defense (DOD) for the bulk of their work should be provided the assistance they need to diversify into commercial products and compete successfully in the commercial marketplace on a global basis.

Monitoring of Best Manufacturing Practices

Every day in the most competitive companies worldwide, manufacturing professionals are developing new practices and improving existing practices in attempts to enhance their competitiveness. These best manufacturing practices should be monitored, information about them should be deposited in a central location, and a streamlined process for disseminating this information or otherwise making it available should be developed.

Investment in High-Technology Infrastructure

Infrastructure is typically viewed as consisting of roads, bridges, and other physical structures that help move people and goods. In the modern world, infrastructure also includes technology and systems for moving information, such as telecommunication and computer networks. A U.S. industrial policy should provide incentives that encourage investment in information-moving infrastructure.

Technology Transfer

Technology is the physical manifestation of knowledge. Every day across the United States, scientists work in private and public research laboratories to create new knowledge and, as a result, new technologies. The process by which the new technologies become the consumer products of tomorrow is known as *technology transfer.* It has two components: commercialization and diffusion. Figure 2–15 illustrates how the process works. A U.S. industrial policy should provide incentives to public and private laboratories that encourage them to push new technologies out of the laboratories and into the hands of entrepreneurs, and to entrepreneurs that encourage them to pull new technologies out of the laboratories. In addition, a good policy should provide for the identification and elimination of barriers that inhibit technology transfer.

Industrial Exports

As other countries have increased their exports of such key goods as automobiles and consumer electronics products, the United States lost market share in industrial exports. An industrial policy that simplifies export financing and increases the amount of attention focused on helping small and midsized manufacturers become exporters is needed.

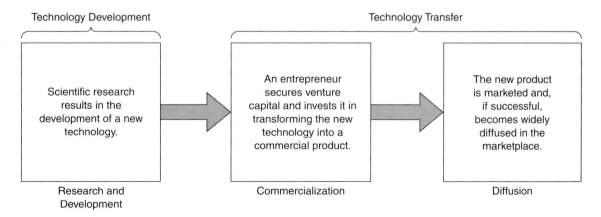

Figure 2–15
Relationship of R&D to Technology Transfer

Education Reform and Investment

Part of an industrial policy should be aimed at providing direction and incentives to stimulate education reform and investment, particularly in the areas of math, science, and engineering.

Tax Incentives

The tax incentive is one of the most powerful tools government can use to stimulate activity. All industrial policy should put tax incentives in place that stimulate the activity needed in all of the other components of the policy with emphasis on promoting long-range planning and development.

Criticism of an Industrial Policy

The need for an industrial policy is not universally accepted. According to *Business Week:*

> The critics will chorus that government shouldn't interfere with the marketplace. An industrial policy, they argue, puts government in the position of picking winners and losers. An industrial policy costs billions of dollars. An industrial policy smacks of central planning, the critics charge, citing the catastrophic failure of the Soviet command economy.[14]

In fact, a poorly conceived and poorly deployed industrial policy would do exactly what the critics suggest. However, a well-conceived and properly deployed policy can have a positive effect on enhancing the relative competitive status of the United States in the international community.

For businesspeople to claim that government should stay out of business is like a football coach saying the team's management should stay out of his way. Management shouldn't interfere with the coach's game plan, starting roster, play selection, and so on, but management should involve itself in ways that make it easier for the coach to get, keep, and properly train the best players and assistant coaches. In this analogy, a football team's management plays the role government can and should play regarding industrial policy. It gives its players (individual businesses) the support they need to do the best job possible.

QUALITY TIP

Math and Science Education

"The government should address the glaring weakness in science and math in primary and secondary schools. It should also subsidize the education of more engineers and scientists. As companies and government hike research, development, and investment spending, the demand for these graduates will soar."

Source: "Industrial Policy," *Business Week,* 6 April 1992, 74.

Can an industrial policy be crafted that will help rather than hurt business? Are counterproductive bureaucratic entanglements inevitable by-products of government initiatives? These are questions that should be debated openly and objectively before deciding whether the United States should adopt an industrial policy. Readers must draw their own conclusions.

TECHNOLOGY AND COMPETITIVENESS

Technology is the physical manifestation of knowledge. Therefore, in a competitive environment where knowledge is king, technology that is well designed to extend human capabilities can enhance an organization's competitiveness. In the early 2000s, the United States lags behind Japan and Germany in the use of such technologies. This is true across the board with both large and small manufacturers, and it is especially substantial for small manufacturers. Not only does a technology gap exist between the United States and its primary foreign competitors, but there is also a technology gap between large and small manufacturers within the United States.

For a country to succeed in the global marketplace, it must export manufactured goods. To be an exporter, a manufacturer must be able to outperform foreign competitors in terms of both quality and productivity. When U.S. manufacturers, particularly small manufacturers (those with fewer than 500 employees), lag behind in the use of key competitiveness-enhancing manufacturing technologies, they are not able to export their products. Figure 2–16 summarizes these technologies.

Because more than 80% of all manufacturers in the United States are small by definition, it is critical that small manufacturers become exporters. For this to happen, small manufacturers must begin using technologies that are essential to competitiveness (see Discussion Assignment 2–5).

Figure 2–17 shows the relative use of several competitiveness-enhancing technologies by small manufacturers in the United States as compared with those in Japan. The technologies used as the basis of comparison are those that have proven most effective in enhancing the competitiveness of small manufacturers. This figure shows that small manufacturers in the United States have lagged behind their counterparts in Japan in the adoption of key technologies. Not shown in this figure are automated warehouse equipment and assembly robots. The ratio for the former is 24.1 to 1.0 in Japan's favor. The ratio for the latter is 10.6 to 1.0 in Japan's favor.

The previous section of this chapter stressed the importance of an industrial policy that provides technical assistance to small manufacturers. In other countries, such assistance is part of an organized system of industrial extension similar to the agricultural extension system in the United States. Although the United States leads the world in agricultural extension services, it lags behind in providing industrial extension services (see Discussion Assignment 2–6). However, this is beginning to change.

Trend in Industrial Extension Services

The state of industrial extension in the United States has begun to change for the better. A noticeable trend exists in the development of public, private, and public/private

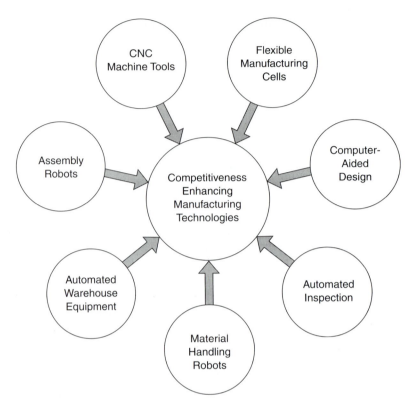

Figure 2–16
Competitiveness-Enhancing Manufacturing Technologies

industrial extension initiatives. Economic development is the principal driver of these initiatives.

In the last decade of the 20th century, industrialized states began to suffer as foreign competition eroded the manufacturing base, driving up unemployment costs while simultaneously shrinking the tax base. The problem was compounded by Department of Defense downsizing that further eroded the manufacturing base.

The federal government and several states responded by appropriating funds to help small and medium-sized manufacturers overcome hurdles that were preventing them from adopting competitiveness-enhancing technologies. An example of a federal government initiative is the establishment of manufacturing technology centers (MTCs). Dr. Stuart Rosenfeld, in his landmark book *Smart Firms in Small Towns,* describes this initiative as follows:

> The Omnibus Trade and Competitiveness Act, for instance, authorizes the National Institute of Standards and Technology to fund regional manufacturing technology centers (MTCs). The MTCs are aimed at helping SMEs [small and medium-sized enterprises] overcome their reluctance to take advantage of advanced technologies developed at NIST in

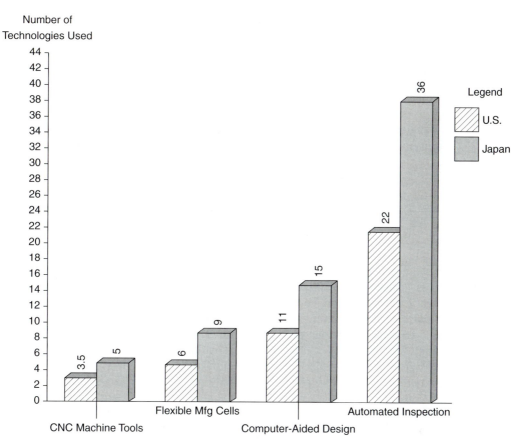

Number of
Technologies Used

Figure 2–17
Technology Adoption Comparisons: Small Manufacturers in Japan and the United States
Source: The Management Institute, *Global Update* (May 2001), 8.

Gaithersburg, Maryland. In practice, however, MTCs seek to promote technology from a range of federal and commercial sources. Although the MTCs have not been funded at anywhere near the level anticipated in the authorizing legislation, they are stimulating states' interest in modernization and leveraging other resources and NIST's technology extension awards have stimulated new programs in more than a dozen states.[15]

The MTCs work with small and medium-sized manufacturers to help them overcome barriers to the adoption and use of off-the-shelf competitiveness-enhancing technologies. These technologies include but are not limited to CNC machine tools, flexible manufacturing cells, computer-aided design, automated inspection, material-handling robots, automated warehouse equipment, and assembly robots. Services provided by MTCs include pilot testing of new equipment and processes, demonstrations of best manufacturing practices, customized on-site training, and brokering of solutions to company-specific problems.

QUALITY TIP

Defense Conversion Defined

"Redeploying defense resources toward the challenge from our economic competitors. Using the technologies, companies, and people we used to win the cold war. Redirecting some of the research in our laboratories, small businesses, and universities to meet the economic challenge of today."

Source: U.S. Air Force, Eglin Air Force Base, Wright Laboratory, Technology Transfer Training Document.

In addition to the MTCs, the federal government now funds a variety of initiatives under the general heading of defense conversion. The DOD became concerned that defense cutbacks would reduce the military–industrial complex to the point that the United States would not be able to gear up quickly should the need arise in the future. The point of defense conversion is to maintain a strong research, development, and manufacturing sector that can convert quickly from commercial products to defense products should military need emerge. To maintain a strong research, development, and manufacturing sector that can convert quickly, manufacturers that have historically depended on defense contracts—particularly small and medium-sized manufacturers— must convert to the production of commercial products that are closely related to military products.

To this end, the federal government sponsors a variety of initiatives to promote defense conversion, with special emphasis on the technologies shown in Figure 2–18. These are dual-use technologies or critical technologies that are in high demand in both the defense and commercial sectors. These initiatives take the form of public/private partnerships that promote technology transfer, including the following prominent initiatives.

Defense Dual-Use Critical Technology Partnerships

The objective of defense dual-use critical technology partnerships is to provide for the establishment of cooperative arrangements, also referred to as partnerships, between the DOD and other entities to encourage and provide for research, development, and application of dual-use technologies. Typical projects that may be included in these partnerships are as follows:

- Digital communications and processing methods
- Optical electronics
- Lightweight, low-clearance multipassenger ground vehicles
- Advanced materials, including precision forging technologies to meet high-strength, low-weight design criteria

- Interferometric synthetic aperture radar technology
- Electrical propulsion of ground vehicles for reduced signature emissions
- Marine biotechnology
- Environmentally compliant manufacturing technologies for the production of computers for military and nonmilitary use as may be identified by a partnership
- Fuel cell and high-density energy storage
- Unexploded ordnance disposal technology and microchip module integration
- Robotics application to defense environmental restoration activities
- Integrated telecommunications technologies for advanced manufacturing
- Advanced automatic control systems technology

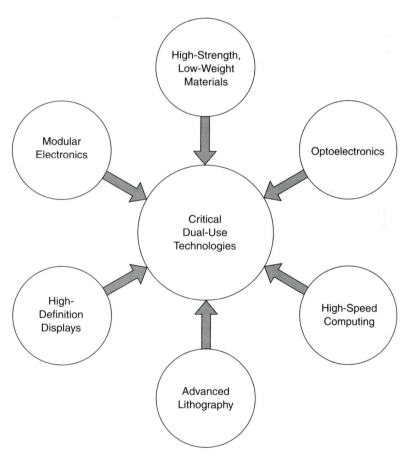

Figure 2–18
Critical Dual-Use Technologies

Commercial–Military Integration Partnerships

The objective of commercial–military integration partnerships is to provide for the development of viable commercial technologies that can also meet future reconstitution requirements and other needs of the DOD.

Regional Technology Alliances Assistance Program

The objective of the regional technology alliances assistance program is to facilitate the use of one or more defense-critical technologies for defense and commercial purposes by an industry in the region served by that center to maintain domestic industrial capabilities that are vital to the national security of the United States. This program was formerly called the critical technology application centers assistance program.

Defense Advanced Manufacturing Technology Partnerships

The objective of defense advanced manufacturing technology partnerships is to encourage and provide for research and development of advanced manufacturing technologies with the potential of having military and dual-use applications. The statute does not specify what constitutes advanced manufacturing technologies. However, to the extent practicable, partnerships will be directed to efforts in manufacturing technologies that would significantly reduce the health, safety, and environmental hazards of existing manufacturing processes.

Manufacturing Extension Programs

The objective of manufacturing extension programs is to accomplish the following:

- Increase the involvement of appropriate segments of the private sector in activities that improve the manufacturing quality, productivity, and performance of United States–based small manufacturing firms
- Promote the development of a broad range of such programs that will benefit both the national security and the economic prosperity of the United States
- Operate through existing manufacturing extension programs

Defense Dual-Use Assistance Extension Program

The objective of the defense dual-use assistance extension program is to further defense investment, diversification, and conversion. This program will assist businesses economically dependent on DOD expenditures to acquire dual-use capabilities.

In addition to initiatives sponsored by the federal government, there are also a number of regional, state, and local initiatives. One of the best examples of an innovative regional approach to industrial extension is the Consortium for Manufacturing Competitiveness (CMC) (see Discussion Assignment 2–7). The CMC includes 14 states in the United States with *ex officio* members from two European countries (see Figure 2–19). It consists of colleges and universities that award two-year degrees and are actively involved in such industrial extension activities as technology transfer, modernization-

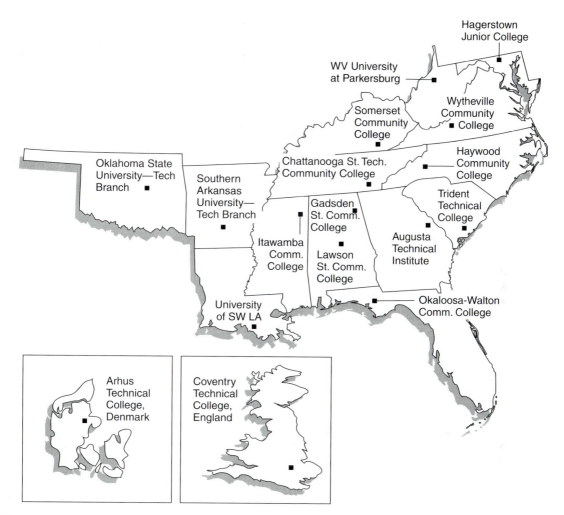

Figure 2–19
Consortium for Manufacturing Competitiveness

related training, customized technical assistance, and demonstrations of best manufacturing practices.

Individual states are becoming active in the promotion of industrial extension services as a way to save and restore their respective manufacturing bases. Discussion Assignment 2–8 and 2–9 describe industrial extension efforts in Pennsylvania and Maine. Similar efforts are under way in Florida, Kansas, North Carolina, Arkansas, Oklahoma, Texas, and several other states. In fact, the trend toward providing industrial extension as a way to reclaim a competitive edge in the global marketplace is being led by individual states.

Local communities are also getting involved, particularly those that have historically depended in large part on the Department of Defense. One of the best-known local efforts is the Technology Coast Manufacturing and Engineering Network (TeCMEN). This is a group of small manufacturing firms that came together to gain economy of scale through the following types of activities (see Figure 2–20):

- Sharing marketing costs
- Sharing costs of education and training
- Jointly undertaking manufacturing contracts that are larger than any individual member company could undertake on its own

HUMAN RESOURCES AND COMPETITIVENESS

The point is made continually throughout this book that the most valuable resources for enhancing competitiveness are human resources. The truth of this point becomes apparent if one studies the approach taken by Germany and Japan to rebuild from the rubble of World War II. Both countries were devastated: "Neither country has much in the way of natural resources. All that they have is their people and the means they have chosen to develop and organize their people for economic effort."[16]

Faced with only one real resource, Germany and Japan were forced to adopt an approach that used this resource to the greatest possible advantage. Discussing the system adopted by Germany, Marshall and Tucker say:

The whole Germany system, then, is self-reinforcing at all the vital points. Its irresistible tendency is to push the decisions of private employers constantly toward business strategies that emphasize competition on productivity and quality rather than price, that enable them to pay even higher wages and improve their competitiveness.[17]

The German and Japanese systems are not perfect, nor are they infallible. They are examples of approaches that work as well as any two systems can in a continually changing and unsure global marketplace. Marshall and Tucker describe these systems as follows:

All these factors combine to produce an economy that is booming in good times, recovers quickly in bad times, and provides a high level of income to a society in which income is far more evenly distributed than in the United States. The unmistakable foundation of success is careful attention to human resources, paralleled by measures that induce employers to make the most of those resources through worker participation in management and high-performance work organizations that both demand high worker skills and use them to good effect.[18]

Business, government, and labor leaders in the United States could learn a great deal from Germany and Japan. People often respond to suggestions that such study might be helpful by claiming that the culture of the United States is so different that what works in these countries won't work in the United States. Such thinking misses

---

**The Technology Coast Manufacturing & Engineering Network
(TeCMEN)**
1170 Martin Luther King, Jr. Boulevard
Fort Walton Beach, Florida 32547
904/243-5812

The Technology Coast Manufacturing & Engineering Network is a group of high-technology companies located in Okaloosa County, Florida. TeCMEN's goal is to identify and provide opportunities for member firms to improve their individual business by working directly with other local firms on projects of mutual interest.

TeCMEN companies offer a vast and diverse array of expertise, resources, and experiences in such areas as engineering, analysis, testing, manufacturing, facility operation, and support services. In addition, TeCMEN will facilitate the formation of teams of 2 to 30 companies based on the needs, demands, and specifications of the contract. TeCMEN consists of companies that can either become prime or subcontractors depending on the specific contractual opportunity, including companies that can prime as an 8(a), Small Disadvantaged Business (SDB), or big business in a full and open competition.

Education and training services are provided to TeCMEN by Okaloosa-Walton Community College and the Graduate Engineering Research Center of the University of Florida. Services provided range from customized training offered on-site to graduate engineering courses.

Okaloosa-Walton
Community College
904/863-6501

University of Florida
Graduate Research Center
904/882-5614

Aeromech Industries, Inc.	L.W. Looney & Son	Sanders Manufacturing
ARINC	Magna Manufacturing	S T Keltec Florida, Inc.
Computer Science & Applications	Manufacturing Technology, Inc.	Suncoast Scientific, Inc.
Crestview Aerospace Corporation	McCowan Corporations	SVERDRUP Technology, Inc.
Datacom, Inc.	Merriwether Circuit Design, Inc.	TASC
Delta Research	Micro Systems	Transistor Devices
E.J. Mlynarczyk & Co.	Orlando Technologies, Inc.	Tybrin Corporation
Florida Industrial Machinery	QMS Circuits	Ver-Val Enterprises, Inc.
Fort Walton Machining	RMS Technologies	Vitro Services Corporation
ISN	SAIC	

Figure 2–20
Description of the Technology Coast Manufacturing and Engineering Network

the point entirely: few countries could be more different from one another than Japan and Germany, yet the approaches to competitiveness adopted by these countries are strikingly similar. The basic philosophical constructs underlying the human resource aspects of the competitiveness of both Japan and Germany are explained in the following list (see also Figure 2–21).

QUALITY CASE

Ford Motor Company and Global Success

Key decision makers in the home offices of multinational companies must understand that they have no monopoly on strategic and innovative thinking. Quality is not a uniquely American trait. One of the best ways to succeed as a multinational company is to have multinational decision makers in top-level positions. This is a philosophy of doing business that Ford Motor Company has used to good advantage over the years. In fact, Ford prides itself on being not just a multinational company, but a multicultural, multiexperienced company that makes international experience a high priority in nurturing and selecting its executive leaders.

Ford Motor Company is the world's largest manufacturer of trucks and the second largest manufacturer of automobiles. Ford cars and trucks are driven in most developed countries of the world. In fact, approximately one third of Ford's overall market is outside of the United States. In addition to the "Ford" brand, Ford manufactures automobiles under the brands of Lincoln, Mercury, Aston Martin, Jaguar, Land Rover, and Volvo. Ford is even a part owner of the Japanese truck and automobile manufacturer, Mazda.

Many multinational companies fail to perform to their full potential when operating in foreign countries because they don't trust the natives. All key decision makers come from the corporate office in the United States regardless of their country of operation. Alex Trotman, chairman and CEO of Ford Motor Company from 1993 to 1998, (himself a native of Middlesex, England) dealt effectively with this misguided strategy during his tenure at the top. According to Trotman, there is no such thing as an American business leader. There are good business leaders, mediocre business leaders, and bad business leaders; nationality has nothing to do with it. There are excellent business executives of all shades, colors, backgrounds, and citizenship. Correspondingly, just being from a given culture does not necessarily mean an individual will be a good business leader in that culture. Cultural knowledge is important, but so are courage, durability, teamwork, integrity, and commitment. Following these principles has helped make Ford Motor Company a multinational success story.

Source: Thomas J. Neff and James M. Citrin, *LESSONS FROM THE TOP* (New York: Doubleday, 2001), 299–305.

■ *Cooperation among business, labor, and government.* According to Marshall and Tucker:

The term *social partners* is used in Japan just as it is in Germany. Prior to the war, the prevailing view among top business executives was very like that of the American rugged individualist. As in Germany, however, the crisis provoked by defeat shifted the balance toward

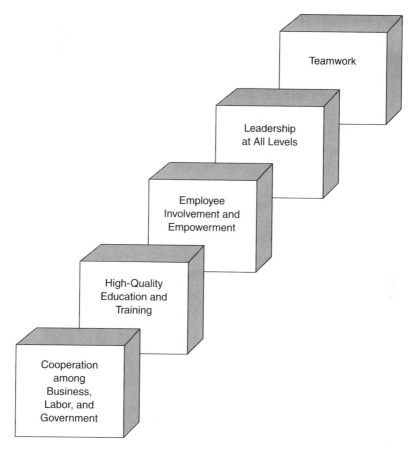

Figure 2–21
Strategies for Human Resource Competitiveness in Japan and Germany

those who believed that business must join with the other major economic actors in the task of rebuilding the society, tying its own goals in with the larger interests of the nation as a whole.[19]

■ *High-quality education and training.* Germany and Japan take different routes, but they arrive at the same place regarding education and training. Germany uses a highly structured apprenticeship program that emphasizes both skills development and academic achievement. Japan relies on excellent primary and secondary education supplemented by industry-based training to prepare front-line employees.

■ *Employee involvement and empowerment.* In both Germany and Japan, employees are involved in functions traditionally viewed as management functions in the United States. These functions include setting working hours, introducing new technologies, establishing compensation levels, human resource planning, work design, and the provision of training.

- *Leadership at all levels.*   In both Germany and Japan, leadership occurs at all levels, and leadership training is provided not just for managers but also for front-line employees. This is important in that it tends to improve the quality of employee involvement in continual improvement efforts.
- *Teamwork.*   In Germany and Japan, not only is work done by teams of employees, but the planning and designing of work, introduction of new technologies, and establishment of compensation levels are also done by teams that involve representatives from labor and management.

The economies of Germany and Japan are not perfect. Establishment of the European Economic Community (EEC) introduced uncertainties into the German economy. Japan's unprecedented annual economic growth rate of 6% to 7% slowed to just 1.5% in 1992 and has stayed there since. One of the reasons for this is that Japanese companies, thinking that the 6% annual growth rate would last forever, made manufacturing infrastructure improvements with money that was borrowed at unrealistically high interest rates. These companies have since found it difficult to service their debt.

Japanese and German companies are as susceptible to unpredictable economic trends, errors in judgment, and unexpected socioeconomic events as U.S. companies. The primary difference is that Japan and Germany adopted approaches that allow them to adapt quickly to changing conditions.

The end of the 20th century will be remembered as a time of turmoil and change for U.S. manufacturers, but in the early 2000s, a new trend has emerged. U.S. firms are showing increased interest in the total quality approach, and leaders from government and business have begun serious discussions about how they can form partnerships that will help U.S. manufacturers compete and win in the global marketplace. Movement in this direction is sporadic and disjointed. It may not reach the point of critical mass until well into the new century, but the trend in the United States is toward a new approach that works better in the age of global competition.

QUALITY TIP

Codetermination in Germany and Japan

"Because workers under codetermination have a strong voice in the management of the firm, they are far more willing than American workers to take the long view, foregoing short-term benefits for the long range. That makes it possible for management, in turn, to concentrate on the long term too. Both management and labor are in this way significantly isolated from some of the short-term pressures that operate on American employees."

Source: Ray Marshall and Marc Tucker, *Thinking for a Living* (New York: Basic Books, 1992), 49.

CHARACTERISTICS OF WORLD-CLASS ORGANIZATIONS[20]

It is often said that only "world-class" organizations can compete in the global marketplace. But what is a world-class organization? In an attempt to answer this question, the American Management Association (AMA) conducted a global survey of more than 2,000 managers in 36 different countries. The AMA received 1,797 usable responses from 36 countries. According to this survey, the following are the top 15 areas in which organizations are concerned about doing well as they attempt to compete in the global marketplace:

1. Customer service
2. Quality control and assurance
3. Research and development/new product development
4. Acquiring new technologies
5. Innovation
6. Team-based approach (adopting and using effectively)
7. Best practices (study and use of)
8. Manpower planning
9. Environmentally sound practices
10. Business partnerships and alliances
11. Reengineering of processes
12. Mergers and acquisitions
13. Outsourcing and contracting
14. Reliance on consulting services
15. Political lobbying

Of the 15 areas listed in the survey, several are directly associated with the larger issue of quality. Customer service, quality control and assurance, innovation, team-based approach to work, partnerships and alliances, and reengineering of processes are all topics that figure prominently in any discussion of total quality.

In addition to these issues, the AMA survey also found that respondents were concerned about several human resources topics. The ten most important of these are:

1. Worker productivity (improvement)
2. Employee training and development
3. Open communication between management and employees
4. Employee benefits and perquisites
5. Codes of workplace conduct
6. Conflict resolution
7. Employee satisfaction
8. Flextime arrangements
9. Management–employee–union relations
10. Child care

Once again the AMA survey identified numerous quality-related concerns and functions that organizations must do well if they hope to compete globally. Worker productivity, employee training and development, codes of workplace conduct, conflict resolution, employee satisfaction, and management–employee–union relations are all total quality–related topics that are addressed at various points in this text.

World-Class Manufacturing: What It Takes[21]

Organizations in business sectors ranging from banking to commercial transportation attempt to compete on a global scale. The most prominent of these come from the manufacturing sector. In an attempt to determine what it takes for manufacturers to compete globally, the American Management Association (AMA) conducted a survey to identify what it calls "ultimate manufacturers." According to the AMA, ultimate manufacturers turned out to be those that lead the way in competitive analysis, performance measurement, involvement of customers in production and the supply chain, customization, doing business via the Internet, and compensation systems. Out of 766 respondents, the AMA was able to identify only 13 that qualify as ultimate manufacturers.

Competitive Analysis Strategies

In the area of competitive analysis, ultimate manufacturers use the following methods to compare themselves with the competition for the purpose of improving their own performance: cost efficiencies in operations, speed to market, research and development supremacy, rapid delivery from suppliers, first-class delivery logistics, zero defects, real-time order management, seamless integration with sales and marketing, close to zero inventory, and networked or collaborative operations. By applying these criteria to themselves and their competitors, ultimate manufacturers determine where their performance is and where it needs to be in order to compete globally.

Production and Supply Chain

In the area of production and supply chain strategies, ultimate manufacturers use the following methods to stay ahead of the competition: collaborative planning, forecasting, and replenishment; collaborative manufacturing and product design; direct delivery of materials to point of use; supplier-managed inventory; and use of channel-assembly distributors. Other manufacturers also use these strategies to varying degrees. Ultimate manufacturers stay ahead of the competition by using them extensively.

Customization Strategies

In the area of customization strategies, ultimate manufacturers use the following methods: building to order, mass production that is configured for individual customers, configuring to order (linking sales operations to production schedules), one-to-one customization for customers in real time, and global sourcing and manufacturing. As with the other strategies, it is not just the fact that ultimate manufacturers use these customization methods that makes them world class; it is the extent to which they use them.

Electronic Commerce Strategies

In the area of electronic commerce strategies, ultimate manufacturers use the following methods: supply management, buying, auctioning, Internet ordering, status and availability tracking by Internet, placing orders by Internet, and accepting Internet orders from customers. Ultimate manufacturers use electronic commerce strategies almost twice as often as their competitors. In addition, these world-class organizations are on track to increase their use of electronic commerce over the next five years at a rate well beyond the projected rates of competitors.

Compensation Systems

In the area of compensation systems, ultimate manufacturers use the following methods as benchmarks for rewarding and recognizing managers and employees: product profitability, inventory levels, manufactured/delivered costs per unit, worker productivity, level of customer satisfaction, manufacturing cycle time, cost efficiencies in operations, employee retention rates, speed of response to market demands, percent of revenues from new products, total delivered cost per unit, zero defects, percent of costs saved from strategic outsourcing, integration of functions across the organization, economic value added, and percent of products from strategic alliances. Figure 2–22 contains a brief checklist of minimum performance benchmarks manufacturers must be able to meet in order to compete in the global marketplace.

E-COMMERCE, INFORMATION QUALITY, AND COMPETITIVENESS

The use of electronic commerce as a business strategy clearly has its advantages. This fact was made apparent in the previous discussion on ultimate manufacturers. However, like any business strategy, electronic commerce can also have a downside if not handled properly.

> Clearly, e-commerce will be a major business strategy of many companies ... But many will fail because they don't realize the new demands on their quality systems and won't make the changes necessary to manage in this highly competitive world. In a world where customers can compare prices, quality, and service offerings from several competitors in a matter of minutes, companies are rethinking their entire quality strategy and quality systems. It is not enough to take an order and pass it off to the supplier or distributor; the company must manage the entire process to assure its promises are met. When a business partner fails, the e-commerce company must quickly correct the problem."[22]

A. Blanton Godfrey gives the example of Dell Computers.[23] Dell developed an intelligent Web site that allows customers to design and order their own customized computers, an innovation that accounts for approximately $10 million in business per day. But developing the Web site was the easy part of the process. With the site in place, Dell must now live up to all of the promises it makes to customers via the site or the entire strategy will lose its value. An example of the lengths Dell will go to in order to keep its electronic promises is instructive. A customer used Dell's Web site to design her own computer and placed her order. The computer was delivered by Dell's

Performance Measures	World-Class Benchmarks
Quality Rejects per Million Parts	< 10
Utilized Capacity	90%
Breakdown Losses	1%
On-Schedule Production	Targeted
Design Producibility	100%
Design Meets Cost Target	95%
Engineering Change Process Response Time	1 day
Annual Training Days per Employee	20

Figure 2–22
To compete in the global marketplace, manufacturers must consistently exceed these benchmarks.
Source: International Finance Center, Washington, D.C.

distributor to the wrong address. To correct this problem, Dell immediately constructed another computer and sent it to the customer by overnight express. Dell took responsibility not just for the design and production of the computer, but also for the error of its distributor. This is the type of commitment that is necessary in order to gain the full benefit of electronic commerce.

Information Quality Program[24]

Organizations that hope to gain the benefits of electronic commerce must make a full commitment to information quality. Poor information quality can be costly. For example, the American Medical Association estimates that poor information in the diagnosis and treatment of medical problems accounts for 120,000 unnecessary deaths in the United States every year. Business leaders are not dealing with life-or-death situations, but they are dealing with the life or death of their organizations. Quality information can contribute to a long and prosperous business life, while poor information can contribute to a slow and painful business death. Establishing an effective information quality program involves the following four steps:

Assessment. All key business processes should be systematically reviewed to identify potential information quality problems.
Prioritizing change. Once potential problems have been identified, they should be prioritized. Those that have the greatest potential payoff should be given the highest priority.
Redesign and reeducation. Information systems and processes should be redesigned as necessary and personnel should be trained to properly use the newly redesigned processes.
Reintegration. An ongoing process of smoothing out the boundaries among the various redesigned processes. This never-ending process should involve stream-

lining, simplifying, and integrating processes and behaviors to ensure that the right way becomes the accepted way.

MANAGEMENT BY ACCOUNTING: ANTITHESIS OF TOTAL QUALITY

The Harvard Business School Press released a book by H. Thomas Johnson and Robert S. Kaplan entitled *Relevance Lost: The Rise and Fall of Management Accounting*. The basic premise of this book is that the concept of managerial accounting has contributed to the loss of competitiveness experienced by businesses in the United States since the end of World War II.[25]

Managerial accounting in the years after World War II evolved into the concept of *management by accounting*. In essence, this means managing an organization's financial results instead of managing the people and processes that produce those results. It is sometimes referred to as *managing by the numbers*. This concept is the equivalent of a tennis player focusing exclusively on the scoreboard rather than on his or her game. Johnson describes this situation as follows:

> Managers before World War II seemed to act intuitively as though well-run processes and satisfied customers made their companies competitive and profitable. While they did not eschew the use of accounting systems to track and budget financial returns, they did not seem to succumb to the belief that focusing everyone in the organization on costs and net income was the way to become competitive and profitable. By 1970, however, we see businesses define process in terms of the financial results. They define process—i.e., the things you control to achieve results—in terms of a cost function, a revenue function, or the accounting determinants of return on investment.[26]

While future managers in the United States were learning to focus on financial results, future managers in Japan and Europe were learning to focus on customers, employees, and processes. While management-by-accounting advocates in the United States were treating customers as "objects of persuasion and employees as cogs in the gears of a deterministic machine,"[27] their counterparts in other industrialized countries were "empowering employees to control processes and remove constraints that stood in the way of profitably satisfying customer wants."[28]

Management by accounting puts organizations at a competitive disadvantage in the following ways:

- ■ It creates an analytically detached approach to decision making in which managers study financial printouts instead of gaining the insight that comes from firsthand knowledge of the situation.

- ■ It promotes a focus on short-term cost reduction to the exclusion of long-term improvements to people and processes.

- ■ It makes for narrowly focused managers who view every problem from a finance and accounting perspective when what is needed by organizations that must compete globally is a broadly focused view that integrates many functions, such as marketing, production, finance, and human resources.

The modern organization's ability to compete in the global marketplace will be determined in large measure by its ability and willingness to replace the management-by-accounting approach with the total quality approach. Johnson made this point in a speech given at Portland State University when he said:

> I now believe that accounting abstractions—no matter how free of distortions, no matter how timely—can tell a business nothing about the processes required to be profitable and competitive. To be competitive and profitable, businesses always must understand how to lead people and satisfy customers. TQM focuses attention on the competitive power that resides in building relationships and empowering people to solve problems—specifically, problems that impede profitable satisfaction of customer wants.[29]

The master's of business administration degree, or MBA, is an excellent credential. So are the various other undergraduate and graduate degrees available from colleges and universities in the United States. It is the concept of focusing excessively on the score rather than the game—management by accounting—that is being questioned by quality advocates, not any specific degree. Management by accounting is an approach to management, not an academic credential.

As anyone knows, both the game and the score are important. We advocate a blending of the principles of quality management with the curricula of business, engineering, technology, and management programs. Students pursuing a degree in any of these disciplines should learn the principles of quality management set forth in this book as well as their traditional curriculum content. This will ensure that they know how to continually improve both performance and the score.

KEY GLOBAL TRENDS

The student of total quality and its potential impact on the global competitiveness of an organization should be familiar with key global trends as well as the geopolitical and economic forces behind the trends. According to Murray Weidenbaum, the following geopolitical and economic forces will shape the global trends of the 21st century:[30]

1. No major traditional hot or cold war for the foreseeable future. There will continue to be friction, trouble spots, and increasing acts of terrorism throughout the world, but no nation will be able to effectively challenge the United States militarily in the traditional sense. This stability sets the stage for increased global trade (although such trade will be intermittently affected by the acts of international terrorists).
2. Although there will be serious banking challenges, there will be no major disruptions as occurred throughout the world during the Great Depression of the 1930s. The absence of both a major depression and high levels of inflation bodes well for international business growth.
3. There is an increasing potential for an interruption in the world's oil supply by terrorist acts aimed at the Western world. The Middle East is the "Achilles' heel" of economic growth around the globe.

4. A growing return to pro-union sentiments in the United States could mean another fundamental shift in labor–management relations that could cause an increasing number of jobs to be moved offshore.

Current Global Trends[31]

With the geopolitical and economic forces already discussed as the backdrop, there are three major global trends underway that will affect businesses throughout the world. These trends all serve to promote the increased globalization of business. Murray Weidenbaum makes this point using the shipping label on a box used to ship the product of a major electronics firm in the United States. The label reads:

> "Made in one or more of the following countries: Korea, Hong Kong, Malaysia, Singapore, Taiwan, Mauritius, Thailand, Indonesia, the Philippines. The exact country of origin is unknown."[32]

All of the following trends have the same potential effect: increased levels of globalization in the world of business.

Growing Irrelevance of Distance

The importance of distance in the world economy has been steadily reduced by advances in technology. Access to effective, low-cost telecommunications technologies has given businesses anywhere on the globe instant access to markets. The cost of a 3-minute call from New York to London dropped from $245 in 1920 to $9 in 1950 and less than 80 cents in 2002. During this same period, the cost of air travel dropped from 68 cents per passenger mile to just 12 cents.

Shifts in the Rates of Growth of Countries

Countries that were economic powerhouses at the end of the 20th century may see themselves challenged by faster growing countries in the 21st century. Currently, the fastest growing nations (5 percent growth or more) are found in Asia, Eastern Europe, and Latin America. These countries are experiencing a rise in the number of middle-class citizens and the benefits of entering participating in the modern economy. Moderately growing nations (3 to 5 percent growth) include the United States and Canada. These countries are still the pacesetters in the modern economy and are especially strong in the areas of technology and services. Slower growing nations (3 percent or less) include Western Europe, Africa, and Japan. Japan and the Western European countries are undergoing difficult institutional changes, and African countries face serious economic problems. Japan, however, has managed to maintain its strength in manufacturing and finance.

Rise of Megacities

The world is experiencing a sharp rise in the number of megacities (urban centers with populations of 10 million or more). Megacities alter the economic face of society by changing the location of both employees and customers and by concentrating them in specific locations. Megacities in the Western hemisphere include Rio de Janeiro, New York, Los Angeles, Mexico City, Sao Paulo, and Buenos Aires. The only megacity in

Europe is London. Africa has two megacities—Cairo and Lagos. The largest number of megacities is in Asia. These cities are Seoul, Bombay, Shanghai, Calcutta, Beijing, Tokyo, Osaka, Manila, Delhi, Karachi, Dhaka, and Tianjin.

All three of these trends feed the larger trend of globalization of business. As location continues to become less and less relevant, companies will increasingly do business across distance. As the rates of growth continue to shift, new competitors will make their way into the global marketplace. Megacities will have a major effect on globalization because they represent such heavy concentrations of both employees and customers. Businesses will learn that if they want these employees and they want these customers, they will have to go, at least electronically, to where they are—in the megacities.

U.S. COMPANIES: GLOBAL STRENGTHS AND WEAKNESSES[33]

As business continues the current trend toward globalization, how are companies in the United States faring? A business trying to compete in the global marketplace is like an athlete trying to compete in the Olympics. Nowhere is the competition tougher. Correspondingly, no country in the world gives its businesses such a solid foundation from which to work. According to Weidenbaum, the following six factors account for a country's ability to compete in the international marketplace:[34]

1. An economy that is open to foreign investment and trade.
2. A government that minimizes controls on business, but does a good job of supervising financial institutions.
3. A judicial system that works well and helps reduce corruption.
4. Greater transparency and availability of economic information.
5. High labor mobility, which enhances productivity and thus living standards.
6. Ease of entry by new businesses.

In varying degrees, the United States meets all six of these criteria. Of course, how well these criteria are fulfilled is a matter of debate between and among various interest groups and stakeholders. Nonetheless, when compared with other countries competing in the global marketplace, the United States fares well in all six of these key areas. This being the case, a key advantage of American firms trying to compete in the global marketplace is these six factors working in their favor. Other advantages and disadvantages are summarized in the following sections.

Global Advantages of U.S. Companies

In the global marketplace, the United States is the world leader in the following industries: aerospace, airlines, beverages, chemicals, computer services, electrical products, entertainment, general merchandise, motor vehicles, office equipment, paper products, pharmaceuticals, photographic and scientific equipment, semiconductors, soap and cosmetics, and tobacco. Some of the reasons the United States is able to lead the world in these key industries are the following advantages:

1. Strong entrepreneurial spirit
2. Presence of a "small capitalization" stock market for small and midsized companies

3. Rapidly advancing technologies
4. Comparatively low taxes
5. Low rate of unionization
6. World-class system of higher education (colleges and universities)

The United States leads the world in new business start-ups. This is because the entrepreneurial spirit is an integral part of the American persona. The presence of a small capitalization stock market allows small and midsized companies to start up and expand without having to use all of their own capital or to take out higher interest loans from banks, as is often the case in other countries. The United States leads the world in the development, transfer, diffusion, and use of technology. This helps ensure a continual stream of new products on the one hand and improved productivity on the other. Americans complain constantly about taxes (as they are entitled to do in practicing their rights as free citizens). But when compared with other industrialized nations, the United States has a low tax burden. Tension between labor and management can harm productivity and, in turn, decrease a company's ability to compete in the global marketplace. The amount of tension that exists between labor and management can typically be demonstrated by the level of union activity: the more tension, the more union activity. Compared with other industrialized nations, union activity in the United States is low.

The United States also provides the world's higher education system. The number of top-notch colleges and universities in the United States is so much greater than those in other countries that comparisons are irrelevant. The cost of higher education in America, although viewed as high by U.S. citizens, is inexpensive when compared with that of other industrialized nations. In addition, financial aid is so readily available that almost any person with the necessary academic ability can pursue a college education in the United States.

Global Disadvantages of U.S. Companies

In spite of the many strengths companies in the United States can bring to the global marketplace, and in spite of this country's world-leading position in several key industries, there are still some disadvantages with which companies have to deal. The primary global disadvantages of U.S. companies are:

1. Expanding government regulation
2. A growing "underclass" of have-nots
3. A weak public school system (K–12)
4. A poorly skilled labor force and poor training opportunities
5. An increasing protectionist sentiment (to restrict imports)
6. Growing public alienation with large institutions (public and private)

Regardless of which major political party has controlled Congress over the past 25 years, the general trend has been toward increasing government regulation of business. Regulating business is a difficult balancing act. On the one hand, businesses cannot be allowed to simply pursue profits, disregarding the potential consequences to the environment and other national interests. On the other hand, too much regulation or unnecessary regulation can make it impossible to compete globally. The growing divide between "haves" and "have-nots" in the United States might lead to the establishment and perpetuation of a permanent economic and social underclass. This is precisely what happened in Russia when Czar Nicholas II was overthrown by the Communists in the early 1900s. People who lose hope might very well respond in ways that threaten the peace, stability, and social fabric of the United States. One of the key factors in the establishment of a social and economic underclass is the failure of America's public school system (K–12). Even with the best system of higher education in the world, America cannot overcome the shortcomings of its K–12 system. In fact, if drastic improvements are not made, over time those shortcomings will begin to erode the quality of our higher education system.

The most fundamental problem with the public school system from the perspective of global competition is that most of the jobs in companies that need to compete globally require less than a college education. These jobs must be performed by high school graduates who, if they cannot read, write, speak, listen, think, and calculate better than their counterparts in other countries, will be outperformed. Poorly skilled workers are an outgrowth of the failure of the nation's public school system, in which the overwhelming majority of Americans are educated. Ideally, every high school graduate should be fully prepared to either go to work or go on to college. When this is not the case, as it certainly is not, American companies must try to compete with a less skilled labor force. This is like a baseball coach trying to win with a team of players who cannot pitch, catch, run, or hit.

One of the factors that contributed to the Great Depression of the 1930s was global protectionism. Americans wanted their farmers and their manufacturers to be "protected" from their counterparts in other countries. Protectionism hurts everyone and never really protects anyone. But as other countries have entered U.S. markets (principally Japan, Korea, and China), the jobs of American workers have been threatened. A natural but ill-informed response is to call for protectionist measures and to adopt slogans such as "Buy American." Economists are quick to point out, however, that the only valid reason to "buy American" is that American products are the best made. If they are not, buying them makes little sense and is nothing more than misguided patriotism. The better approach is to ask why the American products are not the best and then to do what is necessary to make them the best.

The final factor that gives U.S. companies a disadvantage is the growing tendency of the public to see big organizations as the "bad guys." This is displayed in many different ways. Disgruntled employees will sometimes pretend injuries and file fraudulent workers' compensation claims. Employees will cheat and steal from their employers. Of course the most common way animosity toward big business is acted out is by employees giving less than their best on the job. Another expression is when the public at large supports antibusiness legislation and unnecessary regulations.

According to Weidenbaum,

> On balance, it is quite accurate to describe the American economy as the world's strongest, albeit with serious weak spots. Some subjective factors help produce an upbeat assessment: People risk their lives to come to the United States, and American English is the Latin of our time, the nearly universal language of business, science, and transportation. In a broader

sense—one that transcends current issues of economics and business—American culture, warts and all, is the pacesetter for a great portion of the world's population.[35]

SUMMARY

1. The relationship between quality and competitiveness can be summarized as follows: In a modern global marketplace, quality is the key to competitiveness.
2. The costs of poor quality include the following: waste, rejects, testing, rework, customer returns, inspection, recalls, excessive overtime, pricing errors, billing errors, excessive turnover, premium freight costs, development cost of the failed product, field service costs, overdue receivables, handling complaints, expediting, system costs, planning delays, late paperwork, lack of follow-up, excess inventory, customer allowances, and unused capacity.
3. The United States came out of World War II as the only major industrialized nation with its manufacturing sector completely intact. Germany and Japan were devastated by damage during the war. They rebuilt their manufacturing bases on the assumption that to compete globally they would have to produce goods of world-class quality. That strategy helped them recover and become world leaders in manufacturing.
4. Several factors can inhibit competitiveness, including those related to business and government, family, and education.
5. When making comparisons among internationally competing countries, the following indicators are used: standard of living, trade/export growth, investment, and manufacturing productivity.
6. Historically, the United States has shunned the concept of an industrial policy. If such a policy is developed, it should contain the following components: investment in research and development, industrial extension, monitoring of best manufacturing practices, investment in high-technology infrastructure, technology transfer, industrial exports, education reform and investment, and tax incentives. Criticism of industrial policy has centered around the fear of government interference in business.
7. Technology is the physical manifestation of knowledge. Technology that is properly designed to extend human capabilities can enhance an organization's competitiveness. Small and medium-sized manufacturers in the United States have been slow to adopt competitiveness-enhancing technologies.
8. Federal government initiatives to promote technology transfer and industrial competitiveness in the United States include the following: defense dual-use critical technology partnerships, commercial-military integration partnerships, regional technology alliances assistance programs, defense advanced manufacturing technology partnerships, manufacturing extension programs, and defense dual-use assistance extension programs.
9. The most important key in maximizing competitiveness is the human resource. Following World War II, this was the only resource that Germany and Japan had to draw on. Consequently, they built economic systems that encourage private

employers to make business decisions that emphasize improved productivity and quality rather than price.

10. The basic philosophical constructs underlying the human resource aspects of the competitiveness of both Japan and Germany are as follows: cooperation among business, labor, and government; high-quality education and training; employee involvement and empowerment; leadership at all levels; and teamwork.

11. Ultimate manufacturers are those that perform at world-class levels in in the following areas: competitive analysis, production and supply chain management, customization, electronic commerce, and compensation systems.

12. Key global trends that are increasing the level of globalization in business are the growing irrelevance of distance, shifts in the rates of growth in certain countries throughout the world, and the rise of megacities.

 KEY TERMS AND CONCEPTS

Best manufacturing practices

Competitive analysis strategies

Competitiveness

Competitiveness-enhancing manufacturing technologies

Customization strategies

Cost of poor quality

Defense dual-use critical technologies

Education-related factors

Electronic commerce strategies

Export growth

Family-related factors

Global integration

High-technology infrastructure

Human resources

Industrial exports

Industrial extension

Industrial policy

Information quality program

Investment

Manufacturing productivity

Megacities

Product development cycle

Quality of life

Research and development

Standard of living index

Tax incentives

Technology

Technology adoption

Technology gap

Technology transfer

 FACTUAL REVIEW QUESTIONS

1. Explain the relationship between quality and competitiveness.
2. Explain how the costs of poor quality can affect competitiveness.
3. Describe the evolution of the rebuilding effort undertaken by Japan and Germany following World War II.

4. Explain the actions of U.S. manufacturers during the same period in which Japan and Germany were rebuilding following World War II.
5. How does a nation's ability to compete affect its quality of life?
6. Describe how business/government factors can inhibit competitiveness.
7. How can family-related factors inhibit competitiveness?
8. Explain how education-related factors can inhibit competitiveness.
9. Compare investment and manufacturing productivity in the United States with investment and manufacturing productivity in Japan.
10. List and explain the essential components of an industrial policy.
11. Describe five different ways government can help enhance competitiveness in the manufacturing sector.
12. Explain the role technology can play in the enhancement of competitiveness.
13. What are defense dual-use critical technologies?
14. List and briefly explain the basic philosophical constructs underlying the human resource aspects of the competitiveness of Japan and Germany.
15. What is an "ultimate manufacturer" and what does it take to be one?
16. Describe three important current trends that are increasing the level of globalization in business.

 CRITICAL THINKING ACTIVITY

Debate over Industrial Policy

Two quality managers meet for lunch every Friday to discuss common problems, compare notes, and make suggestions to each other. Today, their discussion has turned into a debate. The topic is the lack of a comprehensive industrial policy in the United States. "A government policy on competitiveness is the last thing we need," said one quality manager. "It would just be more government intrusion in business, and we already have enough of that." "I don't agree," said the other manager. "In fact, a good industrial policy could actually get rid of some of the bureaucratic red tape that gets in the way of quality and makes it hard to compete." Join this debate. Take one side or the other and complete one of the following activities:

1. Debate the issue in class with your fellow students.
2. Make a list of potential problems typically associated with "big government." How do these issues affect your opinion in this debate?
3. Write a position paper explaining your opinions.

DISCUSSION ASSIGNMENT 2–1

Education and Competitiveness in the United States

"As American business has suffered increasingly from the competitive assault of foreign companies, government has responded with economic policies based in part on theories that were blind to the most serious problems posed by the changing shape of the international economy, and business executives have pursued strategies that have worked in the past, but are counterproductive in the present. Educational institutions, for their part, proud of past successes and largely unaware of the changes in the world's economic structure, and therefore in domestic skill requirements, have continued to pursue policies that are wholly inadequate for the present and future.

"But the countries that have had the most economic success in recent years have pursued a different course. Committed to the broad goals of full employment and high wages, these nations have built policies firmly grounded on the assumption that they would be able to provide high incomes for everyone only if the quality of their human resources was very high and employers were organized to use those highly educated people effectively in the workplace. What is striking is how similar these policies are in countries as dissimilar in other ways as Japan and Germany, Singapore and Sweden. It is vital for us as a nation to understand how our competitors link economic, labor market, and education policy into one integrated strategy, how the practices of firms and education and training institutions are affected by this strategy, and what the outcomes are for their people."

Discussion Assignment
Discuss the following questions in class or with your fellow students outside of class:

1. Does the private sector play any role in the formulation of education policy in the United States?
2. How can the private sector help improve the quality of K–12 education in the United States?

Source: Ray Marshall and Marc Tucker, *Thinking for a Living* (New York: Basic Books, 1992), xvii.

DISCUSSION ASSIGNMENT 2–2

How U.S. Business Incurred Nonproductive Debt

"In the business world, much of the borrowing that went on in the 1980s was related to takeover threats. As firms were threatened by takeovers, they went deep into debt, using the proceeds to buy back their stock, and thereby driving stock prices up to levels that made the companies unattractive to the raiders. Some managers used the same techniques in the form of leveraged buyouts, borrowing enough money to buy back all the stock of the company. Still others bought back their stock to keep share prices and per-share earnings up because their own tenure and income depended on these indicators. Thus enormous debt was accumulated without many new assets being added to America's productive capacity."

Discussion Assignment

Discuss the following questions in class or with your fellow students outside of class:

1. How does debt affect a company's ability to compete? Why?
2. How does the national debt affect the ability of the United States to compete globally? Why?

Source: Ray Marshall and Marc Tucker, *Thinking for a Living* (New York: Basic Books, 1992), xiv.

DISCUSSION ASSIGNMENT 2–3

The Status of the Family in the United States

"The main family problems are due to the fact that too many households have members who work full-time for wages below the poverty level, too many children are born into poor households, too many women have inadequate prenatal care for themselves and their babies, too many fathers do not support their children at all, too many children are born to unwed mothers who are unable to care for them, and too many mothers transmit drug addition to their children. On all of these indicators, the American experience is much worse than that of any other major industrial country.

"Divorce rates have doubled since the 1960s, and the United States has by far the highest divorce rate in the world. About half of all first marriages and 60 percent of all second marriages end in divorce."

Discussion Assignment

Discuss the following questions in class or with your fellow students outside of class:

1. Can a high rate of single-parent households affect a country's ability to compete in the global marketplace? If so, how?
2. Can a high rate of infant drug addiction affect a country's ability to compete in the global marketplace? If so, how?

Source: Ray Marshall and Marc Tucker, *Thinking for a Living* (New York: Basic Books, 1992), 168.

DISCUSSION ASSIGNMENT 2–4

A Comparison of Western and Japanese Business Practices

Dr. Myron Tribus pointed out that an important difference between Japan and the Western world lies in the way a corporation is regarded by its stockholders and its employees. In Japan, the physical assets of a company are not thought to belong exclusively to the stockholders, who are regarded as silent partners of the managers and workers. Profits are allocated, often equally, to the shareholders, to reinvestment, and to management and workers. Japanese stocks seldom pay dividends of more than 3% since Japanese investors are interested in long-term capital growth more than in dividends.

If there is a downturn in Japanese business, the dividend is decreased, then decreased again. Then, management takes a cut in pay and finally, the hourly workers are given a cut. No one is turned out. A business downturn in most U.S. companies would usually precipitate actions in the reverse order: layoffs first and cuts in dividends last.

When top management is obsessed with figures relating to short-term financial gains, the goal communicated to the production workers tends to be to produce as much as possible without consideration of quality of output. But others in the company—quality assurance inspectors, for example—may have goals that conflict with production by the numbers. The result is dissension, disagreement, and an increase in the normal discontent found in employees who work in a "production goal" environment.

Discussion Assignment

Discuss the following questions in class or outside of class with your fellow students:

1. Which approach to doing business (Japanese or U.S.) makes the most sense from the perspective of global competitiveness?
2. Is the Japanese approach possible in the United States? Why or why not?

Source: Nancy R. Mann, *The Keys to Excellence* (Los Angeles: Prestwick, 1989), 153.

DISCUSSION ASSIGNMENT 2–5

Technology Adoption by Small Manufacturers

"The Office of Technology Assessment (OTA) reports that only 11 to 15 percent of all machine tools in the U.S. are automated; the majority of those automated are found in large companies.

"A Census Bureau study found that half of the small manufacturers surveyed don't use any of 17 technologies cited by experts as critical to competitiveness and didn't plan to do so in the next five years. Among firms that used one of the technologies, 60 percent had no plans to adopt others."

Discussion Assignment
Discuss the following questions in class or outside of class with your fellow students:

1. Why don't more small companies in the United States adopt modern technologies?
2. How can modern technologies be justified from a cost perspective in a small company?

Source: National Center for Manufacturing Sciences, "Smaller Players Struggle with Unique Issues, *FOCUS* (November 1992), 8.

DISCUSSION ASSIGNMENT 2–6

Industrial Extension Services in the United States

The federal government spends approximately $80 million on manufacturing extension every year while spending $1.4 billion on agricultural extension. This, in spite of the fact that manufacturing contributes 10 times more than agriculture to the gross national product.

Discussion Assignment
Discuss the following questions in class or with your fellow students outside of class:

1. The United States produces more agricultural crops every year than any other country. Why, then, are so few people employed in farming as compared with manufacturing, construction, and processing?
2. What role(s) might a college or university play in providing manufacturing extension services?

DISCUSSION ASSIGNMENT 2–7

The Consortium for Manufacturing Competitiveness

"Among the educational institutions with the capacity to support modernization, perhaps the most promising and most underutilized are the two-year colleges. Created in many states to support economic development, community and technical colleges are quietly rising to preeminence in technical education. With universities oriented toward cutting-edge research and high schools toward basic skills, the two-year college stands alone with a primary mission of encouraging economic development. As local institutions, they are more likely to be trusted by and accessible to small, rural manufacturers.

"Some of the colleges with strong industrial technology programs and sophisticated technology centers have begun to realize a potential impact that extends farther than education and training, using their experience and expertise with new technologies to influence and support modernization efforts through extension and demonstration. Attesting to their technical expertise, IBM's CIM-Higher Education Consortium is directing more of its equipment to two-year colleges than to four-year colleges. In an attempt to capitalize on this potential and to develop it further, the Southern Technology Council organized a demonstration project, the Consortium for Manufacturing Competitiveness (CMC), comprised of colleges with outstanding industrial technology resources and programs from each of 14 southern states. The original design called for the colleges to work cooperatively to develop innovative approaches and to share their ideas and experiences with other colleges in their state. They were also charged with achieving greater economies of scale by offering services collectively rather than one-on-one. Due to the success of CMC and its companion program, the National Alliance for Manufacturing Productivity (which was started at about the same time as CMC by Autodesk, Inc., a CAD software producer), NIST's manufacturing technology centers have come to rely heavily on the technical colleges to reach and assist SMEs.

"The Consortium for Manufacturing Competitiveness effectively demonstrates the ability of technical colleges to assume new and expanded responsibilities for industrial modernization. But the true test of an effective demonstration is the extent to which it affects practice elsewhere. To be replicated on a large scale, states will have to acknowledge and provide support for this function. To date, most states will only fund colleges on the basis of full-time equivalent enrollments, and the institutions that want to provide technology extension and demonstration for their manufacturing base are forced to find additional revenues through various entrepreneurial schemes. A few states are adopting the CMC model. The Florida legislature appropriated funds to four additional colleges to become manufacturing technology centers, and Oklahoma is redefining the role of selected faculty in its vocational-technical colleges as technology and innovation brokers."

Discussion Assignment

Discuss the following questions in class or with your fellow students outside of class:

1. If you were hired to manage a manufacturing extension center, what type of services would you want to provide?
2. In your opinion, why are there so few for-profit manufacturing extension centers?

Source: Stuart Rosenfeld, *Smart Firms in Small Towns* (Washington, DC: Aspen Institute, 1992), 66–69.

DISCUSSION ASSIGNMENT 2–8

Pennsylvania's Industrial Resource Centers

"[T]he Pennsylvania Department of Commerce established nine industrial resource centers (IRCs) as nonprofit corporations managed and operated by private industry but supported by the commonwealth. . . . Each was given the mission of helping small and medium-sized manufacturers, both individually and collectively, learn about and adopt new production technologies, techniques, and philosophies in ways that respond to total business needs. Each IRC has considerable leeway in how it was organized, administered, and governed, and how it chose to work with firms.

"The Northwest Pennsylvania Industrial Resource Center (NPIRC) reaches from its home port of Erie into 13 primarily rural counties to help modernize the 2,000 manufacturers based there. Given the large territory, NPIRC has opened an outreach office at Clarion University, about 90 miles south of Erie. This office focuses on powdered metals and related industries. The IRC is governed by a board that consists of three representatives of each member of the coalition it formed. Members include two local universities, two economic development agencies, the manufacturers' association, the chamber of commerce and the National Institute of Flexible Manufacturing. . . .

"The most common services, which account for more than half the projects, have been market expansion and technology improvement, manufacturing strategies, and production planning and inventory control. All of the IRCs are evaluated yearly and funded according to availability of funds and evaluation results."

Discussion Assignment
Discuss the following questions in class or with your fellow students outside of class:

1. What kind of help that is not technology oriented might be needed by a small manufacturer?
2. How could the Pennsylvania model be applied to service sector and other nonmanufacturing companies?

Source: Stuart Rosenfeld, *Smart Firms in Small Towns* (Washington, DC: Aspen Institute, 1992), 32–34.

DISCUSSION ASSIGNMENT 2–9

Maine's Center for Technology Transfer

"[T]he Maine Science and Technology Commission established the Center for Technology Transfer to address the needs of small and medium-sized metals and electronics manufacturers throughout the state. This was one of three state technology transfer centers, the other two targeted for biotechnology and aquaculture. . . . CTT's host organization is the University of Maine system, but its advisory board is drawn from both the public and private sectors. Given that the budget supports only two full-time staff, the advisory board and committees are active participants. . . .

"CTT's six services are: 1) demonstrating advanced technologies, 2) stimulating productivity improvement, 3) disseminating technical information, 4) monitoring emerging issues, 5) brokering industry needs and services, and 6) facilitating interfirm networks and strategic alliances. Its market is the entire state, although practically all manufacturing there is south of Bangor. Lacking field staff, CTT reaches out through its publications and faculty at university branches.

"CTT also is working to organize the state's electronics industry through an association that will eventually provide real services, perhaps modeled on the Maine Metal Products Association.

"Maine's Center for Technology Transfer represents an attempt by a very rural and sparsely populated state to focus on a couple of important industries, to utilize available resources, and to be a catalyst for change. Its funding is not sufficient for the task at hand, but its design and procedures are noteworthy."

Discussion Assignment

Discuss the following questions in class or outside of class with your fellow students:

1. The Maine Center has no field staff to go on-site and see the difficulties of small companies firsthand. Do you see this as a problem? Why or why not?
2. The Maine Center focuses on just two industry sectors—metals and electronics. Do you see this as a strength or a weakness? Why or why not?

Source: Stuart Rosenfeld, *Smart Firms in Small Towns* (Washington, DC: Aspen Institute, 1992), 32–34.

 ENDNOTES

1. DeFeo, Joseph A. "The Tip of the Iceberg," *Quality Progress*, Vol. 34, No. 5 (May 2001), 29–37.
2. Ibid., 30.

3. Ibid., 30.

4. Ibid., 31.

5. Ibid., 37.

6. Ray Marshall and Marc Tucker, *Thinking for a Living* (New York: Basic Books, 1992), xiv.

7. U.S. Bureau of Labor Statistics, Employment and Earnings, Howard Fullerton, Jr., "New Labor Force Projections, Spanning 1988 to 2000," *Monthly Labor Review* 112 (November 1939), 4.

8. Marshall and Tucker, xvi.

9. Peter R. Scholtes, *The Team Handbook* (Madison, WI: Joiner Associates, 1992), 2–7.

10. Marshall and Tucker, 168.

11. Marshall and Tucker, 165.

12. Associated Press, 22 June 1992.

13. "Second Annual Index Shows Competitiveness Gains, Losses," *Challenges* (July 1989), 5–7.

14. "Industrial Policy," *Business Week*, 6 April 1992, 104.

15. Stuart Rosenfeld, *Smart Firms in Small Towns* (Washington, DC: Aspen Institute, 1992), 62–65.

16. Marshall and Tucker, 43.

17. Marshall and Tucker, 48.

18. Marshall and Tucker, 49.

19. Marshall and Tucker, 49.

20. American Management Association. "AMA Global Survey on Key Business Issues," *Management Review* (December 1998), 27–38.

21. American Management Association. "Survey on Manufacturing," *Management Review* (September 1999), 18–21.

22. Godfrey, A. Blanton. "E-QUALITY: Success in the World of E-Commerce Requires a Commitment to Quality," *Quality Digest*, Vol. 19, No. 6 (June 1999), 22.

23. *Ibid.*

24. Albrecht, Karl. "Information: The Next Quality Revolution?," *Quality Digest*, Vol. 19, No. 6 (June 1999), 32

25. H. Thomas Johnson, "Managerial Accounting: The Achilles Heel—A Synopsis of Relevance Lost after Five Years," speech presented on 28 July 1992 to the Corporate Associates of Portland State University's School of Business Administration.

26. Johnson, 5.

27. Johnson, 6.

28. Johnson, 6.

29. Johnson, 12.

30. Weidenbaum, Murray. "All The World's a Stage," *Management Review* (October 1999), 44.

31. Ibid., 44–45

32. Ibid., 45

33. Ibid., 45–48

34. Ibid., 45–46

35. Ibid., 48.

CHAPTER THREE

Strategic Management: Planning and Execution

A mission statement is "a long, awkward sentence that demonstrates management's inability to think clearly."

Scott Adams (Author of *Dilbert*)

MAJOR TOPICS

- What Is Strategic Management?
- Components of Strategic Management
- Strategic Planning Overview
- Creative Thinking in Strategic Planning
- Conducting the SWOT Analysis
- Developing the Vision
- Developing the Mission
- Developing the Guiding Principles
- Developing Broad Strategic Objectives
- Developing Specific Tactics (Action Plan)
- Executing the Strategic Plan
- Strategic Planning in Action: A "Real-World" Case
- Revolutionary Thinking in Strategic Planning

To survive and thrive in a globally competitive marketplace, organizations must adopt a broad strategy that gives them a sustainable competitive advantage. All such strategies fall into one or more of the following categories:

- *Cost/leadership strategies.* Strategies in this category seek to improve efficiency and control costs throughout an organization's activity-cost chain (supplier activity costs, in-house activity costs, and distribution activity costs).
- *Differentiation strategies.* Strategies in this category seek to add value, as defined by customers, to the organization's products or services. Such strategies typically

involve gaining technological superiority over competitors, continually outperforming competitors in the area of quality, providing more and better support services to customers, and/or providing customers more value for their money.

■ *Market-niche strategies.*   Strategies in this category focus on a narrowly defined segment of the market (market niche) and attempt to make the organization in question the market leader in that niche. Leadership can be achieved by adopting cost leadership or differentiation strategies or both designed to appeal specifically to the target market.

Total quality relates to strategic management in that it enhances an organization's ability to gain a sustainable competitive advantage in the marketplace. Handled properly, total quality can be the most effective cost leadership and/or differentiation strategy an organization can adopt. This is because the total quality approach is the best way to continually improve efficiency and cut costs throughout an organization's activity-cost chain, while simultaneously continually improving the features of the product or service that differentiates it in the marketplace. Total quality can also improve an organization's chances of becoming a leader in a given market niche.

WHAT IS STRATEGIC MANAGEMENT?

To understand strategic management, one must first understand the concept of organizational strategy. Strategies are defined as follows:

> Organizational strategies are the approaches adopted by organizations to ensure successful performance in the marketplace. These approaches are typically set forth in a comprehensive document called the *strategic plan.*

Strategic management is management that bases all actions, activities, and decisions on what is most likely—within an ethical framework—to ensure successful performance in the marketplace. From the strategic manager's perspective, resources are wasted unless they contribute to success in the marketplace, and the more direct the contribution, the better.

COMPONENTS OF STRATEGIC MANAGEMENT

Strategic management consists of two interrelated activities: (a) strategic planning and (b) strategic execution. These two primary components of strategic management are described in the following sections.

Strategic Planning

Strategic planning is the process by which an organization answers such questions as the following: Who are we? Where are we going? How will we get there? What do we hope to accomplish? What are our strengths and weaknesses? What are the opportunities and threats in our business environment? Strategic planning involves developing a written plan that has the following components: an organizational vision; an organizational mission; guiding principles; broad strategic objectives; and specific tactics, projects, or activities for achieving the broad objectives. Specific tactics, projects, and activities are often referred to as the "action plan."

Strategic Execution

Strategic execution involves implementing strategies set forth in strategic planning, monitoring progress toward their achievement, and adjusting as necessary. Strategic execution is implementation that achieves maximum efficiency and effectiveness.

Monitoring involves constantly checking actual performance against performance benchmarks. Strategic monitoring answers such questions as these: Are we achieving our objectives? This is the *effectiveness* question. Are we performing as well as we need to perform? This is the *efficiency* question. Adjusting as necessary involves making corrections when the specific strategies or tactics adopted are not producing the desired results. Such adjustments can involve a minor tweaking of plans, finding ways to overcome unexpected barriers that are encountered, or even adopting a whole new set of specific strategies.

STRATEGIC PLANNING OVERVIEW

Strategic planning, as described previously is the process whereby organizations develop a vision, a mission, guiding principles, broad objectives, and specific strategies for achieving the broad objectives. Before even beginning the planning process, an organization should conduct a SWOT analysis. SWOT is the acronym for *strengths, weaknesses, opportunities,* and *threats.* A SWOT analysis answers the following questions: What are this organization's strengths? What are this organization's weaknesses? What opportunities exist in this organization's business environment? What threats exist in this organization's business environment?

The steps in the strategic planning process (Figure 3–1) should be completed in order, because each successive step grows out of the preceding one. The SWOT analysis provides a body of knowledge that is needed to undertake strategic planning. The mission grows out of and supports the vision. The guiding principles, which represent the organization's value system, guide the organization's behavior as it pursues its mission. The broad objectives grow out of the mission and translate it into measurable terms. Specific strategies tie directly to the broad objectives. Typically there will be two to five strategies for each objective, but this is a general guideline, not a hard and fast rule.

CREATIVE THINKING IN STRATEGIC PLANNING

"No one saw corporate life as clearly or expressed its values so eloquently as the late William H. Whyte. In his definitive book, *The Organization Man,* Whyte depicted post-Depression American men who repressed their individuality, creativity, and entrepre-

Figure 3–1
The Strategic Planning Process

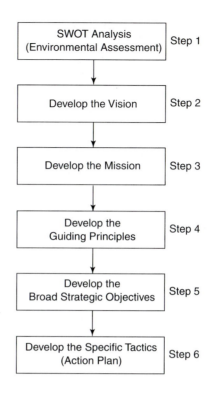

SWOT Analysis (Environmental Assessment)	Step 1
Develop the Vision	Step 2
Develop the Mission	Step 3
Develop the Guiding Principles	Step 4
Develop the Broad Strategic Objectives	Step 5
Develop the Specific Tactics (Action Plan)	Step 6

neurial spirit in exchange for lifetime employment and a sense of belonging. When these men retired from General Motors or IBM after 40 years of service, the standard prize offered by the conformist organization was the gold watch. This type of corporate life no longer exists. The late 20th century free-for-all—every man for himself—has emerged in counterpoint to Whyte's view of the organization man who sold his creative soul for a paycheck."[1]

According to Marlene Piturro, in companies that have achieved world-class status, the "organization man" has been replaced by the "mindshifter."[2] Mindshifters are creative thinkers who look at their organization's business environment from a different perspective, challenge outdated paradigms that no longer work, and prod defenders of the status quo. Mindshifters are mavericks in organizations who constantly seek gaps in the marketplace their companies can fill.

"The mindshifter thrives in an environment where there is room to try new things and explore new avenues—even if some lead to dead ends."[3] Mindshifters are bright people who stay on top in their business and are willing to take calculated risks. They are also smart enough to learn from their mistakes. "Hence, mindshifters break new ground by building companies on ideas so deviant from their industry's norms that the notion of success seems preposterous."[4]

Mindshifters, as mentioned, are creative thinkers. Consequently, they can be invaluable resources in the strategic planning process. For example, when considering the organization's strengths, weaknesses, opportunities, and threats (SWOT analysis), creative thinkers are more likely to see opportunities where others might see only threats. They might also see organizational strengths where others see only weaknesses.

Piturro uses the example of ACCION International to show how creative thinking can add to the strategic planning process.[5] In this case, creative thinkers saw an opportunity where other competing banks saw only a threat. Most banks lend money to people and companies who are so financially solvent they don't really need it. This means that many entrepreneurs, particularly small entrepreneurs who lack a successful investment track record, are unable to secure the capital they need to turn their ideas into businesses. ACCION International made the strategic decision to target financially strapped but promising entrepreneurs as a new niche market for loans. Using the concept of microlending (making small loans to many entrepreneurs), ACCION International began making loans to small entrepreneurs in the United States and in several small countries. The results have exceeded the highest expectations of even the most creative thinkers in the company. In 13 Latin American countries and six cities in the United States, ACCION International has awarded microloans totaling more than $1.7 billion. The loans average only $600 each, but the repayment rate has been an unheard-of 98 percent. These loans have helped almost 1.5 million small entrepreneurs both start and grow businesses. These businesses, in turn, have created more than a million jobs in low-income and rural communities.

Creative thinking is critical in strategic planning. Organizations undertaking strategic planning should not just welcome creative thinking; they should promote, encourage, and protect it. John Davis, CEO of Pegasus Systems, Inc., recommends the following rules for encouraging creative thinking:[6]

1. Bring creative people together in the same room, even if they drive each other crazy.
2. Listen to every idea, even if 90 percent seem ridiculous.
3. Make it safe to express even bad ideas.
4. Handle argumentative people who put down the ideas of other people.
5. Support champions of ideas by reinforcing their vision.
6. Keep an open door and put drawing boards everywhere so that creative thinking will become a normal, expected, ongoing process.
7. If you are not creative, acknowledge and support creativity in others.

CONDUCTING THE SWOT ANALYSIS

The rationale for conducting a SWOT analysis before proceeding with the development of the strategic plan is that the organization's plan should produce a good fit between its internal situation and its external situation. An organization's internal situation is defined by its strengths and weaknesses. An organization's external situation is defined by the opportunities and threats that exist in its business environment. The strategic plan should be designed in such a way that it exploits an organization's strengths and opportunities, while simultaneously overcoming, accommodating, or circumventing weaknesses and threats.

Identifying Organizational Strengths

An organizational strength is any characteristic or capability that gives the organization a competitive advantage. The following are examples of common organizational strengths:

- Financial strength
- A good reputation in the marketplace
- Strategic focus
- High-quality products/services
- Proprietary products/services
- Cost leadership
- Strong management team
- Efficient technological processes
- Talented workforce
- Faster time to market

These are just some of the strengths an individual organization may have; many others are possible. The key is accurately defining an organization's strengths before beginning to develop its strategic plan.

Identifying Organizational Weaknesses

An organizational weakness is any characteristic or capability that is lacking to the extent that it puts the organization at a competitive disadvantage. These are examples of common organizational weaknesses:

- Strategic confusion/lack of direction
- Obsolete facilities
- Obsolete processes
- Weak management team
- Insufficient skills/capabilities in the workforce
- Poorly defined operating procedures
- Too narrow a product line
- Products with decreasing demand
- Too diverse a product line
- Poor image in the marketplace
- Weak distribution system
- Weak financial position
- High unit costs compared with those of competitors
- Poor quality in products/services

These are just a few of many weaknesses an organization may have. The main thing is to identify an organization's weaknesses accurately before undertaking the strategic planning process.

Identifying External Opportunities

External opportunities are opportunities in the organization's business environment that represent potential avenues for growth and/or gaining a sustainable competitive advantage. The following are examples of external opportunities that organizations may have:

- Availability of new customers
- An expanding market for existing or potential/planned products
- Ability to diversify into related products/services
- Removal of barriers that inhibit growth
- Failures of competitors
- New on-line technologies that enhance productivity or quality

Of course, other external opportunities might be available to an organization besides these. You need to identify all such opportunities accurately before undertaking the strategic planning process.

Identifying External Threats

An external threat is a phenomenon in an organization's business environment that has the potential to put the organization at a competitive disadvantage. Such external threats might include the following:

- Entry of lower cost competitors
- Entry of higher quality competitors
- Increased sales of substitute products/services
- Significant slowdown in market growth
- Introduction of costly new regulatory requirements
- Poor supplier relations
- Changing tastes and habits of consumers
- Potentially damaging demographic changes

Many other external threats might confront an organization. Accurately identifying every potential external threat before you begin the strategic planning process is a must.

DEVELOPING THE VISION

An organization's guiding force, the dream of what it wants to become, and its reason for being should be apparent in its vision. A vision is like a beacon in the distance toward which the organization is always moving. Everything about the organization—its structure, policies, procedures, and allocation of resources—should support the realization of the vision.

In an organization with a clear vision, it is relatively easy to stay appropriately focused. If a policy does not support the vision, why have it? If a procedure does not support the vision, why adopt it? If an expenditure does not support the vision, why make it? If a position or even a department doesn't support the vision, why keep it? An organization's vision must be established and articulated by executive management and understood by all employees. The first step in articulating an organizational vision is writing it down. This is called the *vision statement*.

Writing the Vision Statement

A well-written vision statement, regardless of the type of organization, has the following characteristics (Figure 3–2). It:

- is easily understood by all stakeholders
- is briefly stated, yet clear and comprehensive in meaning
- is challenging, yet attainable
- is lofty, yet tangible
- is capable of stirring excitement for all stakeholders
- is capable of creating unity of purpose among all stakeholders
- is not concerned with numbers
- sets the tone for employees

From these characteristics it can be seen that crafting a worthwhile vision statement is a challenging undertaking. What follows are three vision statements—two for service providers and one for a manufacturer—that satisfy the criteria set forth in Figure 3–2.

- The Institute for Corporate Competitiveness will be recognized by its customers as the provider of choice for organizational development products that are the best in the world.
- Business Express Airlines will be recognized by customers as the premiere air carrier in the United States for business travelers.
- Pendleton Manufacturing Company will be the leading producer in the United States of fireproof storage cabinets.

These vision statements illustrate the practical application of the criteria set forth in Figure 3–2. Are these statements easily understood? Yes. Any stakeholder could read the vision statements and understand the dreams of the organizations they represent. Are they briefly stated, yet clear and comprehensive in meaning? Yes. Each of the statements consists of one sentence, but the sentence in each case clearly and comprehensively conveys the intended message. Are these vision statements challenging, yet attainable? Yes. Each vision presents its respective organization with the challenge of being the best in a clearly defined market and a clearly defined geographic area. Being the best in the United States or in the world is a difficult challenge in any field, but it is an attainable challenge. It can be done. Are these visions lofty, yet tangible? Yes. Trying to be the best is a lofty challenge, but, still, it is achievable and therefore tangible. Pick a field, and some organization is going to be the best in that field. It could be this organization. Are these visions capable of stirring excitement among stakeholders? Yes.

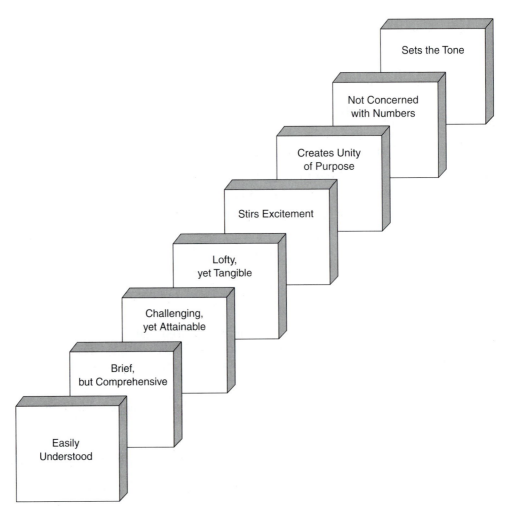

Figure 3–2
Characteristics of a Well-Crafted Vision Statement

Trying to be the best in any endeavor is an exciting undertaking, the kind in which people want to take part.

Are these visions capable of creating unity of purpose? Yes. All three give stakeholders a common rallying cry. This is the type of thing that happens when a sports team sets its sights on the championship. The players, coaches, fans, and management all rally around the vision, pulling together as one in an attempt to achieve it. Do these statements concern themselves with numbers? No. Numbers are left for later in the strategic planning process. Do these visions set the tone for employees? Yes. Clearly, the organizations in question are going somewhere, and employees are expected to do their part to ensure that the organizations get there expeditiously.

DEVELOPING THE MISSION

We have just seen that the vision statement describes what an organization would like to be. It's a dream, but it's not "pie in the sky." The vision represents a dream that can come true. The mission takes the next step and describes *who* the organization is, *what* it does, and *where* it is going. Figure 3–3 contains the mission statements for the three organizations introduced in the previous section.

Assess these mission statements using the three Ws—*who, what,* and *where*—as the criteria. In the first example, the ICC describes *who* it is as follows: "a business development company dedicated to helping organizations continually improve their ability to compete in the global marketplace." This description of who the ICC is also describes who its customers are. Regardless of whether both "who's" can be explained in one sentence, both should be explained in the mission. *What* ICC does is described as follows: "provides high-quality, competitiveness-enhancing products and services." From this statement an outsider with no knowledge of ICC could determine what the company does. *Where* ICC is going is described as reaching "an ever-increasing number of organizations in the United States." Clearly, ICC wants to grow as much as possible within the geographic boundaries of the United States.

In the second example, BEA describes *who* it is as a "domestic air carrier dedicated to providing business travelers with air transportation." This simple statement describes both BEA and who its customers are. BEA is a domestic air carrier, and its customers are business travelers. *What* BEA does is described as "providing business travelers with air transportation that exceeds their expectations." *Where* BEA is going can be seen in the following portion of the mission statement: "BEA provides air carrier service to and from a steadily increasing number of major hub airports." Like ICC, BEA wants to grow continually in the United States.

In the third example, Pendleton Manufacturing describes *who* it is as a "hazardous-materials storage company dedicated to making your work environment safe and healthy." From this statement one can easily discern who Pendleton Manufacturing and its customers are. Any company that either produces hazardous waste or uses toxic materials is a potential customer. *What* Pendleton Manufacturing does is described as

QUALITY TIP

Strategic Vision

"Effective strategic leadership starts with a concept of what the organization should and should not do and a vision of where the organization needs to be headed."

Source: Arthur A. Thompson, Jr., and A. J. Strickland III, *Strategic Management* (Boston: Irwin, 1993), 215.

- The Institute for Corporate Competitiveness (ICC) is a business-development company dedicated to helping organizations continually improve their ability to compete in the global marketplace. To this end, ICC provides high-quality, competitiveness-enhancing products and services to an ever-increasing number of organizations in the United States.
- Business Express Airlines (BEA) is a domestic air carrier dedicated to providing business travelers with air transportation that exceeds their expectations in terms of cost, convenience, service, and dependability. To this end, BEA provides air carrier service to and from a steadily increasing number of major hub airports in the United States.
- Pendleton Manufacturing Company is a hazardous materials storage company dedicated to making your work environment safe and healthy. To this end, Pendleton produces high-quality fireproof cabinets for safely storing toxic substances and hazardous materials for an ever-broadening market in the United States.

Figure 3–3
Sample Mission Statements

follows: "produces high-quality fireproof cabinets for safely storing toxic substances and hazardous wastes." *Where* Pendleton Manufacturing is going can be seen in that part of the final sentence of the mission statement that says it wants to serve "an ever-broadening market in the United States."

All three of the companies in these examples want to grow continually, but only in domestic markets. No interest is expressed in international markets. This is a major strategic decision that will determine the types of actions taken to achieve their respective missions.

In developing the mission statement for any organization, one should apply the following rules of thumb:

- Describe the *who, what,* and *where* of the organization, making sure the *who* component describes the organization and its customers.
- Be brief, but comprehensive. Typically one paragraph should be sufficient to describe an organization's mission.
- Choose wording that is simple, easy to understand, and descriptive.
- Avoid *how* statements. How the mission will be accomplished is described in the "Strategies" section of the strategic plan.

DEVELOPING THE GUIDING PRINCIPLES

An organization's guiding principles establish the framework within which it will pursue its mission. Each guiding principle encompasses an important organizational value. Together, all of the guiding principles represent the organization's value system—the foundation of its corporate culture.

Freedom through control might be one such guiding principle. It is one of the cornerstones of total quality. It is a concept that applies at all levels, from line employees through executive managers. It means that once parameters have been established for a given job, level, or work unit, all employees to which the parameters apply are free to operate innovatively within them. In fact, they are encouraged to be innovative and creative within established parameters. This means that as long as they observe applicable controls, employees are free to apply their knowledge, experience, and judgment in finding ways to do the job better. Once a method is established that is better than the existing one, that new procedure should become the standard throughout the organization.

An organization's guiding principles establish the parameters within which it is free to pursue its mission. These principles might be written as follows:

- XYZ Company will uphold the highest ethical standards in all of its operations.
- At XYZ Company, customer satisfaction is the highest priority.
- XYZ Company will make every effort to deliver the highest-quality products and services in the business.
- AT XYZ Company, all stakeholders (customers, suppliers, and employees) will be treated as partners.
- At XYZ Company, employee input will be actively sought, carefully considered, and strategically used.
- At XYZ Company, continual improvement of products, processes, and people will be the norm.
- XYZ Company will provide employees with a safe and healthy work environment that is conducive to consistent peak performance.
- XYZ Company will be a good corporate neighbor in all communities where its facilities are located.
- XYZ Company will take all appropriate steps to protect the environment.

From this list of guiding principles, the corporate values of XYZ Company can be discerned. This company places a high priority on ethics, customer satisfaction, quality, stakeholder partnerships, employee input, continual improvement, a safe and healthy work environment, consistent peak performance, corporate citizenship, and environmental protection.

With these values clearly stated as the organization's guiding principles, employees know the parameters within which they must operate. When ethical dilemmas arise, as they inevitably will in business, employees know they are expected to do the right thing. If safety or health hazards are identified in the workplace, eliminating them will be a top priority. If employees spend their own time participating in community activities, they know it will reflect positively in their performance appraisals, because XYZ Company values corporate citizenship.

Developing guiding principles is the responsibility of an organization's executive management team. However, the recommended approach in a total quality organization is for executive managers to solicit input from all levels before finalizing the guiding principles.

DEVELOPING BROAD STRATEGIC OBJECTIVES

Broad strategic objectives translate an organization's mission into measurable terms. They represent actual targets the organization aims at and will expend energy and resources trying to achieve. Broad objectives are more specific than the mission, but they are still broad. They still fall into the realm of *what* rather than *how*. The *how* aspects of strategic planning come in the next step: developing specific tactics, projects, and activities for accomplishing broad objectives. Well-written broad organizational objectives have the following characteristics (Figure 3–4). They:

- Are stated broadly enough that they don't have to be continually rewritten
- Are stated specifically enough that they are measurable, but not in terms of numbers
- Are each focused on a single issue or desired outcome

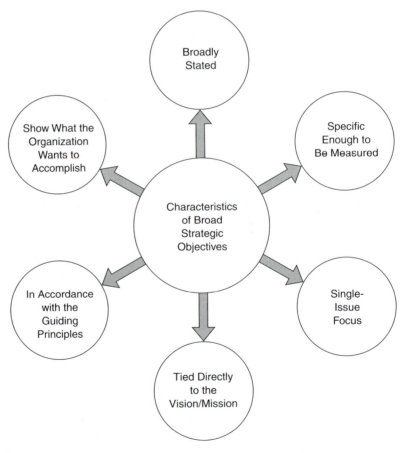

Figure 3–4
Characteristics of Well-Written Broad Strategic Objectives

- Are tied directly into the organization's mission
- Are all in accordance with the organization's guiding principles
- Clearly show what the organization wants to accomplish

In addition to having these characteristics, broad objectives apply to the overall organization, not to individual departments within the organization. In developing its broad objectives, an organization should begin with its vision and mission. A point to keep in mind is that broad strategic objectives should be written in such a way that their accomplishment will give the organization a sustainable competitive advantage in the marketplace. What follows is an organizational vision presented earlier as an example and its corresponding broad objectives:

> The Institute for Corporate Competitiveness will be recognized by its customers as the provider of choice for organizational development products that are the best in the world.

> The broad objectives that translate this vision into measurable terms are :

1. To produce organizational development products of world-class quality that are improved continually
2. To provide organizational development services of world-class quality that are improved continually
3. To establish and maintain a world-class workforce at all levels of the organization
4. To continually increase the organization's market share for its existing products/services
5. To continually introduce new products/services to meet emerging needs in the organizational development market

Five Steps for Writing Broad Strategic Objectives

In actually writing broad objectives for an organization, the following five steps should be observed (Figure 3–5):

1. *Assemble input.*   Circulate the mission widely throughout the organization, and ask for input concerning objectives. Ask all stakeholders to answer the following question: "What do we have to accomplish as an organization in order to fulfill our mission?" Assemble all input received, summarize it, and prepare it for further review.
2. *Find the optimum input.*   Analyze the assembled input, at the same time judging how well individuals' suggestions support the organization's vision and mission. Discard those suggestions that are too narrow or that do not support the vision and mission.
3. *Resolve differences.*   Proposed objectives that remain on the list after step 2 should be discussed in greater depth in this step. Allow time for participants to resolve their differences concerning the objectives.
4. *Select the final objectives.*   Once participants have resolved their differences concerning the proposed objectives, the list is finalized. In this stage the objectives are rewritten and edited to ensure that they meet the criteria set forth in Figure 3–4.
5. *Publicize the objectives.*   All stakeholders need to know what the organization's objectives are. Employees, managers, suppliers, and even customers have a role to play in accomplishing the organization's objectives. These stakeholders cannot play

Figure 3–5
Steps in Writing Broad Strategic Objectives

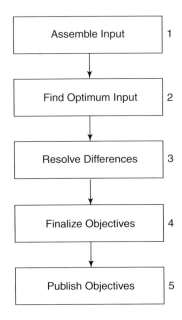

their respective roles unless they know what the objectives are. Publicizing the organization's objectives can be done in a number of ways. Variety and repetition are important when trying to communicate. Wall posters, wallet-sized cards, newsletters, personal letters, company-wide and departmental meetings, videotaped presentations, and annual reports can all be used to publish and communicate the organization's objectives. A rule of thumb to follow is *the more different communication vehicles used, the better*. It's also a good idea to publish the objectives along with the vision, mission, and guiding principles.

Cautions Concerning Broad Strategic Objectives

Before actually developing broad objectives for an organization, it is a good idea to become familiar with several applicable cautions. These cautions are as follows (Figure 3–6):

- Restrict the number of objectives to just a few—from five to eight. This is a rule of thumb, not an absolute. However, if an organization needs more than eight objectives, it may be getting too specific.
- Keep the language simple so that the objectives are easily understood by all employees at all levels of the organization.
- Tie all objectives not just to the mission but also to the vision. All resources and efforts directed toward achieving the broad objectives should support the mission and the vision.
- Make sure objectives do not limit or restrict performance. This is best accomplished by avoiding numerical targets when writing them.

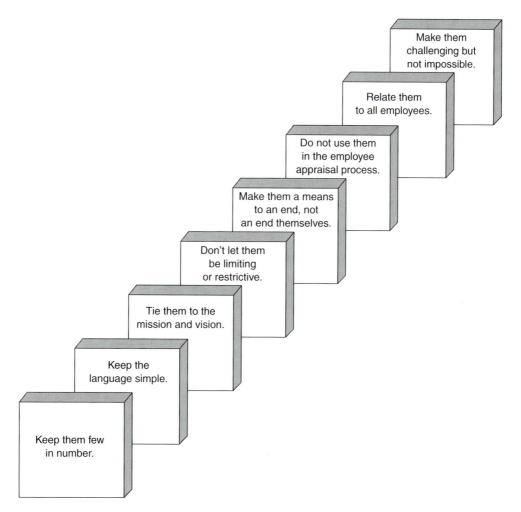

Figure 3–6
Cautions Concerning Broad Strategic Objectives

- Remember that achieving objectives is a means to an end, not an end in itself (the vision is the end).
- Do not use broad objectives in the employee appraisal process. The only aspect of the overall strategic plan that might be used in the employee appraisal process is the specific-tactics component. This is because only the specific tactics in the strategic plan are assigned to specific teams or individuals and given specific time frames within which they should be completed. Broad objectives, on the other hand, are everyone's responsibility.

- Relate broad objectives to all employees. This means there should be objectives covering the entire organization. Employees should be able to see that their work supports one or more of the broad objectives.
- Make broad objectives challenging, but not impossible. Good objectives will challenge an organization without being unrealistic.

DEVELOPING SPECIFIC TACTICS (ACTION PLAN)

The action plan consists of specific tactics that are well-defined, finite projects and activities undertaken for the purpose of achieving a specific desired outcome. They are undertaken for the purpose of accomplishing an organization's broad strategic objectives. Tactics have the following characteristics (Figure 3–7). They:

- Are specific in nature
- Are measurable
- Can be quantifiable
- Can be accomplished within a specified time frame
- Can be assigned to a specific individual or group
- Are tied directly to a broad objective

Drafting the Individual Tactics

In drafting tactics, an organization should begin with its broad strategic objectives. Each objective will have at least one, but typically three or four, tactics accompanying it. Figure 3–8 is a tool that can be used for drafting tactics. Notice that it contains the broad strategic objective to which the tactics relate. This is not necessarily a complete list of tactics for this objective; these are just examples of tactics that might be developed in support of the objectives. The nature of tactics is such that they are accomplished and then replaced by new tactics relating to the relevant objective.

QUALITY TIP

Making the Strategy Happen

"Implementing strategy is a tougher, more time-consuming challenge than crafting strategy. Practitioners emphatically agree that it is a whole lot easier to develop a sound strategic plan than it is to make it happen."

Source: Arthur A. Thompson, Jr., and A. J. Strickland III, *Strategic Management* (Boston: Irwin, 1993), 215.

Figure 3–7
Characteristics of Specific Tactics

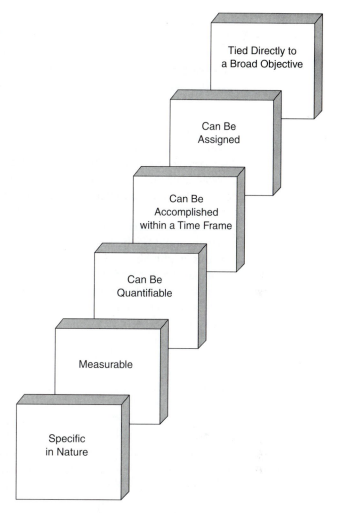

Evaluate the five tactics in Figure 3–8 by applying the criteria set forth in Figure 3–7. Are these tactics specific in nature? Yes. They are finite and limited in scope. Are the tactics measurable? Yes. In each case, the organization can easily determine whether each activity was completed within the specified time frame. Are the tactics quantifiable? Only tangentially, in that one can determine whether *all* employees received the desired service within the specified time frame. It is not necessary for all tactics to be quantifiable. This is an optional criterion. Can the tactics be accomplished within a specified time frame? Yes. A time frame is specified for the completion of each tactic. Can the tactics be assigned to a specific individual or group? Yes. In each case, a responsible party is named by position. Do the tactics tie directly to a specific broad objective? Yes. The related objective is shown on the form in the example. All of the tactics meet the applicable criteria.

QUALITY CASE

Executing Strategies and Measuring Performance at Procter & Gamble

The Procter & Gamble Company is the leading American manufacturer of household products and one of the largest advertisers in the world. Procter & Gamble is also a multinational company worth $38 billion, the products of which are widely recognized and used worldwide. These products include Tide detergent, Folgers coffee, Pampers diapers, Cover Girl makeup, and many other brands.

When he took over as chairman and CEO of Procter & Gamble, John Pepper knew that it would take more than just a good strategic plan to keep his company on top in an intensely competitive global marketplace. Pepper knew that performance would be the key. No matter how effective the company's advertising campaigns might be, in the end Procter & Gamble would have to outperform the competition if it wanted to stay ahead. John Pepper knew that the keys to performance would be *execution and measurement.* Simply put, a strategy is no good unless it is effectively executed, and performance will not improve unless it is measured.

Pepper uses several yardsticks to gauge his company's performance in the international arena. His most critical measure of quality, value, and performance is market share. It tells him immediately how a given product is doing when compared with the competition. Another important measure is percentage of product sales that have clear-cut superiority when compared to the competition. This is measured by conducting blind consumer testing ("try this product compared with that product and tell me which you prefer and why").

Execution, according to Pepper, begins with selecting the right people. He looks for individuals who will demonstrate a strong sense of ownership in terms of their responsibilities. Pepper wants people who, if trusted and properly supported, will exercise personal leadership and feel like they "own" the mission they have been assigned. With these types of individuals responsible for Procter & Gamble's various product lines, the company's strategic plan is more than just a dusty document on a shelf. It is a living, breathing road map for international success.

Source: Thomas J. Neff and James M. Citrin, *LESSON FROM THE TOP.* (New York: Doubleday, 2001), 247–252.

Making Tactics Detailed and Specific

One of the most difficult aspects of developing the action plan component of a strategic plan is making tactics sufficiently detailed and specific. There is a tendency to stop short of the explicitness needed and write tactics that, although more detailed than the broad strategic objective to which they correspond, are still not as specific as they should be. Developing specific, detailed tactics is a process that takes practice. One of the best ways to learn how to develop well-written tactics is to ask the following three questions about the broad strategic objective to which the tactics relate:

1. What are the individual, specific activities that will have to be completed in order to achieve this broad strategic objective?
2. Who should be responsible for completing each individual activity identified?
3. By what date must each activity be completed?

Answering these questions is how the Tactics Development Form in Figure 3–8 was completed.

Perhaps the best way to begin learning to develop specific, detailed tactics is to work through an example to which most students taking a class in quality management/engineering can easily relate. The authors have used the following example numerous times in college classes on quality management/engineering. First, suppose that you are a college student who wants to have a successful career in the field of quality. Assume that the vision for your career is as follows:

To have a long and successful career as a quality manager or engineer.

Now suppose that in support of this career vision you adopt the following mission statement pertaining to your college preparation:

I am an ambitious person who would like to begin a career in quality management/engineering and rise to the top of my profession. Consequently, my current mission is to complete a college degree in a field directly related to quality management/engineering.

You have both a vision and a mission. Now suppose that you develop your broad strategic objectives and that one of them reads as follows:

To successfully complete a course in quality management/engineering as part of my college studies.

Since you are already enrolled in college, we can skip the processes of admissions and academic advising. Let's assume you are at the point of wanting to enroll in the course referred to in your broad strategic objective (quality management/engineering). What are the specific and detailed activities you will have to complete? For the purpose of this exercise, assume that all of the activities will be your responsibility and that deciding on completion dates is not an issue. Focus all of your attention on identifying the specific activities that will be necessary in order to complete a course in quality management/engineering. What follows is a list developed by students in a class taught by the author:

1. Go to the counseling and advising office and make sure that the course in question is part of your approved program of study.
2. Go to the registrar's office and register for the course.
3. Go to the business office and pay the tuition and fees for the course.
4. Go to the bookstore and purchase the required textbook and other related materials.
5. Begin attending classes and participate fully in all class discussions.

6. Complete all assignments for the course.
7. Prepare for all tests and examinations required for the course.
8. Complete all tests and examinations for the course with passing grades.
9. Create a permanent file of your course materials and text to use for reference once you begin your first job in quality management/engineering.

This is an adequately specific and detailed list. However, upon close scrutiny you will find that each of the activities could have been broken down even further and made even more specific. Deciding how specific to be when developing tactics requires practice and experience. Until you become well versed in the art and science of developing action plans, it is better to err on the side of more specificity and more detail.

EXECUTING THE STRATEGIC PLAN

The old saying, "The best-laid plans of mice and men often go astray" is, unfortunately, all too true. Many organizations devote time, energy, and money to developing comprehensive, thorough, detailed plans, only to see them come apart at the seams shortly after they begin to be executed. Execution is a critical component of strategic management, but for some reason it rarely receives the attention it deserves. To put less energy and thought into execution than into planning is a major error because even the best strategic plan won't help an organization if it is poorly executed.

Broad Strategic Objective

To establish and maintain a world-class workforce at all levels of the organization.

Tactic	Responsible Individual/Unit	Timeframe/Deadline
1. Arrange TQ training for all executive managers.	CEO	Completed by January 15
2. Arrange teamwork training for all executive managers.	CEO	Completed by January 30
3. Give all employees training in the use of problem-solving/ quality tools.	Department Managers	January 15–February 20
4. Give all employees training in continual improvement methods.	Department Managers	March 15–May 15
5. Establish a company-supported off-duty education program for all employees.	Human Resources Department	In place by February 28

Figure 3–8
Tactics Development Form

Picture the following scenario. A family plans a vacation to a national park. The parents envision a relaxing, fun-filled two weeks of camping, hiking, swimming, and biking. Their mission (that of the parents) is to get away together and share some quality time as a family. Certain guiding principles concerning behavior, spending, and work sharing are established by the parents. They also set up some broad objectives concerning the various activities the family wants to pursue. Wisely, the parents involve the children in this step. Also working with the children, parents draft strategies for accomplishing their objectives.

The family had an excellent plan for an enjoyable vacation, but as soon as the plan went into execution, the problems began:

- Disagreements among the children concerning destinations and activities
- Problems when the children did not know how to perform some of their assigned duties (e.g., setting up the tent at the campsite, building a proper campfire, monitoring daily gas mileage while the parents drove)
- Attitude problems concerning various aspects of the trip, including distance to cover daily, how many rest stops to make, and where to eat while driving to the eventual destination

As a result of these difficulties, the planned vacation of shared family fun and relaxation turned into an emotionally draining two weeks of stress, anger, and frustration. The family's problems were the results of faulty execution of the vacation plan. The parents in this example failed to apply the following steps, all of which are critical to successful execution:

- *Communicate.* Make sure all stakeholders understand the plan and where they fit into it.
- *Build capabilities.* Make sure all stakeholders have the skills needed to carry out their assignments and responsibilities in the plan.
- *Establish strategy-supportive stimuli.* People in the workplace respond to stimuli. When trying to execute a plan, it is important to ensure that strategy-supportive stimuli are in place. Typically, in a work setting the most effective stimuli are reward and recognition incentives. It is not uncommon to find that an organization's strategic plan expects people to move in one direction, while its incentives encourage them to move in another.
- *Eliminate administrative barriers.* Every organization establishes administrative procedures for accomplishing its day-to-day work. If executing a new strategy changes the intended direction of the organization, administrative procedures may need to be changed correspondingly. A mistake commonly made by organizations attempting to execute a plan is leaving outdated administrative procedures in place. Administrative procedures put in place when the organization was moving in one direction can become inhibitors when the organization decides to move in another direction.
- *Identify advocates and resisters.* In any organization there will be advocates and resisters when it comes to executing the strategic plan. This is natural and should be expected. As the plan unfolds, if it is successful, resisters will become advocates.

If it fails, advocates will try to distance themselves from the plan, and resisters will say, "I told you so." This is just human nature and should be expected. Consequently, it is important to give the plan the best possible chance of succeeding. One way to do this is to assign all initial activities to advocates. Giving initial assignments to resisters when executing a new strategic plan is likely to ensure failure of the plan. Eventually all employees must play a role in executing the strategic plan, but in the critical early stages, stick with advocates and avoid resisters.

■ *Exercise strategic leadership.*   It is important that managers at all levels set a positive example by (a) showing that they believe in the strategic plan, (b) ensuring that all decisions are based on that action that best supports the strategic plan, and (c) allocating resources based on priorities established in the strategic plan.

■ *Be flexible and improvise.*   According to General Colin Powell, "No battle plan survives contact with the enemy."[7] General Powell's statement could easily be misconstrued. After all, if he is right, why should an organization go to the trouble of planning in the first place? Powell's statement does not advocate against planning. Rather, it advocates flexibility. Plans, when they are developed, are based on assumptions that may not be accurate. They are also based on a presumed set of circumstances, circumstances that even if they were accurate when the plan was developed might change before the plan is executed. This is why strategic plans must be viewed as a set of flexible guidelines rather than a hard-and-fast road map from which it is impossible to deviate. Every traveler knows there will be detours on even the best-planned trip. Managers who want to make their organizations competitive must be willing to improvise when necessary.

During the 1970s, teachers of grades K–12 had to learn how to develop lesson plans. They were taught by seminar leaders and college professors to plan their daily lessons right down to the minute, with each activity assigned a specific amount of time. Of course, the wiser, more experienced teachers were able to predict the outcome of this approach, but sometimes it is best to let reality make your point for you. As teachers attempted to hold rigidly to schedules built into their plans, they found themselves falling farther and farther behind as problems cropped up they had not foreseen when developing their lesson plans.

Before long, lesson plans were scrapped and replaced by planned lessons. The difference between a lesson plan and a planned lesson can be summed up in one word: *flexibility*. Planned lessons give teachers a general direction, some expected outcomes, and loose guidelines on how to get where they are going. They contain no rigid time constraints but, more important, give teachers the flexibility to change directions and pursue a whole new set of activities if an opportunity for learning presents itself.

This example holds a lesson for managers who develop and execute strategic plans: be flexible. The assumptions on which the plan was built might not be accurate. The circumstances in which the plan was supposed to be implemented might change. Planning is necessary so that resources can be properly allocated and so that employees on whom the organization depends for progress can get a better picture of where the organization is trying to go. But if somebody moves the target, don't continue to shoot at the spot where it used to be.

This gets back to the point made at the beginning of this section by General Powell's quote. Organizations should plan thoroughly and carefully, based on the most accurate assumptions possible at a given point in time. However, if on implementing the plan, it becomes obvious that the assumptions are not valid or that circumstances have changed, organizations should not rigidly adhere to steps that no longer make sense. If the plan, or any part of it, is no longer relevant, improvise and move on.

■ *Monitor and adjust as needed.* Developing and executing a strategic plan is an example of the *plan-do-check-adjust* cycle in action. The *plan* component of the cycle involves developing the strategic plan. The *do* component is the execution phase. The *check-adjust* component involves monitoring progress toward completion of specific strategies and making the necessary adjustments when roadblocks are encountered. Will it take longer than you thought to complete a project or activity? Adjust the time frame. Have unexpected barriers been encountered? Decide what needs to be done to overcome the barriers, and do it. Did you complete the project only to find it didn't yield the expected results? Develop a substitute tactic and try again.

Strategic Planning in Action: A "Real-World" Case

This section takes the reader through the strategic planning process undertaken by a company that contracted with the authors to facilitate it. The process involved the company's executive management team and took place over a three-day period at an off-site location away from the company's facility. The strategic planning process always varies from organization to organization in specific details. However, this particular company was selected as an example because the process it underwent is representative of what most companies go through in developing a comprehensive strategic plan. This case should help the reader see how the various components of the strategic plan fit together as well as how each respective component is actually developed.

Developing a Strategic Plan: Delcron Manufacturing Company

Delcron Manufacturing Company (DMC) started as a small, minority-owned Department of Defense contractor. When the company was classified as an "8A" firm, it became eligible for government set-asides. Set-asides under the government's "8A" program are contracts awarded outside of the bidding process to minority-owned firms. The idea is to give such firms an opportunity to gain a foothold in business while learning how to compete without the set-asides. DMC entered this arena as a manufacturer of low-voltage power supplies for military aircraft. Its "8A" status lasted for five years. During this time, DMC grew from a small shop in a garage to a company that employed almost five hundred people.

When the company was just over a year from having to make the transition from "8A" status to the competitive marketplace, its executive management team decided the company needed a strategic plan. For several years these executives had been so busy establishing the company and helping it grow that they had given little thought to what would happen to DMC once it graduated from the "8A" program. Would the company be able to compete successfully in the open marketplace? Should it attempt to diversify into other markets? Should the executives simply sell the company and move on to other endeavors? These questions had begun to weigh heavily on the minds of DMC's executives when the CEO suggested they hire a strategic planning consultant, go through the process, and see what transpires. DMC's executives hired the authors and went through the strategic planning process. The remainder of this section documents the process and its outcomes.

Overview of the Process

The authors set up the strategic planning process in the conference center of a resort about 75 miles from DMC's facility. The idea, as they explained it to the company's executives, was to conduct the strategic planning process at a location that would guarantee both privacy and focus. No cellular telephones or visits to the office or home were allowed. In addition, family members were not included. The authors explained that in order to come away with an acceptable draft of a strategic plan, the executives would need three entire days of uninterrupted, fully focused work. The group would be given a morning and afternoon break each day to make telephone calls and check e-mail messages. Beyond that, their administrative assistants and family members knew how to reach them in the event of an emergency.

The first hour of the first day was devoted to learning the process and how it works. The authors explained all of the various components of the strategic plan (the SWOT analysis, vision, mission, guiding principles, and broad strategic goals). This first three-day session would conclude with the development of broad strategic goals. Another session would be scheduled during which the executives would develop the action-plan component. The action-plan session would involve a broader group that, in addition to the executive management team, would include other management and supervisory personnel. The session would be conducted by the authors after each member of the company's executive management team had solicited input from all these employees directly reporting to him or her.

The SWOT Analysis

Prior to beginning the SWOT analysis, the authors placed four large flip charts in the room. The flip charts were labeled *strengths, weaknesses, opportunities, and threats.* Beginning with strengths, the authors led participants through a brainstorming session. Each time a strength, weakness, opportunity, or threat was identified by a participant and agreed on by the group, the authors recorded it on the corresponding flipchart. Every time a flip chart page would fill up, the authors tore it off and taped it to the wall so that participants had a visible record of their work.

Discussion was intense at times, and there were disagreements among participants. One executive would identify a weakness in another executive's area of responsibility, and arguments would ensue. It took the authors a while to convince the executives to drop their defenses and to be open and frank without getting their feelings hurt or being territorial. Another dynamic was that what one executive saw as a strength, another perceived as a weakness. Once the authors worked through these and other issues that inevitably occur during strategic planning sessions, a cohesive group emerged and began to cooperate well as a team. The results of the SWOT analysis follow.

Strengths

Participants identified and reached consensus on the following strengths: strong manufacturing capability approaching the Six Sigma level, solid business contacts in the Department of Defense industry, a proven track record of excellent performance in completing contracts on time, low turnover rate with regard to critical employees, comparatively low labor rates (most employees of DMC are retired military personnel who view their salary as a second income), and an up-to-date facility equipped with modern technology.

Weaknesses

Participants identified and reached consensus on the following weaknesses: no experience outside of the Department of Defense market, no marketing component, no experience being the lead contractor on a major project (all of DMC's work up to this point had been subcontracted to it by larger Department of Defense contractors such as Boeing and Lockheed-Martin), no experience in the international marketplace, no design component (DMC had been a build-to-print operation up to this point), and no research and development component.

Opportunities

Participants identified and reached consensus on the following opportunities: expansion into commercial aircraft markets, expansion into international commercial markets, expansion into foreign military markets (military aircraft of America's allies), and availability of a strong international marketing team that can work both commercial and military markets (one of DMC's potential competitors in the commercial marketplace had just been purchased by a larger company and its entire marketing team had been eliminated as part of the buyout).

Threats

Participants identified and reached consensus on the following threats: DMC's pending loss of its "8A" status, potential cutbacks in the development of new military aircraft in the United States, a tight labor market that could inflate labor costs, and the potential for ever-increasing levels of competition from foreign and domestic sources.

Developing the Vision

Before developing a vision for DMC, the company's executive management team had to decide if there would even be a DMC after graduation from the "8A" program. Going ahead with the company would mean risking the investment of both their time and money in DMC. On the other hand, these executives could probably sell their shares in the company, walk away with a handsome profit, and find high-level positions with other firms in their respective fields or even help start another "8A" company. After a lively discussion, the executives decided they had invested too much of their money and themselves in DMC to walk away from the company now. Consequently, they turned to the task of developing a post–"8A" vision for DMC.

The authors led participants through a lengthy discussion that revolved around the following questions (which came from the SWOT analysis): Should DMC stick with just domestic military markets or expand into the commercial marketplace too? Should DMC consider pursuing contracts with foreign militaries? Should DMC pursue international commercial contracts? Should DMC stick with low-voltage power supplies as its principal product or diversify into other product lines? Should DMC add a design function or continue as just a build-to-print company? Should DMC add a research and development function to develop new product lines?

Answering these questions was the most difficult part of the strategic planning process for DMC's executives. How they answered these questions would determine everything else about the future of the company and, correspondingly, about their professional futures. After an intense discussion, participants decided that DMC would need to expand into both commercial and foreign markets while retaining its Department of Defense base. They also agreed that the company's expertise is in the production of low-voltage power supplies. Consequently, they ruled out adding a research and development function, but they did decide to expand into design. The rationale of DMC's executives was that the company would always be at the mercy of other, larger contractors unless it could design power supplies in addition to just manufacturing them.

With these questions answered, the authors were able to lead participants through the process of developing a vision statement that would encompass their dreams for DMC. The vision statement developed reads as follows:

> Delcron Manufacturing Company will be an international leader in the production of low-voltage power supplies for aircraft.

Developing the Mission

The executives at DMC found that developing a vision answers a lot of important questions. With the vision in place, developing a mission statement was not overly difficult; it was just a matter of following the criteria set forth by the authors for well-written mission statements. Most of the discussion focused on wording as opposed to concepts. The mission statement the participants finally decided on reads as follows:

> Delcron Manufacturing Company (DMC) is a design and manufacturing firm dedicated to providing high-quality products for the aircraft industry. To this end, DMC designs and manufactures low-voltage power supplies for military and commercial aircraft in the United States and abroad.

Developing Guiding Principles

The authors described "guiding principles" to participants as written statements that convey DMC's corporate values. They encouraged the company's executives to mentally put the following sentence before the guiding principles in order to better understand what they represent: "While pursuing our vision and mission, we will apply the following guiding principles in everyday operations and in all decisions made."

The authors asked participants to brainstorm important corporate values without concern, for the moment, about wording. Participants were encouraged to simply offer up value-laden terms (e.g., ethics, quality, customer satisfaction, etc.) that the authors recorded on flip charts. Once the terms had all been selected, the authors asked participants to select the six to ten most critical on the list. With the most important corporate values identified, participants worked with the authors to develop more explicit wording for each one. The guiding principles developed are as follows:

1. *Ethics.* All of DMC's employees and management personnel are expected to exemplify the highest ethical standards in doing their jobs.
2. *Customer delight.* In dealing with customers, DMC will go beyond customer satisfaction to achieve customer delight.
3. *Continual improvement.* Continually improving its products, processes, and people is a high priority for DMC.
4. *Quality.* DMC is committed to delivering the highest-quality products possible on time, every time.
5. *Employee empowerment.* DMC is committed to seeking, valuing, and using employee input and feedback.
6. *Partners.* DMC is committed to treating its customers, suppliers, and employees as partners.

Developing Broad Strategic Goals

The final component of the first strategic planning session involved developing broad strategic goals. These goals had to represent actions that, if accomplished, would move the company ever closer to the full realization of its corporate vision. Before beginning development of the goals, the authors gave participants typed copies of the results of the SWOT analysis. They explained to the executives that the broad goals developed should all satisfy one or more of the following criteria: 1) exploit one or more of the organization's strengths, 2) correct one or more of the organization's weaknesses, 3) take advantage of one or more of the opportunities available to the organization, or 4) prepare against one or more of the threats facing the organization. With this guidance given, participants developed the following broad strategic goals for DMC:

1. Expand the company's business base to include both military and commercial markets in the United States and abroad.
2. Strengthen all functional units in the company in the area of commercial products and markets.
3. Expand the company's core capabilities to include both design and manufacturing.
4. Fully achieve a Six Sigma quality level in manufacturing low-voltage power supplies.
5. Develop and implement a supplier certification program to create a reliable group of dependable, high-quality supplier partners.
6. Establish a comprehensive training program to maximize the capabilities of all employees at all levels in the company.

With the broad strategic objectives established, the "strategic" portion of the plan was completed. Developing an action plan for carrying out the strategic portion of the plan would be scheduled later after DMC's executive management team had communicated the vision, mission, guiding principles, and broad strategic goals to all employees. The action planning session took place two months later and DMC began implementation of the new strategic plan began immediately thereafter. Just two years after DMC graduated from the "8A" program, it had expanded from under five hundred employees to almost 700, and its business base was growing steadily. The company now is well established in both commercial and military markets in the United States and abroad. The plan developed during a three-day, off-site session is working.

REVOLUTIONARY THINKING IN STRATEGIC PLANNING

Writing for the *Harvard Business Review*, Gary Hamel makes a case for revolutionary thinking as a business strategy.[8] He describes three types of companies that can be found in any industry:

- *Rule makers*—These are the companies that built the industry in question. IBM, Sears, and Coca-Cola are examples of rule makers.
- *Rule takers*—These are the companies that follow the rules made by the industry leaders. J. C. Penney, Fujitsu, and U.S. Air exemplify rule takers.
- *Rule breakers*—These are the maverick companies that break the rules, ignore precedent, and cast aside convention. IKEA, Charles Schwab, and Southwest Airlines are rule breakers.

Before beginning strategic planning, executives must decide which of these three types of companies they want. Rule makers adopt one type of vision, rule takers adopt another, and rule breakers yet another. Rule makers will adopt a vision in which they dominate and set the rules in a given market. Rule takers will adopt a vision in which they are industry leaders but perennially less than first-place finishers; their strategies will focus on continuous incremental improvement to work their way ever closer to the market leader. Rule breakers will adopt a vision that focuses on carving out a market niche that because of its characteristics may not be well served by the market leaders. Such companies will then seek to dominate their niche market.

To be a rule maker or a rule breaker, a company must go against the grain of traditional thinking. Hamel sets forth 10 principles for helping executives "step out of the box" and take a revolutionary approach to strategic planning and management:

- *Strategic planning is not strategic.* For most companies, strategic planning is ritualistic, reductionist, calendar driven, and easy. It consists primarily of attempting to better position a company within boundaries and conventions that are simply accepted when they should be challenged. To be *strategic*, strategic planning must be inquisitive, expansive, inventive, and demanding. It must challenge convention and reject boundaries.

■ *Strategy making must be subversive.*   Think of subversive in this context as unorthodox. Strategic planning should involve identifying the 10 or 20 most fundamental beliefs in the industry in question and challenging every one of them.

■ *The bottleneck is at the top.*   Executive managers tend to be the people in an organization who have the greatest investment in orthodoxy and the status quo. The danger here is that the market is changing so rapidly that years of experience can actually be a detriment when it comes to revolutionary thinking.

■ *Revolutionaries exist in every company.*   Every company has revolutionary thinkers, although they may not be at the top. But the midmanagers and employees may have strategic ideas that can propel a company into the future. For this reason, strategic planning must be *inclusive*.

■ *Change is not the problem; engagement is.*   Too many senior executives assume that midmanagers and employees are opposed to change. Consequently, change must be achieved by far-sighted executives who drag the company kicking and screaming into the next decade. In reality, change is not what employees resist. Rather, they resist how change is engaged. Too often *change* is just a code word when something undesirable is about to happen. Change that is inflicted on employees will be resisted, of course. But change that is engendered by employees who are involved in and prepared for it will be accepted and promoted.

■ *Strategy making must be democratic.*   Strategy making should involve employees at all levels, including new employees. Executives must truly involve the employees if the process is going to work. Inviting employees to the table without giving them a real voice is worse than not inviting them at all. (We prefer the term *inclusive* here to Hamel's term *democratic*. Democratic might be misconstrued to imply voting and majority rule, both of which work against revolutionary thinking.)

■ *Anyone can be a strategy activist.*   In too many companies, middle managers and employees are inclined to see themselves as unwilling victims. They view themselves as having no role in shaping strategy. But in reality, any employee can be a strategy activist by asking difficult questions, persistently garnering support for ideas, and making strategic suggestions.

■ *A new perspective can be better than intelligence.*   Executives cannot make themselves much smarter than they already are, but they can gain a better viewpoint. Sometimes the only way to find a solution is to look at the problem from a different perspective. For example, choosing to view a customer's complaint as an opportunity for improvement rather than a threat is adopting a new perspective. There can be no change in thinking without a change of perspective.

■ *Top down and bottom up are not the alternatives.*   Senior executives control the means for shaping change, but they do not control the thinking. Every employee in an organization has a brain. Consequently, strategy making should involve bringing together the top and the bottom in a broadly participative process in which all parties play an active role. Executives who do not truly buy into a coming together of the top and bottom will quickly drown out the voices of midmanagers and employees. The key is to give employees a voice and then listen to what they have to say.

■ *You cannot see the end from the beginning.*   Revolutionary strategy making is open ended and unpredictable. Once it starts, no one can predict where it will end up. A new product line, a new service, or a whole new market might result from a broad cross-section of employees brought together and given license to think. Some of the most successful products and some of the most effective innovations are not what they started out to be. Senior executives must learn to let the strategy-making process stumble its way toward the possible.

 SUMMARY

1. Strategies that organizations can adopt for gaining a sustainable competitive advantage are cost leadership, differentiation, and market-niche strategies.
2. Strategies are approaches adopted by organizations to ensure successful performance in the marketplace.
3. Strategic management is management that bases all actions, activities, and decisions on what is most likely to ensure successful performance in the marketplace. The two major components of strategic management are strategic planning and strategic execution.
4. Mindshifters are creative thinkers in organizations who constantly seek gaps in the marketplace their companies can fill. Consequently, they can be invaluable resources in the strategic planning process.
5. Strategic planning is the process whereby organizations develop their vision, mission, guiding principles, broad objectives, and tactics for accomplishing the broad objectives.
6. An organization's vision is its guiding force, the dream of what it wants to become, and its reason for being.

 QUALITY TIP

Creativity in Strategic Planning

"Any company intent on creating industry revolution has four tasks. First, the company must identify the unshakable beliefs that cut across the industry—the industry's conventions. Second, the company must search for discontinuities in technology, lifestyles, working habits, or geopolitics that might create opportunities to rewrite the industry's rules. Third, the company must achieve a deep understanding of its core competencies. Fourth, the company must use all of this knowledge to identify the revolutionary ideas, the unconventional strategic options that could be put to work in its strategic domain."

Source: Gary Hamel, "Strategy as Revolution," *Harvard Business Review* (July–August 1996), 80.

7. An organization's mission describes *who* an organization is, *what* it does, and *where* it is going.

8. An organization's guiding principles establish the framework within which it will pursue its mission. Together, the guiding principles summarize an organization's value system, the things it believes are most important.

9. An organization's broad strategic objectives translate its mission into more specific terms that represent actual targets at which the organization aims. The objectives are more specific than the mission, but they are still broad.

10. Tactics are well-defined, finite projects and activities undertaken for the purpose of specific desired outcomes in support of the broad objectives.

11. Even the best strategic plan will serve no purpose unless it is effectively executed. To promote successful execution of strategies, organizations should undertake the following activities: communicate, build capabilities, establish strategy-supportive stimuli, eliminate administrative barriers, identify advocates and resisters, exercise strategic leadership, and monitor and adjust as needed.

12. To ensure revolutionary strategy making, senior executives must be willing to include employees from all levels and give them an active voice to which the executives truly listen.

KEY TERMS AND CONCEPTS

Advocates	Plan-do-check-adjust cycle
Broad objectives	Resisters
Cost leadership strategies	Specific strategies
Differentiation strategies	Strategic execution
Guiding principles	Strategic management
Market-niche strategies	Strategic planning
Mission	Vision

FACTUAL REVIEW QUESTIONS

1. What is strategic management?
2. List the steps in the strategic planning process.
3. What is a "mindshifter" and what role can one play in strategic planning?
4. What is SWOT analysis?
5. Write a sample vision for a hypothetical organization.
6. Write a mission statement for the hypothetical organization in question 4.
7. Draft a set of guiding principles for the hypothetical organization in question 4.
8. Establish two or three broad objectives for the hypothetical organization in question 4.
9. Describe the steps you would apply in executing your strategic plan developed in questions 4–7.
10. List and explain 10 principles for achieving revolutionary strategy making.

CRITICAL THINKING ACTIVITY

Locked Out of the Process

Alex Parker is the quality director at CompuTech, Inc. His is a middle-management position. CompuTech's senior executives have announced that they intend to hire a strategic planning consultant to develop a plan for the company. Planning will begin in about two months. It will take place in the conference center of a local resort and will involve only executive-level managers and the consultant.

Parker has several ideas he would like to propose. In addition, he wants to make sure that quality is treated as a strategic issue during the planning process. Unfortunately, Parker and his fellow middle managers have not been invited to participate. They are locked out of the process.

What do you think of CompuTech's approach to strategic planning? Would you change the process in any way? If so, how? If the process remains as it is, how can Parker get his ideas included?

 ENDNOTES

1. Piturro, Marlene. "MINDSHIFT," *Management Review*, May 1999, 46.
2. Ibid., 47.
3. Ibid.
4. Ibid.
5. Piturro, 48.
6. Ibid., 50.
7. Colin Powell, as quoted in "Good/Bad News about Strategy" by Oren Harari, *Management Review* 84, no. 7 (July 1995), 29.
8. Gary Hamel, "Strategy as Revolution," *Harvard Business Review* (July–August 1996), 69–82.

CHAPTER FOUR

Quality Management and Ethics

"It is easier to fight for one's principles than to live up to them."
   Alfred Adler

MAJOR TOPICS

- Ethics Defined
- Trust and Total Quality
- Values and Total Quality
- Integrity and Total Quality
- Responsibility and Total Quality
- Manager's Role in Ethics
- Organization's Role in Ethics
- Handling Ethical Dilemmas
- Ethics Training and Codes of Business Conduct
- Models for Making Ethical Decisions
- Beliefs versus Behavior: Why the Disparity?
- Ethical Dilemmas: Cases

Insider trading, accepting a higher commission for selling an inferior product, accepting gifts from suppliers, hiring a friend or relative instead of a more qualified applicant— these are all examples of ethics violations that are common in today's workplace. Yet, nearly universal agreement exists that business practices in the modern workplace should be above reproach with regard to ethical behavior. Few people are willing to defend unethical behavior, and, for the most part, business and industry operate within the scope of acceptable legal and ethical standards. Ethical behavior is particularly important in a total quality setting in which trust, integrity, and values figure prominently in everyday human interactions.

QUALITY TIP

Ethics as a Key Business Indicator

Financial measures, customer service, and quality are no longer the only key business indicaters. Social responsibility must be added to the list. The entire community, not just customers, expects and desires fair and ethical treatment.

Nevertheless, unethical behavior does occur frequently enough that modern managers need to be aware of the types of ethical dilemmas they might face. They should also be knowledgeable about how to deal with these dilemmas. This chapter explains the importance of ethics in a total quality setting and how ethical behavior in the workplace can be promoted.

ETHICS DEFINED

According to Edward Stead, Dan Worrell, and Jean Garner Stead:

> Managing ethical behavior is one of the most pervasive and complex problems facing business organizations today. Employees' decisions to behave ethically or unethically are influenced by a myriad of individual and situational factors. Background, personality, decision history, managerial philosophy, and reinforcement are but a few of the factors which have been identified by researchers as determinants of employees' behavior when faced with ethical dilemmas.[1]

There are many definitions of the term *ethics*. No one definition has emerged as universally accepted. According to Paul Taylor, the concept can be defined as "inquiry into the nature and grounds of morality where morality is taken to mean moral judgments, standards, and rules of conduct."[2] According to Peter Arlow and Thomas Ulrich, ethical dilemmas in the workplace are more complex than ethical situations in general.[3] Ethical dilemmas in the workplace involve societal expectations, competition, and social responsibility, as well as the potential consequences of an employee's behavior on customers, fellow workers, competitors, the organization, and the public at large. The result of the often conflicting interests of workers, customers, competitors, and the general public is a natural propensity for ethical dilemmas in the workplace.

Any time ethics is the topic of discussion, such terms as *conscience, morality, legality, trust, values, responsibility,* and *integrity* will be frequently heard. Although these terms are closely associated with ethics, they do not, by themselves, define it. For the purpose of this book, *ethics* is defined as follows:

> Ethics is the study of human behavior within a moral context.

Morality refers to the values that are subscribed to and fostered by society in general and by individuals within society. Ethics attempts to apply reason in determining rules of human behavior that translate morality into everyday behavior. *Ethical behavior* is that which falls within the limits prescribed by widely accepted moral values. The concepts of *trust, responsibility*, and *integrity* are part of the value system of total quality.

How does one recognize ethical behavior? Ethical questions are not always black and white; they often fall into a gray area between the two extremes of clearly right and clearly wrong. This gray area is often clouded by personal experience, self-interest, point of view, and external pressure.

Guidelines for Determining Ethical Behavior

This section explains some guidelines that can be used in sorting out ethical and unethical behavior. However, before presenting the guidelines, we must distinguish between the concepts of *legal* and *ethical*; they are not the same thing. Just because a choice made is legal does not necessarily mean it is ethical. A person's behavior can be well within the prescribed limits of the law and still be unethical. The tests for determining ethical behavior shown in Figure 4–1 assume the behavior in question is legal. By applying any one of these tests, a person should be able to see through the gray area surrounding an issue and determine the ethical route to take.

Kenneth Blanchard and Norman Vincent Peale suggest their own test for determining what is the ethical choice in a given situation.[4] Their test consists of the following three questions:

1. Is it legal?
2. Is it balanced?
3. How will it make me feel about myself?

Morning-After Test
How will you feel about this behavior tomorrow morning?

Front-Page Test
How would you like to see this behavior written up on the front page of your hometown newspaper?

Mirror Test
How will you feel about this behavior when you look in the mirror?

Role Reversal Test
How would you feel about being on the receiving end of this behavior?

Commonsense Test
What does everyday common sense say about this behavior?

Figure 4–1

Tests of Ethical Behavior

Adapted with permission from David L. Goetsch, *Effective Supervision* (Upper Saddle River, NJ: Prentice Hall, 2001).

Naturally, if a potential course of action is not legal, no further consideration of it is in order. If an action is not legal, it is also not ethical.

A course of action that is balanced will be fair to all involved. This means that managers and employees have responsibilities that extend beyond their unit, organization, or company.

A course of action that leaves people feeling good about themselves is one that is consistent with their personal value system.

Blanchard and Peale also describe what they call the Five P's of Ethical Power:[5]

- *Purpose.*   Individuals see themselves as ethical people who let their conscience be their guide and in all cases want to feel good about themselves.
- *Pride.*   Individuals apply internal guidelines and have sufficient self-esteem to make decisions that may not be popular with others.
- *Patience.*   Individuals believe right will prevail in the long run, and they are willing to wait when necessary.
- *Persistence.*   Individuals are willing to stay with an ethical course of action once it has been chosen and see it through to a positive conclusion.
- *Perspective.*   Individuals take the time to reflect and are guided by their own internal barometers when making ethical decisions.

These tests and guidelines can help managers and employees make ethical choices in the workplace. In addition to internalizing the guidelines themselves, managers should share them with their workers.

Factors That Influence Ethical Behavior

Managers in a total quality setting need to understand the factors that can influence a person's behavior. Research by L. K. Trevino indicates that ethical behavior in organizations is influenced by both individual and social factors.[6] Trevino suggests three personality measures that can influence a person's ethical behavior: ego strength, Machiavellianism, and locus of control.

A person's *ego strength* is his or her ability to undertake self-directed tasks and to cope with tense situations. *Machiavellianism* is the extent to which a person will attempt to deceive and confuse others. *Locus of control* is the perspective of people concerning who or what controls their behavior. People with an internal locus of control feel they control their own behavior. People with an external locus of control feel their behavior is controlled by external factors (rules, regulations, their supervisors, etc.).

J. E. Prebel and P. Miesing suggest that certain social factors can also influence ethical behavior in organizations, including gender, role differences, religion, age, work experience, nationality, and the influence of other people who are significant in an individual's life.[7] F. Luthans and R. Kreitner state that people learn appropriate behavior by observing the behavior of significant role models (parents, teachers, public officials, etc.).[8] Because managers are role models in organizations, it is critical that they exhibit ethical behavior that is beyond reproach in all situations.

TRUST AND TOTAL QUALITY

The total quality approach cannot be successfully implemented in an organization that does not subscribe to high standards of ethical behavior. This is because ethical behavior builds trust, and trust is an essential ingredient in total quality. Consider the various elements of total quality that depend on trust: communication, interpersonal relations, conflict management, problem solving, teamwork, employee involvement and empowerment, and customer focus (see Figure 4–2).

In human communication, receivers do not accept messages from senders they don't trust. In interpersonal relations, trust is the most fundamental element. People who trust each other will be able to get along and work well together even in the worst of circumstances. On the other hand, people who do not trust each other will not be able to get along and work well together even in the best of circumstances. Trust is also a critical element in conflict management. A manager who is not trusted by both sides in a human conflict cannot help resolve the conflict.

For people to put aside their personal agendas and work together as a team, they must trust each other. If even one team member is concerned that another team member is promoting his or her self-interests over those of the team, teamwork will not succeed. Managers will not involve and empower employees unless they trust them. Involving employees and giving them a real voice in decisions that affect them is a clear demonstration of trust.[9]

Ethics plays a critical role in the successful application of total quality. Ethical behavior on the part of the organization is just as important as the behavior of managers and employees. A company that does not honor warranties, treats employees poorly, or pollutes its community cannot expect employees to disregard the example it sets and promote a trusting environment in the workplace.

If the trust that results from ethical behavior is so important to total quality, then it follows that modern managers need to be good trust builders. Although it is important that managers be able to establish themselves as trustworthy, that by itself is not enough. Managers in a total quality setting must also be able to build trust in the organization and among its employees.

One of the best ways managers can help build trust is "to be loyal to those who are not present."[10] When a manager speaks up for someone who is not present but is being questioned or attacked, employees get two simultaneous messages:

- Talking behind a colleague's back is not acceptable behavior.
- If this manager doesn't let me talk about fellow employees who are absent, he or she won't let others talk about me when I'm absent.

Knowing that they will be included in any conversation about them or that affects them builds trust among employees. A sincere apology can also build trust. Managers sometimes make mistakes or do things that hurt employees. Making excuses, pointing the finger of blame at someone else, or ignoring the situation only exacerbates it. By simply and sincerely saying, "This is my fault; I'm sorry," managers can build trust even when they have made mistakes.

Keeping promises is another way managers can build trust. Dependability builds trust. It is human nature to want to be able to depend on what others tell us. Promises in the

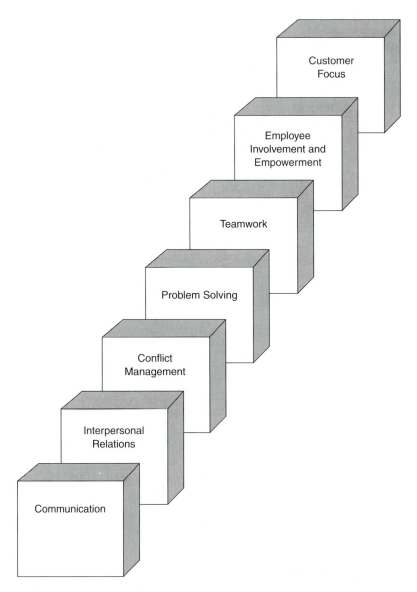

Figure 4–2
Elements of Total Quality That Depend on Trust

workplace often take the form of deadlines. A deadline promised should be a deadline kept. Regardless of the type of promises made, managers and employees in a total quality setting should keep them and expect others to do the same. It is easier to trust people who are dependable, even when we don't agree with them, than it is to trust someone who is not dependable. "Be a person of your word" is a good rule of thumb to follow when trying to build trust.

QUALITY TIP

Role of Trust in an Organization

"Trust is the lubrication that makes it possible for organizations to work. It's hard to imagine an 'organization' without some semblance of trust operating somehow somewhere. An organization without trust is more than an anomaly, it's a misnomer, a dim creature of Kafka's imagination. Trust implies accountability, predictability, reliability."

Source: Warren Bennis and Burt Nanus, *Leaders: The Strategies for Taking Charge* (New York: Harper & Row, 1985), 43.

In attempting to build trust in the workplace, managers should take the initiative, even though in a total quality setting trust building and other tasks necessary for success in the intensely competitive global marketplace are the responsibility of everyone. Managers should not sit back and expect trust building to just happen. Their role is to get things moving and to "stir the pot" as necessary to keep things moving.

Motivating employees and continually developing their job skills are important responsibilities of managers in a total quality setting. Managers who are not trusted will not be effective at fulfilling these responsibilities. This is because employees must trust that they, as well as the organization, will benefit from new skills before they are willing to apply themselves to developing the skills.

VALUES AND TOTAL QUALITY

Values are those deeply held beliefs that form the very core of who we are. A person's conscience or internal barometer is based on his or her values. Our values guide our behavior. This statement also applies to organizations. An organization will not produce

QUALITY TIP

Keeping Promises Builds Trust

"We need to generate a feeling of confidence by creating an environment where trust can exist. People don't trust other people that they don't know. Without recognition, there can be no trust. How do you build trust with others? Simply by making promises and keeping them."

Source: James A. Belohlav, *Championship Management: An Action Model for High Performance* (Cambridge, MA: Productivity Press, 1980), 113.

a quality product or provide a quality service unless the organization values quality. Knowledge and skills are important, but by themselves they do not guarantee results. This is because individual employees and organizations as a whole will most willingly apply their knowledge and skills to what they value, what they believe in, what they feel is important.

Ethical behavior begins with values. Values that lead to ethical behavior include fairness, dependability, integrity, honesty, and truthfulness. Values that lead to peak performance and excellence include achievement, contribution, self-development, creativity, synergy, quality, and opportunity.[11] These different sets of values are not mutually exclusive. In fact, they tend to support and supplement each other in a work environment that involves, empowers, values, and nurtures people; one that holds employees responsible but also gives them the support, leeway, and resources needed to fulfill their responsibilities.

Discussion Assignment 4–2, presented at the end of this chapter, shows the dramatic effects values can have in the workplace. In this discussion assignment, the pride of a Japanese individual was hurt by the shoddy reputation of products produced in his country. By making quality a nationally ingrained value, this situation was reversed, and, as a result, Japan emerged as an international leader in quality. This dramatic reversal took more than 20 years to accomplish, which shows the importance of persistence as a value.

INTEGRITY AND TOTAL QUALITY

Another aspect of ethical behavior is integrity. *Integrity*, as a personal and organizational characteristic, combines honesty and dependability. When an individual or an organization has integrity, ethical behavior automatically follows.

It is important for managers in a total quality setting to understand that although honesty is fundamental to it, integrity is more than just honesty. People with integrity can be counted on to do the right thing, do things correctly, accomplish tasks thoroughly and completely, finish work on time, and keep promises. The same is true of organizations.

According to Peter Drucker, one essential step in developing a reputation for integrity is to observe the part of the physician's Hippocratic oath that reads *primum non nocere*, or "above all, do not knowingly do harm."[12] Drucker explains his advice on this subject as follows:

> No professional, be he doctor, lawyer, or manager, can promise that he will indeed do good for his client. All he can do is try. But he can promise that he will not knowingly do harm. And the client, in turn, must be able to trust the professional not knowingly to do him harm. Otherwise he cannot trust him at all. And *primum non nocere* . . . is the basic rule of professional ethics.[13]

Managers who follow this rule will keep the best interests of their organization and employees in mind when making decisions and in all other aspects of their jobs. Committing to not knowingly harm an employee, a customer, the organization, or the public at large forces managers to think about the consequences of their actions before taking action. This rule also applies to employees and the organization as a whole.

QUALITY CASE

Integrity Is the Key at Computer Associates

Computer Associates International, Inc., is sometimes known in the world of business as "the other computer company,"as in *it's not Microsoft, it's the other company*. Computer Associates is the third largest independent software company in the world (behind Microsoft and Oracle). It is a $7 billion company that offers more than 500 different software products in several broad market categories, including security and storage management; network, event, and performance management; and application development.

Some business analysts attribute the success of Charles Wang, founder and chairman of Computer Associates, to smart, aggressive acquisitions and mergers, but Wang attributes his success in building a world-class company to one thing—*integrity*. According to Wang, who founded the company in 1976 to provide software for IBM mainframes, business leaders must have integrity in order to succeed in the long run. Their word must be their bond, a bond that can be trusted implicitly, and they must have a moral compass. Customers, suppliers, and other stakeholders will quickly learn if the executive leaders of a company are trustworthy. If they are not, the company will suffer.

As an example of ethical business dealings, Wang—who has guided his company through numerous acquisitions and mergers—says that some companies use legal technicalities to wiggle out of obligations made by companies they acquire. Computer Associates honors the obligations, good and bad, of the companies it takes over. According to Wang, legal loopholes do not change what is right. Executives at Computer Associates know they are expected to do the right thing even when it hurts. Doing the right thing has given Wang and his company a well-deserved reputation for integrity; a reputation that pays off in more than $6 billion in sales every year.

Source: Thomas J. Neff and James M. Citrin, *LESSONS FROM THE TOP,* (New York: Doubleday, 2001), 327–332.

RESPONSIBILITY AND TOTAL QUALITY

Part of ethical behavior is accepting responsibility. This is critical in the modern workplace because employees are drawn from a society that, as a rule, shuns responsibility—which is why ours has become such a litigious society.

People want to blame others for their own shortcomings and failures. Students graduate from high school unable to read and immediately file lawsuits against the school board as if they had no part in their own failure. A burglar trips on a skateboard after robbing a house, sues the homeowner, and wins! Inmates upset over the quality of

QUALITY TIP

Individual and Organizational Integrity

"Integrity has been the hallmark of the superior organization through the ages. Be that as it may, today's accelerating uncertainty gives the issue new importance. People on the front line must be able to deal quickly across traditional functional barriers; sole source arrangements must be made with suppliers in the face of uncertain future demands. Successful organizations must shift from an age dominated by contracts and litigiousness to an age of handshakes and trust."

Source: Tom Peters, *Thriving on Chaos* (New York: Harper & Row, 1987), 626.

food take their guards hostage and burn down an entire wing of the prison, saying their rights have been violated. Modern society has evolved into one that focuses on rights but ignores the responsibilities that must accompany those rights.

Passing blame has become commonplace. Employees often refer to their employer as "they" rather than "we." Go to a fast-food restaurant or a retail store and complain to a salesclerk. Chances are good that the salesclerk will pass the blame on to an unseen "they." This is not ethical behavior. In a total quality setting, people are responsible for their actions and accountable for their performance.

Accepting responsibility helps build trust, integrity, and all the other elements of ethics that are so important in a total quality environment. W. Stephen Brown claims that there are only two routes that managers, employees, and organizations can take: performance or excuses.[14] He describes these approaches as follows:

> Two distinct and entirely different attitudinal approaches exist, based on these actions, and only one manages successfully. Internalists, those who are performance oriented, accept personal accountability for their actions, successes, and failures. They know that if they feel unhappy with their results, they have only to look into a mirror to stare the culprit in the eye.
>
> Others refuse to accept their responsibility for their position in life and hide behind excuses. Because they constantly blame some external source, condition, or other people for their personal failures, we'll call them externalists. We'd rather not call them managers.[15]

Brown describes the latter approach, failing to accept responsibility, as a formula for failure.[16] This is an example of why ethical behavior is not just something that is adhered to for the sake of being nice. In the long run, it is also the profitable thing to do.

MANAGER'S ROLE IN ETHICS[17]

By applying the information set forth so far in this chapter, managers can make ethical decisions. Unfortunately, deciding what is ethical is much easier than actually doing what is ethical. In this regard, trying to practice ethics is like trying to lose weight. It is

QUALITY TIP

Integrity Is More than Honesty

"A reputation for integrity is earned only through doing what one has agreed to do, doing it on time, and with completeness. Just being honest is not enough. Honesty is mostly not doing things that are dishonest and is more or less expected of respectable people. Integrity though, is built up block by block through planned employee and management actions based on processes and procedures that are completely understood and agreed upon."

Source: Phillip B. Crosby, *The Externally Successful Organization* (New York: McGraw-Hill, 1988), 32.

not so much a matter of knowing you should exercise and cut down on eating as it is a matter of following through and actually doing it.

This fact defines managers' roles with regard to ethics. Managers have three main responsibilities: (a) they are responsible for setting an example of ethical behavior, (b) they are responsible for helping employees make ethical choices, and (c) they are responsible for helping employees follow through and exhibit ethical behavior after the appropriate choice has been made. Figure 4–3 shows three approaches managers can use in carrying out their responsibilities relating to ethics.

Figure 4–3
Approaches to Ethics

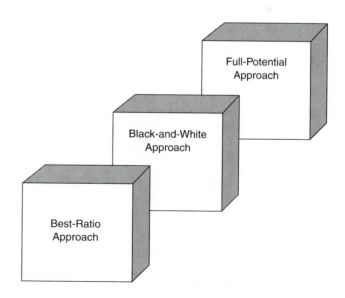

Best-Ratio Approach

The best-ratio approach is a pragmatic approach based on the belief that people are basically good, that in the right circumstances they will behave ethically, and that under certain conditions they can be driven to unethical behavior. Therefore, managers should do everything possible to create conditions that promote ethical behavior and try to maintain the best possible ratio of good choices to bad choices and ethical behavior to unethical behavior. When hard decisions must be made, managers should make the choice that will do the most good for the most people. This approach is sometimes called *situational ethics*.

Black-and-White Approach

With the black-and-white approach, right is right, wrong is wrong, and conditions are irrelevant. The manager's job is to make ethical decisions and carry them out. It is also to help employees behave ethically regardless of circumstances. When difficult decisions must be made, managers should make fair and impartial choices regardless of the outcome and do the right thing without concern for short-term circumstances.

Full-Potential Approach

With the full-potential approach, decisions made are based on how they will affect the ability of those involved to achieve their full potential. The underlying philosophy is that people are responsible for realizing their full potential within the confines of morality. Choices that can achieve this goal without infringing on the rights of others are considered ethical.

The values of the organization and the manager will determine which approach is used. Which one is best is a philosophical question that could be debated at length without being resolved and can only be discussed within the context of a values system.

QUALITY TIP

Managers Must Take Responsibility

"In business, everything begins and ends with management. And in order to work effectively, management must be accountable. When Harry Truman was President of the United States, he had a sign in the Oval Office: The Buck Stops Here. Each manager should adopt the same dictum."

Source: W. Stephen Brown, *13 Fatal Errors Managers Make and How You Can Avoid Them* (New York: Berkley, 1985), 2.

ORGANIZATION'S ROLE IN ETHICS

Organizations have a critical role to play in promoting ethical behavior among their employees. Managers cannot establish ethical standards without support from all levels above them in the organization. An organization's tasks in ethics can be summarized as these:

- Creating an internal environment that promotes, expects, and rewards ethical behavior
- Setting an example of ethical behavior in all external dealings[18]

Creating an Ethical Environment

An organization creates an ethical environment by establishing policies and practices that ensure that all employees are treated ethically, and then by enforcing those policies. Do employees have the right of due process? Do employees have access to an objective grievance procedure? Are appropriate health and safety measures in place to protect employees? Are employees protected from harassment based on race, sex, or other reasons? A company that establishes an environment that promotes, expects, and rewards ethical behavior can answer "yes" to all of these questions.

One effective way to create an ethical environment is to develop an ethics philosophy with specific guidelines for putting the philosophy into operation, to put it in writing, and to share it with all employees. Martin Marietta Corporation of Orlando, Florida, has a *Code of Ethics and Standards* that is shared with all employees. The *Code* begins with the following statement of philosophy:

> Martin Marietta Corporation will conduct its business in strict compliance with applicable laws, rules, regulations, and corporate and operating unit policies, procedures, and guidelines, with honesty and integrity, and with a strong commitment to the highest standards of ethics. We have a duty to conduct our business affairs within both the letter and the spirit of the law.[19]

QUALITY TIP

A Formula for Failure

"We may predict and calculate the amount of failure an individual will experience with this formula: 'People fail in direct proportion to their willingness to accept socially acceptable excuses for failure.'"

Source: W. Stephen Brown, *13 Fatal Errors Managers Make and How You Can Avoid Them* (New York: Berkley, 1985), 3.

This statement sets the tone for all employees at Martin Marietta. It lets them know that upper management not only supports ethical behavior but expects it. This approach makes it less difficult for managers when they find themselves caught in the middle, between the pressures of business and the maintenance of ethical behavior in their departments.

In addition to its corporate ethics policy, Martin Marietta also published a more specific *Credo and Code of Conduct*. The Martin Marietta *Credo* is summarized as follows:

In our daily activities we bear important obligations to our country, our customers, our owners, our communities, and to one another. We carry out these obligations guided by certain unifying principles:

- Our foundation is INTEGRITY.
- Our strength is our PEOPLE.
- Our style is TEAMWORK.
- Our goal is EXCELLENCE.[20]

This *Credo* tells employees that they have obligations that extend beyond the workplace and that how they perform their work can have an impact, negative or positive, on fellow employees and on their organization, customers, and country. Key concepts set forth in the *Credo* are integrity, people, teamwork, and excellence. Managers who stress, promote, and model these concepts will make a major contribution to ethical behavior in the workplace.

The concept of ethics in the workplace is not new. According to David Shanks, Robert Wood Johnson, the leader who built Johnson & Johnson into a major international corporation, developed the following ethics credo for his company in the mid-1940s:

- To customers and users: quality and service at reasonable prices.
- To suppliers: a fair opportunity.
- To employees: respect, equal opportunity, and a sense of job security.
- To communities: a civic responsibility.
- To the environment: protection.
- To shareholders: a fair return.[21]

Written philosophies and guidelines such as those developed by Martin Marietta Corporation and Johnson & Johnson are the first step in creating an ethical environment in the workplace. Managers can play a key role in promoting ethical behavior on the job by encouraging upper management to develop written ethics philosophies/credos/guidelines and then by modeling the behavior they encourage.

Setting an Example

Organizations that take the "Do as I say, not as I do" approach to ethics will not succeed. Employees must be able to trust their employers to conduct all external and internal dealings in an ethical manner. Companies that do not pay their bills on time, companies that pollute, companies that do not live up to advertised quality standards, companies that do not stand behind their guarantees, and companies that are not good neighbors in their communities are not setting a good ethical example. Such companies can expect employees to mimic their unethical behavior. Finally, in addition to creating an ethical internal environment and handling external dealings in an ethical manner, organizations must support managers who make ethically correct decisions—not just when such decisions are profitable but in all cases.

HANDLING ETHICAL DILEMMAS

Every manager will eventually confront an ethical dilemma. When this happens, the manager's response is very important. Figure 4–4 provides guidelines managers can use in such situations. Later, Discussion Assignment 4–4 describes an ethical dilemma of the type a modern manager might face in a total quality setting.

ETHICS TRAINING AND CODES OF BUSINESS CONDUCT

Ethical behavior and the rationale for it can be taught. In fact, almost 40% of the organizations in the United States with 100 or more employees provide ethics training. A survey by the Ethics Resource Center in Washington, D.C., revealed that "28 percent of the 711 responding companies provide specific training on ethics. That number jumps much higher if you include those companies providing training only on narrow ethics topics particularly important to specific industries."[22]

What are these industry-specific issues? The Ethics Resource Center identified the following as topics that are widely addressed to in corporate-sponsored ethics training programs:[23]

- Drug and alcohol abuse
- Employee theft

Before taking any action in situations involving ethical dilemmas, answer the following questions:

- What solution is most likely to build trust among those involved?
- What solution fits best into the company's value system?
- What solution will pass the morning-after, front-page, mirror, role reversal, and commonsense tests?
- What solution is most likely to enhance the organization's integrity?
- What solution is the most responsible option?

Figure 4–4
Guidelines for Handling Ethical Dilemmas

- Conflicts of interest
- Quality control
- Misuse of proprietary information
- Abuse of expense accounts
- Plant closings and layoffs
- Misuse of company property
- Environmental pollution
- Methods of gathering competitors' information
- Inaccuracy of books and records
- Receiving excessive gifts and entertainment
- False or misleading advertising
- Giving excessive gifts and entertainment
- Kickbacks
- Insider trading
- Relations with local communities
- Antitrust issues
- Bribery
- Political contributions and activities
- Improper relations with local government representatives
- Improper relations with federal government representatives
- Inaccurate time charging to government
- Improper relations with foreign government officials
- Making exaggerated advertising claims

Codes of Business Conduct

Ethics training should not take place in a vacuum, nor should it be nothing more than a list of "thou shalt nots." Rather, organizations should develop codes of business conduct written from a positive perspective that encourage employees to do the right thing. Figure 4–5 is an example of one organization's code of business conduct.

Ethics training is becoming increasingly important as the pressures of succeeding in an intensely competitive global marketplace grow. Based on survey responses from 1,073 Columbia University Business School alumni, former associate professors John Delaney and Donna Sockell estimate that the average business school graduate faces about five major ethical dilemmas annually.

> Among the dilemmas reported were incidents involving insider trading, solicitation of bribes, income tax evasion, the marketing of unsafe products, discrimination against women and minorities, kickbacks, discharge of untreated toxic wastes into the environment, and the falsification of company invoices to support a request for a larger budget.[24]

In discussing what works and what does not work when providing ethics training, Brad Lee Thompson makes the following recommendations:[25]

Code of Business Conduct
INTERNATIONAL SERVICE CORPORATION

The owners and management of International Services Corporation (ISC) are committed to the highest ethical standards in the conduct of the company's business. All employees at all levels are expected to conduct their dealings with customers, suppliers, colleagues, and each other with honesty and integrity. At ISC, the ethical way is the right way. When making decisions, ISC personnel will endeavor to achieve fairness to all stakeholders. Our reputation is our most important strategic advantage in the global marketplace.

Figure 4–5
Example of a Company's Code of Business Conduct

- *Stimulate discussion.* Talking through ethical dilemmas and how they might be responded to gives people the opportunity to consider various opinions and gain different perspectives. It also allows people who have confronted ethical dilemmas to explain how they handled them and what the results were.

- *Facilitate, don't preach.* Lectures that resemble sermons are more likely to turn people off than to help them develop an ethical point of view. Facilitating discussions that help people come to their own conclusions will be more effective than lecturing them on what they should do and how they should behave.

- *Integrate ethics training.* Ethical decisions are not made in a vacuum, but ethics training is often provided in one. Separate courses tend to make ethics appear to be something that is itself separate and apart from other workplace issues. Almost everything done in the workplace has a moral dimension; ethical questions are integrated throughout everything done in the workplace. Consequently, ethics training should be incorporated into an organization's overall workplace training program including orientation, employee development, and management training.

- *Highlight practical applications.* The study of ethics is typically viewed as lofty and philosophical and, by necessity, does have a philosophical foundation. However, ethical dilemmas in the workplace are very practical in nature, and the consequences of decisions made regarding these dilemmas typically have tangible consequences. For this reason, it is important to use case studies that allow participants to relate to and personally understand the consequences of their opinions.

MODELS FOR MAKING ETHICAL DECISIONS

Ethics as a concept exists within a framework defined by organizational values. Just as the values of organizations can differ, so can the decision-making models used for determining the ethical course of action in a given situation. This section briefly describes several such models that can be used for making ethical decisions. The actual model chosen will depend on the values of the organization and of the larger community in which the organization does business.

QUALITY TIP

Importance of Ethics Training

"Ethics training becomes important when organizations need to respond to challenges to their behavior. Questions about fairness, equal treatment, morality . . . may look simple at first, but they can quickly become mired in circular discussions. An ethics training program cannot transform personal values, cure the world's ills, or provide a bully pulpit for self-styled corporate philosophers. But it can provide a mechanism for discussing ethical issues and dilemmas so optimal choices can be made."

Source: Brad Lee Thompson, "Ethics Training Enters the Real World," *Training* (October 1990), 84.

In a total quality organization all stakeholders—external and internal—have a say in establishing and judging the corporate values. Consequently, it is important to select a model that will withstand the scrutiny of all stakeholders. Having done so, an organization should stick with its model and apply it consistently. The most widely used models are as follows:

■ *Categorical imperative model.* The categorical imperative model is also known as the *black-and-white model*. With this model, right is right, wrong is wrong, and there are no gray areas.

■ *Full-disclosure model.* With this model, the functional criterion is a simple question: Could the organization explain its actions to the satisfaction of a broad cross section of stakeholders? Only when this question can be answered in the affirmative is an action considered ethical. This model has the advantage of applying the values of stakeholders in deciding what is ethical.

■ *Doctrine of the mean model.* In this model, the *mean* refers to the average or middle point between two extremes. Translated for practical application, this model suggests that in any situation, a moderate middle-ground option is likely to be an ethical option. Said another way, this model suggests that moderation is ethical.

■ *Golden Rule model.* This model is based on the Golden Rule that says, "Do unto others as you would have them do unto you." It is one of the most popular models in Western society. Like the full-disclosure model, it takes the viewpoints of stakeholders into account in deciding what is ethical.

■ *Market-ethic model.* This model is based on the belief that any legal action that promotes profitability is ethical. Proponents of this model profess that the purpose of a business is to make a profit. Consequently, what is ethical should be decided within a framework of profit and loss. They argue that in the long run the market will reject unethical corporate behavior, making it thereby unprofitable.

■ *Organizational ethic model.* This model is based on loyalty to the organization. Its underlying premise is that the most ethical decision is the one that best serves the organization's interests. Unless an organization has adopted a set of guiding principles that ensure ethical behavior, the organizational ethic model is difficult to defend.

■ *Equal freedom model.*   The underlying principle of this model is that organizations have the freedom to behave as they wish unless their actions infringe on the rights of stakeholders. This is a more confining model than it might appear at first glance. For example, suppose an organization decides to use a new chemical that improves product quality and costs substantially less than the one it will replace. Before making a decision, the organization learns that the community's current water safety technology may not be able to screen out the chemical. The worst-case scenario is that it could contaminate the local water supply. Applying the equal freedom model, adopting the new chemical would be unethical because it might violate the rights of stakeholders (everyone who depends on the local water supply).

■ *Proportionality ethic model.*   This model is based on the assumption that the world is so complex that decisions are seldom clearly right or wrong. Consequently, the best an organization can do is make sure that the good outweighs the bad when making decisions.

■ *Professional ethic model.*   This model is based on the principle of peer review. It states that a decision is ethical if it can be explained to the approval of a broad cross section of professional peers. Professions that subscribe to this model typically adopt a professional code of ethics.

BELIEFS VERSUS BEHAVIOR: WHY THE DISPARITY?

Ethics in the workplace manifests itself through the application of such values as honesty, loyalty, fairness, caring, respect, tolerance, and duty. Most, but not all, people in the workplace subscribe to these values. Why, then, is there so often a gap between what people believe and what they do? In other words, if people believe in honesty, why are they sometimes dishonest? If people believe in fairness, why are they sometimes unfair? These are questions that could be debated at length in a broad philosophical context. However, several reasons explain the disparity, at least on a practical level.

Self-Interest/Self-Protection

People are, by their very nature, self-interested and, as a result, self-protective. The driving force behind such slogans as "Me First" and "Looking Out for Number One" is a self-centered attitude. Most people work against human nature to put someone else's needs ahead of their own. Yet, this is precisely what one must do to deal ethically with fellow employees and the public at large.

Being honest sometimes means putting yourself at a disadvantage or having to admit something you would rather not. In such situations, the natural instinct for self-protection takes over and the inclination is to evade the truth. Consider the example of John.

John had some minor repair work to do on his truck but didn't have a certain tool he needed. Consequently, at the end of his shift at Autoworld, Inc., he borrowed the needed tool from the company's tool locker. That night he completed the work on his truck, but the next morning he forgot about the tool and left it at home. By the time he remembered the tool, it was too late to go back home without missing the beginning of his shift. John drove on to work hoping that the tool wouldn't be needed that day. If nobody needed it, the tool wouldn't be missed and he could return it the next day.

Everything went well until midmorning when a job came in requiring the missing tool. When it wasn't in its place in the tool locker, the shift supervisor began asking around to determine who was using it or who had used it last. Finally, when a thorough search of the shop failed to produce it, the supervisor asked if anyone had borrowed it.

John knew that Autoworld had a strict policy against just this sort of thing. He also knew that if he didn't speak up, the last person to have used the tool would be required to pay for it. That was the rule. Unfortunately, if he did admit having borrowed the tool, he would be reprimanded and fined, even if he returned it. John did not want to see a colleague forced to pay for his mistake. But, on the other hand, he had already been reprimanded and fined once this month. He certainly didn't want to go through that again. John faced a common dilemma: tell the truth and bear the consequence, or give in to his natural instinct for self-protection. Faced with such a dilemma, people will go one way or the other, depending on which factor—conscience or self-protection—has the strongest pull on them at the time.

It is not uncommon for self-interest or self-protection to win out under the pressure of the moment, only to have the conscience take over once there has been time for quiet reflection. This is why stories abound about thieves who later return the money or successful people who make a large financial contribution to assuage their guilt over an earlier transgression.

Conflicting Values

People who believe in ethical values (honesty, loyalty, fairness, etc.) sometimes find themselves in situations where these values seem to conflict. For example, consider the dilemma faced by Mary Ann, a sales representative for Construction Products, Inc. (CPI).

CPI is having a bad year and desperately needs every contract it can get. Mary Ann has an opportunity to win a contract to supply all of the kitchen cabinets for a 56-house subdivision that is going to be built in the near future. But there is a problem; CPI cannot deliver the cabinets by the required date. Failure to deliver on time will throw off the customer's entire production schedule. Mary Ann's boss is pressuring her to agree to the delivery dates, even though they both know that the deadline cannot be met. Mary Ann wants to give an honest reply with a more realistic delivery date, but she is in a quandary over what to do.

This morning her boss took her aside and said, "You are so worried about honesty that you've forgotten about loyalty. What about loyalty to this company and your friends who work here? If we don't bring in some work soon, we are all going to be out looking for jobs!"

Honesty versus loyalty—what does one do when ethical values seem to conflict? People obviously choose one over the other based on their interpretation of the situation, the facts as they know them, and contributing personal influences. However, rather than ask what to do when ethical values seem to conflict, it might be better to ask whether the values do, in fact, truly conflict.

What kind of loyalty would require one to be dishonest, unfair, or disrespectful? Although ethical values sometimes appear to be in conflict, a closer look will usually reveal a different story. For example, the loyalty issue in Mary Ann's case was false loyalty. True loyalty would rarely, if ever, require dishonesty.

Tangible/Intangible–Immediate/Deferred

One of the main reasons that people make decisions that run counter to their beliefs is that the benefits of ethical decisions are often intangible or deferred. Put another way, the consequences of unethical behavior are often intangible or deferred while the perceived benefits are usually both tangible and immediate.

Take the case of Mary Ann previously described. If she is willing to deceive the contractor by submitting a false delivery date, there will be a direct benefit that is both immediate and tangible. Her company will win a badly needed contract, and she will be the author of the victory. The downside is that at some point in the future the company will lose the trust and, as a likely result, the future business of the contractor she deceived. The benefit in this situation is immediate, the downside is deferred.

Making Ethics Tangible and Immediate

Because the benefits of ethical behavior can be perceived as being intangible and deferred, people will sometimes choose the unethical option—even people who believe in ethical values. The challenge to management is to help employees see that the benefits of ethical behavior are tangible and that even when deferred they still accrue.

Periodic focus groups conducted by an outside facilitator in which employees discuss ethics-related issues can be an effective way to make ethics tangible. During these meetings employees discuss very specific situations that include ethical dilemmas. The facilitator asks questions such as, "What would you do in this situation? Why? What is the right thing to do? What would keep you from doing the right thing? What are the consequences of choosing an unethical option?"

Group members discuss the issue and respond to the facilitator's questions. The facilitator's job is to guide the discussion toward the tangible and immediate consequences of unethical behavior, or, put another way, the tangible and immediate benefits of ethical behavior.

Sometime during the meeting, the facilitator will ask the following types of questions and lead participants through discussion and debate:

- How does management reward ethical behavior?
- How does management unknowingly promote unethical behavior?
- Does management unknowingly reward unethical behavior?

Discussing these questions will sometimes reveal that management expects ethical behavior but does not reward it or, worse yet, unknowingly rewards unethical behavior. For example, is ethical behavior a part of the performance evaluation process? Does management publish ethics guidelines and reward employees who follow them? Is management's commitment to ethical values real or just lip service? For example, in the earlier case of Mary Ann, would upper management have supported the supervisor or Mary Ann?

ETHICAL DILEMMAS: CASES

This section contains ethical dilemmas that are representative of these faced by managers every day in the world of business. While studying these dilemmas, the reader is encouraged to consider the various factors such as pressure from superiors or peers, personal interest, ambition, financial need, job security, and others that tend to promote unethical behavior on the part of people who are normally honest and trustworthy. While reading these cases, ask yourself, "What would I do in the same situation if I were facing the same pressures?"

Let's consider some examples. Certain models of sport utility vehicles manufactured by Ford begin to show a pattern of high-speed accidents. The similarity in these accidents leads investigators to suspect tire defects. Ford quickly points to its supplier, Firestone. Firestone defends itself and points back at Ford. Union Carbide establishes a processing plant in Bhopal, India, where the laws protecting the safety of employees and the environment are less rigid than those in the United States. The processes at Union Carbide's Bhopal plant involve the use of extremely toxic chemicals and gases. When various emergency protection systems either are not working or fail, more than 40 tons of lethal gas is released into the atmosphere, killing more than 3,000 people.

Most of the discussion surrounding the Ford–Firestone and Union Carbide cases focused on litigation and who would eventually be held responsible for financial damages. However, very little attention was paid to the ethical questions involved. Did these multinational corporations behave in ways that were fair to all stakeholders? Should companies locate plants in developing nations to take advantage of less restrictive safety and environmental protection laws than the U.S.? Should the company that actually sells the product to the consumer pass responsibility along to its supplier, or accept responsibility itself?

The Ford–Firestone and Union Carbide cases received a great deal of media attention worldwide. However, these "big name" cases represent just a few of the thousands of similar situations that arise in the corporate world every year. From small "mom-and-pop" operations to large multinationals, dealing with ethical dilemmas is an everyday part of doing business. In order to succeed in the long term, managers must know how to deal with these dilemmas, and they must understand that just knowing what is right is not enough. Most people who commit ethical violations know what is morally correct.

Typically, people intuitively know the difference between right and wrong. Since this is the case, why, then, do basically "good" people still sometimes behave unethically? This is a valid question and one that has been debated and discussed by philosophers for thousands of years. After all, if we know what is right, will we not do what is right? The answer to this question, unfortunately, is "not always." Even people who have a strong sense of right and wrong can be pressured to behave unethically. This is because an individual's sense of right and wrong can sometimes be overpowered by a stronger sense of ambition, need, fear of the consequences of making the ethical choice, peer pressure, pressure from superiors, and numerous other human factors.

Case 1: "I Need This Promotion"

Janice Carlson had always seen herself as an ethical person. She took pride in always telling the truth, even when doing so was uncomfortable. She also insisted that those she supervised at Comstock Engineering Company (CEC) do the same. Carlson frequently admonished her employees to be "straight" with her. She was fond of saying, "I can accept mistakes. They happen. I can even overlook an occasional bad day. But I will not put up with lying." Close friends knew that Carlson's distaste for lying grew out of an unhappy marriage she had endured for years with a husband who lied to her as a matter of course. When she could take her husband's dishonesty no more, Carlson had filed for divorce.

Her commitment to honesty is why Carlson now feels, as she quietly admits to herself, "lower than a snake in the grass." What makes things even worse is that this is a day on which Carlson should be overjoyed. After 15 years of loyal and effective service to CEC, several of which were spent as the only female engineer in the company, Carlson has just been promoted to director of the civil engineering department. Her promotion means a substantial salary increase, and Carlson needs it. Her daughter has just started college at a private institution. It is an excellent school, but the tuition rate is sky high and her ex-husband, true to form, has refused to help. Why then, on this day of all days, does Carlson feel so bad? The answer is simple: she got the promotion because she lied.

The process for selecting CEC's new Director of Civil Engineering had been difficult. The competition had been especially tough. One of Carlson's long-time colleagues and friends had also been a leading candidate. Since Carlson and her friend were equally qualified and equally experienced, the ultimate selection had come down to solving a complex engineering problem developed by the outgoing director, who was retiring.

A couple of days before the candidates were scheduled to take the promotion test, Carlson had gone to the director's office to return a file she had borrowed. The problem she would have to solve on the promotion test was on the director's desk. The director was out of the office for the day. Carlson saw the problem and knew immediately what it was. She started to turn away, but felt herself drawn to it. Almost without realizing what she was doing, Carlson leaned over the director's desk and looked at the solution. It was a really tough problem.

When the two candidates had completed the test, Carlson's friend and colleague asked how she had done. "I think I solved it," was her response. "Not me," said her friend. "That was the trickiest engineering problem I've ever seen. I've heard that the director had this really complicated problem that no one has ever been able to solve, except him, of course. I don't suppose you had ever seen this problem before, had you?" Janice Carlson could not look her friend in the eye when she said, "No. I've heard about it too. But that was the first time I had ever seen it. I guess I just got lucky." Her friend had smiled and held out his hand, saying, "Anyway, congratulations. It looks like you get the promotion."

Carlson is a person who prides herself on honesty, but in this case her personal interest overcame her commitment to the truth. On the one hand, she needs the promotion in order to help pay her daughter's college costs. On the other hand, the way she received it was dishonest. Put yourself in Carlson's shoes. What would you have done?

Case 2: To Pay or Not to Pay? That Is the Question

John Hingas didn't know what to do, but he did know that he would have to make a recommendation, and soon. He had been the leading marketing representative for Government Products, Incorporated (GPI), for years. In fact, he was practically a legend in the company. That's why GPI's president had given him the current assignment to "break into the Mexican market." GPI produces various office products ranging from desks and chairs to filing cabinets and shelves. The company's major customers are local, state, and federal government organizations.

With the passage of the North American Free Trade Agreement (NAFTA), GPI's executives had decided to expand into Mexico. Unfortunately, they were getting nowhere. After nine months of concerted effort, GPI had nothing to show for its attempts to gain a foothold in Mexican markets except a stack of invoices for airline tickets, motel rooms, and restaurants. Finally, GPI's executives decided to send in the "A Team." That is when John Hingas received the call. While meeting with GPI's executive management team, Hingas quickly showed why he had always been so effective. After analyzing the company's marketing plan for Mexico, Hingas told the executives, "I have just one question. How many of the marketing representatives we send to Mexico actually speak Spanish?" There was an embarrassed silence before Hingas said, "Why don't we step back from the Mexico initiative for a while and give me a chance to look into it? I'll then come back with recommendations."

"How long do you need?" asked the company's CEO.

"Six months," said Hingas

"Why so long?"

"Because before I go down to Mexico to look into things, I need to learn to speak the language."

Eight months and many trips to Mexico later, Hingas knew exactly what would be necessary to succeed in the Mexican markets. By learning to speak Spanish and by getting to know a number of key contact people, Hingas had learned precisely what GPI would have to do in order to compete in its targeted market in Mexico. In a word, the answer was "bribery." GPI could make the best products in the world at the most reasonable prices, but unless its marketing representatives became adept at playing the bribery game, the company would never sell one piece of furniture to a government organization in Mexico. GPI's competitors had already figured this out and were using it to their advantage.

Hingas knew GPI could "play the game" as well or even better than its competitors, but should it? On the one hand, bribery is simply a way of life, a part of the culture in the markets GPI is trying to reach. The hard truth is clear to Hingas; no bribes, no contracts. However, with just a few well-placed bribes, GPI could increase its annual sales by more than 15 percent in less than two years. On the other hand, GPI enjoys a well-deserved reputation for integrity with its customers, and nobody in the company wants to damage that reputation. Hingas has a recommendation to make, and he will have to make it soon. Put yourself in his place. What would you recommend?

Case 3: The Product Is Inferior, but the Profits Are Good

The executive management team of Athletic Footwear, Incorporated (AFI), faces both a threat and an opportunity. The threat is that unless it can find a buyer for a large production run of soccer shoes, the company is going to lose a lot of money. The opportunity is that the vice-president of marketing has found a buyer. The problem is that, although this batch of shoes is the company's best-selling, most popular model, the shoes are defective.

Several months earlier, AFI's management team had decided to save on production costs by using a different glue provided by a new supplier. The glue came highly recommended, and it was much less expensive than that previously used. Consequently, AFI's management team had jumped at the opportunity to save money without first running in-house tests on the glue. Much to their dismay, the new glue turned out to be inferior to that normally used when securing the sole of the shoe. Now the company is stuck with a warehouse full of defective shoes.

Normally the company would simply write off the defective shoes and absorb the loss. However, the company has just gone through a year-long battle to stave off a hostile takeover. As a result, its coffers are practically empty and its debt has nearly doubled. Nobody seated around the table in the executive conference room is in a mood to just absorb the potential loss they face. Legal action against the supplier has already been ruled out for fear of permanently damaging the company's image and credibility. Nobody wants the company's regular customers to know that a defective batch of shoes was produced. Management doesn't want customers thinking, "If AFI produces one large batch of defective shoes, maybe it will produce another."

The potential buyer is a distributor that has retail outlets throughout South America. This company is even willing to pay more than the market price for the shoes in order to be the first distributor in South America to carry the AFI brand. No sport in South America is more favored than soccer, and the AFI soccer shoe is very popular in the United States, Canada, and Europe. The shoe has a reputation for being comfortable and durable. It lasts a long time in even the most demanding conditions. But the defective batch in question won't; in fact, based on initial trial runs, the soles will probably begin to separate after less than 20 hours of use. What should AFI's executives do? Should they sell the shoes, knowing they are defective, or destroy them and find a way to take a loss they really don't need at this point in the company's history? If you were an AFI executive, what would you suggest?

SUMMARY

1. *Ethics* is the study of morality. As applied in the workplace, morality is translated into standards of conduct. If a certain behavior is illegal, it is also unethical. However, conduct may be legal but unethical. In determining whether a certain action or

decision is ethical, managers can apply the following tests: morning-after, front-page, mirror, role reversal, and commonsense. The Five P's of Ethical Power are purpose, pride, patience, persistence, and perspective. Ethical behavior can be influenced by an individual's ego strength, Machiavellianism, and locus of control.

2. Trust is a critical element of ethics, which, in turn, makes ethics critical in total quality. Many of the fundamental elements of total quality depend on trust and ethical behavior, including communication, interpersonal relations, conflict management, problem solving, teamwork, employee involvement and empowerment, and customer focus. Trust can be built by being loyal to those not present, keeping promises, and sincerely apologizing when necessary.

3. *Values* are those core beliefs that guide our behavior. Individuals and organizations apply their knowledge and skills most willingly to efforts in which they believe. Managers should work to establish an environment in which values that lead to ethical behavior and values that lead to peak performance are the same.

4. Integrity requires honesty, but it is more than just honesty. *Integrity* is a combination of honesty and dependability. People with integrity can be counted on to do the right thing, do it correctly, and do it on time.

5. Accepting responsibility is part of ethical behavior. People who pass blame are not behaving ethically. In a total quality setting, people are responsible for their performance. When speaking of their organization, ethical people say "we" instead of "they."

6. Managers play a key role in ethics in an organization. They are responsible for setting an example of ethical behavior, helping employees make ethical choices, and helping employees follow through and behave ethically after making an ethical choice. In carrying out these responsibilities, managers can use the best-ratio approach, black-and-white approach, and full-potential approach.

7. The organization's role in fostering ethical behavior includes creating an ethical environment and setting an ethical example. Key in creating an ethical environment is having a comprehensive ethics policy. Key in setting an example is following the policy, expecting all employees to follow the policy, and rewarding those who do.

8. In handling ethical dilemmas, managers should select the option that is most likely to build trust, integrity, and a sense of responsibility and that is most likely to pass the various ethics tests (i.e., front-page, morning-after, etc.).

9. People who believe in ethical values will sometimes make unethical decisions because of self-interest, self-protection, conflicting values, or because they see the benefits as being intangible or deferred.

KEY TERMS AND CONCEPTS

Best-ratio approach	Ethics
Black-and-white approach	Front-page test
Commonsense test	Full-potential approach
Conflicting values	Integrity
Ego strength	Locus of control
Ethical dilemma	Machiavellianism

Mirror test	*Primum non nocere*
Morality	Purpose
Morning-after test	Responsibility
Patience	Role reversal test
Persistence	Trust
Perspective	Values
Pride	

 FACTUAL REVIEW QUESTIONS

1. Define the term *ethics*.
2. What is morality?
3. Explain how a certain behavior could be legal but not ethical.
4. List and explain five tests that can be used to determine if a choice or a certain behavior is ethical.
5. Describe the Five P's of Ethical Power.
6. Explain the three personality measures that can influence a person's ethical behavior.
7. What role does trust play in a total quality setting?
8. Describe how managers can build trust.
9. What role do values play in a total quality setting?
10. What role does integrity play in a total quality setting?
11. Explain the concept *primum non nocere*.
12. What role does responsibility play in total quality?
13. Describe and differentiate among the following approaches to ethics: best-ratio approach, black-and-white approach, and full-potential approach.
14. What is the manager's role in ethics?
15. Explain the organization's role in promoting ethical behavior.
16. Why, in your own words, would an otherwise ethical person make an unethical decision?

 CRITICAL THINKING ACTIVITY

An Ethical Dilemma

Image Products, Inc., and Lovan Corporation are major producers of shampoos, conditioners, and other bath products for women. They are also intense competitors. Recently Lovan has surged ahead in the marketplace on the strength of a new shampoo that is outselling the leading products of all competitors, including Image Products. This new product is based on a supersecret formula that produces a shampoo with a silky texture and a pleasant aroma. Consumers love it.

John Parker, supervisor of the research and development department for Image Products, is under intense pressure to replicate Lovan's formula. Unfortunately for Parker, all of his attempts to break the code have failed. As a result, Image Products has lost so much market share that company-wide layoffs are imminent. Parker's wife and their daughter, who is 3 months pregnant, both work at Image Products. Both will lose their jobs in the first round if layoffs are necessary. Because the firm is located in a rural one-company town, the chances of his wife and daughter finding another job are slim at best.

A potential solution has fallen into Parker's lap. A disgruntled chemist from Lovan has applied for a vacant position in Parker's department. He has made it clear to Parker that, if hired, he can bring Lovan's coveted formula with him. However, he will do so only if Parker brings him in at top dollar—a demand that is not justified by the chemist's credentials.

Assume that John Parker is a friend and that he asks your advice about how to handle this situation. What will you recommend to him, and why? What model would you apply in dealing with this dilemma if you were in Parker's position? Why?

DISCUSSION ASSIGNMENT 4–1

Trust Building at Motorola

"Trust is accomplished in two ways: line of sight and equity. Line-of-sight management at Motorola means that individuals can see the impact of their actions and the actions of others on the success of the organization. In actual practice, line of sight revolves around the use of teams so that individuals have an opportunity to see that their actions can affect their team's performance. Because PMP teams are small and structured around naturally separate work functions, line of sight is readily apparent. Line of sight, however, goes farther than just the immediate work group. The operating environment is set up so that employees can judge policies, procedures, and managers based on what is actually done, rather than just on what is said. The second element in developing trust is equity. Equity is addressed primarily through a corporate bonus structure. Rewards are based on skill levels, experience, and education and are made as close in time as possible to actual performance. Thus, a sense of sharing and sharing fairly is achieved throughout the organization."

Discussion Assignment
Discuss the following questions in class or outside of class with fellow students:

1. In your own words, can you explain how Motorola's line-of-sight system builds trust?
2. Would a system such as Motorola's corporate bonus structure cause you to trust the company's senior managers?

Source: James A. Belohlav, *Championship Management: An Action Model for High Performance* (Cambridge, MA: Productivity Press, 1980), 66.

DISCUSSION ASSIGNMENT 4–2

Values and Quality in Japan

"We have focused on individuals, yet there are some dramatic examples of changes for entire nations that have come from conscious application of values. As a young Japanese businessman visiting the West in the 1950s, Akio Morita was deeply humiliated to learn that Made in Japan was an international synonym for shoddiness, a phrase that produced jokes. Morita returned to Japan determined to change that, and as chief executive of Sony, he is one of the business leaders who have made that determination a reality. Over the past thirty years, Morita says, 'We have been striving to be the Picassos and Beethovens of electronics.' Is this as good as the ones made in Japan?"

Discussion Assignment

Discuss the following questions in class or outside of class with your fellow students:

1. In your opinion, how did Japan become the world's leading producer of consumer electronics products?
2. What will the United States have to do to regain its leadership in this critical market?

Source: Charles Garfield, *Peak Performers: The New Heros of American Business* (New York: Morrow, 1986), 267.

DISCUSSION ASSIGNMENT 4–3

Ethics at Lockheed Martin

Lockheed Martin Corporation (formerly Marietta Corporation) is one of the largest government contractors in the United States. As a producer of military material and equipment, Lockheed Martin is paid by public tax dollars. Consequently, it is critical that the company and its employees follow a high ethical standard. To ensure that this happens, Lockheed Martin developed a *Code of Ethics and Standards of Conduct* booklet that is given to all employees. Employees must acknowledge receipt of the booklet in writing. Contained in this booklet is the following philosophy statement:

> Marietta Corporation will conduct its business in strict compliance with applicable laws, rules, regulations, and corporate operating unit policies, procedures, and guidelines, with honesty and integrity, and with a strong commitment to the highest standards of ethics. We have a duty to conduct our business affairs within both the letter and the spirit of the law.

The company's corporate ethics booklet goes into great detail in such areas as bidding, negotiations, performance on contracts, conflict of interest, acceptance of gifts, political contributions, and other ethics-related areas of concern.

Discussion Assignment

Discuss the following question in class or outside of class with your fellow students:

1. If you were a Lockheed Martin employee, would you feel comfortable pointing out unethical behavior in the company? Why, or why not?

Source: Martin Marietta Corporation, *Code of Ethics and Standards of Conduct* (Orlando, FL: Martin Marietta Corporate Ethics Office, 1990), 2–12.

DISCUSSION ASSIGNMENT 4–4

An Ethical Dilemma

Vanessa Jones is the manager of the phenolics department of PlastiTech, a manufacturer of industrial plastics and composite materials. She is facing a dilemma. There is an opening in her department, and her brother has applied for it. Because her brother has been out of work for several months, he is having a great deal of trouble supporting his wife and two children. The family pressure on Vanessa to hire her brother is intense. But there is a problem: although she has no doubt her brother could do the job well, two other applicants are better qualified. Further, Vanessa, her brother, and the two better qualified applicants are all minorities, so her company's equal employment opportunity policies will not help in making the decision, nor will the company's other employment policies, which allow the hiring of family members. Vanessa feels a strong sense of responsibility for the productivity of her department. She also loves her brother and wants to help him. This is the first real ethical dilemma she has ever faced.

If Vanessa doesn't hire her brother, she might be ostracized by her family, and her brother's already precarious financial condition might get even worse. On the other hand, she is responsible for hiring the best new team member she can identify. This type of ethical dilemma is not uncommon in the modern workplace.

Discussion Assignment

Discuss the following question in class or outside of class with your fellow students:

1. If you found yourself facing a similar dilemma, what would you do? Why?

ENDNOTES

1. Edward W. Stead, Dan L. Worrell, and Jean Garner Stead, "An Integrative Model for Understanding and Managing Ethical Behavior in Business Organizations," *Journal of Business Ethics* 9 (1990), 233.
2. Paul Taylor, *Principles of Ethics: An Introduction* (Encino, CA: Dickson, 1975), 78.
3. Peter Arlow and Thomas A. Ulrich, "Business Ethics, Social Responsibility, and Business Students: An Empirical Comparison of Clark's Study," *Akron Business and Economic Review* 3 (1980), 17–23.
4. Kenneth Blanchard and Norman Vincent Peale, *The Power of Ethical Management* (New York: Ballantine, 1988), 10–17.
5. Blanchard and Peale, 79.
6. L. K. Trevino, "Ethical Decision-Making in Organizations: A Person-Situation Interactionist Model," *Academy of Management Review* 11 (1986), 601–17.
7. J. E. Prebel and P. Miesing, "Do Adult MBA and Undergraduate Business Students Have Different Business Philosophies?" *Proceedings*, National Meeting of the American Institute for the Decision Sciences, November 1984, 346–348.
8. F. Luthans and R. Kreitner, *Organizational Behavior Modification and Beyond: An Operant and Social Learning Approach* (Glenview, IL: Scott, Foresman, 1985).
9. The same is true of customer focus. When a company asks its external customers to define quality, it is saying, "We trust your judgment and want your input."
10. Stephen R. Covey, *The 7 Habits of Highly Effective People* (New York: Simon & Schuster, 1989), 196.
11. Charles Garfield, *Peak Performers: The New Heros of American Business* (New York: Morrow, 1986), 266.
12. Peter F. Drucker, *Management: Tasks, Responsibilities, Practices* (New York: Harper & Row, 1974), 368–369.
13. Drucker, 369.
14. W. Stephen Brown, *13 Fatal Errors Managers Make and How You Can Avoid Them* (New York: Berkley, 1985), 3.
15. Brown, 3.
16. Brown, 3.
17. David L. Goetsch, *Effective Supervision* (Upper Saddle River, NJ: Prentice Hall, 2001), 119.
18. Goetsch, 137.
19. Martin Marietta Corporation, *Code of Ethics and Standards of Conduct* (Orlando, FL: Martin Marietta Corporate Ethics Office, 1990), 1.
20. Martin Marietta, 2–3.
21. David C. Shanks, "The Role of Leadership in Strategy Development," *Journal of Business Strategy* (January/February 1989), 32.
22. Brad Lee Thompson, "Ethics Training Enters the Real World," *Training* (October 1990), 84.
23. Thompson, 85.
24. Thompson, 86.
25. Thompson, 91–94.

CHAPTER FIVE

Partnering and Strategic Alliances

"A friendship founded on business is better than a business founded on friendship."
John D. Rockefeller, Jr.

MAJOR TOPICS

- Partnering/Strategic Alliances
- Innovative Alliances and Partnerships
- Internal Partnering
- Partnering with Suppliers
- Partnering with Customers
- Partnering with Potential Competitors
- Global Partnering
- Education and Business Partnerships

Partnering for mutual benefit is fundamental to total quality. In an intensely competitive marketplace, where quality is defined by the customer, such practices as low-bid contracts, antagonistic internal relationships, and attempting to operate as an island are being replaced by *partnering*. Working together for mutual benefit sounds like a nice thing to do, and it is. However, being nice has little to do with this contemporary approach to doing business. On the contrary, the partnering philosophy is solidly grounded in the practical demands of the marketplace. This chapter provides the information needed to facilitate partnering relationships with suppliers, customers, internal units, and potential competitors.

PARTNERING/STRATEGIC ALLIANCES

The simplest way to understand the concept of *partnering* or the strategic alliance is to think of it as working together for mutual benefit. Those who work together may be

Figure 5–1
Potential Partnership Participants

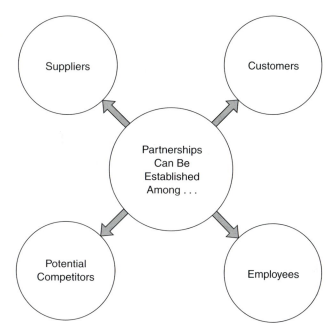

suppliers, fellow employees, customers, and even businesses that are potential competitors (see Figure 5–1). Charles C. Poirier and William F. Houser, in their book *Business Partnering for Continuous Improvement*, describe partnering as

> the creation of cooperative business alliances between constituencies within an organization and between an organization and its suppliers and customers. Business partnering occurs through a pooling of resources in a trusting atmosphere focused on continuous, mutual improvement.[1]

The maximum benefits of partnering are realized when all parties in the chain of partners cooperate. In the traditional supplier–customer chain shown in Figure 5–2, each link in the chain operates independently. Invisible walls exist between each one. The manufacturer in the middle of the chain produces a product used by the customer (end user). For this example, assume that the product is an upscale running shoe. The manufacturer receives leather, fabric, synthetic rubber, glue, and other materials from its suppliers. However, because there is no partnering among the three links in the chain, the manufacturer does not fully understand who buys its shoes and why, what the end users like and dislike about the shoes, or what changes end users think would improve the shoes. Because the manufacturer doesn't know its market and because it doesn't partner with suppliers, the suppliers can't help it better meet the needs of end users.

Now look at the contemporary supplier–customer chain (Figure 5–3). The walls are removed, and the overlapping portions represent partnering. In this model, the manufacturer knows who buys its shoes and why. Further, by involving its customers in the ongoing product development process, the manufacturer designs in the features important to end users and eliminates problems or undesirable characteristics. Because the

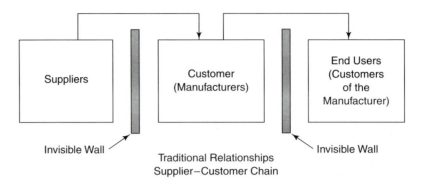

Figure 5–2
Traditional Relationships: Supplier–Customer Chain

manufacturer in the contemporary model knows its customers and their needs, it can work with its suppliers to enlist their help in meeting those needs.

Figures 5–2 and 5–3 depict relationships between external suppliers and customers. These same models are repeated numerous times within companies, in which employees are internal suppliers and internal customers to each other. The types of partnering relationships explained earlier must also occur within total quality companies. In traditional supplier–customer relationships there are invisible walls that block out communication and cooperation (Figure 5–2). With contemporary supplier–customer relationships, these walls are broken down. There is communication, input, feedback, and cooperation (Figure 5–3).

Benefits of Partnering

Several benefits can be derived from partnering (see Figure 5–4). Partnering can lead to continual improvements in such key areas as processes and products, relationships between customers and suppliers, and customer satisfaction. Internal partnering can

QUALITY TIP

Partnering Leads to Important Benefits

"In its simplest form, business partnering can be characterized as a process of improvement that brings an organization and its constituent parts to the point where special benefits not found in competing networks can be created."

Source: Charles C. Poirier and William F. Houser, Business Partnering for Continuous Improvement (San Francisco: Berrett-Koehler, 1993), 57.

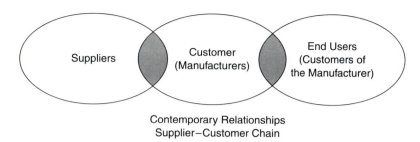

Figure 5–3
Contemporary Relationships: Supplier–Customer Chain

improve relationships among employees and among departments within an organization. When taken as a whole, these individual benefits add up to enhanced competitiveness.

Partnering Model

Establishing partnering relationships with suppliers and/or customers is a process that should be undertaken in a systematic way. Haphazardly formed relationships based on halfhearted commitments are worse than having no partnerships at all. Figure 5–4 provides a model that can be used as a guide in the formulation of partnering relationships with both external and internal suppliers and customers.

Develop a Partnering Briefing

Partnering is about creating cooperative alliances. Before trying to establish such an alliance, make sure everyone involved understands partnering as a concept. A briefing in which the concept is explained and employees who will be involved are given opportunities to question and discuss should be the first step.

Identify Potential Partners

Any external or internal supplier or customer is a potential partner. Choose partners in an order determined by how much value the partnership can have toward enhancing quality, productivity, and competitiveness. Internal partnerships between the design and manufacturing departments have considerable potential in this regard. Partnerships between the manufacturing department and major external suppliers also have excellent potential.

Identify Key Decision Makers

In every organization (unit, department, etc.), there are key people whose support is needed to make an initiative involving their organization work. Identify these key decision makers in any organization considered a potential partner. Their support must to be won if a successful partnership is to be formed.

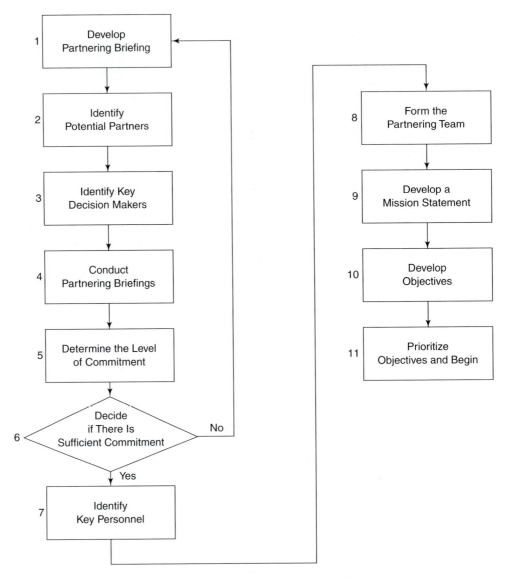

Figure 5–4
Partnering Model

Conduct a Partnering Briefing

Call a meeting of the key decision makers in both organizations, yours and the potential partner's. Present a briefing explaining the partnering concept, with time built in for discussion and questions. This briefing should answer such questions as the following:

- How can we mutually benefit from a partnership?
- What is expected of each partner?

Determine the Level of Commitment

After the key decision makers have been briefed, gauge their level of commitment. Are they willing to commit to the partnership for the long term? Are they willing to make any and all procedural and/or philosophical changes that may be necessary for the partnership to work?

Decide Whether There Is Sufficient Commitment

If the key decision makers show noticeable reluctance, they are not likely to make a full commitment to the partnership. There is no need to proceed any further with potential partners who seem reluctant. The better course of action in such a case is to break off further involvement and begin the process again with another potential external partner. However, if the level of commitment is sufficient, proceed to the next step in the process.

Identify Key Operational Personnel

If the level of commitment is sufficient to proceed with the partnership, who are the key people from both organizations needed to put it into operation? Are personnel needed from marketing? Purchasing? Engineering? Manufacturing? Receiving? Accounting? Identify the people who will be needed to put into action the commitment made by executive-level decision makers.

Form the Partnership Team

The key people identified as necessary to putting the partnership into operation should be formed into a team. This means more than just naming them to the team. They must

QUALITY TIP

The Partnership Team Should Involve All Constituencies

"The alliances that form the central part of the partnering process must supersede normal business arrangements. They need to go beyond the terms of a collective bargaining agreement or traditional interdepartmental relationships. The alliances necessary to make partnering a success have to extend to all constituencies affecting the organization and must bind them to a mutually beneficial improvement effort."

Source: Charles C. Poirier and William F. Houser, Business Partnering for Continuous Improvement (San Francisco: Berrett-Koehler, 1993), 72.

be given opportunities to get to know and trust each other. The success of the partnership will depend in great measure on the willingness and ability of these team members to work together in a mutually supportive and trusting manner. (Teamwork and team building are discussed in Chapter 10.)

Develop a Mission Statement

The partnership team needs a clear and concise mission statement so that everyone involved understands what the team is supposed to do. The mission statement should be developed by executive-level decision makers from both organizations. Figure 5–5 is an example of a mission statement for a supplier–customer partnership.

Develop Objectives

The mission statement is written in general terms. It is translated into more specific terms by objectives. These objectives should be developed by the partners and ratified by the executive-level decision makers of both partnership organizations. Well-written objectives are stated in measurable terms, such as these:

- Each week 100 low-voltage power supplies will be delivered according to a just-in-time schedule.
- All power supplies delivered will be free of defects.

These objectives are specific and measurable. In the first one, the expected quantity (100) and the delivery schedule can both be easily checked. In the second objective, if even one power supply is rejected, the objective has not been accomplished. All such objectives must be agreed to by both partners before being sent forward for ratification by executive-level personnel.

This mission statement is the guiding vision of the customer/supplier partnership that exists between Keltron Electronics (the customer) and Precision Machining, Inc., (the supplier). The mission of this partnership is to promote a mutually supportive working relationship that will help maximize the quality, productivity, competitiveness, and profitability of both partners. In carrying out this mission the following agreements apply:

- **Precision Machining, Inc.,** agrees to deliver quality discrete components to Keltron Electronics just in time and at the best possible price.
- **Keltron Electronics** agrees to purchase discrete machine components from Precision Machining, Inc., at a negotiated price as a sole source provider without requests for competitive bids.

Figure 5–5
Mission Statement: Keltron Electronics–Precision Machining, Inc., Customer–Supplier Partnership

Prioritize Objectives and Begin

It will typically take several objectives to completely translate the mission statement into measurable action. The importance of these objectives is relative. Although all are important, the objectives should be prioritized and listed in order from the most important to the least. After priorities have been established and confirmed by executive-level personnel, the work necessary to accomplish them begins. Results should be monitored and appropriate action taken when problems arise.

INNOVATIVE ALLIANCES AND PARTNERSHIPS

Partnering between and among companies can take many forms. A group of small and medium-sized companies might form a partnership to save money through *consortium buying*. This is a concept wherein two or more companies get together to purchase common items in bulk; by doing so, they gain the cost benefits of size. Another innovative type of partnership involves suppliers and their customers. Major customers agree to welcome an *in-house supplier representative* who works with the customer's personnel to continually improve the supplier–customer relationship. Having an in-house supplier representative gives the customer an advocate who sees firsthand what is needed from the supplier, when, where, and why. In turn, it allows the supplier to gain first-hand knowledge of how to better serve the customer. Another innovative partnership takes the form of the *customer focus group.* The customer focus group is an example of a partnership between a supplier and the users of its products or services. Such a focus group consists of customers who are pulled together by a supplier to provide feedback concerning the quality of an existing product or service or input concerning a proposed product or service. There are many examples of innovative partnerships in today's highly competitive global marketplace.

Coca-Cola and Nestlé formed a research partnership to develop a line of ready-to-drink teas and coffees. Procter & Gamble and Wal-Mart formed a partnership to better serve their shared customers through improved shipping and receiving procedures. IBM has a formal partnering program called the Business Partner Program in which IBM and more than 1,000 partners share information of mutual benefit and develop strategies to better serve mutual customers. Partnerships among automobile manufacturers are now common; Ford partners with Mazda, and General Motors with Suzuki.

There are no limits on the types of partnerships and alliances that businesses can form for mutual benefit, and there are no limits as to the types of companies that might form partnerships; even competitors may do so. Anything that can be done better through cooperation represents a potential basis for a partnership.

INTERNAL PARTNERING

Partnering should begin at home. This means an organization should initiate its partnering efforts internally. Internal partnering occurs at three levels:

- Management-to-employee partnerships
- Team-to-team partnerships

QUALITY TIP

Internal Partnering and Trust

"What is required is confidence in the power of the average worker and management to raise performance by magnitudes of 400–1,000 percent. Also required is a level of trust and honesty on a world-class scale that will induce workers to give freely of their knowledge and skills so higher performance can be achieved."

Source: Charles C. Poirier and William F. Houser, Business Partnering for Continuous Improvement (San Francisco: Berrett-Koehler, 1993), 173.

■ Employee-to-employee partnerships

The overall purpose of internal partnering is to harness the full potential of the workforce and focus it on the continuous improvement of quality.

Internal Partnering Defined

Internal partnering goes by a number of different names. It has been called *employee involvement* (see Chapter 8), *employee empowerment*, and various other terms. Regardless of what it is called, the concept can be defined as follows:

Internal partnering is creating an environment and establishing mechanisms within it that bring managers and employees, teams, and individual employees together in mutually supportive alliances that maximize the human resources of an organization.

The key concepts in this definition are as follows:

■ Environment
■ Mechanisms
■ Mutually supportive alliances
■ Human resources

Does an organization have an environment that is conducive to internal partnerships? If it does, partnering is welcomed, encouraged, and rewarded. Providing a conducive environment is important, but by itself it is not enough. Within the environment, mechanisms must exist through which employees are able to channel their ideas for improvement. Mutually supportive alliances among management and employees, teams, and individual employees are relationships in which each partner helps the other do better. Mutual support within an organization is a much more effective way to achieve continuous improvement than the traditional approach of internal competition among individuals and teams.

With internal competition, somebody within the organization loses. When this happens, the organization also loses. But with mutually supportive internal partner-

QUALITY CASE

Internal Partnerships at Citigroup

Strategic alliances and partnerships should start at home, according to Sandy Weill, chairman and CEO of Citigroup, Inc. Citigroup was formed in 1998 through the merger of Citicorp and the Travelers Group, a move that made it the largest financial services company in the United States. Citigroup does business, and lots of it ($82 billion annually) in more than 100 countries with product lines that include credit cards, consumer banking, commercial banking, insurance, and investment services. One of the most effective strategies used by this successful multinational company is internal partnerships. Weill contends that Citigroup gains a competitive advantage in the global marketplace by forming internal partnerships with the company's employees.

The internal partnerships at Citigroup are based on Weill's philosophy that employees should be paid well, but their pay should be based on performance. According to Weill, many companies make the mistake of failing to differentiate employee earnings based on performance. Citigroup, on the other hand, makes the majority of an employee's earnings contingent on his performance and the performance of his or her division within the company. Weill started a program of stock-option grants that allows employees to receive earnings in the form of stock. This strategy did more than just give the employees a more secure future or a better reason to continually improve performance; it made them partners in the company. As a result of the stock-option program, employees at Citigroup are true partners in the success of their company. As partners, they feel a strong sense of ownership over their individual performance as well as that of the company. The company, in turn, has gained the strategic advantage of a dedicated, determined, motivated workforce.

Source: Thomas J. Neff and James M. Citrin, *LESSONS FROM THE TOP.* (New York: Doubleday, 2001), 333–338.

ships, all internal partners can win and the organization's competitive energy is directed outward against other competing organizations. This is when the organization is truly victorious.

The definition speaks to the importance of human resources. Maximizing human resources is essential in a total quality setting. This is how the most significant workplace improvements are made and maintained.

Ford Motor Company identified the following ways that employees and managers can work together as internal partners to continually improve quality and productivity:[2]

■ Mutually identifying opportunities
■ Working together in problem-solving teams

QUALITY TIP

Human Resources Are Critical Resources

"What an enormous potential must be residing within the minds and hearts of American workers! Leaders must tap that reserve to gain a stronger global position for our businesses. Our successes have been confined to distinct islands of progress instead of being a nationwide characteristic. The missing ingredient is utilization of the total, national talent pool."

Source: Charles C. Poirier and William F. Houser, Business Partnering for Continuous Improvement (San Francisco: Berrett-Koehler, 1993), 159.

■ Mutually analyzing problems and isolating causes
■ Arriving at mutually acceptable alternatives for solutions
■ Mutually selecting the best solution for implementation

Involving both managers and employees as internal partners in these types of activities is an excellent way to maximize human resources and promote mutually supportive alliances. Such alliances will tap the creativity of all internal partners, allowing good ideas to be turned into improvements.

PARTNERING WITH SUPPLIERS

Relationships between an organization and its suppliers have traditionally been characterized by adversarial activities such as the low-bid process, in which at least one and often both parties lose. Rather than working together to find ways for both to win, buyers use their leverage to force suppliers to absorb costs to win the low bid, and suppliers look for ways to minimize their losses by barely meeting the buyer's specifications. Such relationships will not help either party succeed in the long run in a competitive marketplace. Poirier and Houser describe the concept of partnering with suppliers as follows:

> True supplier partnering requires an understanding of each party's needs and capabilities to establish a clear vision for focusing the efforts of people who work for buyer and supplier. Necessary improvement can then be dedicated to those areas identified as requiring attention and improvement. If that focus is placed, as it has been in most cases, solely on the buyer's perceived needs, nothing happens to establish overall improvement to the buyer–seller network. Long-term commitments then become nebulous and difficult to achieve. Continuous improvement is often reduced to a short-term project instead of an ongoing process.[3]

To understand the rationale for partnering with suppliers, one must first understand the goal.

The goal is to create and maintain a loyal, trusting, reliable relationship that will allow both partners to win, while promoting the continuous improvement of quality, productivity, and competitiveness.

The traditional adversarial relationship between suppliers and buyers is not likely to contribute much to the accomplishment of this goal.

Not all suppliers can participate in such relationships. In fact, suppliers should be required to qualify to participate. Qualifying a supplier shows that it can guarantee that its products will be delivered when and where they are needed in the specified quantities and without defects. Suppliers who can meet these criteria all of the time meet the technical requirements to qualify as a partner. Whether they will actually become a partner depends on their level of commitment and the synergism and trust that develops between buyer and supplier personnel.

Mandatory Requirements of Supplier Partnerships

Successful supplier partnerships require commitment and continual nurturing. Poirier and Houser identify the following points as mandatory requirements of supplier partnerships:[4]

- Supplier personnel should meet with buyer personnel beyond those in the purchasing office. It is particularly important for them to meet with personnel who actually use their products so that needed improvements can be identified and made.
- The price-only approach to buyer–supplier negotiations should be eliminated. Product features, quality, and delivery concerns should also be part of the negotiations. The goal of the negotiations should be to achieve the optimum deal when price, features, quality, and delivery issues are all factored in.
- The quality of supplier products should be guaranteed by the supplier's quality processes. The buyer should have no need to inspect the supplier's products.
- The supplier should fully understand and be able to practice just-in-time (JIT). Buyers should not need to maintain inventories.
- Both partners should be capable of sharing information electronically so that the relationship is not inhibited by paperwork. Electronic data exchange is particularly important for successful JIT.

Stages of Development in Supplier Partnerships

Successful supplier partnerships don't just happen overnight; they evolve over time. According to Poirier and Houser, this evolution occurs in the stages depicted in Figure 5–6 and detailed in the following sections.

Uncertainty and Tentativeness

In the uncertainty and tentativeness stage, the buyer and seller are like two people on their first date. There is interest, but it is tentative and prefaced with uncertainty. Neither party knows exactly what to expect of the other. At this point, there is no trust between the partners.

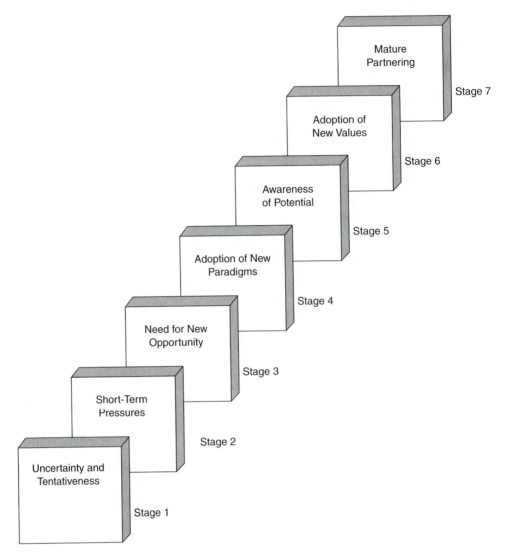

Figure 5–6
Evolution of Supplier Partnerships

Short-Term Pressures

The typical short-term pressures that apply in a traditional business setting manifest themselves in this step. The buyer will be under the usual pressure to cut costs. The supplier will be under the usual pressure to increase sales volume. Both partners will be cautious, and initial attempts to begin putting some substance to the partnership will be probing and vague.

Need for New Opportunity

In the need for new approaches stage, traditional negotiations will inevitably occur. The buyer will press for price discounts, improved payment terms, freight allowances, and other concessions that save it money but cost the supplier. The supplier will press for higher volume to offset concessions made to the buyer. Then, if either partner has involved total quality–conscious personnel, it will dawn on them that quality is not being served by this traditional negotiating. Both will begin to realize that a new approach is needed. If this awareness does not occur, the partnership will fail.

Adoption of New Paradigms

In the adoption of new paradigms stage, both partners explore ways to move toward the concept of *mutual benefit*. The key is for both partners to accept the principle that absorbing costs within the partnership (by either the supplier or the buyer) gives neither an advantage. The best way to promote competitiveness is for both partners to work together to lower costs. This new way of thinking (paradigm) will give the partnership a competitive advantage over other organizations that produce the same product.

Awareness of Potential

In the awareness of potential stage, both partners become fully aware of the possible benefits that can be realized from the partnership. The potential for a true win–win relationship can now be seen. Rather than negotiating price concessions and volume increases, both partners realize that by working together they can exceed any short-term advantages that might have been realized from these traditional negotiating strategies.

QUALITY TIP

World-Class Competitiveness Requires New Paradigms

"The paradigm change we seek is from the traditional emphasis on price, quantity, and delivery to quality, just-in-time, and *kaizen* (continuous improvement). Such a shift will challenge the most dedicated of organizations. To meet these new objectives, both sides need to alter their perspective. The supplier has to have its international house in order, people who understand quality, and a system to sustain and improve current performance. There must be a meaningful improvement policy reflected in everything that is done. It should be possible to select someone from any level of this type of organization who knows how to help the customer."

Source: Charles C. Poirier and William F. Houser, Business Partnering for Continuous Improvement (San Francisco: Berrett-Koehler, 1993), 190.

Discussion Assignment 5–2, presented later in the chapter, illustrates how a supplier and a buyer can work together to cut costs. In a traditional supplier–buyer relationship, it would be unprecedented to allow a team to visit the workplace in the manner shown in this assignment. It would have been considered a violation of proprietary rights. This is why trust is so important in a supplier–buyer relationship.

Adoption of New Values

In the adoption of new values stage, both partners adopt the new values inherent in a true supplier–buyer partnership. These values include trust, openness, and sharing. Each party trusts the other to protect the confidentiality of what they learn about one another. Both parties accept that the more information they share, including financial information, the better prepared they will be to help one another.

Mature Partnering

In the mature partnering stage, the partnership has solidified. A high level of trust and cooperation has been established between the partners. Continuous interfacing between pertinent employees at all levels of both organizations exists as fact. Each partner has a strong self-interest in the success of the other partner.

PARTNERING WITH CUSTOMERS

The term *customer* as used in this section means the end user of the product in question and any buyer of a supplier's products. There are other uses of the term, of course. Internal customers (see Chapter 7) exist in every organization, and organizations that buy from suppliers are customers of those suppliers. However, in this section, the term will be used to mean end users and customers of suppliers. In this context, for example, the customer of an automobile manufacturer might be a consumer who buys one of its models or a car rental agency that purchases its fleet from the manufacturer.

The rationale for forming partnerships with customers in this context is simple: it is the best way to ensure customer satisfaction, which is, in turn, the best way to be competitive. To understand this rationale, answer the following questions:

1. Who knows better what the customer wants, your organization or the customer?
2. What makes more sense, guessing what customers want or asking them?
3. Can a producer benefit from seeing how its product is used by customers?
4. What costs more, making design changes early in the product development cycle or recalling faulty products that have already been produced and purchased by customers?

The answers to these questions form the rationale for partnering with customers. No organization can possibly know better than its customers what the customers want. Customer-defined quality is a fundamental part of the total quality philosophy. Whether it is best to guess what a customer wants or to ask is obvious. No organization can afford to squander its resources and, in turn, its competitive edge guessing what customers want.

Ask. Any organization that produces a product can benefit from observing how it is used by customers.

By involving customers early in the product development cycle, a manufacturer can make changes inexpensively and with relative ease. The further along a product is in the development cycle, the more costly such changes become. If modifications are needed after the product is being used by customers, such as in the case of product recalls, the cost cannot be measured in just dollars and cents. Additional costs accrue in the form of lost consumer confidence, diminished trust, and a tarnished corporate image. Product defects can be corrected much more easily than consumer confidence and trust can be restored.

The customer feedback survey conducted by mail and telephone is a widely used strategy for measuring customer satisfaction. Such surveys have a place. However, their after-the-fact nature limits their usefulness. Unless they are just one part of a much broader set of strategies implemented a great deal earlier in the process, customer satisfaction surveys will have only a limited effect on an organization's ability to compete.

The key to success in partnering with customers is to get them involved early in the product development cycle. Let them preview the design. Allow them to observe and even try prototype models. Get their feedback at every stage in the product development cycle, and make any needed changes as soon as they are identified. When this approach is used, customer satisfaction surveys can solicit feedback from a broader audience to verify the input given earlier in the product development cycle. Discussion Assignment 5–3 illustrates this approach.

PARTNERING WITH POTENTIAL COMPETITORS

Partnering with potential competitors sounds like an odd strategy on the surface. Why would organizations that compete for business in the same markets want to form partnerships? The rationale for partnering with potential competitors is the same as that for partnering with suppliers and customers: competitiveness. This is a strategy that applies more frequently to small and medium-sized firms, but it can also be used by even the largest organizations, and sometimes is. For example, the leading computer companies in the United States may form a partnership to develop the next technological breakthrough before a similar team in Japan, Germany, or some other country beats them to it.

Small and midsized enterprises (SMEs) don't typically develop major technological breakthroughs. However, there are many ways in which SMEs can work together to enhance their competitiveness in spite of being competitors in the same markets. The most widely practiced type of partnership among SMEs is the manufacturing network.

Manufacturing Networks of SMEs

A manufacturing network is a group of individual SMEs that cooperate in ways that increase their quality, productivity, and resultant competitiveness to levels beyond what the individual member companies could achieve by themselves. The concept originated in Europe after World War II. It is still practiced extensively in the Emilia-Romagna region of Italy and in Denmark.

These countries applied the concept to rebuild their manufacturing bases after the devastation of World War II, when resources were insufficient to allow manufacturers to rebuild independent of one another. Consequently, rather than trying to completely retool and restaff independently, companies formed networks and shared both human and technological resources. Manufacturing networks were originally conceived as a way to rebuild. They have since evolved into a way to compete—particularly for SMEs.

What makes a manufacturing network succeed are two concepts: interdependence of member companies and mutual need. The member companies depend on each other in developing mutual solutions to common problems. Gregg Lichtenstein describes this concept as follows:

> Manufacturing networks are characterized by relationships of collaboration or interdependence among firms. The interactions between participants are neither dominated by any one firm nor are they simple economic exchanges motivated by calculations of price and quantity. Firms in these networks are partners—they must rely on each other in order to accomplish their objectives.[5]

Manufacturing networks began to appear in the United States in the 1980s. The earliest known network in the United States is the Garment Industry Development Center, established in New York City in 1984. Figure 5-7 shows the broad industrial clusters in which manufacturing networks can be found in the United States. Of these, the greatest number of networks can be found in the metalworking, woodworking, and textile industries. Table 5-1 lists some of the established manufacturing networks in the United States by location.

QUALITY TIP

Manufacturing Networks: A Definition

"Manufacturing networks are groups of firms that come together to gain competitive advantages that no individual company could achieve working alone. The activities of such networks can be quite diverse. They can engage in joint production, collective marketing, worker training, new product development, technology transfer, and the adoption of total quality management practices."

Source: Gregg A. Lichtenstein, H. Catalogue of U.S. Manufacturing Networks, United States Department of Commerce, National Institute of Standards and Technology, NIST GCR 92-616 (Washington, D.C.: Government Printing Office), 1.

Figure 5–7
Industrial Clusters with Networks
in the United States

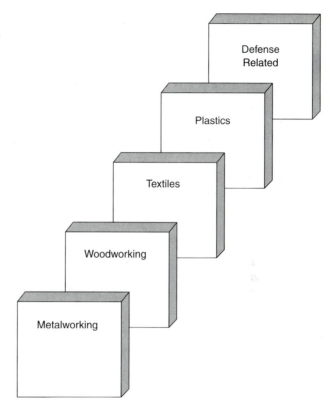

Network Activities

The joint activities in which networks participate vary a great deal, depending on local objectives. Figure 5–8 shows some of the most widely practiced joint activities of manufacturing networks in the United States.

Production

Networked SMEs are able to pursue production contracts larger than any individual member company could undertake alone. Through teaming arrangements, the work and the financing of it are divided among network members as appropriate. The Flex-Cell Group, a manufacturing network in Columbus, Indiana, undertakes joint contracts to produce metal and plastic products. It consists of seven companies that, together, have the following capabilities: pattern making, precision machining, plastics molding and tooling, mechanical engineering, and contract management. Joint projects include the development and production of prototypes for original equipment manufacturers.

Table 5–1
Examples of Manufacturing Networks in the United States

Network	State
The Metalworking Connection, Inc.	Alabama
Technology Coast Manufacturing & Engineering Network	Florida
The Metalworking Consortium	Illinois
The Flex Cell Group	Indiana
Kentucky Wood Manufacturers Network	Kentucky
Louisiana Furniture Industry Association	Louisiana
Mechtech, Inc.	Maryland
Needle Trades Action Project	Massachusetts
Wood Products Manufacturing Network, Inc.	Minnesota
Greater Syracuse Metalworking Industry	New York
Heat Treating Network, Inc.	Ohio
Manufacturing Innovation Network Plastics Initiative	Pennsylvania
Wood Products Competitiveness Corporation	Oregon
WoodNet	Washington

Education and Training

Education and training are often a problem for SMEs. On the one hand, employees need ongoing education and training to continually improve quality, productivity, and competitiveness. On the other hand, they face the following problems because they have a limited employee base:

- Difficulty giving employees time off for education and training and still meeting production schedules
- Difficulty convincing educational institutions that typically need 15 to 20 students to form a class to bring courses on-site
- High expense for education because they do not get the substantial registration discounts large firms get when participating in training provided by private training firms

By partnering, SMEs can solve all three of these problems through economy of scale. Although giving employees time off for training will always be difficult, it can be made easier through the sharing of employees by network members on a reciprocal basis. By bringing together all employees from member companies who need a certain type of training, networks can produce classes large enough to attract educational institutions and to qualify for discounts from private training providers. The Garment Industry Development Corporation (GIDC) in New York City is an example of a network that provides joint training opportunities for its members (see Figure 5–9).

Marketing

Marketing is the most widely practiced joint activity among manufacturing networks of SMEs. Typically the joint capabilities of the network are what is marketed. Member com-

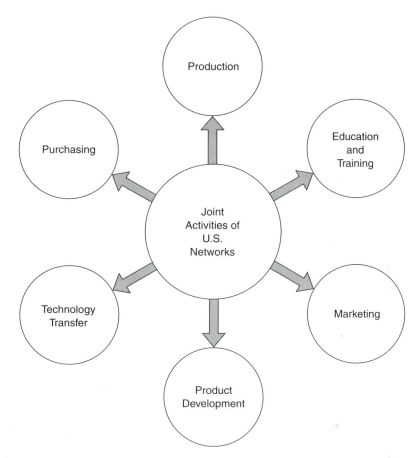

Figure 5–8
Joint Activities of U.S. Networks

panies share the costs of producing marketing tools such as brochures, videos, and promotional materials; of attending trade shows; and of marketing personnel and related expenses such as travel. The Technology Coast Manufacturing and Engineering Network (TeCMEN) in Fort Walton Beach, Florida, conducts a joint marketing program. Members share the costs of participating in selected trade shows, production of a joint marketing brochure, and other miscellaneous related costs (e.g., telephone, travel, postage).

Product Development

Developing new products can be too expensive an undertaking for SMEs. It typically involves such activities as research, design, market analysis, competition analysis, prototype production, performance testing, and test marketing. The costs associated with these activities can be prohibitive for an individual SME. However, when the costs can be divided among network members, product development becomes a more feasible concept.

Figure 5–9
Joint Training Programs:
Garment Industry Development
Corporation (GIDC)

- Sewing machine maintenance and repair
- Computer-assisted shop-floor management
- Manual working and grading operations
- Computerized working and grading operations
- Advanced training (whole-garment construction and quality control)

The Wood Products Manufacturing Network, Inc., in Bemidji, Minnesota, sponsors joint product development projects for its members. The products fall into three categories: craft items, furniture, and building materials.

Technology Transfer

Technology is the physical manifestation of knowledge. *Technology transfer* is the movement of technology from one arena to another. The form of technology transfer that is most readily and widely recognized is the transfer of a new technology from a research laboratory to a production setting. This is often referred to as *technology commercialization*. Another aspect of technology transfer is the movement of a commercialized technology into the hands of users. This aspect is often referred to as *technology diffusion*.

After it has been diffused, technology must be properly used to realize its potential benefits. Proper use of technology requires knowledge. Knowledge sharing is the approach networks use to promote effective technology transfer. An example of how knowledge sharing is applied is the Heat Treating Network, Inc. (HTN), of Cleveland, Ohio. One of the partners in this network is the Edison Material Technology Center (EMTEC). The EMTEC provides a free hotline that HTN members can access for assistance in solving heat treatment problems. The hotline is the primary vehicle for information sharing.

When a technology-related problem is confronted, the member calls EMTEC's hotline and explains it in detail. EMTEC personnel provide a solution within 72 hours or guide the company to another source that can help. The latter approach is known as *brokering*.

Purchasing

One of the most productive applications of the economy of scale gained from networking is in purchasing. SMEs working alone are not able to enjoy the cost savings that large firms achieve by purchasing bulk quantities of expendable materials and other necessities, such as insurance. However, by jointly purchasing necessities, networked SMEs can achieve similar cost savings. For example, members of the Metalworking Connection, Inc., in Arkansas jointly purchase casualty insurance, thereby saving between 20% and 30% of the cost of premiums. (The names, addresses, and telephone numbers of the most active manufacturing networks in the United States are listed in the Appendix.)

The numbers practically sizzle off the page. Just look at this: U.S. direct investment abroad skyrocketed from $467.8 billion in 1991 to more than $796 billion in 1996, according to the

U.S. Department of Commerce. U.S. corporations have invested more than $10 billion in China alone, another $23.6 billion in Brazil, $16 billion in Thailand, and $12.5 billion in Indonesia. Many of these hot markets are growing at a robust GDP (gross domestic product) of 7 percent, or 8 percent a year.

One after another, American companies seek—and seize—new opportunities in rapidly growing yet unfamiliar markets. China, India, Brazil, Russia, and many others offer untold opportunities with their megamillion populations and growing consumer classes. And U.S. firms are right there, offering everything from lightbulbs to power plants and selling their expertise about manufacturing, marketing, and managing global operations.[6]

GLOBAL PARTNERING

The partnering concept, like all contemporary business concepts, has a global aspect. Companies that market to customers worldwide should examine the possibility of partnering with suppliers worldwide. In some cases the government of the host country will actually mandate supplier partnerships as an economic *quid pro quo*. This arrangement is often the case in nondemocratic countries. Government-mandated partnerships are not recommended here because they might be driven more by political than business considerations.

In countries where businesses are free to develop partnerships based on sound business principles rather than politics, the same rules apply as those set forth earlier for domestic partnerships. Modern transportation and telecommunication technologies make geographic separation a manageable issue. But these technologies do not take the place of on-site visits to the facilities of partners, nor do they replace face-to-face interaction with global customers.

A one-size-fits-all product will not suffice in the global marketplace. Local on-the-ground interaction with both suppliers and customers is critical in gearing up for product design. No designer, planner, or manufacturer in the United States can possibly understand all the cultural nuances or country-specific preferences of people from other nations. Put another way, product attributes that are popular in the United States may not be in another country. Consequently, access to suppliers and customers in these countries is critical. Partnering is the best way to gain the necessary access.

EDUCATION AND BUSINESS PARTNERSHIPS

Two of the most important factors in continually improving the performance of an organization are the quality of employees and the quality of human interaction with technology. To improve performance, organizations must first improve their people and the interaction of their people with process technologies. Individuals who lack fundamental work skills cannot perform at globally competitive levels, and people who lack process skills cannot get the most out of technologies available to them. For example, a person who cannot solve general algebraic equations will not be able to learn statistical process control, and a person who uses a word processing system as if it were a typewriter will not get the most out of this technology.

The need to continually improve employees' work skills is the primary force driving business and education partnerships. In such partnerships, educational institutions provide on-site customized training, technical assistance, and consulting services to help organizations continually improve their people and their processes. They also provide workshops and seminars and facilitate focus groups.

Partnering with business and industry has become a common practice for institutions of higher education. Discussion Assignment 5-4 contains two examples of the approaches educational institutions are taking to promote business and industry partnerships.

 SUMMARY

1. *Partnering* means working together for mutual benefit. It involves pooling resources, sharing costs, and cooperating in ways that mutually benefit all parties involved in the partnership. Partnerships may be formed internally (among employees) and externally with suppliers, customers, and potential competitors. The purpose of partnering is to enhance competitiveness. The formation of partnerships should be a systematic process involving such steps as development of a partnering briefing, identification of potential partners, identification of key decision makers, implementation of the partnering briefing, determination of the level of commitment, identification of key personnel, formation of the partnering team, development of a mission statement, development of objectives, prioritization of the objectives, and implementation of the partnership.

2. Internal partnering operates on three levels: management-to-employee partnerships, team-to-team partnerships, and employee-to-employee partnerships. The purpose of internal partnering is to harness the full potential of the workforce and focus it on the continuous improvement of quality. Internal partnering is also called *employee involvement* and *employee empowerment*. Successful internal partnering requires a supportive environment, structured mechanisms, and mutually supportive alliances.

3. The goal of a supplier partnership is to create and maintain loyal, trusting relationships that will allow both partners to win while promoting the continuous improvement of quality, productivity, and competitiveness. The requirements for success in supplier partnerships include the following: supplier personnel should interact with employees who actually use their products, the price-only criteria in the buyer–supplier relationship should be eliminated, the quality of products delivered should be guaranteed by the supplier, the supplier should be proficient in JIT, and both parties should be capable of sharing information electronically. Supplier partnerships typically develop in the following stages: uncertainty and tentativeness, short-term pressure, realization of the need for new approaches, adoption of new paradigms, awareness of potential, adoption of new values, and mature partnering.

4. The rationale for forming customer partnerships is customer satisfaction. The best way to ensure customer satisfaction is to involve customers as partners in the product development process. Doing so is, in turn, the best way to ensure competitiveness. Customer-defined quality is a fundamental aspect of total quality.

5. Small and medium-sized enterprises or SMEs, even those that compete in the same markets, can benefit from partnering. The most widely practiced form of partnership among SMEs is the manufacturing network. A manufacturing network is a group of SMEs that cooperate in ways that enhance their quality, productivity, and competitiveness. Mutual need and interdependence are the characteristics that make manufactur-

ing networks succeed. Widely practiced network activities include joint production, education and training, marketing, product development, technology transfer, and purchasing.

6. Education and business partnerships are formed to help organizations continually improve their people and how well they interact with process technologies. Services provided include on-site customized training, workshops, seminars, technical assistance, and consulting.

KEY TERMS AND CONCEPTS

Brokering	Partnering briefing
Environment	Partnering model
Internal partnering	Partnering with customers
Level of commitment	Partnering with potential customers
Manufacturing network	Partnering with suppliers
Mature partnering	Partnership team
Mechanisms	Price-only approach
Mutually supportive alliances	Short-term pressures
New paradigms	SME
New values	Technology transfer
Partnering	Uncertainty and tentativeness

FACTUAL REVIEW QUESTIONS

1. Define the term *partnering*.
2. What are the benefits of partnering?
3. Describe each step in the partnering model.
4. Define the term *internal partnering*.
5. What is partnering with suppliers?
6. Explain the mandatory requirements of supplier partnerships.
7. List and explain the stages of development in supplier partnerships.
8. Explain the rationale for partnering with customers.
9. What is a manufacturing network?
10. What role does mutual need play in manufacturing networks?
11. List and explain the most widely practiced network activities.
12. What types of services do educational institutions typically provide to business and industry partners?

CRITICAL THINKING ACTIVITY

Does Training Cost or Pay?

John Andrews and Martha Stevens are supervisors in the production department of ATV, Inc., a manufacturer of various types of all-terrain vehicles. ATV is beginning to feel the pressure of global competition. To continue the growth it has enjoyed over the last 5 years, ATV is going to have to improve quality, productivity, customer service, and supplier relationships.

One task force has been formed to investigate the feasibility of supplier partnerships. John and Martha have been appointed to another task force to determine whether a partnership with a local college is feasible as a way to improve employee performance. Representatives from the college have interviewed employees, supervisors, and managers to determine what types of assistance ATV needs.

The college's director of business services made a presentation to the Education Partnership Task Force this morning. He made the following recommendations: (a) all production employees should receive training in bench marking, continuous process improvement, use of the quality tools, and problem solving; (b) the training should be provided on-site and on company time; and (c) the company should pay all costs associated with the training. Andrews and Stevens are discussing these recommendations.

"There is no way ATV is going to pay what the college wants to charge," said Andrews. "And I guarantee there won't be any training conducted on company time. All of this is too expensive. Management will turn the college down flat."

"I don't think so," said Stevens. "If the training improves our performance enough to keep us competitive, it will be worth every penny and more."

Join this debate. Does training for business and industry cost, or does it pay? What is your opinion?

DISCUSSION ASSIGNMENT 5–1

Winter Sports, Inc.

For years, Winter Sports, Inc. (WSI), controlled a substantial share of the snow-ski market. It made good skis, had an excellent marketing program, and invested a respectable percentage of profits into continually modernizing its production facility. However, market research was not a part of its marketing program, supplier relationships were based on low-bid arrangements, and customer feedback was not used in the product development process.

Its failure to form partnerships with suppliers and customers eventually caught up with WSI, resulting in a 42% decrease in sales in just 2 years. Because WSI had antagonistic relationships with its suppliers, it was not aware of a newer, stronger, lighter, and less expensive composite material that would have allowed the company to reduce the cost of its most popular line of skis by 23%. To make matters worse, it turned out that the stronger, lighter material was very popular with customers.

WSI was not aware of this new material or how customers would react to it, but its competitors were. Consequently, the competition's product development cycle got a 2-year head start on WSI. To complicate matters, WSI also failed to recognize the growing popularity of ski boards, an oversight that customer partnering could have prevented. Its competitors, recognizing the trend, got into that market before WSI and established themselves before WSI was able to react. Eventually WSI was purchased by a company that 5 years earlier had been only a minor competitor.

Discussion Assignment
Discuss the following questions in class or outside of class with your fellow students:

1. Is partnering with suppliers a better approach than the low-bid approach? Why, or why not?
2. Is there a downside to partnering with suppliers?

DISCUSSION ASSIGNMENT 5–2

A Supplier–Buyer Partnership That Works

Fort Walton Precision Machining supplies aluminum consoles for various types of electronic measurement devices used in hospitals and clinics. The devices are manufactured by Health Care Electronics, Inc. During the awareness stage of the development of their supplier–buyer partnership, both companies agreed to allow teams to visit their production facilities to identify cost-cutting strategies. A team of managers and employees from Fort Walton Precision Machining spent a week in the production facility of Health Care Electronics observing how its consoles were used and talking to production personnel. At the same time, a team from Health Care Electronics visited Fort Walton Precision Machining.

Fort Walton Machining personnel discovered that two simple design changes in its consoles would allow Health Care Electronics personnel to eliminate two entire steps in its assembly process. Correspondingly, Health Care Electronics personnel found that Fort Walton Precision Machining was spending an inordinate amount of time achieving a surface finish on a part of its consoles that did not require such a finish. In all, the two teams were able to identify ways to cut costs by 13%.

Discussion Assignment
Discuss the following questions in class or outside of class with your fellow students:

1. Have you ever thought, "I wish they would have asked me," about some design features in an automobile or a piece of electronic equipment?
2. Do you see any advantages to be gained by asking for customer input *before* manufacturing a product?

DISCUSSION ASSIGNMENT 5–3

Partnering with Customers Pays Off

Newspaper Concessions Corporation (NCC) manufactures vending machines for newspaper chains nationwide. The machines are placed in conspicuous locations and accessed using the appropriate combination of coins. A great deal of research has gone into continually improving this aspect of the vending machines. Even with this, NCC noticed a sharp drop-off in sales to newspapers serving large urban centers.

As part of its new customer partnership effort, NCC invited representatives from established customers in ten large cities to spend time in its production facility and participate in all phases of the development of the latest model of its vending machine. Input from these representatives resulted in a major design change.

NCC had put a great deal of effort into developing a stronger coin box that could stand up to vandals and thieves. As it turned out, theft of money was only part of the problem. More important to the newspaper representatives was theft of their newspapers. In their cities, people were putting in the correct combination of coins to pay for one newspaper but taking all of the newspapers in the box. The newspaper sellers speculated that drug users were selling the newspapers and keeping the money to support their habits.

What was needed was a vending machine that would drop just one newspaper at a time rather than opening its door to the entire supply. With the necessary design change made, NCC quickly regained its lost sales.

Discussion Assignment
Discuss the following question in class or outside of class with your fellow students:

1. Is there a store, restaurant, or other establishment with which you do business that could improve its service or products? If given the opportunity, what would you tell the owner or manager about improvements that are needed?

DISCUSSION ASSIGNMENT 5–4

Business and Education Partnerships

The Quality Institute (TQI)
The Quality Institute (TQI) is a partnership of Okaloosa-Walton Community College, the University of West Florida, and the Economic Development Council of Okaloosa County. TQI is dedicated to the continual improvement of quality, productivity, and competitiveness in the private and public sectors. To this end, TQI offers education, training, consulting, and technical assistance in a wide range of areas,

including quality management, supervision, manufacturing improvement, workplace health and safety, human resources development and management, and several others.

Institute for Professional Development

The Institute for Professional Development of Okaloosa-Walton Community College was established to help business, industry, and government agencies as well as individuals continually improve their performance in the global marketplace. A knowledgeable and well-trained workforce can provide a formidable competitive advantage. The Institute for Professional Development (IPD) provides seminars, short courses, workshops, and customized contract training to help continually improve the performance of people, processes, products, and organizations. The IPD's services are designed to help individuals and organizations not just survive in today's intensely competitive marketplace but prevail.

Discussion Assignment

Discuss the following questions in class or outside of class with your fellow students:

1. Do you know of any companies that partner with education institutions for training and/or technical assistance?
2. Does the institution you are attending partner with business and industry?
3. In both cases, what is the nature of the partnerships (e.g., what kinds of courses, technical assistance, etc.)?

 ENDNOTES

1. Charles C. Poirier and William F. Houser, *Business Partnering for Continuous Improvement* (San Francisco: Berrett-Koehler, 1993), 56.
2. D. Ephlin, "How to Get Union Leaders and Members Involved in TQM," paper delivered at the Fall Conference of the Quality & Productivity Management Association, St. Louis, MO, September 1991, 2.
3. Poirier and Houser, 201.
4. Poirier and Houser, 183.
5. Gregg Lichtenstein, *H. Catalogue of U.S. Manufacturing Networks*, United States Department of Commerce, National Institute of Standards and Technology, NIST GCR 92-616 (Washington, D.C.: Government Printing Office), 1.
6. Charlene Marmer Solomon, "Don't Get Burned by Hot New Markets," *Global Workforce*, vol. 3, no. 1 (January 1998), 13.

CHAPTER SIX

Quality Culture

On changing the culture: "Don't ever take a fence down until you know the reason why it was put up."

    G. K. Chesterton

MAJOR TOPICS

- Understanding What a Quality Culture Is
- Quality Culture versus Traditional Cultures
- Activating Cultural Change
- Changing Leaders to Activate Change
- Laying the Groundwork for a Quality Culture
- Learning What a Quality Culture Looks Like
- Countering Resistance to Cultural Change
- Establishing a Quality Culture
- Maintaining a Quality Culture

One of the greatest obstacles faced by organizations attempting to implement total quality is the cultural barrier. Many organizations do an excellent job of committing to total quality, involving employees in all aspects of planning and implementation, and providing the training needed to ensure that employees have the necessary skills, only to have their efforts fall flat. The culprit in many of these cases is organizational inertia. No effort has been made to overcome the comfort employees at all levels feel in doing things the way they have always been done. In other words, no effort has been made to change the organization's culture. This chapter explains the concept of organizational culture and how to go about changing it.

UNDERSTANDING WHAT A QUALITY CULTURE IS

To understand what a quality culture is, one must first understand the concept of *organizational culture*. Every organization has one. An organization's culture is the everyday manifestation of its underlying values and traditions. It shows up in how employees behave at work, their expectations of the organization and each other, and what is considered normal in terms of how employees approach their jobs. Have you ever shopped at a store or eaten in a restaurant in which the service was poor and the employees surly or disinterested? Such organizations have a cultural problem. Valuing the customer is not part of their culture. No matter what slogans or what advertising gimmicks they use, the behavior of their employees clearly says, "We don't care about customers."

An organization's culture has the following elements:

- Business environment
- Organizational values
- Cultural role models
- Organizational rites, rituals, and customs
- Cultural transmitters

The business environment in which an organization must operate is a critical determinant of its culture. Organizations that operate in a highly competitive business environment that changes rapidly and continually are likely to develop a *change-oriented* culture. Organizations that operate in a stable market in which competition is limited may develop a *don't-rock-the-boat* culture.

Organizational values describe what the organization thinks is important. Adherence to these values is synonymous with success. Consequently, an organization's values are the heart and soul of its culture.

Cultural role models are employees at any level who personify the organization's values. When cultural role models retire or die, they typically become legends in their organizations. While still active, they serve as living examples of what the organization wants its employees to be.

Organizational rites, rituals, and customs express the organization's unwritten rules about how things are done. How employees dress, interact with each other, and approach their work are all part of this element of an organization's culture. Rites, rituals, and customs are enforced most effectively by peer pressure.

Cultural transmitters are the vehicles by which an organization's culture is passed down through successive generations of employees. The grapevine in any organization is a cultural transmitter, as are an organization's symbols, slogans, and recognition ceremonies.

What an organization truly values will show up in the behavior of its employees, and no amount of lip service or advertising to the contrary will change this. If an organization's culture is its value system as manifested in organizational behavior, what is a quality culture?

> A quality culture is an organizational value system that results in an environment that is conducive to the establishment and continual improvement of quality. It consists of values, traditions, procedures, and expectations that promote quality.

How do you recognize an organization with a quality culture? It is actually easier to recognize a quality culture than to define one. Organizations with a quality culture,

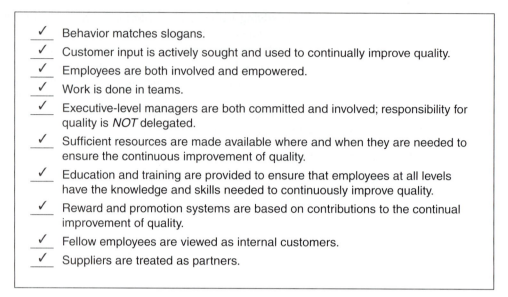

√   Behavior matches slogans.

√   Customer input is actively sought and used to continually improve quality.

√   Employees are both involved and empowered.

√   Work is done in teams.

√   Executive-level managers are both committed and involved; responsibility for quality is *NOT* delegated.

√   Sufficient resources are made available where and when they are needed to ensure the continuous improvement of quality.

√   Education and training are provided to ensure that employees at all levels have the knowledge and skills needed to continuously improve quality.

√   Reward and promotion systems are based on contributions to the continual improvement of quality.

√   Fellow employees are viewed as internal customers.

√   Suppliers are treated as partners.

Figure 6–1
Characteristics Shared by Organizations with a Quality Culture

regardless of the products or services they provide, share a number of common characteristics, presented in Figure 6–1.

How Are Organizational Cultures Created?

Many factors contribute to the creation of an organization's culture. The value systems of executive-level decision makers are often reflected in their organization's culture. How managers treat employees and how employees at all levels interact on a personal basis also contribute to the organizational culture. Expectations are important determinants of organizational culture. What management expects of employees and what employees, in turn, expect of management both contribute to an organization's culture. The stories passed along from employee to employee typically play a major role in the establishment and perpetuation of an organization's culture. All of these factors can either help or hurt an organization.

If managers treat employees with trust, dignity, and respect, employees will be more likely to treat each other in this way, and trust, dignity, and respect in everyday interaction will become part of the organization's culture. On the other hand, if management treats employees poorly, employees are likely to follow suit. Both situations, if not changed, will become ingrained as traditions. These traditions will be perpetuated both by the behavior of employees and by the stories they pass along to one another. This is why it is so important to establish a quality culture. If mistrust is part of the organizational culture, it will be difficult to build partnerships between internal and external customers. It will also be difficult to establish an environment of mutually supportive teamwork. Organizations that have these problems are not likely to be world-class competitors (see Discussion Assignment 6–1, presented later).

Commitment to quality cannot be faked. Employees know when management is just going through the motions. Changing an organization's culture requires a total commitment and a sustained effort at all levels of the organization.

QUALITY CULTURE VERSUS TRADITIONAL CULTURES

Organizations that develop and maintain a quality culture will differ significantly from those with a traditional culture. The differences will be most noticeable in the following areas:

- Operating philosophy
- Objectives
- Management approach
- Attitude toward customers
- Problem-solving approach
- Supplier relationship
- Performance-improvement approach

Operating Philosophy

In an organization with a traditional culture, the primary focus is return on investment and short-term profits. Often the methods used to maximize profits in the short term have a negative effect in the long run. In order to improve the organization's bottom line on the next quarter's profit/loss statement, executives might decide to "unload" a defective product on customers, put off critical technology upgrades, or eliminate training programs for employees. An organization might cut back on equipment maintenance, employee benefits, or performance-incentive programs. All of these shortsighted methods are common in organizations with traditional cultures, and while they might prop up the bottom line temporarily, they invariably lead to disaster in the long run. A short-term operating philosophy is the reason traditional organizations often experience so much turnover at the top. CEOs who apply this short-term operating philosophy are often "cut-and-run" managers who maximize short-term profits by eliminating essential functions, activities, and personnel. They then take their percentage of the resulting profits and leave, only to repeat the charade at another organization.

In an organization with a quality culture, the core of the operating philosophy is customer satisfaction. Quality organizations focus on doing what is necessary to exceed the reasonable expectations of customers. Such an approach can lower profits in the short run, but is the key to long-term survival and prosperity. For example, making a major investment in an expensive technology upgrade can cause the next quarter's profit/loss statement to be flat. Over time, however, the benefits of the new technology will take hold and will be reflected in profit/loss statements for years to come. Organizations that adopt a quality culture typically have less turnover at the top. This is because such a philosophy encourages decision makers to stay in their positions long enough to either enjoy or suffer the consequences of their decisions.

Objectives

Organizations with traditional cultures typically adopt short-term objectives. The focus is on what the organization should accomplish over the next several weeks and months. Organizations that adopt a quality culture plan strategically. They develop both long- and short-term objectives, and they do so within the context of an organizational vision.

Management Approach

In organizations with traditional cultures, managers think and employees do. In fact, employees don't just do, they do what they are told. Managers are seen as "bosses" who give orders and enforce policies, procedures, and rules. In organizations with quality cultures, managers are seen as coaches of the team. They communicate the vision, mission, and goals; provide resources; remove barriers; seek employee input and feedback; build trust; provide training; and reward and recognize performance.

Attitude toward Customers

Organizations with traditional cultures tend to look inward. They are more concerned about their needs than those of customers. Customer relations might actually be adversarial. Organizations with a quality culture are customer focused. Customer satisfaction is the highest priority and is the primary motivation driving continual improvement efforts.

Approach to Problem Solving

There is a lot of finger pointing in organizations with a traditional culture. When problems occur, decision makers and employees tend to expend more energy on deflecting or assigning blame than on identifying the root cause of the problem, which must occur before the problem can be solved. Traditional organizations suffer from the "most valuable player (MVP) syndrome" in which problem solving is viewed as an individual undertaking wherein independent "heroes" operating all alone jump into the breach to put things right just in the nick of time. This approach is erratic at best.

Another phenomenon that occurs in traditional cultures is the "waiting game." With this strategy, decision makers hold back until someone appears to have the problem almost solved; then they jump on board and act as if the idea was theirs all along. Such an approach encourages manipulation and subterfuge rather than innovation and creative thinking.

When difficulties occur in organizations with a quality culture, the focus is on identifying and isolating the root cause so that the problem, and not just its symptoms, can be eliminated. Problem solving is typically a systematic process undertaken by teams, with input solicited from all stakeholders. The goal is to create solutions, not "heroes."

Supplier Relationships

In organizations with a traditional culture, suppliers are kept at arm's length in relationships that are often adversarial. As much pressure as possible is exerted on suppliers to bring down prices and speed up delivery, even when such an approach is likely to drive the supplier out of business. In organizations with a quality culture, suppliers are viewed as partners. Supplier and customer work together cooperatively for the good of both. Each gets to know the other's processes, problems, strengths, and weaknesses and they collaborate, using this information to continually improve the relationship and the performance of both.

Performance-Improvement Approach

In organizations with a traditional culture, performance improvement is an erratic, reactive undertaking that is typically triggered by problems. In organizations with a quality culture, continual improvement of processes, people, products, the working environment, and every other factor that affects performance is at the very core of the operating philosophy.

ACTIVATING CULTURAL CHANGE

To attempt the implementation of total quality without creating a quality culture is to invite failure. Organizations in which the prevailing culture is based on traditional management practices are not likely to succeed in the implementation of total quality. Successful total quality requires cultural change. Several primary reasons cultural change must either precede or at least parallel the implementation of total quality are described here.

Change Cannot Occur in a Hostile Environment.   The total quality approach to doing business may be radically different from what management and employees are accustomed. Managers who are used to sitting in their lonely towers at the top of the pecking order and issuing edicts from on high are likely to reject the concept of employee involvement and empowerment.

Employees who are used to competing against their fellow employees for promotions and wage increases may not be open to mutually supportive internal partnerships and teamwork. Situations such as these can create an environment that is hostile toward change, no matter how desirable that change is. Change can be difficult, even when people want to do so. It can be impossible in a hostile environment.

Moving to Total Quality Takes Time.   The nature of total quality is such that the organization may have to go down somewhat before it can turn things around and start to come up. In a conversion to total quality, positive results are rarely achieved in the short run. This characteristic gives nonbelievers and people who just don't want to change (and such people are often in the majority at first) the opportunity to promote the "I told you it wouldn't work" syndrome.

It Can Be Difficult to Overcome the Past.   Employees who have worked in an organization for any period of time have probably seen a variety of management fads come and

QUALITY TIP

Management by Walking Around (MBWA)

Managers cannot lead while hidden away in their offices. They need to know what is going on. "Management by Walking Around," or MBWA is a technique in which managers get out of their offices and interact directly with employees to exchange information. It is an excellent way for managers to "keep their fingers on the pulse" of the organization. Managers who practice MBWA are burdened with few unpleasant surprises. Managers who do not practice MBWA run the risk of becoming aloof, isolated, and out of touch.

Source: Thomas J. Peters and Robert H. Waterman, *In Search of Excellence* (New York: Harper & Row, 1982).

go. Promoting the latest management gimmick and then letting it die for lack of interest may be part of the existing organizational culture. If this is the case, it will be difficult to overcome the past. Employees will remember earlier fads and gimmicks and characterize total quality as being just the latest one; they may take a "This too shall pass" attitude toward it. The past is not just an important part of an organization's culture; it can be the most difficult part to leave behind.

CHANGING LEADERS TO ACTIVATE CHANGE

Cultural change is one of the most difficult challenges an organization will ever face. It is hard to achieve under even the best of circumstances. Leadership from the top is essential. Consequently, sometimes an organization's culture simply cannot be changed without a change in leadership.

This possibility arises when the staunchest defenders of the status quo are the most senior managers. Senior managers are likely to be the individuals in an organization with the greatest investment in the past and, as a result, the greatest loyalty to orthodoxy. If it is true, as the old adage proclaims, that "an organization is the lengthened shadow of one person," then the CEO must be the key player in changing an organization's culture.

How does one know or how can one tell when it will be necessary to change leaders to change the organization? What follows are several questions that can be used by senior executives for self-assessment or by the organization in making its own assessment of the need for new leadership:

1. Are the current leaders fully knowledgeable of the need to change and the ramifications of not changing?
2. Are the current leaders able to articulate a vision for the new organization?
3. Have the current leaders set the tone for change and established an organization-wide sense of urgency?

4. Are the current leaders willing to remove all obstacles to cultural change?
5. Do the current leaders have a history of following through on change initiatives?
6. Are the current leaders willing to empower employees at all levels of the organization to make cultural change?

In an organization that needs to make a major cultural change, the answer to all of these questions must be yes. Senior executives who fail to comprehend the need to change and the ramifications of not changing cannot lead an organization through a major cultural change. Senior executives who cannot picture in their minds (envision) the new organization and cannot articulate what they see will be unable to lead an organization through the change. If they fail to set the tone for cultural change, they will inhibit rather than lead the organization.

Senior executives who fail to create a sense of urgency will see cultural change fall victim to complacency. Senior executives who are unwilling to remove obstacles that inhibit cultural change have the wrong set of priorities. This sometimes happens when the obstacles are perquisites such as corporate aircraft or luxurious office suites to which executives have grown accustomed.

Senior executives who have a history of starting change initiatives but failing to follow through on them are poor candidates to lead an organization through a major cultural shift. With such executives in leadership roles, employees at all levels are likely to adopt an attitude of "This too shall pass." Finally, senior executives who are unwilling to empower employees at all levels to help lead change will actually ensure that it fails. Cultural change requires support, ideas, and leadership from employees at all levels. Senior executives who are unwilling to empower employees to *think* and *do* will block cultural change.

LAYING THE GROUNDWORK FOR A QUALITY CULTURE

Establishing a quality culture is a lot like constructing a building. First, you must lay the groundwork, or foundation. According to Peter Scholtes, management should begin by developing an understanding of what he calls the "laws" of organizational change.[1] These laws are explained in the following paragraphs.[2]

Understand the History behind the Current Culture. Organizational cultures don't just happen. Somebody wrote the policy that now inhibits competitiveness. Somebody started the tradition that is now such a barrier. Times and circumstances change. Don't be too quick to criticize. Policies, traditions, and other aspects of the existing culture that now seem questionable may have been put in place for good reason in another time and under different circumstances. Learn the history behind the existing culture before trying to change it.

Don't Tamper with Systems—Improve Them. Tampering with existing systems is not the same as improving them. Tampering occurs when changes are made without understanding why a given system works the way it does and without fully comprehending what needs to be changed, and why. To improve something, you must first understand what is wrong with it, why, and how to go about changing it for the better.

QUALITY TIP

Be Sensitive to the Needs of People Involved

"The transformation to quality leadership is often dramatic, and almost always traumatic. Change is seldom easy. It is unlikely anyone will figure out how to change an organization without requiring its people to change. Therefore, we must all be sensitive to the problem that people will have with the transformation."

Source: Peter R. Scholtes, *The Team Handbook* (Madison, WI: Joiner Associates, 1992), 1–20.

Be Prepared to Listen and Observe.   People are the primary inhibitors of change in any organization. Consequently, it is easy to become frustrated and adopt an attitude of "We could get a lot done if it weren't for the people in this organization." The problem with such an attitude is that people are the organization. For this reason, it is important to pay attention to both people and systems. Be prepared to listen and observe. Try to hear what is being said and observe what is not being said. Employees who are heard are more likely to participate in changes than those who are not.

Involve Everyone Affected by Change in Making It.   People will resist change. To do so is normal human behavior. What people really don't like is being changed. It can be difficult to effect change even when people want to do so. It can be impossible when people feel that changes are being imposed on them. The most effective way to ensure that employees will go along with changes is to involve them in planning and implementing the changes. Give them opportunities to express their concerns and fears. Getting problems into the open from the outset will allow them to be dealt with forthrightly and overcome. Shoving them aside or ignoring them will guarantee that even little problems become big ones.

LEARNING WHAT A QUALITY CULTURE LOOKS LIKE

Part of laying the groundwork for a quality culture is understanding what one looks like. This is a lot like a person who wants to lose weight taping a picture of someone whom that person wants to look like to the mirror. The picture serves not only as a constant reminder of the destination but also as a measurement device that indicates when a goal has been met. If a picture of a company with a strong quality culture could be taped to an organization's wall for all employees to see, it would have the following characteristics:[3]

- Widely shared philosophy of management
- Emphasis on the importance of human resources to the organization
- Ceremonies to celebrate organizational events

QUALITY CASE

Culture Is the Key at Southwest Airlines

It is critical for a CEO and every employee to understand the key to a company's success. At Southwest Airlines, the key to success is the company's culture. Consequently, every employee from the CEO to the most junior ticketing agent is actively involved in maintaining and nurturing the culture that has made Southwest Airlines a $5 billion success story. Southwest Airlines not only pioneered the low-cost, no-frills approach to air travel, the company turned it into an art form. Southwest provides more than 2,500 flights daily to more than 50 cities in 29 states. Due in large part to its corporate culture, the airline has enjoyed 27 strike-free years of profits.

In order to maintain its leadership position in one of the most competitive businesses in the world, Southwest works hard to maintain its strategic advantage—a participative corporate culture that emphasizes world-class customer service for both internal and external customers. The company looks for employees who like the kind of work they do and whose enjoyment of the job will show up in excellent customer service—the factor that in the long run is most likely to separate one airline from another. Southwest's CEO believes that every airline can go to Boeing and purchase the same airplanes. The airlines can lease the same type of space in airports and offer all of the same amenities to travelers. Consequently, the best strategic advantage an airline can have is the quality of its customer service.

To maintain its customer service culture, Southwest goes to great pains to hire only people who will "fit in." By "fit in," the company means having the right attitude toward customers, fellow employees, the company, and the work. The company once interviewed more than 100 applicants in an attempt to select the right ramp agent. Not only do managers conduct multiple interviews; before anyone is hired employee input is solicited to make sure the new hire will fit into a fun-loving, nonbureaucratic environment in which customer satisfaction is paramount. Southwest's CEO likes to quote Winston Churchill's famous adage about success never being final. It must be earned over and over again, every day. Southwest Airlines believes the key to success over time is excellence in customer service.

Thomas J. Neff and James M. Citrin, *Lessons from the Top*, (New York: Doubleday, 2001), 187–192.

- Recognition and rewards for successful employees
- Effective internal network for communicating the culture
- Informal rules of behavior
- Strong value system

■ High standards for performance
■ Definite organizational character

Knowing the laws of organizational change and understanding the characteristics of organizations that have strong quality cultures is important to any executive team that hopes to change the culture of its organization. Before implementing any of the specific strategies for establishing a quality culture that are explained later in this chapter, every person who will be involved in the change or affected by it should be familiar with these laws and characteristics.

COUNTERING RESISTANCE TO CULTURAL CHANGE

Change is resisted in any organization. Resistance to change is normal organizational behavior. In this regard, an organization is similar to a biological organism. From the perspective of organizational culture, the alien is change, and the organism is the organization to be changed. Continuous improvement means continuous change. To ensure continuous improvement, one must be able to facilitate continuous change.

Why Change Is Difficult

Most people understand and accept that organizational change will be resisted. However, to be an effective agent of change, one must understand why it is resisted. Joseph Juran describes organizational change as a "clash between cultures."[4] As Figure 6–2 shows, any organization has two separate cultures relating to change: the advocates and the resisters.

Advocates focus on the anticipated benefits of the change. Resisters, on the other hand, focus on perceived threats to their status, beliefs, habits, and security. Often, both advocates and resisters are wrong in how they initially approach change. Advocates are often guilty of focusing so intently on benefits that they fail to take into account the perceptions of employees who may feel threatened by the change. Resisters

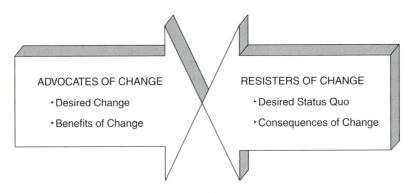

Figure 6–2
Change Causes a Classic Confrontation: Irresistible Force versus Immovable Object

are often guilty of focusing so intently on threats to the status quo that they refuse to acknowledge the benefits. These approaches typically divide an organization into warring camps that waste energy and time instead of focusing resources on the facilitation of change. Table 6–1 shows how advocates and resisters can have different perceptions of the same proposed change.

How to Facilitate Change

The responsibility for facilitating change necessarily falls to its advocates. Figure 6–3 illustrates the broad steps in facilitating change.

Begin with a New Advocacy Paradigm

The first step in facilitating change is to adopt a facilitating paradigm. Juran summarizes the traditional paradigm of change advocates as follows:[5]

■ Advocates of change tend to focus solely on expected results and benefits.

■ Advocates are often unaware of how a proposed change will be perceived by potential resisters.

■ Advocates are often impatient with the concerns of resisters.

If change is to happen, advocates must begin with a different paradigm. When a change is advocated, ask such questions as the following:

■ Who will be affected by this change, and how?

■ How will the change be perceived by those it affects?

■ How can the concerns of those affected be alleviated?

Table 6–1
Same Change, Different Perceptions

Proposed Change	Perception of Advocates	Perception of Resisters
Automate production processes	Improve productivity	Threat to job security
Initiate employee involvement and empowerment	Focus more mental resources on continuous improvement	Loss of authority
Establish a supplier partnership	Mutually beneficial business alliances	Disruption of established purchasing networks
Establish an employee education and training program	More knowledgeable, more highly skilled workforce	Costs too much
Join a manufacturing network	Enhanced competitiveness, shared costs, and shared resources	Competitors will take advantage of what they learn about us

QUALITY TIP

Organizations Have Immune Systems

"All large organizations exhibit some of the features of a biological organism. Invariably one of these features is the immune reaction when something alien is introduced. The organism senses the intrusion of the alien and mobilizes to reject it."

Source: Joseph M. Juran, *Juran on Leadership for Quality* (New York: Free Press, 1989), 192.

Understand Concerns of Potential Resisters

The second step in facilitating change is to understand the concerns of potential resisters—to put yourself in their place. Philip E. Atkinson suggests that people resist change for the following reasons:[6]

- *Fear.* Change brings with it the unwanted specter of the unknown, and people fear the unknown. Worst-case scenarios are assumed and compounded by rumors. In this way, fear tends to feed on itself, growing with time.

Figure 6–3
Steps in Facilitating Change

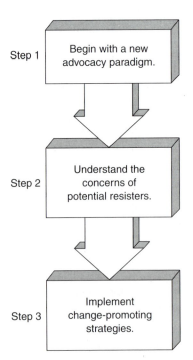

Step 1 — Begin with a new advocacy paradigm.

Step 2 — Understand the concerns of potential resisters.

Step 3 — Implement change-promoting strategies.

■ *Loss of control.*   People value having a sense of control over their lives. There is security in control. Change can threaten this sense of security and cause people to feel as if they are losing control of their lives, jobs, areas of responsibility, and so on.

■ *Uncertainty.*   It is difficult to deal with uncertainty. For better or worse, people like to know where they stand. Will I be able to handle this? What will happen to me if I can't? These are the types of questions people have when confronted with change.

■ *More work.*   Change sometimes means more work, at least at first. This concern includes work in the form of learning. To make the change, people may have to learn more information or develop new skills. For an undefined period, they may have to work longer hours.

Implement Change-Promoting Strategies

The third step in facilitating change is implementing change-promoting strategies. These are strategies that require an advocacy paradigm and take into account the concerns people typically have when confronted with change. Juran recommends the following strategies for handling and overcoming resistance to change.[7]

Involve Potential Resisters

At some point in the process, those affected by change (potential resisters) will have to take ownership of the change or it will fail. By involving them from the outset in planning for the change, organizations can ensure that potential resisters understand it and have adequate opportunities to express their views and concerns about it. This type of involvement will help potential resisters develop a sense of ownership in the change that can, in turn, convert them to advocates.

Avoid Surprises

Predictability is important to people. This is one of the reasons they resist change. Change is unpredictable; it brings with it the specter of the unknown. For this reason, it is better to bring potential resisters into the process from the outset. Surprising potential resisters will turn them into committed resisters.

Move Slowly at First

To gain the support of potential resisters, it is necessary to let them evaluate the proposed change, express their concerns, weigh the expected benefits, and find ways to alleviate problems. This can take time. However, if advocates are perceived as rushing the change through, potential resisters will become distrustful and "dig in their heels."

Start Small and Be Flexible

Change will be more readily accepted if advocates start small and are flexible enough to revise strategies that are not working as planned. This approach offers several benefits, including the following:

1. Starting with a small pilot test or experiment is less threatening than a broad-based, all-encompassing implementation.

2. Conducting a small pilot test can help identify unanticipated problems with the change.
3. The results of a pilot test can be used to revise the plans for change so that valuable resources are not wasted moving in the wrong direction.

Create a Positive Environment

The environment in which change takes place is determined by reward and recognition systems and examples set by managers. A reward and recognition system that does not reward risk taking or that punishes employees for ideas that don't work will undermine change. Managers that take "Do as I say, not as I do" attitudes will also undermine change. Well thought out, sincere attempts to make improvements should be recognized and rewarded even when they fail. Managers should "roll up their sleeves" and do their share of the work associated with change. This approach will create a positive environment that is conducive to change.

Incorporate the Change

Change will be more readily accepted if it can be incorporated into the existing organizational culture. Of course, this is not always possible. However, when it can be done, it should be done. An example might be using an established equipment maintenance schedule to make major new equipment adaptations (e.g., retrofitting manually controlled machine tools for numerical control).

Provide a *Quid Pro Quo*

This strategy could also be called *require something, give something*. If, for example, change will require intense extra effort on the part of selected employees for a given period of time, offer these employees some paid time off either before or immediately after the change is implemented. Using a *quid pro quo* can show employees that they are valued.

Respond Quickly and Positively

When potential resisters raise questions or express concerns, advocates should respond quickly and positively. Making employees wait for answers magnifies the intensity of their concerns. A quick response can often eliminate the concern before it becomes a problem, and it will show employees that their concerns are considered important. A quick response does not mean a surface-level or inaccurate response made before having all the facts. Rather, it means a response made as soon as one can be made thoroughly and accurately. It is also important to respond positively. Advocates should not be offended by or impatient with the questions of potential resisters. A negative attitude toward questions and concerns only serves to magnify them.

Work with Established Leaders

In any organization, some people are regarded as leaders. In some cases, those people are in leadership positions (supervisors, middle managers, team captains, etc.). In other cases, they are informal leaders (highly respected employees whose status is based on their experience or superior knowledge and skills). The support of such leaders is critical. Other employees will take cues from them. The best way to get their support is to involve them in planning for the change from the outset.

Treat People with Dignity and Respect

This strategy is fundamental to all aspects of total quality. It requires behavior that acknowledges the human resource as the organization's most valuable asset. Without this strategy, the others won't matter.

Be Constructive

Change is not made simply for the sake of change. It is made for the sake of continual improvement. Consequently, it should be broached constructively from the perspective of how it will bring about improvements.

ESTABLISHING A QUALITY CULTURE

Establishing a quality culture involves specific planning and activities for every business or department. This section identifies the steps involved, but first it outlines the emotional processes employees go through as the steps are being taken. Managers need to recognize and accommodate the emotional transition required not only of employees but also of themselves while the steps toward making the conversion to quality take place.

Phases of Emotional Transition

A great deal of research has been done about how people undergo transitions from one state of being to another. Most of this research has focused on the stages of transition or recovery that people go through when they confront a major unexpected and unwanted change in their lives. The types of changes that have been studied most include divorce, the death of a loved one, a life-threatening illness, and the loss of a job. Figure 6–4 illustrates the transition process people go through when confronted by one of these major traumatic changes in their lives.

The first emotional response to any type of change is shock. A person is living from day to day, comfortable with the predictability of his or her life. Suddenly an unexpected change intrudes. A typical response to the shock it produces is denial. The change is so unwanted that the natural human response is to simply deny that it has happened. This serves to level the state of mind somewhat from the low experienced during the shock phase. The length of the denial phase varies from person to person. Regardless of its length, the denial phase is temporary.

Events force the issue, and the realization of reality begins to set in. As this happens, the person's state of mind begins to fall. Depression is common during the realization phase. People need a lot of support during this phase. When realization bottoms out, acceptance occurs. Acceptance does not mean the person agrees with what has happened. Rather, it means that he or she is ready to say, "I have this problem; now what can I do about it?"

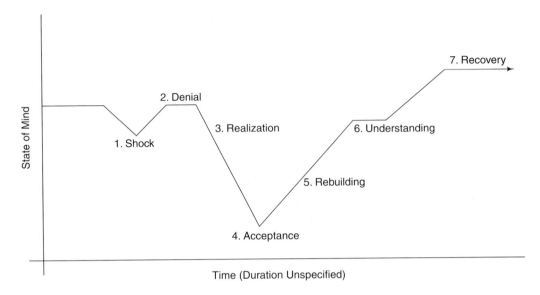

Figure 6–4
Emotional Transition

This attitude allows the rebuilding process to begin. During this phase, people need as much support as they did during the realization phase. As the rebuilding phase is accomplished, understanding sets in. In this phase, people have come to grips with the change, and they are dealing with it successfully. This phase blends into the final phase, recovery. In this phase, people are getting on with their lives.

Managers hoping to instill a quality culture should understand this transitional process. The change from a traditional organizational culture to a quality culture can be traumatic enough to trigger the process. Knowing this and understanding the process will help managers who are trying to instill a quality culture.

Steps in the Conversion to Quality

Figure 6–5 provides a checklist managers can use to guide their organizations through the conversion to a quality culture. The various strategies contained in the checklist are explained in the following subsections.

Identify the Changes Needed

An organization's culture dictates how people in it behave, respond to problems, and interact with each other. If the existing culture is a quality culture, it will have such characteristics as the following:

- Open, continual communication
- Mutually supportive internal partnerships

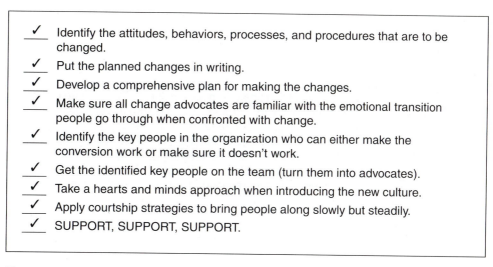

✓ Identify the attitudes, behaviors, processes, and procedures that are to be changed.

✓ Put the planned changes in writing.

✓ Develop a comprehensive plan for making the changes.

✓ Make sure all change advocates are familiar with the emotional transition people go through when confronted with change.

✓ Identify the key people in the organization who can either make the conversion work or make sure it doesn't work.

✓ Get the identified key people on the team (turn them into advocates).

✓ Take a hearts and minds approach when introducing the new culture.

✓ Apply courtship strategies to bring people along slowly but steadily.

✓ SUPPORT, SUPPORT, SUPPORT.

Figure 6–5
Quality Culture Conversion Checklist

- Teamwork approach to problems and processes
- Obsession with continual improvement
- Broad-based employee involvement and empowerment
- Sincere desire for customer input and feedback

Does the organization's culture have these characteristics? The best way to answer this question is to involve the entire workforce from bottom to top in a systematic assessment that is stratified by level (i.e., executive management, middle management, first-line employee, etc.). Figure 6–6 is an example of an assessment instrument that can be used for collecting information on the perceptions of employees at all levels in an organization.

Put the Planned Changes in Writing

A comprehensive assessment of an organization's existing culture will usually identify improvements that need to be made. These improvements will require changes in the status quo. These changes should be listed without annotation or explanation. For example, if the assessment reveals that customer input is not part of the product development cycle, the change list would contain an entry such as the following: *The product development cycle should be changed so that it includes the collection and use of customer input.*

Develop a Plan for Making the Changes

The plan for effecting change is developed according to the who-what-when-where-how model. Each of these elements represents a major section of the plan, as follows:

- Who will be affected by the change? Who will have to be involved in order for the change to succeed? Who is likely to challenge the change?

Position (Type) _____

Date _____

Instructions

The purpose of this survey is to assess the existing culture of our organization. The findings will be compared with what is known about a quality culture for the purpose of identifying the cultural changes needed in our organization to continually improve quality, productivity, and competitiveness. Respond to each of the criteria by circling the number you think best describes our organization as it is today. *Zero* (0) means that we do not meet this criterion at all. *Five* (5) means that we completely satisfy the criterion. Do not respond to items that don't apply or about which you are unsure.

1. All employees know the mission of the organization   0 1 2 3 4 5
2. All employees know their role in helping the organization accomplish its mission   0 1 2 3 4 5
3. Executive management is committed to the continual improvement of quality, productivity, and competitiveness   0 1 2 3 4 5
4. Management treats the workforce as a valuable asset   0 1 2 3 4 5
5. Open, continual communication exists at all levels of the organization   0 1 2 3 4 5
6. Mutually supportive internal partnerships exist between management and employees   0 1 2 3 4 5
7. Mutually supportive internal partnerships exist among employees   0 1 2 3 4 5
8. Quality is defined by customers, internal and external   0 1 2 3 4 5
9. Customers participate in the product development cycle   0 1 2 3 4 5
10. Employees are involved in the decision-making process   0 1 2 3 4 5
11. Employees are empowered to contribute their ideas for promoting continual improvement   0 1 2 3 4 5
12. Performance of processes is measured scientifically
13. Scientific data are used in the decision-making process   0 1 2 3 4 5
14. Employees receive the education and training they need to continually improve their performance   0 1 2 3 4 5
15. All employees at all levels are expected to maintain high ethical standards   0 1 2 3 4 5

   0 1 2 3 4 5

Figure 6–6
Organizational Culture Employee Assessment Worksheet

■ What tasks must be accomplished? What are the most likely barriers? What are the related processes and procedures that will be affected by the change?

■ When should the change be implemented? When should progress be measured? When should the various tasks associated with the change be accomplished? When should implementation be completed?

■ Where will the change be implemented? Where are the people and processes that will be affected?

■ How should the change be made? How will it affect existing people and processes? How will it improve quality, productivity, and competitiveness?

The plan should contain all five elements, and each element should be dealt with comprehensively. However, the plan should be brief. Be comprehensive and thorough, but keep it as brief as possible.

Understand the Emotional Transition Process
Advocates of the change will play key roles in its implementation. The success of the implementation will depend to a large extent on how well advocates play their roles. It is essential that they understand the emotional transition people go through when forced to deal with change, particularly unwanted change (see Figure 6–4).

The transition consists of seven steps: shock, denial, realization, acceptance, rebuilding, understanding, and recovery. People who confront a change they don't want to make may have to go through all seven steps in the transition. Advocates should understand this and proceed accordingly.

Identify Key People and Make Them Advocates
Key people are those who can facilitate and those who can inhibit implementation of the change. These people should be identified, brought together, and given the plan. Allow advocates and inhibitors opportunities to state their cases. Record all concerns and deal with them. This is the step in which a *quid pro quo* might be used to bring inhibitors around. Executive managers must use their judgment in applying the right amount of the "carrot", the "stick", and peer pressure (from advocates) to turn inhibitors into advocates.

Take a Hearts and Minds Approach
Advocates should be conscious of human nature as they work to implement change. On an intellectual level, people may understand and even agree with the reasons behind a change. But understanding intellectually is rarely enough. People tend to react to change more on an emotional (hearts) level than on an intellectual (minds) level, at least initially. Therefore, it is important to take the time to deal with the inevitable emotional response that occurs in the early stages of implementation.

Frequent, open communication—preferably face to face—is the best strategy. Advocates should allow even the most negative opponents to voice their concerns and objections in open forums. Then these concerns should be answered in an impartial, patient, nondefensive manner. When the majority of employees accept the change, critical mass will set in and peer pressure will begin to work on the side of the advocates.

Apply Courtship Strategies

Courtship is a phase in a relationship that moves slowly but deliberately toward a desired end. During the courtship, the partner hoping to move the relationship forward listens carefully to the other partner and patiently responds to any concerns expressed. This partner is on his or her best behavior. If advocates think of their relationship with potential resisters as a courtship, they will be better able to bring them along and eventually win them over.

Support, Support, Support

This final strategy is critical. It means that the material, moral, and emotional support needed by people undergoing change should be provided. Undergoing change is a lot like walking a tightrope for the first time. It will work out a lot better if you have someone to help you get started, someone waiting at the other end to encourage progress, and a safety net underneath in case you fall. Planning is important. Communication is critical. But support is essential.

MAINTAINING A QUALITY CULTURE

Establishing a quality culture is a challenging undertaking for any organization. It is even more challenging to maintain a quality culture over time. The easiest thing in the world is to become complacent and let the organization's culture begin to slip back into its old mold. According to Frank Gryna, in order to maintain a quality culture organizations must foster the following critical behaviors:[8]

1. Maintain an awareness of quality as a key cultural issue. This is accomplished through the regular dissemination to all personnel of quality goals and the corresponding results relating to these goals. Managers should "keep score" and let all stakeholders know what the score is.

2. Make sure that there is plenty of evidence of management's leadership. Cheerleading is good, but it's not enough. Managers should provide leadership in strategic planning for quality, serve on quality councils, and be actively involved in the implementation of quality initiatives. Employees need to see managers "walking the walk" as well as "talking the talk."

3. Empower employees and encourage self-development and self-initiative. Managers should make sure that jobs are designed for as much self-control as possible, continually seek and use employee input, and encourage self-directed teamwork.

4. Keep employees involved. Do not just seek their involvement through empowerment; structure the organization and its processes in ways that ensure it. This means making employees fully empowered members of the quality council; maintaining a sys-

tem that makes it easy, convenient, and nonthreatening for them to recommend improvements; and involving employees in areas such as product or process design review.

5. Recognize and reward the behaviors that tend to nurture and maintain the quality culture. Recognition involves various forms of public acknowledgment. Rewards are tangible benefits such as salary increases, bonuses, incentives, and promotions.

SUMMARY

1. A quality culture is an organization value system that results in an environment that is conducive to the establishment and continual improvement of quality. It consists of values, traditions, procedures, and expectations that promote quality.
2. Implementing total quality necessitates cultural change in an organization, for the following reasons:
   - Change cannot occur in a hostile environment.
   - Moving to total quality takes time.
   - It can be difficult to overcome the past.
3. The laws of organizational change are as follows:
   - Understand the history behind the current culture.
   - Don't tamper with systems—improve them.
   - Be prepared to listen and observe.
   - Involve everyone affected by change in making it.
4. Change can be difficult because resisting change is natural human behavior. In any organization there will be advocates of change and resisters. Sometimes advocates focus so intently on the expected benefits of change that they fail to realize how the change will be perceived by potential resisters. People resist change for the following reasons: fear, loss of control, uncertainty, and more work. To overcome resistance to change, advocates can apply the following strategies:
   - Involve potential resisters.
   - Avoid surprises.
   - Move slowly at first.
   - Start small and be flexible.
   - Create a positive environment.
   - Incorporate the change.
   - Provide a *quid pro quo*.
   - Respond quickly and positively.
   - Work with established leaders.
   - Treat people with dignity and respect.
   - Be constructive.
5. Strategies for establishing a quality culture include the following:
   - Identify the changes needed.
   - Put the planned changes in writing.
   - Develop a plan for making the changes.

- Understand the emotional transition process.
- Identify key people and make them advocates.
- Take a hearts and minds approach.
- Apply courtship strategies.
- Support.

6. At times, it might be necessary to change an organization's leadership team to ensure needed cultural change. This situation arises when the organization's senior executives have a great deal invested in the status quo and therefore are staunch defenders of orthodoxy.

 KEY TERMS AND CONCEPTS

Advocates	More work
Avoid surprises	Move slowly at first
Be constructive	Organizational culture
Courtship strategies	Quality culture
Create a positive environment	*Quid pro quo*
Emotional transition process	Resisters
Facilitating change	Respond quickly and positively
Fear	Start small and be flexible
Hearts and minds approach	Support
Hostile environment	Treat people with dignity and respect
Incorporate the change	Uncertainty
Involve potential resisters	Who-what-when-where-how model
Laws of organizational change	Work with established leaders
Loss of control	

 FACTUAL REVIEW QUESTIONS

1. Define the expression *quality culture*.
2. Explain why the implementation of total quality requires cultural change.
3. List and describe the steps involved in laying the groundwork for a quality culture.
4. What are the characteristics shared by companies that have a quality culture?
5. Why is change so difficult for people?
6. Describe the paradigm that should be adopted by advocates of change.
7. Explain four reasons why people resist change.
8. List and describe the strategies that can be used to overcome resistance to change.
9. What strategies would you use to establish a quality culture in organizations?
10. Explain which strategy from the two previous questions is the most important, and why.
11. Why is it sometimes necessary to change leaders to ensure cultural change?

CRITICAL THINKING ACTIVITY

Why Is Cultural Change So Hard?

Max Cutter is the envy of his fellow seniors at Stanfield Institute of Technology. He and his classmates all majored in business or technology disciplines that emphasized quality management. They all hope to begin careers soon as quality professionals, and Max just got himself a head start. In his very first interview, arranged by SIT's Career Center, Max was offered a job as quality manager for an old and well-established electric power company.

"I can't believe I got the job!" Max shouted to his classmates. "I spent 10 minutes trying to convince the personnel manager that I could do the job in spite of my lack of experience. When I finally stopped rambling on he said, 'We want to hire you *because* you have no experience, not in spite of the fact. We need someone with fresh ideas who is not wedded to the status quo. Frankly, we need to make major changes, and the sooner, the better. You will come in with no loyalties or biases.'"

"They are going to expect you to make major culture changes," said one of Max's classmates. "Do you think you're up to it? I remember studying the issue in class. Cultural change comes hard, if it comes at all."

"You've got a point there," said Max. "It does concern me. Why is cultural change so hard? I don't know why people can't just get with the program."

What about this question? Why is cultural change so hard? What is your opinion?

DISCUSSION ASSIGNMENT 6–1

The Indifferent Manager

The efforts of Public Communications, Inc., to implement total quality had exceeded expectations at two of its three plants. However, the third plant just didn't seem to be able to get things off the ground. The plant manager, Merrill Stephens, was under a lot of pressure because his colleagues in the other two plants were succeeding and he was floundering. He didn't know what the problem was, and he didn't know how to find out.

Merrill decided to call together a group of line employees and ask for their input. Immediately he sensed their reticence. Clearly, they had something to say but didn't want to say it. Finally, an employee who had been with the company for more than 20 years spoke up. "Mr. Stephens, we're just going through the motions to keep corporate off your back," he said. "We know you don't buy this total quality nonsense." Merrill had to admit that the employees were reading him like a book. Sure, he had followed the implementation guidelines to the letter. His executive team was the quality council. Policies had been developed and deployed. Employees were working in teams, and training in the use of quality tools was being provided. The problem was that Merrill himself was just going through the motions. He didn't really believe in total quality and had hoped it would turn out to be just one more corporate-mandated initiative that would fizzle and eventually go away.

He had only 4 more years until retirement and didn't need this in his life right now. His managers, middle managers, supervisors, and employees knew him and sensed his indifference and halfheartedness. Part of the organizational culture at this plant was that the employees took their lead from the plant manager. If Merrill was really behind an effort, they got behind it. If he wasn't, they didn't. When they accurately sensed his indifference to total quality, they responded accordingly.

Discussion Assignment

Discuss the following question in class or outside of class with your fellow students:

1.   Have you ever been involved in an effort that was halfhearted because the leader of the effort didn't seem to be enthusiastic about it? If so, what happened? If not, discuss what it means for the person in charge in a given situation to set the tone.

DISCUSSION ASSIGNMENT 6–2

One Size Does Not Fit All

His efforts to establish a quality culture at AMDEX, Inc., are not working, and the company's CEO, Calvin Forrester, does not know why. He is doing everything just as he had done it in his last position. Forrester had made a name for himself in the world of electronics fabrication by transforming his previous company from a bankrupt market loser to a globally competitive winner. He had achieved what most had said was impossible, and he had done so the hard way; by completely transforming the company's culture from top to bottom and bottom to top. Unfortunately, the model that had worked so well at his previous company was falling flat at AMDEX. In his previous position the entire workforce had been not just willing, but anxious to change. It had been either change or shut the doors. But with AMDEX things are different. The company is far from perfect and has a long way to go if it is to be globally competitive, but it has a lot of good points and is not even close to going out of business. Consequently, AMDEX employees do not feel the urgency to change that Forrester enjoyed in his last position.

Discussion Assignment

Discuss the following questions in class or outside of class with your fellow students:

1. How would the different levels of motivation in these two companies affect the approach taken to establish culture?
2. If you were Calvin Forrester, how would you proceed?

ENDNOTES

1. Peter R. Scholtes, *The Team Handbook* (Madison, WI: Joiner Associates, 1992), 1-20.
2. Scholtes, 1-20–1-21.
3. Joseph M. Juran, *Juran on Leadership for Quality* (New York: Free Press, 1989), 184.
4. Juran, 316.
5. Juran, 317.
6. Philip E. Atkinson, *Creating Culture Change: The Key to Successful Total Quality Management* (San Diego, CA: Pfeiffer, 1990), 48–49.
7. Juran, 318–319.
8. Frank M. Gryna, *Quality Planning and Analysis*, 4th ed. (New York: McGraw Hill, 2001), 68–75.

CHAPTER SEVEN

Customer Satisfaction and Retention

"The competitor to be feared is the one who never bothers about you at all but goes on making his own business better all the time."

Henry Ford, Sr.

MAJOR TOPICS

- Understanding Who Is a Customer
- Understanding Customer-Defined Quality
- Identifying External Customer Needs
- Identifying Internal Customer Needs
- Communicating with Customers
- Instituting Quality Function Deployment
- Customer Satisfaction Process
- Customer-Defined Value
- Customer Value Analysis
- Customer Retention
- Establishing a Customer Focus
- Recognizing the Customer-Driven Organization
- Value Perception and Customer Loyalty
- Guaspari on Customer Satisfaction

In a total quality setting, customers define quality and employees produce it. Historically, organizations have viewed customers as people who buy and use their products. These are external customers. There are also internal customers within any organization—the staff. With this background, an accurate recasting of the first sentence is as follows: In a total quality setting, external customers define quality and internal customers produce it. This chapter provides the information modern managers need to establish in their organizations a customer focus that encompasses both internal and external customers.

UNDERSTANDING WHO IS A CUSTOMER

Historically, the concept of suppliers and customers has been interpreted as shown in Figure 7–1. An organization uses certain processes by which it produces its products. People who interact with the company prior to the processes taking place have been considered suppliers. Those who interact with the company after the processes have produced the product have been viewed as customers. From this traditional perspective, customers and suppliers are both external entities. Figure 7–2 illustrates a more contemporary view of suppliers and customers.

In a total quality setting, customers and suppliers exist inside and outside the organization. Any employee whose work precedes that of another employee is a supplier for that employee. Correspondingly, any employee whose work follows that of another employee and is dependent in some way on it is a customer. For example, say Employee A attaches several components to a printed circuit board and then hands the board to Employee B to connect the components. In this relationship, Employee A is a supplier for Employee B, and Employee B is a customer of Employee A. Employee B cannot do her job correctly unless Employee A has done his correctly. The quality of Employee A's work affects that of Employee B. This concept of *dependency* is critical in the supplier–customer relationship. A customer, whether internal or external, depends on suppliers to provide quality work and produce quality products.

Figure 7–1
Traditional View of Suppliers and Customers Showing That Customers and Suppliers Are Strictly External Entities

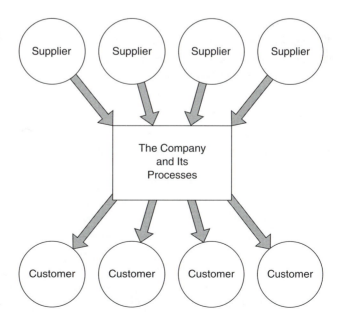

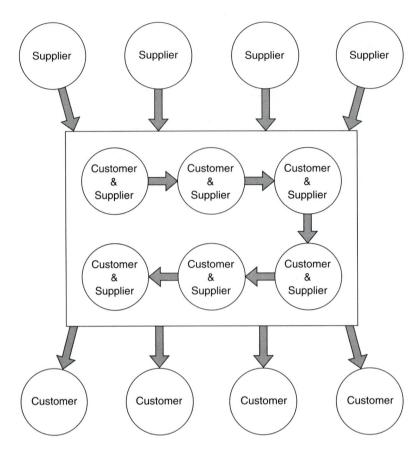

Figure 7–2
Contemporary View of Suppliers and Customers Showing That Employees Are Suppliers and Customers to Each Other

UNDERSTANDING CUSTOMER-DEFINED QUALITY

In a total quality setting, quality is defined by the customer. According to Peter R. Scholtes:

> Only once you understand what processes and customers are will you be able to appreciate what quality means in the new business world. If customers are people who receive your work, only they can determine what quality is, only they can tell you what they want and how they want it. That's why a popular slogan of the quality movement is "quality begins with the customer." You must work with internal and external customers to determine their needs, and collaborate with internal and external suppliers.[1]

In his book *Total Manufacturing Management*, Giorgio Merli makes the following points about customer-defined quality:[2]

- The customer must be the organization's top priority. The organization's survival depends on the customer.
- Reliable customers are the most important customers. A reliable customer is one who buys repeatedly from the same organization. Customers who are satisfied with the quality of their purchases from an organization become reliable customers. Therefore, customer satisfaction is essential.
- Customer satisfaction is ensured by producing high-quality products. It must be renewed with every new purchase. This cannot be accomplished if quality, even though it is high, is static. Satisfaction implies continual improvement. Continual improvement is the only way to keep customers satisfied and loyal.

If customer satisfaction is the highest priority of a total quality organization, then it follows that such an organization must have a customer focus. Traditional management practices that take the management-by-results approach are inward looking. An organization with a customer focus is outward looking. Scholtes describes this concept as follows:

> Whereas Management by Results begins with profit and loss and return on investment, Quality Leadership begins with the customer. Under Quality Leadership, an organization's goal is to meet and exceed customer needs, to give lasting value to the customer.[3]

The key to establishing a customer focus is putting employees in touch with customers and empowering those employees to act as necessary to satisfy the customers. There are a number of ways to put employees in touch with customers. Actual contact may be in person, by telephone, or through reviewing customer-provided data. Identifying customer needs and communicating with customers are covered later in this chapter. At this point, it is necessary to understand only that employee–customer interaction is a critical element in establishing a customer focus.

An excellent example of how a company can establish a customer focus is Zytec Corporation of Eden Prairie, Minnesota. Zytec manufactures power supplies. Writing about Zytec's innovative customer focus, the *Total Quality Newsletter* said this:

> At the heart of its corporate commitment to customer satisfaction, for example, is this message to each of its 700 employees: To solve problems for our customers, you are empowered to spend up to $1,000 on their behalf. No questions asked. There is no paperwork or authorization needed to spend money for a customer. Employees can requisition a check or—if more immediate action is necessary—pay cash for the purchase and be reimbursed. What kind of actions have Zytec employees taken for customers? One chartered a plane to fly some missing equipment pieces from a Zytec plant in southeastern Minnesota to a customer in a central part of the state; another bought new tables for a conference room where customers frequently visited.[4]

IDENTIFYING EXTERNAL CUSTOMER NEEDS

Historically, customers were excluded from the product development process. When this approach is used, the organization producing the product is taking a chance that it will satisfy the customer. In a competitive marketplace that is global in scope, such an approach can be disastrous. In a total quality setting, customer needs are identified clearly as a normal part of product development. According to Scholtes:

> The goal should be to exceed customer expectations, not merely meet them. Your customers should boast about how much they benefit from what you do for them. To attain this goal, you must collect reliable information on what they need and want from your product or service. In doing so, you will find out whether your processes are on target. This strategy can be used to identify potential improvement projects or just to clarify a project's goals.[5]

Scholtes recommends the six-step strategy for identifying customer needs that is illustrated in Figure 7–3 and described in the following subsections.[6]

Speculate about Results

Before gathering information about customer needs, it is a good idea to spend some time speculating about what might be learned. Write down what you think customers will say, so that you can compare your expectations with what is actually said. The purpose of this step is to help representatives of the organization determine whether they are in touch with customer needs.

Develop an Information-Gathering Plan

Information gathering should be systematically undertaken and well organized. Before gathering information, develop a plan. Decide what types of information are needed and who will be asked to provide it. Whenever possible, structure the plan so that information is collected in face-to-face interviews. When personal visits are not possible, use the telephone. Written surveys sent out through the mail cannot produce a level of feedback equal to that gained from the nonverbal messages, impressions, and follow-up questions that are a part of person-to-person interviews.

QUALITY TIP

Satisfying Customers and Profits

"Profit in business comes from repeat customers, customers that boast about your product and service, and that bring friends with them. Fully allocated costs may well show that the profit in a transaction with a customer that comes back voluntarily may be 10 times the profit realized from a customer that responds to advertising and other persuasions."

Source: W. Edwards Deming, *Out of the Crisis* (Cambridge: Massachusetts Institute of Technology, Center for Advanced Engineering Study, 1991), 141.

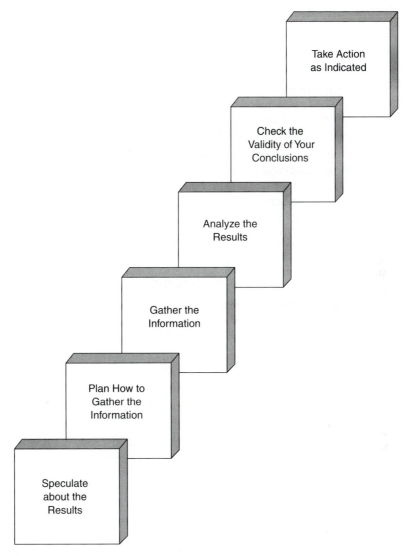

Figure 7–3
Scholtes's Six-Step Strategy for Identifying Customer Needs

Figure 7–4 is an example of an information-gathering plan. Notice that it is structured for collecting information either face to face or by telephone interviews.

Gather Information

Before implementing the entire information-gathering plan, it is a good idea to conduct a smaller pilot study involving just a few customers. This will identify problems with

---

Power Pac Panels of Florida
19 Industrial Parkway, West
Fort Walton Beach, Florida 32548

- **Customers to Be Contacted**
  Twenty manufacturers of low-voltage power supplies make up 90% of our business. These customers are located in five southeastern states. Feedback will be collected from all twenty of these customers.

- **What Information Will Be Sought?**
  Each customer will be asked the following questions:
  1. Rate the performance of our product from 1 to 10 with 10 representing a perfect score.
  2. What specific problems do you have with our product?
  3. What can we do to our product that would raise its performance score to a 10?
  4. What is the priority ranking of your suggested improvements?

- **Information-Collection Method**
  Half of the customers will be contacted in person and half by telephone.

- **Information Analysis**
  Information collected will be analyzed by a team with representatives from the following departments: Design, Production, Quality, Marketing, and Accounting.

Figure 7–4
Information-Gathering Plan

the information-gathering methodology that should be corrected before you proceed on a larger scale. After the methodology has been appropriately refined, gather information in a timely manner.

Analyze the Results

Results should be analyzed carefully and objectively. Do they match the speculated results from the first step? How do they agree and disagree? What problems did customers identify? What strong points? Were there trends? How many customers complained of the same problem? What changes in the product or services relating to it were suggested?

Check the Validity of Conclusions

Having drawn conclusions based on the information gathered, the next step is to check the validity of those conclusions. Customers can be a valuable source of help. Select several customers and share the conclusions with them. Do they agree with the conclusions? Also share the conclusions with other people in the organization and get their feedback. Adjust your conclusions as needed based on this external and internal feedback.

Take Action

Based on the final conclusions, what changes need to be made? Which of these changes are short term in nature, and which are long term? Which can be made immediately, and which will require a longer time? Take any corrective action that can be made immediately, and lay out a plan for completing any that is long term in nature. Meet with customers and let them know what is going to be done and when. Make sure that changes are made, to the extent possible, in the same order of priority as that dictated by customer needs.

Obtaining Customer Information from the Internet

Obtaining customer information has become much easier with the advent of the World Wide Web. Most businesses now have extensive Web sites containing volumes of information about themselves. This information can be obtained by going on-line and accessing the Web site of the company in question. Careful scrutiny of a customer's Web site can reveal useful information about what is important to that customer. It can also reveal who the key players are, who is in charge of what, and who the best contact might be in a given situation.

A word of caution is in order, however. There are ethical issues to be considered when using the Internet to obtain customer information. Only information available for public consumption from the customer's Web site should be used. In the age of technology, Web surfers and computer hackers are sometimes able to access data about customers that is not intended to be public. This type of information has value for the hackers if they can sell it or use it in any other way for personal gain. Before deciding to use the Internet as a source for customer knowledge, organizations should develop guidelines for the ethical use of information obtained electronically. These guidelines should also specify the types of information that will be pursued electronically and how to respond when offered customer data that might have been obtained unethically.

IDENTIFYING INTERNAL CUSTOMER NEEDS

Identifying the needs of internal customers is a matter of ensuring that employees who depend on one another as individuals, as well as departments that depend on each other as units, communicate their needs to one another continually. However, one should not assume that communication will just happen. As important as it is, communication rarely just happens in any setting. Rather, it must be encouraged and facilitated.

Quality circles, self-managed teams, cross-departmental teams, and improvement teams are all examples of mechanisms for improving communication and, in turn, quality. These mechanisms serve to facilitate communication among internal customers and suppliers. However, they are not the only mechanisms available. Communication that occurs over a cup of coffee in the break room or during lunch can be equally effective. Training that promotes communication and helps improve communication skills is also important (this is discussed in greater depth in Chapter 11). Teamwork and how it can improve communications is discussed in Chapter 10.

QUALITY TIP

Employees Treat Customers as Management Treats Them

"It never ceases to amaze me. A company will have signs and banners promoting customer satisfaction. Management will lecture employees at length on the subject. But employees are perceptive. They watch managers. They know how the managers treat customers, and that is how they, the employees, will treat customers. Managers who want good customer service must set the example first and lecture second."

Source: David L. Goetsch. From speech to a conference of business leaders, January 2000.

COMMUNICATING WITH CUSTOMERS

Continual communication with customers is essential in a competitive marketplace. Establishing effective mechanisms for facilitating communication and then making sure the mechanisms are used are critical strategies in establishing a customer focus. One of the main reasons continual communication is required is that customer needs change, and, at times, they can change rapidly. Quality pioneer Joseph M. Juran explains this concept as follows:

> Customers' needs do not remain static. There is no such thing as a permanent list of customers' needs. We are beset by powerful forces that keep coming over the horizon and are ever changing directions: new technology, market competition, social upheavals, international conflicts. These changing forces create new customer needs or change the priority given to existing ones.[7]

Communication with customers must extend to both external and internal customers. What applies on the outside also applies within the organization. According to Juran:

> Most companies do a good job of communicating essential quality information to workers on matters such as specifications, standards, procedures, and methods. This information, although necessary, is not sufficient with respect to willful errors. We should also (a) provide means for workers to communicate their views and ideas, and (b) explain to workers those management actions that on their face are antagonistic to quality.[8]

Communication with customers is sometimes misunderstood as one of the basic strategies used in a total quality setting. It does not mean asking customers what new products should be invented. Customers don't typically think in these terms. In his book *Out of the Crisis*, W. Edwards Deming writes:

> The customer is not in a good position to prescribe a product or service that will help him in the future. The producer is in a far better position than the consumer to invent new design and new service. Would anyone that owned an automobile in 1905 express a desire for

QUALITY TIP

Contact with Customers Is Critical

"Everyone, whether he sees the customer or not, has a chance to build quality into the product or into the service offered. The people that see customers have a role that is not usually appreciated by supervisors and other management. Many customers form their opinions about the product or about the service solely by their contacts with the people they see."

Source: W. Edwards Deming, *Out of the Crisis* (Cambridge, MA: Institute of Technology, Center for Advanced Engineering Study, 1991), 192.

pneumatic tires, had you asked him what he needed? Would I, carrying a precise pocket-watch, have suggested a tiny calculator and quartz timepiece?"[9]

All of the market research in the world won't spare the entrepreneur the anxiety of dealing with the inescapable element of risk. However, having taken the risk to produce a product, communicating with customers about that product can ensure that it gets the best possible reception in the marketplace and that it changes as the needs of customers change.

Know Your Customer's Operations

As a supplier to other companies (customers), it is important to know their operations. The more that is known about a customer's operations, the easier it will be to provide products that meet its needs.

What does the customer do with our product? How is it used? Is our product part of a larger assembly? Does the customer use our product in the way we expect or in some different way? Does the customer modify our product in any way? What processes does the customer use in working with our product? Knowing the answers to questions such as these can help a supplier improve customer satisfaction. The answers to these types of questions can lead to such benefits as the following:

- *Product enhancements.* By knowing a customer's operations, suppliers might be able to modify their products to better fit in with these operations. They might also be able to add attributes that will make the product even more attractive to the customer.
- *Improved productivity.* By knowing a customer's operations, suppliers might be able to propose process modifications that will improve their productivity.
- *Internal improvements.* By knowing a customer's operations, suppliers might learn facts that lead to internal improvements in quality, productivity, and design in their own organizations.

Customers don't always use a product in the way a supplier assumes. By getting to know customers and their operations, suppliers have not just made process and product improvements; some have actually started new product lines. In any case, the better suppliers know their customers' operations, the better they can serve them. The better suppliers serve their customers, the greater the likelihood of satisfied, long-term customers.

INSTITUTING QUALITY FUNCTION DEPLOYMENT

It is important to know that a product will meet the needs of customers before you put it into production. This is the main reason for conducting the research necessary to identify customer needs and for communicating with internal and external customers. Quality function deployment (QFD) was developed with this in mind. In his book *Total Manufacturing Management*, Merli describes QFD as follows:

> Quality Function Deployment (QFD) was developed to ensure that products entering production would fully satisfy the needs of their customers by building in the necessary quality levels as well as maximum suitability at every stage of product development.[10]

QFD is an integrated approach to product development and quality in all preproduction activities. It was introduced in the United States in 1983 by Professor Y. Akao of the University of Tamagawa as part of a paper he presented at a quality conference in Chicago. QFD is actually a model for incorporating customer input and feedback into product development. In effect, it establishes an operational structure for the concept of building in quality. The philosophy underlying QFD is that even a perfectly manufactured product may not satisfy the customer because it may be nothing more than a perfect example of what the customer doesn't want.

QFD allows for the systematic incorporation of customer needs, production capabilities and capacity, and all other relevant parameters into product development. According to Merli, QFD consists of the following basic activities:[11]

- Deployment of customer requirements (quality needs)
- Deployment of measurable quality characteristics
- Determination of the correlation between quality needs and characteristics
- Assignment of numerical values to each quality characteristic
- Integration of quality characteristics into the product
- Detailed design, production, and quality control of the product

QFD is discussed in greater detail in Chapter 17. It is introduced here only to show that it can have a positive role to play in making sure customer needs are satisfied or exceeded.

CUSTOMER SATISFACTION PROCESS

Customer focus is more than just sending out surveys. Customer focus is part of a process that leads to continual improvements in the organization that, in turn, result in customer satisfaction. Resources are limited; consequently, they must be applied where they will do the most to improve customer satisfaction and customer retention. The process described in the following list will help meet all these goals:[12]

- Determine who your customers are.
- Determine what attributes of your product/service are most important to your customers.
- Arrange these attributes in the order of importance indicated by your customers.
- Determine your customers' level of satisfaction with each of these attributes.
- Tie results of customer feedback to your processes.
- Develop a set of metrics (measurements) that tell how you are performing and which areas within the process are having the greatest impact on performance.
- Implement measurements at the lowest possible level in the organization.
- Work on those processes that relate to attributes that have high importance, but low customer satisfaction ratings.
- Work on those areas within the process that offer the greatest opportunity to improve.
- Update customer input and feedback on a continual basis. Then, as process improvements correspondingly increase customer satisfaction, move on to the process improvements that are next in importance.
- Maintain open, continual communication with all stakeholders on what is being done, why, what results are expected, and when.
- Aggregate metrics organization-wide into a format for management review on a continual basis. Adjust as necessary.

CUSTOMER-DEFINED VALUE

It is important for organizations to understand how customers define value. The value of a product or service is the sum of a customer's perceptions of the following factors:

- Product/service quality
- Service provided by the organization
- The organization's personnel
- The organization's image
- Selling price of the product/service
- Overall cost of the product/service

All of these factors are important to customers. The product or service must have the attributes customers want, and those attributes must be of the quality expected. The customer's interaction with the organization and how this interaction is measured are important. Just making a good product/service available is not enough. Customer satisfaction will also be affected by how effectively, courteously, and promptly customers are served. The appearance, knowledge, and attitudes of an organization's personnel also affect the level of satisfaction customers experience. Customers will build relationships with personnel in the organization who are knowledgeable, professional in appearance, and positive. Such relationships promote loyalty. On the other hand, no matter how satisfied customers are with a product or service, if they don't like an organization's people, they are likely to defect to the competition.

An organization's image is important to customers. Consequently, it is vital not just to have quality products, service, and personnel but also to project an image that is consistent with these quality characteristics. Think of the adage that one should "not just talk the talk but walk the walk." In establishing and nurturing an image, it is important to do both those things. The key is that organizations must be concerned with both substance and appearances. An organization's image is defined by what customers *believe* to be true about it.

Selling price is important to customers, of course. It is the easiest characteristic to compare. The point to understand here is that customers have become so sophisticated that they no longer confuse selling price and cost. In other words, they know the difference between *cheap* and *inexpensive*. A competitive selling price is a must in the modern workplace, but it should not be achieved by sacrificing quality or service.

Most customers know that the selling price is just the beginning of the actual cost of a product. Only when maintenance, upkeep, replacement parts, warranty issues, and service are factored in does one know the product's real cost. Customers who don't understand the difference between price and cost soon learn—the hard way. The organization that teaches this difficult lesson is not likely to retain its customers.

Whether customers are satisfied will depend on the sum of their perceptions relative to all of these factors. The issue of customer satisfaction is complicated even more by the fact that different customers place a different priority on these factors. That fact makes it even more critical that organizations maintain close, personal, and continuous contact with their customers.

Customer-Defined Value at Federal Express

Federal Express is one of the most successful package delivery companies on the globe. Many businesspeople think Federal Express sets the worldwide standard in customer service. One of the keys to the success of this company is its commitment to customer-defined value. Part of the operating philosophy of Federal Express is that customers are the best judges of quality.

In order to capture customers' input in a meaningful and useful manner, Federal Express developed a system of Service Quality Indicators (SQI). These indicators reflect the customers' views concerning their satisfaction with the performance of Federal Express. The list of SQIs is as follows:

1. Delivery on the right day, but after the promised time
2. Delivery on the wrong day

3. Unsuccessful trace of a package
4. Customer complaints
5. Proof of performance is missing
6. Missed pickups from customers
7. Damaged packages
8. Lost packages
9. Unanswered calls from customers (not answered within 20 seconds)

Federal Express continually collects data for all of these criteria and uses the data to identify service problems and their root causes. Of course, identifying the problems is just the first step. Correcting them and eliminating the root causes follows quickly once a cause has been identified.

CUSTOMER VALUE ANALYSIS

What is it that customers want from our organization? What is it about our products or services that customers value? A total quality organization must know the answers to these questions. Organizations that don't know what their customers value run the risk of wasting valuable resources improving the wrong things. The process used to determine what is important to customers is called *customer value analysis* (CVA). The CVA process consists of the following five steps:

1. *Determine what attributes customers value most.* Ask customers to describe the attributes of the product/service. At this point, no priority value is assigned to the attributes. This can be done using a written survey, a telephone survey, one-on-one interviews, or focus groups.

2. *Rate the relative importance of the attributes.* Ask customers to prioritize the list of attributes identified in the first step. The ranking should run from *most important* to *least important*.

3. *Assess your organization's performance relative to the prioritized list of attributes.* Is the most important attribute on the list the strongest attribute of the product or service in question? Ideally, the relative strength of the attributes of a given product or service will match the priorities established by customers.

4. *Ask customers to rate all attributes of your product/service against the same attributes of a competitor's product/service.* To consistently beat the competition, an organization's product/service must have more value for customers on an attribute-by-attribute basis. By asking customers to rate attributes, the organization can determine how they perceive the value, on a relative basis, of its product/service attributes. This gives the organization the information it needs to improve the attributes of its products/services in accordance with customer preferences.

5. *Repeat the process periodically.* Over time, customer preferences might change, as might the attributes of competing products/services. Consequently, it is important to periodically repeat the CVA process.

QUALITY CASE

Customer Trust Is Re-Earned Every Day at Meredith

Work hard to build brand loyalty, but don't ever take it for granted. Customer trust, once earned, must be re-earned every day. This is the operating philosophy at Meredith Corporation, the media giant located in Des Moines, Iowa. Meredith is a one billion dollar corporation that publishes 20 leading consumer magazines for the home and family market. Its flagship publications are *Better Homes and Gardens* and *Ladies' Home Journal.*

Perhaps the fastest way to lose customers is to take them for granted. This happened to thousands of American businesses in the 1980s when customers migrated to products made in Japan and other foreign countries, products they saw as being superior in both quality and price. According to Bill Kerr, Meredith's CEO, the best way to hang on to customers is to re-earn their trust and loyalty every day. To do this, companies must first stay focused internally. This means that businesses should figure out what they do well and focus on that. They should commit to doing this core business better than anyone else. Second, businesses should find out what customers want and give it to them.

Kerr knows that customer trust is essential to the success of Meredith Corporation. Consequently, he insists that all of Meredith's personnel be vigilant in producing products that exemplify the family values the company's customers expect. For example, Kerr personally vetoed a proposal for Meredith to carry the hugely profitable *Howard Stern* program on the company's CBS–affiliated television stations. His rationale was simple: the program would offend Meredith's traditional customer base.

Kerr believes that building trust can also build profits, and he proves his point by continuing to carve out new market niches within the broader home and family market. When Meredith produces a new product, customers will give it a try and a chance because they trust Meredith's publications. Kerr is committed to re-earning this trust every day.

Source: Thomas J. Neff & James M. Citrin, *Lessons from the Top,* (New York: Doubleday, 2001), 193–198.

CUSTOMER RETENTION

Customer satisfaction is a fundamental cornerstone of total quality. An organization develops a customer focus to be better able to satisfy its customers. Consequently, forward-looking organizations use customer satisfaction data to measure success. But measuring customer satisfaction alone is not enough. Another important measure of success is customer retention.

Certainly customer satisfaction is the critical component in customer retention, but the two factors are not necessarily synonymous. A customer satisfied is not always a customer retained. Frederick F. Reichheld makes the point that although "it may seem intuitive that increasing customer satisfaction will increase retention and therefore profits, the facts are contrary. Between 65 and 85 percent of customers who defect say they were satisfied or very satisfied with their former supplier."[13] Reichheld's findings suggest that there is more to customer retention than just customer satisfaction.

Many business leaders assume that having acquired customers they need only provide high-quality products and services to retain them. Michael W. Lowenstein calls this the *"myth of customer satisfaction."*[14] According to Lowenstein, "Conventional wisdom of business, academia, and the consulting community is that . . . if satisfied, the customer will remain loyal. Reality proves that customer loyalty or retention is a more complex, yet more definitive indicator of quality performance."[15]

It is important to understand what Lowenstein is saying here. Is he saying that customer satisfaction is not important? No, of course not. Customer satisfaction is critical, but it is a means to an end, not an end in itself. The desired end is customer retention. What Lowenstein is saying is that organizations should measure success based on customer retention data rather than on customer satisfaction data. The issue is not whether customers are satisfied with the organization's products or services, it is whether they are satisfied enough to be retained. Satisfied customers will sometimes defect in spite of their satisfaction, if for no other reason than curiosity about a competitor or the ever-present lure of variety. How, then, can an organization go beyond just satisfying its customers to retaining them? The short answer to this question is as follows:

To retain customers over the long term, organizations must turn them into partners and proactively seek their input rather than waiting for and reacting to feedback provided after a problem has occurred.

The following strategies can help organizations go beyond just satisfying customers to retaining them over the long term. These strategies will help organizations operationalize the philosophy of turning customers into partners.

Be Proactive—Get Out in Front of Customer Complaints

Many organizations make the mistake of relying solely on feedback from customers for identifying problems; the most widely used mechanism in this area is the customer complaint process. Feedback-based processes, although necessary and useful, have three glaring weaknesses. First, they are activated by problems customers have already experienced. Even if these problems are solved quickly, the customer who complains has already had a negative experience with the organization. Such experiences are typically remembered—even if only subconsciously—no matter how well the organization responds.

The second weakness of feedback-oriented processes is that they are based on the often invalid assumption that dissatisfied customers will take the time to lodge a complaint. Some will, but many won't. Some people are just too busy to take the time to complain. Others provide their feedback by simply going elsewhere. In a survey of retail customers conducted by the Institute for Corporate Competitiveness (ICC), 67% of respondents said they would simply go elsewhere if dissatisfied rather than take the time to complain.[16] Retail customers don't necessarily have the same characteristics as customers of service or production organizations. However, the ICC survey still points to a fundamental weakness with customer complaint processes that rely on information collected *ex post facto*.

The third weakness of customer complaint processes is that the information they provide is often too sketchy to yield an accurate picture of the problem. This situation can result in an organization wasting valuable resources chasing after symptoms rather than solving root causes. The weaknesses associated with after-the-fact processes do not mean that organizations should stop collecting customer feedback. On the contrary, customer feedback can be important when used to supplement the data collected using input-based processes.

Customer input is customer information provided *before* a problem occurs. An effective vehicle for collecting customer input is the focus group. *Focus groups* consist of customers who agree to meet periodically with representatives of the organization for the purpose of pointing out issues before they become problems. Focus groups can provide a mechanism for overcoming all three of the weaknesses associated with feedback systems. Participants point out weaknesses or potential issues to the organization's representatives so that they can be dealt with preemptively; focus group input does not depend on the willingness of customers to lodge complaints. Participants agree to provide input at periodic meetings before becoming members of the group.

A variation on the focus group concept is the input group. The purpose of both types of groups is to provide input the organization can use to improve its processes and products and services. The difference between the two is that focus group participants meet together for group discussion. Input group participants provide their data individually, usually by mail, telephone, or facsimile machine. They do not meet together for group interaction.

The focus group approach can also solve the problem of sketchy information. In a focus group, there is discussion, debate, and give-and-take. This type of interaction provides the organization's representatives with opportunities to dig deeper and deeper until they get beyond symptoms to root causes. Input provided by one participant will often trigger input from another.

To be effective, the focus group must consist of participants who understand what they are being asked to do. The organization is well served by neither sycophants nor witch hunters. What is needed is information that is thoroughly thought out and objective, given in the form of open, honest, constructive criticism. Members of the focus group should change periodically to bring in new ideas and a broader cross section of input.

Other methods for collecting customer input include hiring test customers and conducting periodic surveys of a representative sample of the customer base. *Test customers* are individuals who do business with the organization and report their perceptions to designated representatives of the organization. This method can backfire unless employees are fully informed that it is a method the organization employs. This does not mean that employees should know who the test customers are; they shouldn't, or this method will lose its value. However, they should know that any customer they interact with might be a test customer.

Customer surveys conducted periodically can help identify issues that may become problems. If this method is used, the survey instrument should be brief and to the point. One of the surest ways to "turn off" customers is to ask them to complete a lengthy survey instrument. Some type of reward should be associated with completing the survey that says, "Thank you for your valuable time and assistance." Each time a survey is conducted, care should be taken to select a different group of customers. Asking the same people to complete surveys over and over is sure to alienate even the most loyal customers.

Collect Both Registered and Unregistered Complaints

Many organizations make the mistake of acting solely on what customers say in complaints instead of going beyond what is said to include what is unspoken. Lowenstein calls this phenomenon the *Iceberg Complaint Model.*[17] In other words, registered complaints from customers are just the tip of the iceberg that is seen above the surface of the water. A much larger portion of the iceberg floats quietly beneath the surface. For this reason it is important for organizations to collect both registered and unregistered complaints.

Focus groups—already discussed—are an excellent way to solicit unregistered complaints. Customer surveys and test customers can also serve this purpose. Another way to get at that part of the iceberg that floats beneath the surface is the *follow-up interview.* With this method, customers who have registered complaints are contacted either in person or by telephone to discuss their complaints in greater depth. This approach gives representatives of the organization the opportunity to ask clarifying questions and to request suggestions.

Another way to get at unregistered complaints is to use the organization's sales representatives as collectors of customer input. Sales representatives are the employees who have the most frequent face-to-face contact with customers. If properly trained concerning what to look for, what to ask, and how to respond, sales personnel can bring back invaluable information from every sales call. In addition to providing sales personnel with the necessary training, organizations should also provide them with appropriate incentives for collecting customer input. Otherwise they may fall into the trap of simply agreeing with the customer about complaints received, thereby undermining the customer relationship even further.

ESTABLISHING A CUSTOMER FOCUS

Companies that have successfully established a customer focus share a number of common characteristics. In his book *The Customer Driven Company*, Richard C. Whitely suggests that these characteristics can be divided into seven clusters:[18]

- ■ *Vision, commitment, and climate.* A company with these characteristics is totally committed to satisfying customer needs. This commitment shows up in everything the company does. Management demonstrates by deeds and words that the customer is important, that the organization is committed to customer satisfaction, and that customer needs take precedence over internal needs. One way such organizations show their commitment to customers and establish a climate in which customer satisfaction prevails is by making the goal of being customer focused a major factor in all promotions and pay increases.

■ *Alignment with customers.*   Customer-driven companies align themselves with their customers. Customers are included when anyone in the organization says "we." Alignment with customers manifests itself in several ways, including the following: customers play a consultative role in selling, customers are never promised more than can be delivered, employees understand what product attributes the customers value most, and customer feedback and input are incorporated into the product development process.

■ *Willingness to find and eliminate customers' problems.*   Customer-driven companies work hard to continually identify and eliminate problems for customers. This willingness manifests itself in the following ways: customer complaints are monitored and analyzed; customer feedback is sought continually; and internal processes, procedures, and systems that create no value for customers are identified and eliminated.

■ *Use of customer information.*   Customer-driven companies not only collect customer feedback but also use it and communicate it to those who need it to make improvements. The use of customer information manifests itself in the following ways: all employees know how the customer defines quality, employees at different levels are given opportunities to meet with customers, employees know who the "real" customer is, customers are given information that helps them develop realistic expectations, and employees and managers understand what customers want and expect.

■ *Reaching out to customers.*   Customer-driven companies reach out to their customers. In a total quality setting, it is never enough to sit back and wait for customers to give evaluative feedback. A competitive global marketplace demands a more assertive approach. Reaching out to customers means doing the following:

> Making it easy for customers to do business
> Encouraging employees to go beyond the normal call of duty to please customers
> Attempting to resolve all customer complaints
> Making it convenient and easy for customers to make their complaints known

■ *Competence, capability, and empowerment of people.*   Employees are treated as competent, capable professionals and are empowered to use their judgment in doing what is necessary to satisfy customer needs. This means that all employees have a thorough understanding of the products they provide and the customer's needs relating to those products. It also means that employees are given the resources and support required to meet the customer's needs.

■ *Continuous improvement of products and processes.* Customer-driven companies do what is necessary to continuously improve their products and the processes that produce them. This approach to doing business manifests itself in the following ways: internal functional groups cooperate to reach shared goals, best practices in the business are studied (and implemented wherever they will result in improvements), research and development cycle time is continually reduced, problems are solved immediately, and investments are made in the development of innovative ideas.

These seven clusters of characteristics can be used as a guide in establishing a customer focus. The first step is a self-analysis in which it is determined which of these characteristics are present in the organization and which are missing. Characteristics that are missing form the basis for an organization-wide implementation effort.

RECOGNIZING THE CUSTOMER-DRIVEN ORGANIZATION

Is a given organization customer driven? In today's competitive business environment, the answer to this question must be yes. Since this is the case, it is important for quality professionals to be able to recognize a customer-driven organization and to be able to articulate its distinguishing characteristics.

According to Whitely, a customer-driven organization can be recognized by the following characteristics:[19]

■ *Reliability*—an organization that dependably delivers what is promised on time every time
■ *Assurance*—an organization that is able to generate and convey trust and confidence
■ *Tangibles*—an organization that pays attention to the details in all aspects of its operations
■ *Empathy*—an organization that conveys a real interest in its customers
■ *Responsiveness*—an organization that is willingly attentive to customer needs

In addition to these characteristics, Thomas Cartin and Donald Jacoby recommend that quality professionals look for the following management factors:[20]

■ *Internal support.* Has the organization developed the right structure management and employee skills (organizational development)?
■ *Use of knowledge.* Does the organization systematically collect customer feedback and continually assess changing customer expectations and needs?
■ *Use of metrics.* Has the organization established market-based performance measures that are tied to customer needs rather than internal factors?
■ *Communication.* Does the organization regularly communicate progress toward meeting customer needs to its employees?

By applying the characteristics and management factors set forth in this section, quality professionals can recognize customer-driven organizations. They can also quickly surmise where problems and shortcomings exist and develop a clearly focused plan of improvement.

VALUE PERCEPTION AND CUSTOMER LOYALTY

Companies work hard to build customer loyalty. Think of Coke and Pepsi, for example. The goal is to keep the customer on board for the long term. The theory is that a *loyal* customer is a customer forever. It is easy to see what drives the desire to create customer loyalty. Companies spend so much in marketing to attract customers that they must keep them for the long term to recoup their investments.

Writing for the *Harvard Business Review*, Robert Blattberg and John Deighton cite the following example of why companies want to build customer loyalty:

> Lands' End, the second-largest clothing catalog retailer in the United States, operates with unusually high inventory levels. Its manager would rather inflate inventory than fail to fill an order and risk losing a customer. "If we don't keep the customer for several years, we don't make money," said the company's CEO. . . . "We need a long-term payback for the expense of coming up with a buyer."[21]

In spite of all the attention given to the concept, Karl Albrecht claims that customer loyalty is a myth. In fact, he believes, "Aiming for customer satisfaction is a prescription for mediocrity."[22] To prove his point, Albrecht uses the basic Theory of Service Relativity formula, which is as follows:

$V = R - E$
$V =$ Value
$R =$ Results
$E =$ Expectations

This formula means that value (V) as perceived by the customer is equal to actual results (R) minus expectations (E). Consequently, when results and expectations are equal, the perceived value is zero. In other words, there is no perceived value when an organization simply meets customer expectations.

Albrecht's point is that the goal should go beyond just meeting expectations to creating value in the eyes of the customer. "This means that the battle for future business has to be fought at every transaction and at every moment of truth."[23]

Albrecht also challenges the concept of customer loyalty. "There is no such thing as customer loyalty. The term . . . encodes a flawed concept, just as customer satisfaction does."[24] Customer loyalty implies a personal bond that transcends the business relationship. Customer preference, not customer loyalty, is the goal. Customers stay with an organization not out of loyalty but because they prefer to do business with it. And they prefer to do business with the organization that creates the most value for them.

GUASPARI ON CUSTOMER SATISFACTION

> "Now I see it!" he exclaimed. . . . Our customers tell us we have a quality problem, and we turn to our specs and our tolerances to see if they are right. Of course they're right! Customers aren't interested in our specs. They're interested in the answer to one simple question: did the product do what I expected it to do? If the answer is yes, then it's a quality product. If the answer is no, then it isn't."[25]

This passage comes from *I Know It When I See It: A Modern Fable about Quality* by John Guaspari. Guaspari uses the creative approach of the fable to show that customer satisfaction is a function of pleasing the customer, not meeting specifications. This is an important point because a product that does not please the customer even though it meets the customer's specifications is just as bad as one that does not meet specifications. Heresy? Not really, when you think about it. A company that makes a product to a specification that is poorly thought out, incomplete, or insufficient in any other way is still going to have an unhappy customer.

The customer-supplier relationship is a partnership. If a customer provides a set of specifications that can be improved on, the supplier–partner should point out the problems and resolve them before proceeding. That is what a partner does. Using inferior specifications as an excuse for producing an inferior product will not produce satisfied customers. It is like the driver of the car who has a green light at a busy intersection and so speeds through it without even bothering to look either way. When his passenger asks why he isn't concerned about another car running a red light and slamming into him, he responds, "I have the right-of-way." His passenger responds, "Sure, you would be right— *dead* right." Meet the specifications, but lose the customer—this is a formula for failure.

 # SUMMARY

1. Historically customers were considered outsiders who used a company's products, and suppliers were outsiders who provided the materials needed to produce the products. A more contemporary view is that every organization has both internal and external customers. An *external customer* is the one referred to in the traditional definition. An *internal customer* is any employee whose work depends on that of employees whose work precedes his or hers.

2. In a total quality setting, customers define quality. Therefore, customer satisfaction must be the highest priority. Customer satisfaction is achieved by producing high-quality products that meet or exceed expectations. It must be renewed with each purchase. The key to establishing a customer focus is to put employees in touch with customers so that customer needs are known and understood.

3. Scholtes's six-step strategy for identifying customer needs is as follows: speculate about results, develop an information-gathering plan, gather information, analyze the results, check the validity of conclusions, and take action.

4. Customer needs are not static. Therefore, constant contact with customers is essential in a total quality setting. Whenever possible, this contact should be in person or by telephone. Written surveys can be used, but they will not produce the level of feedback that personal contact can generate.

5. Quality function deployment (QFD) is a mechanism for putting into operation the concept of building in quality. It makes customer feedback a normal part of the product development process, thereby improving customer satisfaction.

6. Measuring customer satisfaction alone is not enough. Many customers who defect are satisfied. Organizations should, in addition, measure customer retention.

7. Organizations should go beyond satisfying customers to creating value for them in every supplier–customer interaction.

KEY TERMS AND CONCEPTS

Communicating with customers	Internal customers
Customer	Myth of customer satisfaction
Customer-defined quality	Perceived value
Customer focus	Quality function deployment
Customer satisfaction	Registered complaints
Customer value analysis	Reliable customers
External customers	Supplier
Iceberg Complaint Model	Test customers
Information-gathering plan	Unregistered complaints

 FACTUAL REVIEW QUESTIONS

1. Explain the contemporary concepts of *customer* and *supplier*.
2. How does the contemporary view of customers and suppliers differ from the more traditional view?
3. Explain the role of the customer in a total quality setting.
4. What is a reliable customer?
5. What role does customer satisfaction play in the development of reliable customers?
6. How is customer satisfaction ensured?
7. Briefly describe what is meant by an organization that has customer focus.
8. How does an organization go about establishing a customer focus?
9. Explain the six-step strategy for identifying customer needs.
10. Describe how organizations should go about communicating with their customers.
11. Explain briefly the concept of QFD and how it relates to customer satisfaction.
12. Explain why it is important to measure customer retention.
13. Explain why just meeting customer specifications might not produce customer satisfaction.

 CRITICAL THINKING ACTIVITY

Are Customers Really Loyal?

"I'll stick with you through thick and thin," or "What have you done for me lately?" Which best describes the attitude of customers in today's marketplace? Two quality managers are debating the concept of customer loyalty. One of them, Jack Hayes, claims that customer loyalty does exist, that it can be won, and that winning a customer's loyalty should be every organization's goal. According to Jack, "If you have a history of satisfying a customer, he will be loyal enough to overlook an occasional bad experience."

"No way," says Anna Cage. "It takes only one bad experience to lose a customer."

Join this debate. You are a customer. Are you loyal to any organizations? If so, how many bad experiences will it take to overcome your loyalty? Have you ever decided to withhold your business from a store, restaurant, or other service provider based on poor service? Do you usually give an organization more than one chance to win your business, or is one bad experience all it takes to turn you off?

DISCUSSION ASSIGNMENT 7–1

Motorola's Commitment to Customer Satisfaction

Motorola is a world-class leader in the manufacture of consumer electronics products. As a corporation, its fundamental objective is total customer satisfaction, and achieving this objective is every employee's responsibility. Motorola sets the tone for its customer focus by stating its key beliefs, key goals, and key initiatives as follows:[26]

Key Beliefs—How We Will Always Act
- Constant Respect for People
- Uncompromising Integrity

Key Goals—What We Must Accomplish
- Increased Global Market Share
- Best in Class
- People
- Marketing
- Technology
- Product
- Manufacturing
- Service
- Superior Financial Results

Key Initiatives—How We Will Do It
- Six-Sigma Quality
- Total Cycle Time Reduction
- Product and Manufacturing Leadership
- Profit Improvement
- Participative Management within and Cooperation between Organizations

Discussion Assignment
Discuss the following question in class or outside of class with your fellow students:

1. Given the opportunity, is there anything you would add to Motorola's statement of commitment? If so, what, and why?

Source: James A. Belohlav, *Championship Management* (Cambridge, MA: Productivity Press, 1990), 63.

DISCUSSION ASSIGNMENT 7–2

Federal Express Knows Customers Want It Now!

"You know you have hit the big time when the purple-and-orange-lettered FedEx truck driven by a smiling courier pulls into your driveway or walks up your stairwell and hands you the purple-and-orange envelope. First off, it's always a blessed relief to get what's inside, because it absolutely, positively had to get there.

"Fred Smith, a 20-year-old sophomore at Yale University, understood this characteristic of American culture when he thought up his topic for his business class project. We all know the legend, how he wrote a term paper describing his idea for overnight delivery of documents anywhere in the country by means of a fleet of airplanes flying to a hub system converging on Memphis of all places. Fred got a C on the paper, and most of his buddies thought he was lucky.

"But it took more than full faith in a sophomore's brainchild to make FedEx what it is today. The company serves one of the most important aspects of American culture and business: the need for a NOW response. And FedEx knows that the only thing it has to sell is NOW—without fail."

Discussion Assignment

Discuss the following questions in class or outside of class with your fellow students:

1. Americans have developed what has been called a "microwave mentality." We like quick service, instant food, and fast results. Does this describe you as a customer?
2. What American companies besides Federal Express base their success on NOW service?

Source: Josh Hammond and James Morrison, *The Stuff Americans Are Made Of* (New York: Macmillan, 1996), 171–172.

DISCUSSION ASSIGNMENT 7–3

Winning Back Lost Customers Is a Costly Challenge

Perhaps the most famous case illustrating the enormous cost of winning back lost customers is that of the "Tylenol Murders." In 1982, seven people in the Chicago area died suddenly after taking Tylenol capsules. An investigation revealed that the capsules had been laced with cyanide, a deadly poison. When the story ran on the nightly news programs, a nationwide panic ensued that caused Tylenol's sales to plummet overnight. Many business analysts predicted that Tylenol's manufacturer, Johnson & Johnson, would not survive the tragedy. Johnson & Johnson surprised the analysts by undertaking one of the most successful campaigns in history to win back customers. It worked, but the cost was huge. This case led not just Johnson & Johnson, but all major drug manufacturers, to develop the tamperproof bottle. Having done so, Johnson & Johnson undertook an intensive public relations campaign to win back the trust of its customers.

Discussion Assignment
Discuss the following questions in class or outside of class with your fellow students:

1.   If the Tylenol incident were to happen today and you were a user of Tylenol, would Johnson & Johnson be able to win back your trust?
2.   What would it take for Johnson & Johnson to win back your trust if one of the victims was a friend of yours?

 ENDNOTES

1. Peter R. Scholtes, *The Team Handbook* (Madison, WI: Joiner Associates, 1992), 2-6.
2. Giorgio Merli, *Total Manufacturing Management* (Cambridge, MA: Productivity Press, 1990), 6–7.
3. Scholtes, 1-11.
4. "Zytec Employees Can Spend up to $1,000 to Satisfy Customers," *Total Quality Newsletter* 3 (July 1992), 4.
5. Scholtes, 5-48.
6. Scholtes, 5-48–5-49.
7. Joseph M. Juran, *Juran on Leadership for Quality* (New York: Free Press, 1989), 101.
8. Juran, 314.
9. W. Edwards Deming, *Out of the Crisis* (Cambridge: Massachusetts Institute of Technology, Center for Advanced Engineering Study, 1991), 167.
10. Merli, 222.
11. Merli, 228–238.
12. Michael Butchko, Mike Butchko Consulting, "Customer Satisfaction Process," a white paper dated July 25, 1995. Used with permission.
13. Frederick F. Reichheld, "Loyalty-Based Management," *Harvard Business Review* (March–April 1993), 71.
14. Michael W. Lowenstein, *Customer Retention—An Integrated Process for Keeping Your Best Customers* (Milwaukee: ASQC Quality Press, 1995), 1.
15. Lowenstein, 10.
16. "Customer Complaint Survey," Final Report, Institute for Corporate Competitiveness, 1994.
17. Lowenstein, 37–39.

18. Richard C. Whitely, *The Customer Driven Company: Moving from Talk to Action* (New York: Addison-Wesley, 1991), 221–225.

19. Whitely, 109–111.

20. Thomas J. Cartin and Donald J. Jacoby, *A Review of Managing Quality and a Primer for the Certified Quality Manager Exam* (Milwaukee, WI: ASQC Quality Press, 1997), 106.

21. Robert C. Blattberg and John Deighton, "Manage Marketing by the Customer Equity Test," *Harvard Business Review* (July–August 1996), 136.

22. Karl Albrecht, "Evaluating the Customer Loyalty Myth," *Quality Digest*, vol. 18, no. 4 (April 1998), 64.

23. Albrecht, 64.

24. Albrecht, 64.

25. John Guaspari, *I Know It When I See It* (New York: American Management Association, 1991), 70.

26. James A. Belohlav, *Championship Management* (Cambridge, MA: Productivity Press, 1990), 68.

CHAPTER EIGHT

Employee Empowerment

"Light is the task where many share the toil."
  Homer

MAJOR TOPICS

- Employee Empowerment Defined
- Rationale for Empowerment
- Inhibitors of Empowerment
- Management's Role in Empowerment
- Implementing Empowerment
- Management's Role in Suggestion Systems
- Improving Suggestion Systems
- Evaluating Suggestions
- Handling Poor Suggestions
- Achieving Full Participation
- How to Recognize Empowered Employees
- Avoiding Empowerment Traps
- Beyond Empowerment to Enlistment

Involving people in decisions made relating to their work is a fundamental principle of good management. With total quality, this principle is taken even further. First, employees are involved not only in decision making but also in the creative thought processes that precede decision making. Second, not only are employees involved; they are empowered. This chapter explains the concepts of involvement and empowerment, their relationship, and how they can be used to improve competitiveness.

EMPLOYEE EMPOWERMENT DEFINED

Employee involvement and empowerment are closely related concepts, but they are not the same. In a total quality setting, employees should be empowered:

James Monroe, CEO of a midsized electronics manufacturing firm, decided more than a year ago to get his employees involved as a way to improve work and enhance his company's

QUALITY TIP

Involve Employees

"Involve all personnel at all levels in all functions in virtually everything; for example, quality improvement programs and 100 percent self-inspection; productivity improvement programs; measuring and monitoring results; budget development, monitoring, and adjustment; assessment of new technology; recruitment and hiring; making customer calls and participating in customer visit programs."

Source: Tom Peters, *Thriving on Chaos: A Handbook for a Management Revolution* (New York: Harper & Row, 1987), 342.

competitiveness. He called his managers and supervisors together, explained his idea, and had suggestion boxes placed in all departments. At first, the suggestion boxes filled to overflowing. Supervisors emptied them once a week, acted on any suggestions they thought had merit, and discarded the rest. After a couple of months, employee suggestions dwindled down to one or two a month. Worse, recent suggestion forms have contained derisive remarks about the company and its suggestion system. Productivity has not improved, and morale is worse than before. Monroe is at a loss over what to do. Employee involvement was supposed to help, not hurt.

In this example, the CEO involved his employees, but he failed to empower them. To understand the difference, it is necessary to begin with an understanding of employee involvement as a concept. Peter Grazier, author of *Before It's Too Late: Employee Involvement . . . An Idea Whose Time Has Come*, describes employee involvement as follows:

What is Employee Involvement? It's a way of engaging employees at all levels in the thinking processes of an organization. It's the recognition that many decisions made in an organization can be made better by soliciting the input of those who may be affected by the decision. It's an understanding that people at all levels of an organization possess unique talents, skills, and creativity that can be of significant value if allowed to be expressed.[1]

What, then, is empowerment? Stated simply, *empowerment* is employee involvement that matters. Another good definition of empowerment is the *controlled transfer of authority to make decisions and take action*. It's the difference between just having input and having input that is heard, seriously considered, and followed up on whether it is accepted or not.

Most employee involvement systems fail within the first year, regardless of whether they consist of suggestion systems, regularly scheduled brainstorming sessions, daily quality circles, one-on-one discussions between an employee and a supervisor, or any combination of the various involvement methods. The reason is simple: they involve but do not empower employees. Employees soon catch on to the difference between having input and having input that matters. Without empowerment, involvement is just another management tool that doesn't work. This is what went wrong in the example of the electronics manufacturing firm at the beginning of this section. The CEO implemented a system that involved but did not empower his employees.

Management Tool or Cultural Change

The management strategies developed over the years to improve productivity, quality, cost, service, and response time would make a long list. Is empowerment just another of these management tools, another strategy to add to the list? This is an important question, and it should be dealt with in the earliest stages of implementing empowerment.

Employees who have been around long enough to see several management innovations come and go may be reluctant to accept empowerment if they see it as just another short-lived management strategy. This is known as the WOHCAO syndrome (pronounced WO-KAY-O). WOHCAO is short for "Watch out, here comes another one."

Successful implementation of empowerment requires change in the corporate culture—a major new direction in how managers think and work. The division of labor between managers and workers changes with empowerment. Grazier describes the cultural change required by empowerment as follows:

> It's not something "nice" we do for our employees to make them feel better. It's an understanding that it's everyone's obligation—part of the job—to constantly look for better ways of doing things. It's part of the job to ask questions and raise issues of concern, to get them out on the table so they can be resolved. How else can we get better? Above all, employee involvement is not simply another management tool, but a major change of direction in the way we lead our workforce. It's a change that affects the culture of the workplace as we've come to know it in this century, and, therefore, it's a change that must be implemented with great care and attention.[2]

QUALITY TIP

Let Employees Conceive and Implement

"Many companies emphasize programs that encourage first-line workers to develop ideas for improvement, but the best companies also foster in their employees the ability to execute those improvement ideas. Ideas conceived by the workers should also be executed by the workers."

Source: Iwao Kobayashi, *20 Keys to Workplace Improvement*, trans. Warren W. Smith (Cambridge, MA: Productivity Press, 1990), 163.

Empowerment Does Not Mean Abdication

It is not uncommon for traditional managers to view empowerment as an abdication of power. Such managers see involvement and empowerment as turning over control of the company to the employees. In reality, this is hardly the case. Empowerment involves actively soliciting input from those closest to the work and giving careful thought to that input.

Pooling the collective minds of all people involved in a process, if done properly, will enhance rather than diminish managers' power. It increases the likelihood that the information on which decision makers base their decisions is comprehensive and accurate.

An example from medicine illustrates this point. Suppose that a child is gravely ill. Every established medical procedure known has been used without success. The only remaining option is a rarely employed and unapproved procedure that has been performed by just 10 surgeons. There are variations of the procedure, each carrying its own specific risks as well as its own probability of success.

Because the procedure is not approved, the child's parents must decide which variation is to be applied. They may simply close their eyes, make a choice, and hope for the best, or they may take another approach in arriving at a decision. That other approach involves soliciting input from the 10 surgeons who have performed the operation in its several variations. The surgeons are willing and eager to give their advice.

Which approach is most likely to produce the best decision? Most people would want to pool the minds of the surgeons who have performed the operation. They are closest to the job and therefore likely to have valuable insight concerning how it should be done. Responsibility for ultimately making the decision still rests with the parents. However, these parents will probably solicit the input of the 10 surgeons in question and give it careful consideration. In other words, although they will ultimately make the decision, the parents in this example will involve and empower the surgeons.

The same concept applies in the workplace. Managers do not abdicate their responsibility by adopting empowerment. Rather, like the parents in this example, they increase the likelihood of making the best possible decisions and thereby more effectively carry out their responsibility.

RATIONALE FOR EMPOWERMENT

Traditionally, working hard was seen as the surest way to succeed. With the advent of global competition and the simultaneous advent of automation, the key to success became not just working hard but also working smart. In many cases, decision makers in business and industry interpreted working smart as adopting high-tech systems and automated processes. These smarter technologies have made a difference in many cases. However, improved technology is just one aspect of working smarter, and it's a part that can be quickly neutralized when the competition adopts a similar or even better technology.

An aspect of working smart that is often missing in the modern workplace is involving and empowering employees in ways that take advantage of their creativity and promote independent thinking and initiative on their part. In other words, what's missing is empowerment. Creative thinking and initiative from as many employees as possible will increase the likelihood of better ideas, better decisions, better quality, better productivity, and, therefore, better competitiveness. The rationale for empowerment is that it represents the best way to bring the creativity and initiative of the best employees to bear on improving the company's competitiveness.

Human beings are not robots or automatons. While working they observe, think, sense, and ponder. It is natural for a person to continually ask such questions as the following:

- Why is it done this way?
- How could it be done better?
- Will the customer want the product like this?

Asking such questions is an important step in making improvements. As employees ask questions, they also generate ideas for solutions, particularly when given the opportunity to regularly discuss their ideas in a group setting that is positive, supportive, and mutually nurturing.

Grazier describes the process as follows:

> The human brain stores billions of thought impulses that are not always easily retrieved. A "best solution" to a problem may be locked away in someone's mind, but the person either doesn't know it or can't retrieve it. The person needs a catalyst. In a group of people, the catalyst is provided by someone else's thought. One thought triggers another, releasing more and more ideas. And since people in a group tend to build on each other's thoughts, the ideas and ultimate solutions of the group are better than the individual's idea alone. In addition to mind-pooling, the group process serves an even greater value: it builds trust, confidence, skills, and teamwork.[3]

Empowerment and Motivation

According to Dr. Isobel Pfeiffer and Dr. Jane Dunlop of the University of Georgia:

> Empowerment is the key to motivation and productivity. An employee who feels he or she is valued and can contribute is ready to help and grow in the job. Empowerment enables a person to develop personally and professionally so that his or her contributions in the workplace are maximized.[4]

Empowerment is sometimes seen by experienced managers as just another name for participatory management. However, there is an important distinction between the two. Participatory management is about managers and supervisors asking for their employees' help. Empowerment is about getting employees to help themselves, each other, and the company. This is why empowerment can be so effective in helping maintain a high level of motivation among employees. It helps employees develop a sense of ownership of their jobs and of the company. This, in turn, leads to a greater willingness on the part of employees to make decisions, take risks in an effort to make improvements, and speak out when they disagree.

QUALITY TIP

Empowerment and Global Competition

"[P]owerlessness is the root cause of many of the problems that management is concerned with in the workplace today. There are both economic and psychological reasons why empowerment has become a hot topic in today's business world. In the new era of stiff global competition, American corporations need to respond much more quickly to market forces. . . . As a result, the pressure on employees to produce or perish is increasing. Companies are looking for ways to improve their own bottom lines and help employees cope with the added stress."

Source: Peter Kizilos, "Crazy about Empowerment?" *Training* (December 1990), 47–56.

INHIBITORS OF EMPOWERMENT

The primary inhibitor of empowerment, resistance to change, is an indigenous characteristic of human nature. According to Grazier:

> Resistance to employee involvement is real. Since 1970, when the concept of employee involvement was first introduced in America, resistance has caused the failure of many employee involvement efforts. It is puzzling that a concept that produces both tangible benefits for the organization and a great deal of satisfaction for the workforce creates so much resistance. Resistance to empowerment, when it occurs, comes from three different groups: employees, unions, and management. Although the greatest resistance to empowerment traditionally comes from management, sometimes consciously and sometimes unconsciously, resistance from unions and employees should not be overlooked.[5]

Resistance from Employees and Unions

Earlier in this chapter, the WOHCAO syndrome was discussed. In this syndrome, employees have experienced enough flash-in-the-pan management strategies that either did not work out or were not followed through on that they have become skeptical.

In addition to skepticism, there is the problem of inertia. Resistance to change is natural. Even positive change can be uncomfortable for employees because it involves new and unfamiliar territory. However, when recognized for what they are, skepticism and inertia can be overcome. Strategies for doing so are discussed later in this chapter.

Unions can be another source of resistance when implementing empowerment. Because of the traditional adversarial relationship between organized labor and management, unions may be suspicious of management's motives in implementing empowerment. They might also resent an idea not originated by their own organization. However, unions' greatest concern is likely to be how empowerment will affect their future. If union leaders think it will diminish the need for their organization, they will throw up roadblocks.

Discussion Assignment 8–1, at the end of the chapter, describes a situation in which a union fought implementation of empowerment, at least at first. The company, which now has a nationally recognized quality improvement program, initially failed to involve union leaders in developing and implementing the program and, as a result, suffered the predictable consequences. However, after union leaders were involved and through their involvement learned that empowerment represented no threat to their organization, resistance and suspicion were replaced by support and ownership.

Clear evidence exists that resistance to empowerment on the part of labor unions is becoming less and less an issue. Leaders of some of the largest unions in the United States—including the Communications Workers of America, International Chemical Workers Union, International Brotherhood of Electrical Workers, and the Amalgamated Clothing and Textiles Workers Union—participated in a 2-year study as members of the government-sponsored Economic Policy Productivity Panel. The *Philadelphia Inquirer* carried the following excerpt from the panel's report on the findings of this study:

> By involving personnel at all levels of an enterprise in problem solving and decision making, by enriching the working lives of employees, by helping resolve personal concerns that affect their job performance, and by creating a climate in which people can achieve job satisfaction through directing their creativity and talents toward improving their work and their environment, worker participation programs can provide a foundation for sustained productivity growth.[6]

Resistance from Management

Even if employees and labor unions support empowerment, it will not work unless management makes a full and wholehearted commitment to it. Peter Kizilos says that "too many companies are attempting only halfheartedly to empower employees."[7] What he means is that companies are attempting to implement empowerment without first making the necessary fundamental changes in organizational structure or management style.

The importance of management commitment cannot be overemphasized. Employees take their cues from management concerning what is important, to what the company is committed, how to behave, and all other aspects of the job. Discussion Assignment 8–2 shows what can result when either a real or a perceived lack of commitment exists on the part of management. Because of a lack of management support, the company referred to in this assignment lost a program that had made substantial improvements in its ability to compete.

Grazier summarizes the reasons behind management resistance to empowerment as follows: insecurity, personal values, ego, management training, personality characteristics of managers, and exclusion of managers.[8]

Insecurity

An old adage states, "Knowledge is power." By controlling access to knowledge as well as the day-to-day flow of knowledge, managers can maintain power over employees. Managers who view the workplace from an us-against-them perspective tend to be insecure about any initiative they perceive as diminishing their power.

Another source of management insecurity is accountability. Pooling the minds of employees for the purpose of making workplace improvements is a sure way to identify problems, roadblocks, and inhibitors. Some managers fear they will be revealed as the culprit in such a process. The natural reaction of an individual who feels threatened is to resist the source of the threat. Managers are no different in experiencing this feeling.

Grazier describes the typical response of an insecure manager to empowerment as *subtle sabotage.*[9] He describes the phenomenon as follows:

> Maybe the supervisors don't attend any of the team's meetings, or maybe they make it difficult for a member to attend. Maybe they increase the workload of the members and then make them feel guilty when it's time for their meeting. Maybe there's a subtle put-down of the team's ideas and work. Maybe they approve team recommendations, then drag their feet on implementation. Subtle sabotage takes many forms, and has the cumulative effect of demoralizing the team. Much of the time, a demoralized team will disband, leaving a bad taste in everyone's mouth about employee involvement.[10]

Personal Values

Many of today's managers have a dogmatic mind-set when it comes to working with employees. This means they think employees should do what they are told, when they are told, and how they are told. Such a value system does not promote empowerment. Managers who feel this way will resist involvement and empowerment as being inappropriate. They are likely to think, "There can be only one boss around here, and that boss is me."

Ego

People who become managers may be understandably proud of their status and protective of the perquisites that accompany it. Status appeals to the human ego, and ego-focused managers may project an "I am the boss" attitude. Such managers may have difficulty reining in their egos enough to be effective participants in an approach they view as an encroachment on territory that should be exclusively theirs.

QUALITY TIP

Believing in Employees

Writing about the failure of some companies in the United States to seek the input, initiative, and creative ideas of its employees, Tom Peters says, "In my view this mind-set has proven disastrous. . . . Indeed, the chief reason for our failure in world-class competition is our failure to tap our workforce's potential."

Source: Tom Peters, *Thriving on Chaos: A Handbook for a Management Revolution* (New York: Harper & Row, 1987), 345.

Management Training

Many of today's managers were educated and trained by modern disciples of Frederick Taylor, the father of scientific management. Taylor's followers, whether university professors or management trainers, tend to focus on applying scientific principles to the improvement of processes and technology. Less attention is given to people-oriented improvements. Writing about Taylor, Grazier says:

> Management was continuously looking for methods of dealing with the "labor problem" and Taylor's scientific approach permitted them to deal with work flow procedures and equipment improvements rather than the more complex issues of employee commitment and morale. Taylor believed that it was the "experts" who solved problems in organizations. And the culmination of his work was a philosophy that clearly defined management's role as the "thinker" and labor's role as the "doer"; that is, management does the thinking, and labor does what management says.[11]

Attitudes like the one reflected in this quote are inappropriate in the modern workplace. Nevertheless, vestiges of Taylor's school of thought remain in college management programs throughout the United States. Actually, much of what Taylor professed about applying scientific principles in improving the workplace is valid and has gained new status with the advent of total quality. Function analysis, statistical process control (SPC), and just-in-time manufacturing (JIT)—all total quality tools—are examples of science applied to workplace improvement. However, unlike Taylor's followers, proponents of total quality involve employees in the application of these scientific methods. Management schools that still cling, even subtly, to the management-as-thinkers and labor-as-doers philosophy produce graduates who are likely to resist empowerment. Experienced managers who were schooled in this philosophy in years past and have practiced it throughout their careers are also likely to resist empowerment.

Personality Characteristics of Managers

Old-school managers are often found to be more task oriented than people oriented. They tend to focus more on the task at hand and getting it done than on the people who actually perform the task. Discussion Assignment 8–3, presented later, involves such a manager.

Wanda Brown in Discussion Assignment 8–3 is a task-oriented manager. Before her promotion, she did most of the work associated with task accomplishment herself. Her dependence on and interaction with other employees was minimal. As a result, Wanda's task-oriented personality served her and the company well. Now, as a manager, she is responsible for organizing work and getting it done by others.

In this new setting, she has found that the quality and quantity of employees' work can be affected by problems they are having. She has found that, in spite of her well-earned reputation for getting the job done, other employees have ideas of their own about workplace improvements, and they want their ideas to be heard and given serious consideration. Finally, Wanda Brown has learned that employees have feelings, egos, and personal agendas and that these things can affect their work. Managers such as Wanda Brown who have a strong task orientation are not likely to support efforts such as empowerment that call for a balanced attitude in which the manager is concerned with both tasks and people (see Figure 8–1).

Figure 8–1
Orientation of Managers

Exclusion of Managers

Empowerment is about the total involvement of all personnel who will be affected by an idea or a decision. This includes the first level of management (supervisors), midmanagement, and executive management. Any manager or level of management excluded from the process can be expected to resist. Even with a full commitment from executives and enthusiastic support from employees, empowerment will not succeed if midmanagers and supervisors are excluded. Those who are excluded, even if they agree conceptually, may resist simply because they feel left out.

Workforce Readiness

An inhibitor of employee empowerment that receives little attention in the literature is workforce readiness. Empowerment will fail quickly if employees are not ready to be empowered. In fact, empowering employees who are not prepared for the responsibilities involved can be worse than not empowering them at all. On the other hand, a lack of readiness—even though it may exist—should not be used as an excuse for failing to empower employees. The challenge to management is twofold: (a) determine whether the workforce is ready for empowerment; and (b) if it is not ready, get it ready.

How, then, does one know whether the workforce is ready for empowerment? One rule of thumb is that the more highly educated the workforce, the more ready its members will be for empowerment. Because well-educated people are accustomed to critical thinking, they are experienced in decision making, and they tend to make a point of being well informed concerning issues that affect their work. This does not mean, however, that less educated employees should be excluded. Rather, it means that they may need to be prepared before being included.

In determining whether employees are ready for empowerment, ask the following questions:

- Are the employees accustomed to critical thinking?
- Are the employees knowledgeable of the decision-making process and their role with regard to it?
- Are employees fully informed of the "big picture" and where they fit into it?

Unless the answer to all three of these questions is yes, the workforce is not ready for empowerment. An empowered employee must be able to think critically. It should be second nature for an employee to ask such questions as the following: Is there a better way to do this? Why do we do it this way? Could the goal be accomplished some other way? Is there another way to look at this problem? Is this problem really an opportunity to improve things?

These are the types of questions that lead to continuous improvement of processes and effective solutions to problems. These are the sorts of questions that empowered employees should ask all the time about everything. Employees who are unaccustomed to asking questions such as these should be taught to do so before being empowered.

Employees should understand the decision-making process, both on a conceptual level and on a practical level (e.g., how decisions are made in their organization). Being empowered does not mean making decisions. Rather, it means being made a part of the decision-making process. Before empowering employees, it is important to show them what empowerment will mean on a practical level. How will they be empowered? Where do they fit into the decision-making process? They also need to be aware of the boundaries. What decisions are they able to make themselves or within their work teams? Employees should know the answers to all of these questions before being empowered.

An employee who does not know where the organization is going will be unable to help it get there. Before empowering employees, it is wise to educate them concerning the organization's strategic plan and their role relative to it. When employees can see the goal, they are better able to help the organization reach it.

Organizational Structure and Management Practices

Most resistance to empowerment is attitudinal, as the inhibitors explained so far show. However, a company's organizational structure and its management practices can also mitigate against the successful implementation of empowerment. Before attempting to implement empowerment, the following questions should be asked:

- How many layers of management are there between workers and decision makers?
- Does the employee performance-appraisal system encourage or discourage initiative and risk taking?
- Do management practices encourage employees to speak out against policies and procedures that inhibit quality and productivity?

Employees, like most people, will become frustrated if their ideas have to work their way through a bureaucratic maze before reaching a decision maker. Prompt feedback on suggestions for improvement is essential to the success of empowerment. Too many layers of managers who can say no between employees and decision makers who can say yes will inhibit and eventually kill risk taking and initiative on the part of employees.

Risk-taking employees will occasionally make mistakes or try ideas that don't work. If this reflects negatively in their performance appraisals, initiative will be replaced by a play-it-safe approach. This also applies to constructive criticism of company policies and management practices. Are employees who offer constructive criticism considered problem solvers or troublemakers? Managers' attitudes toward constructive criticism will determine whether they receive any. A positive, open attitude in such cases is essential. The free flow of constructive criticism is a fundamental element of empowerment.

MANAGEMENT'S ROLE IN EMPOWERMENT

Management's role in empowerment can be stated simply. It is to do everything necessary to ensure successful implementation and ongoing application of the concept. The three words that best describe management's role in empowerment are *commitment*, *leadership*, and *facilitation* (see Figure 8–2). All three functions are required to break down the barriers and overcome the inherent resistance often associated with implementation of empowerment or with any other major change in the corporate culture.

Grazier describes the manager's role in empowerment as demonstrating the following support behaviors:

- Exhibiting a supportive attitude
- Being a role model
- Being a trainer
- Being a facilitator
- Practicing management by walking around (MBWA)
- Taking quick action on recommendations
- Recognizing the accomplishments of employees[12]

Figure 8–2
Management's Role in
Empowerment

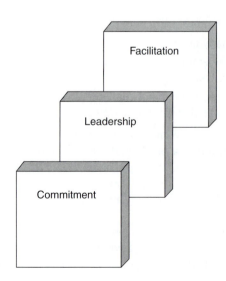

Facilitation

Leadership

Commitment

IMPLEMENTING EMPOWERMENT

Figure 8–3 shows the four broad steps in the implementation of empowerment. Creating a workplace environment that is positive toward and supportive of empowerment so that risk taking and individual initiative are encouraged is critical. Targeting and overcoming inhibitors of empowerment is also critical. These two steps were discussed earlier in this chapter. The third and fourth steps are dealt with in this section, after an explanation of the history and development of suggestion systems and a discussion of how to consider the employee's point of view.

History and Development of Suggestion Systems

A suggestion system is any vehicle through which employees can channel their ideas for workplace improvements. Suggestion systems actually preceded the Industrial Revolution (see Discussion Assignment 8–4). However, suggestion systems relating specifically to workplace improvements developed parallel to the Industrial Revolution and are still associated with it.

Around 1880, a Scottish shipbuilder named William Denny solicited suggestions from his employees concerning how to build ships better and at less cost. In 1898, the Eastman Kodak company paid an employee named William Connors an incentive prize of two dollars for his suggestion that windows be washed regularly to keep the workplace well lighted and bright.[13]

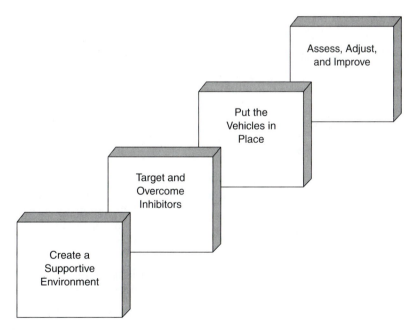

Figure 8–3
Steps in Implementing Empowerment

U.S. companies led the way in the use of suggestion systems from the late 1800s through the mid-1960s. By the late 1960s, however, U.S. industrial leaders had, as a rule, begun to look to technology rather than to their employees for productivity improvements. Japan, on the other hand, with leadership from U.S. consultants, began to promote the use of employee suggestion systems as part of the effort to rebuild their industrial base.

According to the Japanese Human Relations Association (JHRA):

> TWI (Training Within Industries), introduced to Japanese industry in 1949 by the U.S. occupation forces, had a major effect in expanding the suggestion system to involve all workers rather than just a handful of the elite. Japanese executives who traveled to the United States after the war were impressed by U.S. suggestion systems. Many Japanese companies introduced suggestion systems to follow up on the job modification movement begun by TWI. This occurred at Toshiba in 1946, at Matsushita Electric in 1950, and at Toyota in 1951. The Japanese media reported on the "blossoming suggestion systems," and suggestion boxes were set up in many offices during the 1950's. Although the suggestion system was pushed vigorously, it was still a direct copy of the Western suggestion system.[14]

Making suggestions for workplace improvements was historically a highly individualistic undertaking in the United States. An employee with an idea would typically write it down and drop it in a suggestion box. If the suggestion was accepted and worked, the employee was rewarded appropriately and given recognition as an individual. Such an approach fit well into the individualistic personality of the typical American worker.

Efforts to mimic this approach failed in Japan, where group consensus was the norm and individualism was scorned. According to the JHRA:

> During the mid-1950's to the mid-1960's, most Japanese companies that had suggestion systems averaged less than one suggestion per worker per year. Those who made suggestions were considered "fanatics" by their co-workers.[15]

QUALITY TIP

Breaking Down Barriers to Empowerment

"Numerous scholarly monographs and case studies offer complex theories about ingrained stereotypical assumptions and resistance to change, the principal bases for the incessant battles that take place between major functions in any organization. But the most effective tactic I have observed for overcoming these stereotypes is the manager at the local level taking the initiative and offering, in a nonthreatening way, to sit down and chat with everyone involved; that is, from all functions. It takes a while—perhaps a long time—for any noticeable change to occur. But with lots of patience, the results can be startling."

Source: Tom Peters, *Thriving on Chaos: A Handbook for a Management Revolution* (New York: Harper & Row, 1987), 530.

Japanese companies did not begin to experience success with suggestion systems until the mid-1960s, when the Japan Union of Scientists and Engineers (JUSE) proposed quality control, or QC circles and the Zero Defects (ZD) movement.

> Spurred by the ZD movement and the QC circles, other production floor activities to improve quality and reduce errors spread like wildfire among large Japanese companies. It was natural, therefore, for small groups to become the core units of activity in a participative suggestion system.[16]

In the years between 1965 and the present, Japanese companies continued to develop and refine the group approach to suggestion making. U.S. companies persisted with the individualistic approach as a rule, until the late 1980s and early 1990s when total quality concepts began to take root in the United States as well as in other industrialized countries throughout the world.

Considering the Employee's Point of View

Employee risk taking and initiative require a certain environment and a specific set of conditions. Richard Hamlin describes this phenomenon from the employee's perspective by saying, "As an employee, I am willing to learn new thinking and doing skills if the following conditions exist: 1) I see that I will be better off for the learning, and 2) I perceive that a nonpunitive pathway is available to me."[17] Hamlin goes on to describe what managers can do to facilitate risk taking and initiative on the part of employees as follows:[18]

- Believe in their ability to be successful.
- Be patient and give them time to learn.
- Provide direction and structure.
- Teach them new skills in small, incremental steps.
- Ask questions that challenge them to think in new ways.
- Share information with them sometimes to just build rapport.
- Give timely, understandable feedback and encourage them throughout the learning process.

QUALITY TIP

Believing in Employees

"All great leaders have possessed the capacity of believing in the capabilities and talents of others. Those who are always disdainful of subordinates, who constantly denigrate their work, who always compare their efforts unfavorably with their own will wind up leading no one but themselves."

Source: *Bits and Pieces*, July 25 1992, 19.

- Offer alternative ways of performing tasks.
- Exhibit a sense of humor and demonstrate care for them.
- Focus on results and acknowledge personal improvement.

Having managers who are able to view the workplace from the point of view of employees can help build their trust, and trust is critical to empowerment. Trust shows employees that management believes in them.

Putting Vehicles in Place

A number of different types of vehicles can be used for soliciting employee input and channeling it to decision makers. Such vehicles range from simply walking around the workplace and asking employees for their input, to periodic brainstorming sessions, to regularly scheduled quality circles. Widely used methods that are typically most effective are explained in the following subsections.

Brainstorming

With brainstorming, managers serve as catalysts in drawing out group members. Participants are encouraged to share any idea that comes to mind. All ideas are considered valid. Participants are not allowed to make judgmental comments or to evaluate the suggestions made. Typically, one member of the group is asked to serve as a recorder. All ideas suggested are recorded, preferably on a marker board, flip chart, or another medium that allows group members to review them continuously.

After all ideas have been recorded, the evaluation process begins. Participants are asked to go through the list one item at a time, weighing the relative merits of each. This process is repeated until the group narrows the choices to a specified number. For example, managers may ask the group to reduce the number of alternatives to three, reserving the selection of the best of the three to themselves.

Brainstorming can be an effective vehicle for collecting employee input and feedback, particularly if managers understand the weaknesses associated with it and how they can be overcome. Managers interested in soliciting employee input through brainstorming should be familiar with the concepts of groupthink and groupshift. These two concepts can undermine the effectiveness of brainstorming and other group techniques.

Groupthink is the phenomenon that exists when people in a group focus more on reaching a decision than on making a good decision.[19] A number of factors can contribute to groupthink, including the following: overly prescriptive group leadership, peer pressure for conformity, group isolation, and unskilled application of group decision-making techniques. Mel Schnake recommends the following strategies for overcoming groupthink:[20]

- Encourage criticism.
- Encourage the development of several alternatives. Do not allow the group to rush to a hasty decision.
- Assign a member or members to play the role of devil's advocate.
- Include people who are not familiar with the issue.

■ Hold last-chance meetings. When a decision is reached, arrange a follow-up meeting a few days later. After group members have had time to think things over, they may have second thoughts. Last-chance meetings give employees an opportunity to voice their second thoughts.

Groupshift is the phenomenon that exists when group members exaggerate their initial position hoping that the eventual decision will be what they really want.[21] If group members get together prior to a meeting and decide to take an overly risky or unduly conservative view, this can be difficult to surmount. Managers can help minimize the effects of groupshift by discouraging reinforcement of initial points of view and by assigning group members to serve as devil's advocates.

Nominal Group Technique

The nominal group technique (NGT) is a sophisticated form of brainstorming involving five steps (see Figure 8–4). In the first step, the manager states the problem and provides clarification if necessary to make sure all group members understand. In the second step, each group member silently records his or her ideas. At this point, there is no discussion among group members. This strategy promotes free and open thinking unimpeded by judgmental comments or peer pressure.

Figure 8–4
Steps in Nominal Group
Technique (NGT)

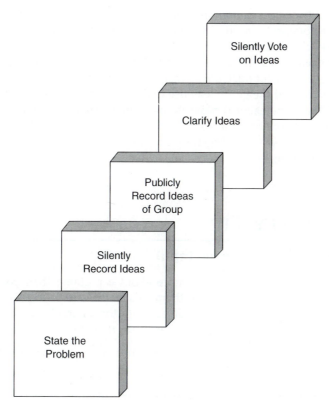

In the third step, the ideas of individual members are made public by asking each member to share one idea with the group. The ideas are recorded on a marker board or flip chart. The process is repeated until all ideas have been recorded. Each idea is numbered. There is no discussion among group members during this step. Taking the ideas one at a time from group members ensures a mix of recorded ideas, making it more difficult for members to recall what ideas belong to which individual.

In the fourth step, recorded ideas are clarified to ensure that group members understand what is meant by each. A group member may be asked to explain an idea, but no comments or judgmental gestures are allowed from other members. The member clarifying the ideas is not allowed to make justifications. The goal in this step is simply to ensure that all ideas are clearly understood.

In the final step, the ideas are voted on silently. There are a number of ways to accomplish this. One simple technique is to ask all group members to record the numbers of their five favorite ideas on five separate 3 × 5 cards. Each member then prioritizes his or her five cards by assigning them a number ranging from 1 (worst idea) to 5 (best idea). The cards are collected and the points assigned to ideas are recorded on the marker board or flip chart. After this process has been accomplished for all five cards of all group members, the points are tallied. The idea receiving the most points is selected as the best idea.

Quality Circles

A *quality circle* is a group of employees that meets regularly for the purpose of identifying, recommending, and making workplace improvements. A key difference between quality circles and brainstorming is that quality circle members are volunteers who convene themselves and conduct their own meetings. Brainstorming sessions are typically convened and conducted by a manager. A quality circle has a team leader who acts as a facilitator, and the group may use brainstorming, NGT, or other group techniques; however, the team leader is typically not a manager and may, in fact, be a different group member at each meeting. Quality circles meet regularly before, during, or after a shift to discuss their work, anticipate problems, propose workplace improvements, set goals, and make plans.

QUALITY TIP

Soliciting Employee Input

"Your people often know more about a problem than you do—but you can't know that unless you ask them. It's unrealistic to expect workers to volunteer opinions. Some workers underestimate the value of their opinions or knowledge. Others may believe you don't care what they think."

Source: Joseph T. Straub, "Ask Questions First to Solve the Right Problems," *Supervisory Management* (October 1991), 7.

Suggestion Boxes

This vehicle is perhaps the oldest method used for collecting employee input and feedback. It consists of placing receptacles in convenient locations into which employees may put written suggestions.[22] Figure 8–5 is an example of a form used for making written suggestions that would be put in a suggestion box. By examining this form, you can see that suggestions made at Manufacturing Technologies Corporation are logged in, acknowledged, and tracked. They may be made by individuals or teams, and they require an explanation of the current situation, proposed improvements, and benefits expected from the improvements.

Walking and Talking

Simply walking around the workplace and talking with employees can be an effective way to solicit input. As mentioned earlier, this approach is sometimes referred to as management by walking around (MBWA). An effective way to prompt employee input is to ask questions. This approach may be necessary to get the ball rolling, particularly when empowerment is still new and not yet fully accepted by employees. In such cases,

QUALITY CASE

Employee Empowerment at Tyco International

In order to survive and prosper, Tyco International must compete on a global scale. This $22 billion corporation is a world leader in such diversified markets as security and fire protection products, undersea fiber-optic cable, printed circuit boards, pipe, pipe fittings, valves, steel tubing, and a variety of health care products. Maintaining its status as a world leader in these highly diversified markets is a challenge for Tyco. One of the strategies Tyco uses to meet this challenge is hiring smart people and empowering them to think, act, and improve continually.

Dennis Kozlowski, Tyco's CEO, says his job is to hire the best people in the world, let them know what is expected of them, and get out of their way. He operates a decentralized organization, never looking over the shoulders of managers or employees. Kozlowski says he believes in putting his trust in the people Tyco hires, helping them during difficult times, and rewarding them in good times. He tells employees that they are actively engaged in creating value at Tyco, and that they will share in the rewards of the value created. Kozlowski puts no limits on Tyco's incentive programs. Consequently, there are no upper limits on the performance of employees and, in turn, the company.

Source: Thomas J. Neff & James M. Citrin, *Lessons from the Top,* (New York: Doubleday, 2001), 204–208.

it is important to ask the right questions. According to Joseph T. Straub, managers should "ask open-ended unbiased questions that respect . . . workers' views and draw them out."[23]

Regardless of the vehicles used for soliciting employee input, organizations need to continually improve the process. Improving suggestion systems is discussed later in this chapter.

Manufacturing Technologies Corporation
Two Industrial Park
Fort Walton Beach, Florida 32549

Name _____
(Individual or team making the suggestion)

Date Submitted _____

Department _____

Telephone Extension _____

Suggestion (Explain current situation, proposed change, and expected benefits.)

_____

_____

_____

_____

_____

Date Received _____

Date Logged In _____

Date Suggestion Was Acknowledged _____

Current Status _____

_____

_____

(Attach sketches or other illustrative material.)

Figure 8–5
Suggestion Form

MANAGEMENT'S ROLE IN SUGGESTION SYSTEMS

Management's role in the implementation and operation of suggestion systems can be divided into several steps, as shown in Figure 8–6.

Establishing Policy   This step involves developing the policies that will guide the suggestion system. Such policies should clearly describe the company's commitment to the suggestion system, the types of rewards that will be used, how suggestions will be evaluated, and how the suggestion system itself will be evaluated.

Setting Up the System   This step involves putting the system in place to do the following:

1. Solicit and collect employee input.
2. Acknowledge and log in suggestions.
3. Monitor suggestions.
4. Implement or reject suggestions.

Figure 8–6
Management's Role in
Suggestion Systems

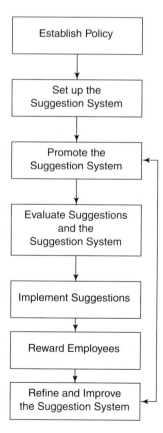

Promoting the Suggestion System   This step involves generating interest and participation in the suggestion system on the part of employees. The following strategies are effective in promoting suggestion systems:

- Sharing the company's policies concerning suggestion systems in frank and open group meetings that encourage questions and discussion
- Sponsoring suggestion competitions
- Asking employees for their input concerning how to increase participation

Evaluating Suggestions and the Suggestion System   This step involves teaching supervisors and managers how to evaluate individual suggestions and the overall system. Both of these topics are explained later in this chapter.

Implementing Suggestions   This is a critical step. If good suggestions are not promptly implemented, the system will lose credibility no matter how well the other components work.

Rewarding Employees   Rewards for suggestions that are implemented can take many forms. Cash awards, public recognition, and paid vacations are a few of the most widely used types of rewards. In addition to rewarding individuals, it is important to reward teams and/or departments that make the most accepted suggestions or that save the largest total dollar amount for the company over a specified period of time. Discussion Assignment 8–5 describes a well-known company's suggestion/reward system.

Refining and Improving the Suggestion System   As with any system, weaknesses will show up in the day-to-day operation of the suggestion system. It is important to identify and correct these weaknesses. Continual improvements to the system in such key areas as soliciting input, tracking and monitoring suggestions, evaluating suggestions, and decreasing the time required to implement suggestions are important. Employee input should be solicited concerning refinements to the suggestion system itself, just as it is with the processes and systems used to do the company's work.

IMPROVING SUGGESTION SYSTEMS

U.S. companies are beginning to accept employee input as not only desirable but necessary to compete. Writing for *Nation's Business*, Skip Berry says:

> The National Association of Suggestion Systems (NASS), a Chicago-based, not-for-profit group that represents companies regarding their suggestion programs, . . . reported that its 900 members received nearly a million suggestions from employees. More than 32 percent of those suggestions were adopted. And for the third consecutive year the member companies said they had achieved at least $2 billion in cost savings as a result of those suggestions.[24]

The growing realization of the importance of employee input is encouraging. However, if only 32% of the suggestions received in a given year are adopted, a clear need exists to improve both individual suggestions and suggestion systems.

Improving Suggestion Processing

The *suggestion system* is the collection of processes used to solicit, collect, evaluate, and adopt or turn down suggestions. According to Bob Scharz, author of *The Suggestion System: A Total Quality Process*, a good suggestion system meets all of the following criteria:[25]

- All suggestions receive a formal response.
- All suggestions are responded to immediately.
- Performance of each department in generating and responding to suggestions is monitored by management.
- System costs and savings are reported.
- Recognition and awards are handled promptly.
- Good ideas are implemented.
- Personality conflicts are minimized.

From these criteria, it can be seen that operating a suggestion system involves more than having employees toss ideas into a box, accepting some, and throwing the rest away. The best suggestion systems require that ideas be submitted in writing on a special form (Figure 8–5). Such forms make it easier for employees to submit suggestions and for employers to give immediate and formal responses. The forms ease the job of logging in suggestions and tracking them through the entire life cycle of the idea until it is either adopted or rejected.

An approach that is being used increasingly is the computer-aided suggestion system. With such a system, employees make suggestions in writing on a form such as the one in Figure 8–5. Immediate acknowledgment is provided by the manager of the department in question or by the employee who administers the computer-aided system. Acknowledged suggestions are entered into a personal computer and monitored from that point forward, creating a database of suggestions that can be updated continually. As the status of a suggestion changes, the information is passed on to the employee who operates the computer-aided system and the database is updated.

Such a system makes it easy for management to assess the performance of individual departments in generating and responding to suggestions and to monitor system costs and savings. Figure 8–7 is an example of a monthly report prepared from the database of a computer-aided suggestion system. In this example, only suggestions that receive a rating of 5 or higher on a scale of 1 to 10 are included. This is a wise approach on the part of the company because it will deter employees from submitting frivolous suggestions in an attempt to show higher numbers on the monthly report.

Improving Individual Suggestions

To make good suggestions, employees need to know how to do two things:

- Identify problems and formulate ideas for improvement.
- Clearly and concisely communicate their ideas in written and graphic form.

Monthly Report
Suggestions Submitted by Department*
GLOBAL MANUFACTURING, INC.
Crestview Plant

Department	Number of Suggestions for Month of __March__
	5 10 15 20
Administration	
Sales/Marketing	
Purchasing	
Engineering	
Electronics Mfg.	
Metal Fabrication	
Wire Harness Fab.	
Quality Assurance	
Shipping/Receiving	
Accounting	

* Only suggestions receiving a rating of 5 (out of a possible 10) or higher are included in this report.

Figure 8–7
Suggestion Summary Report

QUALITY TIP

Creativity in Problem Solving

"A thirsty man came to a well that had a hand pump to get a drink of water. He would pump vigorously, then try to catch the water from the spigot in his palms. Only a few droplets were coming out of the spigot, however, by the time he got his hands in position. The man did this several times before he left, discouraged. Another man came to the well. Instead of repeating what the first man had done, he held the pump with one hand, sealed the spigot opening with the other and drove the pump two or three times. Then he would release his hand and drink the water held back by his palm. There are many ways to do something and some ways work better than others!"

Source: Japan Human Relations Association, ed., *The Idea Book* (Cambridge, MA: Productivity Press, 1988), 37.

Yes	No	Potential Problem Areas
		Are there defects?
		Can costs be reduced?
		Are deadlines missed?
		Are there safety problems?
		Is there wasted material?
		Is there wasted time?
		Is your work *hurry up and wait* in nature?
		Are interpersonal relationships with other employees good?
		Are work results inconsistent?
		Are sufficient materials available when needed?
		Are there defects in raw materials?
		Are work processes well organized?
		Do work processes flow smoothly?
		Do machines and equipment operate properly?
		Do any required tasks detract from quality?
		Do any required tasks detract from productivity?

Comments:

_____

_____

_____

Figure 8–8
Problem Identification Checklist

Teaching employees to use the following three steps in formulating ideas for workplace improvements can increase the number of ideas they generate and the quality of those ideas:

1. *Problem identification* involves identifying situations that differ from the desired result. At this point, no attempt is made to determine the cause of the problem, nor should a cause be too quickly assumed. Employees should be helped to learn to

approach problems systematically and in a step-by-step manner. In this step, the task is to record problem situations that are candidates for improvements. This is the "What is wrong?" step. Figure 8–8 provides a sample of the kind of checklist that employees can use to help identify problems.

2. *Research* comes next. Whereas the preceding step (problem identification) deals with the issue of what is wrong, research deals with why it is wrong. Research into why is important because, if done properly, it can prevent the expenditure of resources to treat symptoms rather than causes. The JHRA recommends asking why five times.[26] According to the JHRA:

> Careful analysis of an occurrence will usually show there is more than one cause. You will also find a cause of a cause, and a cause of a cause of a cause. It is therefore imperative to consider whether the first "cause" you detect for a particular phenomenon is really the true cause. What you intuitively feel is the cause is usually not the whole story; the true cause is usually hidden. If you go back at least five "generations" of cause you will be likely to find the true cause.[27]

To help determine causes, numerous total quality tools can be used, including Pareto charts, cause-and-effect diagrams, bar graphs, check sheets, scatter diagrams, and histograms. These tools are discussed later in this book.

3. *Idea development* involves the development of ideas for solving the problem identified in the first step by eliminating the cause identified in the second step. Figure 8–9 on the previous page is a checklist that can be used as an aid in the development of ideas. This checklist corresponds with the JHRA's recommendation that the

Broad Question	Examples of More Specific Questions
What?	• What is the task in question? • What would happen if the task were eliminated?
Why?	• Why is the task performed? • Why haven't alterations been attempted?
Where?	• Where is the task performed? • Where else might the task be performed?
When?	• When is the task performed? • When else could the task be performed?
Who?	• Who performs the task? • Who else might perform the task?
How?	• How is the task performed? • How else could the task be performed?
How much?	• How much does the task cost now? • How much will it cost after improvements are implemented?

Figure 8–9
Idea Development Checklist

following factors be considered when developing ideas: subject matter, purpose, location, sequence, people, method, and cost.[28]

When ideas for improvement have been formulated, employees should write them down using graphics (sketches, diagrams, graphs, charts, and/or photos) as appropriate to illustrate their ideas. One of the essential steps in improving individual suggestions is to help employees learn to write clear, concise, definitive descriptions of their ideas for improvement.

Figure 8–10 shows a good idea but a poorly written suggestion. It recommends a major rearrangement of the plant but does not explain how or why the company might benefit. Figure 8–11a is the same suggestion rewritten with supportive illustrations attached (Figure 8–11b). The difference between the two suggestions is that the second one does a better job of communicating the problem, the proposed solution, and the expected benefits. The accompanying diagram (Figure 8–11b) shows how the machines will be arranged after the suggestion is implemented.

Managers can help employees improve their individual suggestions by coaching them to do the following:

- Briefly, succinctly, and clearly explain the current situation that creates the problem.
- Get to the heart of the proposed change with no preliminaries or rationalizations, and be specific.
- Prepare illustrations to clarify the proposed change, in every case where this is appropriate.
- Explain the expected benefits using quantifiable terms (percentages, dollars, time, numbers, or whatever is appropriate).
- Take nothing for granted. Assume that the decision makers who will read the suggestion know nothing about the situation.

EVALUATING SUGGESTIONS

Perhaps the most critical and most difficult aspect of operating an effective suggestion system is evaluating suggestions. How does one recognize a good idea? How can one respond to a poor idea without discouraging the employee who submitted it? These questions are dealt with in this section.

Evaluating and Rating Suggestions

Employees are often in the best position to make suggestions for workplace improvements, but in the final analysis it is still the manager who must decide whether the proposed idea is feasible. It's a lot like a loan officer in a bank trying to decide which loan applicants will pay off and which will be deadbeats. There is an art and a science to the process, and those who succeed are the ones who are best able to select the winners most consistently. This also applies to managers who evaluate employee suggestions.

Manufacturing Technologies Corporation
Two Industrial Park
Fort Walton Beach, Florida 32549

Name Roy Jones and Tuwanda Morris

Date Submitted September 10

Department Metal Fabrication

Telephone Extension 496

Suggestion (Explain current situation, proposed change, and expected benefits.)

We need to rearrange the shop floor for better work flow.

Date Received September 10

Date Logged In September 10

Date Suggestion Was Acknowledged September 10

Current Status Suggestion being rewritten with more descriptive detail.

(Attach sketches or other illustrative material.)

Figure 8–10
Poorly Written Suggestion

```
┌────────────────────────────────────────────────────────────────┐
│                                                                  │
│              Manufacturing Technologies Corporation              │
│                       Two Industrial Park                        │
│                  Fort Walton Beach, Florida 32549                │
│                                                                  │
│   Name   Roy Jones and Tuwanda Morris                            │
│          (Individual or team making the suggestion)              │
│                                                                  │
│   Date Submitted   September 10                                  │
│                                                                  │
│   Department   Metal Fabrication                                 │
│                                                                  │
│   Telephone Extension   496                                      │
│                                                                  │
│   Suggestion (Explain current situation, proposed change, and    │
│   expected benefits.)                                            │
│                                                                  │
│   The shop floor is currently arranged with machines grouped by  │
│   type. This causes                                              │
│   extra material handling and wasted time. Rearrange the shop    │
│   floor into flexible                                            │
│   cells, each containing a lathe, mill, and drill. This will     │
│   allow one part to be completely                                │
│   machined in each cell without the constant movement of         │
│   partially manufactured                                         │
│   parts back and forth among machine groups. A sketch of a       │
│   proposed typical                                               │
│   machine cell is attached. The suggested rearrangement will     │
│   reduce throughput                                              │
│   time by an estimated 23 percent.                               │
│                                                                  │
│                                                                  │
│   Date Logged In   September 11                                  │
│                                                                  │
│   Date Suggestion Was Acknowledged   September 11                │
│                                                                  │
│   Current Status   Suggestion being implemented.                 │
│                                                                  │
│                                                                  │
│                                                                  │
│   (Attach sketches or other illustrative material.)              │
│                                                                  │
└────────────────────────────────────────────────────────────────┘
```

Figure 8–11a
Rewritten Suggestion

Figure 8-11b
Supporting Illustration for
Rewritten Suggestion: Proposed
Typical Machining Cell

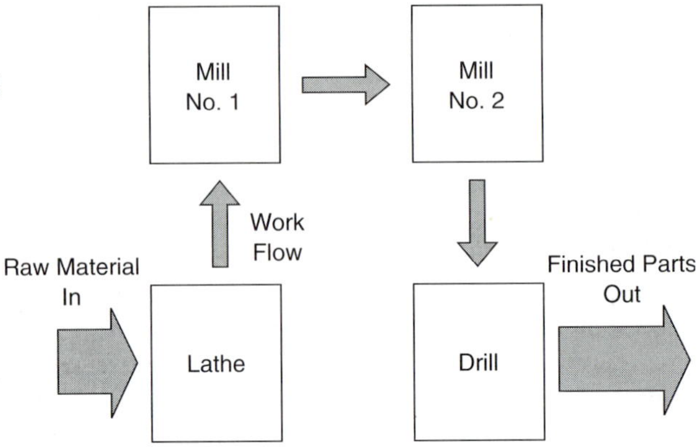

The Japan Human Relations Association recommends the following points be remembered when evaluating employee suggestions:[29]

■ Not all suggestions can be evaluated fairly while sitting at your desk. Questions about a suggestion should be resolved by going to the source and discussing them with the suggester. Be patient and find out what the suggester is trying to say. If necessary, help the suggester improve his or her suggestion.

■ A suggestion that looks bad or outdated on the surface may still contain good ideas. Study all suggestions carefully. Don't jump to conclusions and discount the whole suggestion if part of it is bad.

■ Be generous with first-time suggesters to build up their confidence. If possible, try to accept the suggestion, or at least part of it, making revisions to the suggestion if necessary.

■ With employees who are experienced in suggestion making, evaluate their suggestions carefully and challenge them to set higher goals.

■ Consider the level of the suggester when evaluating a suggestion. If you underestimate a person's ability, he or she will not be challenged and grow. If you expect too much, you will discourage the worker's creativity and initiative.

■ Suggesters will be anxious for feedback. Evaluate suggestions promptly. If a delay is unavoidable, notify the suggester of the reason for it.

■ If possible, notify the suggester of evaluation results in person. Make sure to add a few words of encouragement, especially if the idea is not accepted. When providing a written notice, always add a positive comment along with the result.

■ A suggestion is often the product of the suggester and his or her supervisor. Remember to compliment all contributors for their efforts.

Establish a formal rating system for evaluating suggestions to ensure consistency. The rating system should provide a means for quantifying the results of evaluations and have at least the following components:

■ *Criteria and criteria rating.* Criteria are the factors considered most important in assessing the feasibility of employee suggestions. The Evaluation and Rating Form in Figure 8–12 contains three such criteria. They are the expected benefit of the suggestion, time that will elapse before benefits will begin to be realized, and how successful suggestions made by the individual or team submitting this suggestion have been in the past (track record). The actual criteria might differ from company to company. However, within a company they should be the same for all departments, to ensure consistency and fairness. Each criterion must be assigned a numerical score or rating. This is the key step in the entire process. It requires judgment, common sense, a thorough knowledge of the situation in question, and an open mind. The example in Figure 8–12 uses a scale of 0 to 10 for each criterion.

■ *Weight factors and weighted scores.* Weight factors accommodate the fact that some criteria are more important than others. The individual rating for each criterion is multiplied times its assigned weight factor to determine its weighted score. In Figure 8–12, the expected benefit criterion is assigned a weight factor of 5.0; the time criterion, 3.0; and the final criterion, 2.0. A criterion rating of 7.0 multiplied by a weight factor of 5.0 results in a weighted score of 35.0 points.

■ *Total points and conversion scale.* The weighted scores are added together to determine the total score for the suggestion. Just as numerical scores are converted

Poultry Processing, Inc. Highway 90 East DeFuniak Springs, FL 32816			Suggestion Identification No. _____	
Rating Criteria	Criteria Rating (0–10)	Weight Factor	Weighted Score	Comments
Expected benefit of suggestion		5.0		
Time until benefits will be realized		3.0		
Successful suggestions submitted in past		2.0		
TOTAL SCORE				
Criteria Rating × Weight Factor = Weighted Criteria Score				
Conversion Scale (For Total Score)				
Category 1 .... 10 points Category 2 .... 20 points Category 3 .... 30 points Category 4 .... 40 points Category 5 .... 50 points			Category 6 .... 60 points Category 7 .... 70 points Category 8 .... 80 points Category 9 .... 90 points Category 10 .... 100 points	

Figure 8–12
Employee Suggestion Evaluation and Rating Form

to letter grades in a college class, the total score for the suggestion is converted to a level, category, or grade of suggestion. The conversion scale in Figure 8–12 converts numerical scores to categories. Company policy will typically set the minimum level or category of suggestion that will be considered beyond the supervisory level. For example, Poultry Processing, Inc., might establish a company policy that only suggestions categorized as Level 6 or higher (see Figure 8–12) will be considered for implementation.

HANDLING POOR SUGGESTIONS

Even the best suggestion system will not completely eliminate bad suggestions. In such cases, suggestions must be rejected. "Never adopt a bad idea because you feel sorry for someone or feel you 'owe' him a break,"[30] advises George Milite in an article with the poignant title "When an Employee's Idea Is Just Plain Awful." However, when rejecting a suggestion, precautions should be taken to maintain the employee's interest and morale.

Milite recommends the following strategies for rejecting poor suggestions in a positive manner:[31]

- ■ *Listen carefully.* Give employees an opportunity to explain their suggestions in greater detail. Aspects of the suggestion that have merit or could be developed to make the suggestion more viable may not have been completely explained on the suggestion form.
- ■ *Express appreciation.* Be sure to let employees know that their suggestions are appreciated. Encourage future suggestions. The message to leave them with is this: "This suggestion didn't work out, but your effort is appreciated and your ideas are valued. I am looking forward to your next suggestion."
- ■ *Carefully explain your position.* Don't make excuses or blame the company, higher management, or anyone else. Explain the reasons the suggestion is not feasible, and do so in a way that will help employees make more viable suggestions in the future.
- ■ *Encourage feedback.* Encourage feedback from employees. You may have overlooked an aspect to the suggestion, or the suggester may not fully understand the reasons for the rejection. Solicit sufficient feedback to ensure that you understand the employees and they understand you.
- ■ *Look for compromise.* It may be possible to use all or a portion of a suggestion if it is modified. Never adopt a bad idea, but if a suggestion can be modified to make even part of it worthwhile, a compromise solution may be possible.

ACHIEVING FULL PARTICIPATION

To achieve optimum potential, a suggestion program must have a broad base of participating employees. The closer a company comes to full participation, the better its chances are of having a successful suggestion system. When an effective suggestion system is in place and working, managers can focus on the next level of concern: achieving full participation. This goal amounts to removing hidden barriers, encouraging new employees to get involved, and coaching reluctant employees.

Removing Hidden Barriers

Hidden barriers are the less obvious factors that inhibit participation. They can vary widely from company to company and can be difficult to detect. However, the hidden factors that are most often present are these:

- *Negative behavior.* Attitudes and behaviors of personnel to whom suggestions are submitted can inhibit participation. This point applies regardless of whether the recipient is a supervisor or a clerk at a central suggestion submittal point. Negative facial expressions, harsh voice tones, or nonsupportive comments by such persons can turn away tentative employees.

- *Poor writing skills.* Employees with poor writing skills are not unusual in an age when illiteracy has become commonplace. Because attempts at writing suggestions will call attention to the difficulty, employees with poor writing skills are not likely to participate in the suggestion program. This problem can be overcome through coaching, counseling, and training.

- *Fear of rejection.* Most human beings fear rejection and, as a result, avoid situations that might subject them to it. Consequently, some employees will not participate in the suggestion program for fear of having their suggestions rejected. This problem can be avoided by working with such employees to help them formulate their initial suggestions and build confidence, which is the key to overcoming fear of rejection.

- *Inconvenience.* Employees are reluctant to participate in suggestion programs that are inconvenient. Are suggestion forms complicated? Is the central submittal point too far away or available only at limited times? Do suggestions need the approval of several people before they can be submitted? Inconveniences such as these will discourage participation.

QUALITY TIP

Poor Suggestions

"No matter how much you believe that everyone has the potential to be creative, or that everyone can unlock inner resources by expressing ideas, there's one fact you can't escape: Even the most brilliantly creative person can come up with an idea that has no redeeming value."

Source: George Milite, "When an Employee's Idea Is Just Plain Awful," *Supervisory Management* (October 1991), 3.

These are just four examples of hidden barriers that can mitigate against full participation. Talking with employees, observing, and listening can help identify others. Perhaps the most effective way to discover hidden barriers is to use the suggestion program to improve the suggestion program. Encourage employees to submit their ideas concerning factors that discourage participation. Allowing anonymous suggestions in such cases is a good strategy for encouraging honest constructive criticism of the system.

Encouraging New Employees

New employees may experience a natural reluctance to participate in the suggestion program. Encouragement, support, and coaching can overcome this reluctance. The JHRA recommends the following measures for dealing with reluctant new employees:[32]

- Teach new employees why suggestions are important (i.e., quality improvements, efficiency, cost reduction, productivity improvements, better safety, etc.).
- Make a small improvement in the new employee's job, write it up as an example of a suggestion, and use the example to show how the process works.
- Have the new employee work cooperatively with an experienced employee to develop one or more suggestions.
- Give positive feedback on the new employee's first several tries. Be careful to point out ways the suggestions might be improved.

Coaching Reluctant Employees

Coaching is an important ingredient in all aspects of total quality. In fact, with total quality the middle manager's role is more like that of the coach than that traditionally associated with managers. We typically associate coaching with developing, encouraging, training, and monitoring the performance of athletes. These tasks also apply in the

QUALITY TIP

Employees with Poor Writing Skills

"People tend to think they can forget about writing once they leave school. Writing is a very tedious task for some people. Many workers are embarrassed about their writing ability. . . . To convince someone like this to write a suggestion, the leader must empathize with the worker. Logical arguments are not the answer—the leader must understand what the worker is feeling and offer a helping hand. This kind of caring inspires the worker and nurtures creativity."

Source: Japan Human Relations Association, ed., *The Idea Book* (Cambridge, MA: Productivity Press, 1988), 111.

modern workplace where managers concentrate on developing, encouraging, training, and monitoring employees who are members of teams.

Coaching reluctant employees in an attempt to promote participation is similar to coaching athletes in that it involves applying both "the carrot and the stick," or, put another way, it entails both encouraging and challenging employees. However, before doing either, it is important to identify the cause of the reluctance.

If an employee wants to make improvements but will not make suggestions, hidden barriers to his or her participation must exist. If the employee doesn't care, he or she may not be in the right job. Occasionally an employee's personal goals will be at odds with those of the work team or the company. If an employee has the requisite knowledge and skills needed to make a contribution but isn't doing so, he or she may be playing the wrong position on the work team. This is a common occurrence on athletic teams: a seemingly talented athlete is not performing well in his or her current position but when moved to another position suddenly comes to life and starts helping the team. A similar result can sometimes be achieved in the workplace.

In such cases, managers must play the role of career coach. Career coaching is a way of getting the right employees into the right jobs and having them take responsibility for their performance, for participating in continual improvement efforts, and for contributing to the work team's performance.

G. M. Sturman recommends the following five-point approach to career coaching that managers can use to better match employees and jobs and to transform employees from noncontributors to self-starters who participate actively in workplace improvement efforts:[33]

- *Assess.* Help employees form a clear picture of their interests and aptitudes as well as their strengths and weaknesses.
- *Investigate.* Help employees investigate all possible opportunities within the company. Would they be better suited to another department, another team, or even another job?
- *Match.* Having assessed interests, aptitudes, strengths, and weaknesses and investigated opportunities, help employees find the optimum match available to them in the company.
- *Choose.* Work with employees to help them make the choice that is best for them.
- *Manage.* Help employees develop a personal career management plan that will result in the accomplishment of their goals. Do they need additional education and training to get the job they really want? Would another job within the company more fully meet their needs? Their plan should be a road map that guides reluctant employees from where they are to where they can make a contribution to the company while simultaneously meeting their personal goals.

HOW TO RECOGNIZE EMPOWERED EMPLOYEES[34]

There will always be managers and supervisors who resist the concept of empowerment out of fear of losing control or losing their authority. Some will give lip service to empowerment while continuing to do things the same way they always have. How can an organization's leaders know that their empowerment efforts are working? In other words, how does one recognize an empowered employee? The following comparisons will help leaders in an organization determine whether they have empowered employees or employees who are required to do things as they have always been done, by supervisors who talk about empowerment but do not really believe in it.

Taking Initiative Empowered employees will take the initiative in ambiguous situations and define problems in ways that enable further analysis and lead to positive action. Employees who are not empowered will wait for someone in authority to define the problem and initiate action.

Identifying Opportunities Empowered employees will identify opportunities for improvement in the problems that arise. Employees who are not empowered may solve the immediate problem, but fail to go beyond just solving it to identifying ways to improve processes and prevent future occurrences.

Thinking Critically Empowered employees feel free to think critically, question the status quo, and challenge assumptions. Employees who are not empowered are more likely to take information at face value without testing its validity.

Building Consensus Empowered employees build consensus among all stakeholders within groups and across functional areas. Employees who are not empowered are more likely to simply look to a higher authority to mandate a decision.

No matter what an organization's supervisors and leaders say about empowerment, and no matter how elaborate the systems put in place to promote empowerment, employees are not empowered until they are willing and able to take the initiative when action is needed, identify opportunities for continual improvement in the problems that occur, build consensus for a given action or decision, and think critically when considering actions, decisions, and assumptions.

AVOIDING EMPOWERMENT TRAPS[35]

According to Kyle Dover, despite their potential, "empowerment programs often fall victim to the very structural and cultural problems that made them desirable in the first place. Many managers view empowerment as a threat and continue to measure their value by the authority they wield. Meanwhile, some employees mistake empowerment for discretionary authority—the power to decide things unilaterally—and lack the collaborative skills that management neglects, or refuses, to teach them. Others resist the need to assume more power and cling to a comfortable dependence on authority."[36] Organizations should avoid falling into the following empowerment traps:

Defining Power as Discretion and Self-Reliance Empowerment gives employees the authority to think critically, make decisions within controlled parameters, and participate fully. It does not give them the authority to act unilaterally on the basis of their own discretion. Empowerment is about being a full participant in a team process. It is not about being a self-reliant loner.

Failing to Properly Define Empowerment for Managers and Supervisors When managers and supervisors view empowerment as a loss of authority, status, or power, they tend to resist it. They can feel as if they worked long and hard to gain a position of authority only to be asked to hand that authority over to employees. Ensuring that managers and supervisors understand the issue of controlled delegation of authority is critical if empowerment is to achieve its potential as a performance-improving concept.

Assuming Employees Have the Skills to Be Empowered You should not ask an employee who does not know how to think critically, make decisions, identify improvement opportunities, or build consensus to do so. Before implementing an empowerment program, it is important to assess the skill levels of employees and provide them with the training and mentoring needed to make effective use of their status as empowered employees.

Getting Impatient Making the transition from the traditional approach to management/employee interaction to empowerment represents a major change, a change that can take time. Processes have to be changed, employees have to be trained, fears have to be worked through, and sufficient time must pass to allow people to acclimate to a new culture. Expecting immediate results is unrealistic. Leaders of empowerment efforts should be prepared to deal with and overcome organizational impatience.

BEYOND EMPOWERMENT TO ENLISTMENT

Involvement and empowerment focus the experience, knowledge, creativity, and ideas of a broad cross section of stakeholders on a problem. By involving and empowering stakeholders, organizations find better solutions to problems. Empowerment is now widely accepted and practiced in competitive organizations. What, then, is next? How can organizations go beyond involvement and empowerment?

To answer this question, it is necessary to view the issue as a continuum, as shown in Figure 8–13. The extreme left-hand position (zero) on the continuum represents the old management philosophy that *managers think and employees work*. Organizations that practice this philosophy neither seek nor allow employee input.

The next position (1) on the continuum is employee involvement. Organizations that practice employee involvement provide various mechanisms that allow employees to submit input concerning decisions that affect them. Involvement is a passive approach that *allows* employees to submit input.

The next position (2) on the continuum is employee empowerment. Organizations that practice employee empowerment don't just allow employee input—they actively

QUALITY TIP

Managers as Coaches

The most fundamental goal of coaching in the workplace is improved performance. When an employee's performance improves, he or she wins, the work team wins, the manager wins, and the company wins. This is the rationale behind the philosophy that modern managers should be more like coaches than bosses.

seek it. Empowerment is an active approach in which employee input is sought and given serious consideration. Empowered employees provide input concerning decisions that affect them and can apply their own ingenuity in seeking improvements themselves within specified limits. Like involvement, empowerment *allows* employees to be part of the decision-making process.

The extreme right-hand position (3) on the continuum is employee *enlistment*. Enlistment goes beyond empowerment in that it not only allows employees to provide input and to innovate but *expects* them to do so. Mechanisms that allow employees to be part of the decision-making process also let them not be a part. In other words, with involvement and empowerment, employees can choose not to participate; they can simply opt out. Employees who do this deny organizations the benefit of their knowledge, experience, point of view, and ingenuity.

Organizations trying to survive and thrive in a competitive environment need their employees to bring all of their intellectual tools to bear on continual improvement every day. To do this, they must go beyond empowerment to enlistment. Employee enlistment means not simply empowering employees to participate in the decision-making process but expecting them to do so.

Every employee is a valuable resource. Consequently, organizations need to make full use of employees. This cannot be done if employees opt out of participating in the decision-making process, which is what employees do when they don't provide input. Some strategies that will help organizations move beyond empowerment to enlistment are as follows:

■ Make it clear to all employees that their participation is not just wanted and needed but that it is expected.

Figure 8–13
Involvement–Empowerment–
Enlistment Continuum

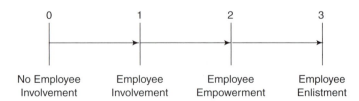

- Make participation a criterion in the performance appraisal process. However, make it clear that what counts is well thought out input. Don't reward frivolous participation.
- In meetings, call on the wallflowers. Don't let employees just be present in the room; expect them to be engaged.
- Make enlistment a guiding principle in the organization's strategic plan and an organizational value that becomes part of the corporate structure.

 SUMMARY

1. Empowerment means engaging employees in the thinking processes of an organization in ways that matter. *Involvement* means having input. *Empowerment* means having input that is heard and seriously considered. Empowerment requires a change in organizational culture, but it does not mean that managers abdicate their responsibility or authority.
2. The rationale for empowerment is that it is the best way to increase creative thinking and initiative on the part of employees. This, in turn, is an excellent way to enhance an organization's competitiveness. Another aspect of the rationale for empowerment is that it can be an outstanding motivator.
3. The primary inhibitor of empowerment is resistance to change. Resistance might come from employees, unions, and management. Management-related inhibitors include insecurity, personal values, ego, management training, personality characteristics, exclusion, organizational structure, and management practices.
4. Management's role in empowerment is best described as commitment, leadership, and facilitation. The kinds of support managers can provide include having a supportive attitude, role modeling, training, facilitating, employing MBWA, taking quick action on recommendations, and recognizing the accomplishments of employees.
5. The implementation of empowerment has four broad steps: creating a supportive environment; targeting and overcoming inhibitors; putting the vehicles in place; and assessing, adjusting, and improving. Vehicles include brainstorming, nominal group technique, quality circles, suggestion boxes, and walking and talking.
6. Management's role in suggestion systems includes establishing policy, setting up the system, promoting the suggestion system, evaluating suggestions, evaluating suggestion systems, implementing suggestions, rewarding employees, and refining suggestion systems.
7. Improving suggestion systems involves improving both the system and individual suggestions. Systems can be improved by ensuring that all suggestions receive a formal response, all suggestions are responded to immediately, participation is monitored, system costs and savings are monitored, recognition and awards are handled promptly, good ideas are implemented, and personality conflicts are minimized. Improving individual suggestions involves helping employees learn problem identification, research, and idea development.
8. A rating system for evaluating suggestions should include rating criteria, a scoring mechanism, weight factors and weighted scores, a point total, and a conversion scale.

9. Poor suggestions can be handled without damaging morale by listening carefully, expressing appreciation, carefully explaining your position, encouraging feedback, and looking for compromise.
10. Full participation in a suggestion system can be achieved by removing hidden barriers (for example, negative behavior, poor writing skills, fear of rejection, and inconvenience), encouraging new employees, and coaching reluctant employees.
11. A workforce that is ready for empowerment is accustomed to critical thinking, understands the decision-making process, and knows where it fits into the big picture.

 KEY TERMS AND CONCEPTS

Assess	Initiative
Brainstorming	Investigate
Choose	Leadership
Coaching	Manage
Commitment	Match
Conversion scale	Negative behavior
Empowerment	Nominal group technique
Enlistment	Poor writing skills
Facilitation	Problem identification
Fear of rejection	Rating criteria
Groupshift	Research
Groupthink	Risk taking
Hidden barriers	Suggestion boxes
Idea development	Suggestion system
Inconvenience	Weight factors

 FACTUAL REVIEW QUESTIONS

1. Define the term *empowerment*, being sure to distinguish between *involvement* and *empowerment*.
2. Explain the following statement: "Successful implementation of empowerment requires change in the corporate culture."
3. Give a brief rationale for empowerment.
4. What is the relationship between empowerment and motivation?
5. List three inhibitors of empowerment and how they can be overcome.
6. Explain the various root causes of management resistance to empowerment.
7. In what ways can an organization's structure and management practices inhibit empowerment?
8. Describe management's role in empowerment.
9. Give a brief explanation of the historical development of suggestion systems.

10. Describe how to use brainstorming to promote empowerment.
11. What is a quality circle?
12. Describe the concept of MBWA.
13. Summarize management's role with regard to suggestion systems.
14. Describe how a suggestion system might be improved.
15. Describe how individual suggestions can be improved.
16. Explain how suggestions should be evaluated.
17. How would you handle an obviously misguided suggestion?
18. List three examples of hidden barriers to full participation in a suggestion system and how each barrier can be overcome.
19. What is meant by the term *career coaching*?
20. Explain the concept of workforce readiness as it relates to empowerment.

 CRITICAL THINKING ACTIVITY

Empowerment Can Be a Tough Sell

"We are the market leader in our field," said Mark Hansen, CEO of Gosport Shipbuilding, Inc. (GSI). "I built this company from the ground up. I know more about constructing gambling ships than anybody in the business. That's why we are number one. My motto is 'I think and employees work.' This empowerment nonsense you're selling is just that—nonsense. If I want an employee's opinion, I'll give it to him!"

Luke O'Hara, GSI's new quality director, listened respectfully as his boss ranted on. But he had to admit that Hansen had a point. GSI's CEO could do every job in the yard better than the best employees on the payroll. He was also right about GSI's position of market leadership. He thought, "Employee empowerment is going to be a tough sell with Hansen. After all, strip away the bombast and what the CEO is saying is 'Why fix what isn't broken?'"

Put yourself in Luke O'Hara's place. You're the new quality director and want to convince your new boss of the benefits of employee empowerment. How would you persuade Mark Hansen to change his mind?

DISCUSSION ASSIGNMENT 8–1

Union Resistance to Empowerment

Florida Power and Light Company is well known for its commitment to employee empowerment. FP&L's Quality Improvement Program is one of the best in the country. However, what works well now was not universally popular at first. In fact, the idea was originally rejected by the union leadership of FP&L. This resistance on the part of the union resulted in a difficult beginning for FP&L's Quality

Improvement Program. Why did the union resist at first? Because union leaders were not involved in the development and implementation of the program. What overcame their resistance? Getting them involved.

Discussion Assignment
Discuss the following questions in class or outside of class with your fellow students:

1. Assume you are a supervisor. For the sake of discussion, what objections might you have to employee empowerment?
2. If you were a senior manager and a supervisor raised these objections, how would you respond?

Source: Peter B. Grazier, *Before It's Too Late: Employee Involvement . . . An Idea Whose Time Has Come* (Chadds Ford, PA: Teambuilding, 1989), 89.

DISCUSSION ASSIGNMENT 8–2

A Lack of Management Commitment

The employees of a midsized printed circuit–board manufacturer had been excited about their empowerment program during its first several months of operation. A number of solid suggestions for improvement had been made, accepted, and implemented, saving the company substantial amounts of money by reducing throughput time by 19%. During this period, management was very supportive. Because of its new employee-driven competitiveness, the company became a hot item and was sold at a sizable profit. The new management team voiced agreement with the empowerment program, but it soon became apparent that management support was hollow and halfhearted at best. Within three months of the sale, employee interest in the program had died, and the program had been dissolved.

Discussion Assignment
Discuss the following question in class or outside of class with your fellow students:

1. If you were the quality manager for this company and the employee empowerment effort had been your idea, how would you have gained a commitment to it from the new management team?

DISCUSSION ASSIGNMENT 8–3

The Task-Oriented Manager

Wanda Brown had worked hard to achieve her rapid advancement from shipping/receiving clerk to shipping/receiving manager. She had an uncanny ability to focus on a task, break it into its component parts, arrange the parts in a logical sequence, and tackle each part in order until the entire task has been accomplished—usually well ahead of schedule. She used this ability to quickly climb from the bottom of the ladder to the top in her department.

Now, as manager, things seem to be falling apart for Wanda. Rather than focusing exclusively on tasks, she is finding it necessary to deal with people. Often her subordinates don't agree with her concerning how best to do the job. They have ideas, problems, and feelings—none of which Wanda wants to hear about. Her attitude is "Forget your ideas, problems, and feelings; just focus on your work and do it my way."

Discussion Assignment
Discuss the following questions in class or outside of class with your fellow students:

1. Why might it be difficult for a manager who used to be a talented technician to let employees do their jobs?
2. What personal inhibitors will such an individual have to overcome to empower his or her employees?

DISCUSSION ASSIGNMENT 8–4

An Early Suggestion System

"In August 1721, a small box called the meyasubako was placed at the Takinoguchi entrance to Edo Castle by the order of Yoshimune Tokugawa, the eighth shogun. All citizens, regardless of their social standing, were allowed to drop written suggestions, requests, and complaints into the box. The meyasubako was the shogun's way of finding out how people felt about his policies and what people were thinking in general. Good suggestions were rewarded, and a man named Sensen Ogawa wrote a suggestion that led to the opening of a health-care facility for the poor. A suggestion dropped in the meyasubako also led to the development of a firefighting policy for the city of Edo."

Discussion Assignment

Discuss the following questions in class or outside of class with your fellow students:

1. Suggestion systems don't usually work very well in American organizations. Why?
2. If you were an employee of an organization, how would you want its suggestion system to work?

Source: Japan Human Relations Association, ed., *The Idea Book* (Cambridge, MA: Productivity Press, 1988), 201.

DISCUSSION ASSIGNMENT 8–5

Patagonia's Opportunity for Improvement Program

Patagonia is a world-leading textile manufacturer that specializes in clothing for children and adults. Employees are the primary source of workplace improvements at this company, where empowerment is the norm and the suggestion system is called the Opportunity for Improvement Program. Patagonia employees submit written suggestions on a form that asks three questions: "What needs improvement?" "Why?" and "How should the improvements be implemented?" Employees keep a copy of their suggestion, send one to their supervisor, and send one to a central office where it is entered into a suggestions database and tracked. Rewards for suggestions that are implemented range from token gifts such as movie tickets to paid adventure holidays.

Discussion Assignment

Discuss the following questions in class or outside of class with your fellow students:

1. How does an organization know whether its suggestion system is worth the time and effort needed to make it work?
2. If a suggestion system is costing more to operate than it is generating in improvements, how would you respond?

 ENDNOTES

1. Peter B. Grazier, *Before It's Too Late: Employee Involvement . . . An Idea Whose Time Has Come* (Chadds Ford, PA: Teambuilding, 1989), 8.
2. Grazier, 14.
3. Grazier, 9.

4. Isobel Pfeiffer and Jane Dunlop, "Increasing Productivity through Empowerment," *Supervisory Management* (January 1990), 11–12.

5. Grazier, 90–91.

6. *Philadelphia Inquirer*, cited by Grazier, 90–91.

7. Peter Kizilos, "Crazy about Empowerment?" *Training* (December 1990), 56.

8. Grazier, 87.

9. Grazier, 97.

10. Grazier, 97–98.

11. Grazier, 103–104.

12. Grazier, 129–142.

13. Japan Human Relations Association, ed., *The Idea Book* (Cambridge, MA: Productivity Press, 1988), 201, 202.

14. Japan Human Relations Association, 202.

15. Japan Human Relations Association, 203.

16. Japan Human Relations Association, 203.

17. Richard Hamlin, "A Practical Guide to Empowering Your Employees," *Supervisory Management* (April 1991), 8.

18. Hamlin, 8.

19. D. G. Myers and H. Lamm, "The Group Polarization Phenomenon," *Psychological Bulletin* 85 (1976), 602–627.

20. Mel E. Schnake, *Human Relations* (New York: Macmillan, 1990), 285–286.

21. R. D. Clark, "Group-Induced Shift toward Risk: A Critical Appraisal," *Psychological Bulletin* 80 (1971), 251–270.

22. J. R. Jablonski, *Implementing TQM* (London: Pfeiffer, 1992), 98.

23. J. T. Straub, "Ask Questions First to Solve the Right Problems," *Supervisory Management* (October 1991), 7.

24. S. Berry, "Ideas That Pay Off," *Nation's Business* (April 1991), 34.

25. B. Scharz, *The Suggestion System: A Total Quality Process* (Cambridge, MA: Productivity Press, 1991), 92.

26. Japan Human Relations Association, 21.

27. Japan Human Relations Association, 21.

28. Japan Human Relations Association, 33.

29. Japan Human Relations Association, 140.

30. George Milite, "When an Employee's Idea Is Just Plain Awful," *Supervisory Management* (October 1991), 3.

31. Milite, 3.

32. Japan Human Relations Association, 108.

33. G. M. Sturman, "The Supervisor as a Career Coach," *Supervisory Management* (November 1990), 6.

34. Kyle Dover, "Avoiding Empowerment Traps," *Management Review* (January 1999), 53.

35. Ibid., 51–55.

36. Dover, 51–52.

CHAPTER NINE

Leadership and Change

"There is no way to make people like change. You can only make them feel less threatened by it."

   Frederick Hayes

MAJOR TOPICS

- Leadership Defined
- Leadership for Quality
- Leadership Skills: Inherited or Learned?
- Leadership, Motivation, and Inspiration
- Leadership Styles
- Leadership Style in a Total Quality Setting
- Building and Maintaining a Following
- Leadership versus Management
- Leadership and Ethics
- Leadership and Change
- Employees and Managers on Change
- Restructuring and Change
- How to Lead Change
- Lessons from Distinguished Leaders
- Servant Leadership and Stewardship

Leadership is an intangible concept that produces tangible results. It is referred to sometimes as an art and at other times as a science. In reality, leadership is both an art and a science.

   The impact of good leadership can be readily seen in any organization where it exists. Well-led organizations, whether they are large companies or small departments within a company, share several easily identifiable characteristics:

- High levels of productivity
- Positive, can-do attitudes

- Commitment to accomplishing organizational goals
- Effective, efficient use of resources
- High levels of quality
- Mutually supportive teamwork approach to getting work done

Where good leadership exists, work is accomplished by teams. These teams are built deliberately, nurtured carefully, and improved continually. This chapter explains the concepts of leadership and leadership during times of change—and how they are applied in a total quality setting.

LEADERSHIP DEFINED

Leadership can be defined in many different ways, partly because it has been examined from the perspective of so many different fields of endeavor. Leadership has been defined as it applies to the military, athletics, education, business, industry, and many other fields. For the purpose of this book, *leadership* is defined as it relates specifically to total quality:

> Leadership is the ability to inspire people to make a total, willing, and voluntary commitment to accomplishing or exceeding organizational goals.

This definition contains a key concept that makes it particularly applicable in a total quality setting: the concept of inspiring people. Inspiring people is a higher order of human interaction than motivating them, which is a concept more frequently used in defining leadership. *Inspiration*, as used here, means motivation that has been internalized and therefore comes from within employees, as opposed to motivation that is simply a temporary response to external stimuli. Motivated employees commit to the organization's goals. Inspired employees make those goals their own. When employees are inspired, the total, willing, and voluntary commitment described in the definition follows naturally.

QUALITY TIP

How Leaders Are Alike

"At their best, . . . leaders—a fairly disparate group in many superficial ways—commit themselves to a common enterprise and are resilient enough to absorb the conflicts; brave enough, now and then, to be transformed by its accompanying energies; and capable of sustaining a vision that encompasses the whole organization."

Source: Warren Bennis and Burt Nanus, *Leaders: The Strategies for Taking Charge* (New York: Harper & Row, 1985), 217.

What Leaders Must Be Able to Do

In their book *Leaders: The Strategy for Taking Charge*, Warren Bennis and Burt Nanus describe three lessons for leadership that summarize what leaders must be able to do:[1]

- *Overcome resistance to change.*   Some people in management positions attempt to do this using power and control. Those who are leaders overcome resistance by achieving a total, willing, and voluntary commitment to shared values and goals.
- *Broker the needs of constituency groups inside and outside of the organization.* When the needs of the company and one of its suppliers appear to conflict, leaders must be able to find ways of bringing the needs of both organizations together without shortchanging either of them.
- *Establish an ethical framework within which all employees and the company as a whole operate.*   This is best accomplished by doing the following:

    Setting an example of ethical behavior

    Choosing ethical people as team members

    Communicating a sense of purpose for the organization

    Reinforcing appropriate behaviors within the organization and outside of it

    Articulating ethical positions, internally and externally

What Is a Good Leader?

Good leaders come in all shapes, sizes, genders, ages, races, political persuasions, and national origins. They do not look alike, talk alike, or even work alike. However, good leaders do share several common characteristics. These are the characteristics necessary to inspire people to make a total, willing, and voluntary commitment. Regardless of their backgrounds, good leaders exhibit the characteristics shown in Figure 9–1.

Good leaders are committed to both the job to be done and the people who must do it, and they are able to strike the appropriate balance between the two. Good leaders project a positive example at all times. They are good role models. Managers who pro-

QUALITY TIP

Leadership and Ethics

"In the end, trust, integrity, and positioning are all different faces of a common property of leadership—the ability to integrate those who must act with that which must be done so that it all comes together as a single organism in harmony with itself and its niche in the environment."

Source: Warren Bennis and Burt Nanus, *Leaders: The Strategy for Taking Charge* (New York: Harper & Row, 1985), 186.

Figure 9–1
Characteristics of Good Leaders

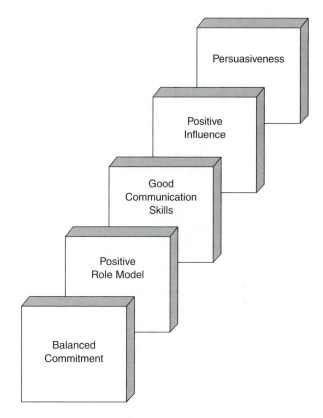

ject a "Do as I say, not as I do" attitude will not be effective leaders. To inspire employees, managers must be willing to do what they expect of workers, do it better, do it right, and do so consistently. If, for example, dependability is important, managers must set a consistent example of dependability. If punctuality is important, a manager must set a consistent example of punctuality. To be a good leader, a manager must set a consistent example of all characteristics that are important on the job.

Good leaders are good communicators. They are willing, patient, skilled listeners. They are also able to communicate their ideas clearly, succinctly, and in a nonthreatening manner. They use their communication skills to establish and nurture rapport with employees. Good leaders have influence with employees and use it in a positive manner. *Influence* is the art of using power to move people toward a certain end or point of view. The power of managers derives from the authority that goes with their jobs and the credibility they establish by being good leaders. Power is useless unless it is converted to influence. Power that is properly, appropriately, and effectively applied becomes positive influence.

Finally, good leaders are persuasive. Managers who expect people to simply do what they are ordered to do will have limited success. Those who are able to use their communication skills and influence to persuade people to their point of view and to help people make a total, willing, and voluntary commitment to that point of view can have unlimited success.

Leaders versus Misleaders

In his book *Managers for the Future: The 1990's and Beyond*, Peter Drucker makes the point that leadership is not a function of charisma.[2] Too many managers have been led to believe that dressing for success and developing a charismatic personality are the keys to being a good leader. Although there is something to be said for personal appearance, and charisma is certainly not a negative quality, one should not make the mistake of confusing image with substance.

Some of the world's most effective leaders have had little or no charisma:

> Dwight Eisenhower, George Marshall, and Harry Truman were singularly effective leaders, yet none possessed any more charisma than a dead mackerel. No one had less charisma than that of Lincoln of Illinois, the raw-boned, uncouth, backwoodsman of 1860. John F. Kennedy may have been the most charismatic person ever to occupy the White House, yet few presidents got as little done.[3]

Those who place image above substance and try to lead by charisma alone are misleaders, not leaders. What follows are several criteria Drucker uses to distinguish leaders from misleaders:[4]

- Leaders define and clearly articulate the organization's mission.
- Leaders set goals, priorities, and standards.
- Leaders see leadership as a responsibility rather than a privilege of rank.
- Leaders surround themselves with knowledgeable, strong people who can make a contribution.
- Leaders earn trust, respect, and integrity.

Myths about Leadership

Over the years a number of myths have grown up about the subject of leadership. Managers in a total quality setting should be aware of these myths and be able to dispel

QUALITY TIP

Leadership Means Substance over Image

"Books, articles, and conferences on leadership and on the qualities of the leader abound. Every CEO, it seems, has to be made to look like a dashing Confederate cavalry general or a boardroom Elvis Presley. Leadership does matter, of course. But, alas, it is something different from what is now touted under this label. It has little to do with 'leadership qualities' and even less to do with 'charisma.' It is mundane, unromantic, and boring. Its essence is performance."

Source: Peter F. Drucker, *Managing for the Future: The 1990's and Beyond* (New York: Truman Talley Books/Dutton, 1992), 119.

them. Bennis and Nanus describe the most common myths about leadership as follows:[5]

■ *Leadership is a rare skill.* Although it is true that few great leaders of world renown exist, many good, effective leaders do. Renowned leaders such as Winston Churchill were simply good leaders given the opportunity to participate in monumental events (World War II in Churchill's case). Another example is General Norman Schwarzkopf. He had always been an effective military leader. That's how he became a general. But it took a monumental event—the Gulf War—coupled with his leadership ability to make General Schwarzkopf a world-renowned leader. His leadership skills didn't appear suddenly; he had them all along. Circumstances allowed them to be displayed on the world stage.

Most effective leaders spend their careers in virtual anonymity, but they exist in surprisingly large numbers, and there may be little or no correlation between their ability to lead and their relative positions in an organization. The best leader in a company may be the lowest-paid wage earner, and the worst may be the CEO. In addition, a person may be a leader in one setting and not in another. For example, a person who shows no leadership ability at work may be an effective leader in his or her church. One of the keys to success in a total quality setting is to create an environment that brings out the leadership skills of all employees at all levels and focuses them on continually improving competitiveness.

■ *Leaders are born, not made.* This myth will be addressed later in this chapter. Suffice it to say here that leadership, attitudes, and behaviors can be learned, even by those who do not appear to have inborn leadership potential.

■ *Leaders are charismatic.* This myth was dispelled in a previous section. Some leaders have charisma and some don't. Some of history's most renowned leaders have had little or no charisma. Correspondingly, some of history's greatest misleaders have been highly charismatic. Generals Dwight Eisenhower and Omar Bradley are examples of great but uncharismatic leaders. Adolf Hitler and Benito Mussolini are examples of great misleaders who relied almost exclusively on charisma to build a following.

■ *Leadership exists only at the top.* Total quality would not work if this myth were true. Total quality relies on the building of teams at all levels in an organization and teaching employees in these teams to be leaders. In reality, the opposite of this myth is often true. Top managers may be the least capable leaders in a company. Leadership is about producing results and generating continual improvement, not one's relative position within the organization.

■ *Leaders control, direct, prod, and manipulate.* If practice is an indicator, this myth is the most widely believed. The "I'm the boss, so do what I say" syndrome is rampant in business and industry. It seems to be the automatic fallback position or default approach for managers who don't know better. Leadership in a total quality setting is about involving and empowering, not prodding and manipulating.

■ *Leaders don't need to be learners.* Lifelong learning is a must for leaders. One cannot be a good leader without being a good learner. Leaders don't learn simply for the sake of learning (although to do so is a worthwhile undertaking). Rather, leaders continually learn in what Bennis and Nanus call an "organizational context."[6] This means they approach learning from the perspective of what matters most to their organizations. A manager who is responsible for the metal fabrication department in a manufacturing firm might undertake to learn more about the classics of European literature. Although this would certainly make her a better educated person, studying European literature is not learning in an organizational context for the manager of a metal fabrication department. Examples of learning in an organizational context for such a manager include learning techniques to improve speed and feed rates, statistical process control (SPC), team-building strategies, computer numerical control programming, information about new composite materials, total productive maintenance, and anything else that will help improve the department's performance.

LEADERSHIP FOR QUALITY

Leadership for quality is leadership from the perspective of total quality. It is about applying the principles of leadership set forth in the preceding section in such a way as to continually improve work methods and processes. Leadership for quality is based on the philosophy that continually improving work methods and processes will, in turn, improve quality, cost, productivity, and return on investment.

This is the philosophy articulated in the Deming Chain Reaction developed by quality pioneer W. Edwards Deming. Deming's philosophy is that each improvement in work methods and processes initiates a chain reaction that results in the following:[7]

- Improved quality
- Decreased costs
- Improved productivity
- Decreased prices
- Increased market share
- Longevity in business
- More jobs
- Improved return on investment

Principles of Leadership for Quality

The principles of leadership for quality parallel those of total quality. In his book *The Team Handbook*, Peter R. Scholtes summarizes these principles as described in the following sections.[8]

Customer Focus

Leadership for quality requires a customer focus. This means the organization's primary goal is to meet or exceed customer expectations in a way that gives the customer lasting value. In a total quality setting, there are both internal and external customers.

QUALITY TIP

Quality Leadership

"Give customer concerns top priority . . . so that the final product or service exceeds customer expectations. Simply producing a certain number of widgets is replaced with producing widgets that consistently and exactly suit customer needs, and through a process that creates no scrap, rework, nor lost time. This can only be done by building excellence into every aspect of the company. Quality leadership, therefore, focuses on creating a workplace that encourages everyone to contribute to the company."

Source: Peter R. Scholtes, *The Team Handbook* (Madison, WI: Joiner Associates, 1992), 1–9.

Internal customers are other employees within the organization whose work depends on the work that precedes theirs. External customers are people who purchase and/or use the organization's products.

Obsession with Quality

Obsession with quality is an attitude that must be instilled and continually nurtured by leaders in an organization. It means that every employee aggressively pursues quality in an attempt to exceed the expectations of customers, internal and external.

Recognizing the Structure of Work

Leadership for quality requires that work processes be analyzed to determine their appropriate structural makeup (organization, order of steps, tools used, motion required, etc.). When the optimum structure is in place, work processes should be analyzed, evaluated, and studied continually in an attempt to improve them.

Freedom through Control

Control in a total quality setting refers to human control of work methods and processes. All too often in the age of high technology, the "tail wags the dog" in that machines run people instead of people running machines. Leaders must ensure that managers and employees take control of work processes and methods by collaborating to standardize them. The goal is to reduce variations in output by eliminating variations in how work is done.

Unity of Purpose

One of the most important responsibilities of a leader is to articulate the organization's mission clearly and accurately so that all employees understand it, believe in it, and commit to it. When there is unity of purpose, all employees pull together toward the same end.

QUALITY TIP

The Deming Chain Reaction

"With each improvement, processes and systems run better and better. Productivity increases as waste goes down. Customers get better products, which ultimately increases market share and provides better return on investments."

Source: W. Edwards Deming, quoted in Peter Scholtes, *The Team Handbook* (Madison, WI: Joiner Associates, 1992), 1–9.

Looking for Faults in Systems

Quality pioneers W. Edwards Deming and Joseph M. Juran believed that 85% of an organization's failures are failures of management-controlled systems. In their opinion, employees who do the work can control only 15% of what causes failures. Leadership for quality requires a change in focus from assessing blame for problems to assessing systems in an attempt to ferret out and correct systemic problems.

Teamwork

Rugged individualism has long been a fundamental element of the American character. The strong, silent stranger who rides into town and single-handedly runs out the bad guys (the character typified by Clint Eastwood over the years) has always had popular appeal in the United States. Individual performance has been encouraged and rewarded in the American workplace since the Industrial Revolution. Not until competition among companies became global in nature did it become necessary to apply a principle that has been known for years—that a team of people working together toward a common goal can outperform a group of individuals working toward their own ends. Leadership for quality requires team building and teamwork. These critical topics are covered in Chapter 10.

Continuing Education and Training

In the age of high technology, the most important machine in the workplace is the human mind. This is the premise of the book *Thinking for a Living* by Ray Marshall, secretary of labor during the Carter administration, who now serves as professor of economics at the University of Texas and is cochair of the Commission on the Skills of the American Workforce.[9] Continued learning at all levels is a fundamental element of total quality. Working hard no longer guarantees success. In the age of high technology, it is necessary to work hard and work smart.

The Juran Trilogy[10]

In his book *Juran on Leadership for Quality*, Joseph M. Juran sets forth his trilogy on leadership for quality: planning, control, and continuous improvement. The Juran Trilogy is composed of the following elements:[11]

■ *Quality planning.*   Quality planning consists of the following steps: identify customers, identify the needs of customers, develop products based on customer needs, develop work methods and processes that can produce products that meet or exceed customer expectations, and convert the results of planning into action.

■ *Quality control.*   Quality control consists of the following steps: evaluate actual performance, compare actual performance with performance goals, and take immediate steps to resolve differences between planned performance and actual performance.

■ *Quality improvement.*   Continual improvement of quality is a fundamental element of total quality. The steps involved are these: establish an infrastructure for accomplishing continual quality improvement; identify specific processes or methods in need of improvement; set up teams responsible for specific improvement projects; and provide improvement teams with the resources and training needed to diagnose problems and identify causes, decide on a remedy, and establish appropriate improvements after they have been made.

Planning, control, and improvement of quality do not happen automatically in any organization. They happen as the result of leadership. Leaders in a total quality setting must ensure that these principles are applied daily at all levels of their organizations.

LEADERSHIP SKILLS: INHERITED OR LEARNED?

Perhaps the oldest debate about leadership revolves around this question: "Are leaders born or made?" Can leadership skills be learned, or must they be inherited? This debate has never been settled and probably never will be. There are proponents on both sides of the debate, and this polarity is not likely to change because, as is often the case in such controversies, both sides are partially right.

The point of view set forth in this book is that leaders are like athletes: some athletes are born with natural ability, whereas others develop their ability through determination and hard work. Inborn ability, or the lack of it, represents only the starting point. Success from that point forward depends on the individual's willingness and determination to develop and improve. Some athletes born with tremendous natural ability never live up to their potential. Other athletes with limited natural ability do, through hard work, determination, and continuous improvement, perform beyond their apparent potential.

This phenomenon also applies to leadership. Some managers have more natural leadership ability than others. However, regardless of their individual starting points, managers can become good leaders through education, training, practice, determination, and effort.

LEADERSHIP, MOTIVATION, AND INSPIRATION

One of the characteristics shared by effective leaders is the ability to inspire and motivate others to make a commitment. The key to motivating people lies in the ability to relate their personal goals to the organization's goals. The key to inspiring people lies in the ability to relate what they believe to the organizational goals. Implicit in both cases is the leader's need to know and understand workers, including both their individual goals and their personal beliefs.

Understanding Individual Needs

Perhaps the best model for explaining individual human needs is that developed by psychologist Abraham H. Maslow. Maslow's Hierarchy of Needs (Figure 9–2) arrays the basic human needs on five successive levels. The lowest level in the hierarchy encompasses basic survival needs. All people need air to breathe, food to eat, water to drink, clothing to wear, and shelter in which to live. The second level encompasses safety/security needs. All people need to feel safe from harm and secure in their world. To this end, people enact laws, pay taxes to employ police and military personnel, buy insurance, try to save and invest money, and install security systems in their homes.

The third level encompasses social needs. People are social animals by nature. This fact manifests itself through families, friendships, social organizations, civic groups, special clubs, and even employment-based groups such as company softball and basketball teams. The fourth level of the hierarchy encompasses esteem needs. Self-esteem is a key ingredient in the personal happiness of individuals. All people need to feel self-worth, dignity, and respect. People need to feel that they matter. This fact manifests itself in a variety of ways. It can be seen in the clothes people wear, the cars people drive, and the behavior people exhibit in public.

The highest level of Maslow's hierarchy encompasses self-actualization needs. Complete self-fulfillment is a need that is rarely satisfied in people. The need for self-actualization manifests itself in a variety of ways. Some people seek to achieve it through their work; others through hobbies, human associations, or leisure activities.

Figure 9–2
Maslow's Hierarchy of Needs

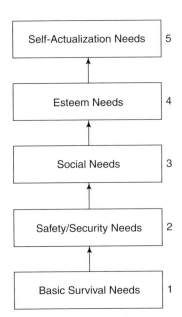

Self-Actualization Needs	5
Esteem Needs	4
Social Needs	3
Safety/Security Needs	2
Basic Survival Needs	1

Leaders need to understand how to apply Maslow's model if they hope to use it to motivate and inspire workers. Principles required for applying this model are as follows:

1. Needs must be satisfied in order from the bottom up.
2. People focus most intently on their lowest unmet need. For example, employees who have not met their basic security needs will not be motivated by factors relating to their social needs.
3. After a need has been satisfied, it no longer works as a motivating factor. For example, people who have satisfied their need for financial security will not be motivated by a pay raise.

Understanding Individual Beliefs

Each person has a basic set of beliefs that, together, forms that individual's value system. If leaders know their fellow employees well enough to understand those basic beliefs, they can use this knowledge to inspire them on the job. Developing this level of understanding of employees comes from observing, listening, asking, and taking the time to establish trust.

Leaders who develop this level of understanding of workers can use it to inspire employees to higher levels of performance. This is done by showing employees how the organization's goals relate to their beliefs. For example, if pride of workmanship is part of an employee's value system, a leader can inspire the person to help achieve the organization's quality goals by appealing to that value.

Inspiration, as a level of leadership, is on a higher plane than motivation. Managers who become good enough leaders to inspire their workers will achieve the best results.

LEADERSHIP STYLES

Leadership styles have to do with how people interact with those they seek to lead. Leadership styles go by many different names. However, most styles fall into the categories shown in Figure 9–3.

Autocratic Leadership

Autocratic leadership is also called *directive* or *dictatorial leadership*. People who take this approach make decisions without consulting the employees who will have to implement them or who will be affected by them. They tell others what to do and expect them to comply obediently. Critics of this approach say that although it can work in the short run or in isolated instances, in the long run it is not effective. Autocratic leadership is not appropriate in a total quality setting.

Democratic Leadership

Democratic leadership is also called *consultive* or *consensus leadership*. People who take this approach involve the employees who will have to implement decisions in making them. The leader actually makes the final decision, but only after receiving the input and recommendations of team members. Critics of this approach say the most popular

Figure 9–3
Leadership Styles

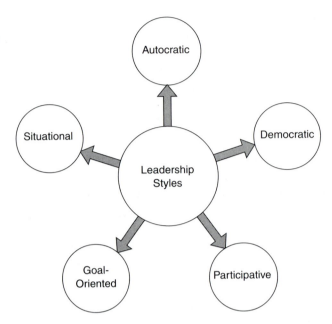

decision is not always the best decision and that democratic leadership, by its nature, can result in the making of popular decisions as opposed to right decisions. This style can also lead to compromises that ultimately fail to produce the desired result.

Participative Leadership

Participative leadership is also known as *open, free-reign,* or *nondirective leadership.* People who take this approach exert little control over the decision-making process. Rather, they provide information about the problem and allow team members to develop strategies and solutions. The leader's job is to move the team toward consensus. The underlying assumption of this style is that workers will more readily accept responsibility for solutions, goals, and strategies that they are empowered to help develop. Critics of this approach say consensus building is time consuming and works only if all people involved are committed to the best interests of the organization.

Goal-Oriented Leadership

Goal-oriented leadership is also called *results-based* or *objective-based leadership.* People who take this approach ask team members to focus solely on the goals at hand. Only strategies that make a definite and measurable contribution to accomplishing organizational goals are discussed. The influence of personalities and other factors unrelated to the specific goals of the organization is minimized. Critics of this approach say it can break down when team members focus so intently on specific goals that they overlook opportunities and/or potential problems that fall outside of their narrow focus. Advocates of total quality say that results-oriented leadership is too narrowly focused and often centered on the wrong concerns.

QUALITY TIP

Leaders Focus on People, Not Technology

Talking about corporate executives Bob and Jim Swiggett of Kollmorgen Corporation, James A. Belohlav says, "Indeed, what the Swiggetts realized was that the basis for success in their industry, innovation and customer satisfaction, was correlated with how well they utilized the people within their company. What the Swiggetts also came to appreciate was what successful military leaders have always known: Armies are made up of individual people. Technology changes the way battles are fought, but people determine who the ultimate winners and losers are."

Source: James A. Belohlav, *Championship Management: An Action Model for High Performance* (Cambridge, MA: Productivity Press, 1990), 46.

Situational Leadership

Situational leadership is also called *fluid* or *contingency leadership*. People who take this approach select the style that seems to be appropriate based on the circumstances that exist at a given time. In identifying these circumstances, leaders consider the following factors:

- Relationship of the manager and team members
- How precisely actions taken must comply with specific guidelines
- Amount of authority the leader actually has with team members

Depending on what is learned when these factors are considered, the manager decides whether to take the autocratic, democratic, participative, or goal-oriented approach. Under different circumstances, the same manager would apply a different leadership style. Advocates of total quality reject situational leadership as an attempt to apply an approach based on short-term concerns instead of focusing on the solution of long-term problems.

LEADERSHIP STYLE IN A TOTAL QUALITY SETTING

The appropriate leadership style in a total quality setting might be called participative leadership taken to a higher level. Whereas participative leadership in the traditional sense involves soliciting employee input, in a total quality setting it involves soliciting input from empowered employees, listening to that input, and acting on it. The key difference between the traditional participative leadership and participative leadership from a total quality perspective is that, with the latter, employees providing input are empowered.

Collecting employee input is not new. However, collecting input, logging it in, tracking it, acting on it in an appropriate manner, working with employees to improve weak suggestions rather than simply rejecting them, and rewarding employees for improvements that result from their input—all of which are normal in a total quality setting—extend beyond the traditional approach to participative leadership.

Discussion Assignment 9–2 illustrates the concept of participative leadership as applied in a total quality setting at a U.S.–based electronics company. This assignment illustrates how important freedom and respect for the individual are in today's intensely competitive world of business and industry.

BUILDING AND MAINTAINING A FOLLOWING

Managers can be good leaders only if the people they hope to lead will follow them willingly and steadfastly. Followership must be built and, having been built, maintained. This section is devoted to a discussion of how managers can build and maintain followership among the people they hope to lead.

Popularity and the Leader

Leadership and popularity are not the same thing. However, many managers confuse popularity with leadership and, in turn, followership. An important point to understand in leading people is the difference between popularity and respect. Long-term followership grows out of respect, not popularity. Good leaders *may* be popular, but they *must* be respected. Not all good leaders are popular, but they are all respected.

Managers occasionally have to make unpopular decisions. This is a fact of life for leaders, and it is why leadership positions are sometimes described as lonely ones. Making an unpopular decision does not necessarily cause a leader to lose followership, provided the leader is seen as having solicited a broad base of input and given serious, objective, and impartial consideration to that input. Correspondingly, leaders who make inappropriate decisions that are popular in the short run may actually lose followership in the long run. If the long-term consequences of a decision turn out to be detrimental to the team, team members will hold the leader responsible, particularly if the decision was made without first collecting and considering employee input.

Leadership Characteristics That Build and Maintain Followership

Leaders build and maintain followership by earning the respect of those they lead. Here are some characteristics of leaders that build respect, as shown in Figure 9–4:

■ *Sense of purpose.* Successful leaders have a strong sense of purpose. They know who they are, where they fit in the overall organization, and the contributions their areas of responsibility make to the organization's success.

■ *Self-discipline.* Successful leaders develop discipline and use it to set an example. Through self-discipline, leaders avoid negative self-indulgence, inappropriate displays of emotion such as anger, and counterproductive responses to the everyday pressures of the job. Through self-discipline, leaders set an example of handling problems and pressures with equilibrium and a positive attitude.

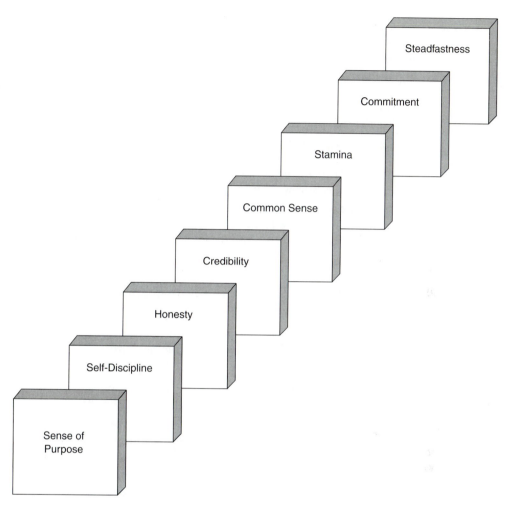

Figure 9–4
Characteristics That Build and Maintain Followership

- *Honesty.* Successful leaders are trusted by their followers. This is because they are open, honest, and forthright with other members of the organization and with themselves. They can be depended on to make difficult decisions in unpleasant situations with steadfastness and consistency.
- *Credibility.* Successful leaders have credibility. Credibility is established by being knowledgeable, consistent, fair, and impartial in all human interaction; by setting a positive example; and by adhering to the same standards of performance and behavior expected of others.

- *Common sense.*   Successful leaders have common sense. They know what is important in a given situation and what is not. They know that applying tact is important when dealing with people. They know when to be flexible and when to be firm.
- *Stamina.*   Successful leaders must have stamina. Frequently they need to be the first to arrive and the last to leave. Their hours are likely to be longer and the pressures they face more intense than those of others. Energy, endurance, and good health are important to those who lead.
- *Commitment.*   Successful leaders are committed to the goals of the organization, the people they work with, and their own ongoing personal and professional development. They are willing to do everything within the limits of the law, professional ethics, and company policy to help their team succeed.
- *Steadfastness.*   Successful leaders are steadfast and resolute. People do not follow a person they perceive to be wishy-washy and noncommittal. Nor do they follow a person whose resolve they question. Successful leaders must have the steadfastness to stay the course even when it becomes difficult.

Pitfalls That Can Undermine Followership

The previous section explained several positive characteristics that will help managers build and maintain the respect and, in turn, followership of those they hope to lead. Managers should also be aware of several common pitfalls that can undermine followership and the respect managers must work so hard to earn:

- *Trying to be a buddy.*   Positive relations and good rapport are important, but leaders are not the buddies of those they lead. The nature of the relationship does not allow it.
- *Having an intimate relationship with an employee.*   This practice is both unwise and unethical. A positive manager–employee relationship cannot exist under such circumstances. Few people can succeed at being the lover and the boss, and few things can damage the morale of a team so quickly and completely.
- *Trying to keep things the same when supervising former peers.*   The supervisor–employee relationship, no matter how positive, is different from the peer–peer relationship. This can be a difficult fact to accept and a difficult adjustment to make. But it is an adjustment that must be made if the peer-turned-supervisor is going to succeed as a leader.

Paradigms of Human Interaction

In his book *The 7 Habits of Highly Effective People*, Stephen R. Covey describes the following paradigms of human interaction:[12]

- *Win/win* is an approach to human interaction that seeks mutual benefit. Rather than pursuing a your-way or my-way solution, win/win proponents seek best-way solutions.
- *Win/lose* is an approach to human interaction that says, "Go ahead and have things your way. I never get what I want anyway." This approach results in a definite winner and a definite loser.

■ *Lose/lose* is an approach to human interaction in which both parties are so stubborn, ego driven, and vindictive that, ultimately, they both lose regardless of what decision is made.

■ *Win* is an approach to human interaction that says, "I don't necessarily want you to lose, but I definitely want to win." It is the result of a *"You take care of yourself and I'll take care of myself"* attitude.

QUALITY CASE:

Leading by Example at Autodesk

Autodesk, Inc., was one of the pioneers of computer-aided design and drafting (CAD) software for use on a personal computer. The company is now a leading developer of PC software for engineering, architecture, construction, and manufacturing. Autodesk's flagship product, AutoCAD, is one of the most widely used software packages for design, modeling, drafting, mapping, and rendering in business. To achieve world-class status, Autodesk needed world-class leadership, and the company found it in CEO Carol Bartz. Bartz is one of the most-studied CEOs in America. As a woman who not only broke through the glass ceiling but led her company to the top in a globally competitive industry, Bartz is widely respected as an effective leader.

When Bartz took over at Autodesk in 1992, she saw immediately that changes were needed. She raised the quality standards in manufacturing, hired a stronger sales force, and began developing new products. Bartz's approach to leadership is simple: she leads by example and holds everyone accountable, including and especially herself. Bartz models the behavior she wants to see in Autodesk employees. To Bartz, accountability implies measurement. If you want something to improve, you have to measure it. Consequently, she leads her managers in developing realistic but challenging performance goals, then expects them to commit to achieving them. Autodesk's managers know that Bartz expects them to deliver.

Bartz makes a point of letting employees at Autodesk know what is important and, again, she does this by example. Bartz knew her leadership style was working and that she was setting the right example when she began to hear managers at Autodesk say, "Carol wouldn't do that." Bartz thinks a leader must be so excited about her vision that people cannot wait to join her in pursuing that vision. Under Bartz's leadership, Autodesk is approaching the $1 billion sales level. Hers is a leadership style that is successful.

Source: Thomas J. Neff and James M. Citrin., *Lessons from the Top* (New York: Doubleday, 2001), 41–44.

Of the four paradigms just presented, the win/win approach is the one that will most help leaders build and maintain a following. Unlike the other paradigms, win/win places value on the opinions of both parties and requires them to work together to find solutions.

Discussion Assignment 9–3 describes a company that applies the win/win approach to build followership among its employees. S. C. Johnson values its employees and demonstrates its commitment to them by removing barriers, both physical and psychological, between managers and employees. The following is one example of how positive interaction between management and employees is encouraged:

> As one talks to Roger Mulhollen, retired vice-president of corporate personnel, about S. C. Johnson, he will recount how, in his first days with the corporation, he went to eat at the cafeteria. At a company that previously employed him, there were several cafeterias each serving a different segment of the employee population (workers, managers, etc.). The amazing sight for Roger was that everyone ate together at S. C. Johnson. That first day he shared his lunch with a vice-president and an R&D technician.[13]

LEADERSHIP VERSUS MANAGEMENT

Leadership and management, although both are needed in the modern workplace, are not the same thing. To be a good leader and a good manager, one must know the difference between the two concepts. According to John P. Kotter, leadership and management "are two distinctive and complementary systems of action."[14] Kotter lists the following differences between management and leadership:[15]

■ Management is about coping with complexity; leadership is about coping with change.

■ Management is about planning and budgeting for complexity; leadership is about setting the direction for change through the creation of a vision.

■ Management develops the capacity to carry out plans through organizing and staffing; leadership aligns people to work toward the vision.

■ Management ensures the accomplishment of plans through controlling and problem solving; leadership motivates and inspires people to want to accomplish the plan.

Bennis quotes Field Marshall Sir William Slim, who led the British Army's brilliant reconquest of Burma during World War II, on drawing the distinction between management and leadership:

> Managers are necessary; leaders are essential . . . . Leadership is of the Spirit, compounded of personality and vision . . . . Management is of the mind, more a matter of accurate calculation, statistics, methods, timetables, and routine.[16]

Writing for *Training* magazine, Bennis compares leaders and managers:[17]

■ Managers administer; leaders innovate.

■ Managers are copies; leaders are originals.

■ Managers maintain; leaders develop.

■ Managers focus on systems and structure; leaders focus on people.

- Managers rely on control; leaders inspire.
- Managers take the short view; leaders take the long view.
- Managers ask how and when; leaders ask what and why.
- Managers accept the status quo; leaders challenge it.
- Managers do things right; leaders do the right thing.

Bennis takes a more critical view of management than does Kotter, and the points he makes deserve consideration. In reality, the most successful managers in a total quality setting will be those who can appropriately combine the characteristics of both managers and leaders.

Trust Building and Leadership

Trust is a necessary ingredient for success in the intensely competitive modern workplace. It means, in the words of D. Zielinski and C. Busse, "employees who can make hard decisions, access key information, and take initiative without fear of recrimination from management, and managers who believe their people can make the right decisions."[18]

Building trust requires leadership on the part of managers. Trust-building strategies include the following:

- *Taking the blame but sharing the credit.*   Managers who point the finger of blame at their employees, even when the employees are at fault, do not build trust. Leaders must be willing to accept responsibility for the performance of people they hope to lead. Correspondingly, when credit is due, leaders must be prepared to spread it around appropriately. Such unselfishness on the part of managers builds trust among employees.
- *Pitching in and helping.*   Managers can show leadership and build trust by rolling up their sleeves and helping when a deadline is approaching. A willingness to "get their hands dirty" when circumstances warrant it helps managers build trust among employees.
- *Being consistent.*   People trust consistency. It lets them know what to expect. Even when employees disagree with managers, they appreciate consistent behavior.
- *Being equitable.*   Managers cannot play favorites and hope to build trust. Employees want to know that they are treated not just well, but as well as all other employees. Fair and equitable treatment of all employees will help build trust.

LEADERSHIP AND ETHICS

In his book *Managing for the Future: The 1990's and Beyond*, Drucker discusses the role of ethics in leadership.[19] Setting high standards of ethical behavior is an essential task of leaders in a total quality setting. Drucker summarizes his views regarding ethical leadership in the modern workplace as follows:

> What executives do, what they believe and value, what they reward and whom, are watched, seen, and minutely interpreted throughout the whole organization. And nothing

QUALITY TIP

An Expectation of Ethics

"Both the heads of large corporations and the few 'tycoons' have come to be seen as society's leaders. And leaders are expected to set an example. They are not supposed to behave as we know that we behave. They are expected to behave as we know that we ought to behave. The more cynical we have become about the behavior of earlier leadership groups—politicians, preachers, physicians, lawyers, and so on—the more we have come to expect virtue from business and businesspeople."

Source: Peter F. Drucker, *Managing for the Future: The 1990's and Beyond* (New York: Truman Talley Books/Dutton, 1992), 113.

is noticed more quickly—and considered more significant—than a discrepancy between what executives preach and what they expect their associates to practice. The Japanese recognize that there are really only two demands of leadership. One is to accept that rank does not confer privileges; it entails responsibilities. The other is to acknowledge that leaders in an organization need to impose on themselves that congruence between deeds and words, between behavior and professed beliefs and values, that we call personal integrity.[20]

Drucker speaks to ethical behavior on the part of corporate executives who hope to lead their companies effectively. The concept also applies to managers at all levels and any other person who hopes to lead others in a total quality setting.

LEADERSHIP AND CHANGE

In a competitive and rapidly changing marketplace, industrial companies are constantly involved in the development of strategies for keeping up, staying ahead, and/or setting new directions. What can managers do to play a positive role in the process? David Shanks recommends the following strategies:[21]

- Have a clear vision and corresponding goals.
- Exhibit a strong sense of responsibility.
- Be an effective communicator.
- Have a high energy level.
- Have the will to change.

Shanks developed these strategies to help executives guide their companies through periods of corporate stress and change, but they also apply to other personnel.

These characteristics of good leaders apply to any manager at any level who must help his or her organization deal with the uncertainty caused by change.

Facilitating Change as a Leadership Function

The following statement by management consultant Donna Deeprose carries a particularly relevant message for modern managers:

> In an age of rapidly accelerating technology, restructuring, repositioning, downsizing, and corporate takeovers, change may be the only constant. Is there anything you can do about it? Of course there is. You can make change happen, let it happen to you, or you can stand by while it goes on around you.[22]

Deeprose divides managers into three categories, based on how they handle change: driver, rider, or spoiler.[23] People who are drivers lead their organizations in new directions as a response to change. Managers who just go along, reacting to change as it happens rather than getting in front of it, are riders. Managers who actively resist change are spoilers. Deeprose gives examples of how a driver would behave in a variety of situations:[24]

- In viewing the change taking place in an organization, drivers stay mentally prepared to take advantage of the change.
- When facing change about which they have misgivings, drivers step back and examine their own motivations.
- When a higher manager has an idea that has been tried before and failed, drivers let the boss know what difficulties were experienced earlier and offer suggestions for avoiding the problems this time.
- When a company announces major changes in direction, drivers find out all they can about the new plans, communicate what they learn with their employees, and solicit input to determine how to make a contribution to the achievement of the company's new goals.
- When permission to implement a change that will affect other departments is received, drivers go to these other departments and explain the change in their terms, solicit their input, and involve them in the implementation process.
- When demand for their unit's work declines, drivers solicit input from users and employees as to what modifications and new products or services might be needed and include the input in a plan for updating and changing direction.
- When an employee suggests a good idea for change, drivers support the change by justifying it to higher management and using their influence to obtain resources for it while countering opposition to it.
- When their unit is assigned a new, unfamiliar task, drivers delegate the new responsibilities to their employees and make sure that workers get the support and training needed to succeed.

These examples show that a driver is a manager who exhibits the leadership characteristics necessary to play a positive, facilitating role in helping workers and organizations successfully adapt to change on a continual basis.

EMPLOYEES AND MANAGERS ON CHANGE

One of the difficulties organizations face when attempting to facilitate change is the differing perceptions of employees and managers concerning change. Employees often view change as something done to them. Managers often regard it as something done in spite of employees who just won't cooperate. The following quote is representative of how employees feel about change:

"I'm struggling with this new business environment of ours. I'm doing the best I can, but I'm scared. A little more sensitivity and patience on your part will go a long way toward helping me cope."[25]

Managers who listen to employees can learn a valuable lesson. It's not that they dislike change so much. Rather, it's that they don't like how it's done. The key to winning the support of employees for change is *involvement*. Make them part of the process from the beginning. Give them a voice in how change is implemented. Make sure that change is something done *with* employees rather than *to* them.

From the perspective of employees, managers are often viewed as the "bad guys" when changes are made. This viewpoint is just as unfair and counterproductive as the one that sees employees as inhibitors of change. This quote is typical of how managers often feel about change:

"Maybe you see me as the instigator or 'perpetrator' of change. If you do, to a degree, you're right. Sponsoring and supporting change is one of my responsibilities—and it's an absolute necessity in order to keep our organization successful and protect our jobs. But besides being a source of change, I am also a victim of it. And when it comes to 'rolling downhill,' I end up having to make as many adjustments as anyone else."[26]

To respond effectively to change, organizations must continually apply at least the following strategies:

■ Promote a "we are in this together" attitude toward change.
■ Make sure all employees understand that change is driven by market forces, not management.
■ Involve everyone who will be affected by change in planning and implementing the response to it.

RESTRUCTURING AND CHANGE

Few words can strike as much fear into the hearts of employees at all levels as *restructuring*. The term at one time was synonymous with *reorganization*. However, because of the way so many organizations have used the word, it has become a euphemism for layoffs, terminations, plant closings, and workforce cuts. As a result, the typical employee responses to the term are uncertainty and panic.[27]

Because of the ever-changing conditions of the global marketplace, few organizations will escape the necessity for restructuring, and few people will complete a career without experiencing one or more restructurings. Acquisitions, mergers, buyouts, and downsizing—common occurrences in today's marketplace—all typically involve corporate restructuring. This fact is market driven and can be controlled by neither individuals nor organizations. However, organizations and individuals can control how they respond to the changes brought by restructuring, and it is this response that will determine the effectiveness of the restructuring effort. The remainder of this section is devoted to explaining strategies for effectively handling the changes inherent in restructuring.

Be Smart and Empathetic

Restructuring can be traumatic for employees. Managers should remember this point when planning and implementing the changes that go with restructuring. The challenge to management is to be both smart and empathetic. The following strategies can help maintain employee loyalty and calm employee fears during restructuring:

- Take time to show employees that management cares and is concerned about them on a personal level.
- Communicate with employees about why the changes are necessary. Focus on market factors. Use a variety of tools to ensure effective communication (e.g., face-to-face meetings, newsletters, videotaped messages, posted notices, etc.).
- Provide formal outplacement assistance to all employees who will lose their jobs.
- Be fair, equitable, and honest with employees. Select employees to be laid off according to a definite set of criteria rather than as the result of a witch-hunt.
- Remember to provide support to those individuals who will be the primary change agents.

Have a Clear Vision

One of the best ways to minimize the disruptive nature of change is to have a clear vision of what the organization is going to look like after the change. A good question to ask is "What are we trying to become?" Managers should have a clear vision and be able to articulate that vision. This will give the organization a beacon in the distance to guide it through the emotional fog that can accompany change.

Establish Incentives That Promote the Change

People respond to incentives, especially when those incentives are important to them on a personal level. Managers can promote the changes that accompany restructuring by establishing incentives for contributors to those changes. Incentives can be monetary or nonmonetary, but they should motivate employees on a personal level.

An effective way to identify incentives that will work is to form an *ad hoc* task force of employees and discuss the issue of incentives. List as many monetary and nonmonetary incentives as the group can identify. Then give the members a week to discuss the list with their fellow employees. Once a broad base of employer input has been collected, the task force meets again and ranks the incentives in order of preference. The team then establishes a menu of incentives management can use to promote change. The menu concept allows employees to select incentives from among a list of options. This increases the likelihood that the incentives will motivate on a personal level.

Continue to Train

During times of intense change, the tendency of organizations is to put training on hold. The idea is "we'll get back to training again when things settle down." In reality, putting off training during restructuring is the last thing an organization should do.

One of the primary reasons employees will oppose change is that it will require skills they don't have. Training should actually be increased during times of intense change to make sure that employees have the skills required during and after the transition period.

HOW TO LEAD CHANGE

Kotter opens the first chapter of his book *Leading Change* with the following statement:

> By any objective measure, the amount of significant, often traumatic, change in organizations has grown tremendously over the past two decades. Although some people predict that most of the reengineering, restrategizing, mergers, downsizing, quality efforts, and cultural renewal projects will soon disappear, I think that is highly unlikely. Powerful macroeconomic forces are at work here, and these forces may grow even stronger over the next few decades. As a result, more and more organizations will be pushed to reduce costs, improve the quality of products and services, locate new opportunities for growth, and increase productivity.[28]

A critical aspect of leadership in today's globally oriented organization involves leading change. To survive and thrive in a competitive environment, organizations must be able to anticipate and respond to change effectively. However, successful organizations don't just respond to change—they get out in front of it.

Change is a constant in today's global business environment. Consequently, organizations must structure themselves for change. In other words, organizations must *institutionalize* the process of change. The model in Figure 9–5 is an effective tool for doing so. The following sections explain the various steps in this model.

Reality of Continual Change

Change is not something organizations do because they want to or because they get bored with the status quo. Rather, it is something they do because they must. Every organization that has to compete in an open marketplace is forced by macroeconomic conditions to constantly reduce costs, improve quality, enhance product attributes, increase productivity, and identify new markets. None of these things can be accomplished without changing the way things are currently done. Employees at all levels in an organization need to understand this point.

If it is true that change is the only constant on the radar screen of the modern organization, why do so many miss the point? A typical reaction is that human beings don't

Figure 9–5
Change Facilitation Model
This model is adapted from John P. Kotter, *Leading Change* (Boston: Harvard Business School Press, 1996), 21.

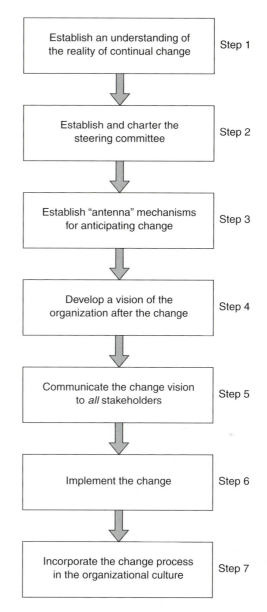

Establish an understanding of the reality of continual change — Step 1

Establish and charter the steering committee — Step 2

Establish "antenna" mechanisms for anticipating change — Step 3

Develop a vision of the organization after the change — Step 4

Communicate the change vision to *all* stakeholders — Step 5

Implement the change — Step 6

Incorporate the change process in the organizational culture — Step 7

like change. But in reality this is not the case. Employees who object to change typically object to how it is handled, not the fact that it's happening. Kotter identifies the following reasons that employees may not understand the reality of and need for change:[29]

- Absence of a major crisis
- Low overall performance standards
- No view of the big picture

- Internal evaluation measures that focus on the wrong benchmarks
- Insufficient external feedback
- A "kill the messenger" mentality among managers
- Overfocus among employees on the day-to-day stresses of the job
- Too much "happy talk" from executive managers

An organization's senior management team is responsible for eliminating any of these factors except, of course, a major crisis caused by external forces. Creating an artificial crisis is a questionable strategy, but correcting the remainder of these factors is not just appropriate but also advisable.

Performance standards should be based on what it takes to succeed in the global marketplace. Every organization should have a strategic plan that puts the "big picture" into writing, and every employee should know the big picture. Internal evaluation measures should mirror overall performance standards in that they should ensure globally competitive performance.

Organizations cannot succeed in a competitive environment without systematically collecting, analyzing, and using external feedback. This is how an organization knows what is going on. External feedback is the most effective way to identify the need for change.

Every employee in an organization is a potential for change. Employees attend conferences, read professional journals, participate in seminars, browse the Internet, and talk to colleagues. If managers welcome feedback from employees, they can turn them collectively into an effective mechanism for anticipating change. On the other hand, managers who "kill the messenger" will quickly extinguish this invaluable source of continual feedback.

As a rule, employees will focus most of their attention on their day-to-day duties, which is how it should be. Consequently, management must make a special effort to communicate with employees about market trends and other big-picture issues. All such communications with employees should be accurate, thorough, and honest. Managers must make sure they do not go overboard and create panic, but to sugarcoat important issues with "happy talk" is the same as deceiving employees.

Establish and Charter the Steering Committee

> Major transformations are often associated with one highly visible leader. Consider Chrysler's comeback from near bankruptcy . . . and we think of Lee Iacocca. Mention Wal-Mart's ascension from small-fry to industry leader, and Sam Walton comes to mind. Read about IBM's efforts to renew itself, and the story centers on Lou Gerstner. After a while, one might easily conclude that the kind of leadership that is so critical to any change can come only from a single larger-than-life person.[30]

Although a visible and visionary individual can certainly be a catalyst for organizational change, the reality is that one person does often change an organization.

The media like to create the image of the knight on a charging steed who single-handedly saves the company. This story makes good press, but it rarely squares with reality. Organizations that do the best job of handling change have what Kotter calls a "guiding coalition." The *guiding coalition* is a team of people who are committed to the change in question and who can make it happen. Every member of the team should have the following characteristics:

■ *Authority.*   Members should have the authority necessary to make decisions and commit resources.

■ *Expertise.*   Members should have expertise that is pertinent in terms of the subject change so that informed decisions can be made.

■ *Credibility.*   Members must be well respected by all stakeholders so that they will be listened to and taken seriously.

■ *Leadership.*   Members should have the leadership qualities necessary to drive the effort. These qualities include those listed here plus influence, vision, commitment, perseverance, and persuasiveness.

Establish Antenna Mechanisms

Leading change is about getting out in front of it. It's about driving change rather than letting it drive you. To do this, organizations must have mechanisms for sensing trends that will generate future change. These "antenna" mechanisms can take many forms, and the more, the better. Reading professional journals, attending conferences, studying global markets, and even reading the newspaper can help identify trends that might affect an organization. So can attentive marketing representatives.

For example, a computer company that markets primarily to colleges and universities learned that two large institutions had adopted a plan to stop purchasing personal computers. Instead, they intend to require all students to purchase their own laptops. The potential for a major change in the organization's business was identified by a marketing representative in the course of a routine call on these universities.

This information allowed the computer company quickly to develop and implement a plan for getting out in front of what is likely to be a nationwide trend. The company now markets laptop computers directly to students through the bookstores at both universities and is promoting the idea to colleges and universities nationwide.

Every employee in an organization should have his or her antennae tuned to the world outside and bring anything of interest to the attention of the steering committee.

Develop a Vision

What will we look like after the change in question has been made? The organization's vision answers this question. Kotter calls a vision a "sensible and appealing picture of the future."[31] The following scenario illustrates how having a cogent vision for change can help employees accept the change more readily.

Two tour groups are taking a trip together. Each group has its own tour bus and group leader. The route to the destination and stops along the way have been meticulously planned for maximum tourist value, interest, and enjoyment. All members of both groups have agreed to the itinerary and are looking forward to every stop. Unfortunately, several miles before the first scheduled attraction, the tour group leaders receive word that a chemical spill on the main highway will require a detour that will, in turn, require changes to the itinerary.

The tour group leader for group A simply acknowledges the message and tells the bus driver to take the alternate route. To the members of group A he says only that an unexpected detour has forced a change of plans. With no more information than this to go on, the members of group A are confused and quickly become unhappy.

The tour leader for group B, however, is a different sort. She acknowledges the message and asks the driver to pull over. She then says to the members of group B:

> Folks, we've had a change of plans. A chemical spill on the highway up ahead has closed down the route on which most of today's attractions are located. Fortunately, this area is full of wonderful attractions. Why don't we just have lunch on me at a rustic country restaurant? It's up the road about 5 miles. While you folks are enjoying lunch, I'll hand out a list of new attractions I know you'll want to see. We aren't going to let a little detour spoil our fun!

Because they could see how things would look after the change, the members of group B accepted it and were satisfied. However, the members of group A, because they were not fully informed, became increasingly frustrated and angry. As a result, they went along with the changes only reluctantly and, in several cases, begrudgingly. Group A's tour leader kept his clients in the dark. Group B's leader gave hers a vision.

The six characteristics of an effective vision are as follows:

- *Imaginable.*   It conveys a picture that others can see of how things will be after the change.
- *Desirable.*   A vision that points to a better tomorrow for all stakeholders will be well received even by those who resist change.
- *Feasible.*   To be feasible, a vision must be realistic and attainable.
- *Flexible.*   An effective vision is stated in terms that are general enough to allow for initiative in responding to ever-changing conditions.
- *Communicable.*   A good vision can be explained to an outsider who has no knowledge of the business.

Communicate the Vision to All Stakeholders

People will buy into the vision only if they know about it. The vision must be communicated to all stakeholders. A good communication package will have at least the following characteristics:

- Simplicity
- Repetitiveness
- Multiple formats
- Feedback mechanisms

The simpler the message, the better. Regardless of the communication formats chosen, keep the message simple and get right to it. Don't beat around the bush, lead with rationalizations, or attempt any type of linguistic subterfuge—what politicians and journalists call "spin."

Repetition is critical when communicating a new message. Messages are like flies. If a fly buzzes past your face just once and moves on, you will probably take no notice of it. But if it keeps coming back persistently, buzzing, and refuses to go away, before long you will take notice of it. Repetition forces employees to take notice.

It is also important to use multiple formats, such as small-group meetings, large-group meetings, newsletter articles, fliers, bulletin board notices, videotaped messages, E-mail, and a variety of other methods. A combination of visual, reading, and listening vehicles will typically be the most effective.

Regardless of the communication formats used, one or more feedback mechanisms must be in place. In face-to-face meetings, the feedback can be spontaneous and direct. This is always the best form of feedback. However, telephone, facsimile, E-mail, and written feedback can also be valuable.

Implement the Change

Implementing the change is a step that is usually composed of numerous substeps. It involves doing everything necessary to get the change made, which might include some or all of the following substeps:

- Removing structural inhibitors to change
- Enabling employees through training
- Confronting managers and supervisors who continue to resist change
- Planning and generating short-term wins to get the ball rolling
- Eliminating unnecessary interdependencies among functional components of the organization

Incorporate the Change Process

Once an organization has gone through the transformational process of change, both the change itself and the process of change should be incorporated as part of the organization's culture. In other words, two things need to happen. First, the major change that has just occurred must be anchored in the culture so that it becomes the normal way of doing business. Second, the process explained in this section for facilitating change must be institutionalized.

Anchoring the new change in the organizational culture is critical. If this does not occur, the organization will quickly begin to backslide and retrench. The following strategies will help an organization anchor a major change in its culture:

- *Showcase the results.* In the first place, a change must work to be accepted. The projected benefits of making the change should be showcased as soon as they are realized and, of course, the sooner, the better.
- *Communicate constantly.* Don't assume that stakeholders will automatically see, understand, and appreciate the results gained by making the change. Talk about results and their corresponding benefits constantly.
- *Remove resistant employees.* If key personnel are still fighting the change after it has been made and is producing the desired results, give them the "get with it or

get out" option. This approach might seem harsh, but employees at all levels are paid to move an organization forward, not to hold it back.

Institutionalizing the process of change is an important and final element of this step. Change is not something that happens once and then goes away. It is a constant in the lives of every person, in every organization. Consequently, the change facilitation model explained here (Figure 9–5) must become part of normal business operations. Antenna mechanisms continue to anticipate change all the time, forever. They feed what they see into the model, and the organization works its way through each step explained here. The better an organization becomes at doing this, the more successful it will be at competing in the global marketplace.

LESSONS FROM DISTINGUISHED LEADERS

Some of the most distinguished leaders in America's history can be found in fields outside of business and industry. In many cases, their leadership philosophies and methods, though applied in other fields, have direct applications in the world of business. Three distinguished leaders from outside the field of business are profiled in this section. Those chosen for inclusion had to meet the following criteria: 1) recognized widely as a distinguished leader in a specific field, 2) deceased long enough for history to have formed an accurate perspective, and 3) advocated a leadership philosophy that has direct application in today's fast-paced, highly competitive business environment. The leaders chosen for inclusion here are Abraham Lincoln, Harry Truman, and Winston Churchill. Each of these individuals had a leadership style that distinguished him from his contemporaries and set him apart from competitors. No attempt is made to explain every aspect of each leader's philosophy; rather, key aspects have been gleaned from the many for their distinctive application to the contemporary world of global business.

Abraham Lincoln on Leadership

Abraham Lincoln has been called the man who "saved the Union," and deservedly so. He led the United States through four of the most bitter and difficult years in its history, those years when the North and South were embroiled in the American Civil War. In a horrific conflict that pitted brother against brother and friend against friend, Abraham Lincoln prevailed against the forces of secession by clinging steadfastly to his vision of one nation, undivided.

Lincoln was the 16th President of the United States. He gained a national reputation for opposing slavery, which led to his nomination for the presidency by the new Republican Party. He was elected in 1860 shortly before the onset of the Civil War. He led the northern states through the long and deadly years of a war most people expected to last only weeks. In 1863, seeking an issue to rally the North and save the Union, Lincoln issued the Emancipation Proclamation freeing the slaves in the areas under Confederate control. He was reelected in 1864. During its 1864–1865 session, Congress passed the 13th Amendment abolishing slavery.

With slavery ended and the war over, Lincoln looked forward to healing the country's wounds and bringing the North and South back together "with charity for all and malice toward none." He never got the chance. In April 1865, shortly after the war ended, Lincoln was assassinated by John Wilkes Booth while attending a play at Ford's Theater in Washington, D.C. With him died the hopes of a charitable reconciliation with the South. Instead, a period of "reconstruction" ensued that was punitive at best and, in many cases, brutal. In some ways the country is still scarred by the Civil War and the reconstruction period that followed. But the nation survived, prevailed, and remains undivided, thanks to the steadfast determination of Abraham Lincoln to preserve the vision of America's founding fathers.

Lincoln's vision for the United States grew out of the Declaration of Independence. To this self-educated country lawyer, the words of Thomas Jefferson represented what modern business leaders call the corporate vision. The Constitution represented the strategic plan for achieving the vision. Lincoln articulated this vision over and over to anyone who would listen and to many who wouldn't. So determined was Lincoln to preserve what the founding fathers had established that he was not above temporarily rescinding the very rights he was so committed to protecting (e.g., the writ of *habeas corpus*).

Lincoln exemplified many leadership strategies that have direct applications in today's global business environment. According to Donald T. Phillips, some of Lincoln's strongest beliefs about leading were as follows:[32]

- *Get out of the office and circulate among the troops.* Lincoln spent as much time as he could in the field with his commanders and troops. Apparently Lincoln knew about management by walking around more than 100 years before Tom Peters made the concept part of the quality lexicon.

- *Persuade rather than coerce.* Lincoln used amusing stories and country-bumpkin humor to persuade people to his way of thinking. Building consensus rather than just dictating is fundamental to leadership in a quality management setting.

- *Honesty and integrity are the best policies.* Lincoln developed a reputation for telling the truth even when it hurt. Even people who did not like Lincoln, and many did not, usually trusted him. Trust is the cornerstone of quality leadership. People will follow only those they trust.

- *Have the courage to handle unjust criticism.* Lincoln was the most criticized president in our nation's history. Abolitionists criticized him for moving too slowly in freeing the slaves, while proslavery advocates criticized him for moving too fast. By its very nature, leadership involves promoting and facilitating change. Consequently, leaders are subject to the unjust criticism of those who oppose change.

- *Have a vision and continually reaffirm it.* Lincoln's vision for the country could be found in the words of the Declaration of Independence, and he took every opportunity to share his vision. Leaders must show those they would lead what is important, what they believe in, and where they want the organization to go. This is accomplished by articulating a clear, concise vision that is worthy of their commitment.

These are just a few of the leadership strategies exemplified by Abraham Lincoln in a life cut short by an assassin's bullet. There are many more, but it is the last strategy—*have a vision and continually reaffirm it*—that more than any of the others sets Lincoln apart as being worthy of emulation by today's business leaders.

It's well known and documented that during the Civil War Abraham Lincoln, through his speeches, writings, and conversations, "preached a vision" of America that has never been equaled in the course of American history. Lincoln provided exactly what the country needed at that precise moment in time: a clear, concise statement of the direction of the nation and justification for the Union's drastic action in forcing civil war.[33]

Many examples could be cited, but just two will adequately show how Lincoln continually articulated a clear and concise vision around which the North could rally. Speaking about the Civil War in a speech delivered to a special session of Congress on July 4, 1861, Lincoln said,

> This is essentially a people's contest. On the side of the Union, it is a struggle for maintaining in the world that form and substance of government whose leading object is to elevate the condition of men—to lift artificial weights from all shoulders—to clear the paths of laudable pursuit for all—to afford all an unfettered start, and a fair chance, in the race of life.[34]

Perhaps the best example of Lincoln articulating the national vision came during his address in November 1863, during ceremonies dedicating the national cemetery at Gettysburg, Pennsylvania. Standing on the site of a great and terrible battle that left more than 50,000 dead and wounded in its bloody aftermath, Lincoln said:

> Four score and seven years ago our fathers brought forth on this continent a new nation, conceived in Liberty, and dedicated to the proposition that all men are created equal. Now we are engaged in a great civil war, testing whether that nation, or any nation so conceived and so dedicated, can long endure. We are met on a great battlefield of that war. We have come to dedicate a portion of that field, as a final resting place for those who gave their lives that that nation might live. It is altogether fitting and proper that we do this. But in a larger sense, we cannot dedicate—we cannot consecrate—we cannot hallow—this ground. The brave men, living and dead, who struggled here have consecrated it, far above our poor power to add or detract. The world will little note, nor long remember what we say here, but it can never forget what they did here. It is for us the living, rather, to be dedicated here to the unfinished work which they who fought here have thus far so nobly advanced. It is rather for us to be here dedicated to the great task remaining before us—that from these honored dead we take increased devotion to that cause for which they gave the last full measure of devotion—that we here highly resolve that these dead shall not have died in vain—that this nation, under God, shall have a new birth of freedom—and that government of the people, by the people, and for the people, shall not perish from the earth."[35]

Lincoln's message was simple, yet inspiring, as any good vision must be. It told the people of the North, a people worn down and weary of war, that what they were fighting for was worthy of the terrible price being paid, and that preserving the concepts of freedom and liberty for which the founding fathers had fought and died was the issue at stake. Business leaders struggling to keep their organizations focused on the difficult challenge of competing globally can profit from studying the lessons of Abraham Lincoln.

Harry Truman on Leadership

In his early years, Harry Truman's was not a name that came to mind when the topic of conversation was leadership. Physically unimposing, he was small, nonathletic, and wore thick glasses that magnified his eyes. As a boy, Harry Truman might have been called a "nerd" had the word existed at the time. He was inept at farming, had no profession, and showed little promise when World War I intervened. It was while serving in the army as a captain of artillery that Harry Truman first displayed evidence of the leadership ability for which he is now famous. Truman was put in charge of an artillery company made up of some tough characters who had already run off two previous commanders. But Harry Truman, they would soon find, was a different kind of leader. He quickly applied that leadership adage, "When you are put in charge, take charge." Before long, Captain Truman had won both the respect and admiration of his men, some of whom remained lifelong friends.

Truman was America's 33rd President, serving from 1945 to 1953. He succeeded to the office on the death of Franklin D. Roosevelt just as World War II was entering its final and most crucial phase. Later, he was the president who had to deal with the Korean War at a time when the last thing Americans wanted was another conflict. Although Harry Truman exemplified many important leadership strategies, he is best remembered for the following:

1. Making hard decisions and sticking by them
2. Taking responsibility
3. Believing in yourself when no one else does

Making the Hard Decisions

As vice president Harry Truman was not part of President Franklin D. Roosevelt's inner circle. In fact, he knew very little of what was going on as Roosevelt made the momentous decisions of a wartime president. Roosevelt died while resting at his "Little White House" in Warm Springs, Georgia, and Harry Truman was sworn in as president on April 12, 1945. Just hours after being sworn in, Truman was informed by Secretary of War Henry Stimson that the United States had successfully developed the most destructive bomb in the history of the world and, that Truman would have to decide whether to drop it on Japan. After weighing the facts presented to him by his top military advisors, Truman decided to use the atomic bomb to bring the war to a speedy conclusion rather than risk the additional 1 million American casualties projected should the United States have to invade the Japanese homeland. In making this decision, Harry Truman knew that he was sentencing thousands of Japanese people—military and civilian; men, women, and children—to a fiery death. Balancing this terrible knowledge against the almost certain 1 million American deaths, Truman made the tough decision to use the bomb. One could argue that no leader in history has had to make so monumental a choice and with so little time to consider it.

Another difficult decision Truman had to make involved the hugely popular military leader, General Douglas MacArthur. By the time Truman became president, MacArthur was a living legend. One of the most charismatic and highly decorated military leaders in America's history, MacArthur had served with distinction in World War I and again in World War II, where he commanded all military forces in the southwest Pacific theater of war. MacArthur was awarded the Medal of Honor for leading the courageous but doomed garrison of American troops that held out at against the Japanese army and navy at Bataan and Corregidor in the Phillipines long enough to allow the United States to recover from the devastation of Pearl Harbor. When President Roosevelt ordered MacArthur to turn over his command to General Jonathan Wainwright and vacate the Philippines, MacArthur escaped by undertaking a perilous journey through Japanese-held waters to Australia. On arrival he gave the famous speech in which he said, "I shall return," and return he did, freeing the Philippines from years of Japanese rule. In doing so, MacArthur garnered for himself a place not just in history, but also in the hearts of the American people.

After the war MacArthur served as military governor of Japan, helping rebuild the devastated country and draft its new constitution. He was serving in this capacity when the communist North Korean Army crossed the 38th parallel, caught South Korean forces off guard, and nearly overran the entire Korean peninsula. In a short time, the United States was at war again and General MacArthur was in charge. He added to an already brilliant military career by rallying the demoralized American and South Korean armies and pulling off an incredibly daring and risky invasion behind enemy lines at the port of Inchon, Korea. Soon he had the communist North Korean Army on the run. In fact, his forces pushed them back across the 38th parallel and kept pushing them almost to the border of China. This is where his problems with President Truman began.

MacArthur wanted permission to pursue the North Korean army into China, which was giving the Koreans not just support, but sanctuary. Truman wanted to limit the war to Korea and prevent the tragedy of an all-out war with China. While MacArthur and Truman disagreed over the conduct of the war—sometimes publicly, much to the chagrin of Truman—the Chinese decided the matter of a broader war themselves by coming to the aid of the communist North Korean army. Soon the most advanced forces under MacArthur found themselves surrounded by 38 divisions of the army of the People's Republic of China. MacArthur advocated a strong response, including a nuclear attack against China, if necessary. Many Americans agreed with him. Truman refused, and his disagreements with MacArthur became heated and public. Eventually the feud reached the point where Truman thought MacArthur had crossed the line and become insubordinate. Consequently, Truman did the unthinkable; he fired one of the most popular, most highly decorated military heroes in America's history. The response of Americans was immediate and volatile. MacArthur returned home to a hero's welcome, complete with ticker-tape parades and the opportunity to address a joint session of Congress. Truman, on the other hand, found himself isolated and vilified. He quickly became one of the most unpopular presidents ever to hold office. He would later decide to forego running for reelection.

Firing MacArthur made Harry Truman a pariah for several years, but with the passing of time people began to reassess the situation. Before he died, Harry Truman's firing of Douglas MacArthur had come to be viewed as not just the right thing to do, but one of the most politically courageous acts ever undertaken by an American president. It came to be seen for what it really was: the civilian commander in chief asserting his authority under the United States Constitution when a military leader presumed to challenge that authority. It could be said that Harry Truman sacrificed his political career to protect the integrity of the Constitution.

Taking Responsibility

A leader must be willing to share the credit and take the blame. Having the courage to take responsibility for one's decisions and behavior is an absolute necessity for a leader in any field. As president, Harry Truman became famous for his willingness to do what he thought was right and take responsibility for the consequences; a characteristic all too often missing in today's political leaders. He had a sign on his desk that read, "The buck stops here." In other words, everyone else might be able to take the politically expedient way out and "pass the buck" (meaning pass responsibility or blame to someone else), but as president, Truman would not.

He willingly accepted responsibility for the decision to drop the atomic bomb on Japan, and he took responsibility for firing General Douglas MacArthur, paying for the latter with his political career. This is an example that every business leader facing difficult decisions should seek to emulate.

Believing in Yourself

Every leader will at one time or another be faced with a "no confidence" situation. The leader has made a decision or outlined a course of action only to be met with opposition, dissension, and negativity. But leaders who have considered the dissent, weighed the facts, and still think they are right must have the strength of their own convictions to go forward, even if they are alone in believing in themselves.

Having to believe in himself when no one else would was a lifelong burden for Harry Truman. As a child, Truman was not one to garner the confidence of others. Nobody looked at Harry Truman and thought, "Here is a boy who might grow up to be president of the United States." Truman's bookish appearance and plain-spoken personality belied the fact that he was bright, well read, had a will of iron, and possessed great depth of character. Because of this, even when no one else believed in Harry Truman, Harry Truman did. Never was this more apparent as when Truman ran for reelection to the presidency in 1948 against the Republican candidate, Thomas Dewey.

Dewey was everything that Truman wasn't: handsome, well educated, and urbane. Political professionals and media figures gave Truman little or no chance of winning the election. Even Truman's own supporters did not think he could beat Dewey, but Truman did. Harry Truman believed he was the right man for the job, that his ideas for moving the country forward were the best ideas, and that the American people would support him if they heard the truth rather than the biased reporting of the media. In order to get around the media and directly to the American people, Truman undertook a nationwide "whistle-stop" campaign in which his train stopped at every little city, town, or community along the line and Truman spoke to the people from the back of the caboose (a car specially renovated for his use). The political professionals and the media gave Truman no chance. So sure were they of a Truman defeat that one newspaper printed its front-page headline proclaiming Dewey the victor even before the votes had been counted. When the votes were tallied, Truman shocked everyone but himself by winning convincingly. To this day, the most famous photograph of Harry Truman is one showing him holding up that newspaper and pointing to the erroneous headline. When the chips were down, Harry Truman believed in himself. This is an example that leaders in today's global world of business would do well to copy.

Winston Churchill on Leadership

Business leaders in need of a role model who exemplified perseverance in the face of adversity can learn much by studying the life of Sir Winston Churchill, especially his years as Great Britain's prime minister during World War II. Winston Churchill had already amassed a long record of public service to the British Crown when Adolph Hitler first came to power. Seeing the future more accurately than many of his fellow citizens, Churchill began urging the British government to rearm and prepare to defend itself against the rise of Nazism. Unfortunately, few paid Churchill any mind. Consequently, Great Britain was caught unprepared when in September 1939, Hitler's troops quickly overran Poland, a British ally. Two days later, honoring its alliance with Poland, Britain declared war on Nazi Germany and Churchill was elevated to the position of prime minister, a position he held from 1940 until the end of World War II in 1945. France joined Great Britain in declaring war on Germany.

In the early months of the war, nothing went right for Churchill's tiny island nation. Rather than join Britain and France in an alliance against Hitler's Germany, the Soviet Union shocked the world by entering into a pact with the Nazis. In short order, Germany won a series of victories in Norway, Denmark, the Netherlands, Belgium, and Luxembourg. Then, on May 17, 1940, the German army and *Luftwaffe* (air force) swept into France, quickly brushing aside that nation's army and simply bypassing the vaunted Maginot Line erected after World War I to prevent just such an invasion. British troops sent to help stop the Nazi *blitzkrieg* (lightning warfare) were quickly thrown back, along with their French counterparts. By May 26, 1940, more than 200,000 British and 100,000 French troops had been pushed all the way to the coast of France. They stood on the beaches of Dunkirk, their backs to the English Channel, surrounded by superior German forces. Only the incredible resourcefulness of the British people in organizing a cross-channel evacuation involving nearly every craft on the British coast that could float, coupled with bad weather that stymied the German Luftwaffe, saved the British army and the remnants of the French forces.

In June 1940, Italy declared war on Great Britain and before the month was out France had surrendered and agreed to German occupation. Every day brought more and more bad news for Churchill and his beleaguered compatriots. Then, when it seemed to the people of Great Britain that things could not possibly get any worse, they did. On July 10, 1940, Germany began a bombing assault on Great Britain that would continue into the summer of 1941. In addition, Germany declared a complete blockade of this tiny island nation. By the end of 1940, Britain stood practically alone in the world against the Nazi onslaught, cut off from help by German U-boats (submarines) that patrolled the Atlantic ocean, sinking any ship that might carry much-needed supplies for the British.

Churchill's challenge was to hold his nation together—a nation surrounded by hostile forces, bombed mercilessly every night, and starved of badly needed provisions—until he could convince the United States to come to Britain's assistance. It was a challenge that Churchill accepted with courage, optimism, and unshakable resolve. In the words of Maurice Hankey,

> We owed a good deal in those early days to the courage and inspiration of Winston Churchill who, undaunted by difficulties and losses, set an infectious example....His stout attitude did something to hearten his colleagues.[36]

According to Steven Hayward,

> The key to Churchill's courage was his unbounded optimism. Only an optimist can be courageous, because courage depends on hopefulness that dangers and hazards can be overcome by bold and risky acts.[37]

Churchill combined an optimistic spirit and a bulldog tenacity into a "can-do" attitude that was contagious. He convinced his beleaguered compatriots that if they would hang on and do their duty, the forces of good would overcome the forces of evil in due course. Churchill's favorite phrase was "All will come right." He repeated this phrase over and over again in speeches given during the darkest hours of World War II. Churchill never ended a speech on anything but an optimistic note, even during the worst of times. But his messages to the British people and to the world were not pie-in-the-sky cheerleading. He never flinched in telling the British people just how bad things were; after all, they knew. The bombs were falling on them every night. Their sons, husbands, and brothers were coming home from the war wounded or in coffins. The comforts of peacetime no longer existed. Churchill's message was not that "everything is fine." Rather, he told the people that things were bad and would probably get even worse, but in due course the tide would turn. Britain would eventually prevail because it stood, even if at the time it stood alone, for what is right and good and decent in the world. His was a powerful message, and it worked.

Because of Churchill's steadfast courage, optimism, and perseverance in the face of adversity, Great Britain was able to hold on until the Japanese attack on Pearl Harbor (December 7, 1941) brought the United States into the war as an ally. With any less a leader than Sir Winston Churchill at the helm during those dark early years of the war, Great Britain might not have resisted. Had Britain fallen, one can only speculate as to how the world of today might look. Leaders of organizations going through difficult times, organizations that are barely holding on while trying to survive, can benefit from studying the life of Sir Winston Churchill.

SERVANT LEADERSHIP AND STEWARDSHIP

Leadership in business is about ensuring that organizations operate at peak performance levels on a consistent basis. It is about getting the best out of the organization in terms of efficiency and effectiveness. A concept that has this same goal at its core, but challenges the traditional approaches to leadership, is servant leadership and stewardship. Like the traditional approaches to leadership, service leadership and stewardship must pass the tests of competitiveness in the global marketplace. Like any concept that seeks to ensure the optimum performance of an organization, servant leadership and stewardship seeks to do a better job of serving both external and internal customers than do traditional approaches. In other words, the concept differs from conventional leadership ideas not so much in its overall goal, but in its approach to achieving that goal.

Servant Leadership and Stewardship Defined

Advocates of servant leadership believe those who serve best lead best. According to Professor Sean Aland, servant leaders set an example of putting their employees, customers, organization, and community ahead of their own personal needs.[38] Being a servant leader is being a good steward in terms of the organization and its various stakeholders. Employees who see managers being good stewards are more likely to buy into the concept themselves. Advocates of this philosophy believe that employees at all levels should be committed to being good stewards; that is, they should, of their own volition and without coercion, do what is necessary to improve the organization because they feel an intense and personal responsibility for its performance. The servant leadership/stewardship philosophy is an approach to organization and management that seeks to go beyond employee empowerment to employee autonomy, while still meeting all the demands of a competitive marketplace.

Proponents of servant leadership and stewardship claim that this level of commitment cannot be achieved in a traditionally led organization. In an organization in which the philosophy of servant leadership and stewardship is fully accepted and practiced, employees are given the autonomy to think and act for the greater good of the larger group (service and stewardship) rather than just themselves, a team, or some other individual unit. In order to do this, employees must feel that they are in control of their safety and security.

Criticism of Traditional Leadership Approaches

Advocates of servant leadership/stewardship believe that traditional approaches to leadership are inherently limiting and restrictive. According to Peter Block,

> The strength in the concept of leadership is that it connotes initiative and responsibility. It carries the baggage, however, of being inevitably associated with behaviors of control, direction, and knowing what is best for others. The act of leading cultural and organizational change by determining the desired future, defining the path to get there, and knowing what is best for others is incompatible with widely distributing ownership responsibility on an organization. Placing ownership and felt responsibility close to the core work is the fundamental change we seek. To state it bluntly, strong leadership does not have within itself the capability to create the fundamental changes our organizations require. It is not the fault of the people in these positions, it is the fault of the way we have framed the role. Our search for strong leadership in others expresses a desire for others to assume the ownership and responsibility for our group, our organization, our society. The effect is to localize power, purpose, and privilege in the one we call leader.[39]

Block summarizes his criticisms of traditional leadership philosophies as follows:

1. traditional leaders often have more impact in the news than on our lives
2. traditional leaders reinforce the idea that accomplishments can come only from great individual acts
3. traditional leaders cause our attention to be focused on the top
4. traditional leaders who succeed tend to start believing their own press[40]

Potential Benefits of Servant Leadership/Stewardship

Any organization that must compete in the global marketplace faces three challenges:

1. doing more with less
2. learning to adapt to customers and the marketplace
3. creating passion and commitment in employees.

Advocates of servant leadership/stewardship claim that their philosophy is the best way to meet these three challenges.

> Only by rediscovering what it means to commit ourselves to acts of service will these business demands be met. Each of us needs to believe the organization is ours to create if any shift is to take place in how customers are served. Cost control and quality improvement are questions of individual accountability and ownership. Strategies of control and consistency, for all their strengths, tend to be expensive, are slow to react to a marketplace, and drain passion from human beings. With the element of service at its core, stewardship creates a form of governance that offers choice and spirit to core workers so they, in turn, can offer the same to the marketplace. When governance has the texture of service, it calls for a like response from those governed. Leadership-based governance, no matter how loving the leader, swims upstream in giving choice and optimism to those at the bottom.[41]

 SUMMARY

1. Leadership is the ability to inspire people to make a total, willing, and voluntary commitment to accomplishing or exceeding organizational goals. Good leaders overcome resistance to change, broker the needs of constituent groups inside and outside the organization, and establish an ethical framework. Good leaders are committed to both the job to be done and the people who must do it. They are good communicators, and they are persuasive.
2. Leadership for quality is based on the following principles: customer focus, obsession with quality, recognition of the structure of work, freedom through control, unity of purpose, looking for faults in the systems, teamwork, and continuing education and training.
3. Common leadership styles include the following: democratic, participative, goal oriented, and situational. The appropriate leadership style in a total quality setting is participative taken to a higher level. Leadership characteristics that build and maintain followership are a sense of purpose, self-discipline, honesty, credibility, common sense, stamina, steadfastness, and commitment.
4. Leaders can build trust by applying the following strategies: taking the blame but sharing the credit, pitching in and helping, being consistent, and being equitable.

5. To facilitate change in a positive way, leaders must have a clear vision and corresponding goals, exhibit a strong sense of responsibility, be effective communicators, have a high energy level, and have the will to change.

6. When restructuring, organizations should show that they care, let employees vent, communicate, provide outplacement services, be honest and fair, provide for change agents, have a clear vision, offer incentives, and train.

7. The change facilitation model contains the following steps: establish the reality of change, charter the steering committee, develop a change vision, establish antenna mechanisms, communicate, implement, and incorporate change.

8. Servant leadership/stewardship goes beyond employee empowerment to employee autonomy and seeks to create an environment in which employees perform out of a spirit of ownership and commitment.

 KEY TERMS AND CONCEPTS

Alignment	Mission statement
Assessment/action	Needs assessment
Autocratic leadership	Obsession with quality
Celebration	Participative leadership
Coaching	Problem-solving skills
Common sense	Restructuring
Communicable	Self-discipline
Credibility	Sense of purpose
Customer focus	Servant leadership
Deming Chain Reaction	Shared leadership/followership
Democratic leadership	Situational leadership
Desirable	Stamina
Ethics and leadership	Steadfastness
Feasible	Steering committee
Juran Trilogy	Stewardship
Leadership	Stretching tasks
Leadership for quality	Trust/respect
Lose/lose	Win/lose
Maslow's Hierarchy of Needs	Win/win
Misleaders	

 FACTUAL REVIEW QUESTIONS

1. Define the term *leadership*.
2. Explain the concept of a good leader.

3. How can one distinguish between leaders and misleaders?
4. Describe and debunk three common myths about leadership.
5. List and briefly explain the principles of leadership.
6. What is the Juran Trilogy?
7. Describe Maslow's Hierarchy of Needs and how it can be used in a total quality setting.
8. What leadership style is most appropriate in a total quality setting? Why?
9. Explain the leadership characteristics that build and maintain followership.
10. Explain the pitfalls that can undermine followership.
11. Compare and contrast leadership and management.
12. List the strategies leaders can use to play a positive role in facilitating change.
13. Explain what organizations must do to respond effectively to change.
14. What can organizations do to promote a positive response to restructuring?
15. Explain each step in change facilitation.
16. Explain the main leadership lessons that can be learned by studying the lives of Abraham Lincoln, Harry Truman, and Winston Churchill.
17. How does the concept of servant leadership/stewardship differ from traditional leadership philosophies?.

 CRITICAL THINKING ACTIVITY

How Do You Change a Complacent Organization?

Mark Bolten, CEO of Trans-Tech Corporation, is frustrated. Trans-Tech is the market leader in the manufacture of avionics components for commercial airliners and has been for years. But looking to the future, Mark sees problems. Not now, but within 5 years Trans-Tech's situation could change drastically for the worse. Mark sees this and wants to get his company started right away making major but necessary changes.

The challenge he faces is organizational inertia based on complacency. Not even one member of his management team sees the need to change. The collective attitude of Trans-Tech's senior managers seems to be, "We are the market leaders—why rock the boat?"

What is especially frustrating for Mark is the fact that his senior managers are solid, talented professionals. Together with him they built Trans-Tech into a leading company. He can't simply replace them with more future-minded managers. They need to be part of the solution.

Put yourself in Mark's place. What can he do to break through the inertia and get Trans-Tech started on making the necessary changes? How would you handle this dilemma?

DISCUSSION ASSIGNMENT 9–1

Leadership for Quality at Lincoln Electric

Lincoln Electric Company in Cleveland, Ohio, manufactures arc welding equipment. Lincoln has the highest-paid workers in this extremely competitive market, and it is protected neither by patents nor price supports. In spite of this, Lincoln Electric controls 40% of the arc welding market. How is this possible? Lincoln Electric outperforms its competitors in both quality and productivity. By way of comparison, Lincoln Electric has sales in excess of $167,000 per employee while the industry average is $70,000 per employee.

This is accomplished by applying, in the age of high technology, the following leadership principles set forth by James F. Lincoln in 1895:

- Viewing people as the company's most valuable asset
- Practicing Christian ethics
- Making decisions based on principles
- Observing simplicity in all things
- Competing
- Focusing on the customer

Source: Adapted from James A. Belohlav, *Championship Management: An Action Model for High Performance* (Cambridge, MA: Productivity Press, 1990), 27–29.

DISCUSSION ASSIGNMENT 9–2

Leadership for Quality at Kollmorgen Corporation

Kollmorgen Corporation is a diversified technology company that operates in the highly competitive electronics industry. Kollmorgen relies on its employees pulling together to outperform the competition worldwide in the areas of quality and productivity. The focus of leaders at Kollmorgen is on individuals and helping them want to achieve peak performance levels.

Kollmorgen achieves this by applying the following strategies:

- Managers make sure that employees can focus on their work rather than paperwork.
- The distance between people, both physical and psychological, is reduced to promote effective communication.
- Positive personal relationships are stressed.

Source: Adapted from James A. Belohlav, *Championship Management: An Action Model for High Performance* (Cambridge, MA: Productivity Press, 1990), 45–51.

DISCUSSION ASSIGNMENT 9–3

Building Followership at S. C. Johnson and Son, Inc.

"Most people recognize S. C. Johnson and Son, Inc., by its hallmark product line—Johnson Wax. This $2 billion consumer products giant is made up of 46 companies spread across 45 countries. It employs about 11,000 people worldwide. Even though it is quite large, S. C. Johnson never evolved into the kind of entangled morass of people that is literally strangling many of its contemporaries. What magic formula does S. C. Johnson use to create an environment that allows both the company and its people to prosper? S. C. Johnson can be described as a classless society where everyone is allowed to take part in the success of the organization. But probably a better or more accurate way to describe S. C. Johnson is as the Johnson Wax family."

Source: James A. Belohlav, *Championship Management: An Action Model for High Performance* (Cambridge, MA: Productivity Press, 1990), 34.

 ENDNOTES

1. W. Bennis and B. Nanus, *Leaders: The Strategies for Taking Charge* (New York: Harper & Row, 1985), 184–186.

2. Peter F. Drucker, *Managing for the Future: The 1990's and Beyond* (New York: Truman Talley Books/Dutton, 1992), 112–123.

3. Drucker, 120.

4. Drucker, 122.

5. Bennis and Nanus, 222–226.

6. Bennis and Nanus, 189.

7. Peter Scholtes, *The Team Handbook* (Madison, WI: Joiner Associates, 1992), 1-9.

8. Scholtes, 1-11–1-13.

9. Ray Marshall and Marc Tucker, *Thinking for a Living* (New York: Basic Books, 1992).

10. The Juran Trilogy® is a registered trademark of Juran Industries, Inc.

11. Joseph M. Juran, *Juran on Leadership for Quality: An Executive Handbook* (New York: Free Press, 1989), 20–26.

12. Stephen R. Covey, *The 7 Habits of Highly Effective People* (New York: Simon & Schuster, 1989), 204–234.

13. James A. Belohlav, *Championship Management: An Action Model for High Performance* (Cambridge, MA: Productivity Press, 1990) 36.

14. John P. Kotter, "What Leaders Really Do," *Harvard Business Review* 2(3): 103–104.

15. Kotter, 105.

16. W. Bennis, "Leadership in the 21st Century," *Training* (May 1990), 44.

17. Bennis, 44.
18. D. Zielinski and C. Busse, "Quality Efforts Flourish When Trust Replaces Fear and Doubt," *Total Quality* (December 1990), 103.
19. Drucker, 113–117.
20. Drucker, 116–117.
21. D. C. Shanks, "The Role of Leadership in Strategy Development," *Journal of Business Strategy* (January/February 1989), 36.
22. D. Deeprose, "Change: Are You a Driver, a Rider, or a Spoiler?" *Supervisory Management* (February 1990), 3.
23. Deeprose, 3.
24. Deeprose, 3.
25. Eric Harvey and Steve Venture, *Walk Awhile in My Shoes* (Dallas: Performance Publishing, 1998), 1.
26. Harvey and Venture, 1.
27. This section is based on Alastair Rylatt, "Coping with Restructuring," *American Management International* 1 (May 1998), 4–5.
28. John P. Kotter, *Leading Change* (Boston: Harvard Business School Press, 1996), 3.
29. Ibid., 40.
30. Kotter, 51.
31. Kotter, 71.
32. Donald T. Phillips, *Lincoln on Leadership*, (New York: Warner Books, 1992), 13–137.
33. Ibid., 163.
34. Phillips, 163–164.
35. Phillips, 167–168.
36. Steven F. Hayward, *Churchill on Leadership*, (Rocklin, California: Forum, 1998), 115.
37. Ibid.
38. Sean Aland, servant leadership seminar, "(Panama City, Florida), July 9, 2001.
39. Peter Block, *Stewardship*, (San Francisco: Berrett-Koehler Publishing, Inc: 1996), 13.
40. Ibid., 40.
41. Peter Block, 21–22.

CHAPTER TEN

Team Building and Teamwork

"No member of a crew is praised for the rugged individuality of his rowing."
   Ralph Waldo Emerson

MAJOR TOPICS

- Overview of Team Building and Teamwork
- Building Teams and Making Them Work
- Four-Step Approach to Team Building
- Character Traits and Teamwork
- Teams Are Not Bossed—They Are Coached
- Handling Conflict in Teams
- Structural Inhibitors of Teamwork
- Rewarding Team and Individual Performance
- Recognizing Teamwork and Team Players

OVERVIEW OF TEAM BUILDING AND TEAMWORK

Teamwork is a fundamental element of total quality. The reason for this is simple and practical, as can be seen in the following quote:

> Someone may be great at his or her job, maybe even the best there ever was. But what counts at work is the organization's success, not personal success. After all, if your organization fails, it does not matter how great you were; you are just as unemployed as everyone else.[1]

What Is a Team?

A team is a group of people with a common, collective goal. The collective goal aspect of teams is critical. This point is evident in the performance of athletic teams. For example, a basketball team in which one player hogs the ball, plays the role of the prima donna, and pursues his or her own personal goals (a personal high point total, MVP status, publicity, or something else) will rarely win against a team of players, all of whom pull together toward the collective goal of winning the game.

An example of teamwork succeeding over individualism is the No Name Defense of the Miami Dolphins of the National Football League (NFL) during the early years of the franchise. In a sport that promotes and spawns media stars, no member of the Dolphin defense stood out above the others. In fact, although it was arguably the best defense in the NFL at the time, individual members of the team were not well known—hence the nickname No Name Defense.

Taken separately, the individual team members were not particularly impressive when compared with the league's best at their respective positions. Bigger defensive ends, stronger nose tackles, quicker linebackers, and swifter defensive backs played on competing teams. Where the Dolphin's defense excelled was in working together as a team. Its members were a perfect example of the fact that a team's ability is more than just the sum of the ability of each individual member.

Rationale for Teams

In the example of the No Name Defense, the team's ability was more than the sum of the abilities of individual members. This is one of the primary reasons for advocating teamwork. Perry L. Johnson describes the rationale for teams as follows:[2]

- Two or more heads are better than one.
- The whole (the team) is greater than the sum of its parts (individual members).
- People in teams get to know each other, build trust, and as a result, want to help each other.
- Teamwork promotes better communication.

It is well established that teams can outperform individuals, provided they are properly handled. A team is not just a group of people. A group of people becomes a team when the following conditions exist:

QUALITY TIP

Individuals versus Teams

"If we act as a group of individuals, concerned with individual goals, we waste time, motion, effort, and resources—frequently missing the group goal or even making it impossible to achieve. When we act as a team, we get the job done."

Source: Perry L. Johnson, Rob Kantner, and Marcia A. Kikora, *TQM Team-Building and Problem-Solving* (Southfield, MI: Perry Johnson, Inc., 1990), 1-1.

■ Agreement exists as to the team's mission. For a group to be a team and a team to work effectively, all members must understand and agree on the mission.

■ Members adhere to team ground rules. A team must have ground rules that establish the framework within which the team's mission is pursued. A group becomes a team when there is agreement as to mission and adherence to ground rules.

■ Fair distribution of responsibility and authority exists. Teams do not eliminate structure and authority. Football teams have quarterbacks, and baseball teams have captains. However, teams work best when responsibility and authority are shared and team members are treated as equals.

■ People adapt to change. Change is not just inevitable in a total quality setting—it is desirable. Unfortunately, people typically resist change. People in teams should help each other adapt to change in a positive way.

Types of Teams

According to Johnson, Rob Kantner, and Marcia Kikora, most teams can be classified as being one of three types:[3]

■ *Department improvement team.* This is the most common type of team. It is composed of the personnel who make up a specific unit, department, or function within the organization and is sometimes called a *quality circle.*

■ *Process improvement team.* This type of team has improving an entire process as its mission. Consequently, it is made up of people from each individual phase of the process.

■ *Task force.* A task force, sometimes called a *project team,* is a temporary team formed for a specific, well-defined mission. Special project and problem solution teams fall into this category. Task forces are composed of the people who are most likely to be able to carry out their specific missions and are disbanded once the mission in question has been accomplished.

 # QUALITY TIP

Teamwork Skills Must Be Learned

"It is folly to assume that a group assigned to a task will simply find a way of working cooperatively. Learning to work together is as hard as learning to make improvements. The ordinary project team is a complicated creature. Members must work out personal differences, find strengths on which to build, balance commitments to the project against the demands of their everyday jobs, and learn how to improve quality."

Source: Peter R. Scholtes, *The Team Handbook* (Madison, WI: Joiner Associates, 1992), x, 6-1.

Learning to Work Together

A group of people does not a team make. People in a group do not automatically or magically find ways to work together. Concerning the pressures that can keep a group from becoming a team, teamwork expert Peter R. Scholtes says:

> Dealing with internal group needs that arise from these pressures is as important as the group's external task of making improvements. Yet even teams that grasp the importance of improving quality often underestimate the need for developing themselves as teams. . . . A team that fails to build relationships among its members will waste time on struggle for control and endless discussions that lead nowhere.[4]

One of the reasons teams don't always work as well as they might is certain built-in human factors that, unless understood and dealt with, can mitigate against success. Scholtes describes these factors as follows:[5]

■ *Personal identity of team members.*   It is natural for people to wonder where they fit into any organization. This tendency applies regardless of whether the organization is a company or a team within a company. People worry about being an outsider, getting along with other team members, having a voice, and developing mutual trust among team members. The work of the team cannot proceed effectively until team members feel as if they fit in.

■ *Relationships among team members.*   Before people in a group can work together, they have to get to know each other and form relationships. When people know each other and care about each other, they will go to great lengths to support one another. Time spent helping team members get acquainted and establish common ground among themselves is time invested well. This is especially important now that the modern workforce has become so diverse; common ground among team members can no longer be assumed.

■ *Identity within the organization.*   This factor has two aspects. The first has to do with how the team fits into the organization. Is its mission a high priority in the company? Does the team have support at the highest management levels? The second aspect of this factor relates to how membership on a given team will affect relationships with nonteam members. This concern is especially important in the case of task forces and project teams on which team members will want to maintain relationships they have already established with fellow employees who are not on the team. They may be concerned that membership on the team might have a negative impact on their relationships with fellow workers who aren't included.

How to Be a Leader, How to Be a Member

Managers in the modern workplace will be called on to lead teams in some situations and to be members of teams in other cases. Mary Walsh Massop recommends the following strategies for being an effective team leader:[6]

■ *Be clear on the team's mission.*   The team's first organizational meeting should be used to draft a mission statement. This task, although guided by the team leader, should involve all team members. The statement should explain the reasons for the team's existence and define the limits of its authority. The mission statement will be the yardstick against which team performance is measured.

■ *Identify success criteria.*   The team must define what constitutes success and put it in writing. Remember, in a total quality setting, success should be defined in terms of the customer—internal and external. This means that team members must understand the needs and expectations of its customers before identifying success criteria.

■ *Be action centered.*   For every success criterion, the team should develop an action statement or plan that specifies exactly what must be done to satisfy the criterion, the time frame within which it must be done, and by whom.

■ *Establish the ground rules.*   The team needs to decide how it will operate. Team leaders should work to achieve consensus on such issues as these:

> Calling meetings only when necessary
> Making sure all team members come to meetings well briefed and fully prepared
> Determining how much time to allocate for agenda items
> Encouraging participants to be brief
> Determining who will serve as the recorder during meetings
> Deciding how and when to subdivide into subgroups
> Keeping disturbances and interruptions out of meetings
> Finishing an agenda item before moving on to the next item
> Allowing time for informed interaction among members before and/or after meetings

■ *Share information.*   Information should be shared freely inside and outside the team. Communication is a fundamental element of total quality; everyone should know what is going on. Teams do not operate in a vacuum. They are all part of a larger team, the organization. Keeping everyone informed of team activities will eliminate idle, nonproductive, and typically inaccurate speculation.

■ *Cultivate team unity.*   During the 1992 Summer Olympic Games in Barcelona, Spain, the United States fielded what many sports experts and fans thought was the best basketball team ever assembled. Aptly named, the Dream Team won a gold medal with ease. But in spite of the enormous talent of individual team members (Magic Johnson, Michael Jordan, Larry Bird, Moses Malone, Scotty Pippen, and others), the gold medal was really won because these incredibly talented athletes developed a sense of unity, a team identity. As a result, they put aside individual goals, left their egos in the locker room, and played as a team—each supporting the other and contributing to the team's score rather than his own individual statistics. This does not mean they gave up their individual identities completely. On the contrary, the more team unity grew, the more supportive the team became of individual performances.

These strategies will help team leaders build a strong and effective team. A manager who leads one team may be a nonleading member of another team. In such cases, it is important to be a positive, contributing member. Massop recommends the following strategies for being a good team member:[7]

QUALITY TIP

Team Leaders and Members Must Know Fundamentals

"Every day thousands of project teams fail because their leaders aren't sure what they're doing and/or team members aren't sure how to work with others. As a manager you'll frequently find yourself in one position or the other. Without knowing the fundamentals of teamwork, you risk failure before you even get started."

Source: Mary Walsh Massop, "Total Teamwork: How to Be a Member," in *Management for the 90s: A Special Report from* Supervisory Management (Saranac Lake, NY: American Management Association, 1991), 8.

- *Gain entry.*   Get acquainted with all fellow team members as soon as possible. Let them know who you are and what you can contribute, but more important, find out who they are and what they can contribute.
- *Be clear on the team's mission.*   Team members cannot contribute to accomplishing the team's mission unless they know what it is. Learn the mission statement; know the goals; understand tasks, time frames, and assignments; and communicate problems, progress, and other important information freely.
- *Be well prepared and participate.*   Good team members never wing it at team meetings. Before attending a meeting, review previous minutes and your personal notes, familiarize yourself with the agenda, do any reading or research necessary relating to agenda items, and write down any concerns you may have about agenda items. During the meeting, participate. Share what you have to offer, but do so as briefly, accurately, and succinctly as possible.
- *Stay in touch.*   Good team members stay in touch between meetings and communicate frequently. Keep fellow team members up to date on your progress and ask for their help with problems.

Following these strategies helps managers and employees be effective team members. In addition to applying the strategies for team leaders set forth earlier in this section, leaders should teach all team members the strategies for being a good team member.

Team Excellence and Performance

Teamwork is not a magic cure-all. Poorly run teams can do more damage to an organization's performance and corresponding competitiveness than having no teams at all. For this reason, it is critical that excellence in team performance be an overriding goal of the organization.

Quality Tip

Building Team Unity and Identity

"Cultivating an 'esprit de corps' doesn't preclude individuality. In fact, the more the members become a team, the more individuality is permitted. Differences of opinion are respected and people agree to disagree."

Source: Mary Walsh Massop, "Total Teamwork: How to Be a Member," in *Management for the 90s: A Special Report from* Supervisory Management (Saranac Lake, NY: American Management Association, 1991), 8.

Dennis King, personnel manager for Procter & Gamble, recommends the following strategies, which he calls the Ten Team Commandments:[8]

- *Interdependence.* Team members should be mutually dependent on each other for information, resources, task accomplishment, and support. Interdependence is the glue that will hold a team together.

- *Stretching tasks.* Teams need to be challenged. Responding to a challenge as a team builds team spirit and instills pride and unity.

- *Alignment.* An aligned team is one in which all members not only share a common mission but are willing to put aside individualism to accomplish it.

- *Common language.* Teams often consist of members from different departments (manufacturing, marketing, accounting, etc.). These different departments typically have their own indigenous languages that may be foreign to people from other departments. Consequently, it is important for team leaders to ensure that department-specific terms and phrases are used minimally and fully explained in common terms when employed.

- *Trust/respect.* For team members to work well together, there must be trust and respect. Time and effort spent building trust and respect among team members is time invested well.

- *Shared leadership/followership.* Some group members tend to emerge as more vocal, while others sit back and observe. If group leaders are allowed to dominate, the group will not achieve its full potential. A better approach is to draw out the special talents of each individual group member so that leadership and followership are shared.

- *Problem-solving skills.* Time invested in helping group members become better problem solvers is time well spent. Much of the business of groups is problem solving.

- *Confrontation-/conflict-handling skills.* Human conflict is inevitable in a high-pressure, competitive workplace. Even the best teams and closest families have disagreements. Learning to disagree without being disagreeable and to air disagreement openly and frankly—attacking ideas, issues, and proposed solutions without attacking the people proposing them—are critical skills in a total quality environment.

- ■ *Assessment/action.*   Assessment is a matter of asking and answering the question "How are we doing?" The yardsticks for answering this question are the group's mission statement and corresponding action plan. The action plan contains goals, objectives, timetables, and assignments of responsibility. By monitoring these continually, group members can assess how the group is doing.

- ■ *Celebration.*   An effective team reinforces its successes by celebrating them. Recognition of a job well done can motivate team members to work even harder and smarter to achieve the next goal.

BUILDING TEAMS AND MAKING THEM WORK

Part of building a successful team is choosing team members wisely. This section describes strategies for selecting team members, naming officers (or otherwise assigning responsibility), creating a mission statement, and developing collegial relations among team members.

Makeup and Size of Teams

Teams should be composed of those people who are most likely to be able to satisfy the team's mission efficiently and effectively. The appropriate makeup of a team depends in part on the type of team in question (whether it is departmental, process improvement, or task force/project oriented). Departmental teams such as quality circles are made up of the employees of a given department. However, process improvement teams and task forces typically cross departmental lines.

The membership of such teams should be open to any level of employee—management, supervisors, and hourly wage earners. A good rule of thumb is that the greater the mix, the better. According to Johnson:

> The bigger the mix, the better the results. As for the size of the team, we want a group large enough to guarantee a mix of people and opinions, yet small enough to make meetings

QUALITY TIP

Excellence in Team Performance

"Bringing a group together and calling it a team doesn't ensure great performance. A successful team effort doesn't just happen. High-performance teams have certain characteristics in common. Knowing what these are is the first step to . . . achieving team excellence."

Source: Dennis King, "Team Excellence," in *Management for the 90s: A Special Report from* Supervisory Management (Saranac Lake, NY: American Management Association, 1991), 15.

comfortable, productive, and brief. So, what's the right number? No fewer than six, no more than twelve. Eight or nine is just about right.[9]

Choosing Team Members

When putting together a team, the first step is to identify all potential team members. This is important because there will often be more potential team members than the number of members actually needed (maximum of 12 members). After the list has been compiled, volunteers can be solicited and actual team members selected from among those who volunteer. However, care should be taken to ensure a broad mix, as discussed in the previous section. This rule should be adhered to even if there are no volunteers and team members must be drafted. The more likely case is that there will be more volunteers than openings on most teams. Johnson recommends handling this by periodically rotating the membership, preferably on an alternating basis so that the team always includes both experienced and new members.[10]

Responsibilities of Team Leaders

Most teams will have members who are managers, supervisors, and hourly employees. However, it should not be assumed that the highest-level manager will automatically be the team leader. Correspondingly, it should not be assumed that the most junior hourly worker cannot be the team leader. The team should select its own leader.

The first step in selecting a team leader is to develop an understanding of the role and responsibilities of this individual. Beyond the description set forth, Scholtes lists the following as responsibilities of team leaders:[11]

- ■ Serve as the official contact between the team and the rest of the organization.
- ■ Serve as the official record keeper for the team. Records include minutes, correspondence, agendas, and reports. Typically, the team leader will appoint a recorder to take minutes during meetings. However, the team leader is still responsible for distributing and filing minutes.

QUALITY TIP

Role of Team Leaders

"The team leader is the person who manages the team: calling and facilitating meetings, handling or assigning administrative details, orchestrating all team activities, and overseeing preparations for reports and presentations. The team leader should be interested in solving the problems that prompted this project and be reasonably good at working with individuals and groups. Ultimately, it is the leader's responsibility to create and maintain channels that enable team members to do their work."

Source: Peter R. Scholtes, *The Team Handbook* (Madison, WI: Joiner Associates, 1992), 3-8.

- Serve as a full-fledged team member, but with care to avoid dominating team discussions.
- Implement team recommendations that fall within the team leader's realm of authority, and work with upper management to implement those that fall outside it.

Other Team Members

In addition to the team leader, most teams will need a team recorder and a quality advisor. The *recorder* is responsible for taking minutes during team meetings and assisting the team leader with the various other types of correspondence generated by the team. The quality advisor is an important part of the team in a total quality environment. In addition to the description set forth, Scholtes lists the following as responsibilities of the *quality advisor*:[12]

- Focus on team processes rather than products and on *how* decisions are made as opposed to *what* decisions are made.
- Assist the team leader in breaking tasks down into component parts and assigning the parts to team members.
- Help the team leader plan and prepare for meetings.
- Help team members learn to use the scientific approach of collecting data, analyzing data statistically, and drawing conclusions based on the statistical analysis.
- Help team members convert their recommendations into presentations that can be made to upper management.

Creating the Team's Mission Statement

After a team has been formed, a team leader selected, a reporter appointed, and a quality advisor assigned, the team is ready to draft its mission statement. This is a critical step in the life of a team. The mission statement explains the team's reason for being. A mission statement is written in terms that are broad enough to encompass all it will be expected to do but specific enough that progress can be easily measured. The following sample mission statement meets both of these criteria:

QUALITY TIP

The Team's Quality Advisor

"A quality advisor is versed in the tools and concepts of quality improvement, including approaches that help a team have effective, productive meetings. The quality advisor is there to help facilitate the team's work—coaching team members in needed skills and tools—but not to participate directly in the team's activities."

Source: Peter R. Scholtes, *The Team Handbook* (Madison, WI: Joiner Associates, 1992), 3-8.

The purpose of this team is to reduce the time between when an order is taken and when it is filled while simultaneously improving the quality of products shipped.

This statement is broad enough to encompass a wide range of activities and to give team members room within which to operate. The statement does not specify by how much throughput time will be reduced nor by how much quality will be improved. The level of specificity comes in the goals set by the team (e.g., reduce throughput time by 15% within 6 months; improve the customer satisfaction rate to 100% within 6 months). Goals follow the mission statement and explain it more fully in quantifiable terms.

This sample mission statement is written in broad terms, but it is specific enough that team members know they are expected to simultaneously improve both productivity and quality. It also meets one other important criterion: simplicity. Any employee could understand this mission statement. It is brief, to the point, and devoid of all esoteric nonessential verbiage.

Team leaders should keep these criteria in mind when developing mission statements: broadness, appropriate specificity, and simplicity. A good mission statement is a tool for communicating the team's purpose—within the team and throughout the organization—not a device for confusing people or an opportunity to show off literary dexterity.

Developing Collegial Relationships

A team works most effectively when individual team members form positive, mutually supportive peer relationships. These are *collegial relationships*, and they can be the difference between a high-performance team and a mediocre one. Scholtes recommends the following strategies for building collegial relationships among team members:[13]

- Help team members understand the importance of honesty, reliability, and trustworthiness. Team members must trust each other and know that they can count on each other.
- Help team members develop mutual confidence in their work ability.

QUALITY TIP

The Team's Mission Statement

"Why was our team formed? That is the question the team mission statement answers. Whatever kind of team we have, it has an overriding purpose, an objective, a reason for being. Before we can start working toward our team's goal, we must decide what it is."

Source: Perry L. Johnson, Rob Kantner, and Marcia A. Kikora, *TQM Team-Building and Problem-Solving* (Southfield, MI: Perry Johnson, Inc., 1990), 2-5.

- ■ Help team members understand the pressures to which other team members are subjected. It is important for team members to be supportive of peers as they deal with the stresses of the job.

These are the basics. Competence, trust, communication, and mutual support are the foundation on which effective teamwork is built. Any resources devoted to improving these factors is an investment well made.

Promoting Diversity in Teams

The American workplace has undergone an unprecedented transformation. Formerly dominated by young to middle-aged white males, the workplace now draws from a labor pool dominated by women and minorities. This means that today's employees come from a variety of different cultures and backgrounds. Consequently, they are likely to have different values and different outlooks. This situation can be good or bad, depending on how it is handled.

Dealing with diversity in a way that makes it a strength has come to be known as *managing diversity*. According to Sharon Nelton:

> Managing diversity meant, and still means, fostering an environment in which workers of all kinds—men, women, white, disabled, homosexual, straight, elderly—can flourish and, given opportunities to reach their full potential and contribute at the highest level, can give top performance to a company.[14]

When diversity is properly managed, the glass ceiling that exists in some workplaces can be eliminated. The glass ceiling consists of "artificial barriers that prevent qualified minorities and women from advancing into middle and senior levels of management."[15] By working together in well-supervised teams that include women and men, young and old, minorities and nonminorities, employees can learn how to realize the full potential of diversity. Diversity in teamwork can be promoted by applying the following strategies:

- ■ *Continually assess circumstances.*   Is communication among diverse team members positive? Do bias and stereotyping exist among team members? Do minorities and nonminorities with comparable jobs and qualifications earn comparable wages? Factors that might undermine harmonious teamwork should be anticipated, identified, and handled.

- ■ *Give team members opportunities to learn.*   Humans naturally tend to distrust people who are different, whether the differences are attributed to gender, culture, age, race, or any other factor. Just working with people who are different can help overcome this unfortunate but natural human tendency. However, it usually takes more than just working together to break down barriers and turn a diverse group of employees into a mutually supportive, complementary team in which the effectiveness of the whole is greater than the sum of its parts. Education and training aimed at promoting sensitivity to and appreciation of human differences should be provided. Such training should also help team members overcome the stereotypical assumptions that society in general seems to promote.

QUALITY CASE

Diversity and Teamwork at Chase Manhattan

Chase Manhattan is the second largest bank in the United States, behind Bank of America, and one of the biggest in the world. Chase Manhattan was created through the merger of Chemical Bank and the old Chase Manhattan Bank. The bank operates in more than 50 different countries and provides a comprehensive package of services including commercial, consumer, and investment banking; credit cards; automobile loans; on-line banking; money management; merger and acquisition consulting; and investment research. Chase Manhattan must compete in one of the most competitive business sectors in the global marketplace. One of its strategies for doing so is diversity. According to CEO Walter Shipley, you have a much stronger organization when a diverse workforce is made up of people who feel valued. He sees his job, as the team leader, as establishing and maintaining an environment in which people can excel.

Shipley thinks one of the keys to successful teamwork is a willingness on the part of employees to accept diversity in people, ideas, and opinions. When team members value diversity, it can be the team's greatest strength. He sees the value of diversity in the new ideas and different perspectives it can bring to the team. Shipley thinks that managers who create an environment that is open to diversity will gain the benefits of different thinking and different problem-solving approaches. He thinks that this kind of environment liberates the creative energies of people and makes their value in a team multiplicative rather than just additive. To Shipley, the way Chase Manhattan promotes diversity of people, ideas, opinions, and solutions gives the bank a competitive advantage in the marketplace.

Source: Thomas J. Neff and James M. Citrin, *Lessons from the Top,* (New York: Doubleday, 2001), 271–275.

For metal to have optimum strength and resiliency characteristics, it must be alloyed with other metals. High-performance, space-age metals are all mixtures of several different component metals, each different from the others and each possessing its own desirable characteristics. In the modern workplace, this analogy can be applied to the team. Diverse employees, properly managed and trained, can make high-performance, world-class teams.

FOUR-STEP APPROACH TO TEAM BUILDING

Effective team building is a four-step process:

1. Assess
2. Plan

QUALITY TIP

Diversity in the Workplace

"Workplace 2000, a landmark report from the Hudson Institute, a policy-research organization in Indianapolis, startled American business with its relations of how the labor pool was changing dramatically.

"The report showed that white males were already in a minority in the workplace, and it forecast that 85% of the net growth in the U.S. labor force throughout the rest of the century will be workers who are minorities, or are white women, or are immigrants."

Source: Sharon Nelton, "Winning with Diversity," *Nation's Business* (September 1992), 19.

3. Execute
4. Evaluate

To be a little more specific, the team-building process proceeds along the following lines: (a) assess the team's developmental needs (e.g., its strengths and weaknesses), (b) plan team-building activities based on the needs identified, (c) execute the planned team-building activities, and (d) evaluate results. The steps are spelled out further in the next sections.

Assessing Team Needs

If you were the coach of a baseball team about which you knew very little, what is the first thing you would want to do? Most coaches in such situations would begin by assessing the abilities of their new teams. Can we hit? Can we pitch? Can we field? What are our weaknesses? What are our strengths? With these questions answered, the coach will know how best to proceed with team-building activities.

This same approach can be used in the workplace. A mistake commonly made by organizations is beginning team-building activities without first assessing the team's developmental needs. Resources are often limited in organizations. Consequently, it is important to use them as efficiently and effectively as possible. Organizations that begin team-building activities without first assessing strengths and weaknesses run the risk of wasting resources in an attempt to strengthen characteristics that are already strong, while at the same time overlooking characteristics that are weak.

For workplace teams to be successful, they should have at least the following characteristics:

- Clear direction that is understood by all members
- "Team players" on the team
- Fully understood and accepted accountability measures

Figure 10–1 is a tool that can be used for assessing the team-building needs of workplace teams. It consists of criteria arranged in three broad categories: *direction and understanding, characteristics of team members,* and *accountability.* Individual team members record their perceptions of the team's performance and abilities relative to the specific criteria in each category. The highest score possible for each criterion is 6; the lowest score possible, 0. The team score for each criterion is found by adding the scores of individual members for that criterion and dividing by the number of team members. For example, a four-person team might produce the following score on a given criterion:

Instructions

To the left of each item is a blank for recording your perception regarding that item. For each item, record your perception of how well it describes your team. Is the statement *Completely True* (CT), *Somewhat True* (ST), *Somewhat False* (SF), or *Completely False* (CF)? Use the following numbers to record your perception.

$$CT = 6$$
$$ST = 4$$
$$SF = 2$$
$$CF = 0$$

Direction and Understanding

_____ 1. The team has a clearly stated mission.
_____ 2. All team members understand the mission.
_____ 3. All team members understand the scope and boundaries of the team's charter.
_____ 4. The team has a set of broad goals that support its mission.
_____ 5. All team members understand the team's goals.
_____ 6. The team has identified specific activities that must be completed in order to accomplish team goals.
_____ 7. All team members understand the specific activities that must be completed in order to accomplish team goals.
_____ 8. All team members understand projected time frames, schedules, and deadlines relating to specific activities.

Characteristics of Team Members

_____ 9. All team members are open and honest with each other at all times.
_____ 10. All team members trust each other.
_____ 11. All team members put the team's mission and goals ahead of their own personal agendas all of the time.
_____ 12. All team members are comfortable that they can depend on each other.
_____ 13. All team members are enthusiastic about accomplishing the team's mission and goals.
_____ 14. All team members are willing to take responsibility for the team's performance.

Figure 10–1
Team-Building Needs Assessment

_____ 15. All team members are willing to cooperate in order to get the team's mission accomplished.

_____ 16. All team members will take the initiative in moving the team toward its final destination.

_____ 17. All team members are patient with each other.

_____ 18. All team members are resourceful in finding ways to accomplish the team's mission in spite of difficulties.

_____ 19. All team members are punctual when it comes to team meetings, other team activities, and meeting deadlines.

_____ 20. All team members are tolerant and sensitive to the individual differences of team members.

_____ 21. All team members are willing to persevere when team activities become difficult.

_____ 22. The team has a mutually supportive climate.

_____ 23. All team members are comfortable expressing opinions, pointing out problems, and offering constructive criticism.

_____ 24. All team members support team decisions once they are made.

_____ 25. All team members understand how the team fits into the overall organization/big picture.

Accountability

_____ 26. All team members know how team progress/performance will be measured.

_____ 27. All team members understand how team success is defined.

_____ 28. All team members understand how ineffective team members will be dealt with.

_____ 29. All team members understand how team decisions are made.

_____ 30. All team members know their respective responsibilities.

_____ 31. All team members know the responsibilities of all other team members.

_____ 32. All team members understand their authority within the team and that of all other team members.

_____ 33. All team goals have been prioritized.

_____ 34. All specific activities relating to team goals have been assigned appropriately and given projected completion dates.

_____ 35. All team members know what to do when unforeseen inhibitors impede progress.

Figure 10–1   continued

Team member 1:   4
Team member 2:   2
Team member 3:   2
Team member 4:   4

$\qquad$ 12: $12 \div 4 = 3$ (team average score)

The lower the team score for each criterion, the more work needed on that criterion.

Team-building activities should be developed and executed based on what is revealed by this assessment. Activities should be undertaken in reverse order of the assessment scores (e.g., lower scores first, higher scores last). For example, if the team score for criterion 1 (clearly stated mission) is the lowest score for all the criteria, the first team-building activity would be the development of a mission statement.

Planning Team-Building Activities

Team-building activities should be planned around the results of the needs assessment conducted in the previous step. Consider the example of a newly chartered team. The highest score for a given criterion in Figure 10-1 is 6. Consequently, any team average score less than 6 indicates a need for team building relating to the criterion in question. The lower the score, the greater the need.

For example, say the team in question had an average score of 3 for criterion 2 ("All team members understand the mission"). Clearly, part of the process of building this team must be explaining the team's mission more clearly. A team average score of 3 on this issue indicates that some members understand the mission and some don't. This solution might be as simple as the responsible manager or team leader sitting down with the team, describing the mission, and responding to questions from team members.

On the other hand, if the assessment produces a low score for criterion 9 ("All team members are open and honest with each other all the time"), more extensive trust-building activities may be needed. In any case, what is important in this step is to (a) plan team-building activities based on what is learned from the needs assessment and (b) provide team-building activities in the priority indicated by the needs assessment, beginning with the lowest scores.

Executing Team-Building Activities

Team-building activities should be implemented on a just-in-time basis. A mistake made by many organizations that are interested in implementing total quality is rushing into team building. All employees are given teamwork training, even those who are not yet part of a chartered team. Like any kind of training, teamwork training will be forgotten unless it is put to immediate use. Consequently, the best time to provide teamwork training is after a team has been formed and given its charter. In this way, team members will have opportunities to apply what they are learning immediately.

Team building is an ongoing process. The idea is to make a team better and better as time goes by. Consequently, basic teamwork training is provided as soon as a team is chartered. All subsequent team-building activities are based on the results of the needs assessment and planning process.

Evaluating Team-Building Activities

If team-building activities have been effective, weaknesses pointed out by the needs assessment process should have been strengthened. A simple way to evaluate the effectiveness of team-building activities is to readminister the appropriate portion of the needs assessment document. The best approach is to reconstitute the document so that it contains the relevant criteria only. This will focus the attention of team members on the specific targeted areas.

If the evaluation shows that sufficient progress has been made, nothing more is required. If not, additional team-building activities are needed. If a given team-building activity appears to have been ineffective, get the team together and discuss it. Use the feedback from team members to identify weaknesses and problems and use the information to ensure that team-building activities become more effective.

CHARACTER TRAITS AND TEAMWORK

A study conducted by the Institute for Corporate Competitiveness (ICC) suggests that organizations may be "missing the boat" unless character building is part of their team-building program.[16] This conclusion is based on the findings of ICC's year-long study in which 10 focus groups were questioned at length concerning which factors contribute most to helping people work well in teams.

Five of the focus groups consisted of participants who were members of successful teams. The other five groups consisted of participants who had been members of unsuccessful teams. Participants in each group were asked to discuss the factors that contributed most to their team's success or its failure, as applicable. ICC facilitators recorded the consensus responses of each group and summarized them in order of importance (as prioritized by the groups).

There was a strong correlation between the composite data of the successful groups and that of the unsuccessful groups. The factors that both types of groups identified as having the greatest impact on the success of teams in the workplace can best be described as character traits. These traits, in the order of importance established by the focus groups, are shown in Figure 10–2. Participants from successful teams identified the presence of selected character traits in team members as the most important determinant of team success. Participants from unsuccessful teams identified the lack of these same character traits as the most important determinant of team failure.

Study subjects were vocal and clear on the contributions that the character traits shown in Figure 10–2 make to successful teamwork and, conversely, on the harmful impact the lack of these traits can have. The input of study participants is summarized as follows:[17]

■ *Honesty.* To build trust, team members must be honest with each other. Honesty is the cornerstone of trust, and trust is the cornerstone of teamwork. Team members

Figure 10–2
Character Traits That Promote Successful Teamwork
Source: Institute for Corporate Competitiveness, *Final Report: Team Success Study* (Niceville, FL: Author, 1995), 2–4.

- Honesty/integrity
- Selflessness
- Dependability
- Enthusiasm
- Responsibility
- Cooperativeness
- Initiative
- Patience
- Resourcefulness
- Punctuality
- Tolerance/Sensitivity
- Perseverance

depend on each other in ways that affect them every day on a personal level (e.g., job performance, job security, wages, promotions). It is difficult at best for people to place their personal interests even partially in the hands of other people. To do so, there must be a high level of trust. Building that trust begins with honesty.

■ *Selflessness.* This character trait means that people are willing to put the team's interests ahead of their own. A team can move only as fast as its slowest member. This means that there will always be team members who will get out in front of the pack unless they rein themselves in. Being willing to do this is critical to the success of a team. Rather than running ahead of the pack, faster team members should help slower team members improve so that the pace of the overall team is improved.

■ *Dependability.* People who are dependable consistently do what they are supposed to do, when they are supposed to do it, and how they are supposed to do it. Because team members must depend on each other, this character trait is critical. The performance of the team depends on the performance of its members. Consequently, the team relies on its members and the members rely on each other.

■ *Enthusiasm.* The concept of team spirit is real. People who are enthusiastic about their work typically do it better. The good news is that enthusiasm is contagious. The bad news is that despondency and negativism are also contagious. Every team will face roadblocks, unexpected barriers, and difficulties. Enthusiasm can help team members persevere when the road gets rocky. Despondency, on the other hand, promotes a defeatist attitude that says, "When times are tough, give up."

■ *Responsibility.* This character trait means that people know what is expected of them and are willing to be held accountable for doing what is expected. Successful teams and team members take responsibility for their actions, decisions, and performance. Failing teams and team members tend to avoid responsibility. They are prone to blame others as well as each other when things go wrong. Taking responsibility holds teams together during difficult times. Pointing the finger of blame breaks teams up when things aren't going well.

■ *Cooperativeness.* People who work together must cooperate with each other. Think of a team that runs relay races. Much of the success of the team depends on how well they cooperate in passing the baton. If the individual members of the team do not cooperate in this critical phase of the process, the team will lose no matter how fast each individual runs his or her portion of the race. The same concept applies to members of work teams. Moreover, in a work setting the race is run again every day, and the baton is passed more frequently.

■ *Initiative.* Initiative means recognizing what needs to be done and doing it without waiting to be told. This character trait means that team members never say, "That's not my job." In a team setting, whatever is necessary to get the job done is everybody's job.

■ *Patience.* The most difficult challenge facing members of teams is learning to work together. It is easier for people not to get along than it is to get along. In a sense, working in a team is contrary to human nature. People by nature tend to be individualistic. Consequently, there are going to be tensions and troubles as people make the difficult transition from individuals to group members. To stay together long enough to make this transition, team members must be patient with and supportive of each other.

■ *Resourcefulness.*   Resourceful people find ways to get the job done in spite of an apparent lack of resources. A resourceful person will make wise use of materials and ideas that others might overlook or even discard. Having such people on a team will make the team more effective.

■ *Punctuality.*   People who are punctual (on time, on schedule) show respect for their team members and their time. A team cannot function fully without all of its members present; members who are tardy or absent impede the performance of their team. Team members who are punctual can be depended on to be where they are supposed to be when they are supposed to be there.

■ *Tolerance/sensitivity.*   The people in teams can be different in many ways. They might be different in terms of gender, race, or religion. They might have cultural differences and different political outlooks. The modern workplace is an increasingly diverse environment. Diversity can strengthen a team, provided that team members are sensitive to and tolerant of individual differences in people. Insensitive team members who can relate only to people like themselves don't make good team players.

■ *Perseverance.*   To persevere is to persist unrelentingly in trying to accomplish a task in spite of obstacles. People who persevere make valuable team members because they serve as beacons of encouragement when the team becomes engulfed in a fog of difficulty. The natural human tendency is to want to give up when problems arise. However, if just one or two team members are willing to persevere, the others will usually buckle down and keep trying.

TEAMS ARE NOT BOSSED—THEY ARE COACHED

If employees are going to be expected to work together as a team, managers and supervisors have to realize that teams are not bossed—they are coached. Team leaders, regardless of their respective titles (manager, supervisor, etc.) need to understand the difference between bossing and coaching. Bossing, in the traditional sense, involves planning work, giving orders, monitoring programs, and evaluating performance. Bosses approach the job from an "I'm in charge—do as you are told" perspective.

Coaches, on the other hand, are facilitators of team development and continually improved performance. They approach the job from the perspective of leading the team in such a way that it achieves peak performance levels on a consistent basis. This philosophy is translated into everyday behavior as follows:

■ Coaches give their teams a clearly defined charter.

■ Coaches make team development and team building a constant activity.

■ Coaches are mentors.

■ Coaches promote mutual respect between themselves and team members and among team members.

■ Coaches make human diversity within a team a plus.

Clearly Defined Charter

One can imagine a basketball, soccer, or track coach calling her team together and saying, "This year we have one overriding purpose—to win the championship." In one simple statement this coach has clearly and succinctly defined the team's charter. All team members now know that everything they do this season should be directed at winning the championship. The coach didn't say the team would improve its record by 25 points, improve its standing in the league by two places, or make the playoffs, all of which would be worthy missions. This coach has a greater vision—this year the team is going for the championship. Coaches of work teams should be just as specific in explaining the team's mission to team members.

Team Development/Team Building

The most constant presence in an athlete's life is practice. Regardless of the sport, athletic teams practice constantly. During practice, coaches work on developing the skills of individual team members and the team as a whole. Team development and team-building activities are ongoing forever. Coaches of work teams should follow the lead of their athletic counterparts. Developing the skills of individual team members and building the team as a whole should be a normal part of the job—a part that takes place regularly, forever.

Mentoring

Good coaches are mentors. This means they establish a helping, caring, nurturing relationship with team members. Developing the capabilities of team members, improving the contribution individuals make to the team, and helping team members advance their careers are all mentoring activities. According to Gordon F. Shea, effective mentors help team members in the following ways:[18]

- Developing their job-related competence
- Building character
- Teaching them the corporate culture
- Teaching them how to get things done in the organization
- Helping them understand other people and their viewpoints
- Teaching them how to behave in unfamiliar settings or circumstances
- Giving them insight into differences among people
- Helping them develop success-oriented values

Mutual Respect

It is important for team members to respect their coach, for the coach to respect his or her team members, and for team members to respect each other. According to Robert H. Rosen, "Respect is composed of a number of elements that, like a chemical mixture, interact and bond together."[19]

■ *Trust made tangible.*   Trust is built by (a) setting the example, (b) sharing information, (c) explaining personal motives, (d) avoiding both personal criticisms and personal favors, (e) handing out sincere rewards and recognition, and (f) being consistent in disciplining.

■ *Appreciation of people as assets.*   Appreciation for people is shown by (a) respecting their thoughts, feelings, values, and fears; (b) respecting their desire to lead and follow; (c) respecting their individual strengths and differences; (d) respecting their desire to be involved and to participate; (e) respecting their need to be winners; (f) respecting their need to learn, grow, and develop; (g) respecting their need for a safe and healthy workplace that is conducive to peak performance; and (h) respecting their personal and family lives.

■ *Communication that is clear and candid.*   Communication can be made clear and candid if coaches will do the following: (a) open their eyes and ears—observe and listen; (b) say what you want, say what you mean (be tactfully candid); (c) give feedback constantly and encourage team members to follow suit; and (d) face conflict within the team head-on; that is, don't let resentment among team members simmer until it boils over—handle it now.

■ *Ethics that are unequivocal.*   Ethics can be made unequivocal by (a) working with the team to develop a code of ethics; (b) identifying ethical conflicts or potential conflicts as early as possible; (c) rewarding ethical behavior; (d) disciplining unethical behavior, and doing so consistently; and (e) before bringing in new team members, making them aware of the team's code of ethics. In addition to these strategies, the coach should set a consistent example of unequivocal ethical behavior.

■ *Team members are assets.*   Professional athletes in the United States are provided the best medical, health, and fitness services in the world. In addition, they practice and perform in an environment that is as safe and healthy as it can be made. Their coaches insist on these conditions because they understand that the athletes are invaluable resources. Their performance determines the ultimate success or failure of the organization. Coaches of work teams should take a similar approach. To protect their assets (team members), coaches can apply the following strategies: (a) form a partnership between the larger organization and the team to promote healthy habits and control; (b) encourage monitoring and screening of high-risk conditions such as high blood pressure, high cholesterol, and cancer; (c) promote nonsmoking; (d) encourage good nutrition and regular exercise; (e) organize classes, seminars, or workshops on such subjects as HIV/AIDS, prenatal care, stress management, stroke and heart attack prevention, workplace safety, nutrition, and ergonomics; (f) encourage upper management to establish an employee-assistance plan (EAP); and (g) stress important topics such as accident prevention and safe work methods.

Human Diversity

Human diversity is a plus. Sports and the military have typically led American society in the drive for diversity, and both have benefited immensely as a result. To list the contributions to either sports or the military made by people of different genders, races, religions, and so on, would be a task of gargantuan portions. Fortunately, leading organizations in the United States have followed the positive example set by sports and the military. The smart ones have learned that most of the growth in the workplace will be among women, minorities, and immigrants. These people will bring new ideas and varying perspectives, precisely what an organization needs to stay on the razor's edge of competitiveness. However, in spite of steps already taken toward making the American workplace both diverse and harmonious, wise coaches understand that people—consciously and unconsciously—tend to erect barriers between themselves and people who are different from them. This tendency can quickly undermine that trust and cohesiveness on which teamwork is built. To keep this from happening, coaches can do the following:

■ *Conduct a cultural audit.* Identify the demographics, personal characteristics, cultural values, and individual differences among team members.

■ *Identify the specific needs of different groups.* Ask women, ethnic minorities, and older workers to describe the unique inhibitors they face. Make sure all team members understand these barriers, then work together as a team to eliminate, overcome, or accommodate them.

■ *Confront cultural clashes.* Wise coaches meet conflict among team members head-on and immediately. This approach is particularly important when the conflict is based on diversity issues. Conflicts that grow out of religious, cultural, ethnic, age-, and/or gender-related issues are more potentially volatile than everyday disagreements over work-related concerns. Consequently, conflict that is based on or aggravated by human differences should be confronted promptly. Few things will polarize a team faster than diversity-related disagreements that are allowed to fester and grow.

■ *Eliminate institutionalized bias.* A company the workforce of which had historically been predominantly male now has a workforce in which women are the majority. However, the physical facility still has 10 men's rest rooms and only 2 for women. This imbalance is an example of institutionalized bias. Teams may find themselves unintentionally slighting members, simply out of habit or tradition. This is the concept of *discrimination by inertia.* It happens when the demographics of a team changes but its habits, traditions, procedures, and work environment do not.

An effective way to eliminate institutional bias is to circulate a blank notebook and ask team members to record—without attribution—instances and examples of institutional bias. After the initial circulation, repeat the process periodically. The coach can use the input collected to help eliminate institutionalized bias. By collecting input directly from team members and acting on it promptly, coaches can ensure that discrimination by inertia is not creating or perpetuating quiet but debilitating resentment.

HANDLING CONFLICT IN TEAMS

The following conversation took place in a meeting the authors once attended. A CEO had called together employees in his company to deal with issues that were disrupting work. Where the company wanted teamwork, it was getting conflict. Where it wanted mutual cooperation, it was getting bickering. The conversation started something like this:

CEO:	We all work for the same company, don't we?
Employees:	[Nods of agreement.]
CEO:	We all understand that we cannot do well unless the company does well, don't we?
Employees:	[Nods of agreement.]
CEO:	Then we want the company to do well, don't we?
Employees:	[Nods of agreement.]
CEO:	Then we are all going to work together toward the same goal, aren't we?
Employees:	[Silence. All employees stared uncomfortably at the floor.]

This CEO made a common mistake. He thought that employees would automatically work together as a team because this approach is so obviously the right thing to do. In other words, just give employees a chance and explain things to them and they'll work together. James R. Lucas calls this belief a "fatal illusion."[20] He lists the following reasons that people might not work well in teams—reasons that, in turn, can lead to conflict:[21]

■ Because we live in an age of rapid change, employees can conclude that the only person they can trust is themselves.
■ Employees who are too proud or too broken and wounded by others might resist relying on others to help achieve their goals.
■ Employees steeped in the traditions of *rugged individuality* and *competition is king* can feel that cooperation is not fitting for a vigorous person or organization.

In addition to these personal inhibitors of teamwork and promoters of conflict, there is the *example* issue. Organizations that espouse teamwork among employees but clearly are not good team players themselves are setting an example that works against teamwork. Poor teamwork on the part of an organization will manifest itself in either or both of the following ways: (a) treating suppliers poorly while advocating a partnership and (b) treating customers poorly while advocating customer satisfaction.

If organizations want employees to be team players, they must set a positive example of teamwork. If organizations want employees to resolve team conflicts in a positive manner, they must set an example of resolving supplier and customer conflicts in a positive manner.

Resolution Strategies for Team Conflicts

Conflict will occur in even the best teams. Even when all team members agree on a goal, they can still disagree on how best to accomplish it. Lucas recommends the following strategies for preventing and resolving team conflict:[22]

■ Plan and work to establish a culture where individuality and dissent are in balance with teamwork and cooperation.

- Establish clear criteria for deciding when decisions will be made by individuals and when they will be made by teams.
- Don't allow individuals to build personal empires or to use the organization to advance personal agendas.
- Encourage and recognize individual risk-taking behavior that breaks the organization out of unhelpful habits and negative mental frameworks.
- Encourage healthy, productive competition, and discourage unhealthy, counterproductive competition.
- Recognize how difficult it can be to ensure effective cooperation, and spend the energy necessary to get just the right amount of it.
- Value constructive dissent, and encourage it.
- Assign people of widely differing perspectives to every team or problem.
- Reward and recognize both dissent and teamwork when they solve problems.
- Reevaluate the project, problem, or idea when no dissent or doubt is expressed.
- Avoid hiring people who think they don't need help, who don't value cooperation, or who are driven by the desire to be accepted.
- Ingrain into new employees the need for balance between the concepts of cooperation and constructive dissent.
- Provide ways for employees to say what no one wants to hear.
- Realistically and regularly assess the ability and willingness of employees to cooperate effectively.
- Understand that some employees are going to clash, so determine where this is happening and remix rather than waste precious organizational energy trying to get people to like each other.
- Ensure that the organization's value system and reward/recognition systems are geared toward cooperation with constructive dissent rather than dog-eat-dog competition or cooperating at all costs.
- Teach employees how to manage both dissent (not let it get out of hand) and agreement.

Quality Tip

Benefits of Conflict

"Conflict, though stressful, can be more of a benefit than a liability if it is channeled to increase members' understanding of one another, and with these new perspectives, to increase group creativity. As members disagree with one another, more ideas than at first come out of the discussion, increasing team productivity."

Source: Florence M. Stone and Randi T. Sachs, *The High-Value Manager* (New York: AMACOM, American Management Association, 1995), 129.

■ Quickly assess whether conflict is healthy or destructive, and take immediate steps to encourage the former and resolve or eliminate the latter.

STRUCTURAL INHIBITORS OF TEAMWORK

One of the primary and most common reasons that teamwork never gains a foothold in certain organizations is that those organizations fail to remove built-in structural inhibitors. A *structural inhibitor* is an administrative procedure, organizational principle, or cultural element that works against a given change—in this case, the change from individual work to teamwork. Organizations often make the mistake of espousing teamwork without first removing the structural inhibitors that will guarantee its failure.

Consider the following example of how a structural inhibitor can undermine the most well-intended plans. A college decided to implement distance learning as a delivery system for students whose personal circumstances made it difficult to attend in a traditional classroom setting. The college had no provision for paying professors extra for especially large classes. Because the physical size of classrooms limited traditional classes to 40 or fewer students, the issue had never come up. But distance learning students attend class at home. Consequently, an instructor might easily have more than 40 students. In fact, with distance learning, academic integrity was the only factor limiting class size.

When the new distance learning program was first established, the faculty showed no interest and the program fizzled. A survey of the faculty soon revealed several structural inhibitors that had guaranteed failure. The first was a salary schedule that would pay a faculty member the same for 15 students (the minimum class size) as for 100. The second was a decision-making process that gave professors no say in setting maximum enrollment limits. The final structural inhibitor was a policy that did not compensate professors for developing courses and the corresponding courseware.

These inhibitors virtually guaranteed that the college's faculty would turn its back on distance learning. However, when they were removed—with maximum input from faculty members—there was immediate buy-in and the program took off.

In an article for *Quality Digest*, Michael Donovan describes some of the structural inhibitors to effective teamwork that are commonly found in organizations:[23]

■ *Unit structure.* Teams work best in a cross-functional environment as opposed to the traditional functional-unit environment. This allows teams to be process or product oriented. Failing to change the traditional unit structure can inhibit teamwork.

■ *Accountability.* In a traditional organization employees feel accountable to management. This perception can undermine teamwork. Teams work best when they feel accountable to customers. Managers in a team setting should view themselves as internal emissaries for customers.

■ *Unit goals.* Traditional organizations are task oriented, and their unit goals reflect this orientation. A task orientation can undermine teamwork. Teams work best when they focus on overall process effectiveness rather than individual tasks.

- *Responsibility.* In a traditional organization, employees are responsible for their individual performance. This individual orientation can be a powerful inhibitor to teamwork. Teams work best when individual employees are held responsible for the performance of their team.

- *Compensation and recognition.* The two most common stumbling blocks to teamwork are compensation and recognition. Traditional organizations recognize individual achievement and compensate on the basis of either time or individual merit. Teams work best when both team and individual achievement are recognized and when both individual and team performance are compensated.

- *Planning and control.* In a traditional organization managers and supervisors plan and control the work. Teams work best in a setting in which managers and teams work together to plan and control work.

Organizations that are serious about teamwork and need the improved productivity that can result from it must begin by removing structural inhibitors. In addition to the inhibitors described earlier, managers should be diligent in rooting out others that exist in their organizations. An effective way to identify structural inhibitors in an organization is to form focus groups of employees and ask the following question: *"What existing administrative procedures, organizational principles, or cultural factors will keep us from working effectively in teams?"* Employees are closer to the most likely inhibitors on a daily basis and can, therefore, provide invaluable insight in identifying them.

REWARDING TEAM AND INDIVIDUAL PERFORMANCE

An organization's attempts to institutionalize teamwork will fail unless it includes implementation of an appropriate compensation system; in other words, if you want teamwork to work, make it pay. This does not mean that employees are no longer compensated as individuals. Rather, the most successful compensation systems combine both individual and team pay.

This matter is important because few employees work exclusively in teams. A typical employee, even in the most team-oriented organization, spends a percentage of his or her time involved in team participation and a percentage involved in individual activities. Even those who work full-time in teams have individual responsibilities that are carried out on behalf of the team.

Figure 10–3
Typical Team Compensation
Components

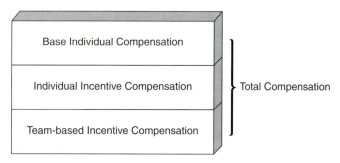

Consequently, the most successful compensation systems have the components shown in Figure 10–3. With such a system, all employees receive their traditional individual base pay. Then there are incentives that allow employees to increase their income by surpassing goals set for their individual performance. Finally, other incentives are based on team performance. In some cases the amount of team compensation awarded to individual team members is based on their individual performance within the team or, in other words, on the contribution they made to the team's performance.

An example of this approach can be found in the world of professional sports. All baseball players in both the National and American Leagues receive a base amount of individual compensation. Most also have a number of incentive clauses in their contracts to promote better individual performance. Team-based incentives are offered if the team wins the World Series or the league championship. When this happens, the players on the team divide the incentive dollars into shares. Every member of the team receives a certain number of shares based on his perceived contribution to the team's success that year.

Figure 10–4 is a model that can be used for establishing a compensation system that reinforces both team and individual performance. The first step in this model involves deciding what performance outcomes will be measured (individual and team outcomes). Step 2 involves how the outcomes will be measured. What types of data will tell the story? How can these data be collected? How frequently will the performance measurements be made? Step 3 involves deciding what types of rewards will be

Figure 10–4
Model for Developing a Team and Individual Compensation System

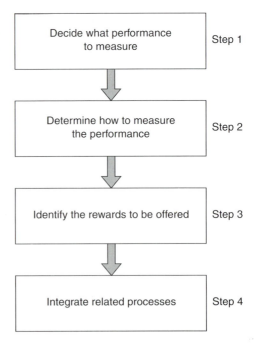

offered (monetary, nonmonetary, or a combination of the two). This is the step in which rewards are organized into levels that correspond to levels of performance so that the reward is in proportion to the performance.

The issue of proportionality is important when designing incentives. If just barely exceeding a performance goal results in the same reward given for substantially exceeding it, just barely is what the organization will get. If exceeding a goal by 10% results in a 10% bonus, then exceeding it by 20% should result in a 20% bonus, and so on. Proportionality and fairness are characteristics that employees scrutinize with care when examining an incentive formula. Any formula that is perceived as unfair or disproportionate will not have the desired result.

The final step in the model in Figure 10–4 involves integrating the compensation system with other performance-related processes. These systems include performance appraisal, the promotion process, and staffing. If teamwork is important, one or more criteria relating to teamwork should be included in the organization's performance-appraisal process.

Correspondingly, the employee's ratings on the teamwork criteria in a performance appraisal should be considered when making promotion decisions. An ineffective team player should not be promoted in an organization that values teamwork. Other employees will know, and teamwork will be undermined. Finally, during the selection process, applicants should be questioned concerning their views on teamwork. It makes no sense for an organization that values teamwork to hire new employees who, during their interview, show no interest in or aptitude for teamwork.

Nonmonetary Rewards

A common mistake made when organizations first attempt to develop incentives is thinking that employees will respond only to dollars in a paycheck. In fact, according to Douglas G. Shaw and Craig Eric Schneier, nonmonetary rewards can be even more effective than actual dollars.[24] Widely used nonmonetary rewards that have proven to be effective include the following: movie tickets, gift certificates, time off, event tickets, free attendance at seminars, getaway weekends for two, airline tickets, and prizes such as electronic and/or household products.

Different people respond to different incentives. Consequently, what will work can be difficult to predict. A good rule of thumb to apply when selecting nonmonetary incentives is "Don't assume—ask." Employees know what appeals to them. Before investing in nonmonetary incentives, organizations should survey their employees. List as many different potential nonmonetary rewards as possible and let employees rate them. In addition, set up the incentive system so that employees, to the extent possible, are able to select the reward that appeals to them. For example, employees who exceed performance goals (team or individual) by 10% should be allowed to select from among several equally valuable rewards on the "10% Menu." Where one employee might enjoy dinner tickets for two, another might be more motivated by tickets to a sporting event. The better an incentive program is able to respond to individual preferences, the better it will work.

RECOGNIZING TEAMWORK AND TEAM PLAYERS

One of the strongest human motivators is *recognition*. People don't just want to be recognized for their contributions; they *need* to be recognized. The military applies this fact very effectively. The entire system of military commendations and decorations (medals) is based on the positive human response to recognition. No amount of pay could compel a young soldier to perform the acts of bravery that are commonplace in the history of the United States military. But the recognition of a grateful nation continues to spur men and women on to incredible acts of valor every time our country is involved in an armed conflict. There is a lesson here for nonmilitary organizations.

The list of methods for recognizing employees goes on *ad infinitum*. There is no end to the ways that the intangible concept of employee appreciation can be expressed. For example, writing in the *Quality Digest*, Bob Nelson recommends the following methods:[25]

■ Write a letter to the employee's family members telling them about the excellent job the employee is doing.

■ Arrange for a senior-level manager to have lunch with the employee.

■ Have the CEO of the organization call the employee personally (or stop by in person) to say, "Thanks for a job well done."

■ Find out what the employee's hobby is and publicly award him or her a gift relating to that hobby.

■ Designate the best parking space in the lot for the "Employee of the Month."

■ Create a "Wall of Fame" to honor outstanding performance.

These examples are provided to trigger ideas but are only a sampling of the many ways that employees can be recognized. Every individual organization should develop its own locally tailored recognition options. When doing so, the following rules of thumb will be helpful:

■ Involve employees in identifying the types of recognition activities to be used. Employees are the best judge of what will motivate them.

■ Change the list of recognition activities periodically. The same activities used over and over for too long will become stale.

■ Have a variety of recognition options for each level of performance. This will allow employees to select the type of reward that appeals to them the most.

 SUMMARY

1. A team is a group of people with a common, collective goal. The rationale for the team approach to work is that "two heads are better than one." A group of people becomes a team when the following conditions exist: there is agreement as to the mission, members adhere to ground rules, there is a fair distribution of responsibility and authority, and people adapt to change.

2. Teams can be classified as department improvement, process improvement, and task force teams. Factors that can mitigate against the success of a team are: per-

sonal identity of team members, relationships among team members, and the team's identity within the organization.

3. To be an effective team leader, one should apply the following strategies: be clear on the team's mission, identify success criteria, be action centered, establish ground rules, share information, and cultivate team unity. One can be a good team member by applying the following strategies: gain entry, be clear on the team's mission, be well prepared and participate, and stay in touch.

4. The Ten Team Commandments are interdependence, stretching tasks, alignment, common language, trust/respect, shared leadership/followership, problem-solving skills, confrontation/conflict-handling skills, assessment/action, and celebration.

5. After a team has been formed, a mission statement should be drafted. A good mission statement summarizes the team's reason for being. It should be broad enough to encompass all the team is expected to do but specific enough to allow for the measurement of progress.

6. Character traits that promote successful teamwork are honesty, selflessness, dependability, enthusiasm, responsibility, cooperativeness, initiative, patience, resourcefulness, punctuality, tolerance/sensitivity, and perseverance.

7. Teams are not bossed—they are coached. Coaches are facilitators and mentors. They promote mutual respect among team members and foster cultural diversity.

8. Employees will not always work well together as a team just because it's the right thing to do. Employees might not be willing to trust their performance, in part, to other employees.

9. Common structural inhibitors in organizations are unit structure, accountability, unit goals, responsibility, compensation, recognition, planning, and control.

10. Team and individual compensation systems can be developed in four steps: (a) decide what performance to measure, (b) determine how to measure the performance, (c) identify the rewards to be offered, and (d) integrate related processes.

KEY TERMS AND CONCEPTS

Alignment	Nonmonetary rewards
Assessment/action	Process improvement team
Celebration	Proportionality
Coaching	Stretching tasks
Collegial relationships	Structural inhibitors
Common language	Task force
Conflict	Team building
Department improvement team	Teamwork
Diversity	Trust building
Interdependence	Trust/respect
Mission statement	Unity of purpose
Needs assessment	

FACTUAL REVIEW QUESTIONS

1. What is a team, and why are teams important?
2. When does a group of people become a team?
3. Explain the strategies for being an effective team leader.
4. List and explain the Ten Team Commandments.
5. What are the characteristics of a good team mission statement?
6. Define the concept of collegial relationships.
7. Describe how to promote diversity in teams.
8. Explain the concept of institutionalized bias.
9. Explain why some employees are not comfortable being team players.
10. List and describe four common structural inhibitors of teamwork in organizations.
11. Explain the concept of nonmonetary rewards.

 CRITICAL THINKING ACTIVITY

Everybody Talks about Teamwork, But Nobody Does It

Teamwork is not working at Southeastern Electric Company (SEC). Juan Morales, quality director for SEC, is growing increasingly frustrated. The company's executive managers all advocate teamwork. A cross-functional team developed a promotional campaign with a teamwork motto, banners, bulletins, and a video. Everybody talks about teamwork but nobody does it. Juan doesn't know what to do. If he asked for your advice, what would you tell him to do? How can he identify the problem, and what steps should he take to resolve it?

DISCUSSION ASSIGNMENT 10–1

It is important to learn how to develop a comprehensive, clearly articulated team charter. Assume that you and your fellow students are a team in an organization (you choose the kind and size of organization). The task is to develop a team charter.

Discussion Questions
Discuss the following questions in class or outside of class with your fellow students:

1. Discuss with your class members how you would go about developing a team charter.

2. Following the discussion, each student should develop a comprehensive team charter. Once this assignment has been completed, exchange charters with your classmates and compare them. Discuss any differences and revise the individual charters to strengthen them, based on the discussion.

ENDNOTES

1. Perry L. Johnson, Rob Kantner, and Marcia A. Kikora, *TQM Team-Building and Problem-Solving* (Southfield, MI: Perry Johnson, Inc., 1990), 1-1.
2. Johnson, Kantner, and Kikora, 1-2.
3. Johnson, Kantner, and Kikora, 1-4, 1-5.
4. Peter R. Scholtes, *The Team Handbook* (Madison, WI: Joiner Associates, 1992), 6-1.
5. Scholtes, 6-1, 6-2.
6. Mary Walsh Massop, "Total Teamwork: How to Be a Member," in *Management for the 90s: A Special Report from* Supervisory Management (Saranac Lake, NY: American Management Association, 1991), 8.
7. Massop, 8.
8. Dennis King, "Team Excellence," in *Management for the 90s: A Special Report from* Supervisory Management (Saranac Lake, NY: American Management Association, 1991), 16–17.
9. Johnson, Kantner, and Kikora, 2-1.
10. Johnson, Kantner, and Kikora, 2-2, 2-3.
11. Scholtes, 3-9, 3-10.
12. Scholtes, 3-8.
13. Scholtes, 3-8.
14. Sharon Nelton, "Winning with Diversity," *Nation's Business* (September 1992), 19.
15. Nelton, 19.
16. Institute for Corporate Competitiveness, *Final Report: Team Success Study* (Niceville, FL: Author, 1995), 2–4.
17. Institute for Corporate Competitiveness, 5, 10.
18. Gordon F. Shea, *Mentoring* (New York: American Management Association, 1994), 49–50.
19. Robert H. Rosen, *The Healthy Company* (New York: Perigee Books, Putnam, 1991), 24.
20. James R. Lucas, *Fatal Illusions* (New York: AMACOM, American Management Association, 1996), 155.
21. Lucas, 159–160.
22. Lucas, 160–161.
23. Michael Donovan, "Maximizing the Bottom Line Impact of Self-Directed Work Teams," *Quality Digest* 16(6) (June 1996), 38.
24. Douglas G. Shaw and Craig Eric Schneier, "Team Measurements and Rewards: How Some Companies Are Getting It Right," *Human Resource Planning* 18(3) (1995), 39.
25. Bob Nelson, "Secrets of Successful Employee Recognition," *Quality Digest* 16(8) (August 1996), 29.

CHAPTER ELEVEN

Communication and Interpersonal Relations

"Communicating is depositing a part of yourself in another person."
   Anonymous

MAJOR TOPICS

- Defining Communication
- Understanding the Role of Communication in Total Quality
- Understanding Communication as a Process
- Recognizing Inhibitors of Communication
- Establishing a Conducive Communication Climate
- Communicating by Listening
- Understanding Nonverbal Communication Factors
- Communicating Verbally
- Communicating in Writing
- Communicating Corrective Feedback
- Improving Communication
- Developing Interpersonal Skills
- Personality and Communication

Of all the many skills needed by managers in a total quality setting, communication skills are the most important. All of the other total quality components and strategies presented in this book depend either directly or indirectly on effective communication. Communication is fundamental to leadership, decision making, problem solving, teamwork, training, employee involvement and empowerment, ethics, and all other areas of concern to modern managers. This chapter seeks to help practicing and prospective managers accomplish the following:

- Improve communication skills
- Improve interpersonal skills

These are critical skills in a total quality setting.

A cover story in *Nation's Business* dealt with 10 threats to success in business.[1] Among the threats explained in this article, two are particularly significant when it comes to implementing total quality: poor interpersonal skills and poor communication skills. These shortcomings were listed as threats 4 and 8, respectively.[2] Effective communication is essential to quality, productivity, and competitiveness.

Defining Communication

Inexperienced managers sometimes make the mistake of confusing telling with communicating and hearing with listening. Then, when a problem arises, they blame everyone else but themselves. The following quotation illustrates the point that telling and hearing don't necessarily result in communication: "I know you believe you understand what you think I said, but I am not sure you realize that what you heard is not what I meant."

What you say is not necessarily what the other person hears, and what the other person hears is not necessarily what you intended to say. What is missing here is understanding. Communication may involve telling, but it is much more than just telling. Communication may involve hearing, but it is much more than just hearing. For the purpose of this book, communication is defined as follows:

> Communication is the transfer of a message (information, idea, emotion, intent, feeling, or something else) that is both received and understood.

A message may be sent by one person and received by another, but until the message is understood by both, no communication has occurred (see Figure 11–1). This applies to all forms of communication, including verbal, nonverbal, and written.

Effective Communication in Total Quality

When the message received is understood, there is communication. However, communication by itself is not necessarily effective communication. Effective communication means that the message is received, understood, and acted on in the desired manner. It is the sender's responsibility to ensure that there is effective communication.

Quality Tip

The Need for Communication and Interpersonal Skills

"To enlist the commitment of others, entrepreneurs must be able to share their vision. They need to sell goods, services, and ideas to customers. They need to hear employees and understand what employees are thinking and feeling. These are just a few of the uses of communication in a successful company."

Source: U.S. Chamber of Commerce, "Ten Key Threats to Success," *Nation's Business*, 80 (June 1992), 18–28.

Figure 11–1
Communication

For example, suppose a manager asks her team members to stay 15 minutes after quitting time for the next week to ensure that an important order goes out on schedule. Each team member receives the message and verifies that he or she understands it. However, two team members decide they are not going to comply. This is an example of ineffective communication. The two nonconforming employees understood the message but decided against complying with it. The manager in this case failed to achieve acceptance of the message.

Effective communication is a higher order of communication. It involves receiving, understanding, and acting on the message. This means that effective communication may require persuasion, motivation, monitoring, and leadership on the part of managers.

Communication Levels

Figure 11–2 shows the various levels at which communication can take place in a company. These levels are explained as follows:

- One-on-one-level communication involves one person communicating with one other person. This might involve face-to-face conversation, a telephone call, or even a simple gesture or facial expression.

- Team- or unit-level communication is communication within a peer group. The primary difference between one-on-one and team-level communication is that, with the latter, all team members are involved in the process at once. A team meeting called to solve a problem or to set goals would be an opportunity for team-level communication.

QUALITY TIP

Listening to Employees and Customers

"To provide support and feedback, the first step must start with listening (really listening to what people are saying) so that you can understand and properly provide what people need. Your subordinates are just as much your customers as are the people who buy your product or service. They are internal customers providing the means to create the product or service for the external customers."

Source: James A. Belohlav, *Championship Management: An Action Model for High Performance* (Cambridge, MA: Productivity Press, 1990), 122.

QUALITY TIP

Effective Communication

"On a peak performing team, everybody knows what is going on. There are no hidden agendas likely to sabotage the team's efforts. Among individuals on the team, effective communication calls for empathy, authenticity, and concreteness."

Source: Charles Garfield, *Peak Performers: The New Heroes of American Business* (New York: Morrow, 1986), 196.

■ Company-level communication is communication among groups. A meeting involving various different departments within a company is an opportunity for company-level communication.

■ Community-level communication occurs among groups inside a company and groups outside the company. Perhaps the most common examples of community-level communication are a company's sales force communicating with clients and the purchasing department communicating with vendors.

UNDERSTANDING THE ROLE OF COMMUNICATION IN TOTAL QUALITY

If total quality is the engine, communication is the oil that keeps it running. Much of what total quality is all about depends on effective communication. Without it, total

QUALITY TIP

Communication at Motorola

"Communication is the third principle. It is seen as a vital process for creating and enhancing meaningful participation. One of the techniques that Motorola uses to set this principle into motion is the 'I recommend' process. Within each work area, there is a readily accessible bulletin board displaying 'I recommend' forms. Individuals wishing to make a recommendation submit one of the forms, signing or not signing their names as they wish. Prompt and accurate answers are required. The name of the person working on the solution and the date are posted, and the answer must be provided within 72 hours. If an answer cannot be given within this time frame, a final date must be provided. All recommendations and answers are posted on the 'I recommend' board."

Source: James A. Belohlav, *Championship Management: An Action Model for High Performance* (Cambridge, MA: Productivity Press, 1990), 66.

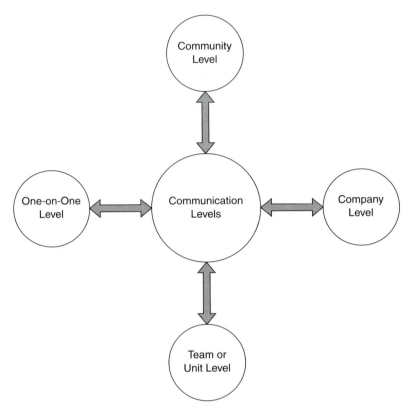

Figure 11–2
Communication Levels

quality breaks down. Some of the key elements of the total quality concept are customer focus (internal and external), total employee involvement and empowerment, leadership, teamwork, decision making, problem prevention, problem solving, and conflict resolution. Each of these elements is dependent on effective communication.

Customer focus means basing decisions and actions on the needs of customers. Determining the needs of customers involves listening, asking, observing, and probing while simultaneously being mindful of not just what is said but how it is said, and additionally, what isn't said. Total employee involvement and empowerment require the establishment of a workplace environment that promotes open, frank communication.

Effective leadership by definition requires effective communication. Effective leaders are those who inspire others to make a full and willing commitment to the organization's goals. To accomplish this, managers must communicate with employees about the organization's goals and how accomplishing these goals will, in turn, help employees accomplish their own personal goals.

Teamwork, by its very nature, depends on communication. To succeed, a team must be comprised of employees who are informed concerning team goals, how they are to be accomplished, who is responsible for what, and how it all fits together. This means that team members must continually communicate among themselves, with managers, and with other teams, and their communication must be effective. This dependence on communication also applies in decision making, problem prevention, problem solving, and conflict resolution. Clearly, communication plays a critical role in total quality. The role it plays is *facilitation*.

UNDERSTANDING COMMUNICATION AS A PROCESS

Communication is a process with several components (see Figure 11–3). These components are the message, the sender, the receiver, and the medium. The *sender* is the originator or source of the message. The *receiver* is the person or group for whom the message is intended. The *message* is the information, idea, feeling, or intent that is to be conveyed, understood, accepted, and acted on. The *medium* is the vehicle used to convey the message.

There are four basic categories of media: verbal, nonverbal, written, and electronic. The *verbal* category includes face-to-face conversation, telephone conversation, speeches, public address announcements, press conferences, and other approaches for conveying the spoken word. The *nonverbal* category includes gestures, facial expressions, and body language. The *written* category includes letters, memorandums, billboards, bulletin boards, manuals, books, and any other method of conveying the written word. The *electronic* category includes the transmission of digital data as well as any other form of electronic transmission that can be converted into a message understood by humans (e.g., the dot and dash impulses of a telegraph).

Technological developments are having a major impact on our ability to convey information. These developments include word processing, satellite communication, computer modems, cordless telephones, cellular telephones, telephone answering machines, facsimile machines, pocket-size dictation machines, and electronic mail. No matter how advanced these communication-enhancing devices have become, as many inhibitors of effective communication as there ever were still exist—maybe even more. Modern managers should be familiar with these inhibitors in order to be able to avoid or overcome them.

Figure 11–3
Communication as a Process

QUALITY TIP

Communication Means Getting Desired Results

"It is not enough that a person be able to get along well with others—that person must be able to get others to do things."

Source: John R. Noe, *People Power* (New York: Berkley, 1988), 142.

RECOGNIZING INHIBITORS OF COMMUNICATION

Managers should be familiar with the various factors that can inhibit effective communication in the workplace. If properly handled, these inhibitors can be overcome or avoided. The most common inhibitors of effective communication are as follows (see Figure 11–4):

- *Differences in meaning.*   Differences in meaning can cause problems in communication. People have different backgrounds, levels of education, and cultures. As a result, words, gestures, and facial expressions can have altogether different meanings to different people. This is why managers should invest time getting to know employees.

- *Lack of trust.*   A lack of trust can inhibit effective communication. If receivers do not trust senders, they may be overly sensitive and guarded. They might concentrate so hard on reading between the lines and looking for hidden agendas that they miss the message. This is why trust building between managers and employees is so important.

- *Information overload.*   Information overload is more of an inhibitor than it has ever been. Computers, modems, satellite communication, facsimile machines, E-mail, and the many other technological devices developed to promote and enhance communications can actually cause communication breakdown.

  Because of advances in communication technology and the rapid and continual proliferation of information, employees and managers often receive more information than they can deal with effectively. This is *information overload*. Managers can guard against information overload by screening, organizing, summarizing, and simplifying the information they convey to employees.

- *Interference.*   Interference is any external distraction that inhibits effective communication. It might be something as simple as background noise or as complex as atmospheric interference with satellite communications. Regardless of its source, interference distorts or completely blocks out the message. This is why managers must be attentive to the environment in which they plan to communicate.

- *Condescending tone.*   A condescending tone when conveying information can inhibit effective communication. People do not like to be talked down to, and they

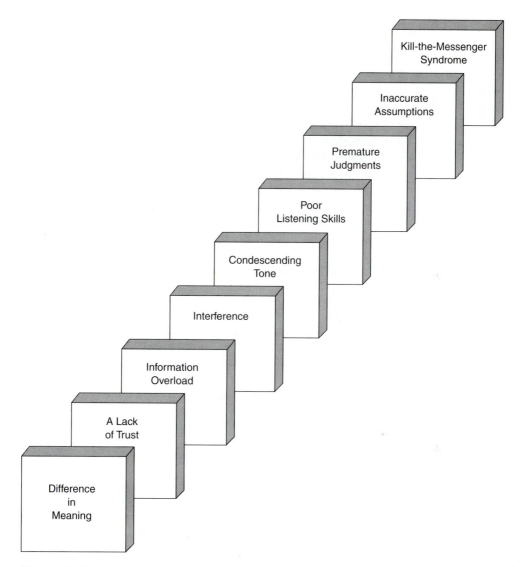

Figure 11–4
Inhibitors of Communication

typically respond to tone of voice as much as or more than the content of the message. It is a mistake to talk down to employees.

■ *Poor listening skills.* Poor listening skills can seriously inhibit effective communication. Problems can result when the sender does not listen to the receiver and vice versa.

■ *Premature judgments.* Premature judgments by either the sender or the receiver can inhibit effective communication. This is primarily because such judgments interfere with listening. When people make a judgment, they are prone to stop listening at that point. One cannot make premature judgments and maintain an open mind. Therefore, it is important for managers to listen nonjudgmentally when talking with employees.

- *Inaccurate assumptions.*   Our perceptions are influenced by our assumptions. Consequently, inaccurate assumptions tend to shut down communication before it has a chance to get started.
- *Kill-the-messenger syndrome.*   In the days when gladiators dueled in Rome's Coliseum, it was common practice to kill the bearer of bad news. A more civilized version of this practice is still very common, particularly in the workplace. Managers who "kill the messenger" when an employee tells the hard truth will eventually hear only what employees think management wants to hear. This dangerous situation quickly leads to uninformed, ill-advised managers.

ESTABLISHING A CONDUCIVE COMMUNICATION CLIMATE

Corwin P. King describes a climate that is conducive to communication as one that "gives people the information they need to do their jobs well and also builds morale and encourages creativity."[3] He describes a bad communication climate as one that "creates doubt and confusion, demotivating people and leading to cynicism."[4] Corwin goes on to describe how managers can guarantee a bad communication climate by doing the following:[5]

- Communicating with peers and employees as little as possible, while, at the same time, being secretive and mysterious
- Being vague and obscure; speaking in generalities
- Communicating with only a select few individuals
- Limiting employee access
- Communicating only when it is personally advantageous to do so
- Ignoring the good ideas of employees

QUALITY TIP

Only the Good News, Please

"Employees shouldn't feel they'll be penalized when the information conveyed isn't to a manager's liking."

Source: Robert A. Glacalone and Stephen B. Kaouse, "Reducing the Need for Defensive Communication," *Management for the 90s: A Special Report from* Supervisory Management (Saranac Lake, NY: American Management Association, 1991), 18.

If these are the steps to creating a bad communication climate, the opposite of each step should create a climate that is conducive to communication. Therefore, managers should take special care to communicate specific, detailed information to as many employees as possible, as frequently as possible, while at the same time soliciting ideas for improvement from employees.

COMMUNICATING BY LISTENING

One of the most important communication skills is listening. It is also the skill people are least likely to have. Are you a good listener? Complete the listening skills assessment in Figure 11–5 to find out.

QUESTION	RESPONSE	
	Yes	No
1. When in a group of people, do you talk more than you listen?		
2. When talking with someone, do you frequently interrupt before he or she completes a statement?		
3. In conversations, do you tune out and think ahead to your response?		
4. In a typical conversation, can you paraphrase what the speaker has said and repeat it?		
5. When talking, do you frequently state your opinion before the other parties have made their case?		
6. Do you continue doing other tasks when someone is talking with you?		
7. Do you ask for clarification when you don't understand what has been said?		
8. Do you frequently tune out and daydream during meetings?		
9. Do you fidget and sneak glances at your watch during conversations?		
10. Do you find yourself finishing statements for people who don't move the conversation along fast enough?		

Figure 11–5
Listening Skills Assessment

If you are a good listener, your answers to the questions in Figure 11–5 will be as follows: 1, no; 2, no; 3, no; 4, yes; 5, no; 6, no; 7, yes; 8, no; 9, no; and 10, no.

Good listeners typically listen more than they talk. It's one of the reasons they are good listeners. Interrupting people before they complete a statement is a sign of impatience and a lack of interest in what is being said. This can have a doubly negative effect on communication. First, interrupting a speaker lessens the listener's chances of properly perceiving what is being said. Second, it can turn the speaker off because it sends the message, "I don't have time to listen to you."

People who tune out and think ahead to their response have more interest in their own message than that of the speaker. Accurate perception is difficult enough when tuned in. It is impossible when tuned out. A tuned-in listener should be able to digest a speaker's message, paraphrase, and repeat it back. Paraphrasing does not mean parroting back the speaker's exact words: it means summarizing the message in your own words. This lets speakers know they have been heard and understood.

Stating an opinion before a speaker has finished his or her message or continuing to do other tasks while someone is speaking both send the same message: "I don't want to hear it." Managers who send this message will get what they ask for. The problem with this is that in a total quality setting, managers *need* to hear it.

Even the best listeners sometimes must ask questions for clarification. When the message isn't clear, managers should ask questions. Not only does questioning improve perception, but it also shows the speaker that the listener is tuned in and wants to understand. There are two ways to handle questions. The first is to wait until the speaker pauses or begins to move on to another train of thought. This is an appropriate time to raise a question; doing so at this point will not cause the speaker to lose his or her original train of thought. However, if it is critical that a point be clarified immediately, stopping the speaker by raising a hand in a gesture that says "Hold on a moment" is acceptable. If it is necessary to stop a speaker in this way, make a mental or written note of where he or she left off, in case a reminder is needed to get the conversation started again.

Daydreaming during meetings and sneaking glances at a watch both make the same statement: "I've got something better to be doing with my time." Time pressures and

QUALITY TIP

Killing a Good Communication Climate

"Trust, confidence, achievement, motivation, and all of the other products of a good communication climate will not last long if you are genuinely devoted to killing them."

Source: Corwin P. King, "Crummy Communication Climate (and How to Create It)," *Management for the 90s: A Special Report from* Supervisory Management (Saranac Lake, NY: American Management Association, 1991), 22.

QUALITY TIP

Becoming a Compulsive Listener

"Today's successful leaders will work diligently to engage others in their cause. Oddly enough, the best way, by far, to engage others is by listening—seriously listening—to them. If talking and giving orders was the administrative model of the last fifty years, listening (to lots of people near the action) is the model of the 1980s and beyond."

Source: Tom Peters, *Handbook for a Management Revolution* (New York: Harper & Row, 1987), 425.

conflicting demands for a manager's time will always exist, but few things a manager does are as important as listening to employees. Managers who find themselves tuning out at meetings and stealing glances at their watches during conversations should give some thought to how they are organizing and managing their time.

Finishing sentences for people who don't move the conversation along fast enough sends the message, "I am in too big a hurry to listen to you." Of course, this might legitimately be the case. When it is, rather than finishing sentences for an employee, managers will get better results by saying, "I don't have time to give you the attention you deserve right now; let's compare schedules and find a time that works for both of us."

What Is Listening?

Hearing is a natural process, but listening is not. A person with highly sensitive hearing abilities can be a poor listener. Conversely, a person with impaired hearing can be an excellent listener. *Hearing* is the physiological decoding of sound waves, but *listening* involves perception. Listening can be defined in numerous different ways. In this book we use the following definition:

> Good listening means receiving the message, correctly decoding it, and accurately perceiving what it means.

Inhibitors of Effective Listening

Effective listening occurs when the receiver accurately perceives the message. Unfortunately, several inhibitors can prevent this from happening. These inhibitors include the following (see Figure 11–6):

- Lack of concentration
- Interruptions
- Preconceived ideas
- Thinking ahead
- Interference

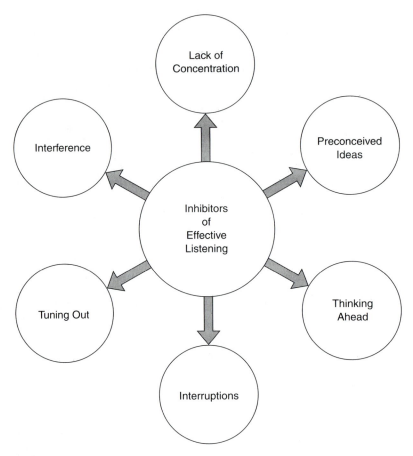

Figure 11–6
Inhibitors of Effective Listening

To perceive a message accurately, listeners must concentrate on what is being said, how it is being said, and in what tone. Part of effective listening is properly reading nonverbal cues (discussed in the next section).

Concentration requires the listener to eliminate as many extraneous distractions as possible and to mentally shut out the rest. J. Lamar Roberts, chief executive officer and president of a successful and dynamic community bank, requires his managers to keep their desks clear of all projects except the one being worked on at the moment. In this way, when an employee enters the office, this project can be easily pushed aside, leaving a clean desk. Work left on the desk can distract the manager and make the employee feel like an intruder. Managers who jump to preconceived notions don't give themselves a chance to listen effectively. Preconceived ideas can cause them to make premature judgments that turn out to be wrong. Even the most experienced managers are better off waiting patiently and listening.

Managers who jump ahead to where they think the conversation is going often get there only to find they are alone. Thinking ahead is typically a response to being hurried, but managers will find that it takes less time to hear an employee out than it does to start over after jumping ahead to the wrong conclusion.

Interruptions not only inhibit effective listening, they can frustrate and confuse the speaker. If clarification is needed during a conversation, it is best to make a mental note and wait for the speaker to reach an interim stopping point. Mental notes are always preferable to written notes. The act of writing notes may distract the speaker or cause the listener to miss a critical point. If managers find it necessary to make written notes, they should keep them short and to a minimum.

Tuning out inhibits effective listening. Some people become skilled at using body language to make it appear they are listening while, in reality, their minds are focused on other areas of concern. Managers should avoid the temptation to engage in such ploys. Skilled speakers may ask the manager to repeat or paraphrase what they just said.

Interference is anything that distracts the listener, thereby impeding either hearing or perception, or both. Background noises, a telephone ringing, and people walking in and out of the office are all examples of interference. Such distractions should be eliminated before beginning a conversation. If they can't be, the conversation should be moved to another location. Figure 11–7 provides a checklist managers can use to improve their listening skills.

Listening Empathically

One way managers can improve their ability to properly perceive the messages they receive is to practice empathic listening.[6] *Empathic listening* means listening with the intent to understand. It does not mean agreeing with what is being said. Rather, it means attempting to fully understand—not just the message but also the messenger—both intellectually and emotionally.[7]

Figure 11–7
Listening Improvement Checklist

☑ Remove all distractions.
☑ Put the speaker at ease.
☑ Look directly at the speaker.
☑ Concentrate on what is being said.
☑ Watch for nonverbal cues.
☑ Make note of the speaker's tone.
☑ Be patient and wait.
☑ Ask clarifying questions.
☑ Paraphrase and repeat.
☑ No matter what is said, control your emotions.

QUALITY TIP

Listening to Understand

"Most people do not listen with the intent to understand; they listen with the intent to reply. They're either speaking or preparing to speak. They're filtering everything through their own paradigms, reading their autobiographies into other people's lives. They're constantly projecting their own home movies onto other people's behavior. They prescribe their own glasses for everyone with whom they interact."

Source: Stephen R. Covey, *The 7 Habits of Highly Effective People* (New York: Simon & Schuster, 1990), 239.

Empathic listening is the highest level of listening, the first three levels being ignoring, pretending, and selective listening. *Ignoring* the message and the messenger can be done overtly or covertly. Overt signs of ignoring include interrupting, fidgeting, and stealing glances at a watch. Covertly ignoring the message means hearing the messenger out but completely disregarding what he or she says. *Pretending* means acting as if you are listening but tuning out. Pretenders are often good at using nonverbal cues such as eye contact, nods, and facial expressions to make people think they are listening.

Selective listening means tuning in only to parts of the message. It's like viewing a painting that is partially covered: the best part may be what you don't see. Empathic listening is attentive listening; it involves focusing intently on the speaker's words while giving equal attention to nonverbal cues, feelings, emotions, intensity, and so on. To perceive some messages fully, it is necessary to do more than just listen to the words.

Listening Responsively

Responsive listening, like the other approaches explained in this chapter, is a way to ensure effective listening.[8] Responsive listening involves seeking to receive and affirm

QUALITY TIP

How to Listen Empathically

"Communications experts estimate, in fact, that only 10% of our communication is represented by the words we say. Another 30% is represented by our sounds, and 60% by our body language. In empathic listening, you listen with your ears, but you also listen with your eyes and with your heart. You listen for behavior. You use your right brain as well as your left. You sense, you intuit, you feel."

Source: Stephen R. Covey, *The 7 Habits of Highly Effective People* (New York: Simon & Schuster, 1990), 241.

Figure 11–8
Descriptors of Responsive
Listening

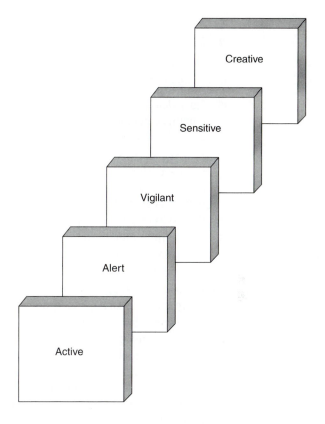

the messenger as well as the message. Figure 11–8 names descriptors associated with responsive listening. To listen responsively, managers and employees must learn to simultaneously use their ears, eyes, brains, and hearts.

Responsive listening can pay substantial dividends for managers in a total quality setting. Personal benefits to managers include the following:

■ Barriers are broken down.

■ Valuable insights are gained.

■ Communication is encouraged.

QUALITY TIP

Responsive Listening

"Responsive listening can present a threat to the insecure because it involves laying aside prejudices and biases.

"I discovered a need I could fill when I quit talking and started listening responsively."

Source: John R. Noe, *People Power* (New York: Berkley, 1988), 87.

■ Misconceptions are cleared up.

■ Learning takes place.

Some pointers that managers can use to become more responsive listeners and to help employees do the same include these:

■ *Slow down.*   Managers who respond, "OK, I can give you 2 minutes," when an employee needs to talk will not be responsive listeners. Hurrying an employee will only inhibit effective communication.

■ *Allocate your listening time.*   There will be times when managers do not have time to listen. This is to be expected. As important as listening is, there is more to management than listening. If an employee wants to talk at a bad time, rather than rush him or begrudgingly stop and listen halfheartedly, set up a time that is more conducive to responsive listening.

■ *Concentrate fully.*   Focus intently and simultaneously on both the verbal and nonverbal aspects of the message. Remove all distractions, physical and mental.

■ *Grant a fair hearing.*   This means laying aside preconceived notions, biases, and prejudices. Managers should be especially attentive to avoiding the tendency to let an employee's lack of speaking ability bias them against the message. A message may have value, even if it isn't delivered well.

■ *Make it easy for the person to talk.*   A friendly smile, a warm handshake, a relaxed attitude, a comfortable chair, and reassuring nonverbal cues will encourage the speaker. These efforts are especially important with employees who are reluctant or who find it difficult to communicate. This strategy applies regardless of the nature of the message. In other words, managers should make it easy for employees to talk even when they bring an unwelcome message.

■ *Understand completely.*   Practice paraphrasing and restating what has been said, and ask questions for clarification to ensure complete understanding. Never allow a conversation to end vaguely. Managers should make sure they understand the message and that employees know they understand before concluding a conversation.

■ *Clarify expectations.*   Before ending a conversation, the wise manager asks, "What would you like me to do?" Even if the message is clear, it's a good idea to ask the question rather than making an assumption. A manager may or may not be able to fulfill the request. However, before concluding the conversation, the manager and employee should both know exactly what is going to happen, and when.

Improving Listening Skills

Most people have room for improvement in their listening skills. Fortunately, listening skills can be enhanced, particularly when there is an awareness of the need to improve. C. Glenn Pearce recommends that managers apply the following strategies for improving their listening skills:[9]

■ *Upgrade your desire to listen.*   Good management requires that managers listen more and talk less. For many managers, this will require a concerted and conscious effort to suppress the natural desire to talk. This is the verbal equivalent of sitting on your hands. Another strategy is to make a conscious effort to learn as much as possible from every conversation. This will force the issue of listening instead of talking.

■ *Ask the right questions.* Two managers can hear the same words but receive different messages. Consequently, it is important to ask questions that will clarify the message. Three types of questions can be helpful in this regard. The first type of question is used to move the speaker on to his or her next point ("I understood your first concern; is there a second?"). The second type of question is used to gain an intermediate summary of the conversation before moving on to a new point ("Can you summarize this concern before we move to the next one?"). The third type of question is used to obtain a summary of the entire conversation ("Before leaving, can you summarize your major concerns for me?").

■ *Judge what is really being said.* This is a matter of listening to more than just words. It also involves observing nonverbal cues, rate of speech, tone of voice, the intensity of the speaker, enthusiasm or a lack of it, and context clues. It involves going beyond what is said to *why* it is being said.

■ *Eliminate listening errors.* Listening errors include failing to concentrate, tuning out, giving in to distractions, and interrupting. To eliminate these errors, keep tabs on how frequently you make each kind. After every conversation for a given week, jot down the listening errors you made during the conversation. This will help focus your efforts on correcting the listening errors you make most frequently.

UNDERSTANDING NONVERBAL COMMUNICATION FACTORS

Communications consultant Roger Ailes explains the importance of nonverbal communication as follows: "You've got just seven seconds to make the right first impression. . . . You broadcast verbal and nonverbal signals that determine how others see you. . . . And whether people realize it or not, they respond immediately to your facial expressions, gestures, stance, and energy, and they instinctively size up your motives and attitudes."[10] Table 11–1 lists the components of nonverbal communication.

QUALITY TIP

Acknowledging Listening Problems

"If managers are to improve their listening skills, they must first acknowledge they have a problem. Only then can they really work on overcoming their listening deficiencies."

Source: C. Glenn Pearce, "Doing Something about Your Listening Ability," *Management for the 90s: A Special Report from* Supervisory Management (Saranac Lake, NY: American Management Association, 1991), 1-6.

Table 11–1
Components of Nonverbal Communication

Body Factors	Voice Factors	Proximity Factors
Posture	Volume	Relative positions
Dress	Pitch	Physical arrangements
Gestures	Tone	Color of the room or environment
Facial expressions	Rate of speech	Fixtures
Body poses		

Body Factors

Posture, body poses, facial expressions, gestures, and dress can convey a message. Even such extras as makeup or the lack of it, well-groomed or unkempt hair, and shined or scruffy shoes can convey a message. Managers should be attentive to these body factors and how they add to or distract from the verbal message.

One of the keys to understanding nonverbal cues lies in the concept of *congruence*. Are the spoken message and the nonverbal message congruent? They should be. *Incongruence* is when words say one thing but the nonverbal cues say another. When the verbal and nonverbal aspects of the message are not congruent, managers should take the time to dig a little deeper. An effective way to deal with incongruence is to confront it gently, with a simple statement such as, "Mary, your words say one thing, but your nonverbal cues say something else."

Voice Factors

Voice factors are an important part of nonverbal communication. In addition to listening to the words, managers should listen for such factors as volume, tone, pitch of voice, and rate of speech. These factors can indicate anger, fear, impatience, unsureness, interest, acceptance, confidence, and a variety of other messages. As with body factors, it is important to look for congruence. It is also advisable to look for groups of nonverbal cues.

Managers can mislead themselves by attaching too much meaning to isolated nonverbal cues. A single cue taken out of context might have little importance, but as part of a group of cues, it can take on significant meaning. For example, if you look through the office window and see a man leaning over a desk pounding his fist on it, it would be tempting to interpret this as a gesture of anger. But what kind of look does he have on his face? Is his facial expression congruent with desk-pounding anger? Or could he simply be trying to knock loose a desk drawer that has become stuck? If he is pounding on the desk with a frown on his face and yelling in an agitated tone, your assumption of anger is probably correct. He might just be angry because his desk drawer is stuck, but he would still be angry.

Proximity Factors

Proximity involves factors ranging from where you position yourself when talking with an employee, to how your office is arranged, to the color of the walls, to the types of fixtures and decorations. A manager who sits next to an employee conveys a different message than one who sits across a desk from the employee. A manager who goes to the trouble to make his or her office a comfortable place to visit is sending a message that invites communication. A manager who maintains a stark, impersonal office sends the opposite message.

To send the nonverbal message that employees are welcome to stop and talk, try using the following strategies:

- Have comfortable chairs available for visitors.
- Arrange chairs so you can sit beside visitors rather than behind your desk.
- Choose soft, soothing colors rather than harsh, stark, or overly bright or busy colors.
- If possible, have refreshments such as coffee, soda, and snacks available for visitors to your office.

COMMUNICATING VERBALLY

Verbal communication ranks close to listening in its importance in a total quality setting. Managers can improve their verbal communication skills by being attentive to the factors shown in Figure 11–9. These factors are explained in the following paragraphs.

Figure 11–9
Improvement of Verbal
Communication Skills

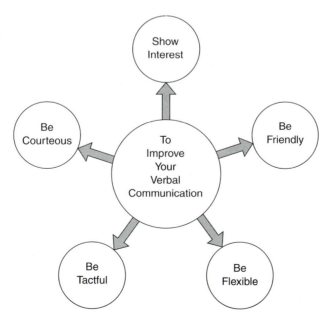

■ *Show interest.* When speaking with employees, show an interest in the topic. Show that you are sincerely interested in communicating your message to them. Also show interest in the receivers of the message. Look them in the eye and, when in a group, spread your eye contact evenly among all receivers.

■ *Be friendly.* A positive, friendly attitude will enhance verbal communication. A caustic; superior, condescending, or disinterested attitude will shut off communication; so will an argumentative attitude. Be patient, be friendly, and *smile*.

■ *Be flexible.* Flexibility can enhance verbal communication. For example, if a manager calls her team together to explain a new company policy but finds they are uniformly focused on a problem that is disrupting their work schedule, she must be flexible enough to put her message aside for the moment and deal with the problem. Until the employees work through what is on their minds, they will not be good listeners.

■ *Be tactful.* Tact is an important ingredient in verbal communication, particularly when delivering a sensitive or potentially controversial message. Tact has been called the ability to "hammer in the nail without breaking the board." The key to tactful verbal communication lies in thinking before talking.

■ *Be courteous.* Courtesy promotes effective verbal communication. Being courteous means showing appropriate concern for the receiver's needs. For example, calling a meeting 10 minutes before quitting time is discourteous and will inhibit communication. Courtesy also means not monopolizing. When communicating verbally, give the receiver ample opportunities to ask questions, seek clarification, and state his or her point of view.

Asking Questions Effectively

In addition to applying the strategies just explained, managers should learn to be skilled questioners. Knowing how and when to question is an important verbal communication skill. It is how managers get at what employees really think and feel. Some general rules of questioning that professional counselors use to draw out their clients' feelings and thoughts can be listed. Modern managers can apply these same rules to enhance their verbal communication with employees (Figure 11–10).

■ *Drop your defenses.* Human interaction is emotional interaction. There is no such thing as fully objective discourse between people. All people have their public and private faces; rarely does what is said completely match what is felt. People learn early in life to build walls and put up defenses. To communicate effectively, it is necessary to get behind the walls and break through the defenses. A strategy counselors use for this is dropping their defenses first. When employees see you open up, they will be more likely to follow suit and respond more openly to your questions.

■ *State your purpose.* The silent question people often ask themselves when asked a question is "Why is he asking that—what does she really want?" You will learn more from your questions by stating your purpose at the outset. This will allow the receiver to focus on your question rather than worrying about why you are asking it.

Figure 11–10
Asking Questions Effectively

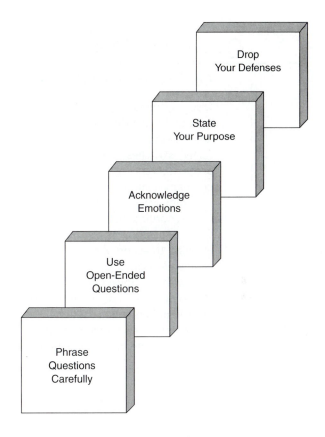

- *Acknowledge emotions.* Avoid what counselors call the "elephant in the living room" syndrome when questioning employees. Human emotions can be difficult to deal with. As a result, some people respond by ignoring them. This is like walking around an elephant in the living room and pretending you don't see it. Ignoring the emotions of people you question may cause them to close up. If a person shows anger, you might respond by saying, "I can see I've made you angry" or "You seem to feel strongly about this." Such nonjudgmental acknowledgments will often draw a person out.

- *Use open-ended questions and phrase questions carefully.* To learn the most from your questions, make them open ended. This allows the person being questioned to do most of the talking, and, in turn, you to do most of the listening. Counselors feel they learn more when listening than when talking. Closed-ended questions force restricted or limited responses. For example, the question "Can we meet our deadline?" will probably elicit a yes-or-no response. However, the question "What do you think about this deadline?" gives the responder room to offer opinions and other potentially useful information.

QUALITY CASE

Communication Is Key at Cisco Systems

Cisco Systems, Inc., is an "enabling" company; that is to say, it produces products that make the Internet work or make it work better. Its principle product lines are network routers, which tell information where to go, and local area network (LAN) switches. Cisco also produces network-management software and dial-up access servers. A $12 billion company, Cisco is the world leader in its field. The company maintains strategic partnerships with other market-leading companies including Microsoft, IBM, and Sun Microsystems.

According to CEO John Chambers, communication is one of the keys to Cisco's success. He claims that every employee knows exactly what the company is trying to accomplish because communication with employees is continuous. It is also electronic. Cisco is a technology company; consequently, all communication with employees—from the time they apply for a job to the time they leave the company—is done through the Internet. Company officials also communicate with suppliers electronically. This commitment to technology, coupled with a commitment to communication, has helped propel Cisco Systems, Inc., to the forefront of the Internet-enabling market.

Source: Thomas J. Neff and James M. Citrin, *Lessons from the Top,* (New York: Doubleday, 2001), 79–83.

Practice using these questioning techniques at home, at work, and even in social settings. It will take practice to internalize them to the point that they become natural. However, with effort managers can become skilled questioners and, as a result, more effective communicators.

COMMUNICATING IN WRITING

The ability to communicate effectively in writing is important for managers in a total quality setting. The types of writing required of managers can be mastered, like any other skill, with the appropriate mix of coaching, practice and genuine effort to improve. This section provides the coaching. Managers must provide the practice and effort for themselves.

Helpful Rules

Several rules of thumb that can enhance the effectiveness of your written communication are shown in Figure 11–11. These rules of thumb are explained in the following sections.

Figure 11–11
Strategies for Improved Written
Communication

> - Plan before you write.
> - Be brief.
> - Be direct.
> - Be accurate.
> - Practice self-editing.

Plan before You Write

One reason some people have trouble writing effectively is that they start writing before deciding what they want to say. This is like getting in a car and driving before deciding where to go. The route is sure to be confusing, as will the message if you write without planning first.

Planning a memorandum or letter is a simple process. It is a matter of deciding to whom you are writing, why, and what you want to say. Figure 11–12 is a planning sheet that will help you plan before you write.

Taking the time to complete such a planning sheet can ensure that the intended message is communicated. After you have used planning sheets for a while, you will be able to complete this step mentally without having to actually fill them out.

Be Brief

One of the negative aspects of modern technology is the potential for information overload. This compounds another negative aspect—the tendency toward shortened attention spans. Modern computer and telecommunication technology have conditioned us to expect instant information with little or no effort on our part.

Written documents can run counter to our expectation of instant information with no effort. Reading takes time and work. Keep this in mind in writing. Be brief. In as few words as possible, explain your purpose, state your points, and tell recipients what you want them to do.

Be Direct

Directness is an extension of brevity. It means getting to the point without "beating around the bush." This is especially important when the message is one readers will not particularly like. No purpose is served by obscuring the message. Come right to the point and state it completely and accurately.

Be Accurate

Accuracy is important in written communication. Be exact. Avoid vague phrases and terms such as *some time ago*, *approximately*, and *as soon as possible*. Take the time to identify specific dates, numbers, quantities, and so on. Then double-check to make sure they are accurate.

Practice Self-Editing

A one-draft writer is rare. But people who send their first drafts are very common. Sending the first draft can cause you to overlook errors that are embarrassing and that

- I am writing to _____

  _____

  _____

  _____

- My purpose in writing is to _____

  _____

  _____

  _____

- I want to make the following points _____

  A. _____

  B. _____

  C. _____

  D. _____

  E. _____

- I want the recipient to do the following: _____

  _____

  _____

  _____

Figure 11–12
Planning Form for Letters and Memorandums

obscure or confuse the message. In your first draft, concentrate on *what* you are saying. In the second draft, concentrate on *how* you say it. These are two different processes that should not be mixed. Even professional writers find it difficult to edit for content and edit for grammar, sentence structure, and spelling simultaneously.

Writing Better Reports

Writing reports is an important task that goes with being a manager. To put it simply, managers must be able to write good reports. Robert Maidment recommends the following steps for writing better reports:[11]

1. *Define the problem.*   Before beginning to write a report, managers should finish the following sentence: "The purpose of this report is. . . . " The problem statement for a report should be brief, to the point, descriptive, and accurate.

2. *Develop a workplan.*   A *workplan* is a list of the tasks to be completed and a projected date of completion for each task. Having one will help keep the development of the report on schedule. An effective way to structure a workplan is to develop a table of contents for the report and then list the tasks that must be accomplished under each major heading and assign a projected completion date to each task (see Figure 11–13).

3. *Gather relevant data.*   This step involves collecting all data pertaining to the problem in question. This is the research step. It might involve searching through files, reading other reports, interviewing employees and/or customers, running tests, or taking any other action that will yield useful data.

4. *Process findings.*   Information is simply data that have been converted into a useful form. Processing findings means converting the raw data collected in the previous step into information on which recommendations can be based. This involves both analysis and synthesis.

5. *Develop conclusions.*   Having identified a problem, gathered all pertinent data relating to the problem, and analyzed and synthesized that data, you next draw conclusions. The conclusions explain, based on data collected, analyzed, and synthesized, what caused the problem. Conclusions should be based on hard facts, stated objectively, and free of personal opinions or editorializing.

6. *Make recommendations.*   This section contains the writer's recommendations for solving the problem. They should grow logically out of the conclusions. Arrange recommendations sequentially in order of priority. Whenever possible and appropriate, give options. Recommendations should be specific and detailed, indicating time frames, the people responsible for carrying them out, costs, and any other pertinent information.

 # QUALITY TIP

Report Writing: An Essential Management Skill

"For managers, report writing is an essential skill. It is the way they make their ideas a reality—and gain organizational visibility. Though managers initially ascend the organizational ladder through draft, design, or divine intervention, how high they climb often depends on the ability to write succinct and persuasive reports."

Source: Robert Maidment, "Seven Steps to Better Reports," *Management for the 90s: A Special Report from* Supervisory Management (Saranac Lake, NY: American Management Association, 1991), 11.

Figure 11–13
Report Workplan

Title of Report: Failure Rate of TP-10 Gear	
Major Report Headings	**Completion Date**
• Problem Statement	
1. Collect a broad base of input	January 15
2. Write problem statement	January 15
• Background	
1. Research all pertinent files	January 17
2. Analyze and synthesize	January 18
3. Develop a background summary	January 18
• Conclusions	
1. Develop conclusions	January 20
2. Summarize conclusions	January 21
• Recommended Solution	
1. Select the best solution	January 22
2. Pilot-test the solution chosen	January 23
3. Summarize the results and the rationale for recommending the solution	January 24

COMMUNICATING CORRECTIVE FEEDBACK

In dealing with employees, managers often must give corrective feedback. To be effective, corrective feedback must be communicated properly. Robert Luke offers the following guidelines for enhancing the effectiveness of corrective feedback:[12]

■ *Be positive.*   For feedback to be corrective, the employee must accept and act on it. This is more likely to happen if it is delivered in a positive "can-do" manner.

■ *Be prepared.*   Focus all feedback specifically on the behavior. Do not get into personality traits. Give specific examples of the behavior you would like to see corrected.

■ *Be realistic.*   Make sure the behaviors you want to change are controlled by the employee. Don't expect an employee to correct a behavior over which he or she has no control.

■ *Don't be completely negative.*   Find something positive to say. Give the employee the necessary corrective feedback, but don't focus wholly on the negative.

Luke also recommends the following two approaches managers can use when giving corrective feedback to employees:

■ *Tell-ask-listen.*   With this approach, the manager tells the employee about the behavior, asks for his or her input, and listens to that input. According to Luke, "The session will have been a success if the employee leaves feeling the issues are important, is appreciative of your input, and is committed to correcting the problem."[13]

■ *Listen–ask–tell.* With this approach, the manager listens first. It may be necessary to ask an open-ended, general question such as, "How are things going with your job?" to get the ball rolling. When the employee starts talking, listen. If the employee talks adequately about the area in which corrective action is needed, reinforce his or her comments. Ask what he or she thinks can and should be done to improve. If the employee does not appear to be fully aware of the problem, move to the tell step. According to Luke, "By communicating, you resolve more problems and eliminate the possibility of hard feelings and mixed messages."[14]

IMPROVING COMMUNICATION

Kim McKinnon, manager of personnel development at the Santa Barbara Research Center, recommends the following strategies for improving communication skills:[15]

■ *Keep up to date.* Managers should make an effort to stay up to date with new information relating to the workplace. You cannot communicate what you don't know. Accurate information is essential in a total quality setting.

■ *Prioritize and determine time constraints.* Communicating does not mean simply passing on everything you learn to your employees. Such an approach might overload them and, as a result, inhibit communication. Analyze your information and decide what your employees need to have. Then prioritize it, from "urgent" to "when time permits," and share the information accordingly.

■ *Decide whom to inform.* After you have prioritized your information, decide who needs to have it. Employees have enough to keep up with without receiving information they don't need. Employees need information that will help them do a better job or that will help them help fellow employees do a better job.

■ *Determine how to communicate.* There are a variety of different ways to communicate: orally, in writing, one on one, in groups, and others. A combination of methods will probably be more effective than any one method by itself. (The next section in this chapter deals with this issue in more depth.)

■ *Communicate and follow up.* Don't just tell employees what you want them to know or write them a memorandum. Follow up. Ask questions to determine whether they have really gotten the message. Encourage employees to ask you questions for clarification. Agree on the next steps (what they should do with the information).

■ *Check understanding and obtain feedback.* Check to see that your communication was understood. Is the employee undertaking the correct next steps? Obtain feedback from employees to ensure that their understanding has not changed and that progress is being made.

Selecting the Appropriate Communication Method

One of the steps to improved communication recommended in the previous section is "Determine how to communicate." Because most workplace communication is either verbal or written, managers need to know when each method is the most and least effective.[16]

Written communication is typically most effective for communicating general information and for information that requires action on the part of employees. For example, general information such as new company policies or announcements of activities that carry dates, times, places, or other specific data are appropriate for written communication. A message that says "Please bring your automobile registration to work no later than noon Friday if you want to have a parking sticker" is appropriately communicated in writing.

Verbal communication is appropriate when reprimanding employees or attempting to resolve conflict between or among employees. In these cases, verbal interaction in private is the best approach.

Written communication is least effective in the following instances:

■ When communicating a message that requires immediate action on the part of employees. The more appropriate approach in such a case is to communicate the message verbally and follow it up in writing.

■ When commending an employee for doing a good job. This should be done verbally, preferably in public, and then followed up in writing.

■ When reprimanding an employee for poor performance. This message can be communicated more effectively if given verbally and in private. This is particularly true for occasional as opposed to repeat offenses.

■ When attempting to resolve conflict among employees about work-related problems. The necessary communication in such instances is more effectively given verbally and in private.

Verbal communication is least effective in the following instances:

■ When communicating a message requiring future action on the part of employees. Such messages are more effectively communicated when given verbally and followed up in writing.

■ When communicating general information such as company policies, personnel information, directives, or orders.

■ When communicating progress reports to an immediate supervisor or upper-level manager.

■ When promoting a safety campaign.

Using Electronic Communication

In the age of high technology, electronic communication has become an important means of sending written messages.[17] Electronic mail (or E-mail) consists of written messages transferred electronically from computer to computer. Electronic communication is doing for written communication what the telephone did for verbal communication, including the following:

■ Messages can be transmitted rapidly. For example, consider the time it might take to transmit a letter to a supplier in another city. Even if overnight express services are used, it will take as long as 24 hours. The same letter sent electronically would be received in a matter of seconds.

■ Messages can be transmitted simultaneously to more than one person. This is particularly helpful when the same notification must be sent to a large number of people. The sender inputs the message and enters the codes of all who are to receive it. The message is sent instantly.

■ Messages can be printed if a hard copy is needed. A hard copy is a paper printout of the information shown on the screen of a computer terminal or personal computer. Electronic messages can be printed if a hard copy is required.

■ Messages can be prompted and acknowledged electronically. Recipients of electronic messages can acknowledge their receipt by simply depressing a key, which allows the sender to know not just that the message was received but when. They can also be *prompted*. This means that some type of visual prompt can inform the receiver that he or she has a message waiting. Typically, the prompt cannot be cleared until the message is read.

Electronic communication has advantages over other forms of communication, but it also has potential disadvantages. The first disadvantage is one inherent in any form of written communication: the inability to transmit body language, voice tone, facial expressions, or eye contact. Each of these aspects of verbal communication are missing with electronic communication.

Another disadvantage is the potential for overuse of electronic communication. Because of the ease of sending a written message by simply pressing a key, users may send more messages than they really need to, send frivolous messages, or send messages that could more appropriately be delivered verbally. Electronic communication systems should be used cautiously and only in situations in which they clearly enhance communication.

DEVELOPING INTERPERSONAL SKILLS

Interpersonal skills are those needed for people to work together in a manner that is conducive to both personal and corporate success. For employees and managers to function effectively in a total quality setting, they must have good interpersonal skills. Positive interpersonal relations among team members, between company representatives and customers, among internal customers, and between company officials and vendors is critical in a total quality setting.

The modern marketplace is international in scope and increasingly competitive. This translates into what Charles, Robert, and John Hobson refer to as a "fast-paced, demanding, and highly stressful" workplace that can result in "emotionless, expressionless, and routinized behavior among employees who initially start out with a great deal of enthusiasm."[18]

Figure 11–14 summarizes the steps managers can take to ensure that members of the workforce have good interpersonal relations:

■ *Recognition of the need.* To have employees with good interpersonal skills, managers must recognize the need for these skills. Historically, the focus of the staffing process has been technical skills and paper credentials. These are important considerations that should remain at the forefront in making staffing decisions. However, to these considerations must be added interpersonal skills.

Figure 11–14
Management Strategies for
Interpersonal
Relationships

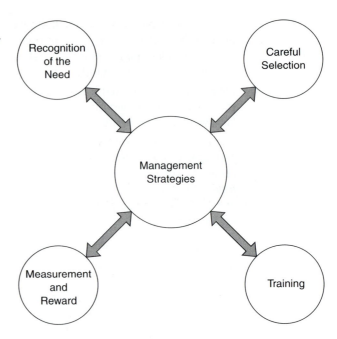

- *Careful selection.* When interpersonal skills are made a part of the selection process, the process changes somewhat. The screening of written credentials and technical skills continues in the normal manner. After the candidates with the best credentials and technical skills have been identified, they are then carefully screened to determine whether they have such interpersonal skills as listening, patience, empathy, tact, open-mindedness, friendliness, and the ability not just to get along in a diverse workplace but to be a positive agent in helping other employees get along with each other.

- *Training.* It is the uncommon individual who possesses inborn interpersonal skills. Some people are naturally good at dealing with others. However, most of us

QUALITY TIP

Positive Interpersonal Skills

"The existence or lack of positive interpersonal skills can mean the difference between business success or failure."

Source: Charles J. Hobson, Robert B. Hobson, and John J. Hobson, "People Skills: A Key to Success in the Service Sector," *Management for the 90s: A Special Report from* Supervisory Management (Saranac Lake, NY: American Management Association, 1991), 17.

have room for improvement in this key area of total quality. Fortunately, interpersonal skills can be acquired. People can learn to listen better, empathize with different types of people, be tactful, and facilitate positive interaction among fellow employees. (Training is the subject of Chapter 12.)

■ *Measurement and reward.* If managers value interpersonal skills, these skills will be measured as part of the normal performance-appraisal process. Correspondingly, the results of such appraisals will be built into the reward system.

Making Human Connections in a World of Technology

Consider the many communication-enhancing technologies that are a normal part of modern life. The television, radio, telephone, facsimile machine, citizens' band radio, computer modem, and cellular telephone were all developed in the name of improved communication, but have they really improved it?

In some ways, technology has actually had a negative effect on communication because it has removed much of the need for face-to-face interaction. But the key to communication is perception. Receivers must be able to perceive feelings, emotions, intent, and other intangible, nonverbal aspects of the message that are missing in a facsimile transmittal, a computer message, or even in a telephone conversation. For example, in certain situations, a simple human touch can communicate a powerful message, a message that cannot be transmitted by technology.

Human connections have become more important than ever because technology has made it so easy to interact without really communicating. John Noe describes this phenomenon as follows:

> The further we go technologically, the more we need to understand and be able to work with other people. It is crucial for us to recognize, make allowances for, and capitalize upon our interrelationship, our interdependence, and our interaction with other people. Even the fictional Lone Ranger felt the need for a sidekick, Tonto. In today's highly connected world, there simply are no Lone Rangers.[19]

QUALITY TIP

Management Focus on Interpersonal Skills

"Until positive interpersonal skills . . . become a focal issue for top management, one cannot expect a great deal of concern on the part of individual employees for the quality of their interpersonal image."

Source: Charles J. Hobson, Robert B. Hobson, and John J. Hobson, "People Skills: A Key to Success in the Service Sector," *Management for the 90s: A Special Report from* Supervisory Management (Saranac Lake, NY: American Management Association, 1991), 21–22.

QUALITY TIP

Has Technology Improved Communication?

"The technological advances we now enjoy—advances that could be used to bring people closer together—are resulting in increased isolation of the individual. It's too easy to enjoy the world vicariously, to sit back and watch it on TV, to read about it in the newspaper, to hear about it on the radio."

Source: John R. Noe, *People Power* (New York: Berkley, 1988), 14.

In a total quality setting, managers should be especially attentive to human interrelationships, interdependence, and interaction and to the negative impact technology can have in each of these areas. The next section explains what managers can do to promote what Noe describes as "responsiveness" among people in the workplace.[20]

Promoting Responsiveness among Employees

Responsive people are those who can perceive the real message from among the verbal, written, and nonverbal cues they receive. Responsive employees and managers are assets in a total quality setting. Responsive employees are more likely to be achievers and to help fellow employees be achievers than unresponsive employees. Consequently, it behooves managers to promote responsiveness in the workplace.

Managers can do this by teaching employees to apply the following strategies and by applying these strategies themselves:

- *Value people.*[21] Valuing people means remembering that the most sophisticated technologies are designed, operated, and maintained by people. Consequently, in the final analysis, improving people and people-oriented processes such as communication is the best way to improve competitiveness. Managers and employees who value people treat all people responsibly, not just those who have something they want.

- *Give people what you want to get back.* People have a natural tendency to mirror the treatment they receive from others. Typically, people who treat others with decency and respect are, in turn, treated with decency and respect. People who are loyal to others are likely to be the beneficiaries of loyalty. Teaching employees to get by giving is a worthwhile task for managers in a total quality setting.

- *Make cooperation a habit.* World-class athletes practice the skills that make them great until these skills become automatic and habitual. Any habit is hard to break; just ask people who have tried to stop smoking or biting their nails. Consequently, employees who practice cooperation until it becomes habitual will practice cooperation for life.

Cooperation in the workplace means learning to use the word *we* instead of *I* and *they*. It means chipping in to get the job done even when what needs to be done is not part of the job description. Cooperation means involving all employees who must do the work in decisions relating to that work.

Cooperation does not mean always saying yes. When the right answer to give is no, that answer should be given and without a sugarcoating. Cooperation in such cases means showing employees they are valued by explaining why the answer must be no.

PERSONALITY AND COMMUNICATION

Few factors affect communication so directly as the personalities of the individuals attempting to communicate. The term *personality* is used to describe a relatively stable pattern of behavior, thought, motives, emotions, and outlook distinctive to a given individual and that characterize that individual throughout life. Every person has a personality, and every individual's personality is different. However, there are certain identifiable central traits that people may share to varying degrees. Psychologists disagree over what these identifiable traits are and how many characteristics should be considered central personality traits. There is, though, something close to consensus among psychologists that the following group of characteristics is central to most people:[22]

Introversion versus extroversion. These traits describe the extent to which an individual is shy. An introvert is considered shy, while an extrovert is considered outgoing. There are, of course, varying degrees of introversion and extroversion. An extrovert is more likely to be talkative, sociable, adventurous, lively, cheerful, and enthusiastic. An introvert is more likely to be silent and reclusive. An extrovert is more likely to be a talker, an introvert is more likely to be a listener. When trying to communicate with an extrovert it might be difficult to get a word in, and you may have to ask the person to paraphrase what you have said and repeat it back. When trying to communicate with introverts, one must remember that they are not likely to volunteer much information; consequently, it can take an extra effort to draw them out.

Neuroticism versus emotional stability. These traits describe the extent to which an individual is emotionally stable. The more neurotic people are, the more anxiety they tend to feel, the less able they are to control their emotions, and the more likely they are to feel negative emotions such as anger, resentment, and scorn. The more emotionally stable people are, the less likely they are to be worriers, complainers, and defeatists. Neurotic people tend to see the bad side of any situation, while emotionally stable people will generally have a more realistic perspective. When trying to communicate with neurotic people, it is necessary to be patient, understated, and calm. Disagreement should be conveyed gently and stated in the most positive light.

Agreeable versus stubborn. These traits describe the extent to which an individual is good natured or irritable, gentle or headstrong, cooperative or abrasive, and secure or suspicious. The communication style of agreeable people is typically characterized by friendliness; the communication style of stubborn people is typically characterized by hostility. When trying to communicate with stubborn people, it is wise to first invest some time earning their trust in order to overcome their inherent suspiciousness.

Conscientious versus undependable. These traits describe the extent to which an individual is responsible or irresponsible, persevering or faint-hearted, and steadfast or fickle. When trying to communicate with undependable people, it is wise to summarize and repeat what was said. It is also important to be confident enough in what you are saying to offset the other person's tendency toward faintheartedness.

Open to experience versus prefers the familiar. These traits describe the extent to which an individual is original, creative, imagining, questioning, artistic, capable of creative thinking, or conforming and predictable. When trying to communicate with "open" people, it is sometimes necessary to rein them in and keep them focused and on task. When trying to communicate with people who prefer the familiar, it is sometimes necessary to take the time to get them to think "outside the box."

One of the most popular tests used in business settings for identifying an individual's central personality traits is the Myers–Briggs Type Indicator. One CEO thought so much of the "Myers–Briggs" that he had all employees take the test and put these results (e.g., introvert or extrovert, etc.) on their desks like a name plate. The idea was that if his employees knew more about their respective personalities, they would be better able to communicate with each other. The author does not recommend this approach, but it is wise to recognize personality traits when trying to communicate with people, particularly those with whom interaction is frequent and ongoing.

 ## SUMMARY

1. Communication is the transfer of a message that is both received and understood. Effective communication is a higher order of communication. It means the message is received, understood, and being acted on in the desired manner.
2. Communication is the oil that keeps the total quality engine running. Without it, total quality breaks down. Communications play the role of facilitation in a total quality setting.
3. Communication is a process that involves a message, sender, receiver, and medium. The message is what is being transmitted (information, emotion, intent, or something else). The sender is the originator of the message, and the receiver is the person to whom it goes. The medium is the vehicle used to transfer the message.
4. Various factors can inhibit communication. Prominent among these are differences in meaning, a lack of trust, information overload, interference, premature judgments, "kill-the-messenger" syndrome, condescending tone, inaccurate assumptions, and listening problems.
5. A climate conducive to communication gives people the information they need to do their jobs, builds morale, and promotes creativity. A bad communication climate creates conflict, confusion, and cynicism. A bad climate can be guaranteed by poor communication.

6. One of the most important communication skills is listening. Good listening means receiving the message correctly, decoding it, and accurately perceiving what it means. Inhibitors to good listening include the following: lack of concentration, preconceived ideas, thinking ahead, interruptions, tuning out, and interference. *Empathic listening* means listening with the intent to understand. *Responsive listening* means seeking to receive and affirm both the messenger and the message.

7. Listening skills may be improved by upgrading the desire to listen, asking the right questions, judging what is really being said, and eliminating listening errors. Body factors and proximity must also be managed carefully to listen well.

8. Verbal communication can be improved by showing interest, being friendly, being flexible, being tactful, being courteous, dropping your defenses, stating your purpose, acknowledging emotions, and using carefully phrased open-ended questions.

9. Written communication can be improved by being brief, being direct, being accurate, and practicing self-editing. The following step-by-step strategy will help managers write better reports: (a) define the problem, (b) develop a workplan, (c) gather relevant data, (d) process findings, (e) develop conclusions, and (f) make recommendations.

10. When it is necessary to communicate corrective feedback, be positive, be prepared, be realistic, and don't be completely negative. There are two approaches to use for communicating corrective feedback: tell–ask–listen and listen–ask–tell.

11. Adopting the following steps can help improve communication skills: (a) keep up to date, (b) prioritize and determine time constraints, (c) decide whom to inform, (d) determine how to communicate, (e) communicate and follow up, (f) check understanding, and (g) obtain feedback.

12. Electronic communication has advantages and disadvantages. The advantages include the following: messages can be transmitted rapidly, messages can be transmitted simultaneously to more than one person, messages can be printed if a hard copy is needed, and messages can be prompted and acknowledged electronically. Disadvantages are the inability to transmit nonverbal messages electronically and the potential for overuse.

13. Interpersonal skills are those needed for people to work together in a positive manner that is conducive to both personal and corporate success. To ensure that employees have good interpersonal skills, managers should recognize the need for them, select personnel carefully, provide training, measure skill, and reward improvement.

KEY TERMS AND CONCEPTS

Bad communication climate

Body factors

Communication

Communication-enhancing technologies

Company-level communication

Condescending tone

Corrective feedback

Effective communication

Empathic listening

Good listening

Inaccurate assumptions

Information overload

Interference

Interpersonal relations

Lack of concentration

Listen–ask–tell

Medium

Message

One-on-one–level communication

Open-ended questions

Preconceived ideas

Premature judgments

Proximity

Responsive listening

Self-editing

Sender

Tact

Team-level communication

Tell–ask–listen

Thinking ahead

Tuning out

Verbal communication

Voice factors

Written communication

 FACTUAL REVIEW QUESTIONS

1. Define the following terms relating to communication: *communication* and *effective communication.*
2. List and explain four levels of communication.
3. Describe the role communication plays in a total quality setting.
4. Explain the process of communication.
5. List and briefly explain six inhibitors of communication.
6. Define *good listening.*
7. List and briefly explain five inhibitors of good listening.
8. What is empathic listening?
9. What is responsive listening?
10. Explain how a person can become a responsive listener.
11. Explain four strategies for improving listening skills.
12. Define the following factors and explain how they affect listening: body factors, voice factors, proximity factors.
13. Describe how a person can improve his or her verbal communication skills.
14. List and explain five rules of thumb for improving written communication.
15. Explain the steps for improving written reports.
16. When it is necessary to communicate corrective feedback, what four guidelines should be applied?
17. Six guidelines to improved communication were set forth in this chapter. Explain all six.
18. Briefly describe the advantages and disadvantages of electronic communication.
19. Define *interpersonal relations.*
20. How can managers ensure that employees have good interpersonal skills?
21. How can a manager promote responsiveness among employees?
22. List five personality traits and describe how they can affect communication.

CRITICAL THINKING ACTIVITY

Are You a Good Listener?

The purpose of this activity is to compare your self-perception with the perceptions of others concerning your listening skills. Figure 11–5 is a listening skills assessment instrument. Complete this instrument for yourself. Then ask at least two other people who you trust to be open and honest to complete the instrument with *you as the subject*. Does their perception of your listening skills match your self-perception? If not, what are the differences? What do you need to do to improve your listening skills?

DISCUSSION ASSIGNMENT 11–1

The Busy Boss

John Gill is a busy man. No matter how fast he works, it seems he's always behind. Consequently, when an employee brings Gill a problem, he is not a good listener. He opens mail, answers the telephone, and constantly glances at his watch while the employee is talking. Because he does not listen well, John keeps being blindsided by new problems that seem to get more and more serious.

Discussion Questions
Discuss the following questions in class or outside of class with your fellow students:

1. What can John do to become a better listener?
2. How might John's poor listening skills be adding to his growing list of problems?

 ENDNOTES

1. U.S. Chamber of Commerce, "Ten Key Threats to Success," *Nation's Business* 80 (June 1992), 18–28.
2. U.S. Chamber of Commerce, 26.
3. Corwin P. King, "Crummy Communication Climate (and How to Create It)," *Management for the 90s: A Special Report from* Supervisory Management (Saranac Lake, NY: American Management Association, 1991), 21.
4. King, 21.
5. King, 21.
6. This section is based on Stephen Covey, *The Seven Habits of Highly Effective People* (New York: Simon & Schuster, 1990), 239–249.
7. Covey, 241.

8. This section is based on John R. Noe, *People Power* (New York: Berkley, 1988), 85–99.

9. C. Glenn Pearce, "Doing Something about Your Listening Ability," *Management for the 90s: A Special Report from* Supervisory Management (Saranac Lake, NY: American Management Association, 1991), 1–6.

10. Roger Ailes, "The Seven-Second Solution," *Management Digest* (January 1990), 3.

11. Robert Maidment, "Seven Steps to Better Reports, *Management for the 90s: A Special Report from* Supervisory Management (Saranac Lake, NY: American Management Association, 1991), 11–14.

12. Robert A. Luke, "How to Give Corrective Feedback to Employees," *Supervisory Management* (March 1980), 7.

13. Luke, 7.

14. Luke, 8.

15. Kim McKinnon, "Six Steps to Improved Communication," Supervisory Management (February 1990), 9.

16. This section is based on D. A. Level, "Communication and Situation," *Journal of Business Communication* 9 (1992), 19–25.

17. This section is based on R. C. Huseman and E. W. Miles, "Organizational Communication in the Information Age: Implications of Computer-Based Systems," *Journal of Business Communications* 12 (2), 181–204.

18. Charles J. Hobson, Robert B. Hobson, and John J. Hobson, "People Skills: A Key to Success in the Service Sector," *Management for the 90s: A Special Report from* Supervisory Management (Saranac Lake, NY: American Management Association, 1991), 20.

19. Noe, 47–70.

20. Noe, 15.

21. Noe, 55–60.

22. Robert R. McCrae and Paul T. Costa, Jr., "Personality Trait Structure as a Human Universal," *American Psychologist*, Vol. 52, 509–516.

CHAPTER TWELVE

Education and Training

"A man can seldom—very, very seldom—fight a winning fight against his training."

Mark Twain

MAJOR TOPICS

- Overview of Education, Training, and Learning
- Rationale for Training
- Training Needs Assessment
- Providing Training
- Evaluating Training
- Managers as Trainers and Trainees
- Workforce Literacy
- Improving Learning
- Why Training Sometimes Fails
- Quality Training Curriculum
- Orientation Training
- Customer Training
- Ethics Training

One of the most fundamental elements of total quality is the ongoing development of personnel, which means education, training, and learning. This chapter provides readers with the information needed to justify, provide, and evaluate education, training, and learning.

OVERVIEW OF EDUCATION, TRAINING, AND LEARNING

It is common to hear the terms *education*, *training*, and *learning* used interchangeably in discussions of employee development. Although common practice is to use the term *training* for the sake of convenience, modern managers should be familiar with the distinctions among them. For purposes of this book, training is defined as follows:

> Training is an organized, systematic series of activities designed to enhance an individual's work-related knowledge, skills, and understanding and/or motivation.

Training can be distinguished from *education* by its characteristics of practicality, specificity, and immediacy. Training should relate specifically to the job performed by those being trained, and it should have immediate practical application on the job. *Education* is a broader term; training is a subset of education. Also, education tends to be more philosophical and theoretical and less practical than training.

The purpose of both education and training is learning. In an educational setting the learning will tend to be more theoretical, whereas in a training setting it will be more practical. However, with both, understanding is implicit in learning. Whether the point is to have the learner understand *why* or *how to*, the point is still to have the learner understand. Understanding is what allows an employee to become an innovator, initiative taker, and creative problem solver in addition to being an efficient and effective performer of his or her job.

Although education typically occurs in a classroom setting while training typically occurs in a less formal environment, there is some overlap in the concepts. Education can certainly occur outside of a classroom, and training can certainly occur in one. For this reason and for the sake of simplicity, the term *training* will be used throughout the remainder of this chapter.

Corporate Training in the United States

Corporate America invests more than $45 billion per year in training.[1] However, there are serious questions about how wisely this money is spent. Is it being used in a way that will bring the best results, or is corporate America spending its training dollars on the wrong people? The following sections describe the status of training in the United

QUALITY TIP

Training and Workplace Improvement

"All employees must understand their jobs and their roles in the company—and how these roles change as quality improves. Such understanding goes beyond the instructions given in manuals or job descriptions. Employees need to know where their work fits into the larger context: how their work is influenced by workers who precede them and how their work influences workers who follow. They must learn new skills for improving work."

Source: Peter R. Scholtes, *The Team Handbook* (Madison, WI: Joiner Associates, 1992), 1-16.

States by job category, sources of training, instructional methods, and types of training in selected industry classifications.

Training Status by Job Category

Industrial firms with more than 100 employees typically have personnel in the following categories of employment:

■ Executive managers
■ Senior managers
■ Middle managers
■ Supervisors
■ Professionals (engineers, scientists, technologists, and technicians)
■ Sales representatives
■ Customer service representatives
■ Production personnel
■ Office personnel

Training provided to personnel in these various categories can be compared in a number of different ways. Two of the most informative comparisons are the percentage of companies that provide training to employees in each category and the average number of hours of training received by employees in each category. Figure 12–1 compares the percentage of companies that provide training in each subject category of employment. The

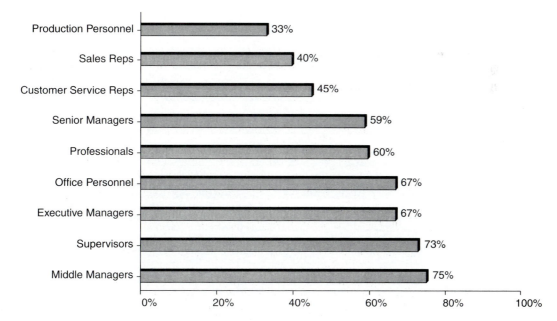

Figure 12–1
Percentage of U.S. Companies That Provide Training to Employees

comparisons represent approximate figures for a typical year for companies with 100 or more employees.

Figure 12–1 shows that in the United States more companies provide training for managers than for any other category of employee. By comparison, only 33% of these companies provide training for production personnel. These figures raise serious questions abut how effectively corporate America's training dollars are being spent.

In a total quality setting, training is a bottom-up enterprise in which the people who do the work receive top priority in the allocation of training dollars. The philosophy of the total quality approach concerning training can be summarized as spending the money where it will do the most good. In practice, this philosophy translates into giving training priority to those employees who are most actively involved in producing products or providing services. The further removed from these processes an employee is, the lower his or her training priority becomes. Followed to its logical conclusion, the total quality philosophy assigns the lowest training priority to managers, a reversal of the figures shown in Figure 12–1.

When training is compared using the average number of hours provided per year per employee, production workers fare only slightly better. Figure 12–2 contains these comparisons for companies in the United States for a typical year. Again, the data are for companies with more than 100 employees. When average hours of training is the criterion, only office personnel receive less training than production personnel.

Figure 12–2 provides a stark illustration of what happened to many of the U.S. firms that managed themselves out of business in the 1980s. By putting more resources into

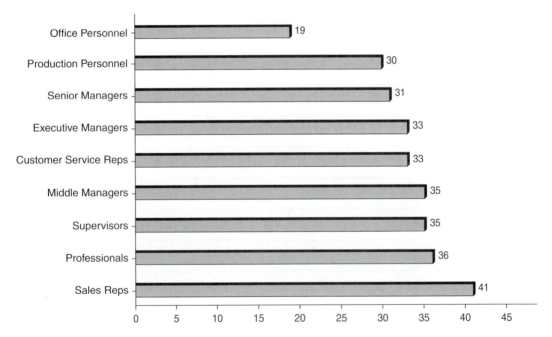

Figure 12–2
Average Training Hours per Typical Year

QUALITY TIP

"Forget Quality—Give Me Better Sales Reps"

"The last two decades saw many companies in the U.S. simply manage themselves to death. Many mistakes were made. Infrastructure was often neglected in favor of short-term profits, managers searched for giant productivity improvements rather than focusing on small but continual improvements, and adversarial relations between management and labor were allowed to persist. Perhaps the worst mistake of all was the neglect of quality in favor of sales gimmicks. Rather than improve the quality of their products, many U.S. companies simply said, 'Forget quality—give me better sales reps.' I suppose these companies thought that a bad product could be sold by a good sales rep. They were wrong. It didn't work. Even the best sales rep can't fool a customer who is determined to get the most for his money."

Source: From a speech by David Goetsch.

training sales representatives than into training production personnel, they made a conscious decision to neglect quality. Such an attitude can be summarized by the statement "Forget quality, just sell harder." The total quality philosophy is the opposite and can be summarized by the statement "Improve quality and you won't have to sell so hard."

Sources of Training

Many sources of training are available to organizations that want to provide training for employees. This section discusses them in terms of broad categories; they are examined in greater detail later in this chapter. All of them fall into one of the following categories: in-house training, external training, or a combination of in-house and external. Figure 12–3 summarizes the percentages provided using each of these approaches.

Most training is provided through a combination of in-house training and external sources, with in-house training the second most widely used option. *In-house training* is a broad heading covering on-the-job training, in-house seminars/workshops, on-site media-based instruction (videotape, audiotape, satellite downlinks, etc.), and on-site computer-assisted instruction. *External training* is a broad category into which fits college or university courses; workshops and seminars provided by colleges, universities, and/or private training organizations; correspondence courses; vendor-sponsored training; and training provided by technical, trade, and professional associations.

Instructional Methods

Numerous instructional methods are used to provide training for employees. Figure 12–4 displays a checklist of instructional methods that are widely employed for both internal

Figure 12–3
Sources of Training

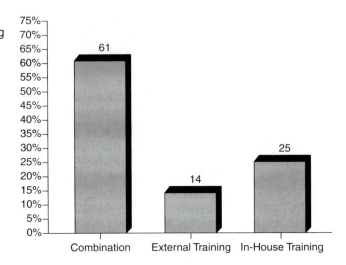

and external instruction. Of the methods listed, the most commonly used are videotaped instruction and lectures. The influence of technology on instruction is apparent from the checklist in Figure 12–4. Of the 12 methods shown, 8 are technology dependent. In addition, technology is often used with the other four to enhance the quality and retention of instruction.

For example, it is now common to supplement a lecture with slides or transparencies. Simulation activities are often computerized, and self-study instruction is frequently centered around videotaped or audiotaped instruction or self-teaching computer software.

Figure 12–4
Instructional Methods Checklist

____ Videotapes

____ Lecture

____ Demonstration (one-on-one)

____ Slides/Transparencies

____ Role-Playing

____ Audiotapes

____ Film

____ Simulation

____ Case Studies

____ Self-Study Instruction

____ Videoconferencing and Teleconferencing

____ Computer Referencing

Types of Training by Industry

Four key industrial sectors in the United States are manufacturing, transportation, communication, and utilities. Figure 12–5 shows the types of training provided for employees in these key industrial sectors. Training topics that are especially important in a total quality setting are technical skills, supervision skills, communication, new work procedures, and customer relations (internal and external customers).

Technical skills training is provided by 86% of the manufacturing, transportation, communication, and utility companies in the United States. Supervision training is important in a total quality setting because supervisors need to learn to be facilitators, coaches, communicators, and leaders rather than bosses in the traditional sense. The need for supervision training is apparently felt in these four key industrial sectors, where 77% of the companies provide this type of training.

Communication skills are taught by 78% of the companies in the subject sectors. Although this figure is more than 75% of the companies, a great many still show no understanding of the importance of communication in workplace improvement and competitiveness. Communication can have a direct influence on an organization's ability to compete.

Because continual process improvement is a fundamental element of total quality, work procedures change frequently in a total quality setting. This means that training in new work procedures should take place continually so that process improvements can

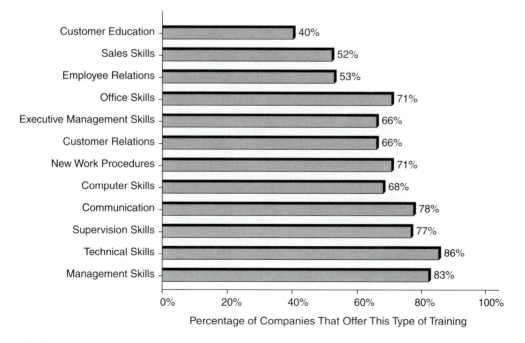

Figure 12–5
Types of Training (Manufacturing, Transportation, Communication, Utilities)

be effectively and efficiently implemented. Figure 12–5 shows that 71% of the companies in the subject sectors offer this type of training.

Changing Role of Training

Although corporate training in the United States is more than 100 years old, it really got its start near the end of World War II. Establishment of the American Society for Training and Development (ASTD) was a milestone development that occurred at this time. According to Dick Schaaf, the ASTD grew out of "wartime manufacturing businesses that placed a premium on rapidly bringing new industrial workers up to speed on large-scale processes and machines that made things."[2]

Since the mid-1940s, interest in corporate training has grown steadily so that today

those that employ more than 100 workers spend some $45 billion a year on developing their human resources, and training directors command comfortably middle-management-level salaries. The field has professionalized, become specialized, developed its own alphabet soup of acronyms and special interest groups, not to mention a handful of formulas to calculate training's organizational role and clout.[3]

Schaaf summarizes the past, present, and current roles of corporate training in terms of such factors as mission, focus, and imperative. For example, he claims that the mission of corporate training has evolved from a focus on developing skills and discipline to a new focus on improving values and motivation. Future focuses, according to Schaaf, will be on enhancing service and quality.[4] Our view is that the mission of corporate training is becoming the maximization of competitiveness through continual improvements in quality, productivity, response time, cost, service, and corporate image.

Attitudes toward Training in the United States and Other Countries

The Commission on the Skills of the American Workforce cochaired by Ray Marshall and William E. Brock—both of whom served as U.S. secretary of labor—issued a report entitled *America's Choice: High Skills or Low Wages*. This report concluded that U.S. businesses place an alarmingly low priority on the skills of their employees. A survey of more than 400 companies yielded the following results:[5]

■ Less than 10% of the companies planned to increase productivity by reorganizing work in ways that call for employees with broader skills.
■ Only 15% expressed concern over the potential for shortage of skilled workers.
■ Less than 30% intended to offer special training programs for women and minorities, although these groups account for 85% of all new workers.
■ More than 80% were more concerned about workers' attitudes than their skills.

Based on these findings, the commission concludes that business is not taking a proactive role in demanding the improvements that are so badly needed in American education and training. Contrasting what is found in the United States with other industrialized countries, the commission concludes further that education and training for non–college-bound students is much better in such countries as Germany, Japan,

Sweden, and Denmark. Consequently, American students rank near the bottom in indicators of school performance when compared with students from these countries. The commission describes the U.S. system for transitioning students from school to work as "the worst of any industrialized country." Many feel the commission's findings are exaggerated; however, clearly there is a problem.

Other industrialized countries are creating what the commission described as a high-performance workplace by reorganizing work in ways that call for multiple abilities and high levels of reading, math, science, and problem-solving skills. As a result, workers are better able continually to develop new skills as technology changes. The ability of such workers to adapt quickly and continually also allows their employers to introduce new products on shorter cycle times and make more frequent changes in production runs.

Traditionally, business and industry in Canada and the United States have had similar attitudes. In the past, those attitudes have been casual at best. However, although interest levels don't yet match those found in European and Pacific Rim countries, attitudes toward training in North America are changing for the better, particularly in Canada. A study completed by Canada's Labor Market and Productivity Center (CLMPC) produced the following findings:

- A majority of the respondents said that both employers and unions should have a direct role in the training of workers, but the main responsibility for training and retraining—including funding—belongs to employers. Labor leaders strongly support the concept of a national training tax on corporations, commonly referred to in Canada as the 1% levy-grant proposal, to fund training programs. According to the CLMPC, three-fourths of the business leaders oppose the idea of a training tax.
- Obstacles to improved training include inadequate facilities for workplace training, low interest in training by many employers, and the practice described by the survey report as interfirm poaching of trained workers.
- International competitiveness is generally regarded as vital to maintaining a high standard of living, but it is valued less by union leaders than by business leaders.
- One-third of the business leaders and one-half of the labor leaders described workers' reading, writing, and mathematical skills as inadequate. Respondents linked these shortcomings to low productivity, high training costs, poor quality control, and difficult recruitment.
- Universities, colleges, and vocational schools do a fair job of preparing people for the working world, according to most of the survey respondents. Elementary and secondary schools were deemed inadequate.
- Although respondents agreed that training was important, they were unhappy with current offerings. Forty percent of business leaders and 86% of labor leaders said workplace programs were inadequate.[6]

RATIONALE FOR TRAINING

The rationale for training can be found in the need to compete. To survive in the modern marketplace, organizations must be able to compete globally. Companies that at one time competed only with their neighbors up the street now find themselves vying against companies from Europe, Asia, Central and South America, and the Pacific Rim.

These companies are like the local high school track star who decides to try out for the Olympic team. Suddenly the competition is much more difficult, and it will be increasingly so at each successive level right up to the Olympic Games. If the athlete is able to make it that far, he or she will face the best athletes in the world. This is the situation in which modern business and industrial firms find themselves every day. Like the Olympic team that must have world-class athletes to win medals, these companies must have world-class employees to win the competition for market share.

Several factors combine to magnify the need for training. The most important of these are:

- Quality of the existing labor pool
- Global competition
- Rapid and continual change
- Technology transfer problems
- Changing demographics

These factors and how they contribute to the rationale for training are explained in the following sections.

Quality of the Existing Labor Pool

The labor pool consists of the people who are available for and wish to have employment. New jobs are filled from the labor pool. For this reason, the quality of the labor pool is critical. Quality in this sense means both preparedness and potential. A high-quality labor pool is one in which its members are well schooled in such fundamental intellectual skills as reading, writing, thinking, listening, speaking, and problem solving. Such people are well prepared in terms of the basics and, as a result, have good potential to quickly learn and adapt when put in a job. The importance of basic knowledge and skills cannot be overstated, as can be seen in Discussion Assignment 12–2.

How does the labor pool from which U.S. organizations draw their employees measure up with regard to quality? Consider the following facts released by the National Center for Manufacturing Sciences:[7]

QUALITY TIP

Importance of a Literate Workforce

"It has been estimated that in 1965, a mechanic who understood about 500 pages of various repair manuals could fix about any car on the road. Today, that same mechanic would need nearly 500,000 pages of manuals—equivalent to roughly 50 New York City telephone books."

Source: National Center for Manufacturing Sciences, *Focus* (August 1992), 8.

■ Youth in the United States spend barely 9% of their first 18 years in school.

■ Approximately 93% of the largest U.S. companies must teach employees the three R's and other basic skills.

■ When compared with their counterparts in Canada, Europe, and Asia, 23-year-old people in the United States place last in math and science.

These deficiencies do not result from the United States spending too little on public education. As a percentage of gross national product (GNP) spent on education, the United States ranks among the highest. Of the most industrialized nations, only Canada (4.1%) devotes a larger percentage of its GNP to education than the United States (3.5%). Japan and Germany spend less than 3%. These figures seem to indicate that the United States spends more on public education but gets less for its money than most other industrialized nations. The implications of this situation in terms of the quality of the labor force are serious.

U.S. companies must invest training dollars in teaching employees basics before they can begin to deal with higher level material that will more directly affect productivity and quality. Compare this situation with the Dutch labor force, which draws its members from high schools where 90% of the students take advanced math courses, or Japan's, where 25% of the time elementary school children spend in school is devoted to math and science.[8] This puts U.S. employers in the position of having to spend more to get the same result.

Global Competition

U.S. companies, even some of the smallest, find themselves competing in a global marketplace, and the competition is intense. A small manufacturer of automotive parts in Michigan might find itself competing with companies from Korea, Japan, Taiwan, and Europe as well as the United States. To win the contest, this small Michigan firm must make its products both better and less expensively than its competitors. Large companies face the same intense global competition.

According to Philip Harkins:

> The Japanese, the Germans, and others are battling fiercely for the lead technological and manufacturing positions. The international business world is being led by managers who are committed to increasing productivity and quality through advanced educational processes and methods. The result: the best-trained workforce will gain the long-term competitive advantage. No longer are we locked in a cold war. Today, the battle is for economic survival and dominance—the most powerful weapons on this new global business battlefield are the education and training of our workforce.[9]

Rapid and Continual Change

Change is a fact of life in the modern workplace. It happens fast and continually. Knowledge and skills that are on the cutting edge today may be obsolete tomorrow. In such an environment, it is critical that employees be updated constantly. The relative literacy of an organization's workforce will determine its ability to keep up. Figure 12–6 provides a

	Yes	No
1. Do you notice employees having trouble reading and spelling at the level required by the job?		
2. Do you notice job applicants who have trouble completing application paperwork?		
3. Do you notice employees having trouble using fractions and decimals?		
4. Do you experience equipment problems because employees cannot read operating manuals?		
5. Do you notice problems in the workplace caused by employees with limited English proficiency?		
6. Do you notice employees who cannot keep up in workplace training programs?		

Figure 12–6
Literacy Assessment Checklist

checklist managers can use to get a feel for whether their organization has a literacy problem.

Rapid and continual change represents an insurmountable barrier to employees who are not functionally literate, meaning their academic performance is between grade levels 4.0 and 8.9. To understand the number of employees in the United States who may not be able to keep pace, consider the following facts:

- Almost 30 million adults in the United States are functionally illiterate.
- Approximately 20% of the workforce in the United States has a reading comprehension level of eighth grade or lower, whereas 70% of the reading material in the modern workplace is at the ninth grade level or higher.
- Approximately 2.5 million people enter the workforce every year with only limited language skills. These include immigrants, high school dropouts, and high school graduates who are still functionally illiterate.

These facts are part of the rationale for workplace training. Organizations that do not provide it may find it difficult, if not impossible, to keep up with the rapid change that is sure to occur.

Technology Transfer Problems

Technology transfer is the movement of technology from one arena to another. The process has two steps. The first step is the commercialization of new technologies developed in research laboratories or by individual inventors. This is a business development issue and does not involve training.

The second step in the process is known as technology diffusion, and it is training dependent. *Technology diffusion* is the process of moving newly commercialized technologies into the workplace where they can be used to enhance productivity, quality, and competitiveness.

This step breaks down unless the workers who will use the technology have been well trained to use it efficiently and effectively. This is critical because new technologies by themselves do not enhance productivity. A word processing system given to an untrained traditional secretary will be nothing more than an expensive typewriter with a screen. To take maximum advantage of the capabilities of new technologies, workers must know how to use them effectively. Knowledge comes from training. Two of the major inhibitors of effective technology transfer are fear of change and lack of know-how. Both of these inhibitors can be overcome by training.

Changing Demographics

Workplace demographics are changing in ways that make training even more important than it was in the past. Figure 12–7 summarizes some of the ways workplace demographics are changing in the United States. One word characterizes the modern workplace— *diversity*. Because teamwork is a fundamental element of total quality, training may be needed to teach employees from vastly different backgrounds how to work together in harmony.

Experienced employees may not adapt easily to an influx of women, minorities, and immigrants. They may even feel threatened and resentful. Women, minorities, and immigrants may not be comfortable in what may, at first, be an alien environment. Simply bringing diverse people together in the workplace does not guarantee that they will work together in harmony. Overcoming cultural, social, and gender differences requires training, commitment, and constant attention.

Benefits of Training

In spite of the fact that billions of dollars are spent on training each year, many employers still do not understand the role or benefits of training in the modern workplace. John Hoerr, writing in *Business Week*, paints a grim picture of the attitudes of U.S. employers toward training.[10] Hoerr cites a study conducted by the National Center on Education and the Economy that compares the education and training of workers in the United States with those of workers in competing countries.

Figure 12–7
Changing Demographics in the Workplace

- More than 80 percent of all new entrants into the workforce are women, minorities, or immigrants.
- The average age of employed people is increasing.
- More than 20 percent of all new entrants into the workforce are immigrants with limited English language skills.

QUALITY TIP

Employer Attitudes toward Training

- Only 1 in 14 employees in the United States has ever received formal training from an employer.
- Employers in the U.S. spend 2% of payroll on training, whereas Japanese companies spend 6%.

Source: *U.S. News & World Report*, June 1998.

This study concludes that less than 10% of U.S. companies use a flexible approach requiring better trained workers as a way to improve productivity. This approach is standard practice in Japan, Germany, Denmark, and Sweden. In addition, less than 30% of U.S. firms have special training programs for women, minorities, and immigrants, in spite of the fact that over 80% of all new workers come from these groups. It is critical to the competitiveness of U.S. industry that employers understand the need for training that results from such factors as intense international competition, rapid and continual change, technology transfer problems, and changing demographics.

Tom Peters claims, "Our investment in training is a national disgrace."[11] According to Peters, 69% of the companies in the United States with 50 or more employees provide training for middle managers, and 70% train executive-level personnel. However, only 30% train skilled personnel. Peters contrasts this with the Japanese and Germans, who outspend U.S. firms markedly in providing training for skilled personnel.

Modern managers should understand the benefits of workforce training and be able to articulate these benefits to upper management. Figure 12–8 is a checklist that summarizes the most important of these. As can be seen from the checklist, the benefits of training build upon themselves. For example, reducing the turnover rate will also contribute to improving safety. Increased safety will, in turn, help minimize insurance costs.

Employers who are not familiar with the benefits of training but are beginning to take an interest in providing it often debate the applicability or job relatedness of training. The argument is often made that "We will provide only the training that relates directly to the job." According to total quality pioneer W. Edwards Deming, focusing too intently on direct applicability is a mistake. Any kind of learning can benefit employees and employers alike in ways that cannot be predicted.

TRAINING NEEDS ASSESSMENT

How do managers know what training is needed in their organizations? The answer is that many don't know. When compared with their competitors from other countries,

Figure 12–8
Checklist of Training Benefits

____ Fewer Production Errors
____ Increased Productivity
____ Improved Quality
____ Decreased Turnover Rate
____ Lower Staffing Costs
____ Improved Safety and Health
____ Fewer Accidents
____ Minimized Insurance Costs
____ Increased Flexibility of Employees
____ Better Response to Change
____ Improved Communication
____ Better Teamwork
____ More Harmonious Employee Relations

U.S. companies appear to spend a great deal of money on the wrong kinds of training. It is a matter of emphasis; by placing the emphasis on management, employers are spending the bulk of their training dollars on those who organize and oversee the work rather than those who actually do it. This is akin to training the coaches instead of the players.

This is not to say that managers don't need ongoing training. In a total quality setting, every employee needs continual training. However, the keys to maximizing the return on training dollars are to place the emphasis on training those who need it most and ensure that the training provided is designed to promote the goals of the organization (quality, productivity, and competitiveness).

Satisfying the first criterion is simply a matter of reversing the emphasis so that it is bottom up in nature rather than top down. Satisfying the second involves assessing training needs. Begin by asking two broad questions:

■ What knowledge, skills, and attitudes do our employees need to have to be world-class level?

■ What knowledge, skills, and attitudes do our employees currently have?

The difference between the answers to these questions identifies an organization's training needs.

Managers may become involved in assessing training needs at two levels: the organizational level and the individual level. Managers who work closely enough with their team members can see firsthand on a daily basis their capabilities and those of the team as a whole. Observation is one method managers can use for assessing training needs. Are there specific problems that persist? Does an individual have difficulties performing certain tasks? Does work consistently back up at a given point in the process? These are indicators of a possible need for training that can be observed.

QUALITY TIP

All Learning Is Good

"People are afraid to take a course. It might not be the right one. My advice is to take it. Find the right one later. And how do you know it is the wrong one? Study, learn, improve.

"You never know what could be used, what could be needed. He that thinks he has to be practical, is not going to be here very long. Who knows what is practical?

"Help people to improve. I mean everybody."

Source: W. Edwards Deming, as quoted in Mary Walton, *The Deming Management Method* (New York: Putnam, 1986), 85.

A more structured way to assess training needs is to ask employees to state their needs in terms of their job knowledge and skills. Employees know the tasks they must perform every day. They also know which tasks they do well, which they do not do well, and which they cannot do at all. A brainstorming session focusing on training needs is another method managers can use. Brainstorming is particularly effective in organizations where employees are comfortable speaking out as part of the continual improvement process.

The most structured approach managers can use to assess training needs is the *job task analysis survey*. With this method, a job is analyzed thoroughly and the knowledge, skills, and attitudes needed to perform it are recorded. Using this information, a survey instrument is developed and distributed among employees who do the job in question. They respond by indicating which skills they have and which they need.

QUALITY TIP

Teaching Employees the Wrong Things

"We teach people about marketing, finance, accounting, operations management, and R&D, but not how to run a business. We teach people about leadership, but not how to lead. We teach people how to make decisions (our quantitative techniques provide us with an unparalleled level of precision), but not how to carry them out. We teach people about strategic management, but not how to compete successfully."

Source: James A. Belohlav, *Championship Management: An Action Model for High Performance* (Cambridge, MA: Productivity Press, 1990), 69.

Before preparing a survey instrument, step back and take in the "big picture". A common mistake is to focus too intently on the finite tasks of a job to the exclusion of the broader, less tangible requirements. For example, while comprehensively breaking down technical process tasks, managers often overlook such criteria as teamwork skills, sensitivity to customer feedback (particularly internal customers), problem solving, and interpersonal skills. For this reason, it is a good idea to involve the employees who will be surveyed in the development of the survey instrument.

Another way to identify training needs is to ask employee groups to convene their own quality circles relating specifically to training. Employee-managed groups such as these that convene without managers and supervisors are often more open to admitting that there are training needs. They will also be less reluctant to identify the training

QUALITY CASE

Top-Down Commitment to Training at Intel

Intel Corporation is the global leader in the production of microprocessors. In addition, Intel produces flash memory chips and microcontrollers, and is the largest e-commerce company in the world. What is the key to Intel's phenomenal success in the global marketplace? According to Andy Grove, Intel's CEO, much of the company's success can be attributed to its commitment to training. Grove believes that many of the problems that rob organizations of their efficiency, effectiveness, and competitiveness could be prevented through proper training. He also believes that managers and executives give more lip service than commitment to training. Most senior managers will profess to believing in training, while in the same breath claiming they don't have time for it.

Grove believes that training is one of the key responsibilities of managers and that those who don't have time to train their personnel don't have time to be managers. According to Grove, managers have only three ways to increase the output of their personnel: 1) change nothing, but expect them to work harder; 2) motivate and inspire them; and 3) increase their individual capabilities through training.

Grove not only advocates training, he conducts training sessions himself on such topics as preparing and delivering performance reviews and conducting productive meetings, as well as giving an orientation seminar on the history, objectives, organization, and management practices of Intel. It is difficult for managers and supervisors at any level to claim they don't have time to train their personnel when the company's CEO is so actively involved in and committed to training. This commitment can be seen in the fact that Intel spends more than $500 million annually on employee and management training.

Source: Thomas J. Neff and James M. Citrin, *Lessons from the Top* (New York: Doubleday, 2001), 163–168.

needs they think supervisors and managers have. The organization's suggestion system should also be used to identify training needs.

Converting Training Needs to Training Objectives

Having identified training needs, the next step is to write training objectives. This responsibility will fall in whole or in part to the manager. Some organizations have training personnel who can assist; others do not. In either case, managers in a total quality setting should be proficient in writing training objectives. The key to writing good training objectives lies in learning to be specific and to state objectives in behavioral terms. For example, suppose a need for training in the area of mathematics has been identified. The manager might write the following training objective:

> Employees will learn mathematics.

This training objective, as stated, lacks specificity and is stated in nonbehavioral terms. Mathematics is a broad concept. What is the need? Arithmetic? Algebra? Geometry? Trigonometry? All of these? To gain specificity, this objective must be broken into several objectives and restated.

To be stated in behavioral terms, these more specific objectives must explain what the employee should be able to do after completing the training. Behavioral objectives contain action verbs. The more clearly training objectives are written, the easier it is to plan training to meet them. The sample training objectives in Figure 12–9 are stated in behavioral terms, they are specific, and they are measurable. Taking the time to write objectives in this way will make it easier to provide and evaluate training.

PROVIDING TRAINING

Ernest Boyer, president of the Carnegie Foundation for the Advancement of Teaching, sets forth four dimensions of the corporate learning enterprise.[12] Boyer credits Nell P. Eurich, a trustee of the Carnegie Foundation, with identifying the following dimensions of corporate education and training:

---

1. Upon completion of this lesson, employees will be able to solve right triangles.
2. Upon completion of this lesson, employees will be able to apply the Law of Sines to the solution of triangles.
3. Upon completion of this lesson, employees will be able to apply the Law of Cosines to the solution of triangles.
4. Upon completion of this lesson, employees will be able to add, subtract, multiply, and divide decimal fractions.
5. Upon completion of this lesson, employees will be able to solve equations containing one unknown variable.

Figure 12–9
Sample Training Objectives

■ In-house education is becoming more frequently used to meet the needs of industry and business, including needs typically referred to as general education.

■ Educational and training facilities owned and operated by corporations are becoming common. Motorola, Xerox, RCA, and Rockwell International have all established their own education and training facilities (see Discussion Assignment 12–3).

■ Degree-granting institutions that are corporate owned and operated and that are accredited by the same associations that accredit both public and private colleges and universities are becoming common. Many community colleges offer degree-granting partnership programs with industrial firms.

■ The Satellite University through which instruction is provided by satellite to local downlink sites is becoming common. The National Technological University (NTU) of Fort Collins, Colorado, transmits instruction up through the master's degree level to corporate classrooms nationwide.

Several firms in the United States have learned the value of education and training and, as a result, are reaping their benefits. IBM, Nissan (in Tennessee), and Motorola are examples. At IBM, training immediately follows each promotion. All IBM employees must complete at least 40 hours of training each year. Today, as IBM faces intense pressure from foreign competitors, training is at the heart of its strategy for confronting the challenge. Before Nissan opened its plant in Smyrna, Tennessee, it spent $63 million training approximately 2,000 employees. This amounts to approximately $30,000 per employee, which is about average for a Japanese company. Motorola stays competitive in the semiconductor business by investing 2.5% of its annual payroll in education and training. In addition to its employees, Motorola also trains its suppliers.

Eurich of the Carnegie Foundation cites IBM and AT&T as leaders among U.S. firms in using education and training to enhance competitiveness.[13] According to Eurich, before divestiture in 1984, AT&T spent $1.7 million annually for education and training. This investment produced 12,000 courses in 1,300 different locations and required 13,000 trainers and training support workers. In a typical year in the late 1980s, IBM invested $700 million in education and training.

Several different methods for providing training are available to organizations. All fall into one of the following three broad categories: internal approaches, external approaches, and partnership approaches. Regardless of the method used, it is important to maximize training resources. Carolyn Wilson recommends five strategies for maximizing training resources:[14]

1. *Build in quality from the start.*   Take the time to do it right from the outset.
2. *Design small.*   Do not try to develop courses that are all things to all people. Develop specific activities around specific objectives.
3. *Think creatively.*   Do not assume that the traditional classroom approach is automatically best. Videotapes, interactive video, or one-on-one peer training may be more effective in some cases.
4. *Shop around.*   Before purchasing training services, conduct a thorough analysis of specific job-training objectives. Decide exactly what you want and make sure the company you plan to deal with can provide it.

QUALITY TIP

Using Stories to Teach

Stories can be an excellent way to enhance learning. The use of stories is one of the oldest and most effective teaching methods.

"Using stories to impact the values of a corporate culture and to teach skills predates by eons the more rational, scientific approach to communication. Cognitive psychologists have been investigating how people process information for 25 years. They have found that stories communicate by affecting memory and, consequently, behavior. Well-told stories hold our attention and allow us to experience the actions and tribulations of others vicariously."

Source: Laurel C. Sneed, "Make Your Video Tell a Story," *Training* (September 1992), 59.

5. *Preview and customize.*   Never buy a training product (videotape, self-paced manual, or something else) without previewing it. If you can save by customizing a generic product, do so.

Internal Approaches

Internal approaches are those used to provide training on-site in the organization's facilities. These approaches include one-on-one training, on-the-job computer-based training, formal group instruction, and media-based instruction. *One-on-one training* involves placing a less skilled, less experienced employee under the instruction of a more skilled, more experienced employee. This approach is often used when a new employee is hired. It is also an effective way to prepare a replacement for a high-value employee who plans to leave or retire.

Computer-based training (CBT) has proven to be an effective internal approach. Over the years, it has continually improved so that now CBT is a widely used training method. It offers the advantages of being self-paced, individualized, and able to provide immediate and continual feedback to learners. Its best application is in developing general knowledge rather than in company-specific job skills.

Formal *group instruction* in which a number of people who share a common training need are trained together is a widely used method. This approach might involve lectures, demonstrations, multimedia use, hands-on learning, question/answer sessions, role-playing, and simulation.

Media-based instruction has become a popular internal approach. Private training companies and major publishing houses produce an almost endless list of turnkey media-based training programs. The simplest of these might consist of a set of audiotapes. A more comprehensive package might include videotapes and workbooks. Interactive laser disk training packages that combine computer, video, and laser disk technology are also effective with the internal approach.

An example of an extensive internal training program is Motorola University, the in-house educational institution operated by Motorola, Inc. Motorola University consists of institutes for manufacturing and engineering personnel, middle managers, and senior managers, as well as an instructional design center. More than 60,000 employees have completed training in Motorola University. The company's goal is to have a minimum of 2% per year of an employee's time spent in training.

External Approaches

External approaches are those that involve enrolling employees in programs or activities provided by public institutions, private institutions, professional organizations, and private training companies. The two most widely used approaches are (a) enrolling employees in short-term training (a few hours to a few weeks) during work hours and (b) enrolling employees in long-term training such as a college course and paying all or part of the costs (i.e., tuition, books, fees). External approaches encompass training methods ranging from seminars to college courses. The external approach is typically used for developing broad, generic skills. However, some institutions will work with employers to develop customized courses.

Partnership Approaches

In recent years, community colleges, universities, and technical schools have begun to actively pursue partnerships with employers through which they provide customized training. These training partnerships combine some of the characteristics of the previous two approaches.

Customized on-site training provided cooperatively by colleges and private companies or associations has become very common. According to Eurich of the Carnegie Foundation, the more than 1,200 community colleges in the United States have built extensive networks and alliances with business and industry.[15] Eurich cites the example of General Motors Corporation, which has contracts with 45 community colleges throughout the country to train service technicians. General Motors contributes to the partnership by training the instructors.

Many universities, community colleges, and technical schools have continuing education or corporate training divisions that specialize in providing training for business and industry. Managers should know the administrator responsible for continuing education at all colleges, universities, and technical schools in their communities.

Partnerships with institutions of higher education offer several advantages to organizations that want to arrange training for their employees. Representatives of these institutions are education and training professionals. They know how to transform training objectives into customized curricula, courses, and lessons. They know how to deliver instruction and have access to a wide range of instructional support systems (libraries, multimedia centers, and instructional design centers). They know how to design application activities that simulate real-world conditions. They know how to develop a valid and reliable system of evaluation and use the results produced to chart progress and prescribe remedial activities when necessary.

In addition to professional know-how, institutions of higher education have resources that can markedly reduce the cost of training for an organization. Tuition costs for continuing education activities are typically much less than those associated with traditional college courses. If these institutions do not have faculty members on staff who are qualified to provide instruction in a given area, they can usually hire a temporary or part-time instructor who is qualified. Other advantages institutions of higher education can offer are credibility, formalization, standardization, and flexibility in training locations. Associating with a community college, university, or technical school can formalize an organization's training program and give it credibility.

This is important because employers sometimes find their attempts at customized training hampered by a lack of credibility. Their employees have been conditioned to expect formal grade reports, transcripts, and certificates of completion. These things formalize training in the minds of employees and make it more real for them. Educational institutions can provide these credibility builders.

Another problem employers sometimes experience when providing their own customized training is lack of standardization. The same training provided in three different divisions might produce markedly different results. Professional educators can help standardize the curriculum and evaluation systems. They can also help standardize instruction to the extent possible by providing train-the-trainer workshops for employees who are serving as in-house instructors.

Regardless of the approach or approaches used in providing training, there is a widely accepted rule of thumb that should be observed and with which managers should be familiar:

■ People learn best when the learning approach used involves them in seeing, hearing, speaking, and doing.

■ Education practitioners hold that the percentages in Figure 12–10 apply regarding what learners remember and retain. Clearly, for learning to be effective, it must involve activity on the part of learners, be interactive in nature, and comprise to the extent possible reading, hearing, seeing, talking, and doing.

Regardless of the approach used to provide training, there are several critical decisions that must be made. Total quality pioneer Joseph M. Juran explains these as follows:[16]

■ *Should training be voluntary or required?* If training is an essential part of total quality and the organization is committed to total quality, training should be mandatory. According to Juran:

Figure 12–10
Learning Retention

- 10 percent of what is read
- 20 percent of what is heard
- 30 percent of what is seen
- 50 percent of what is seen and heard
- 70 percent of what is seen and spoken
- 90 percent of what is said while doing what is talked about

Once it is determined that universal training in managing for quality is essential to meeting the company's quality goals, such training should not be put on a voluntary basis. Instead, training in managing for quality should be mandated by upper management.[17]

■ *How should training be sequenced?*   Although the emphasis in a total quality setting is typically bottom up in terms of how much training is provided, the sequence of training is top down. In other words, managers receive less training than employees, but they receive it first. According to Juran:

Upper management should be the first to acquire the new training. The reasons are quite persuasive: 1) by being first, upper managers become better qualified to review proposals made for training for the rest of the organization; and 2) by setting an example the upper managers change an element of the corporate culture, that is, for lower levels to take the new training is to do what has been done in respected circles.[18]

■ *What subjects should be taught?*   The subject matter of training is dictated by the organization's goals for quality, productivity, and competitiveness. As was discussed earlier in this chapter, training needs are determined by comparing the knowledge, skills, and attitudes needed to accomplish organizational goals and those that are present in the organization. Any gap between what is needed and what is present can be closed by providing appropriate training.

Because the training needs of personnel at different levels are different, it is important to have training tracks (nonmanagement track, middle-management track, executive management track). Training should be arranged so that it is upwardly mobile in nature. This means that training for first-line supervisors should build on that provided for employees. Training provided for middle managers should build on that provided for first-line supervisors, and so on up the line through executive management.

EVALUATING TRAINING

Did the training provided satisfy the training objectives? Are trainees using what they learned? Has the training brought results? Managers need to know the answers to these questions every time training is provided. However, these can be difficult questions to answer. Evaluating training begins with a clear statement of purpose. What is the purpose of the training? This broad purpose should not be confused with the more specific training objectives. The purpose of the training is a broader concept. The objectives translate this purpose into more specific, measurable terms.

The purpose of training is to improve the knowledge, skills, and attitudes of employees and, in turn, the overall quality and productivity of the organization so that it becomes more competitive. In other words, the purpose of training is to improve performance and, in turn, competitiveness. To know whether training has improved performance, managers need to know three things:

■ Was the training provided valid?
■ Did the employees learn?
■ Has the learning made a difference?

Valid training is training that is consistent with the training objectives.

Evaluating training for validity is a two-step process. The first step involves comparing the written documentation for the training (course outline, lesson plans, curriculum framework, etc.) with the training objectives. If the training is valid in design and content, the written documentation will match the training objectives. The second step involves determining whether the actual training provided is consistent with the documentation. Training that strays from the approved plan will not be valid. Student evaluations of instruction conducted immediately after completion can provide information on consistency and the quality of instruction. Figure 12–11 is an example of an instrument that allows students to evaluate instruction.

Determining whether employees have learned is a matter of building evaluation into the training. Employees can be tested to determine whether they have learned, but be sure that tests are based on the training objectives. If the training is valid and employees have learned, the training should make a difference in their performance. Performance on the job should improve; this means quality and productivity should improve. Managers can make determinations about performance using the same indicators that told them training was needed in the first place. Can employees perform tasks they could not perform before the training? Is waste reduced? Has quality improved? Is setup time down? Is in-process time down? Is the on-time rate up? Is the production rate up? Is throughput time down? These are the types of questions managers should ask to determine whether training has improved performance.

Gilda Dangot-Simpkin of Dynamic Development suggests a checklist of questions for evaluating purchased training programs:[19]

- Does the program have specific behavioral objectives?
- Is there a logical sequence for the program?
- Is the training relevant for the trainee?
- Does the program allow trainees to apply the training?
- Does the program accommodate different levels of expertise?
- Does the training include activities that appeal to a variety of learning styles?
- Is the philosophy of the program consistent with that of the organization?
- Is the trainer credible?
- Does the program provide follow-up activities to maintain the training on the job?

Using these questions as a guide, managers can make intelligent choices when selecting commercially produced training objectives.

MANAGERS AS TRAINERS AND TRAINEES

To be good trainers, managers should have such characteristics as a thorough knowledge of the topics to be taught; a desire to teach; a positive, helpful, cooperative attitude; strong leadership abilities; a professional attitude and approach; and exemplary behavior that sets a positive example.[20]

Instructions

On a scale of 1 to 5 (5 = highest rating; 1 = lowest rating), rank your teacher on each item. Leave blank any item which does not apply.

Organization of Course

1. Objectives (Clear to Unclear) 1 2 3 4 5
2. Requirements (Challenging to Unchallenging) 1 2 3 4 5
3. Assignments (Useful to Not Useful) 1 2 3 4 5
4. Materials (Excellent to Poor) 1 2 3 4 5
5. Testing Procedures (Effective to Ineffective) 1 2 3 4 5
6. Grading Practice (Explained to Not Explained) 1 2 3 4 5
7. Student Work Returned (Promptly to Delayed) 1 2 3 4 5
8. Overall Organization (Outstanding to Poor) 1 2 3 4 5

Comments

Teaching Skills

9. Class Meetings (Productive to Nonproductive) 1 2 3 4 5
10. Lectures (Effective to Ineffective) 1 2 3 4 5
11. Discussions (Balanced to Unbalanced) 1 2 3 4 5
12. Class Proceedings (To-the-Point/Wandering) 1 2 3 4 5
13. Provides Feedback (Beneficial to Not Beneficial) 1 2 3 4 5
14. Responds to Students (Positively/Negatively) 1 2 3 4 5
15. Provides Assistance (Always to Never) 1 2 3 4 5
16. Overall Rating of Instructor's Teaching Skills
    (Outstanding to Poor) 1 2 3 4 5

Comments

Substantive Value of Course

17. The course was (Intellectually Challenging to Elementary) 1 2 3 4 5
18. The instructor's command of the subject was
    (Broad and Accurate/Plainly Defective) 1 2 3 4 5
19. Overall substantive value of the course
    (Outstanding to Poor) 1 2 3 4 5

Comments

Figure 12–11
Student Evaluation of Instruction

In addition to having these characteristics, managers should be knowledgeable about the fundamental principles of learning and the four-step teaching method. The principles of learning summarize much of what is known about how people learn best. It is important to conduct training in accordance with these principles. The four-step teaching method is a basic approach to conducting training that has proven effective over many years of use.

Principles of Learning

The principles of learning summarize what is known and widely accepted about how people learn. Trainers can do a better job of facilitating learning if they understand the following principles:

- *People learn best when they are ready to learn.* You cannot make employees learn anything. You can only make them want to learn. Therefore, time spent motivating employees to want to learn is time well spent. Before beginning instruction, explain why employees need to learn and how they and the organization will mutually benefit from their having done so.

- *People learn more easily when what they are learning can be related to something they already know.* Build today's learning on what was learned yesterday and tomorrow's learning on what was learned today. Begin each new learning activity with a brief review of the one that preceded it. Use examples to which all employees can relate.

- *People learn best in a step-by-step manner.* This is an extension of the preceding principle. Learning should be organized into logically sequenced steps that proceed from the concrete to the abstract, from the simple to the complex, and from the known to the unknown.

- *People learn by doing.* This is probably the most important principle for trainers to understand. Inexperienced trainers tend to confuse talking (lecturing or demonstrating) with teaching. These things can be part of the teaching process, but they do little good unless they are followed with application activities that require the learner to do something. Consider the example of teaching an employee how to ride a bicycle. You might present a thorough lecture on the principles of pedaling and steering and give a comprehensive demonstration on how to do it. However, until the employee gets on and begins pedaling, he or she will not learn how to ride a bicycle.

- *The more often people use what they are learning, the better they will remember and understand it.* How many things have you learned in your life that you can no longer remember? People forget what they do not use. Trainers should keep this principle in mind. It means that repetition and application should be built into the learning process.

- *Success in learning tends to stimulate additional learning.* This principle is a restatement of a fundamental principle in management (success breeds success). Organize training into long enough segments to allow learners to see progress but not so long that they become bored.

■ *People need immediate and continual feedback to know if they have learned.* Did you ever take a test and get the results back a week later? If so, that was probably a week later than you wanted them. People who are learning want to know immediately and continually how they are doing. Trainers should keep this principle in mind at all times. Feedback can be as simple as a nod, a pat on the back, or a comment such as "Good job!" It can also be more formal, such as a progress report or a graded activity. Regardless of the form it takes, trainers should concentrate on giving immediate and continual feedback.

Four-Step Teaching Method

Regardless of the setting, teaching is a matter of helping people learn. One of the most effective approaches for facilitating learning is not new, innovative, gimmicky, or high-tech in nature. It is known as the four-step teaching method, an effective approach to use for training. The four steps and a brief description of each follow:

■ *Preparation* encompasses all tasks necessary to get participants prepared to learn, trainers prepared to teach, and facilities prepared to accommodate the process. Preparing participants means motivating them to want to learn. Personal preparation involves planning lessons and getting all of the necessary instructional materials ready. Preparing the facility involves arranging the room for both function and comfort, checking all equipment to ensure it works properly, and making sure that all tools and other training aids are in place.

■ *Presentation* is a matter of presenting the material participants are to learn. It might involve giving a demonstration, presenting a lecture, conducting a question/answer session, helping participants interact with a computer or interactive videodisc system, or assisting those who are proceeding through self-paced materials.

■ *Application* is a matter of giving learners opportunities to use what they are learning. Application might range from simulation activities in which learners role-play to actual hands-on activities in which learners use their new skills in a live format.

■ *Evaluation* is a matter of determining the extent to which learning has taken place. In a training setting, evaluation does not need to be a complicated process. If the training objectives were written in measurable, observable terms, evaluation is simple. Employees were supposed to learn how to do X, Y, and Z well and safely. Have them do X, Y, and Z and observe the results. In other words, have employees demonstrate proficiency in performing a task and see how they do.

Preparing Instruction

Preparing instruction involves the following steps:

1. Preparing (planning) the instruction
2. Preparing the facility
3. Preparing the learners

It is important to accomplish all three steps before attempting to present instruction. The instruction delivered by managers will usually be part of a course, workshop, or seminar. In any case, a course outline must exist that summarizes the major topics covered by the instruction. The outline should state in broad terms the expected outcomes of the instruction or, in other words, what the learner is supposed to be able to do after completing the course, workshop, or seminar. The outline should also have a brief statement of purpose. More specific instructional objectives are developed later when preparing lesson plans. Figure 12–12 is a sample outline for a short course on W. Edwards Deming's Fourteen Points for quality improvement.

The course outline contains just two components: a statement of purpose and a list of intended outcomes. Some instructors prefer to add additional components such as a list of equipment and/or training aids needed, but the components shown in Figure 12–12 are sufficient. A good course outline is a broadly stated snapshot of the scope and sequence of the course. Specific details are typically shown in the lesson plans that are developed next.

Lesson plans are an important part of the planning step. They are road maps or blueprints for the actual instruction that is to take place. In addition, they serve to standardize instruction when more than one person might present the same lesson to different groups. Standardization is particularly important for quality training. If even one member of a work team receives less training than the others, the potential for quality problems is increased by his or her ignorance.

Statement of Purpose

The purpose of this seminar is to familiarize managers with the Fourteen Points set forth by W. Edwards Deming and how they might be used to improve quality in this organization.

Intended Outcomes

Upon completion of this seminar, managers should be able to explain the ramifications for our company of each of the following points:

1. Create constancy of purpose for improvement of product and service.
2. Adopt the new philosophy.
3. Depend less on mass inspection.
4. End the practice of awarding business on the price tag alone.
5. Improve constantly and forever the system of production and service.
6. Institute training.
7. Institute leadership.
8. Drive out fear.
9. Break down barriers between staff areas.
10. Eliminate slogans, exhortations, and targets for the workforce.
11. Eliminate numerical quotas.
12. Remove barriers of pride of workmanship.
13. Institute a vigorous program of education and retraining.
14. Take action to accomplish the transformation.

Figure 12–12
Course Outline: Deming's Fourteen Points for Quality Improvement

Lesson plans can vary in format according to the personal preferences of the trainer. However, all lesson plans should include the components discussed in the following paragraphs:

■ *Lesson title and number.* The lesson title should be as descriptive as possible of the content of the lesson. The number shows where the lesson fits into the sequence of lessons that make up the course.

■ *Statement of purpose.* Like the statement of purpose in the course outline, this component consists of a concise description of the lesson's contents, where it fits into the course, and why it is included.

■ *Learning objectives.* Learning objectives are specific statements of what the learner should know or be able to do as a result of completing the lesson. Objectives should be written in behavioral terms that can be measured or easily observed.

■ *Training aids list.* This component serves as a handy checklist to help trainers quickly and conveniently ensure that all the training aids they will need are present. The list should include every tool, handout, piece of equipment, video, chart, or other item needed to conduct the instruction for that lesson.

■ *Instructional approach.* The instructional approach is a brief action plan for carrying out the instruction. It should begin with a short statement describing the instructional methodology to be used (lecture/discussion, demonstration, computer-assisted instruction, or other method). This statement is followed by a step-by-step summary of the trainer's major activities; for example, deliver lecture on safety regulations, or distribute safety regulations handout.

■ *Application assignments.* The application assignments list details the tasks learners will be required to complete to apply what they are learning.

■ *Evaluation methodology.* The evaluation methodology component explains how learning will be evaluated. Will there be a test? Will performance be observed? Will records be monitored for improvement? Such questions are answered in this section.

Presenting Instruction

As shown earlier in Figure 12–10, educators hold that the following percentages apply regarding what learners retain from instruction they receive:

■ 10% of what is read
■ 20% of what is heard
■ 30% of what is seen
■ 50% of what is seen and heard
■ 70% of what is seen and spoken
■ 90% of what is said while doing what is talked about

Instruction can be presented in several different ways. The most widely used are the lecture/discussion group instruction, demonstration, conference, and multimedia methods. Regardless of the approach used, trainers should keep the percentages listed here in mind. What they indicate is that the trainers should get the learners actively engaged in seeing, saying, listening, and most important, doing.

The Lecture/Discussion Method   The lecture/discussion method of teaching is the oldest, most familiar, most used, and probably most abused method. A lecture is a planned, structured, and sometimes illustrated (using slides, charts, or chalkboard) method of communicating information to a group of people. By itself, the lecture allows for only one-way communication. This serious deficiency is overcome by adding the discussion component. Discussion can be between the instructor and participants or among the participants. During discussion, the instructor's job is to keep the discussion on track and moving in the right direction.

The best justification for using the lecture/discussion method is that it is an effective way to communicate information to groups that are too large to allow for individual interaction between instructor and participants. Another reason for using this method is that it allows the instructor to generate enthusiasm among participants about a topic.

The lecture/discussion method, if used properly, can be an effective teaching technique. However, it does not work in every situation. Trainers need to know when to use it and when to use another method. Use the lecture/discussion method when any of the following things are true:

- The material to be presented deals strictly with data, theory, or information (no skills development).
- Participants need to be motivated before beginning a particular lesson.
- The material to be presented is not available in print.
- Sharing insight or experience in a particular area will enhance learning.
- It is necessary to communicate information to a large group in one session.
- Interaction among participants is desired.

Do not use the lecture/discussion method when any of the following things are true:

- The subject matter deals with skills development or how-to information.
- The group of auditors is small enough to allow individual student–teacher interaction.
- There is no need for interaction among participants.

Participants must be thoroughly prepared prior to the session. If they are not, the session will be all lecture and no discussion. Require participants to approach a lecture/discussion session as if they were entering a debate. This will ensure that they are active, contributing participants rather than passive spectators.

Prepare participants for a lecture/discussion session as follows:

1. Give them a written outline or overview of the lecture.
2. Have participants use the outline as a study guide or a guide to research and reading.
3. Instruct your listeners to note anything that is unclear to them or about which they cannot locate information while reading and/or researching.
4. Have participants put their notes together in an annotated outline, leaving plenty of blank space for taking notes under each item during the lecture/discussion session.

There is a saying in teaching: "When giving a lecture, tell them what you are going to tell them, tell them, and then tell them what you told them." Although it is said with tongue in cheek, this is actually good advice. A well-planned, properly structured lecture contains three distinct components: the *opening*, in which you "tell them what you are going to tell them"; the *body*, in which you "tell them"; and the *closing*, in which you "tell them what you told them."

Discussion may be interspersed within the body or held until after the closing, depending on your preference. The recommended method is to allow discussion during the body of the lecture while questions and concerns are fresh in the participants' minds.

The opening, body, and closing of a lecture all contain specific tasks that should be accomplished in order:

1. Opening
   a. Greet the class.
   b. State the title of the lecture.
   c. Explain the purpose of the lecture.
   d. List the objectives so that participants know exactly what they should be learning.
   e. List and define any new terms that will be used during the session.
   f. Present a general overview of the content of the lecture/discussion.
2. Body
   a. Present the information in the order set forth on the participants' outline.
   b. Accomplish the purpose of the lecture.
   c. Make frequent reference to all visual aids and supportive materials.
3. Closing
   a. Restate the title, purpose, and objectives.
   b. Briefly summarize major points.
   c. State your conclusions.
   d. Answer remaining questions.
   e. Make follow-up assignments to reinforce and apply learning.

These elements are the fundamental or tangible tasks that should be performed in all lecture/discussion sessions. You should also keep in mind a number of intangibles when conducting lecture/discussion sessions:

■ Make sure that the classroom is arranged to accommodate a lecture/discussion.
■ Be enthusiastic. Enthusiasm is contagious.
■ Call on participants by name. They will appreciate the recognition and feel more at ease.
■ Spread your attention evenly. This will make all participants feel that they are part of the lecture/discussion.
■ Maintain eye contact with all participants in the session.
■ Speak clearly, evenly, and slowly enough to be understood, but not in a monotone.
■ Use facial expressions, body language, and movement to emphasize points.

■ It is all right to use an outline or note cards to keep yourself on track, but never read a lecture to the participants.

■ Use carefully prepared visual aids to reinforce major points.

■ Do not dominate—facilitate. Participation is critical. Remember, this method is called lecture/*discussion.*

The Demonstration Method   Demonstration is the process in which the instructor shows participants how to perform certain skills or tasks. While demonstrating, the instructor also explains all operations step by step. The key to giving a good demonstration is preparation. The following checklist contains specific tasks for preparing a demonstration:

1. Decide exactly what the purpose of the demonstration is, why it will be given, what participants should learn from it, what will be demonstrated and in what order, and how long the demonstration will last.
2. Gather all tools, equipment, and instructional aids, and make sure that everything is available and in working order. Never put yourself in the position of being forced to stop a demonstration in the middle because something does not work the way it should or because a necessary part of the demonstration is not on hand.
3. Set up the demonstration so that participants will be able to easily see what is going on and hear what you are saying.
4. Arrange all materials to be used in the demonstration so that they correspond with the order in which the various steps of the demonstration will be presented.
5. Practice the demonstration several times before giving it in order to work out any bugs.

Just as there are specific tasks to be performed in presenting a lecture, there are specific tasks to be performed in giving a demonstration:

1. Orient participants to the demonstration by explaining its purpose and objectives. Give them a brief overview of the content of the demonstration. Explain how the demonstration ties in with what the participants already know.
2. Present the demonstration in a slow, deliberate fashion so that participants can easily follow.
3. Pause between stages to determine whether participants are comprehending or have any questions that should be answered before continuing. Go back over any steps that they did not seem to grasp.
4. Conclude the demonstration with a brief summary and question/answer session.

Activity in skills development by class members is critical. Remember, no matter how well a demonstration is presented, it is really just showing. Showing is important, but participants learn best by doing, so it is vital to provide them with hands-on activities after a demonstration. An effective way to follow up demonstrations is to do the following:

■ Select several participants and ask them to repeat the demonstration you have just given.

■ Assign a number practical application activities in which the participants are required to apply the skills demonstrated.

■ Observe participants on an individual basis as they attempt to perform the practical application activities. Give individual attention and assistance where needed. Be sure to correct mistakes immediately so that the wrong way does not become a habit.

■ Conduct performance evaluations so that skills development can be measured.

Conference Method   The conference method is particularly well suited for corporate training settings. It is less formal than a traditional classroom setting and requires that the trainer serve as a facilitator rather than a teacher. It is best used as a problem-solving teaching method. For example, a manager might use the conference method to make all employees aware of a new quality problem while simultaneously soliciting their input on how to solve the problem.

To be effective facilitators, managers must become adept at defining the problem, soliciting input from participants, drawing out all participants, summarizing and repeating information, and building consensus. The conference teaching method, when effectively used, should result in both well-informed participants and a plan for solving the problem.

Other Presentation Methods   In addition to the presentation methods already explained, several others can be employed. The most widely used are simulation, videotapes, and programmed instruction.

■ *Simulation.*   Simulation, as the name implies, involves structuring a training activity that simulates a live situation. For example, if a manager is teaching a group of workers how to respond when a fellow worker passes on faulty work, he or she might simulate that situation by having a worker role-play this situation. Simulation can also be technology based. Computer simulation activities and those based on interactive laser disc and video technology are becoming more commonly employed. The military has used technology-based simulation for many years to train pilots.

■ *Videotapes.*   The use of videotapes for presenting instruction has become common in corporate training settings. In essence, the videotape takes the place of a lecture or demonstration. The pause or stop functions on the video player can be used to allow for discussion, questions and answers, and group interaction. The playback feature can be used for reviewing material or replaying portions of the tape that are not fully understood by participants.

■ *Programmed instruction.*   Programmed instruction is a technique for individualizing instruction. Traditionally, the programmed medium has been a workbook or text that presents information in segments that proceed as follows:

a. Information presentation

b. Information review

c. Questions, problems, or activities for the participant to work based on the information presented

d. A self-test

Before proceeding to the next lesson, the learner must make a specified score on the self-test for the preceding lesson. Increasingly, programmed instruction is becoming computerized. This enhances the interactive nature of the instruction and, with good software, provides almost immediate feedback for the learner.

■ *Interactive video.*   Interactive video training combines laser disc and personal computer technologies to create an excellent high-tech approach to workplace training. Unlike many other media-based training methods, interactive video is not passive. Rather, it requires the learner to participate actively by making choices, selecting options, and participating in one-on-one simulations of workplace situations. According to John Fisher:

> In operation, an employee sits privately at a TV monitor with (optional) headphones. Those being trained need only touch the TV screen for interaction, so no computer literacy is necessary. Employees become actively engaged in the learning process on a one-on-one basis. Dramatizations of workplace situations actually appear on the screen in live action.[21]

Fisher lists the following advantages of interactive video as a training methodology: "individualizes learning, increases retention, is self-paced, evaluates and records progress, reduces training time, simplifies learning, and is dependable, consistent, and flexible."[22]

Applying Instruction

One of the fundamental principles of learning states that people learn by doing. To the trainer, this means that learners must be given opportunities to apply what they are learning. If the topic of a training session is how to apply statistics to quality, application should involve having the learners actually apply statistics. While the learners practice, the trainer observes, coaches, and corrects. Regardless of the nature of the material, learners should be given plenty of opportunities to apply what they are learning.

The final step in the four-step teaching method is evaluation, which was described in the list at the start of "Four-Step Process" section.

Managers as Trainees

In a total quality setting, some managers will be trainers. However, all managers will be trainees. According to Juran, all managers in a total quality setting should complete training in the following areas:[23]

■ *Quality basics,* including definitions, big Q and little Q, and the Juran Trilogy
■ *Strategic quality management,* including developing a quality strategy and quality policies, and establishing a quality council, goals, deployment, resources, measurement, and rewards
■ *Quality planning,* including how to apply quality planning road maps
■ *Quality improvement,* including infrastructure, cost of poor quality, return on investments in quality, and the project-by-project concept
■ *Quality control,* including establishing quality measures, compiling quality report packages, and conducting quality audits

Several factors can undermine the process of providing training for managers. These include a lack of credibility on the part of trainers and discomfort with training that mixes managers and nonmanagers. Juran recommends that the following common behavior characteristics of managers be taken into account when providing training for them:[24]

- Managers prefer training sessions that enroll only managers.
- Managers are uncomfortable being trained by subordinates.
- Managers prefer to be trained by well-known outsiders.
- Managers enjoy learning of the experiences of other managers from well-managed companies.
- Managers prefer off-site training.
- Managers enjoy visiting companies that have reputations for excellence.

WORKFORCE LITERACY

In recent years, industry has been forced to face a tragic and potentially devastating problem: adult illiteracy, a problem that is having a major impact on the competitiveness of business and industry in the United States.[25] It is estimated that more than 60 million people, or approximately one-third of the adult population in this country, are marginally to functionally illiterate.[26]

People are sometimes shocked to learn that the number of illiterate adults is so high. Ernest Fields and colleagues list several reasons that this number has been obscured in the past and why there is now a growing awareness of the adult illiteracy problem:[27]

- Traditionally the number of low-skill jobs available has been sufficient to accommodate the number of illiterate adults.
- Faulty research methods for collecting data about illiterate adults have obscured the reality of the situation.
- Reticence on the part of illiterate adults to admit they have a problem and to seek help has further obscured the facts.

Illiteracy has become a more compelling problem today because of technological advances and the need to compete in the international marketplace. The problem is compounded in the following ways:

- Basic skill requirements are being increased by technological advances and the need to compete in the international marketplace.
- Broader definitions of literacy that go beyond just reading and writing also include speaking, listening, and mathematics.
- Old views of what constitutes literacy no longer apply.

Impact of Illiteracy on Industry

The basic skills necessary for a worker to be productive in the modern workplace are increasing steadily. At the same time, the national high school dropout rate continues to rise, as does the number of high school graduates who are functionally illiterate in spite of their diplomas. This means that while the number of high-skilled jobs in modern industry is increasing, the number of people able to fill them is on the decline. (Discussion Assignment 12–4 describes a community in which business has worked to improve the population's literacy level.) The impact this problem will have on industry in the United States can be summarized as follows:

- Difficulty in filling high-skill jobs
- Lower levels of productivity and, as a result, a lower level of competitiveness
- Higher levels of waste
- Higher potential for damage to sophisticated technological systems
- Greater number of dissatisfied employees in the workplace

What Industry Can Do

Industry in the United States has found it necessary to confront the illiteracy problem head-on.[28] Companies are doing this by providing remedial education for employees in the workplace. Some companies contract with private training firms, others provide the education themselves, and still others form partnerships with colleges, universities, or vocational schools.

The National Center for Research in Vocational Education conducts research into industry-based adult literacy training programs in the United States. It found that Texas Instruments requires math, verbal and written communication, and basic physics of its employees. Physics skills have not traditionally been viewed as being part of functional literacy. However, to succeed in this high-tech company, employees must have these skills. The approach used by Texas Instruments to provide literacy training can be one of the least expensive; by working in conjunction with public colleges or vocational schools, companies can provide literacy training at little and, in some cases, no cost. Managers should establish a close working relationship with the business and industry representatives in local colleges and vocational schools.

Rockwell International also defines literacy more stringently than has been typical in the past. To function effectively at Rockwell International, employees must be skilled in chemistry and physics. This is another example that is indicative of the need for higher skill levels to be functionally literate. To accomplish this goal for its employees, Rockwell hired its own certified teachers.

Of the companies studied, Polaroid takes the most aggressive approach in defining functional literacy. The skills taught in Polaroid's program are also indicative of the trend, especially in the areas of statistics, problem solving, and computer literacy. The need for these skills is technology driven. Knowledge of statistics is required to use statistical process control (SPC), which is widely employed in total quality settings. Few employees in a modern industrial firm get by without using a computer on the job—hence the need for computer literacy training. Problem-solving skills are critical to companies in a total quality setting. Programs that teach problem-solving skills involve employees in identifying and correcting problems that have a negative impact on quality or do not add value to the company's products. Polaroid considers problem solving part of functional literacy.

What Managers Should Know about Literacy Training

In their study of industry-based literacy training programs, Fields and colleagues made recommendations to assist in the planning and implementation of such programs.[29] Several of these recommendations have relevance for managers in a total quality setting:

- The definition of literacy should be driven by the company's needs.
- Companies should establish an environment in which employees feel comfortable having their literacy skills assessed and in seeking help to improve those skills.
- Whenever possible, companies should establish programs to improve the skills of existing employees rather than laying them off and hiring new employees.
- Whenever possible, companies should collaborate with educational institutions or education professionals to provide literacy training.

As organizations continue to enhance their technological capabilities, the skill levels of the workforce will have to increase correspondingly. Because the number of people in the labor force who have high-level skills is not increasing, the need for workforce literacy training will be a fact of life with which managers will have to deal for some time to come.

IMPROVING LEARNING

One of the difficulties with education and training is that many people don't know *how* to learn. Learning can be improved by teaching all employees good study skills before putting them in a training program, and by using humor to improve teaching in company-sponsored training programs.

Teaching Study Skills

Time spent helping employees learn how to learn will be time invested well. The Channing L. Bete Company recommends teaching employees to use the following strategies:[30]

- *Make a schedule and stick to it.*   Allow 2 hours of study time for each hour of class time. Schedule time for review immediately before and immediately after class. Take short breaks, at least one each hour. Reward yourself for sticking to your schedule.
- *Have a special place to study.*   Designate a quiet place as your study site and equip it with everything you will need (reference books, paper, pens, and whatever else you need).
- *Listen and take notes.*   Concentrate on the presentation and take notes. Don't let your mind wander. In taking notes, don't try to write down everything that is said. Rather, write down key points only, leaving room to expand your notes afterward. Listening for key points will improve your listening ability and, in turn, your retention.
- *Read assertively.*   Just reading course materials will not adequately prepare you for class activities. After reading such materials, go back and make an outline of major points and supportive points.
- *Improve test-taking skills.*   Don't cram for tests; instead, study regularly. When taking tests, skim over the entire document before answering individual items. Answer

the questions you are sure of first. This will allow you to spend more time on those items about which you are less confident.

Using Humor in Training

According to researcher Charles Gruner, a speech and communication professor at the University of Georgia, appropriately used humor can enhance learning.[31] Gruner's research is supported by that of Kellogg fellow Debra Korobkin, who concludes that using humor in training can result in the following benefits:[32]

- Improved retention of material
- Better student–teacher rapport
- Better motivation toward learning
- More satisfaction with learning
- Higher test scores
- More animated class discussion
- More creative thinking
- More divergent, independent thinking
- Less stress and anxiety
- Less monotony

These benefits are derived when appropriate humor is used properly. The qualifiers *appropriate* and *proper* are important here. Humor used improperly can backfire. Gruner suggests the following guidelines for the proper use of humor in training:[33]

- *Properly used, relevant humor can produce a more favorable audience for the trainer.* Relevant humor is humor that relates to the subject matter of the training. Properly used humor is humor that helps make a point, makes learners more open to or motivated toward learning, and/or helps learners relax. Humor at the expense

QUALITY TIP

It's a Classroom, Not a Comedy Club

"Professor Tom Endres of the College of St. Thomas in St. Paul, Minnesota, who has studied the rhetorical impact of the late comedian Lenny Bruce, cautions trainers against adopting a 'tendentious,' or acid, style in the classroom. 'Take Don Rickles,' he says. 'Very pointed, caustic, acerbic. He would be a terrible trainer. He would have no *immediacy*. He would develop no sense of openness, friendliness, or empathy with the trainees. Rickles, Bruce, and Dangerfield purposefully alienate themselves from their audiences: good for a comedy club, terrible for the training room.'"

Source: Ron Zemke, "Humor in Training," *Training* (August 1991), 28.

of the audience should be avoided, as should controversial, sarcastic, and insulting humor.

■ *Self-disparaging humor enhances the trainer's image.* Whereas humor that disparages the audience can backfire on trainers, humor that gently and subtly disparages the trainer can actually enhance his or her stature with learners. It can show that the trainer is human and, although an expert, is fallible.

■ *Humor can enhance the interest level of learners, within limits.* Humor can create and help maintain interest as long as it is an extension of the subject matter and not humor for humor's sake. In other words, humor in training should be like spices in cooking: it should accentuate the meal without becoming the main course.

■ *The use of satire can have unpredictable results.* Satire is a type of humor that presents a point of view. Consequently, the laughs it generates are always at someone's expense. The best rule of thumb to follow in a training setting is avoid satire unless you know the audience and its views very well.

Why Training Sometimes Fails

Training is an essential ingredient in total quality, but training is not automatically good. In fact, training often fails. Training fails when it does for several reasons, such as poor teaching, inadequate curriculum materials, poor planning, insufficient funding, and a lack of commitment.

Some subtle and more serious reasons for training failures are explained by Juran as follows:[34]

■ *Lack of participation in planning by management.* It is important to involve people at the line level in the planning of training. This does not mean management should be excluded, however; in fact, quite the opposite is true. Management must be involved, or the training may become task or technique oriented as opposed to results oriented. It is critical that training be results oriented, or in the long run it will fail.

Quality Tip

Avoid Satirical Humor in Training

"Political affiliation, industry affiliation, open-mindedness, liberal/conservativeness, verbal intelligence, and verbal scores on the Scholastic Aptitude Test (SAT) have all been shown to influence understanding and enjoyment of satire. Unless you know your audience quite well, satirical humor in training is usually best avoided."

Source: Ron Zemke, "Humor in Training," *Training* (August 1991), 29.

QUALITY TIP

Why Training Sometimes Fails

"Training in managing for quality can fail for a variety of rather conventional reasons: inadequacies in facilities, training materials, leaders, and budgets. Such inadequacies are usually obvious enough to generate alarm signals to those directing the program. The more subtle reasons for failure are also the most serious, since they may generate only subtle alarm signals or no signals at all."

Source: Joseph M. Juran, *Juran on Leadership for Quality: An Executive Handbook* (New York: Free Press, 1989), 342.

■ *Too narrow in scope.* Training that is to improve quality should proceed from the broad and general to the more specific. Often organizations jump right into the finite aspects of total quality such as statistical process control, just-in-time manufacturing, or teamwork before employees understand the "big picture" and where these finite aspects fit into it.

Writing for *Training*, Linda Harold makes the point that training sometimes fails because it focuses too specifically on how to more effectively and efficiently perform a task or complete a process instead of focusing on helping employees become independent-thinking, creative problem solvers.[35] According to Harold:

> Over the past few years, many organizations have launched training programs for quality improvement. These programs are designed to teach employees specific processes that will improve quality. And, initially, significant improvements in quality do occur. Once these processes have been implemented, however, many organizations reach a plateau in quality improvement. Why? Employees have been trained in a process. They have not been allowed to think for themselves. When employees complete the process, they don't know how to take the next step because they haven't learned to think through the next step on their own.[36]

QUALITY TRAINING CURRICULUM

For managers to play a leadership role in a total quality setting, they must be well trained in at least what Juran calls the Juran Trilogy: quality planning, quality control, and quality improvement.[37] A curriculum outline for each of these areas is provided in the following sections.

Quality Planning Training

Quality planning is the first component of the Juran Trilogy. According to Juran, quality planning should cover the following topics:[38]

- Strategic management for quality
- Quality policies and their deployment
- Strategic quality goals and their deployment
- The Juran Trilogy
- Big Q and little Q
- The triple-role concept
- Quality planning road map
- Internal and external customers
- How to identify customers
- Planning macroprocesses
- Planning microprocesses
- Product design
- Planning for process control
- Transfer to operations
- Santayana review (lessons learned)
- Planning tools

Quality Control Training

Quality control is the second component of the Juran Trilogy. According to Juran, quality control training should cover the following topics:[39]

- Strategic management for quality
- The feedback loop in quality control
- Controllability (self-control)
- Planning for control
- Control subjects
- Responsibility for control
- How to evaluate performance
- Interpretation of statistical and economic data for significance
- Decision making
- Corrective action
- Quality assurance audits
- Control tools

Quality Improvement Training

Quality improvement is the third component of the Juran Trilogy. According to Juran, quality improvement training should cover the following topics:[40]

- Strategic management for quality
- The Juran Trilogy

QUALITY TIP

Deming on Consistency of Performance

"Another reason Deming focuses so heavily on training is that achieving consistency in the output of employees is as important as reducing the variation in the items produced by two different machines or delivered by two different suppliers. As with everything else in a Deming-oriented process, the goal is to bring workers into statistical control—that is, to have their work be as uniform and predictable as possible. Deming argues that once a group is performing in a stable and predictable manner, defects and problems that occur are the fault not of workers, but rather of the system. Once the performance of the workforce is under control, management and workers can begin to search for more efficient ways to perform a job."

Source: Andrea Gabor, *The Man Who Discovered Quality* (New York: Random House, 1990), 26.

- Quality Council and its responsibilities
- Cost of poor quality: how to estimate it
- Project-by-project concept
- Estimating return on investment
- Nominating, screening, and selecting projects
- Infrastructure for quality improvement
- Macroprocess improvement projects
- Diagnostic journey
- Remedial journey
- Progress review
- Using recognition and reward to motivate
- Quality improvement tools and techniques

By standardizing the curriculum as set forth here and ensuring that all managers complete training in these three broad areas of quality, companies can come closer to achieving consistency of performance, and consistency of performance is critical in a total quality setting because it makes performance easier to measure and improve.

ORIENTATION TRAINING

New employees walk into a new job cold. They don't know the organization, its corporate culture; its rules, regulations, and expectations; or its employees. For this reason, orientation training is important. Although orientation is widely provided, too frequently it is haphazard and of poor quality.[41]

QUALITY TIP

Quality of Orientation Training

"Orientation for new employees consistently turns up as the most prevalent type of formal training in American organizations. . . . Frequency, however, is no guarantee of quality. In most organizations, the content and format of orientation programs are determined by historical precedent, fiat, and whim. The factors that should shape orientation—the message of the organization and the needs of the new employee—only capriciously influence the information and experiences provided to newcomers."

Source: Jeff Brechlin and Allison Rossett, "Orienting New Employees," *Training* (July 1997), 45.

According to Jeff Brechlin of Apple Computers, Inc., and Allison Rossett of San Diego State University, three recurring errors are associated with orientation training:[42]

■ *Insufficient information.* People typically begin new jobs with some trepidation. They want to know where they fit in, what is expected of them, and with whom they will be working. A problem with some orientation programs is that they don't provide enough of the right information to give new employees a good start. The obvious difficulty in such cases is that employees take longer, through no fault of their own, to become productive.

■ *Too much information.* Some organizations determined to give new employees a good start actually give them too much information too fast. The human mind can absorb and act on only so much data at a time. When it tries to take in too much too fast, information overload can occur. The net result is the same as when too little information is given.

■ *Conflicting information.* A common problem with orientation programs is that new employees often receive conflicting information. The orientation provides new employees with the organization's approved approach to a given situation, only to have this information refuted by experienced employees who say, "Ignore that stuff; here is how we really do it."

These three problems can be prevented. Brechlin and Rossett recommend the following principles for providing effective orientation training:[43]

■ *Base orientation topics on a needs assessment.* Before developing an orientation program, assess the needs of both the organization and the new employees. Characteristics shared by good orientation programs are these: they afford new employees privacy as they proceed through the program; they provide new employees with ongoing access to information, people, and resources; and they reflect the culture (tone, feeling, spirit) of the organization.

QUALITY TIP

Orientation Training: Make It Count

"Like any major product or process, a successful orientation depends on establishing a system to ensure adequate resources, planning, communications, development, testing, revision, update, and maintenance. No manager should do it alone. Orientation could mean so much to new employees and organizations, but it rarely does. Too many programs make little or no contribution to the competence, productivity, or satisfaction of new hires. The lesson is clear: You never get a second chance to make a first impression. Make it count."

Source: Jeff Brechlin and Allison Rossett, "Orienting New Employees," *Training* (July 1997), 51.

■ *Establish an organizing framework.*   Should information be organized chronologically, by major functions, or by mission? The first portion of the program might deal with people issues; the second, rules and regulations; the third, work processes; and so on. Regardless of the approach used to organize the training, it is important that it be arranged logically and proceed in a step-by-step manner.

■ *Establish learner control.*   This involves putting the learner in control of his or her learning. This can be accomplished only if instruction is self-paced and individualized. Fortunately, computers and instructional media such as videotapes have made the development and use of individualized instruction easier, which in turn, puts learners in control, letting them pursue learning in their own order of priority.

■ *Make orientation a process, not just an event.*   Typically, orientation programs are front-loaded in the first day or so of employment and treated as a finite event. Such an approach can limit their value. Front-loading only the learning that must be accomplished before beginning work and then spreading the rest appropriately over a period of months can improve results. An important part of the extended orientation process should be regularly scheduled conferences with a supervisor or mentor who can observe progress, identify potential trouble spots, and help facilitate the continued growth of the new employees.

■ *Allow people and personalities to emerge.*   Typically, new employees are keenly interested in the people who make up the human side of the organization. Consequently, information about key personnel should be included in orientation materials. This information should include a brief resume, a one-paragraph job description, and a recent photograph. Including personal information such as hobbies will help orientation trainees see their new colleagues as human beings.

■ *Reflect the organization's mission and culture.*   New employees, regardless of their position in the organization, need to know its mission, goals, and priorities. They also need to know the values on which the organization's culture is based—new employees need to know what their employer thinks is important.

- *Have a system for improving and updating.* Successful organizations are not static. They evolve as the circumstances in which they must compete change. New people, processes, technologies, procedures, and priorities evolve and replace the old. As this happens, the orientation program must be updated accordingly. In addition, ongoing evaluations of the program will reveal weaknesses that should be converted quickly. This means that employees, new and old, and at all levels, should be involved in the continual improvement and updating of the orientation program.

CUSTOMER TRAINING

An old adage states, "The customer is always right." Although the message this saying conveys is a good one, in reality the customer is not always right. In fact, one of the main reasons for consumer product failure is improper use by the customer. According to the customer service research firm, Technical Assistance Research Programs:

> up to one-third of all customer complaints are caused by customers who don't know how to use a product. If you add in people who bought products for the wrong reason and those who had mistaken notions about what a product would do . . . it could be even worse than that.[44]

Customer education has several aspects, including shaping customer expectations, providing user support, and marketing. To be satisfied with a product, customers need to know what to expect from the product. This is important because in a total quality setting, quality is defined in terms of customer expectations. Customers with inaccurate or unrealistic expectations are less likely to be satisfied customers.

Customer expectations are shaped by the promotional literature used in marketing the product and by the user support materials provided with the product. For this reason, it is vital that promotional literature be accurate and that it not contain inflated claims about the product. Accurate customer expectations can also be promoted by the organization's customer service representatives. These employees should be adept at providing one-on-one training for customers in person or by telephone.

Giving customers toll-free access to customer service trainers can be an effective way to promote accurate expectations, as illustrated by the following:

> One notable example of this approach is Armstrong World Industries' tactic of printing an 800 number on the surface of its no-wax floors. Previously, customers didn't care for these floors properly, and the company received many complaints about them as a result. The printing on the flooring tells customers to call an 800 number to find out how to remove the message. Actually the message comes off with warm water, but while Armstrong reps have the new customer on the line, they give them tips on how to take care of the floors correctly.[45]

User support might be provided in the form of user manuals, on-site technical assistance, or training provided at a central company facility. Regardless of the approach, providing user support gives a company an excellent opportunity to train customers in

QUALITY TIP

Customers with Problems Are Trainable

"The good news is that from an educational perspective, the organization is dealing with customers at their most trainable moment: They're in the middle of a problem; they want a solution. Assuming they're given the right information, they will likely retain what they have learned."

Source: Bob Filipczak, "Customer Education (Some Assembly Required)," *Training* (December 1991), 33.

the proper use of its product. To take full advantage of this opportunity, an organization must make sure its user manuals are readable, train its technical representatives to be customer trainers, and give customers immediate access to additional help through a user support telephone number. People who provide user support are in an excellent position to turn a new customer into a satisfied, knowledgeable, loyal customer.

Customer training can also help market a product. The philosophy that joins customer training and marketing can be stated as follows: "You wouldn't buy a car if you did not know how to drive one." To get the full marketing value of customer training, it's a good idea to involve marketing personnel in the development of the training.

ETHICS TRAINING

Ethical behavior and the rationale for it can be taught. In fact, almost 40% of the organizations in the United States with 100 or more employees provide ethics training. A survey by the Ethics Resource Center in Washington D.C. revealed that

> 28% of the 711 responding companies provide specific training in ethics. That number jumps much higher if we include those companies providing training only on narrow ethics topics particularly important to specific industries.[46]

What are these industry-specific issues? The Ethics Resource Center identified the following as topics that are widely addressed to in corporate-sponsored ethics training programs:[47]

- Drug and alcohol abuse
- Employee theft
- Conflicts of interest
- Quality control
- Misuse of proprietary information
- Abuse of expense accounts
- Plant closings and layoffs

QUALITY TIP

Customer Training as a Marketing Strategy

"Companies that make a concerted effort to educate their customers usually have two strategies in mind. Obviously customer service is one, but don't overlook the marketing angle. Customer education that helps market a product, a company, or a company's image is often free and aims to establish the company as an expert in the minds of potential customers."

Source: Bob Filipczak, "Customer Education (Some Assembly Required)," *Training* (December 1991), 32.

- Misuse of company property
- Environmental pollution
- Methods of gathering competitors' information
- Inaccuracy of books and records
- Receiving excessive gifts and entertainment
- False or misleading advertising
- Giving excessive gifts and entertainment
- Kickbacks
- Insider trading
- Relations with local communities
- Antitrust issues
- Bribery
- Political contributions and activities
- Improper relations with local government representatives
- Improper relations with state government representatives
- Improper relations with federal government representatives
- Inaccurate time charging to government entities
- Improper relations with foreign government officials

Ethics training is becoming increasingly important as the pressures of succeeding in an intensely competitive global marketplace grow. The need for such training is clear, as reported by Brad Lee Thompson:

Based on survey responses from 1,073 Columbia University business school alumni, former professors John Delaney and Donna Sockell estimate that the average business school graduate faces about five major ethical dilemmas annually. Among the dilemmas reported were incidents involving insider trading, solicitation of bribes, income tax evasion, the marketing of unsafe products, discrimination against women and minorities, kickbacks, discharge of untreated toxic wastes into the environment, and the falsification of company invoices to support a request for a larger budget.[48]

In discussing what works and what does not work when providing ethics training, Thompson makes the following recommendations:[49]

- *Stimulate discussion.* Talking through ethical dilemmas and how they might be addressed gives people the opportunity to try opinions and gain different perspectives. It also allows people who have confronted ethical dilemmas to explain how they handled them and what the results were.

- *Facilitate, don't preach.* Preachy lectures that resemble sermons are more likely to turn people off than help them develop an ethical point of view. Facilitating discussions that help people come to their own conclusions will be more effective than lecturing them on what they should do and how they should behave.

- *Integrate ethics training.* Ethical decisions are not made in a vacuum, but ethics training is often provided in one. Separate courses tend to make ethics appear to be something that is itself separate and apart from other workplace issues. Almost everything done in the workplace has a moral dimension: ethical questions should be integrated throughout an organization's overall workplace training program including orientation, employee development, and management training.

- *Highlight practical applications.* The study of ethics is typically viewed as lofty and philosophical. By necessity, it does have a philosophical foundation. However, ethical dilemmas in the workplace are very practical in nature, and the consequences of decisions made regarding these dilemmas typically have tangible consequences. For this reason, it is important to use case studies that allow participants to relate to and personally understand the consequences of their opinions.

SUMMARY

1. Training is an organized, systematic series of activities designed to enhance an individual's work-related knowledge, skills, understanding, and/or motivation. Training is distinguished from education by its characteristics of practicality, specificity, and immediacy. Education is a broader concept that is more philosophical and theoretical in nature than training. Corporate training in the United States has historically focused more on managers than on workers. However, with the advent of total quality, the focus is beginning to change.

2. Historically, corporate America has not placed as high a priority on training as have companies from such countries as Germany and Japan. However, with the increased pressure from global competition, this attitude is beginning to change.

3. The rationale for training can be found in the following factors: quality of the existing labor pool, global competition, rapid and continual change, technology transfer problems, and changing demographics.

4. It is important to place the emphasis of training on those who need it most and to ensure that training is designed to promote the organization's goals. These requirements are met by assessing training needs before providing training. Training needs

can be assessed by observing, brainstorming, and/or surveying. Training needs should be converted to training objectives that are stated in behavioral terms.

5. Training can be provided in-house; through corporate-owned education and training facilities; in conjunction with colleges, universities, and professional organizations; or via satellite downlinks. Regardless of the approach used, the following strategies should be applied: build in quality from the start, design small, think creatively, shop around, preview, and customize.

6. Evaluating training begins with a clear statement of purpose. With a statement of purpose drafted, the next step is to ask the following questions: Was the training provided valid? Did the employees learn? Has the training made a difference?

7. Managers who serve as trainers should understand the principles of learning and the four-step teaching method (preparation, presentation, application, and evaluation). In presenting instruction, trainers should remember that people learn by doing. Widely used instructional approaches are lecture/discussion; demonstration; conference; simulation; and videotaped, programmed, and interactive video instruction.

8. Functional illiteracy affects business and industry as follows: difficulty in filling high-skill jobs, lower productivity, higher levels of waste, higher potential for damage to sophisticated equipment, and more dissatisfied employees.

9. Before putting employees in training, it is a good idea to teach them study skills that will enhance their learning. They should learn to make a schedule and stick to it, have a special place to study, listen and take notes, read assertively, and study regularly instead of cramming.

10. When training fails, the reason is often a lack of participation by management or insufficient scope (focusing on the specifics before teaching the "big picture").

11. Quality training should be divided into three broad categories of study: quality planning, quality control, and quality improvement.

12. Orientation training sometimes fails. When it does, the cause is usually one of the following factors: insufficient information, too much information, or conflicting information. To improve orientation training, organizations should base orientation topics on a needs assessment, establish an organizing framework, establish learner control, make orientation a process rather than an event, allow people and personalities to emerge, reflect the organization's mission and culture, and have a system for improving and updating.

13. Topics frequently dealt with in ethics training programs include drug and alcohol abuse, theft, conflicts of interest, abuse of expense accounts, misuse of company property, kickbacks, bribery, improper relations with government officials, and false advertising.

 KEY TERMS AND CONCEPTS

Application	Conference method
Body	Customer training
Changing demographics	Demonstration method
Closing	Design small

Education

Ethics training

Evaluation

External approaches

Four-step teaching method

Global competition

Immediate and continual feedback

In-house training

Internal approaches

Lecture/discussion method

Lesson plan

Opening

Orientation training

Partnership approaches

Preparation

Presentation

Preview and customize

Principles of learning

Programmed instruction

Quality control training

Quality improvement training

Quality of the existing labor pool

Quality planning training

Rapid and continual change

Shop around

Simulation

Study skills

Technology transfer problems

Think creatively

Training

Training aids list

Training objectives

Valid training

Workplace literacy

 FACTUAL REVIEW QUESTIONS

1. Define *training*, and explain how it differs from *education*.
2. Explain the total quality philosophy of training.
3. Name 10 widely used instructional methods.
4. Describe the traditional attitude of corporate America toward training. How does this view compare with that of other industrialized countries?
5. List the five factors that magnify the need for training.
6. What is technology transfer? What effect can it have on an organization's competitiveness?
7. Explain the potential benefits of training.
8. If asked to assess a department's training needs, how would you go about it?
9. What is a training objective? Write a sample training objective in behavioral terms.
10. List and explain five strategies for maximizing training resources.
11. What does a manager need to know to be sure that training satisfied the training objectives? How can these things be determined?
12. Describe how to evaluate a training program before purchasing it.
13. List and explain the principles of learning.
14. Describe the four-step teaching method.
15. What are the minimum recommended contents of a lesson plan?

16. As an instructor, what can you do to help participants retain what they are learning?
17. List and explain the three components of a lecture.
18. Describe the following presentation methods:
    - Simulation
    - Programmed instruction
    - Conference

19. Summarize the status of literacy in the United States and the impact it has on the competitiveness of U.S. companies.

20. What can industry do to deal with workforce illiteracy?

21. Describe what managers in a total quality setting should know about literacy training.

22. List and describe study skills employees should learn before beginning a training program.

23. What are the pitfalls to avoid when using humor in training?

24. Why does training sometimes fail?

25. Juran recommends what three major components in quality training?

26. What are the reasons that orientation training sometimes fails? How can these factors be overcome?

27. List 10 widely taught ethics topics.

 CRITICAL THINKING ACTIVITY

A Training Problem

Amanda Carr is a quality management trouble shooter who has been brought in as a consultant by The Werner Corporation (TWC). TWC is applying the principles of quality management in an attempt to continually improve every aspect of the company's performance—something it must do to compete globally.

TWC's top executive has done everything right as far as he can tell. The company has a good strategic plan, employees are involved and empowered, executive-level commitment is in place, and there is a unity of purpose from top to bottom in the organization. Employees and executives alike want the company's performance to improve, and they all want to do their part. Unfortunately, things are just not improving. Something is wrong, but TWC's management team has been unable to determine what it is. Hence, the contract with Amanda Carr.

Amanda thinks she has found the problem. TWC's managers and employees have plenty of *want-to* but very little *know-how*. Training is needed from top to bottom in the company. How should Amanda proceed? What types of training are needed? How should it be provided? What obstacles will have to be overcome? How can they be overcome?

DISCUSSION ASSIGNMENT 12–1

Education and Training at Motorola

At Motorola, ongoing training for employees at all levels is one of the company's highest priorities and the foundation of its quality and competitiveness effort. The company spends more than $50 million annually on the education and training of employees from top to bottom. A corporate commitment at Motorola is that at least 2% of all employees' time will be spent in training each year. "Being in the highly competitive and changeable electronics industry demands that the company's technical expertise be continually updated. However, Motorola goes far beyond the minimum technical requirements. Motorola has created an advanced learning environment to update technical skills, organizational skills, and reinforce basic values. As changing technology and worldwide competitiveness demand more from companies intent upon being worldclass, the Motorola educational process enables its employees, including senior executives, engineers, and production floor workers, to continually upgrade their skills and remain optimally productive."

Three categories of courses are offered at Motorola: critical path courses, professional development courses, and optional technical courses. First-level managers, middle managers, and senior managers must complete all critical path courses, which include the following: Effective Manufacturing Supervision, Statistical Process Control, Improve the Production System, Workplace Organization, Total Productive Maintenance, Changeover Time Reduction, Group Technology Material Flow, Asset Management, Manufacturing Cycle Management, Leveling Production Schedules, Pull Production System, Project Planning/Project Control, Successful Negotiator, and Motorola Management Institute. Professional development courses are available to first-level and middle managers. They include Interpersonal Skills, Effective Presentations, Managing Engineers' Performance, and Effective Meetings. Optional technical courses are available to managers at all levels. They include Motorola Product Development Design for Manufacturability and Focused Action Study Team (FAST).

Discussion Questions
Discuss the following questions in class or outside of class with your fellow students:

1. Critique the Motorola training commitment and program. Two percent of all employees' time annually amounts to about 40 hours. Is this enough time?
2. What is your opinion of the actual courses provided?

Source: James A. Belohlav, *Championship Management: An Action Model for High Performance* (Cambridge, MA: Productivity Press, 1990), 122.

DISCUSSION ASSIGNMENT 12–2

Importance of the Basics

According to Philip Crosby, Coach Vince Lombardi was known for winning by mastering the basics. He knew that football teams won or lost based on how well they could execute such fundamentals as blocking and tackling.

"The team members were excellent athletes, but not beyond what was to be found on other league football teams. The plays designed by the coaching staff were rather routine; they didn't invent a new, mysterious formation that guaranteed victory. Yet for a few years they were the best in every department. They learned to perform the basics without error. They would block, tackle, run, and otherwise perform their assigned tasks with confidence and reliability. They could depend on each other. A strong discipline, built on respect, held them together and helped them pay attention to the job at hand."

Discussion Questions
Discuss the following questions in class or outside of class with your fellow students:

1. What does the lesson in this case about the importance of the basics say about organizations that must hire employees who are not well prepared?
2. What advantages does an organization gain when all of its employees are well prepared in the basics of reading, writing, calculating, speaking, listening, and critical thinking?

Source: Philip B. Crosby, *The Eternally Successful Organization* (New York: McGraw-Hill, 1988), 26.

DISCUSSION ASSIGNMENT 12–3

Corporate Training at Rockwell International

The Rocky Flats Plant of Rockwell International is committed to hiring hard-core unemployed people for entry-level positions with the intention of migrating them upward. However, many new employees have serious deficiencies in basic education skills. Technological changes compound the problem. As a result, Rockwell has established an in-house training program to help new employees develop basic skills and the more advanced skills needed to migrate upward.

A heavy emphasis is placed on algebra, chemistry, and safety. Rockwell pays all training costs, including the costs of college courses taken off-site. Required courses are taken on company time. Courses for promotion are taken on the employee's time. Self-paced and computer-aided instruction methods are used for teaching basic skills.

DISCUSSION ASSIGNMENT 12–4

Better Education in Dalton, Georgia

"Nestled in the northwestern Georgia hills, the bustling town of Dalton is known as the 'Carpet Capital of the World.' The area's 200 mills manufacture 66% of all the carpet produced in the U.S., but in Dalton and surrounding Whitfield County, 56% of the area's adults have less than a high-school education. The gap between the need and availability of workers who can respond to the demands of modern production technologies is a concern. More than 300 companies in the area signed a pledge to encourage job applicants under 19 to finish high school before seeking a job with their companies and to hire high-school students on a part-time basis only, establishing contact with the school counselor to ensure that attendance and grades are maintained. The other points in the pledge include stressing the value of education to employees and giving special recognition to employees who receive a high school equivalency diploma and to employees' children when they graduate from high school."

Discussion Questions
Discuss the following question in class or outside of class with your fellow students:

1. What other ways can businesses help improve the quality of education in local communities?

Source: Joan C. Szabo, "Schools That Work," *Nation's Business* (October 1991), 20–28.

DISCUSSION ASSIGNMENT 12–5

Improving Learning with Humor

"The best evidence of humor's effect on long-term learning comes from the work of Abner Ziv of Tel Aviv University, one of the most published researchers in the field. In one of Ziv's experiments, some statistics professors were trained to teach an entire statistics course with and without humor. Each professor taught two courses, one intended to include plenty of laughs and the other deliberately unfunny. Students were measured throughout the semester on standardized statistics tests. When the numbers were in, students subjected to the 'humor treatment' scored an average of 15% higher."

Discussion Questions
Discuss the following questions in class or outside of class with your fellow students:

1. Have you ever taken a class in which the professor used humor to make his or her points?
2. Do you like or dislike the use of humor in the classroom?
3. Have you ever had a professor who overused humor to the point that it interfered with learning?

Source: Ron Zemke, "Humor in Training," *Training* (August 1991), 27.

DISCUSSION ASSIGNMENT 12–6

Training at Johnsonville Foods

"Several years ago, Johnsonville Foods changed its management style to allow the people who do the work to make decisions relating to their own jobs. This new way of life yielded great improvements in the commitment and zeal with which its *members* (formerly called *employees*) approached their work.

"These changes have had a measurable effect on the bottom line. From 1982 to 1990, return on assets doubled, sales increased eightfold, rejects were reduced from 5% to less than 1.5%, and the ratio of complaint to compliment letters went from 5-to-1 to 1-to-2.8.

"One of the keys to this transformation was a decision to change the focus of the company from using people to build a great business to using the business to build great people. In other words, Johnsonville became a learning organization. Those in leadership positions had to relinquish their monopoly on decision making and allow people in the plant to make decisions.

"This was not an overnight transformation. One of the first things to change was the name of the training department: It became the member-development department. Its focus broadened from training to all aspects of the mental development of people."

Discussion Questions
Discuss the following questions in class or outside of class with your fellow students:

1. What do you think will happen if employees are allowed to make decisions without first being trained in decision making?
2. What other aspects of employee development are there in addition to training?

Source: Joseph M. Juran, *Juran on Leadership for Quality: An Executive Handbook* (New York: Free Press, 1989), 337.

DISCUSSION ASSIGNMENT 12–7

Is the Customer Really Always Right?

"A software company was having trouble with one of its clients. The customer called repeatedly to complain that the software wouldn't work. Finally the company's support people suggested that the floppy disks might be dusty. The customer assured them this couldn't be the case because the disks had been washed faithfully with soap and water.

"A manufacturer of electronic water-control meters discovered some of its customers were using the meters to control the flow of sulfuric acid. Customers reported the meters were defective. No wonder: Vital parts of the meter had been completely eaten away."

Discussion Question
Discuss the following question in class or outside of class with your fellow students:

1.   Do you think the customer is always right? Defend or refute the following statement: "The customer is not always right, but should always be treated right."

Source: Bob Filipczak, "Customer Education (Some Assembly Required)," *Training* (December 1991), 31.

DISCUSSION ASSIGNMENT 12–8

A Road Map to Competitiveness

"Other nations have pursued creative human-resource policies in part to improve their competitiveness in a rapidly integrating world economy. But becoming more competitive is in itself no challenge at all. There are only two ways to compete—reducing incomes or improving productivity and quality. We can, in other words, become more competitive simply by lowering the value of the dollar in relation to other currencies. That, however, is simply the path to lower income and a diminished standard of living for all Americans—and it is the path we now tread. The challenge is to be competitive while maintaining high incomes and full employment. This is now impossible without high productivity growth rates, and these, in turn, cannot be achieved without very high-quality human resources."

Discussion Assignment
Discuss the following questions in class or with your fellow students outside of class:

1.   What factors are the greatest inhibitors that might prevent an organization from having a world-class workforce?

> 2.   If you were a corporate CEO, how would you increase the quality of your work-force to a world-class level?
>
> Source: Ray Marshall and Marc Tucker, *Thinking for a Living* (New York: Basic Books, 1992), xvi.

↰ ENDNOTES

1. This section is based on Jack Gordon, "Where the Training Goes," *Training* (October 1990), 51–69.
2. Dick Schaaf, "The Changing Role of Training: From Rage to Riches?" *Corporate Development in the 90s: A Supplement to* Training (1991), 5.
3. Schaaf, 5.
4. Schaaf, 5.
5. John Hoerr, "Business Shares the Blame for Workers' Low Skills," *Business Week* (25 June 1990), 71.
6. Brad Lee Thompson and Bob Filipczak, "Training Is Ticket Abroad," *Training* (September 1990), 12.
7. National Center for Manufacturing Sciences, *Focus* (August 1992), 8.
8. National Center for Manufacturing Sciences, 8.
9. Philip J. Harkins, "The Changing Role of Corporate Training and Development," *Corporate Development in the 1990s: Supplement to* Training (1991): 27.
10. Hoerr, 71.
11. Tom Peters, *Thriving on Chaos: Handbook for a Management Revolution* (New York: Harper & Row, 1987), 386.
12. Nell P. Eurich, *Corporate Classrooms: The Learning Business* (Lawrenceville, NJ: Princeton University Press, 1985), x, xi.
13. Eurich, 8.
14. Carolyn Wilson, *Training for Non-Trainers* (New York: AMACOM, 1990), 18–19.
15. Eurich, 16.
16. Joseph M. Juran, *Juran on Leadership for Quality: An Executive Handbook* (New York: Free Press, 1989), 323–327.
17. Juran, 323–324.
18. Juran, 324.
19. Gilda Dangot-Simpkin, "How to Get What You Pay For," *Training* (July 1990), 53–54.
20. This section is based on David L. Goetsch, *Industrial Safety and Health in the Age of High Technology* (Upper Saddle River, NJ: Prentice Hall/Merrill, 1993), 391–402.
21. Quoted in Goetsch, 519.
22. Quoted in Goetsch, 520.
23. Juran, 331–332.
24. Juran, 331–332.
25. This section is based on Juran, 332.

26. Goetsch, 413–415.
27. E. L. Fields, W. L. Hull, and J. A. Sechler, *Adult Literacy: Industry-Based Training Programs* (Columbus, OH: National Center for Research in Vocational Education, 1987), vii.
28. This section is based on Fields, Hull, and Sechler, 5–31.
29. Fields, Hull, and Sechler, 5–31.
30. *About Adult Learners* (South Deerfield, MA: Channel L. Bete, 1991), 8–9.
31. Ron Zemke, "Humor in Training," *Training*, August 1991, 27.
32. Zemke, 27.
33. Zemke, 27.
34. Juran, 342.
35. Linda Harold, "The Power of LEARNING at Johnsonville Foods," *Training* (April 1991), 56.
36. Harold, 56.
37. Juran, 337.
38. Juran, 325.
39. Juran, 336.
40. Juran, 337.
41. This section is based on Jeff Brechlin and Allison Rossett, "Orienting New Employees," *Training* (July 1997), 45–51.
42. Brechlin and Rossett, 46–47.
43. Brechlin and Rossett, 47–51.
44. Bob Filipczak, "Customer Education (Some Assembly Required)," *Training* (December 1991), 31.
45. Filipczak, 32.
46. Brad Lee Thompson, "Ethics Training Enters the Real World," *Training* (October 1991), 84.
47. Thompson, 85.
48. Thompson, 86.
49. Thompson, 91–94.

CHAPTER THIRTEEN

Overcoming Politics, Negativity, and Conflict in the Workplace

"A time-conscious manager can grind genius into gruel."
Thomas A. Stewart

MAJOR TOPICS

- Internal Politics Defined
- Power and Politics
- Organizational Structure and Internal Politics
- Internal Politics in Action
- Internal Politicians and Their Methods
- Impact of Internal Politics on Quality
- Controlling Internal Politics in Organizations
- Overcoming Negativity in Organizations
- Overcoming Territorial Behavior in Organizations
- Managing Conflict in Organizations

Internal politics is the game everybody plays, but nobody talks about. It is one of the most underemphasized but overpracticed barriers to organizational performance that exists. In fact, there are actually books on the market that have the expressed purpose of making people better internal politicians—that is, better practitioners of the art and science of organizational manipulation. The root causes of internal politics are personal insecurity, self-interest, a hunger for power, ambition, and the need for acceptance. In other words, the root cause of internal politics is human nature. Consequently, some people think that internal politics in organizations is inevitable and should be accepted as an unavoidable fact of life. Although it is true that internal politics is an organizational disease that has no known cure, it is equally true that the disease can be controlled, and it should be. Internal politics has no place in a total quality organization. Quality is about *we*. Internal politics is about *me*.

The success of total quality depends, in part, on organization-wide collaboration. One of the fundamental cornerstones of total quality is *unity of purpose*. This means that all employees at all levels understand the vision and are united in their efforts to accomplish it. Consequently, all decisions are based on the same criterion: what option best supports the vision? This requires putting the interests of the overall organization ahead of the interests of individual divisions, departments, units, teams, and employees. Because human beings tend to be self-serving by nature, establishing unity of purpose and collaboration is a challenging undertaking that requires constant attention.

In an organization with unity of purpose, collaboration is the norm. In an organization infected with internal politics, the promotion of self-interest is the norm. Internal politics drains the competitive energy of thousands of organizations every day. To derive the full benefit of total quality, organizations must rid themselves of the debilitating scourge of internal politics.

INTERNAL POLITICS DEFINED

A historic example of internal politics at its worst occurred during World War II. Pearl Harbor and the Philippines had fallen before the Japanese onslaught in the Pacific, and it looked as if New Guinea and Australia would be next. The Allied forces were in a real bind. On the one hand, they couldn't allow these two countries to fall into enemy hands. But on the other hand, they couldn't pull enough troops away from the European and Atlantic theaters of operation to stop Japan.

In desperation, General Douglas MacArthur was ordered by President Franklin D. Roosevelt to leave Bataan, the tiny sliver of land in the Philippines to which MacArthur and his American and Filipino troops had clung tenaciously for several months. MacArthur and his beleaguered troops were holding out against great odds to give the United States time to recover from the devastating blow suffered at Pearl Harbor. This was a monumental decision because if Bataan fell—which it surely would without MacArthur's presence—the Japanese would add the Philippines to their list of conquests. Reluctantly, MacArthur and a small staff slipped away under cover of darkness aboard patrol torpedo (PT) boats of the American navy. Breaking through the Japanese blockade, they eventually reached Australia.

In short order, MacArthur rallied the dangerously understaffed and underequipped military units available to him. Relying on innovative strategy, raw courage, and stubborn determination, MacArthur and his troops made their stand and stopped the Japanese juggernaut on the Owen Stanley mountain range in New Guinea. From that point

QUALITY TIP

"Politics is more dangerous than war for in war you are killed only once."
Winston Churchill

on, MacArthur conducted one of the most brilliant military campaigns in the annals of war, eventually pushing a stronger, healthier, better fed, better equipped Japanese military out of the southwest Pacific.

The Japanese army and navy were well trained, well led, and fanatically determined, but they weren't MacArthur's most difficult foe. Also working against him was the insidious scourge of internal politics. To conduct the type of island-hopping warfare that was necessary in the southwest Pacific, MacArthur needed the U.S. Army, Navy, Marines, and Air Corps, as well as military units from both Australia and New Zealand, to work together in a closely coordinated, collaborative effort. As it turns out, trying to achieve coordination and collaboration among the disparate groups under his command was General MacArthur's greatest challenge and most persistent frustration.

Rivalries among different branches of the U.S. military, as well as within individual branches of the military, were legendary at the time Japan attacked Pearl Harbor. The personal ambitions of military leaders, service loyalties, disagreements over the allocation of resources, and jealousy relating to who was in command of what were all ongoing sources of problems that created almost as much trouble for MacArthur as did the Japanese. In the European theater, General Eisenhower faced the same frustrations daily as he tried to coordinate the military forces of the various Allied nations.

In a meeting with Army Chief of Staff George C. Marshall, MacArthur vented his frustration. Recalling this meeting in his memoirs, MacArthur said, "I felt it fantastic, to say the least, that interservice rivalry or personal ambitions were allowed to interfere with winning the war."[1] This same kind of situation can be found in almost any organization. Internal politics is a natural, if unfortunate, outgrowth of human nature.

Politics, in general, is the art and science of wielding influence in such a way as to gain advantage. We generally think of politics in relation to influencing government. However, politics as a means of influencing outcomes is not limited to government enterprises. In fact, one will find politics practiced with great skill in virtually every type of organization. Internal politics, as practiced in organizations, is defined as follows:

> Internal politics consists of activities undertaken to gain advantage or influence organizational decision making in ways intended to serve a purpose other than the best interests of the overall organization. Internal politics consists of the games people play to promote decisions that are based on criteria other than merit.

Internal politics manifests itself in a number of different behaviors, all of which amount to individuals or groups within an organization putting their interests ahead of those of the overall organization. Internal politics is not necessarily an inherently bad concept. For example, internal politics practiced for the purpose of furthering the interests of the overall organization would be acceptable. The problem with the concept is that it is rarely practiced in a positive way. Even though most invariably claim they are acting in the best interests of the organization, internal politicians typically have their own interests at heart in every action they take.

In a worst-case scenario, internal politicians are people who put self-interest ahead of organizational interests. In a best-case scenario, internal politicians are people who put not self-interest but the interests of their divisions, departments, or teams ahead of those of the overall organization. It is this—the self-serving nature of the concept as it

QUALITY TIP

Root Cause of Internal Politics

Individuals in organizations—frequently the organization's key decision makers—can become so intent on fulfilling their ambitions, satisfying their personal needs, and feeding their individual egos that they lose sight of what's best for the organization that employs them.

Source: David L. Goetsch.

is typically practiced—that makes internal politics such a negative phenomenon in organizations. It undermines collaboration, trust, and unity of purpose—all fundamental elements of total quality.

POWER AND POLITICS

Power is the ability to exert influence. Power is neither inherently positive nor inherently negative. It is a concept that can cut either way, depending on how it is used. Power, properly applied in an organization, is used to move the organization closer to the realization of its vision. Power, improperly applied in an organization, is used to advance an agenda other than that of the overall organization. This is the way in which internal politicians use power. The five different sources of power in an organization are: personal, position, capability, reward, and coercive power. Internal politicians may use any or all of these sources of power to advance their personal agendas.

Personal Power

Personal power is the power of an individual's personality. Individuals with personal power are generally persuasive and/or charismatic. They tend to have strong beliefs, an aura of confidence, and an air of determination. In the military, such people are said to have the intangible attribute of *command presence*. Their personalities appeal to certain people—their followers—on an emotional level.

Position Power

Position power is that bestowed officially by higher authority. People with the authority to hire and fire, allocate resources, evaluate performance, and make decisions that affect the jobs of others have position power. People with position power may or may not be adept at using their power to influence others. Weak leaders in positions of authority often find that their position alone is not enough to ensure their influence in an organization. An important aspect of power is knowing how to use it and being willing to do so.

For example, if a supervisor is unwilling to confront nonproductive employees, he or she will not be able to influence their behavior for the better, in spite of having position power.

Capability Power

Capability power comes from having special knowledge, skills, or talents that are important to an organization and in short supply. In any organization there are critical tasks that are fundamental to the organization's success. People who can perform these tasks often gain power. In some instances their power will extend well beyond that which might be expected of a person in their position. Since automation in the workplace became the norm, people who are knowledgeable about computers have been afforded capability power.

Reward Power

Reward power comes from the authority to control, administer, or withhold something of value to others. Often the something in question is money. People in positions to give or withhold money from others in organizations have reward power. In addition to money and resources, rewards that might be granted or withheld include recognition, promotion, positive feedback, and/or inclusion in a group. The reward of inclusion is one of the favorite tools of the internal politician. People have a natural desire to be part of a group. Consequently, the ability to exclude others gives an individual power over those who want to be associated with a given group.

Coercive Power

People who have the ability to punish others or subject them to unpleasant circumstances have coercive power. Coercive power is founded in fear, and its application is based on threats. The threat—whether implicit or explicit—is to punish those who don't respond as expected.

In a workplace setting, the coercive threat usually has to do with the victim's job security, work schedule, and/or pay. The threat can also be one of physical abuse, although this is less common in the workplace than are threats to job security. The threat of ostracism is also a commonly used coercive tactic. It involves threatening to remove or exclude an individual from a group with which he or she wants to be associated.

ORGANIZATIONAL STRUCTURE AND INTERNAL POLITICS

Internal politics can exist, and usually does, in every type of organization, regardless of organizational structure. In other words, internal politics is not caused by organizational structure; consequently, it cannot be eliminated by simply changing the structure. This point can be illustrated easily enough by considering the most widely used organizational structures: the functional, geographic, decentralized line-of-business, strategic business-unit, and matrix structures.

Each of these structures has its own set of advantages and disadvantages, but they are all susceptible to internal politics. In considering organizational structures, keep the following rule of thumb in mind: factors that promote internal competition—whether intentionally or unintentionally—also promote internal politics. This is because internal politics is just one more tool (or weapon) to be used any time people within an organization compete among themselves. The competition might be about questions of who reports to whom, who gets the lion's share of limited resources, who will fill key positions, what department will be reorganized and how, lines of authority, and any number of other issues in which people have vested interests.

The next few figures exemplify the five most widely used organizational structures. Even a cursory examination of these various structures will reveal built-in characteristics that tend to promote competition and, in turn, internal politics. In the functional structure represented in Figure 13–1, the CEO would want to be sensitive to political machinations among the functional vice presidents. These could grow out of competition for limited resources, personality clashes, competition among the vice presidents for a future promotion to the CEO position, or misguided loyalty to the respective functional areas on the part of employees at all levels within the organization.

The geographic structure (Figure 13–2) is particularly susceptible to internal politics. The Achilles' heel of this structure is distance. Distance has a tendency to promote an "us against them" mentality that can lead to counterproductive internal competition and other forms of negative behavior. The reason for this is simple: it is only natural to be suspicious of people you seldom or never see but with whom you compete for resources. In addition, it is difficult to establish trust, unity of purpose, and collaboration across distance. For this reason, executives in companies that are organized geographically should be especially sensitive to the issue of internal politics.

The decentralized line-of-business structure, as shown in Figure 13–3, tends to promote competition among the various business enterprises, as well as between functional departments within these separate enterprises. This is also true of the strategic business-unit structure (Figure 13–4).

Figure 13–1
Functional Structure

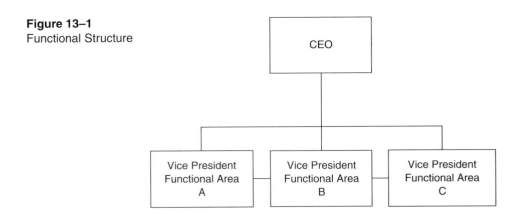

Figure 13–2
Geographic Structure

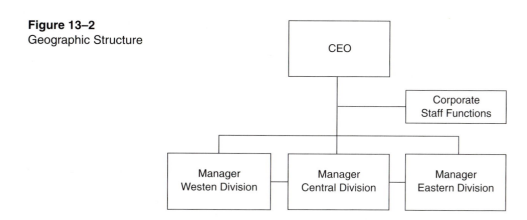

The matrix structure is the least used of the five common organizational structures. With this structure, functional specialists—human resources, accounting, marketing, and so on—are organized under functional areas, with each area having its own manager. Business ventures undertaken by the overall organization then draw on these functional areas as needed. For example, in Figure 13–5 the company is currently pursuing three business ventures. Business Ventures 1 through 3 all draw the functional expertise they need (e.g., marketing, accounting, engineering) from the respective functional areas.

This structure tends to promote internal competition and, in turn, internal politics both vertically and horizontally within the matrix. Vertically there is competition for resources among the various business ventures. For example, say that all business ventures in Figure 13–5 want to draw engineering services at the same time, and the functional vice president of engineering has an insufficient number of engineers. How will available engineers be allocated? Another spin on this same situation occurs when two

Figure 13–3
Decentralized Line-of-Business
Structure

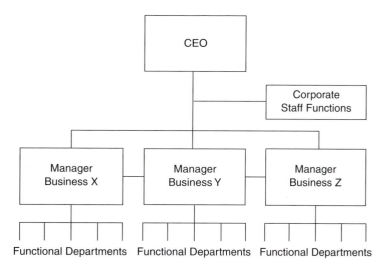

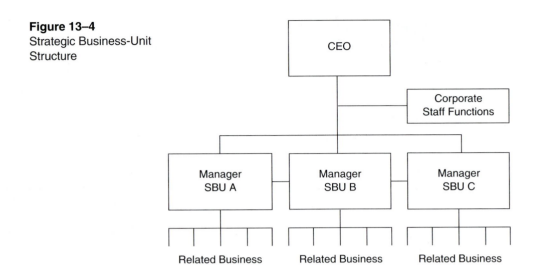

Figure 13–4
Strategic Business-Unit
Structure

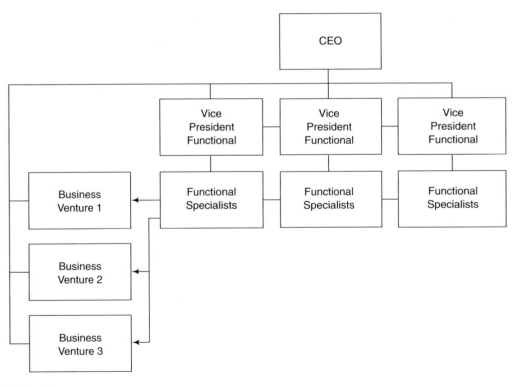

Figure 13–5
Matrix Structure

or more business ventures want the services of the same engineer, a particularly talented individual.

Horizontally, there is competition among the functional areas, certainly over resources, and possibly over personalities and misguided loyalties. For example, say the marketing vice president wants to increase the size of his staff. At the same time, the accounting vice president wants to upgrade her computer system. Resources are limited. Who gets the lion's share?

These few examples, without even considering such other drivers as the need to fit in or personal insecurity, show that regardless of structure, the potential for internal politics exists in any organization. This section is not intended to be an exhaustive description of how organizational structures contribute to the potential for internal politics. Rather, it simply makes the point that regardless of organizational structure, the potential for internal politics exists. Consequently, managers in organizations trying to implement total quality should be sensitive to symptoms of internal politics regardless of how their organizations are structured.

INTERNAL POLITICS IN ACTION

Several motivations—personal insecurity, self-interest, a hunger for power, ambition, ego, and the need for acceptance—are the primary drivers of internal politics. The examples of internal politics provided in this section illustrate these drivers in action. All cases presented are real. The names of people and organizations have been changed, but the facts of the cases have not.

Politics in Reorganization

The two department chairs actually agree with the college administration that a reorganization is in order. Under pressure to cut administrative costs, the college's president has begun merging smaller departments. For example, the history, political science, sociology, and psychology departments have all been merged to form a new social science department. The chair of the old history department is now the chair of the new social science department, and the other three chairs—political science, sociology, and psychology—have returned to the faculty as professors.

A similar scenario has been enacted by combining the art, literature, drama, and music departments to form a new humanities department. Speech and English were combined to form the new communications department. In every case the savings in administrative overhead have been substantial.

The college president is finally down to the last planned merger on her list: the mathematics and statistics departments. These two separate departments are to be merged to form a new mathematics and statistics department with one department chair.

The other mergers have been accomplished with a minimum of discord. In each case, the individual best qualified to become the surviving administrator was obvious and had the support of his or her colleagues. However, this is not the case with the mathematics and statistics departments. For years the chairs of these two departments have been at war with each other. Their ongoing feud began years ago when both were upstart professors struggling to impress their respective supervisors and gain a foothold in the academic world. Since that time their disagreements have become legendary. Now they are competing to see which will emerge as chair of the new mathematics and statistics department and which will become the other's subordinate.

Both are lobbying their dean and the college president persistently—so persistently, in fact, that they are beginning to interfere with the duties of these two administrators. Both are building coalitions, electioneering among the faculty, calling in favors, and even making threats to nontenured faculty members in an attempt to win their support.

In the meantime, their normal duties are being neglected, the faculty is becoming polarized, and the atmosphere on campus has become tense and stressful. These two department chairs have become so intent on satisfying their own personal interests that they are harming the overall organization. Their self-serving political machinations are interfering with the mission of the college: teaching and learning.

The postscript to this case is as follows: fed up with the self-serving internal politics of her mathematics and statistics chairs, the college president returned them both to the classroom and appointed a less senior professor as chair of the new mathematics and statistics department.

Politics in Resource Allocation

There is constant tension at Noonan Computer Center, Inc. (NCCI), between the sales and service departments. John Gates is the manager of the sales department. Amanda Blakely manages the service department. Gates can make a convincing case for hiring additional sales and marketing personnel, and he does so frequently. Blakely can make an equally convincing case for hiring additional service technicians, and she doesn't hesitate to do so.

The two managers are always at odds with each other, feuding constantly over which department should receive the bulk of NCCI's limited resources. They both think their respective departments hold the key to the struggling new company's survival and eventual prosperity. Gates generally states his case as follows: "We need more sales personnel out there knocking on doors, or we will never achieve the volume needed to get over the start-up hump." Blakely typically counters this argument by saying, "Good service is the key. It's how we will gain a reputation that will give us a competitive edge. If we don't service what we sell, we will never be able to sell enough computers to overcome the bad reputation that will result, no matter how many sales reps we hire."

The problem confounding NCCI's owner and CEO, Debbie Parks, is that both Gates and Blakely are right. The company needs more sales reps and more service technicians; there is no question about it. NCCI is still in its infancy. Parks is reinvesting every dollar she earns in the company as fast as she earns it. But it will be some time, even if things continue to go as well as they are now, before NCCI will be able to meet the types of demands Gates and Blakely are making.

Parks needs the cooperation of these two key managers badly. Their constant bickering is dividing the workforce into battling coalitions. Hostility between Gates and

Blakely is subtle, but it is always there. Parks is convinced that they actually covertly undermine each other, even to the point of circulating unfounded rumors, recruiting spies in each other's departments, and promoting vicious gossip about each other.

Parks faces a dilemma. Gates and Blakely are the best there is at what they do, but their internal political maneuvering is distracting employees and draining them of the physical, emotional, and intellectual energy they need to constantly perform at peak levels. Parks needs peak performance from all her employees if she is going to gain a foothold in an intensely competitive marketplace.

The postscript to this case is as follows: NCCI never gained the foothold it needed. After just 18 months in business, the company folded, a victim of internal politics.

Politics in Status

On the company's organizational chart all departmental directors at Payton Temporary Employees, Inc. (PTE), are equal. However, in the minds of the individuals filling these positions, equality among directors at PTE is a concept that looks good on paper but doesn't exist in reality. The more equal among equals are those directors whose offices are located in closest proximity to that of the company's CEO, John Davis. The real winners, according to company folklore, are the directors whose offices are located on the same floor as Davis's office.

If one is to believe company gossip, the directors with offices on the third floor with Davis gain a tangible advantage in that they are more likely to bump into him in the hall, at the water cooler, or, if they are male, even in the rest room. This gives them enhanced access and opportunities for sharing their ideas and proposals. Other directors, without this proximity advantage, must wait for scheduled meetings or make appointments for one-on-one time with the CEO. Right or wrong, at PTE there is a perception of proximity advantage, and this perception creates counterproductive competition, political maneuvering, and ill will among the company's department directors.

The political machinations of the directors take several different forms. Prominent among these are rumormongering, building of coalitions, undermining, electioneering when special committees and task forces are established, and the spreading of gossip. As a result, the company is a balkanized organization of warring factions with an ever-shifting set of loyalties. What it isn't, is a company in which there is a unity of purpose and a concerted effort to have all managers and employees pulling on the same end of the rope toward a common vision.

The postscript to this case is as follows: John Davis's failure to deal with the issue of internal politics at PTE eventually led to his termination. His replacement immediately set about establishing a collaborative culture at PTE; as a result, the company survived and is currently prospering.

Politics and Promotion

Karin Newhouse and Ronald Gorman have a lot in common. Both are plant managers for Northwest Manufacturing Company (NMC), an electronics manufacturer with plants in 10 different locations. Newhouse and Gorman were both gifted college athletes. Newhouse was the best player on a tennis team that won the national championship in her senior year, and Gorman was an Olympic-caliber track star. Both majored in electrical engineering and began their careers with NMC just 15 years ago. Both have been on the fast track to the upper ranks of the company, and both were named managers of their respective plants within a month of each other. There is something else these two fast movers have in common: they both know that NMC's president, Marvin Stanley, plans to retire in 18 months. Newhouse and Gorman both want his job.

Over the years, Newhouse and Gorman have maintained the competitive spirit they developed as champion athletes. Both have used their competitive drive in positive ways to advance their careers. But the moment Marvin Stanley announced his impending retirement, the competition between Newhouse and Gorman intensified. Before long, it began to get out of hand. Eventually it became so bad that both managers, instead of being promoted, were fired, and Marvin Stanley was forced to postpone his retirement to pick up the pieces and get NMC moving forward again.

Here is what happened. Both managers knew they were on the short list to replace Stanley. In fact, they *were* the short list. None of NMC's other eight plant managers were in the running. Newhouse and Gorman decided—independently—that the decision would be made based on whose plant performed best during Stanley's last 18 months on the job. Consequently, both undertook aggressive cost-cutting campaigns. Some of their initial cost-cutting decisions were popular and well received by the company's board of directors. But soon one-upmanship set in. Before long, Newhouse and Gorman were making decisions that—although they paid off in the short run—would actually hurt the company in the long run. And so they did.

Two months before Stanley was scheduled to retire, the employees of Newhouse's plant revolted. All 610 nonmanagement employees took either sick leave or vacation and simply did not show up for work, all on the same day. A spokesperson for the employees said, "We're just trying to get the attention of corporate management. This is a good plant, and we are proud of what we do here, but Newhouse is driving this company into the ground. As a result of her unilateral cost-cutting decisions we can no longer maintain the quality we are known for, nor can we fill orders on time, which has always been one of our hallmarks. We don't know what has been driving the behavior of our plant man-

QUALITY TIP

A Manager's Nightmare

"Interdepartmental politics, attempts at functional empire building, and conflicting functional viewpoints can impose a time-consuming administrative burden on the general manager, who is the only person with authority to resolve cross-functional differences and enforce cooperation."

Source: Arthur A. Thompson, Jr., and A. J. Strickland III, *Strategic Management Concepts and Cases*, 7th ed. (Boston: Irwin, 1993), 224.

ager for the last several months, but somebody from corporate needs to find out before it's too late."

At first, Gorman delighted in the news from Newhouse's plant. "This," he thought, "should clinch the job for me." But what he didn't know was that several of NMC's most important customers had contacted Marvin Stanley complaining about problems with Gorman's plant. The gist of their complaints was that quality had slipped markedly over the past several months and that on-time deliveries were off by 60%.

Unable to believe what was happening, Stanley got his two star plant managers together in his office for a conference. His intentions had been to find out where things had gone awry and get them straightened out. The reports he was hearing from their plants were so out of character for these two go-getters—both of whom he considered protégés—that Stanley simply could not believe them. The meeting had barely gotten started, however, before Newhouse and Gorman were attacking each other with accusations and recriminations. They even made threats.

Stanley was heartbroken. These two were his stars, the two managers he had nurtured and groomed to ensure that NMC would have the leadership it needed to survive and thrive after his retirement. After an hour of unrelenting bitterness between Newhouse and Gorman, Stanley—with reluctance—admitted the obvious to himself. His two stars were so intent on advancing their personal agendas that they had lost sight of what was best for NMC. In his mind, Stanley had just one option: terminate Newhouse and Gorman, and find someone else to take over the reins at NMC.

Stanley undertook a national search for his replacement and was able to find a dynamic leader who not only knew NMC's business but was also well versed in controlling internal politics. The postscript to this case is as follows: under the new CEO's leadership, NMC won back both its employees and its customers and was once again a thriving company.

INTERNAL POLITICIANS AND THEIR METHODS

Internal politicians have many of the characteristics of special interest groups. Special interest groups consist of people who share narrowly focused common goals. Internal politicians are individuals with narrowly focused interests—namely, their own. Special interest groups seek to gain advantage so as to influence governmental decision making. Internal politicians seek to gain advantage so as to influence organizational decision making. With such strong similarities, it should come as no surprise that internal politicians and special interest groups use many of the same methods. The most widely used of these methods are as follows (Figure 13–6):

- Lobbying
- Building coalitions
- Applying harassment and pressure
- Electioneering
- Gossiping and spreading rumors

Figure 13–6
Methods Checklist: Internal
Politicians

> ✓ Lobbying
> ✓ Building coalitions
> ✓ Applying pressure
> ✓ Electioneering
> ✓ Spreading rumors and gossip

Lobbying

According to Alan Gitelson, Robert Dudley, and Melvin Dubnick, "lobbying is the act of trying to influence government decision makers. Named after the public rooms in which it first took place, lobbying now goes on in hearing rooms, offices, and restaurants—any spot where a lobbyist can gain a hearing and effectively present a case."[2] Legend has it that the term *lobbying* is based on the fact that many of the earliest attempts to influence members of Congress occurred in the lobby of the Willard Hotel and other hotels in Washington, D. C.

Governmental lobbyists use favors, financial contributions, and information to influence government officials. Internal politicians, when lobbying, use similar tactics. By doing favors for people in positions to help them, internal politicians hope to curry favor. Ideally, they will establish a *quid pro quo* relationship with someone in a position of influence. Although they don't make financial contributions, internal politicians do have their version of this concept. They might contribute to easing the work load of, or solving a problem for, someone with whom they hope to gain favor.

Sharing of information is another widely practiced lobbying tactic. There is an old saying that knowledge is power. By providing information to carefully selected people, internal politicians attempt to endear themselves. Figure 13–7 contains a list of lobbying tactics commonly used by internal politicians.

Figure 13–7
Lobbying Tactics of Internal
Politicians

- Contacting people formally (by appointment) to present a personal point of view
- Engaging people in informal discussions (over lunch, on the golf course, in the hall, etc.) and presenting a personal point of view
- Providing carefully screened information on a selective basis
- Doing favors to establish *quid pro quo* relationships
- Helping lighten the workload of selected people
- Applying pressure directly to individuals
- Applying pressure through third parties
- Exploiting personal relationships

Doing favors, making contributions, and providing information are not inherently negative activities, quite the contrary. What transforms these otherwise positive activities into negative endeavors is their misuse. If these things are done with the best interests of the organization in mind, they are admirable activities. However, when done for the purpose of advancing a personal agenda at the expense of or without sufficient consideration for the organization's needs, they become negative.

The information-sharing aspect of lobbying is frequently the most misused of the various lobbying tactics. Information provided for lobbying purposes is carefully shaded in favor of the information provider. An internal politician is not going to volunteer information that doesn't serve his or her purpose. This does not mean that internal politicians necessarily lie, nor that they even need to do so. Rather, it means that they carefully control the information they provide, and to whom it is provided, so as to gain the greatest possible benefit. What follows are some examples of how internal politicians can shade information in their favor when lobbying in support of their own personal agendas:

- Myron Conley's boss, John Upfield, is a proud graduate and active alumnus of Centerbury College. Conley has his eye on a promotion. Consequently, he has been looking for an opportunity to mention in casual conversation with Upfield that he, too, attended Centerbury College. When the opportunity finally presented itself, Conley was delighted to see that the association seemed to have a positive effect. What Conley didn't tell Upfield was that he completed less than a year at Centerbury before being expelled for disciplinary reasons and had to finish his degree at another college.

- Myra Gladstone couldn't believe her luck. At the annual corporate banquet she was seated next to her company's corporate vice president, the very person who would decide which unit would get the huge new Johnson account. Gladstone knew that if she got the Johnson account, and if her unit performed well, it would mean a promotion for her. She also knew that corporate would be wiser to give the account to Amanda Perry's unit. Gladstone's unit was good, but for this particular type of account, Perry's unit was better, much better. No matter, Gladstone wanted the account and intended to get it. Taking advantage of her good luck with seating arrangements, Gladstone told the corporate vice president exactly how her unit would handle the Johnson account. Emphasizing strengths and carefully avoiding weaknesses, Gladstone made a convincing case for her unit. The capabilities of Amanda Perry's unit were not mentioned.

- Mack Parmentier knew his division needed more business. In fact, if he didn't increase sales soon, his division would be merged with another larger division and he would be demoted from division vice president back to his old job of product manager in somebody else's division. But what to do? He was struggling with this very question when opportunity knocked—literally. The person knocking on his office door turned out to be Mary Washington, corporate head of marketing. Washington had a new contract, a big multiyear contract, and she was talking with all division vice presidents before recommending which division should get it. Her message was clear: the company had to perform beyond expectations on this contract. If this happened, there would be much more work in the future. If not, the current contract would be the first and last the company would receive from a very important client. Washington told Parmentier she had narrowed the list of divisions down to two, his and the company's eastern division. What appealed to her about Parmentier's division was its strength in the area of precision machining. The contract would involve a great deal of this type of work. Washington told Parmentier, "The eastern division has a strong precision machining unit, but you've got Mike Bates. He is a genius." Seizing on the opportunity, Parmentier pressed his case. "You're right. With Mike and his team, we can outperform anyone on this contract." What Parmentier didn't tell Washington was that Mike Bates and his two best team members had given notice of their intention to resign within 15 days. Parmentier did everything he could to convince Mike Bates and his team to stay. However, their reasons for leaving went beyond just money, and he was unable to hold onto them. As a result, his division did not perform up to par on the contract. Despite assistance provided belatedly by the eastern division, the company did not satisfy expectations and, therefore, lost an important client.

Building Coalitions

A coalition is a group of diverse people brought together by a common interest. In governmental politics, coalitions are formed for the purpose of electing individuals to office, keeping other individuals from being elected, securing budget appropriations, and passing legislation. Organizational coalitions are formed for various reasons, such as getting selected individuals promoted, ensuring that others are not promoted, securing resources, guaranteeing the adoption of favorable policies or procedures, and fostering a favorable organizational structure. The individuals or individual groups that make up a coalition may have nothing in common except the simple cause that brought them together. This fact gave rise to the old adage that "politics makes strange bedfellows." Consequently, once a cause has been satisfied, the coalition typically dissolves. In its place others will form as interests, conditions, and circumstances change.

Perhaps the classic example of a coalition of radically disparate groups working together toward a common goal is the anticrime coalition in the United States. This coalition consists of groups and individuals from both ends of the political spectrum as well as points in between. The fact that they believe that crime has gotten out of hand in the United States may be the only issue on which these groups agree. However, on this particular issue, they do concur. Consequently, a strong anticrime coalition exists and is having a major impact on legislation at the state and federal level.

For the sake of illustration, suppose that the marketing and accounting departments of a hypothetical organization have never gotten along. Marketing personnel in this organization think that their colleagues from accounting are shortsighted, tightfisted, and don't understand that the organization must spend money to make money. Accounting personnel think that their colleagues from marketing are a bunch of high-rolling big spenders who can't work within a budget no matter how large it is. Then, one day, the organization's management team announced its plan to purchase a building across the street from the organization's existing facility. By retaining its existing facility and relocating some of its personnel to the new building across the street, the company will gain badly needed work space. The only downside is that at least one department is going to have to move to the new building.

After analyzing space requirements, the executive team decides that either the engineering department alone or marketing and accounting together must move. It is not a good time for engineering to be disrupted by a relocation because the company has just received a large contract that is engineering intensive and has a "short fuse." There is no problem with marketing and accounting moving except that they are comfortable where they are and don't want to move. The current building that houses their offices has some amenities the new building won't have (e.g., covered parking, a cafeteria). Occupants of the new building will have to park their cars in a lot that is exposed to the weather and walk across the street to the old building to use the cafeteria.

Sensing that they are about to lose some of their valued perquisites, the accounting and marketing vice presidents, along with their respective staffs, form a coalition to lobby against moving. While engineering personnel are busy working on the organization's new contract, the accounting and marketing departments mount an effective lobbying campaign to have engineering relocated to the new building. Their lobbying efforts pay off, and the engineering department is moved across the street. Unfortunately, the disruption causes the department to fall behind in its work, and the organization's new contract goes over schedule. Late fees are assessed, and the relationship with a valuable new customer gets off to a bad start.

Applying Pressure and Harassment

In governmental politics, when pressure is applied there is an implicit threat from voters: "Do what I ask or I won't vote for you." From lobbyists, the threat is more along the lines of "Vote as we ask or lose our financial support." From colleagues in Congress, the unspoken message is "Support my bill or else I won't support yours." In organizations, pressure is applied differently, but the implicit threat is still there. Examples of messages and tactics used by internal politicians to apply pressure are as follows:

- Help me out, or you will be socially ostracized by your peers; or the opposite, help me out, and you will be part of the crowd.
- Help me out, or something you don't want known will be revealed.
- If you help me out, I'll help you when I win. If you don't, you'll be left out when I win.
- Help me out, or something undesirable will happen to someone you care about; or the opposite, help me out, and something good will happen to someone you care about.
- Ongoing harassment that ranges from distracting to threatening behavior

The following scenario illustrates how internal politicians use pressure to serve their self-interests. John Brown is the purchasing agent for Orlando A & M University. Because he has a master's degree in accounting, Brown is able to earn extra income teaching night classes in freshman accounting. With twins on the way, Brown and his wife need the extra income.

In addition to his job as a purchasing agent, Brown chairs the university's staff development committee. This is the committee that allocates the funds used by faculty and staff members to attend professional conferences and to participate in professional development activities. There are always more requests for funds than there are funds available. Consequently, Brown and his committee have established some ironclad rules about the number of activities that will be funded for a given individual within a specified time frame. These rules ensure that the largest possible number of employees of the university get an opportunity to participate in professional development activities.

As chair of the development committee, Brown is accustomed to being in the "hot seat" when someone wants an activity funded and the committee cannot comply. But Brown has never been pressured so hard as he is currently being pressured by Amos Andrews, chair of the Department of Business and Accounting. The committee, sticking to the rules, has turned down a request from Andrews. After Andrews exhausted the list of tactics typically used by people trying to influence the committee, he began to get desperate.

That's when Brown really began to feel the pressure. Working through other faculty members in the business and accounting department, Andrews made sure Brown knew that the extra income he earned by teaching night classes in accounting was in jeopardy. The message given to Brown was clear: If Andrews's request for a waiver of the committee's rules is not approved, Brown has taught his last freshman accounting course. Brown needs the extra income from teaching, but he cannot approve Andrews's request without bumping Maxine Denny from a conference that is very important to the social science department. Brown is in a bind, and he is feeling the pressure.

This is just one example of the many ways that pressure can be applied in the workplace by internal politicians. There are many other ways that internal politicians can and do use personal pressure to advance their individual agendas.

Electioneering

In governmental politics, electioneering means participating in the election process. Participation can take many different forms including raising money for candidates, making contributions, and getting out the vote. Of course, the purpose of electioneering is to

QUALITY TIP

Eliminate Counterproductive Internal Competition

Organizations can devote their time and energy to the battle of the marketplace, or they can devote it to internal battles, but not to both. Resources are finite. Time and energy wasted on internal squabbling are resources that could have been used to improve performance.

Source: David L. Goetsch.

ensure that a certain candidate is elected. Electioneering in an organization is a similar process.

Internal politicians use electioneering tactics to ensure that selected individuals are promoted, that the *right* people are appointed to prestigious committees, and that selected people are chosen to chair important committees and task forces. The following example illustrates electioneering as it might be used in an organizational setting.

Tim Jones is in a bind. In just 6 months he will be promoted to a planning position at the corporate office, unless, that is, someone at corporate learns the truth about his division. Brierfield Products, Inc. (BPI), is undergoing a corporation-wide implementation of total quality. As a division director, Jones is a key player. However, he hasn't played the leadership role the company needs him and its other directors to play.

It is not that Jones is opposed to total quality. It's just that he had hoped to be off to his new position before having to bother with implementing it. "After all," he thought, "why go through all the trouble when I won't be around to enjoy the benefits?" This type of thinking has led Jones to procrastinate. As a result, his division is lagging behind the company's other divisions. Jones has belatedly decided to get started, but he has a long way to go to catch up, and the TQ monitoring visit is scheduled to occur in just one week.

When he made the decision to skate through his last 6 months, leaving the work of the implementation to his successor, Jones had not known that corporate would conduct monitoring visits. Now a monitoring visit was right around the corner. If he didn't do something soon, not only would his promotion be lost, but he'd be lucky to keep his current job. But Jones isn't finished yet. He hadn't gotten to be one of the company's youngest division directors by accident. He possesses considerable skills as an internal politician, and it will be his political skills that will save him.

It was while scanning the list of personnel appointed by corporate to serve as implementation monitors that Jones saw his chance for salvation. The third person on the list was Jake Burns. Jones couldn't believe his luck. He and Burns went way back, and, better yet, Burns owed him. Now all he had to do was make sure that Jake Burns was selected as the monitor for his division. Jones began electioneering in earnest, pulling out all the stops. He made telephone calls, got other people to make telephone calls, applied pressure, made promises, made threats, and called in favors. By the time he was done Jake Burns had been chosen as the TQ monitor for the upcoming visit. Jones could finally relax. Jake Burns would write an appropriately worded, appropriately positive—albeit misleading—monitoring report.

Gossiping and Spreading Rumors

One of the most pernicious weapons in the arsenal of the internal politician is the creation of doubt. Doubt can be created effectively by using gossip to spread rumors about a targeted individual or group. When used by internal politicians, rumors and gossip are not of the harmless chitchat variety. Rather, they are intentional, coldly calculated attempts to advance the agenda of one individual or group at the expense of another.

Rumors and gossip have the greatest impact when they cast doubt on an individual relative to high-priority organizational values. The following scenario demonstrates how rumormongering and gossip can be used by internal politicians to gain an advantage.

Patricia Chitwood is both ambitious and smart. She knows that her company, Drake Services, Inc., places a high priority on ethics, and she plans to use this fact to her advantage. Chitwood and her colleague Pamela McGraw have both applied for the soon-to-be-vacant position of regional sales manager at Drake Services. Like Chitwood, McGraw is good. In terms of both credentials and performance, the two sales reps could be twins. As things stand now, the race for the promotion is dead even—it could go either way. But Chitwood has a plan. At lunch today she will start a rumor that is sure to sow seeds of doubt about McGraw's ethics. Nothing major—just a few whispered comments and well-placed winks concerning McGraw's expense account. Within an hour, the office grapevine will be buzzing. Within a day, the CEO of Drake Services will be wondering about the ethics of one of his best sales reps. Within a week, Chitwood should be the new regional sales manager.

There is no question that Patricia Chitwood is a resourceful internal politician. However, her shortsighted methods may cost her in the long run. The rumor Chitwood started about her rival is likely to make Pamela McGraw's position with Drake Services tenuous at best. Even if McGraw isn't fired, she will probably leave; and with her record, landing a position with a competitor won't be difficult. As regional sales manager, the last thing Chitwood will need is to have her best sales rep, Pamela McGraw, joining forces with the competition.

IMPACT OF INTERNAL POLITICS ON QUALITY

The approach that is the opposite of internal politics is *collaboration*. The rationale for collaboration can be found in the negative impact internal politics can have on an organization. To appreciate just how powerfully negative this impact can be, consider the following scenario:

> The U.S. Marine Corps has an outstanding rifle team that competes in tournaments worldwide. Members of the team are the best marksmen in a branch of the military that places a high priority on marksmanship. The Marine Corps Rifle Team performs well in competition, typically taking top honors. But how well would the team do if its members shot at each other instead of their respective targets?

A far-fetched scenario? Perhaps. But it illustrates exactly what happens in organizations suffering from the scourge of internal politics. Individuals in organizations—frequently the organizations' key decision makers—sometimes become so intent on fulfilling their personal ambitions, satisfying their individual needs, or feeding their own egos that they lose sight of what is best for the organization that employs them.

Internal politics can affect an organization in the same way that cancer affects an individual. Both start covertly inside the victim, often remaining invisible until the damage done is serious, and both can spread quickly. Organizations stricken with the disease of internal politics ultimately suffer the effects shown in Figure 13–8.

Figure 13–8
Effects of Internal Politics on an
Organization

- *Loss of morale* due to infighting, buck passing, and rumormongering
- *Questionable decisions* made for reasons other than what is best for the organization
- *Counterproductive internal competition* that saps the organization of its competitive energy
- *Loss of the best and brightest employees* as they make a statement about their dissatisfaction by leaving
- *Perpetuation of outdated processes, procedures, and technologies* as internal politics is used to promote organizational inertia by those opposed to change
- *Constant conflict* as the political machinations of one group are countered by those of others
- *Loss of quality, competitiveness, and customers* as the organization's focus is diverted from what really matters

An organization's morale suffers when infighting, buck passing, and rumormongering—all of which invariably result from the practice of internal politics—are allowed to become part of the dominant corporate culture. Decisions that are questionable at best and even potentially unsound are not uncommon in organizations that condone internal politics. Any time decisions are made based on criteria other than what is best for the organization, the organization suffers.

Internal politics invariably leads to counterproductive internal competition. Ideally, the only competition in which an organization would engage is market competition. Internal cooperation in the pursuit of a common purpose serves an organization better in the long run than internal competition among its own departments and employees. Organizations can devote their time and energy to the battle of the marketplace, or they can devote it to internal battles, but not to both. Resources are finite; time and energy wasted on internal squabbling are resources that could have been used to improve performance.

Organizations that are suffering the ill effects of internal politics often lose the best and brightest employees. If a work group is low performing and does not want to do better, its members may actually drive out anyone who tries to improve. Their methods range from peer pressure to harassment to outright ostracism. As they become increasingly frustrated by decisions based on politics rather than merit, employees with the most marketable credentials often show their dissatisfaction by leaving. Those who stay tend to be employees that fall into one of two categories. The first category consists of employees who stay because weak credentials make it difficult for them to find a better job. The second category consists of employees who give in to reality, and become internal politicians themselves.

Internal politics tends to perpetuate outdated processes, procedures, and technologies. This is because the tactics of the internal politician are ideally suited for opposing change. Change comes hard for most people. Psychological comfort with the status quo is inherent in the human condition. When internal politics becomes part of the corporate culture, organizations find it even more difficult than usual to make the changes necessary to stay competitive. With a little lobbying, some electioneering, and just the right amount of pressure wisely applied, the natural resistance of people to change can be magnified exponentially by internal politicians opposed to change. When this happens, the employees of an organization gain the psychological comfort associated with the status quo, but the organization loses the competitive edge associated with change.

QUALITY CASE

Preventing Office Politics at Merrill Lynch

Merrill Lynch is the leading investment banking and brokerage company in the United States. The company operates in three market categories: 1) the Private Client Group offers retail brokerage, life insurance, and cash management services; 2) the Corporate and International Group offers investment banking and capital market services to corporations, large organizations, and governments; and 3) Government bonds and derivatives. In addition to being a market leader in the United States, Merrill Lynch has a strong international presence. Together, the company's domestic and foreign markets make Merrill Lynch a $35 billion company.

One of the ways Merrill Lynch maintains its world-class status in the competitive financial services industry is avoiding the debilitating, productivity-sapping effects of negativity, turf wars, and office politics. The way these problems are prevented is through the examples of the company's CEOs, present and past. Merrill Lynch has a history of maintaining a mutually supportive, mutually respectful chain of leadership. The current CEO of Merrill Lynch has a photograph hanging in his office of the company's last five CEOs (including himself). This and other efforts to maintain executive-level continuity and cooperation over time go a long way to explain why Merrill Lynch is so successful.

With many organizations, the arrival of a new CEO sets off a round of turf battles and interoffice politics as new members of the team struggle with existing members for power, position, and recognition. The continuity gained from the long and cooperative chain of leadership has prevented these outbreaks of office warfare at Merrill Lynch. Instead, each successive CEO has insisted that the only question any manager or employee need ask when making decisions is, "What is best for the customer?" rather than "Whose turf is being invaded?" or "Whose office is closest to the CEO's?" As with any other factor that affects competitiveness, employees and managers will take their lead from the top.

Source: Thomas J. Neff and James M. Citrin, *Lessons from the Top,* (New York: Doubleday, 2001), 307–318.

Internal politics invariably multiplies both the frequency and intensity of conflict in an organization. Infighting, backbiting, and ill will are antithetical to quality and competitiveness. Total quality requires unity of purpose and a trusting, mutually supportive work environment. Such environments cannot be maintained in the face of constant conflict that occurs on a personal rather than professional level. All of the individual deleterious effects of internal politics, when taken together, have the cumulative effect of diverting an organization's attention from what really matters. This can occur to such an extent that the organization's quality suffers. When this happens, the organization loses its ability to satisfy and retain customers and, in turn, its ability to survive in a competitive marketplace.

CONTROLLING INTERNAL POLITICS IN ORGANIZATIONS

How does one go about convincing people in organizations—all of whom have their own interests, ambitions, and egos—to put aside the natural inclination to practice internal politics and, instead, practice collaboration? Interestingly, trying to control internal politics in organizations is a lot like trying to prevent overeating in individuals. Both involve finding ways to subdue human nature, both require persistent effort, and both demand constant vigilance. Controlling internal politics in an organization requires a comprehensive effort involving all employees. Such an effort should have at least the components listed in Figure 13–9 and elaborated on in the following sections.

Strategic Planning Component

Controlling internal politics begins with the organization's strategic plan. One of the keys to controlling internal politics over the long term is creating a cultural expectation that all decisions will be based on what is best from the perspective of the organization's strategic plan. If employees are to make all decisions based on this criterion, they have to know the organization's vision, mission, guiding principles, and broad objectives. In other words, if their behavior and decisions are supposed to support the strategic plan, employees have to know the plan. The strategic planning components of an organization's effort to control internal politics should have at least the following elements:

- Explaining the strategic plan to all employees and how it is to be used in guiding all decisions and actions in the organization
- Building a guiding principle into the strategic plan that conveys the message that collaboration is the expected approach in the organization

Explaining the Strategic Plan and Using It in Decision Making

All employees should have a copy of the strategic plan, the plan should be thoroughly explained, and employees should be given ample opportunity to ask questions and seek clarification about the plan. In other words, it's not enough for employees to have a plan—they need to understand it.

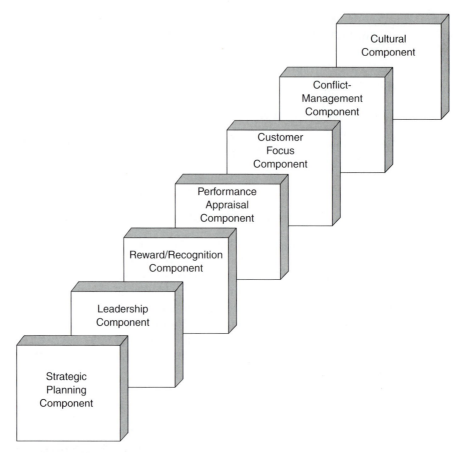

Figure 13–9
Main Components of an Internal Politics Prevention Program

In explaining the strategic plan to employees, managers should make sure that they convey the following message:

> Everything we do in this organization is to be guided by one criteria: support of the strategic plan. A good decision is one that supports accomplishing what is set forth in the strategic plan. A decision that does not meet this criteria is a bad decision. Good policy, good procedures, and good work practices are those that support the strategic plan. Others are unacceptable. Consequently, there is no room in the organization for the advancing of personal agendas or the promotion of self-interest to the detriment of organizational interests. In short, internal politics has no place in this organization.

Conveying this message to all employees, in conjunction with an explanation of the strategic plan, sets the proper tone, establishes the proper expectations, and brings the issue of internal politics out into the open. Setting the proper tone and establishing expectations are essential to developing an organizational culture that does not promote or condone internal politics. Bringing the issue out into the open removes the shroud of secrecy on which internal politicians thrive.

Collaboration as a Guiding Principle

One of the most important components of a strategic plan is the one that contains the organization's guiding principles. The organization's core values are described in its guiding principles. These principles explain in writing what is most important to the organization and how it intends to do business. Consequently, one of the guiding principles of every organization should be *collaboration*. An example of such a guiding principle follows:

> ABC Company places a high priority on collaboration among all employees at all levels. We base all policies, procedures, practices, and decisions on what is best for the organization, rather than what serves the personal interests, agendas, or ambitions of individuals or individual units within the company.

If employees, from the CEO to line workers, know that collaboration is a high priority, it becomes more difficult for them to play the games collectively known as internal politics. This principle, coupled with the remaining strategies in this section, can go a long way to controlling the practice of internal politics in the organization.

Leadership Component

A fundamental premise of leadership is setting a positive example. Managers must be consistent role models of the behavior they expect of employees. If a manager practices internal politics, employees will respond in kind. Consequently, it is important that managers be seen using the organization's strategic plan as the basis for all actions and that they insist on all employees following suit.

Setting an example goes beyond just adopting policies and making decisions based on what best supports the organization's strategic plan. It also involves refusing to condone—explicitly or implicitly—counterproductive behavior on the part of employees (e.g., rumormongering or gossiping). In fact, counterproductive behavior such as rumor mongering and gossiping gives managers excellent opportunities for demonstrating the point that internal politics is not condoned. By openly and consistently refusing to gossip, spread rumors, or respond to either, managers can help take away two of the most potent weapons of internal politicians.

Reward/Recognition Component

If you want to promote a certain type of approach—for example, collaboration—reward it, and recognize employees who practice it. This is a simple but effective management principle. Unfortunately, it's a principle that is preached more than it is practiced. One of the most frequent systemic mistakes made in organizations is failing to match up management expectations with reward/recognition systems. Perhaps the most common example of this failing can be found in organizations that expect teamwork but still maintain a reward system that is based on individual performance. Managers should examine their organization's incentives carefully to identify ways in which internal politics is rewarded, either directly or indirectly. The most obvious question is "What happens to employees at any level who are found to practice internal politics?" Another question is "Does the organization provide incentives that promote employee collaboration, and, if so, what are those incentives?" A well-designed reward/recognition system will simultaneously provide disincentives to internal politics and incentives that promote collaboration.

Disincentives that can work against internal politics include negative performance appraisals, verbal warnings, and written reprimands. Incentives can be both formal and informal, and there are hundreds of both varieties. For an excellent source of help in identifying effective formal and informal incentives, refer to *1,001 Ways to Reward Employees* by Bob Nelson (Workman, 1994).

Performance Appraisal Component

The periodic performance appraisal is how most organizations formally let employees know how they are doing. Consequently, one or more of the criteria in an organization's performance-appraisal instrument should relate to collaboration. Examples of this type of criteria follow:

This employee bases all actions on what is best for the organization.
- Always
- Sometimes
- Usually
- Never

What is this employee's collaboration rating?
- Excellent
- Above average
- Average
- Poor

Making collaboration an issue in performance appraisals ties it directly to pay and promotions. This is critical. Remember that internal politics is driven by self-interest.

QUALITY TIP

Recognition Isn't Expensive

"Some of the most effective sources of recognition cost nothing at all. A sincere word of thanks from the right person at the right time can mean more to an employee than a raise, a formal award, or a whole wall of certificates and plaques."

Source: Bob Nelson, *1,001 Ways to Reward Employees* (New York: Workman, 1994), 28.

Tying pay and promotion to an employee's willingness to practice collaboration means that internal politics no longer serves his or her self-interest.

Customer Focus Component

Customer focus is a fundamental cornerstone of total quality. In organizations with a customer focus, quality is defined by customers, and the organization's strategic plan is written from the perspective of attracting, satisfying, and retaining customers. A customer focus is achieved by partnering with customers.

When an organization partners with its customers, it brings them into the decision-making process by actively seeking their input and feedback. Input, remember, consists of customer recommendations made *before* the decision is made. Feedback is customer information given *after* the decision is made. Input influences the decision that is made. Feedback evaluates the quality of the decision that was made. In organizations that factor customer input and feedback into the decision-making process, it is difficult for internal politicians to play their games. When decisions are driven by customer preference, they cannot be driven by politics.

The full benefit of a customer focus—from the perspective of preventing internal politics—is gained by ensuring that all employees are thoroughly informed concerning customer input and feedback. In this way, customer needs and preferences become the critical criteria by which the viability of policies, procedures, practices, and decisions can be judged. Such criteria make it difficult for internal politicians to play their games. Even the most accomplished internal politician will find it difficult to justify recommending decisions that run counter to customer preferences.

Conflict-Management Component

Internal politics tends to generate counterproductive conflict. This is one of the reasons that managers in organizations should do what is necessary to prevent internal politics. However, it is important to distinguish between conflict and counterproductive conflict. Not all conflict is bad. In fact, properly managed conflict that has the improvement of products, processes, people, and/or the work environment as its source is positive conflict.

Counterproductive conflict—the type associated with internal politics—occurs when people in organizations behave in ways that work against the interests of the overall organization. This type of conflict is often characterized by deceitfulness, vindictiveness, and personal rancor. Productive conflict occurs when right-minded, well-meaning people disagree, without being disagreeable, concerning the best way to support the organization's strategic plan.

Positive conflict leads to discussion, debate, and give-and-take interaction among people whose only goal is to find the best solution or make the best decision. This type of interaction exposes the viewpoints of all participants to careful scrutiny and judges the merits of all arguments by applying criteria that are accepted by all stakeholders. By putting every point of view under the microscope of group scrutiny, weaknesses in arguments are identified, the issue is viewed from all possible angles, and, ultimately, the best solution can be identified. Contrast this approach with that of internal politi-cians whose only goal is to promote self-interest. Internal politicians reveal only that information that serves their interests, while concealing any information that might weaken their case. They lobby, practice electioneering, and apply pressure to influence people to make decisions based on criteria other than merit.

By practicing conflict management, managers in an organization can make it difficult for internal politicians to play their games. Conflict management has the following com-ponents:

■ Establishing conflict guidelines
■ Helping all employees develop conflict prevention/resolution skills
■ Helping all employees develop anger management skills
■ Stimulating/facilitating productive conflict

Establishing Conflict Guidelines

Conflict guidelines establish ground rules for discussing and debating differing points of view, varying ideas, and an array of opinions concerning how best to accomplish the organization's vision, mission, and broad objectives. Figure 13–10 is an example of an organization's conflict guidelines. Guidelines such as these should be developed with a broad base of employee involvement from all levels in the organization.

Developing Conflict Prevention/Resolution Skills

If managers are going to expect employees to disagree without being disagreeable, they are going to have to ensure that all employees are skilled in the art and science of con-flict resolution. The second guideline in Figure 13–10 is an acknowledgement of human nature. It takes advanced human relation skills and constant effort to disagree without being disagreeable. Few people are born with this ability, but fortunately it can be learned. The following strategies are based on a three-phase model developed by Tom Rusk and described in his book *The Power of Ethical Persuasion.*[3]

Explore the other person's viewpoint. A good start on preventing conflict can be made by acknowledging the importance of the other person's point of view. Begin the discus-sion by giving the other party an opportunity to present his or her point of view and, listening carefully, say, "Your viewpoint is important to me, and I'm going to hear you out." The following strategies will help make this phase of the discussion more positive and productive.

1. Establish that your goal at this point in the discussion is mutual understanding.
2. Elicit the other person's complete point of view.
3. Listen nonjudgmentally and do not interrupt.
4. Ask for clarification if necessary.
5. Paraphrase the other person's point of view and restate it to show that you understand.
6. Ask the other person to correct your understanding if it appears to be off-base or incomplete.

Jones Engineering Company encourages discussion and debate among employees at all levels concerning better ways to continually improve the quality of our products, processes, people, and work environment. This type of intellectual interaction, if properly handled, will result in better ideas, policies, procedures, practices, and decisions. However, human nature is such that conflict can easily get out of hand, take on personal connotations, and become counterproductive. Consequently, in order to promote productive conflict, Jones Engineering Company has adopted the following guidelines. These guidelines are to be followed by all employees at all levels:

- The criteria to be applied when discussing/debating any point of contention is as follows: Which recommendation is most likely to move our company closer to realizing the strategic vision?
- Disagree, but don't be disagreeable. If the debate becomes too hot, stop and give all parties an opportunity to cool down before continuing. Apply your conflict-resolution skills and anger management skills. Remember, even when we disagree about how to get there, we are all trying to reach the same destination.
- Justify your point of view by tying it to either the strategic plan or customer input/feedback and require others to follow suit.
- In any discussion of differing points of view, ask yourself the following question: *Am I just trying to win the debate for the sake of winning (ego), or is my point of view really the most valid?*

Figure 13–10
Jones Engineering Company: Conflict Guidelines

Explain your viewpoint.   After you accurately and fully understand the other person's point of view, present your own. The following strategies will help make this phase of the discussion more positive and productive:

1. Ask for the same type of fair hearing for your point of view that you gave the other party.
2. Describe how the other person's point of view affects you. Don't point the finger of blame or be defensive. Explain your reactions objectively, keeping the discussion on a professional level.
3. Explain your point of view accurately and completely.
4. Ask the other party to paraphrase and restate what you have said.
5. Correct the other party's understanding, if necessary.
6. Review and compare the two positions (yours and that of the other party). Describe the fundamental differences between the two points of view and ask the other party to do the same.

Agree on a resolution.   Once both viewpoints have been explained and are understood, it is time to move to the resolution phase. This is the phase in which both parties attempt to come to an agreement. It is also the phase in which both parties may discover that they cannot agree. Agreeing to disagree—in an agreeable manner—is an acceptable solution. The following strategies will help make this phase of the discussion more positive and productive:

1. Reaffirm the mutual understanding of the situation.
2. Confirm that both parties are ready and willing to consider options for coming to an acceptable resolution.
3. If it appears that differences cannot be resolved to the satisfaction of both parties, try one or more of the following strategies:
   - Take time out to reflect and then try again.
   - Agree to third-party arbitration or neutral mediation.
   - Agree to a compromise solution.
   - Take turns suggesting alternative solutions.
   - Yield (this time) once your position has been thoroughly stated and is understood. The eventual result may vindicate your position.
   - Agree to disagree while still respecting each other.

Developing Anger Management Skills

It is difficult, if not impossible, to keep conflict positive when anger enters the picture. If individuals in an organization are going to be encouraged to question, discuss, debate, and even disagree, they must know how to manage their anger. *Anger is an intense emotional reaction to conflict in which self-control may be lost.* Anger occurs when people feel that one or more of their fundamental needs are being threatened. These needs include the following:

1. Need for approval
2. Need to be valued
3. Need to be appreciated
4. Need to be in control
5. Need for self-esteem

When one or more of these needs is threatened, a normal human response is to become angry. An angry person can respond in one of five ways:

1. *Attacking.*   With this response the source of the threat is attacked, usually verbally. For example, when someone disagrees with you (threatens your need for approval), you might attack by questioning their veracity or credentials.
2. *Retaliating.*   With this response, you fight fire with fire, so to speak. Whatever is given, you give back. For example, if someone calls your suggestion ridiculous (threatens your need to be valued), you might retaliate by calling his or her suggestion dumb.
3. *Isolating.*   This response is the opposite of venting. With the isolation response, you internalize your anger, find a place where you can be alone, and simmer. The childhood version of this response was to go to your room and pout. For example, when someone fails to even acknowledge your suggestion (threatens your need to be appreciated), you might swallow your anger, return to your office, and boil over in private.
4. *Coping.*   This is the only positive response to anger. Coping does not mean that you don't become angry. Rather, it means that even when you do, you control your

emotions instead of letting them control you. A person who copes well with anger is a person who, in spite of his or her anger, stays in control. All employees at all levels of a total quality organization need to be able to cope with their anger. The following strategies will help employees manage their anger by becoming better at coping:

- Avoid the use of anger-inducing words and phrases, including the following: *but, you should, you made me, always, never, I can't,* and *you can't.*
- Admit that others don't make you angry but that you allow yourself to become angry. You are responsible for your emotions and your responses to them.
- Don't let pride get in the way of progress. You don't have to be right every time.
- Drop your defenses when dealing with people. Be open and honest.
- Relate to other people as equals. Regardless of position or rank, you are no better than they and they are no better than you.
- Avoid the human tendency to rationalize your angry responses. You are responsible and accountable for your behavior.

Stimulating/Facilitating Productive Conflict

Sycophantic agreement with the boss has no place in the total quality organization. Ideas, suggestions, and proposals should be subjected to careful, even intense scrutiny. Consequently, productive conflict is not only allowed in a total quality organization but also promoted. Productive conflict consists of genuine, harmonious disagreement over the best way to solve a problem.

Productive conflict is productive because the only agenda being advanced is the good of the organization. With productive conflict no hidden agendas or political machinations are at work. All parties are attempting to reach the same destination; the disagreement has to do with how best to get there. Because there are no hidden agendas, all parties are open to questions, challenges, and constructive criticism. In addition, all parties agree on the criteria by which their ideas will be judged.

In a total quality organization, managers actually stimulate discussion and debate (productive conflict) if they think a proposal is moving down the track too fast unimpeded by careful scrutiny. Productive conflict is stimulated using methods such as the following:

■ Openly communicating the message "We want ideas and constructive criticism of ideas. We believe discussion and debate sharpen our ideas."

■ Playing "devil's advocate," and teaching employees to play this role

■ Requiring employees with suggestions to identify the downside when making them

Cultural Component

There are many different definitions for the term *culture*. The *American Heritage College Dictionary* defines it as follows: "The totality of socially transmitted behavior patterns, arts, beliefs, institutions, and all other products of human thought."[4] The key part of this definition reads "socially transmitted behavior patterns." As applied to an organization, the concept means the way things are done in the organization. In other words, an organization's culture is the everyday manifestation of its actual beliefs. It is important to note that the concept grows out of *actual* beliefs as opposed to *written* beliefs. An organization's culture *should be* the everyday manifestation of the guiding principles found in its strategic plan.

However, some organizations are guilty of practicing a set of beliefs that differs from those written down as guiding principles. Culture cannot be mandated. Rather, it develops over time based on actions, not words. This is why it is so important to live out the organization's professed beliefs on a daily basis.

If collaboration is a high priority, it should be promoted, modeled, rewarded, and reinforced on a daily basis at all levels in the organization. Correspondingly, internal politics must be seen to be ineffective and detrimental. For every incentive to collaborate, there should also be a disincentive to play political games. The ultimate disincentive is social pressure. This is why establishing a collaborative culture is so important and beneficial. It ensures that social pressure, which is the most effective enforcer of culture, works *for* collaboration instead of against it.

By applying the various strategies described in this section on a daily basis, organizations can make collaboration a fundamental part of their culture. When this happens, social pressure within the organization will keep the practice of internal politics under control.

OVERCOMING NEGATIVITY IN ORGANIZATIONS

Negativity is any behavior on the part of any employee at any level that works against the optimum performance of the organization. The motivation behind negativity can be as different and varied as the employees who manifest it. However, negative behavior can be categorized. The most common categories of negative behavior are as follows:

- Control disputes
- Territorial disputes (boundaries)
- Dependence/independence issues
- Need for attention/responsibility
- Authority
- Loyalty issues

Recognizing Negativity in the Organization

Managers should be constantly alert to signs of negativity in the workplace because negativity is contagious. It can spread throughout an organization quickly, dampening morale and inhibiting performance. What follows are symptoms of the negativity syndrome that managers should watch for:

- *"I can't" attitudes.* Employees in an organization that is committed to continuous improvement have "can-do" attitudes. If "I can't" is being heard regularly, negativity has crept into the organization.

- *"They" mentality.* In high-performance organizations employees say "we" when talking about their employer. If employees refer to the organization as "they," negativity has gained a foothold.

- *Critical conversation.*   In high-performance organizations, coffee-break conversation is about positive work-related topics or topics of personal interest. When conversation is typically critical, negative, and judgmental, negativity has set in. Some managers subscribe to the philosophy that employees are not happy unless they are complaining. This is a dangerous attitude. Positive, improvement-oriented employees will complain to their supervisor about conditions that inhibit performance, but they don't sit around criticizing and whining during coffee breaks.

- *Blame fixing.*   In a high-performance organization, employees fix problems, not blame. If blame fixing and finger-pointing are common in an organization, negativity is at work.

Overcoming Negativity

Managers who identify negativity in their organizations should take the appropriate steps to eliminate it. What follows are strategies that can be used to overcome negativity in organizations:

- *Communicate.*   Frequent, ongoing, effective communication is the best defense against negativity in organizations, and it is the best tool for overcoming negativity that has already set in. Organizational communication can be made more effective using the following strategies: acknowledge innovation, suggestions, and concerns; share information so that all employees are informed; encourage open, frank discussion during meetings; celebrate milestones; give employees ownership of their jobs; and promote teamwork.

- *Establish clear expectations.*   Make sure all employees know what is expected of them as individuals and as members of the team. People need to know what is expected of them and how and to whom they are accountable for what is expected.

- *Provide for anxiety venting.*   The workplace can be stressful in even the best organizations. Deadlines, performance standards, budget pressures, and competition can all produce anxiety in employees. Consequently, managers need to give their direct reports opportunities to vent in a nonthreatening, affirming environment. This means listening supportively. This means letting the employee know that you will not "shoot the messenger," and then listening without interrupting, thinking ahead, focusing on preconceived ideas, or tuning out.

- *Build trust.*   Negativity cannot flourish in an atmosphere of trust. Managers can build trust between themselves and employees and among employees by applying the following strategies: always delivering what is promised; remaining open-minded to suggestions; taking an interest in the development and welfare of employees; being tactfully honest with employees at all times; lending a hand when necessary; accepting blame, but sharing credit; maintaining a steady, pleasant temperament even when under stress; and making sure that criticism is constructive and delivered in an affirming way.

■ *Involve employees.*   It's hard to criticize the way things are done when you are a part of how they are done. Involving employees by asking their opinions, soliciting their feedback, and making them part of the solution are some of the most effective deterrents to and cures for negativity in organizations.

OVERCOMING TERRITORIAL BEHAVIOR IN ORGANIZATIONS

Territory in the workplace tends to be more a function of psychological boundaries than of physical boundaries. In her book *Territorial Games*, Annette Simmons says this about territory on the job:

> The territorial impulse is deeply rooted in our survival programming. We are territorial because territory helps us survive. It did so thousands of years ago and it still does today. If you look at it backwards, survival needs started the whole concept of territory. The problem now may be that we are still using old territorial behaviors that are no longer appropriate to our new environment.[5]

Manifestations of Territoriality

The territorial instinct shows up in a variety of ways in an organization. Simmons lists the following manifestations of territoriality in an organization:[6]

■ *Occupation.*   These games include actually marking territory as *mine*; playing the *gatekeeper* game with information; and monopolizing resources, information, access, and relationships.

■ *Information manipulation.*   People who play territorial games with information subscribe to the philosophy that information is power. To exercise power they withhold information, bias information to suit their individual agendas (spin), cover up information, and actually give out false information.

■ *Intimidation.*   One of the most common manifestations of territoriality is intimidation—a tactic used to frighten others away from certain turf. Intimidation can take many different forms, from subtle threats to blatant aggression (physical or verbal).

■ *Alliances.*   Forming alliances with powerful individuals in an organization is a commonly practiced territorial game. The idea is to say without actually having to speak the words that "you had better keep off my turf, or I'll get my powerful friend to cause trouble."

■ *Invisible wall.*   Putting up an invisible wall involves creating hidden barriers to ensure that a decision, although already made, cannot be implemented. There are hundreds of strategies for building an invisible wall, including stalling, losing paperwork, forgetting to place an order, and many others.

■ *Strategic noncompliance.*   Agreeing to a decision up front but with no intention of carrying the decision out is called *strategic noncompliance*. This tactic is often used to buy enough time to find a way to reverse the decision.

- *Discredit.* Discrediting an individual as a way to cast doubt on his or her recommendation is a common turf protection tactic. Such an approach is called an *ad hominem* argument, which means if you cannot discredit the recommendation, try to discredit the person making it.

- *Shunning.* Shunning, or excluding an individual who threatens your turf, is a common territorial protection tactic. The point of shunning is to use peer pressure against the individual being shunned.

- *Camouflage.* Other terms that are sometimes used to describe this tactic are *throwing up a smoke screen* or *creating fog.* This tactic involves confusing the issue by raising other distracting controversies, especially those that will produce anxiety such as encroaching on turf.

- *Filibuster.* Filibustering means talking a recommended action to death. The tactic involves talking at length about concerns—usually inconsequential—until the other side gives in just to stop any further discussion or until time to make the decision runs out.

Overcoming Territorial Behavior

Overcoming territorial behavior requires a two-pronged approach: (a) recognizing the manifestations described earlier and admitting that they exist, and (b) creating an environment in which survival is equated with cooperation rather than territoriality. Simmons recommends the following strategies for creating a cooperative environment:[7]

- *Avoid jumping to conclusions.* Talk to employees about territoriality versus cooperation. Ask to hear their views, and listen to what they say.

- *Attribute territorial behavior to instinct rather than people.* Blaming people for following their natural instincts is like blaming them for eating. The better approach is to show them that their survival instinct is tied to cooperation, not turf. This is done by rewarding cooperation and applying negative reinforcement to territorial behavior.

- *Ensure that no employee feels attacked.* Remember that the survival instinct is the motivation behind territorial behavior. Attacking employees, or even letting them feel as if they are being attacked, will only serve to trigger their survival instinct. To change territorial behavior, it is necessary to put employees at ease.

- *Avoid generalizations.* When employees exhibit territorial behavior, deal with it in specifics as opposed to generalizations. It is a mistake to witness territorial behavior on the part of one employee and respond by calling a group of employees together and talking about the issue in general terms. Deal with the individual who exhibits the behavior and focus on specifics.

- *Understand "irrational" fears.* The survival instinct is a powerful motivator. It can lead employees to cling irrationally to their fears. Managers should consider this point when dealing with employees who find it difficult to let go of survival behaviors. Be firm but patient, and never deal with an employee's fears in a denigrating or condescending manner.

■ *Respect each individual's perspective.*   In a way, an individual's perspective or opinion is part of his or her psychological territory. Failure to respect people's perspectives is the same as threatening their territory. When challenging territorial behavior, let employees explain their perspectives and show respect for them, even if you do not agree.

■ *Consider the employee's point of view.*   In addition to giving an appropriate level of respect to employees' perspectives, managers should also try to "step into their shoes." How would you, as a manager, feel if you were the employee? Sensitivity to the employee's point of view and patience with that point of view are critical when trying to overcome territorial behavior.

MANAGING CONFLICT IN ORGANIZATIONS

Human conflict is a normal and unavoidable aspect of the highly competitive modern workplace, even in a total quality setting. One of the human relations skills needed by people in such a setting is the ability to disagree with fellow workers without being disagreeable. However, even if most members of an organization have this skill, there is no guarantee that conflicts will not arise among workers. When people work together, no matter how committed they are to a common goal, human conflict is going to occur. Consequently, managers in a total quality setting must be proficient in resolving conflict. This section contains information managers need to know to be catalysts in resolving human conflict in the workplace.

Causes of Workplace Conflict

The most common causes of workplace conflict are predictable. They include those shown in Figure 13–11.

Limited resources often lead to conflict in the workplace. It is not uncommon for an organization to have fewer resources (funds, supplies, personnel, time, equipment, etc.) than might be needed to complete a job. When this happens, who gets the resources and in what amounts? *Incompatible goals* often lead to conflict, and incompatibility of goals is inherent in the workplace. For example, conflicts between engineering and manufacturing are common in modern industry. The goal of engineering is to design a product that meets the customers' needs. The goal of manufacturing is to produce a high-quality product as inexpensively as possible. In an attempt to satisfy the customer, engineering might create a design that is difficult to manufacture economically. The result? Conflict.

Role ambiguity can also lead to conflict by blurring "turf lines." This makes it difficult to know who is responsible and who has authority. *Different values* can lead to conflict. For example, if one group values job security and another values maximum profits, the potential for conflict exists. *Different perceptions* can lead to conflict. How people perceive a given situation depends on their background, values, beliefs, and individual circumstances. Because these factors are sure to differ among both individuals and groups, particularly in an increasingly diverse workplace, perception problems are not uncommon.

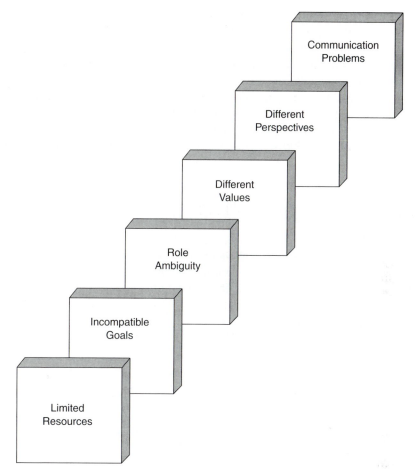

Figure 13–11
Causes of Workplace Conflict

The final predictable cause of conflict is *communication*. Effective communication is difficult at best. Improving the communication skills of employees at all levels is an ongoing goal of management. Knowing that communication will never be perfect, communication-based conflict should be expected.

How People React to Conflict

To deal with conflict effectively, managers need to understand how people react to conflict. According to K. W. Thomas, the ways in which people react to conflict can be summarized as competing, accommodating, compromising, collaborating, and avoiding.[8]

A typical reaction to conflict is *competition* in which one party attempts to win while making the other lose. The opposite reaction to conflict is *accommodation*. In this reaction, one person puts the needs of the other first. *Compromise* is a reaction in which the two opposing sides attempt to work out a solution that helps both to the extent possible. *Collaboration* involves both sides working together to find an acceptable solution for both. *Avoidance* involves shrinking away from conflict. This reaction is seen in people who are not comfortable facing conflict and dealing with it.

In some situations, a particular reaction to conflict is more appropriate than another. Managers who are responsible for resolving conflict need to understand what is and what is not an appropriate reaction to conflict. Thomas has summarized the various situations in which specific reactions to conflict are appropriate:[9]

- Competing is appropriate when quick action is vital or when important but potentially unpopular actions must be taken.
- Collaborating is appropriate when the objective is to learn or to work through feelings that are interfering with interpersonal relationships.
- Avoiding is appropriate when you perceive no chance of satisfying your concerns or when you desire to let people cool down and have time to regain a positive perspective.
- Accommodating is appropriate when you are outmatched and losing anyway or when harmony and stability are more important than the issue at hand.

How Conflict Should Be Handled

Managers have two responsibilities regarding conflicts: *conflict resolution* and *conflict stimulation*. Where conflict is present, managers need to resolve it in ways that serve the organization's long-term best interests. This will keep conflict from becoming a detriment to performance. Where conflict does not exist, managers may need to stimulate it to keep the organization from becoming stale and stagnant.[10] Both of these concepts, taken together, are known as *conflict management*.

D. Tjosvold sets forth the following guidelines that can be used by managers in attempting to resolve conflict:[11]

- Determine how important the issue is to all people involved.
- Determine whether all people involved are willing and able to discuss the issue in a positive manner.
- Select a private place where the issue can be discussed confidentially by everyone involved.
- Make sure that both sides understand that they are responsible for both the problem and the solution.
- Solicit opening comments from both sides. Let them express their concerns, feelings, ideas, and thoughts, but in a nonaccusatory manner.
- Guide participants toward a clear and specific definition of the problem.
- Encourage participants to propose solutions. Examine the problem from a variety of different perspectives and discuss any and all solutions proposed.
- Evaluate the costs versus the gains (cost–benefit analysis) of all proposed solutions and discuss them openly. Choose the best solution.

■ Reflect on the issue and discuss the conflict-resolution process. Encourage participants to express their opinions as to how the process might be improved.

How and When Conflict Should Be Stimulated

Occasionally an organization will have too little conflict. Such organizations tend to be those in which employees have become overly comfortable and management has effectively suppressed free thinking, innovation, and creativity. When this occurs, stagnation generally results. Stagnant organizations need to be shaken up before they die. Managers can do this by stimulating positive conflict or conflict that is aimed at revitalizing the organization. S. P. Robbins says that a yes response to any of the following questions suggests a need for conflict stimulation by the manager:[12]

■ Are you surrounded by employees who always agree with you and tell you only what you want to hear?

■ Are your employees afraid to admit they need help or that they've made mistakes?

■ Do decision makers focus more on reaching agreement than on arriving at the best decision?

■ Do managers focus more on getting along with others than on accomplishing objectives?

■ Do managers place more emphasis on not hurting feelings than on making quality decisions?

■ Do managers place more emphasis on being popular than on high job performance and competitiveness?

■ Are employees highly resistant to change?

■ Is the turnover rate unusually low?

■ Do employees, supervisors, and managers avoid proposing new ideas?

Each time one of these questions is answered in the affirmative, it is an indication that conflict may need to be stimulated. It may be possible to have a vital, energetic, developing, improving organization without conflict, but this isn't likely to happen. Innovation, creativity, and the change inherent in continual improvement typically breed conflict. Therefore, the absence of conflict can also be an indication of the absence of vitality. Because this is the case, managers need to know how to stimulate positive conflict.

According to Robbins, the techniques for stimulating conflict fall into three categories: improving communication, altering organizational structure, and changing behavior.[13]

■ Improving communication will ensure a free flow of ideas at all levels. Open communication will introduce a daily agitation factor that will ensure against stagnation while at the same time providing a mechanism for effectively dealing with the resultant conflict.

■ Altering organizational structure in ways that involve employees in making decisions that affect them and that empower them will help prevent stagnation. Employees in organizations that are structured to give them a voice will use that voice. The result will be positive conflict.

■ Changing behavior may be necessary, particularly in organizations that have traditionally suppressed and discouraged conflict rather than dealing with it. Managers who find themselves in such situations may find the following procedure helpful: (a) identify the types of behaviors you want employees to exhibit, (b) communicate with employees so that they understand what is expected, (c) reinforce the desired behavior, and (d) handle conflict as it emerges using the procedures set forth in the previous section.

Communication in Conflict Situations

The point was made in the previous section that human conflict in the workplace is normal, to be expected, and, in certain instances, to be promoted. In managing conflict—which in essence means resolving conflict when it is having negative effects and promoting conflict when doing so might help avoid stagnation—communication is critical.

James Wilcox, Ethel Wilcox, and Karen Cowart recommend the following guidelines concerning the use of communication in managing conflict:[14]

■ *The initial attitude of those involved in the conflict can predetermine the outcome.* This means if a person enters into a situation spoiling for a fight, he or she will probably get one. Communication prior to such a situation aimed at convincing either or both parties to view it as an opportunity to cooperatively solve a problem can help predetermine a positive outcome.

■ *When possible, conflict guidelines should be in place before conflicts occur.* It is not uncommon for conflict to be exacerbated by disagreements over how it should

QUALITY TIP

Conflict in the Workplace

"Almost any manager could write a book about staff disagreements, interdepartmental squabbles, and traumatic office fights. The common combination of scarce resources, varied personalities, and differing perceptions of what is needed and what should be done provides an explosive mixture that frequently results in hard feelings, withdrawal from colleagues, time spent 'putting out fires' instead of getting the job done, and less than top quality production. Inability to handle conflict effectively may well be the single greatest barrier to managerial job satisfaction and success."

Source: James R. Wilcox, Ethel M. Wilcox, and Karen M. Cowart, "Communicating Creatively in Conflict Situations," *Management Solutions: Special Reports* (Saranac Lake, NY: American Management Association, 1987), 34.

QUALITY TIP

Winning and Losing Trust

"Interpersonal trust is exceedingly difficult to gain and just as easy to lose. Losing trust is especially easy in conflicts where individuals tend to look for reasons to 'condemn the enemy.' In departments characterized by conflict, if a supervisor says that a report will be out tomorrow, and the report is not out by tomorrow, employees are prone to say, 'See, we knew we couldn't trust him to keep his word in getting a simple report out. Obviously he can't be trusted to keep his word.'"

Source: James R. Wilcox, Ethel M. Wilcox, and Karen M. Cowart, "Communicating Creatively in Conflict Situations," *Management Solutions: Special Reports* (Saranac Lake, NY: American Management Association, 1987), 42.

---

be resolved. Before entering into a situation in which conflict might occur, make sure all parties understand how decisions will be made, who has the right to give input, and what issues are irrelevant.

- *Assessing blame should not be allowed.*   It is predictable that two people in a conflict situation will blame each other. If human interaction is allowed to get hung up on the rocks and shoals of blame, it will never move forward. The approach that says, "We have a problem. How can we work together to solve it?" is more likely to result in a positive solution than arguing over who is to blame.

- *"More of the same" solutions should be eliminated.*   When a particular strategy for resolving conflict is tried but proves to be ineffective, don't continue using it. Some managers get stuck on a particular approach and stay with it even when the approach clearly doesn't work. Try something new instead of using "more of the same" solutions.

- *Maintain trust by keeping promises.*   Trust is fundamental to all aspects of total quality. It is especially important in managing conflict. Trust is difficult to win but easy to lose. Conflict cannot be effectively managed by someone who is not trusted. Consequently, managers in a total quality setting must keep their promises and, in so doing, build trust among employees.

Each of the strategies set forth in this section depends on communication. Communication that is open, frank, tactful, continuous, and inclusive can do more than anything else to ensure that conflict is properly managed in the workplace.

 SUMMARY

1. Internal politics consists of activities undertaken to gain advantage or influence organizational decision making in ways intended to serve a purpose other than the

best interests of the overall organization. It is the games people play to promote decisions that are based on criteria other than merit.

2. Organizational structure is not the cause of internal politics. All of the widely used organizational structures are susceptible to internal politics.

3. Several concepts—personal insecurity, self-interest, a hunger for power, ambition, ego, and the need for acceptance—are the primary drivers of internal politics.

4. The most commonly used methods of internal politicians are as follows: lobbying, building coalitions, applying harassment and pressure, electioneering, gossiping, and spreading rumors.

5. The rationale for collaboration is found in the debilitating effect internal politics can have on an organization. Internal politics can drain an organization of its intellectual and physical energy and in the process take away its ability to compete.

6. An organization's effort to control internal politics should have at least the following components: strategic planning, leadership, reward/recognition, performance appraisal, customer focus, conflict management, and culture.

7. The most common categories of negative behavior are control disputes, territorial (boundary) disputes, dependence/independence issues, need for attention/responsibility, authority, and loyalty issues.

8. The following symptoms are indicators of negativity in the workplace: "I can't" attitudes, "they" mentality, critical conversation, and blame fixing among employees.

9. To overcome negativity, organizations should communicate, establish clear expectations, provide opportunities for anxiety venting, build trust, and involve employees.

10. Territorial behavior in organizations manifests itself in the following ways: occupation, information manipulation, intimidation, alliances, invisible walls, strategic noncompliance, discredit, shunning, camouflage, and filibustering.

11. The following strategies will help when trying to overcome territorial behavior: avoid jumping to conclusions, attribute the behavior to instinct rather than people, ensure that employees don't feel attacked, avoid generalizations, understand irrational fears, respect each individual's perspective, and consider the employee's point of view.

12. Causes of workplace conflict include limited resources, incompatible goals, role ambiguity, different perceptions, and poor communication.

13. Managers have two responsibilities regarding conflict in the workplace: conflict resolution and conflict stimulation. Conflict should be stimulated to overcome excessive compliance and complacency.

 KEY TERMS AND CONCEPTS

Alliances	Camouflage
Ambition	Conflict management
Anxiety venting	Critical conversation
Blame fixing	Culture
Building coalitions	Customer focus

Ego	Lobbying
Electioneering	Need for acceptance
Gossiping and spreading rumors	Occupation
Harassment and pressure	Performance appraisal
Hunger for power	Personal insecurity
"I can't" attitudes	Reward/recognition
Information manipulation	Self-interest
Internal politics	Shunning
Intimidation	Strategic noncompliance
Invisible walls	Strategic planning
Leadership	"They" mentality

 FACTUAL REVIEW QUESTIONS

1. Define *internal politics*.
2. Explain the role organizational structure can play in promoting internal politics.
3. Give an example of internal politics in action in an organization.
4. List and briefly describe the most commonly used methods of internal politicians.
5. Describe the impact internal politics can have on the implementation of total quality.
6. Describe how managers can control internal politics in an organization.
7. What are the categories of negativity in the workplace?
8. Explain the strategies for overcoming territorial behavior.
9. When should conflict be encouraged in an organization?

 CRITICAL THINKING ACTIVITY

A Toxic Work Environment

When Maggie Lance took over Air-Tech's packaging and shipping department, it was notorious within the company as a dead-end, high-stress, uncomfortable place to work. Internal politics, territoriality, negativity, and conflict were the rule rather than the exception in the P & S department. Other employees joked that "P & S" stands for "push and shove." The best word Maggie could find to describe the work environment in her new department was *toxic*.

Turnover is high, morale is low, productivity is substandard, and quality is a joke. Clearly, Maggie has her work cut out for her. Put yourself in her place. Where should she start? How should she proceed? What should be the key elements in her turnaround plan?

DISCUSSION ASSIGNMENT 13–1

Conflict over Conflict

"We have too much conflict in this organization," said Larry Parette. "Nobody agrees on anything. It's beginning to get to me."

"Nonsense, Larry," said Mack Porter. "We need conflict. Our best ideas come when smart people disagree, not when everyone goes along just to get along." Parette and Porter are the founders and managing partners of P&P Design, a firm they have grown from just the two of them to more than 300 employees.

"Well, this conflict you like so much has cost us two good employees this month," said Parette. "Both of them told me that working here had become too stressful because of the constant arguing and bickering."

"All right, we lost two good designers," said Porter. "But just last week we saved $250,000 on the Morgandale project because one of our junior designers refused to go along with the crowd and practically forced us to consider his ideas. His ideas caused a lot of grief in several meetings and some of our senior designers even went after his job. But now that his ideas have worked so well, suddenly everyone claims they agreed with him from the beginning. We need this kind of conflict."

Discussion Questions
Discuss the following questions in class or outside of class with your fellow students:

1. Who is right in this case? Is it Parette, Porter, neither, or both?
2. What if anything, should be done about conflict at P&P Design?

DISCUSSION ASSIGNMENT 13–2

Keep Off My Turf!

The best way to describe the relationship between the design and manufacturing departments at Waverly Prestressed Concrete (WPC) is warfare. A typical conversation between the vice president for design and the vice president for manufacturing goes like this: "If you knew anything about manufacturing, you might design a product we can actually make every once in a while," says the vice president for manufacturing.

"If you knew my job better, you could manufacture anything we design," responds the vice president for design.

They are actually both right. The design and manufacturing departments need to work together closely and cooperatively as partners. The designers need to apply the principles of design for manufacturing (DFM) so that the product is not just functional but can be produced economically. Manufacturing personnel need to be a part of the design team so that their input is part of the process from the outset. Unfortunately, both departments have built invisible walls around their domains and adopted attitudes toward each other that clearly say, "Keep off my turf!"

Discussion Questions
Discuss the following questions in class or outside of class with your fellow students:

1. What, if anything, should be done to improve relations at WPC?
2. How can the invisible walls be pulled down?

ENDNOTES

1. Douglas MacArthur, *Reminiscences* (New York: McGraw Hill, 1964), 184.
2. Alan R. Gitelson, Robert L. Dudley, and Melvin J. Dubnick, *American Government* (Boston: Houghton Mifflin, 1993), 225.
3. Tom Rusk, *The Power of Ethical Persuasion* (New York: Penguin, 1994), xvi–xvii.
4. *The American Heritage College Dictionary*, 3rd ed. (New York: Houghton Mifflin, 1993), 337.
5. Annette Simmons, *Territorial Games* (New York: AMACOM, American Management Association, 1998), 6–7.
6. Simmons, 179.
7. Simmons, 187.
8. K. W. Thomas, "Conflict and Conflict Management," *Academy of Management Review* 2 (1977), 163.
9. K. W. Thomas, "Toward Multi-Dimensional Values in Technology," *Academy of Management Review* 2 (1977), 487.
10. S. P. Robbins, "Conflict Management and Conflict Resolution Are Not Synonymous Terms," *California Management Review* 32 (1978), 67–75.
11. D. Tjosvold, "Making Conflict Productive," *Personnel Administrator* 29 (1984), 121–130.
12. Robbins, 67–75.
13. Robbins, 67–75.
14. James R. Wilcox, Ethel M. Wilcox, and Karen M. Cowart, "Communicating Creatively in Conflict Situations," *Management Solutions: Special Reports* (Saranac Lake, NY: American Management Association, 1987), 34–40.

CHAPTER FOURTEEN

ISO 9000 and Total Quality: The Relationship*

"A company that claims that it cannot standardize and must rely on experience is a company without technology."

   Kaoru Ishikawa

MAJOR TOPICS

- ISO 9000: What It Is
- The ISO 9000 Quality Management System: A Definition
- Comparative Scope of ISO 9000 and Total Quality Management
- The Origin of ISO 9000
- Aims of ISO 9000 and Total Quality Management
- Management Motivation for Registration to ISO 9000
- Compatibility of ISO 9000 and Total Quality Management
- Movement from ISO 9000 to TQM and Vice Versa
- ISO 9000 as an Entry into Total Quality Management

ISO 9000: WHAT IT IS

The International Organization for Standardization (ISO) is a federation of the national standards bodies of nations from around the world. The American National Standards Institute (ANSI) is the U.S. member. The American Society for Quality (ASQ), as a member of ANSI, is the body responsible for quality management and related standards. In the United States, ASQ publishes and distributes ISO 9000 documentation as the ANSI/ISO/ASQ Q9000 series.

---

*This chapter has been adapted from David L. Goetsch and Stanley B. Davis, *Understanding and Implementing ISO 9000: 2000* (Upper Saddle River, NJ: Prentice Hall, 2002), Chapters 1 and 10.

ISO 9000 is an international quality standard for goods and services. The term *quality standard* tends to be misleading. For example, ISO 9000 does not set any specifications for quality. Rather, it sets broad requirements for the assurance of quality and for management's involvement. The emphasis is on *prevention* rather than inspection and rework. In fact, this emphasis is placed not only on the production process but also on the product design process. The ISO 9000 approach is completely compatible with the total quality philosophy, though it is not as all encompassing. ISO 9000 is composed of three standards:

ISO 9000: 2000   Quality Management Systems—Fundamentals and Vocabulary
ISO 9001: 2000   Quality Management Systems—Requirements
ISO 9004: 2000   Quality Management Systems—Guidelines for Performance Improvements

ISO 9000 is about standardizing the approach organizations everywhere take to managing and improving the processes that ultimately result in their products and services. Specifically, ISO 9000 establishes the requirements for quality management systems (QMS) that must be employed by all organizations registered to the standard. Registered organizations should enjoy:

- Wider customer acceptance of products and services
- Improved effectiveness and reliability of its processes
- Improved quality of products and services
- Improved organizational performance and competitiveness

Since ISO 9000 was first released in 1987 it has evolved through two revisions, the first in 1994 and the most recent at the end of 2000. This evolution has aligned it more closely with the Total Quality Management philosophy. It seemed to many observers, including the authors, that the 1987 and 1994 versions shied away from association with TQM, or from acknowledging its existence. Even the 2000 version, which borrows heavily from TQM, scarcely acknowledges it. The fact is, of course, that with the tutelage of W. Edwards Deming and Joseph Juran, the Japanese started the development of the management system we now know as TQM in 1950. Over the years several Japanese experts—Kaoru Ishikawa, Shigeo Shingo, Taiichi Ohno, and others—emerged, and by the early 1970s TQM had been widely accepted in Japan. By 1980 the Western world began taking note. By the time ISO 9000:1987 was released, TQM was a mature management system, well understood by many in the West. It is clear that ISO's Technical Committee 176 (TC 176), which was charged with ISO 9000's development, borrowed some TQM elements, most notably its documentation requirements. ISO 9000:1994 moved a bit closer to TQM, at least mentioning (though not requiring) continual improvement. But any acknowledgement of TQM's influence or superiority seemed to be deliberately avoided. ISO 9000:2000 made a giant leap in comparison, especially in the area of continual improvement, which has gone from receiving just cursory treatment to becoming a firm requirement. In addition, the standard now incorporates eight quality management principles that come directly from TQM.[1] They are:

1. Customer focus—understanding customers' needs, striving to exceed their expectations.
2. Leadership—establishing direction, unity of purpose, and a supportive work environment.
3. Involvement of people—ensuring that all employees at all levels are able to fully use their abilities for the organization's benefit.
4. Process approach—recognizing that all work is done through processes, and managing them accordingly.
5. System approach to management—expands on the previous principle in that achieving any objective requires a system of interrelated processes.
6. Continual improvement—as a permanent organizational objective, recognizing and acting on the fact that no process is so good that further improvement is impossible.
7. Factual approach to decision making—acknowledging that sound decisions must be based on analysis of factual data and information.
8. Mutually beneficial supplier relationships—to take advantage of the synergy that can be found in such relationships.

Five of these principles are also principles listed in the primary eleven TQM principles. (See Chapter 1, The Total Quality Approach to Quality Management.) The other three (4, 5, and 8) are also part of the TQM philosophy, the whole of which is now embedded to some degree in ISO 9000:2000.

ISO and TC 176 are to be commended for these positive changes to ISO 9000. Understand, however, that an organization can be certified to ISO 9000 without fully adopting TQM. For those elements of the organization subject to the standard, however, the TQM alignment will be quite close, especially if ISO 9004:2000, which contains much of the TQM philosophy, is followed. (Note: organizations are certified to ISO 9001:2000 only.) ISO 9004 is designed to show the organization how its performance can be further and continually improved.

QUALITY TIP

Implementation of ISO 9000:2000

A three-year transition period went into effect with the December 2000 publication of ISO 9000:2000. Although the intent was for organizations to "upgrade" to the new standard as quickly as possible, ISO permitted new registrations to the 1994 standard during the transition period to accommodate organizations that were nearing readiness for certification. Existing (pre-December, 2000) ISO 9000:1994 certifications retain validity until their next recertification date, no later than December 2003. ISO 9000:1994 certifications issued any time after December 2000 expire December 2003. Thereafter, all certifications will be to ISO 9001:2000.

As a result of ISO 9000, any organization supplying products or services is able to develop and employ a quality management system that is recognized by customers worldwide. Customers around the globe who deal with ISO 9000–registered organizations can expect that purchased goods or services will conform to a set of recognized standards.

ISO 9001's requirements for quality management systems are generic in nature, and are applicable to organizations in any industry or economic sector.[2] Whether the organization manufactures a product or provides a service, whether it is a company or a governmental agency, whether it is large or small, ISO 9000 can apply, and be used to advantage. To be registered the organization must go through a process that includes the following steps:

1. Develop (or upgrade) a quality manual that describes how the company will assure the quality of its products or services.
2. Document procedures (or upgrade existing documentation) that describe how the various processes for design, production, continual improvement, and so forth, will be operated. This must include procedures for management review/audits and the like.
3. The organization must provide evidence of top management's commitment to the QMS and its continual improvement.
4. The organization's top management must ensure that customer requirements are determined and met.
5. The organization must hire an accredited registrar company to examine its systems, processes, procedures, quality manual, and related items. If everything is in order, registration will be granted. Otherwise, the registrar will inform the company of which areas require work (but will not inform the company specifically what must be done), and a second visit will be scheduled.
6. Once registration is accomplished, the company will conduct its own internal audits to ensure that the systems, processes, and procedures are working as intended.
7. Also once registered, the outside registrar will make periodic audits for the same purpose. These audits must be passed to retain registration.

An important point to understand about ISO 9000 is that the organization has to respond to all ISO 9000 requirements and tell the registrar specifically what it is going to do and how. ISO does not tell the organization. Assuming the registrar agrees with the organization's plan, registration is awarded. To retain that registration, the organization *must do what it said it would do.*

THE ISO 9000 QUALITY MANAGEMENT SYSTEM: A DEFINITION

To secure ISO 9000 registration, organizations must develop and use quality management systems conforming to the requirements of ISO 9001:2000. The first question asked by someone unfamiliar with ISO 9000 is, "What is a quality management system?" In ISO's own words, a QMS is a *management system to direct and control an organization with regard to quality.*[3]

QUALITY TIP

Aim of the "Consistent Pair"

The primary aim of the "consistent pair" [ISO 9001:2000 and ISO 9004:2000] is to relate modern quality management to the process and activities of an organization, including the promotion of continual improvement and achievement of customer satisfaction. Furthermore it is intended that the ISO 9000 standards have global applicability. Therefore, the factors that are driving the [ISO 9000: 2000] revision process, among others [include the] provision of a natural stepping-stone towards Total Quality Management.

Source: ISO, *The New Year 2000 ISO 9000 Standards: An Executive Summary,* 1999.

This definition is not as tight as it might be. This being the case, we provide the following definition:

> The quality management system is composed of all the organization's policies, procedures, plans, resources, processes, and delineation of responsibility and authority, all deliberately aimed at achieving product or service quality levels consistent with customer satisfaction and the organization's objectives. When these policies, procedures, plans, and so forth, are taken together, they define how the organization works and how quality is managed.

The quality management system will include the following documentation:

1. *A quality policy.*   This statement describes how the organization approaches quality.
2. *The quality manual.*   This must address each clause of the ISO 9001 standard. It will also typically include an organization chart, or some such device, illustrating management responsibility for operating the quality system. Quality procedures may be part of this manual, or they may be referenced.
3. *Quality objectives.*   Goals sought or aimed for, related to quality, based on the quality policy. Quality objectives are assigned to the relevant organizational functions and levels, and are tracked by top management.
4. *Quality procedures.*   These describe step by step what the company does to meet the quality policies. As a minimum, there will be a procedure for each of the ISO 9001 clauses outlining requirements. There may also be procedures for any processes that can impact quality.
5. *Forms, records, and so on.*   These materials provide proof of activities for the firm and for the auditors.

This documentation is used to assure the necessary consistency in the firm's operations and processes. Auditors use it to verify compliance.

COMPARATIVE SCOPE OF ISO 9000
AND TOTAL QUALITY MANAGEMENT

The two principal quality initiatives at work in the world today are ISO 9000 and total quality management. Consequently, it is helpful to explain the relationship between the two. The following statements outline the relationship. Each statement is explained in the sections that follow in this chapter.

- ISO 9000 and Total Quality Management are not interchangeable.
- ISO 9000 is compatible with, and can be viewed as a subset of TQM.
- ISO 9000 is frequently implemented in a non–TQM environment.
- ISO 9000 can improve operations in a traditional environment.
- ISO 9000 may be redundant in a mature TQM environment.
- ISO 9000 and TQM are not in competition.

ISO 9000 and Total Quality Are Not Completely Interchangeable

Although ISO 9000 has made a great leap toward TQM with the 2000 release, these are not yet the same, and probably never will be. By definition, ISO 9000 is concerned only with quality management systems, for the *design, development, purchasing, production, installation,* and *servicing* of products and services.

On the other hand, Total Quality Management, by definition, encompasses every aspect of the business or organization, not just the systems used to design, produce, and deploy its products and services. This includes all support systems such as human resources, finance, and marketing. Total Quality Management involves every function and level of the organization, from top to bottom.

Total Quality Management also means that management is responsible for developing the organization's vision (its dream), establishing guiding principles (a code of conduct for the organization and all of its employees), and setting the strategy and tactics for achieving the vision within the constraints of the guiding principles. In a Total Quality Management organization the vision is pursued with input from an empowered workforce that cooperates and collaborates with management.

Total Quality Management, based on the teachings of Deming, Juran, Ishikawa, and others, with criteria defined by Deming's Fourteen Points, Juran's Ten Steps to Quality Improvement, and the Malcolm Baldrige National Quality Award, is more pervasive and demanding—literally requiring the transformation of the organization.

Before the advent of the year 2000 release, ISO 9000 was concerned only with the standards which an organization could build its own version of a quality management system. ISO 9000:2000 has closed much of the gap that existed with TQM. The primary remaining difference between ISO 9000 and TQM is in the degree to which the total organization is involved. Whereas TQM requires the involvement of all functions and levels of the organization, ISO 9000 does not require the QMS to include functions and levels that do not play a direct role in the management and execution of the product/service realization processes. Functions that are typically not involved under the QMS include human resources, finance (accounting), sales, and marketing.

Characteristics of Total Quality Management	ISO 9000:2000	TQM
Customer focus (internal and external)	✓	✓
Obsession with quality		✓
Scientific approach to problem solving	✓	✓
Long-term commitment	partial	✓
Teamwork		✓
Continual process and product improvement	✓	✓
Education and training intensive	✓	✓
Freedom through control		✓
Unity of purpose	✓	✓
Employee involvement and empowerment	partial	✓

Figure 14–1
Total Quality Management Characteristics Compared with ISO 9000

Figure 14–1 illustrates how close ISO 9000's evolution has brought it to TQM.

Total Quality is defined as an approach to doing business that attempts to maximize the competitiveness of an organization through the continual improvement of the quality of its products, services, people, and environments by emphasizing the characteristics listed in Figure 14–1.

In comparison, the ISO 9000 quality management system is designed to "provide the framework for continual improvement to increase the probability of enhancing customer satisfaction and the satisfaction of other interested parties. It provides confidence to the organization and its customers that it is able to provide products that consistently fulfill requirements."[4] ISO claims that beyond customer satisfaction, cost and risk-management benefits will also accrue to the organization. These benefits translate to improved competitiveness—the same as TQM's objective. ISO claims these benefits result from emphasizing the eight quality management principles on which the standard is based. See Figure 14–2 for a comparison of ISO's eight quality management principles with Deming's Fourteen Points and TQM.

ISO 9000 Is Compatible with, and Can Be Viewed as a Subset of, TQM

Clearly, TQM and ISO 9000 are not quite the same thing. However, there is nothing inherent in ISO 9000 that would prevent it from becoming part of a larger Total Quality Management environment. There are many examples today of companies that have

ISO 9000's Eight Quality Management Principles	Deming's 14 Points	TQM
1. Customer focus		✓
2. Leadership	#1, #2, #7	✓
3. Involvement of people		✓
4. Process approach		✓
5. System approach to management		✓
6. Continual improvement	#5	✓
7. Factual approach to decision making		✓
8. Mutually beneficial supplier relationships	#4	✓

Figure 14–2
ISO 9000's Quality Management Principles versus Deming's Fourteen Points and TQM

successfully included ISO 9000 as part of a larger total quality effort. Organizations that are already at some level of TQM maturity have typically found it easy to implement ISO 9000. This is because a TQM environment with its infrastructure of top management commitment, documented processes and procedures, continuous improvement, obsession with quality, and so on, easily supports the requirements of ISO 9000.

ISO 9000 Is Frequently Implemented in a Non–TQM Environment

Although total quality is compatible with and may well facilitate an ISO 9000 implementation, it is by no means a prerequisite for ISO 9000. In fact, it is safe to say that the majority of ISO 9000–registered organizations have not fully adopted total quality—at least, not yet.

ISO 9000 Can Improve Operations in a Traditional Environment

By "traditional environment," we mean an organizational environment that has persisted in companies for decades, until the Total Quality Management movement began to change things. A traditional organizational enviroment is one which still operates according to the "old way of doing things" rather than according to the principles of Total Quality Management.

When ISO 9000 is implemented by a traditional organization, the company should be the better for it. We will not go so far as to say it *will* be the better for it, because much depends on the organization's reasons for adopting ISO 9000 and the degree of executive-level commitment to it. Put another way, if ISO 9000 is approached inappropriately and for the wrong reasons, it can become nothing more than a marketing ploy, and the organization's functional departments might develop even more problems than they had before ISO 9000.

ISO 9000 May Be Redundant in a Mature TQM Environment

Just as ISO 9000 should help traditional organizations, it should also benefit TQM organizations. However, in an organization that has achieved a high level of maturity in its total quality journey, say in the 400–600 range on the Baldrige scale of 1,000 points, all ISO 9000 criteria may already be in place. In such a case, the only compelling reason for registration under ISO 9000 would be for marketing purposes. What would a company such as Toyota gain from ISO 9000 registration? Probably nothing. It already does everything required by ISO 9000. Its products and processes are recognized as world class. Consequently, It wouldn't gain even a marketing advantage. However, there are many fine TQM organizations that are not as well known as Toyota. Such organizations, even though they may already meet or exceed the requirements of ISO 9000, may find it necessary to register in order to let potential customers know that their products or services satisfy the international standard.

ISO 9000 and TQM Are Not in Competition

This is not a case of one or the other. Organizations can adopt TQM or ISO 9000, or both. While there may be those who advocate one to the exclusion of the other, in the larger scheme of things, the two concepts fit well with each other. Both have worthwhile and similiar aims. Our view is that not only are TQM and ISO 9000 compatible; they actually support each other and are complementary. There are good reasons for using both in a single management system.

THE ORIGIN OF ISO 9000

ISO 9000 and Total Quality Management originated independently of each other, for different reasons, in different parts of the world, and at different times. You are already

QUALITY TIP

Marketing and ISO 9000

"As a marketing advantage, ISO 9000 may be only temporary. Organizations achieve an advantage if they are registered before their competition. But the advantage lasts only until the competition adopts ISO 9000. *Gaining a marketing advantage is the wrong motivation for adopting ISO 9000.*"

Source: David Goetsch and Stan Davis, *Understanding and Implementing ISO 9000:2000 Standards* (Upper Saddle River, NJ: Prentice Hall, 2002), 316.

familiar from other chapters with the post–World War II origins in Japan of the total quality movement. The ISO 9000 series of standards was originally developed in response to the need to harmonize dozens of national and international quality standards then existing throughout the world. To that end the International Organization for Standardization (ISO), a worldwide federation of national standards organizations from 110 nations, formed Technical Committee 176.

Although sometimes considered to be a European standard (certainly the impetus came from Europe), ISO 9000 was developed by an international team that includes the American National Standards Institute (ANSI), the U.S. member of ISO. ANSI was represented by the American Society for Quality (ASQ), its affiliate responsible for quality management and related standards. The first version of ISO 9000 was released in 1987. By this time, the Total Quality Management movement was more than 35 years old. A revised version of ISO 9000 was released in 1994 and most recently in 2000. As a result of this standard, suppliers of products and services are able to develop and employ a quality management system that is recognized by all their customers, regardless of where on the planet those customers might be. Customers around the world who deal with ISO 9000–registered organizations can expect that their purchases will measure up to a set of recognized standards.

AIMS OF ISO 9000 AND TOTAL QUALITY MANAGEMENT

The aim of ISO 9000 has historically been to assure that the products or services provided by registered organizations are consistently fit for their intended purpose. ISO 9000:2000 has raised the standard's aim to a new level. Customer satisfaction and continual improvement, along with the other six quality management principles that have been incorporated into the standard, seek to make registered organizations more competitive. This is essentially the same objective as that of Total Quality Management.

MANAGEMENT MOTIVATION FOR REGISTRATION TO ISO 9000

Management motivation for adopting either ISO 9000 or TQM can vary widely. There are both appropriate and inappropriate motives. For example, if a company seeks ISO 9000 registration to obtain a marketing advantage, its motive is inappropriate. As a result, the organization will likely give mere lip service to adopting the standard. Appropriate motives for adopting ISO 9000 include the following:

■ To improve operations by implementing a quality management system that satisfies the ISO 9000 requirements for management responsibility; resource management; product realization; and measurement, analysis, and improvement

■ To create or improve a quality management system that will be recognized by customers worldwide

■ To improve product or service quality or the consistency of quality

■ To improve customer satisfaction

- To improve competitive posture
- To conform to the requirements of one or more major customers (although adoption would be better motivated by internal considerations, such as the preceding five)

What we are saying here is that, ideally, management will adopt ISO 9000 as a way to make real improvements in the company's operations, serve its customers in a more responsible way, and, as a result, be more successful. This approach is more likely to assure commitment and participation by top management. Approaching ISO 9000 from a strictly marketing perspective may result in a negative reaction to the amount of work required by the functional departments, and only enough management commitment to do the bare minimum for registration. In other words, if ISO 9000 is viewed as a necessary evil that one must adopt to compete in certain markets, every dollar and every hour spent on ISO 9000 will be seen as a burden to be endured rather than an investment in the organization's future. By definition, a burden is a load that is difficult to bear; the connotation is negative. When negative feelings abound among employees, commitment to ISO 9000 will suffer. It may be possible to fool the ISO 9000 registrar's auditor, but we guarantee that customers will not be fooled—at least not for long. Newfound markets will soon wither and disappear. If ISO 9000 is to have a real and permanent effect, it must be approached with a positive attitude and the unwavering commitment of top management.

COMPATIBILITY OF ISO 9000 AND TOTAL QUALITY

We have discussed the fact that ISO 9000 and TQM are different in scope and were developed from different perspectives, but now have similar requirements and objectives. Now, more than ever, the two concepts are compatible. With the exception of registration and audits, TQM requires everything required by ISO 9000. However, even a mature TQM organization, one that does everything it would do under ISO 9000, and more, will not have the worldwide recognition afforded by ISO 9000 registration. There is no corresponding international certification for TQM. For this reason, even the mature TQM organization may find it necessary to seek ISO 9000 registration as a way to satisfy the demands of its customers. On the other hand, a traditional organization that is registered under ISO 9000 may find that it needs the larger Total Quality Management implementation to become or stay competitive.

ISO 9000 registration can be a good first step into TQM. In a paper published by the European Union entitled *Working Document on a European Quality Promotion Policy, or the European Way toward Excellence*, the authors had this to say:

> ISO 9000 registration often is the most practical route to demonstrate compliance for European regulations with a quality system element. In parallel, the use of ISO 9000 by companies can be regarded as a first step toward a more global management of companies, similar to the one arising from the European model of total quality management.[5]

We are not sure how the "European model" of total quality management differs from that of the rest of the world, but the point is that regardless of geographic location, people who understand both ISO 9000 and total quality have concluded that the two are compatible and that ISO is properly seen as a subset of total quality.

Movement from ISO 9000 to TQM and Vice Versa

An organization that has its processes documented and under control, such as a company involved in Total Quality Management, should find it relatively easy to prepare for ISO 9000. Correspondingly, a traditional organization that has successfully registered under ISO 9000 will have a head start should it decide to implement total quality. The major issues with ISO 9000 are securing top management commitment, focusing on customer requirements and satisfaction, and the work involved in documenting processes and procedures. Total Quality Management requires the same.

ISO 9000 as an Entry into Total Quality Management

How to get started is always an issue for organizations just beginning their total quality journey. In Chapter 22 and in our book *Implementing Total Quality*,[6] we lay out a 20-step process that may be used by any organization for undertaking a total quality implementation (see Figure 22–10). Embedded within the 20 steps are the requirements to assess the organization's strengths, weaknesses, where it wants to go (the vision), and its strategy for achieving the vision. Organizations going through this implementation process may find that a good strategy is registration to the ISO 9000:2000 standard. ISO 9000 preparation projects can be pursued as the entry projects for implementing total quality management.

We have already discussed the fact that ISO 9000 and total quality are compatible, making many of the same demands on the organization, and also that ISO 9000 is, for all practical purposes, a subset of total quality. For an organization attempting to adopt total quality, and one that would also benefit from ISO 9000, our 20-step implementation process should be considered. At the planning phase, steps 12 through 15, the initial implementation approach should be designed to include the steps necessary for ISO 9000 registration. By adopting this strategy, the organization will be engaged in both a Total Quality Management implementation and an ISO 9000 preparation. The ISO 9000 effort will benefit from the total quality preparation phase by having the following components: an executive-level steering committee, a vision with the attendant guiding principles, a set of broad objectives, baselines on employee and customer satisfaction, an objective view of the organization's strengths and weaknesses, and an indication of which employees at all levels can be counted on for support during the implementation. In addition, the organization will have a well thought out means of communicating with employees and all other stakeholders to keep them apprised of the changes taking place, why they are happening, and what they will mean to everyone.

The recommendations in the preceding paragraph apply to organizations that have not yet implemented total quality or ISO 9000, ones that are just thinking about it. But what about the organization that has already started working on ISO 9000 or has already achieved registration? How should such an organization approach the larger task of implementing total quality? The effort expended on ISO 9000 clearly should be seen as a head start, assuming ISO 9000 compliance is approached as a way to improve the organization and not simply as a marketing gimmick. To the degree that the organization has already accomplished the early steps of the 20-step implementation process for Total Quality Management, count it as progress toward the eventual complete implementation of TQM.

This organization should go back and execute any steps that have not already been completed before moving on. For example, there is no requirement in ISO 9000 for forming a steering committee (step 2) composed of the top managers, so it is doubtful one has been established. Similarly, steps 3 (team building for the steering committee) and 4 (total quality training for the steering committee) are not required by ISO 9000 and will not have been done. The same may be said for steps 5, 8, 9, 10, and 11. ISO 9000 does not require its registered organizations to have a vision statement or a set of guiding principles under which the organization will operate.

Nothing in ISO 9000 would require an examination of organizational strengths and weaknesses or the baselining of employee and customer satisfaction. All of these should be done for total quality, and all will benefit ISO 9000. Having gone back to complete steps 1 through 11, the steering committee should start its planning phase by incorporating its ISO 9000 activities into the total quality initiative using steps 12 through 17, then expand beyond ISO 9000 from there.

In summary, the organization that is already involved in ISO 9000 should see itself as having a head start on the larger TQM implementation. One that has started neither, although seeing TQM and ISO 9000 as beneficial, might approach ISO 9000 as a logical part of the initial total quality journey.

If you would like to learn more about ISO 9000 and its registration requirements, refer to our book *Understanding and Implementing ISO 9000:2000* (Upper Saddle River, NJ: Prentice Hall, 2002).

 SUMMARY

1. The following statements describe the relationship between ISO 9000 and TQM: ISO 9000 and TQM are not completely interchangeable; ISO 9000 is compatible with, and can be a subset of, TQM; ISO 9000 is frequently implemented in a non–TQM environment; ISO 9000 can improve operations in a traditional environment; ISO 9000 may be redundant in a mature TQM environment; and ISO 9000 and TQM are not in competition.
2. The origins of ISO 9000 and total quality management are vastly different. ISO 9000 was developed in response to the need to harmonize dozens of national and international standards relating to quality. Total quality got its start in Japan around 1950 as a way to help that nation compete in the international marketplace.
3. The new aim of ISO 9000 is to enable organizations to better serve their customers and to be more competitive through adherence to the standard's eight quality management principles.
4. Appropriate motivations for implementing ISO 9000 are as follows: to improve operations, to improve or create a quality management system, to improve the consistency of quality, to improve customer satisfaction, to improve competitive posture, and to conform to the requirements of customers. The appropriate motivation for implementing TQM is a desire to continually improve all aspects of an organization.

5. ISO 9000 and TQM are compatible in that ISO 9000 can be a complementary subset of TQM. ISO 9000 can give an organization a head start in implementing TQM.

KEY TERMS AND CONCEPTS

Accredited registrar

Aims of ISO 9000 and total quality management

American Society for Quality (ASQ)

American National Standards Institute (ANSI)

Audits (internal and by registrar)

Compatibility of ISO 9000 and TQM

International Organization for Standardization (ISO)

ISO 9000

ISO 9000 and TQM are not interchangeable

ISO 9000's eight quality management principles

Origin of ISO 9000

Quality Management Systems (QMS)

Quality objectives

Quality policy

Registration

Traditional organizational environment

Technical Committee 176 (TC 176)

FACTUAL REVIEW QUESTIONS

1. List six statements that summarize the comparative scope of ISO 9000 and total quality.
2. Explain the origins of ISO 9000 and total quality. How are they different?
3. Contrast the aims of ISO 9000 and total quality.
4. List three appropriate reasons for implementing ISO 9000.
5. What is the most appropriate rationale for implementing total quality?
6. Describe how you would use ISO 9000 as an entry into total quality.

CRITICAL THINKING ACTIVITY

Implementing Total Quality

Create a matrix that shows how ISO 9000 registration fits into a broader total quality implementation.

DISCUSSION ASSIGNMENT 14–1

ISO 9000 or TQM: Which will it be?

Reliance Control Systems is a large manufacturer of control systems for the electrical power generation and distribution industry. Over the past few years Reliance has watched overseas competitors take away market share with products that are priced lower and which, at the same time, have developed a reputation for better reliability. The company is not in a dangerous position yet, but the board of directors wants to see a concerted effort to improve the company's competitive posture. Among the senior management two factions have developed. One, led by the vice president of operations, is pressing the CEO to implement Total Quality Management. After all, the aim of TQM is improved competitiveness, and that is just what is needed. On the other hand, the manufacturing vice president and the director of quality assurance are making the case for ISO 9000:2000.

Discussion Questions

Discuss the following questions in class or outside of class with your fellow students:

1. Assume that you are siding with the VP of Operations, and list the arguments for implementing TQM.

2. Now assume that you are with the heads of Manufacturing and Quality Assurance. List the arguments for ISO 9000:2000.

3. You have heard both sets of arguments, and you believe there might be a third approach that could satisfy both factions. What would you propose?

 ENDNOTES

1. ISO 9000:2000, clause 0.2, Quality Management Principles.
2. ISO 9000:2000, clause 2.2, Requirements for Quality Management Systems and Requirements for Products.
3. ISO 9000:2000, clause 3.2.3, Quality Management System.
4. ISO 9000:2000, clause 2.1, Rationale for Quality Management Systems.
5. "Working Document on a European Quality Promotion Policy, or the European Way toward Excellence," *Quality in Manufacturing* (March 1996), 17.
6. David L. Goetsch and Stanley B. Davis, *Implementing Total Quality* (Upper Saddle River, NJ: Prentice Hall, 1995).

PART TWO

Tools and Techniques

CHAPTER FIFTEEN

Overview of Total Quality Tools

"Think as you work, for in the final analysis your worth to your company comes not in solving problems but in anticipating them."

Herbert H. Ross

MAJOR TOPICS

- Total Quality Tools Defined
- The Pareto Chart
- Cause-and-Effect Diagrams
- Check Sheets
- Histograms
- Scatter Diagrams
- Run Charts and Control Charts
- Stratification
- Some Other Tools Introduced
- Management's Role in Tool Deployment

One of the basic tenets of total quality is *management by facts.* This is not in harmony with the capability so revered in North America and the West in general: the ability to make snap decisions and come up with quick solutions to problems in the absence of input beyond intuition, gut feel, and experience. Management by facts requires that each decision, each solution to a problem, is based on relevant data and appropriate analysis. Once we get beyond the very small business (in which the data are always resident in the few heads involved, anyway), most decision points and problems will have many impacting factors, and the problem's root cause or the best-course decision will remain obscure until valid data are studied and analyzed. Collecting and analyzing data can be difficult. The total quality tools presented in this chapter make that task easy enough for anyone. Their use will assure better decision making, better solutions to problems, and even improvement of productivity and products and services.

Writing about the use of statistical methods in Japan, Dr. Kaoru Ishikawa said:

> The above are the so-called seven indispensable tools . . . that are being used by everyone: company presidents, company directors, middle management, foremen, and line workers. These tools are also used in a variety of [departments], not only in the manufacturing [department] but also in the [departments] of planning, design, marketing, purchasing, and technology."[1]

No matter where you fit into your organization today, you can use some or all of these tools to advantage, and they will serve you well for your future prospects.

This chapter explains the most widely used total quality tools and their applications, provides some insights on the involvement of management and the cross-functional nature of the tools, and issues some cautions.

TOTAL QUALITY TOOLS DEFINED

Carpenters use a kit of tools designed for very specific functions. Their hammers, for example, are used for the driving of nails. Their saws for the cutting of wood. These and others enable a carpenter to build houses. They are *physical* tools. Total quality tools also enable today's employees, whether engineers, technologists, production workers, managers, or office staff, to do their jobs. Virtually no one can function in an organization that has embraced total quality without some or all of these tools. Unlike those in the carpenter's kit, these are *intellectual* tools: they are not wood and steel to be used with muscle; they are tools for collecting and displaying information in ways to help the human brain grasp thoughts and ideas. When thoughts and ideas are applied to physical processes, the processes yield better results. When applied to problem solving or decision making, better solutions and decisions are developed.

The seven tools discussed in the following seven sections of this chapter represent those generally accepted as the basic total quality tools. Some authors would include others, and we discuss some of the others briefly later in this chapter. A case can be made that *just-in-time, statistical process control,* and *quality function deployment* are total quality tools. But these are more than tools: they are complete systems under the total quality umbrella. This book devotes an entire chapter to each of these systems.

A tool, like a hammer, exists to help do a job. If the job includes *continuous improvement,* problem solving, or decision making, then these seven tools fit the definition. Each of these tools is some form of chart for the collection and display of specific kinds of data. Through the collection and display facility, the data become useful information—information that can be used to solve problems, enhance decision making, keep track of work being done, even predict future performance and problems. The beauty of the charts is that they organize data so that we can immediately comprehend the message. This would be all but impossible without the charts, given the mountains of data flooding today's workplace.

THE PARETO CHART

The *Pareto* (pah-ray-toe) *chart* is a very useful tool wherever one needs to separate the important from the trivial. The chart, first promoted by Dr. Joseph Juran, is named after

Italian economist/sociologist Vilfredo Pareto (1848–1923). He had the insight to recognize that in the real world a minority of causes lead to the majority of problems. This is known as the Pareto principle. Pick a category, and the Pareto principle will usually hold. For example, in a factory you will find that of all the kinds of problems you can name, only about 20% of them will produce 80% of the product defects; 80% of the cost associated with the defects will be assignable to only about 20% of the total number of defect types occurring.[2] Examining the elements of this cost will reveal that once again 80% of the total defect costs will spring from only about 20% of the cost elements.

Charts have shown that approximately 20% of the pros on the tennis tour reap 80% of the prize money and that 80% of the money supporting churches in the United States comes from 20% of the church membership.

All of us have limited resources. That point applies to you and to me, and to all enterprises—even to giant corporations and to the government. That means that our resources (time, energy, and money) need to be applied where they will do the most good. The purpose of the Pareto chart is to show you where to apply your resources by revealing the significant few from the trivial many. It helps us establish priorities.

The Pareto chart in Figure 15–1 labels a company's customers A, B, C, D, E, and All Others. The bars represent the percentage of the company's sales going to the respective customers. Seventy-five percent of this company's sales are the result of just two customers. If one adds customer C, 90% of its sales are accounted for. All the other customers together account for only 10% of the company's sales. Bear in mind that "Other" may include a very large number of small customers. Which customers are the ones who should be kept happy? Obviously, A, B, and perhaps C are the most critical. This would suggest that customers A, B, and C are the company's core market and all the other customers represent a marginal business. Decisions on where to allocate resources should be made accordingly.

The Pareto chart in Figure 15–2 shows bars representing the sales of a particular model of automobile by age group of the buyers. The curve represents the cumulative percentage of sales and is keyed to the y-axis scale on the right. The manufacturer has limited resources in its advertising budget, and the chart reveals which age group is the most logical choice to target. Concentrating on the 26–45 age bracket will result in the

QUALITY TIP

Find the Important Causes

"While there are many cause factors, the truly important ones, the cause factors which will sharply influence effects, are not many. If we follow the principle set by Vilfredo Pareto, all we have to do is standardize two or three of the most important cause factors and control them."

Source: Kaoru Ishikawa.

Figure 15–1
Pareto Chart: Percentage of
Total Sales by Customer

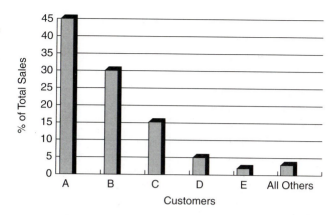

best return on investment because 76% of the Swift V-12 buyers come from the combination of the 36–45 and 26–35 age groups. The significant few referred to in the Pareto principle are in the 26–45 group. The insignificant many are all those under 26 and over 45.

Cascading Pareto Charts

You can cascade Pareto charts by determining the most significant category in the first chart, then making a second chart related only to that category, and repeating this as far as possible, to three, four, or even five or more charts. If the cascading is done properly, root causes of problems may be determined rather easily.

Consider the following example. A company produces complex electronic assemblies, and the test department is concerned about the cost of rework resulting from test failures. They are spending more than $190,000 per year, and that amount is coming directly out of profit. The department formed a special project team to find the cause of the problem and reduce the cost of rework. The Pareto chart in Figure 15–3 showed them that about 80% of the cost was related to just five defect causes. All the others, and there were about 30 more, were insignificant—*at least at that time.*

Figure 15–2
Swift V-12 Sales by Age Group

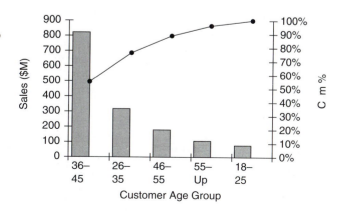

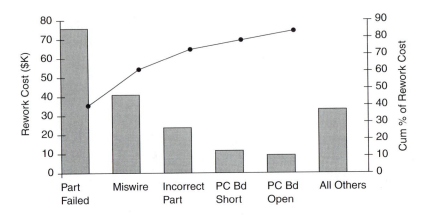

Figure 15–3
Top Five Defects by Rework Cost

The longest bar alone accounted for nearly 40% of the cost. If the problem it represents could be solved, the result would be an immediate reduction of almost $75,000 in rework cost. The team sorted the data again to develop a level 2 Pareto chart, Figure 15–4, to focus on any part types that might be a major contributor to the failures.

Figure 15–4 clearly showed that one type of relay accounted for about 60% of the failures. No other part failures came close. In this case and at this time, the relay was the significant one, and all the other parts were the insignificant many. At this point, another team was formed to analyze the failure modes of the relay in order to determine a course of action for eliminating the relay problem. It was determined that there were a number of failure modes in the relay. They were plotted on the Pareto chart shown in Figure 15–5, which immediately revealed that 66% of all the failures were associated with one failure mode. The second longest bar in Figure 15–5 represented another manifestation of the same root cause. The relay contacts were not switching on *at all* (longest bar) or were not switching on *completely* (next longest bar). With this information known, the relay contacts were carefully examined, and it was determined that the relays were being damaged at incoming inspection where they were tested with a

Figure 15–4
Rework Cost by Top Five Part
Failure Categories

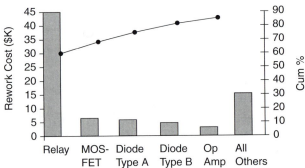

Figure 15–5
Relay Failure Categories

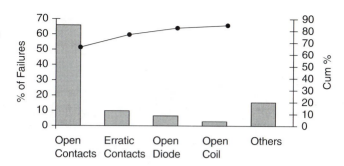

voltage that was high enough to damage the gold plating on the contacts. Changing the incoming test procedure and working with the relay vendor to improve its plating process eliminated the problem.

Earlier in this chapter, we implied that although a particular problem might be insignificant at one point in time, it might not stay that way. Consider what happens to the bars on the cascaded charts when the relay contact problem is solved. The second longest bar on the first chart clearly becomes the longest (assuming it was not being solved simultaneously with the relay problem). At this point, more than $100,000 a year is still being spent from profit to rework product rather than make it properly in the first place. The cycle must continue to be repeated until perfection is approached.

The next cycle of Pareto charts might look like those in Figure 15–6. Starting at the top we see the following points:

1. Miswires (wires connected to the wrong point or not properly attached to the right point) account for 40% of the remaining rework cost.
2. Wires connected with hand-wrapping tools represent more than 70% of all miswires.
3. Of the hand-wrap defects, more than 65% are because of operator error.
4. Of all the operators doing hand-wrap work, operators 33 and 28 contribute more than 80% of the defects.

Attention must be given to those operators in the form of training or, perhaps, reassignment.

The third Pareto chart cascading would break down the Wrong Part problem. For example, perhaps part abc is mistakenly substituted for part xyz on a printed circuit board. The cycle may be repeated over and over, each time dealing with the significant few, while ignoring the trivial many. Eventually perfection is approached. A few companies are getting close with some of their products, but most have ample opportunity for significant improvement. One need not worry about running out of improvement possibilities.

CAUSE-AND-EFFECT DIAGRAMS

A team typically uses a *cause-and-effect diagram* to identify and isolate causes of a problem. The technique was developed by the late Dr. Kaoru Ishikawa, a noted Japanese quality expert, so sometimes the diagram is called an *Ishikawa diagram*. It is also often called a fishbone diagram because that is what it looks like.

Figure 15–6
Second Cascading of Pareto
Charts

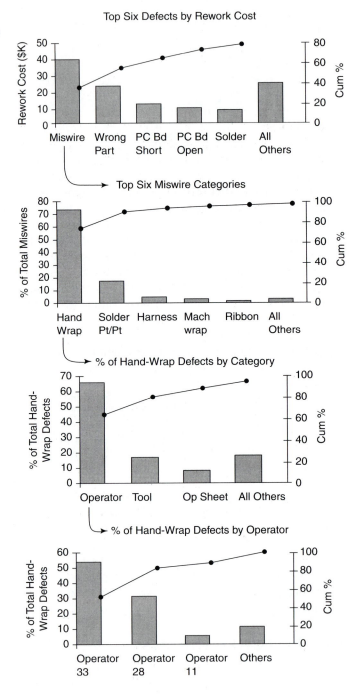

In his book *Guide to Quality Control,* Ishikawa explains the benefits of using cause-and-effect diagrams as follows:

- Creating the diagram itself is an enlightened, instructive process.
- Such diagrams focus a group, thereby reducing irrelevant discussion.
- Such diagrams separate causes from symptoms and force the issue of data collection.
- Such diagrams can be used with any problem.[3]

The cause-and-effect diagram is the only tool of the seven tools that is not based on statistics. This chart is simply a means of visualizing how the various factors associated with a process affect the process's output. The same data could be tabulated in a list, but the human mind would have a much more difficult time trying to associate the factors with each other and with the total outcome of the process under investigation. The cause-and-effect diagram provides a graphic view of the entire process that is easily interpreted by the brain.

Suppose an electronics plant is experiencing soldering rejects on printed circuit (PC) boards. People at the plant decide to analyze the process to see what can be done; they begin by calling together a group of people to get their thoughts. The group is made up of engineers, solder machine operators, inspectors, buyers, production control specialists, and others. All the groups in the plant who have anything at all to do with PC boards are represented, which is necessary to get the broadest possible view of the factors that might affect the process output.

The group is told that the issue to be discussed is the solder defect rate, and the objective is to list all the factors in the process that could possibly have an impact on the defect rate. The group uses brainstorming to generate the list of possible *causes.* The list might look like Figure 15–7.

The group developed a fairly comprehensive list of factors in the PC board manufacturing process—factors that could *cause* the *effect* of solder defects. Unfortunately, the list does nothing in terms of suggesting which of the 35 factors might be major causes, which are minor, and how they relate to each other. This is where the cause-

machine	solderability	operator
solder	conveyer speed	temperature
preheat	materials	parts
operator attitude	operator attention	flux
conveyer angle	wave height	cleanliness
age of parts	age of boards	part preparation
parts vendors	board vendors	type of flux
specific gravity	machine maintenance	training
skill	vibration	storage
instruments	lighting	calibration
handling	wait-time	contamination
air quality	humidity	

Figure 15–7
Brainstormed List of Possible Causes for Solder Defects

and-effect diagram comes into play. Ishikawa's genius was to develop a means by which these random ideas might be organized to show relationships and to help people make intelligent choices.

Figure 15–8 is a basic cause-and-effect diagram. The spine points to the *effect*. The effect is the "problem" we are interested in—in this case, machine soldering defects. Each of the ribs represents a cause leading to the effect. The ribs are normally assigned to the causes considered to be *major factors*. The lower level factors affecting the major factors branch off the ribs. Examine Figure 15–7 to see whether the major causes can be identified. The ribs are assigned to these causes.

Six major groupings of causes are discernible:

1. The solder machine itself is a major factor in the process.
2. The operators who prepare the boards and run the solder machine would also be major factors.
3. The list includes many items such as parts, solder, flux, boards, and so forth, and these can be collected under the word *materials*, which also appears on the list. Materials is a major factor.
4. Temperature within the machine, conveyor speed and angle, solder wave height, and so on, are really the *methods* (usually published procedures and instructions) used in the process. Methods is a major factor.
5. Many of these same items are subject to the plant's methods (how-to-do-it) and measurement (accuracy of control), so measurement is a major factor, even though it did not appear on the list.
6. Cleanliness, lighting, temperature and humidity, and the quality of the air we breathe can significantly affect our performance and thus the quality of output of processes with which we work. We will call this major factor *environment*.

The designated six major factors, or causes, are those that the group thinks might have an impact on the quality of output of the machine soldering process: machine, operator, materials, methods, measurement, and environment. The cause-and-effect fishbone diagram developed from this information has six ribs, as shown in Figure 15–9.

Having assigned the major causes, the next step is to assign all the other causes to the ribs they affect. For example, *machine maintenance* should be assigned to the Machine rib, because machine performance is obviously affected by how well or how poorly the machine is maintained. *Training* will be attached to the Operator rib, because the degree to which operators have been trained certainly affects their expertise in running the machine. In some cases, a possible cause noted on the list may appropri-

Figure 15–8
Basic Cause-and-Effect or
Fishbone Diagram

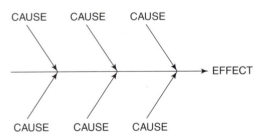

Figure 15–9
Cause-and-Effect Diagram with
Major Causes and Effect
Assigned

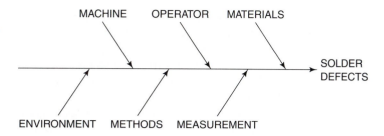

ately branch not from the rib (major cause) but from one of the branches (contributing cause). For example, *solderability* (the relative ease—or difficulty—with which materials can be soldered) would branch from the Materials rib, because it is a contributor to the materials' cause of solder defects. An important cause of poor solderability is age of parts. So *age of parts* will branch not from Materials, but from solderability. Study Figure 15–10 to get a graphic sense of the relationships described in this paragraph.

Figure 15–10 is the completed fishbone diagram. It presents a picture of the major factors that can cause solder defects and in turn the smaller factors that affect the major factors. Examination of the Materials rib shows that there are four factors directly affecting materials in regard to solder defects: the parts themselves, handling of the materials, and the solder and flux used in the process. The chart points out that contamination can affect the solder's performance and also that the big issue affecting the parts is solderability. In this case, the branches go to three levels from the rib, noting that solderability can be affected by the vendor supplying the parts, storage of the parts before use, and age of the parts.

Now you may say, "The diagram didn't configure itself in this way. Someone had to know the relationships before the diagram was drawn, so why is the diagram

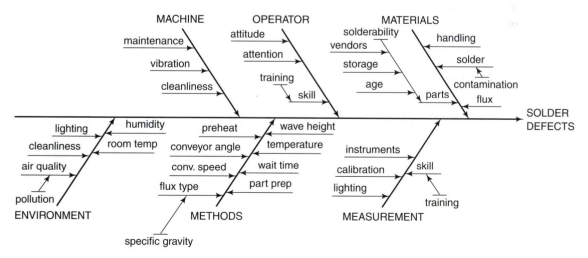

Figure 15–10
Completed Cause-and-Effect Diagram

needed?" First, picture these relationships in your mind—no diagram, just a mental image. If you are not familiar with the process used in the example, pick any process involving more than two or three people and some equipment, such as the process of an athletic event. If you try this, you will probably find it virtually impossible to be conscious of all the factors coming into play, to say nothing of how they relate and interact. Certainly, the necessary knowledge and information already existed before the 35 factors were arranged in the cause-and-effect diagram. The key to the diagram's usefulness is that it is very possible that no *one* individual had all that knowledge and information. That is why cause-and-effect diagrams are normally created by teams of people widely divergent in their expertise.

The initial effort by the team is developing the list of possible factors. This is usually done using brainstorming techniques. Such a list can be made in a surprisingly short time—usually no more than an hour. It is not necessary that the list be complete or even that all the factors listed be truly germane. Missing elements will usually be obvious as the diagram is developed, and superfluous elements will be recognized and discarded. After the list has been compiled, all the team members contribute from their personal knowledge and expertise to assemble the cause-and-effect diagram.

The completed diagram reveals factors or relationships that had previously not been obvious. The causes most likely responsible for the problem (solder defects) will usually be isolated. Further, the diagram may suggest possibilities for action. It is conceivable in the example that the team, because it is familiar with the plant's operation, could say with some assurance that solderability was suspected because the parts were stored for long periods of time. They might recommend that, by switching to a just-in-time system, both storage and aging could be eliminated as factors affecting solderability.

The cause-and-effect diagram serves as an excellent reminder that the items noted on it are the things the company needs to pay attention to if the process is to continually improve. Even in processes that are working well, continual improvement is the most important job any employee or team can have. In today's competitive global marketplace, it is truly the key to survival.

CHECK SHEETS

The *check sheet* is introduced here as the third of the seven traditional tools. The fuel that powers the total quality tools is data. In many companies, elaborate systems of people, machines, and procedures exist for the sole purpose of collecting data. At times, this quest for data has become zealous to the point of obscuring the reason for data collection in the first place. Many organizations are literally drowning in their own data while at the same time not knowing what is actually going on; they are "data rich and information poor." With the advent of powerful desktop computers, information collection has become an end unto itself in many instances.

Having access to data is essential. However, problems arise when trivial data cannot be winnowed from the important and when there is so much of it that it cannot be easily translated into useful information. Check sheets help deal with this problem.

The check sheet can be a valuable tool in a wide variety of applications. Its utility is restricted only by the imagination of the person seeking information. The check sheet can take any form. The only rules are that data collection must be the equivalent of entering a check mark and that the displayed data be easily translated into useful information. For example, it may take the form of a drawing of a product with the check marks entered at appropriate places on the drawing to illustrate the location and type of finish blemishes. An accounts receivable department might set up a check sheet to record the types and number of mistakes on invoices prepared. Check sheets apply to any work environment—not just to the factory floor.

The purpose of the check sheet is to make it easy to collect data for specific purposes and to present it in a way that facilitates conversion from data to useful information. For example, suppose we are manufacturing parts that have a specified dimensional tolerance of 1.120" to 1.130". During the week, each part is measured and the data is recorded. Figure 15–11 is a summary of the week's results.

This figure contains all the data on shaft length for the week of July 11. Without a lot of additional work, it will be difficult to glean much useful information from this list of data. Imagine how much more difficult it would be if, instead of a table, you were presented with a stack of computer runs several inches thick. That is frequently the case in the information age. (The information age should be called the *data age*, in our opinion, reflecting the difference between an abundance of raw, often meaningless data, and the real paucity of *useful* information.)

The computer could be programmed to do something with this data to make it more useful and in some situations that would be appropriate. After all, computers are good at digesting raw data and formatting it for human consumption. But before the computer can do that, some human must tell it exactly what it must do, how to format the information,

Figure 15–11
Weekly Summary of Shaft
Dimensional Tolerance Results
(This is *not* a check sheet.)

Shaft length: Week of 7/11 (Spec: 1.120–1.130")

Date	Length	Date	Length	Date	Length	Rem
11	1.124	11	1.128	11	1.123	
11	1.126	11	1.128	11	1.125	
11	1.119	11	1.123	11	1.122	
11	1.120	11	1.122	11	1.123	
12	1.124	12	1.126	12	1.125	
12	1.125	12	1.127	12	1.125	
12	1.121	12	1.124	12	1.125	
12	1.126	12	1.124	12	1.127	
13	1.123	13	1.125	13	1.121	
13	1.120	13	1.122	13	1.118	
13	1.124	13	1.123	13	1.125	
13	1.126	13	1.123	13	1.124	
14	1.125	14	1.127	14	1.124	
14	1.126	14	1.129	14	1.125	
14	1.126	14	1.123	14	1.124	
14	1.122	14	1.124	14	1.122	
15	1.124	15	1.121	15	1.123	
15	1.124	15	1.127	15	1.123	
15	1.124	15	1.122	15	1.122	
15	1.123	15	1.122	15	1.121	

what to discard, what to use, and so on. If we can't first figure out what to do with the data, no amount of computer power will help. On the supposition that we do know what to do with the data, it is possible that we could *preformat* the data so that it will be instantly useful *as it is being collected*. This is one of the powerful capabilities of the check sheet.

The importance of the data in Figure 15–11 rests in reporting how the work being produced relates to the shaft length specification. The machine has been set up to produce shafts in the center of the range so that normal variation would not spill outside the specified limits of 1.120″ and 1.130″ and thereby create waste. If the raw data could give us a feel for this as it is being collected, that would be very helpful. We would also like to know when the limits are exceeded. The check sheet in Figure 15–12 has been designed to facilitate both data collection and conversion to information.

The check sheet of Figure 15–12 is set up to accept the data very easily and at the same time display useful information. The check sheet actually produces a histogram as the data are entered. (See the following section for information about histograms.) Data are taken by measuring the shafts, just as was done for Figure 15–11. But rather than logging the measured data by date, as in Figure 15–11, the check sheet in Figure 15–12 only requires noting the date (day of month) opposite the appropriate shaft dimension.

Figure 15–12
Check Sheet of Shaft
Dimensional Tolerance Results

Check Sheet
Shaft length: Week of _7 /11_ (Spec: 1.120–1.130″)

1.118**	13
1.119**	11 ** Out of Limits
1.120	11 13
1.121	12 13 15 15
1.122	11 11 13 14 14 15 15 15
1.123	11 11 11 13 13 13 14 15 15 15
1.124	11 12 12 12 13 13 14 14 14 15 15 15
1.125	11 12 12 12 12 13 13 14 14
1.126	11 12 12 13 14 14
1.127	12 12 14 15
1.128	11 11
1.129	14 **Enter day of month for**
1.130	**data point.**
1.131**	
1.132**	

QUALITY TIP

Give the Operator Some Responsibility

The taking of measurements and logging of data on the check sheet should ideally be done by the operator who runs the machine, not a quality control inspector. In a total quality system, the operators are responsible for the quality of their output, for checking it, taking data, responding to it, and so forth. The quality control department is there to audit the processes to make sure they are under control and that procedures are followed.

Source: Stanley B. Davis and David L. Goetsch.

The day-of-month notation serves as a check mark while at the same time keeping track of the day the reading was taken.

This check sheet should be set up on an easel on the shop floor, with entries hand-written. That will make the performance of the machine continuously visible to all—operators, supervisors, engineers, or anyone else in the work area.

The data in Figure 15–11 are the same as the data in Figure 15–12. Figure 15–11 shows columns of sterile data that, before meaning can be extracted, must be subjected to hard work at someone's desk. Assuming it does get translated into meaningful information, it will probably still remain invisible to the people who could make the best use of it—the operators. That can, of course, be overcome by more hard work, but in most cases the data will languish. On the other hand, Figure 15–12 provides a simple check sheet into which the data are entered more easily and once entered, provide a graphic presentation of performance. Should the check sheet reveal that the machine is creeping away from the center of the range, or if the histogram shape distorts, the operator can react immediately. No additional work is required to translate the data to useful information, and no additional work is required to broadcast the information to all who can use it.

To set up a check sheet, you must think about your objective. In this example, we were making shafts to a specification. We wanted to know how well the machine was performing, a graphic warning whenever the machine started to deviate, and information about defects. Setting the check sheet up as a histogram provided all the information needed. This is called a *Process Distribution Check Sheet* because it is concerned with the variability of a process. Other commonly used check sheets include Defective Item Check Sheets (detailing the variety of defects), Defect Location Check Sheets (showing where on the subject product defects occur), Defect Factor Check Sheets (illustrating the factors—time, temperature, machine, operator—possibly influencing defect generation), and many others.

If we wanted to better understand what factors might be contributing to excessive defects on the shop floor, we could set up a Defect Factors Check Sheet. As an example, go back to the section on Pareto charts and look at Figure 15–6. The top chart there revealed that miswires were the most significant defect in terms of cost. To collect some data about the factors that might be contributing to the miswire defects, a reasonable approach would be to set up a Defect Factors Check Sheet and collect data for a week. We are primarily concerned with the operators themselves and the factors that may influence their performance. The check sheet will list each operator's number and bench location within the factory. To determine whether the day of the week or the time of day has anything to do with performance, the data will be recorded by day and by morning or afternoon. We could have included tool numbers as well, but using a tool that produces faulty connections is something the operator must guard against. In other words, we will not consider a tool to be at fault—only the operator if he or she continues to use a defective tool.

In the check sheet, shown in Figure 15–13, five types of miswire defects, covering all types experienced, are coded by symbols, and these symbols are the only raw data entered on the chart. Sums of all defect categories are shown at the bottom of each column, and the weekly total for each operator is shown at the end of each row. A quick glance at the check sheet points to operators 28 and 33 as the sources of the problem. We don't know the *cause* at this point, but we know where to start looking.

In times past, these two people might very well have been summarily fired. In a total quality setting, that decision would be considered the last resort. Most people

Operator No.	Bench No.	11/2 AM	11/2 PM	11/3 AM	11/3 PM	11/4 AM	11/4 PM	11/5 AM	11/5 PM	11/6 AM	11/6 PM	Week Totals
8	A3					●				o □		o - 1 ● - 1 3 □ - 1
10	A2			● ●			□					● - 2 □ - 1 3
11	B1	o						o				o - 2 2
13	A1		o				△				□ □	o - 1 △ - 1 4 □ - 2
28	C2	o ● o	o △	o o o	△ o	o	o o	o △	□ o	o o	o ● o	o - 17 ● - 2 △ - 3 23 □ - 1
33	C3	o o o △ o	● o o o	o △ o o	o o □	o ●	□ o o	△ o o	o o	o o o o △ o	o o o o	o - 28 ● - 2 △ - 4 36 □ - 2
40	B2	+						o ●				o - 1 ● - 1 3 + - 1

Half-day totals		10	7	9	6	4	7	6	6	10	9	39 35
Full-day totals		17		15		11		12		19		74

LEGEND: o = Handwrap
● = Solder point to point
△ = Harness
+ = Ribbon
□ = Other

Figure 15–13
Check Sheet: Defect Factors—Miswires

QUALITY TIP

Statistics Expertise Not Required

We recognize that although much of the following discussion of histograms and control charts is related to statistics, many users of this book will not be expert statisticians. Unfortunately, the scope of this text does not allow for a treatise on statistics, so we have attempted to present the material and mathematical processes in a way that can be followed by the uninitiated who are willing to stay with us. In doing this we have sacrificed nothing in the accuracy of the information presented or the techniques applied. Our objective is that both the statistics novice and the expert will be rewarded with a good understanding of these tools, their applications, and the methodology and significance of the math. For those interested in delving deeper into the tools or statistics, any number of books are dedicated to each.

want to do a good job and will if they are provided with the necessary resources and training. In a case like this, it is not unusual to find that the fault lies with management. The employees were not adequately trained for the job, or some environmental factor (noise, temperature, lighting, or something else) is at fault, or the operators may simply not be equipped for the task (because of vision impairment, impaired motor skills, or some other problem). In any of those scenarios, management is at fault and therefore should do the morally right thing to correct the problem.

Check sheets can be valuable tools for converting data into useful and easy-to-use information. The key is teaching operators how to employ them and empowering them to do so.

HISTOGRAMS

Histograms are used to chart frequency of occurrence. How often does something happen? Any discussion of histograms must begin with an understanding of the two kinds of data commonly associated with processes: *attributes* and *variables* data. Although they were not introduced as such, both kinds of data have been used in the illustrations of this chapter. An *attribute* is something that the output product of the process either has or does not have. From one of the examples (Figure 15–6), an electronic assembly either had wiring errors or it did not. Another example (Figure 15–30) shows that an assembly either had broken screws or it did not. These are attributes. The example of making shafts of a specified length (Figures 15–11 and 15–12) was concerned with *measured data*. That example used shaft length measured in thousandths of an inch, but any scale of measurement can be used, as appropriate for the process under scrutiny. A process used in making electrical resistors would use the scale of electrical resistance in ohms, another process might use a weight scale, and so on. *Variables data* are something that results from measurement.

QUALITY TIP

An Important Distinction to Remember

Attributes Data
- Has or has not
- Good or bad
- Pass or fail
- Accept or reject
- Conforming or nonconforming

Variables Data
- Measured values (dimension, weight, voltage, surface, etc.)

Using the shaft example again, an all-too-common scenario in manufacturing plants would have been to place a Go–No Go screen at the end of the process, accepting all shafts between the specification limits of 1.120" and 1.130" and discarding the rest. Data might have been recorded to keep track of the number of shafts that had to be scrapped. Such a record might have looked like Figure 15–14, based on the original data.

Figure 15–14 tells us what we wanted to know if we were interested only in the number of shafts accepted versus the number rejected. Looking at the shaft process in this way, we are using *attributes data*: either they passed or they failed the screening. This reveals only that we are scrapping between 3% and 4% of all the shafts made. It does not reveal anything about the process adjustment that may be contributing to the scrap rate. Nor does it tell us anything about how robust the process is—might some slight change push the process over the edge? For that kind of insight we need *variables data*.

One can gain much more information about a process when variables data are available. The check sheet of Figure 15–12 shows that both of the rejects (out-of-limits

Shaft Acceptance: Week of ____7/11____ (Spec: 1.120–1.130")

Date	Accepted	Rejected
11.	11	1
12.	12	0
13.	11	1
14.	12	0
15.	12	0
Totals:	58	2

Figure 15–14
Summary Data: Weekly Shaft Acceptance

shafts) were on the low side of the specified tolerance. The peak of the histogram seems to occur between 1.123" and 1.124". If the machine were adjusted to bring the peak up to 1.125", some of the low-end rejects might be eliminated without causing any new rejects at the top end. The frequency distribution also suggests that the process as it stands now will always have occasional rejects—probably in the 2%–3% range at best.

Potential Trap with Histograms

Be aware of a potential trap when using histograms. The histogram is nothing more than a measurement scale across one axis (usually the x-axis) and frequency of like measurements on the other. (Histograms are also called *frequency distribution diagrams*.) The trap occurs when measurements are taken over a long period of time. Too many things can affect processes over time: wear, maintenance, adjustment, material differences, operator influence, environmental influence. The histogram makes no allowance for any of these factors. It may be helpful to consider a histogram to be the equivalent of a snapshot of the process performance. If the subject of a photograph is moving, the photographer must use a fast shutter speed to prevent a blurred image. If the histogram data are not collected over a suitably short period of time, the result will be blurred, just as if the camera's shutter was too slow for the action taking place, because it is possible that the process's performance changes over time. Blurred photographs and blurred histograms are both useless. A good histogram will show a crisp snapshot of process performance as it was at the time the data were taken, not before and not after. This leads some people to claim that histograms should be used only on processes that are known to be *in control*. (See the section on control charts later in this chapter.)

That limitation is not necessary as long as you understand that histograms have this inherent flaw. Be careful that any interpretation you make has accounted for time and its effect on the process you are studying. For example, we do not know enough about the results of the shaft-making process from Figure 15–12 to predict with any certainty that it will do as well next week. We don't know that a machine operator didn't tweak the machine two or three times during the week, trying to find the center of the range. What happens if that operator is on vacation next week? Would we dare predict that performance will be the same? We can only make these predictions if we know the process is statistically in control; thus, the warnings. Taking this into consideration, the histogram in Figure 15–12 provides valuable information.

Histograms and Statistics

Understanding a few basic facts is fundamental to the use of statistical techniques for quality and process applications. We have said that all processes are subject to variability, or variation. There are many examples of this. One of the oldest and most graphically convincing is the bead experiment.[4] This involves a container with a large number of beads. The beads are identical except for color. Suppose there are 900 white beads and 100 red, making 1,000 total. The beads are mixed thoroughly (step 1). Then 50 beads are drawn at random as a sample (step 2). The red beads in the sample are counted. A check mark is entered in a histogram column for that number. All the beads are put back into the container, and they are mixed again (step 3). When you repeat these steps a second time, the odds are that a different number of red beads will be drawn. When a third sample is taken, it will probably contain yet another number of red beads. The process (steps 1, 2, and 3) has not changed, yet the output of the process does change. This is *process variation* or *variability*. If these steps are repeated over and over until a valid statistical sampling has been taken, the resulting histogram will invariably take on the characteristic bell shape common to process variability (see Figure 15–15).

It is possible to calculate the *process variability* from the data. The histogram in Figure 15–15 was created from 100 samples of 50 beads each. The data was as shown in Figure 15–16.

The flatter and wider the frequency distribution curve, the greater the process variability. The taller and narrower the curve, the less the process variability. Even though the variability may change from process to process, it would be helpful to have a common means of measuring, discussing, or understanding variability. Fortunately we do. To express the process's variability we need to know only two things, both of which can be derived from the process's own distribution data: *standard deviation* and *mean*. Standard deviation is represented by the lowercase Greek letter sigma (σ) and indicates a deviation from the average, or *mean*, value of the samples in the data set. The mean is represented by the Greek letter mu (μ). In a normal histogram, μ is seen as a vertical line from the peak of the bell curve to the base, and it is the line from which deviation is measured, minus to the left of μ and plus to the right. Standard deviation (σ) is normally plotted at -3σ, -2σ, -1σ (left of μ), and $+1\sigma$, $+2\sigma$, and $+3\sigma$ to the right (refer to Figure 15–18). Because mean and standard deviation are always derived from data from the process in question, standard deviation has a constant meaning from process to process. From this we can tell what the process can do in terms of its statistical variability (assuming that it remains stable and, no changes are introduced):

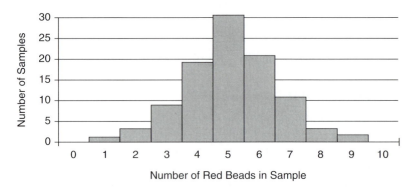

Figure 15–15
Frequency Distribution of Red Beads in Samples

Figure 15–16

Data on Red Beads in Samples

Samples with 0 red beads	0
Samples with 1 red bead	1
Samples with 2 red beads	3
Samples with 3 red beads	9
Samples with 4 red beads	19
Samples with 5 red beads	31
Samples with 6 red beads	21
Samples with 7 red beads	11
Samples with 8 red beads	3
Samples with 9 red beads	2
Samples with 10 red beads	0
Total samples taken	100

- 68.26% of all sample values will be found between $+1\sigma$ and -1σ.
- 95.46% of all sample values will be found between $+2\sigma$ and -2σ.
- 99.73% of all sample values will be found between $+3\sigma$ and -3σ.
- 99.9999998% of all sample values will be found between $+6\sigma$ and -6σ. (Note: As we discussed in Chapter 1, Six Sigma practitioners use 99.99966 percent rather than the actual statistical value.)

This information has a profound practical value, as we shall see as we develop the discussion.

In order to calculate the process mean value (μ) and standard deviation (σ), we must first use the raw process data from Figure 15–16 to develop the information required for those calculations. As we develop the information we will post it in the appropriate columns of Figures 15–17a, b, and c.

Columns 1 and 2 of Figures 15–17a, b, and c contain the measured raw data from the colored bead process from Figure 15–16. Column 1 lists the number of red beads possible to be counted (from 0 to 10) in the various samples. Column 2 lists the number of samples which contained the corresponding number of red beads. The number of samples in column 2 is totaled, yielding $n = 100$.

Calculating the Mean

For a histogram representing a truly normal distribution between $\pm$ infinity, the mean value would be a vertical line to the peak of the bell curve. Our curve is slightly off normal because we are using a relatively small sample, so the mean (μ) must be calculated. The equation for μ is:

$$\mu = \Sigma X \div n$$

where X is the product of the number of beads in a sample times the number of samples containing that number of beads, or for Figure 15–17a, the product of columns 1 and 2.

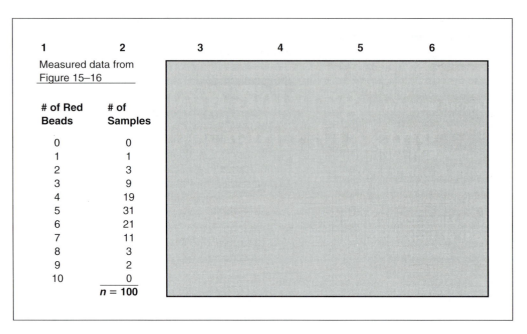

1	2	3	4	5	6
Measured data from Figure 15–16					
# of Red Beads	# of Samples				
0	0				
1	1				
2	3				
3	9				
4	19				
5	31				
6	21				
7	11				
8	3				
9	2				
10	0				
	$n = 100$				

Figure 15–17a
Raw Data from the Colored Bead Experiment (see Figure 15–16)

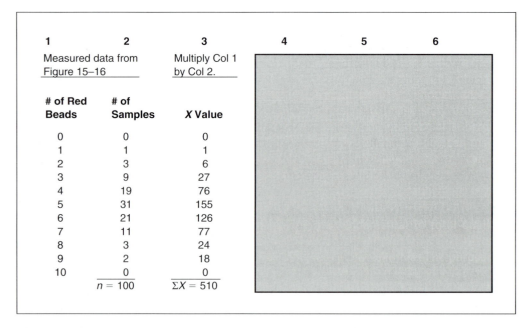

1	2	3	4	5	6
Measured data from Figure 15–16		Multiply Col 1 by Col 2.			
# of Red Beads	# of Samples	X Value			
0	0	0			
1	1	1			
2	3	6			
3	9	27			
4	19	76			
5	31	155			
6	21	126			
7	11	77			
8	3	24			
9	2	18			
10	0	0			
	$n = 100$	$\Sigma X = 510$			

Figure 15–17b
Calculating Values of X and ΣX

1	2	3	4	5	6
Measured data from Figure 15–16	Multiply Col 1 by Col 2.	Deviation from μ $(Col\ 1 - \mu)$	Deviation squared $(Col\ 4)^2$	Sum of Distance^2 $(Col\ 2 \times Col\ 5)$	
# of Red Beads	# of Samples	X Value	d	d^2	
0	0	0	−5.1	26.01	0
1	1	1	−4.1	16.81	16.81
2	3	6	−3.1	9.61	28.83
3	9	27	−2.1	4.41	39.69
4	19	76	−1.1	1.21	22.99
5	31	155	−0.1	0.01	0.31
6	21	126	0.9	0.81	17.01
7	11	77	1.9	3.61	39.71
8	3	24	2.9	8.41	25.23
9	2	18	3.9	15.21	30.42
10	0	0	4.9	24.01	0
	$n = 100$	$\Sigma X = 510$			$\Sigma d^2 = 221$

Figure 15–17c
Completed Deviation Data Table

We calculate Column 3 of Figure 15–17b:

Now that we have the X values in Figure 15–17b, we simply add them up to give us the sum of the X values (ΣX). The figure tells us that n = 100 and $\Sigma X = 510$. Using the equation for μ:

$$\mu = \Sigma X \div n$$
$$\mu = 510 \div 100$$
$$\mu = 5.1$$

The mean (μ) is placed at 5.1 on the histogram's x-axis, and all deviations are measured relative to that. (See Figure 15–18.)

Calculating Standard Deviation (σ)

To understand the process's variability we must know its standard deviation. The formula for standard deviation is

$$\sigma = \sqrt{\Sigma d^2/(n - 1)}$$

where d = the deviation of any unit from the mean
n = the number of units sampled

We already have the value of n (100), but we have not calculated the values of d, d^2, or Σd^2. We will perform these calculations and post the information in the remaining three columns of Figure 15–17c. The values of the deviations (d) are determined by subtracting μ (5.1) from each of the red bead values (0 through 10) of Column 1. The first entry in Column 4 (deviation from μ) is determined by subtracting μ from the value in Column 1, that is, $0 - 5.1 = -5.1$. Similarly, the second entry in Column 4 is the value of column 1 at the 1-bead row minus μ, or $1 - 5.1 = -4.1$. Repeating this process through the 10-bead row completes the deviation column.

Column 5 of Figure 15–17c is simply a list of the Column 4 deviation values squared. For example, in the 0-bead row, Column 4 shows $d = -5.1$. Column 5 lists the square of -5.1, or 26.01. The 1-bead row has $d = -4.1$. Column 5 lists its square, 16.81. This process is continued through the 10-bead row to complete Column 5 of the figure.

Column 6 of Figure 15–17c lists the results of the squared deviations (Column 5) multiplied by the number of samples at the corresponding deviation value (Column 2). For the Column 6 entry at the 0-bead row we multiply 0 (from column 2) by 26.01 (from Column 5). $0 \times 26.01 = 0$, so 0 is entered in Column 6. For the 1-bead row, 1×16.81. 16.81 is the second entry in Column 6. At the 2-bead row, we multiply 3 by 9.61, and enter 28.83 in column 6. This process is repeated through the remaining rows of the figure.

Next we add Column 6's entries to obtain the sum of the squared deviations, Σd^2. Σd^2 for our bead process experiment is 221.

Now we have all the information we need to calculate the standard deviation (σ) for our process.

$$\sigma = \sqrt{\Sigma d^2/(n - 1)}$$

$$\sigma = \sqrt{221 \div 99}$$

$$\sigma = \sqrt{2.23} \quad \text{(2.23 is called the } \textit{mean squared deviation.}\text{)}$$

$$\sigma = 1.49 \quad \text{(1.49 is also called the } \textit{root mean squared deviation.}\text{)}$$

Note: Calculations are to two decimal places.

Next calculate the positions of $\mu \pm 1\sigma$, 2σ, and 3σ.

$$\sigma = 1.49 \qquad 2\sigma = 2.99 \qquad 3\sigma = 4.47$$

These values are entered on Figure 15–15 to create Figure 15–18:

$$\mu - 1\sigma = 5.1 - 1.49 = 3.61$$
$$\mu + 1\sigma = 5.1 + 1.49 = 6.59$$
$$\mu - 2\sigma = 5.1 - 2.99 = 2.11$$
$$\mu + 2\sigma = 5.1 + 2.99 = 8.09$$
$$\mu - 3\sigma = 5.1 - 4.47 = 0.63$$
$$\mu + 3\sigma = 5.1 + 4.47 = 9.57$$

Suppose we have a process that is operating like the curve in Figure 15–18. We have specifications for the product output that requires us to reject any part below 3.6 and above 6.6. It turns out that these limits are approximately $\pm 1\sigma$. We know immediately that about one third of the process output will be rejected. If this is not acceptable, which is highly probable, we will have to improve the process or change to a com-

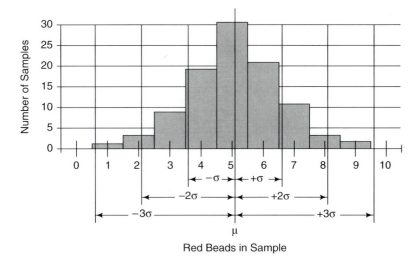

Figure 15–18
Application of Standard Deviation Calculations to Red Bead Histogram

pletely different process. Even if more variation could be tolerated in the product and we took the specification limits out to 2 and 8, about 5 of every 100 pieces flowing out of the process would still be rejected. In a competitive world, this is poor performance indeed. Many companies no longer consider 2,700 parts per million defective ($\pm 3\sigma$) to be good enough. A growing number of organizations are seeking the Motorola version of Six Sigma quality performance. These companies target a defect rate of 3.4 nonconformances per million opportunities for nonconformance (npmo). Technically speaking, 3.4 nmpo is not very close to the statistically pure 6-sigma rate of 0.002 per million opportunities, or 1 nonconformance in 500 million. (We explained this difference in the Six Sigma section of Chapter 1.) Although the popular Six Sigma does not match the true 6 sigma, 3.4 nmpo is a remarkable achievement. Whatever the situation, with this statistical sampling tool properly applied, there is no question about what can be achieved with any process. You will be able to predict the results.

Shapes of Histograms

Consider the shape of some histograms and their position relative to specification limits. Figure 15–19 is a collection of histograms. Histogram A represents a normal distribution. So does B, except it is shallower. The difference between the process characteristics of these two histograms is that process A is much tighter, whereas the looser process B will have greater variances. Process A is usually preferred. Processes C and D are skewed left and right. Although the curves are normal, product will be lost because the processes are not centered. Process E is bimodal. This can result from two batches of input material, for example. One batch produces the left bell curve, and the second batch the curve on the

QUALITY TIP

Six Sigma Process Capability

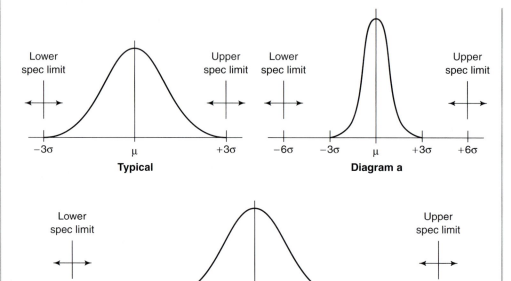

- When specification limits (defining acceptable products) are set to correspond to the $\pm 3\sigma$ capability of the process, npmo will be 2,700. That may be 2,700 unacceptable parts out of every million produced, or it may mean that out of 1 million operations, 2,700 mistakes will be made, and so forth.

- When specification limits correspond to the $\pm 6\sigma$ process capability, a vanishingly small 0.002 npmo will be achieved. See Diagrams a and b. (Note that at $\pm 6\sigma$ the Motorola Six Sigma method will yield a still small 3.4 npmo.)

- One method used in striving for statistical 6σ (or Motorola's Six Sigma) performance involves narrowing the bell curve through the development of superior processes. Compare the typical 3σ diagram above with diagram a. Note that the specification limits have remained constant, but the process variation has been reduced, moving the process's $\pm 6\sigma$ points inward to the specification limits.

- Another method for working toward Six Sigma performance involves designing products that can tolerate wider physical or functional variation in their component parts while still performing to product specifications. Compare the Typical 3σ diagram above with diagram b. This technique is usually referred to as *robust design*.

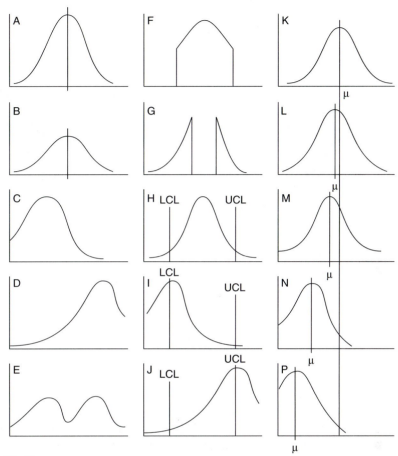

Figure 15–19
Histograms of Varying Shapes

right. The two curves may be separated for a better view of what is going on by stratifying the data by batch. (See the "Stratification" section later in this chapter.)

Histogram F suggests that someone is discarding the samples below and above a set of limits. This typically happens when there is a 100% inspection and only data that are within limits are recorded. The strange Histogram G might have used data from incoming inspection. The message here is that the vendor is screening the parts, and someone else is getting the best ones. A typical case might be electrical resistors which are graded as 1%, 5%, and 10% tolerance. The resistors that met 1% and 5% criteria were screened out and sold at a higher price. You got what was left.

Histogram H shows a normal distribution properly centered between a set of upper and lower control limits. Histograms I and J illustrate what happens when the same normal curve is allowed to shift left or right. There will be a significant loss of product as a control limit intercepts the curve higher up its slope.

Histograms K through P show a normal, centered curve that went out of control and drifted. Remember that histograms do not account for time and you must therefore be careful about making judgments. If all the data that produced Histograms K through P were averaged, or even if all the data were combined to make a single histogram, you could be misled. You would not know that the process was drifting. Plotting a series of histograms over time, such as K through P, clearly illustrates any drift right or left, shallowing of the bell, and the like.

The number of samples or data points has a bearing on the accuracy of the histogram, just as with other tools. But with the histogram there is another consideration: how does one determine the proper number of intervals for the chart? (The intervals are, in effect, the data columns of the histogram.) For example, Figure 15–15 is set up for 11 intervals: 0, 1, 2, and so on. The two outside intervals are not used, however, so the histogram plots data in nine intervals. The rule of thumb is as follows:

Number of Observations (N)	Number of Intervals (k)
<75	5−7
75−300	6−10
>300	10−20

Or you may use the formula $k = \sqrt{N}$

It is not necessary to be very precise with this. These methods are used to get close and adjust one way or the other for a fit with your data.

Suppose we are using steel balls in one of our products and the weight of the ball is critical. The specification is 5 ± 0.2 grams. The balls are purchased from a vendor, and because our tolerance is tighter than the vendor's, we weigh the balls and use only those that meet our specification. The vendor is trying to tighten its tolerance and has asked for assistance in the form of data. Today 60 balls were received and weighed. The data were plotted on a histogram. To give the vendor the complete information, a histogram with intervals every 0.02 grams is established.

Figure 15–20 does not look much like a bell curve because we have tried to stretch a limited amount of data (60 observations) too far. There are 23 active or skipped inter-

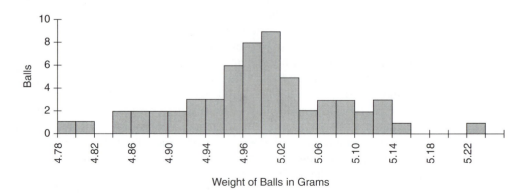

Figure 15–20
Histogram with Limited Amount of Data Stretched

Figure 15–21
Histogram with Appropriate
Intervals for the Amount of Data

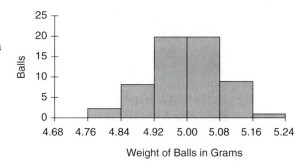

vals. Our rule of thumb suggests 5 to 7 intervals for less than 75 observations. If the same data were plotted into a histogram of 6 intervals (excluding the blank), it would look like Figure 15–21. At least in this version it looks like a histogram. With more data—say, 100 or more observations—one could narrow the intervals and get more granularity. Don't try to stretch data too thin because the conversion to real information can become difficult and risky.

SCATTER DIAGRAMS

The fifth of the seven tools is the *scatter diagram*. It is the simplest of the seven and one of the most useful. The scatter diagram is used to determine the correlation (relationship) between two characteristics (variables). Suppose you have an idea that there is a relationship between automobile fuel consumption and the rate of speed at which people drive. To prove, or disprove, such an assumption, you could record data on a scatter diagram that has miles per gallon (mpg) on the *y*-axis and miles per hour (mph) on the *x*-axis; mpg and mph are the two characteristics.

Examination of the scatter diagram of Figure 15–22 shows that the aggregate of data points contains a slope down and to the right. This is correlation, and it supports the thesis that the faster cars travel, the more fuel they use. Had the slope been upward to the right, as it actually appears to be (for three of the four cars) between 20 and 30 mph,

Figure 15–22
Scatter Diagram: Speed versus
Fuel Consumption for Four
Automobiles

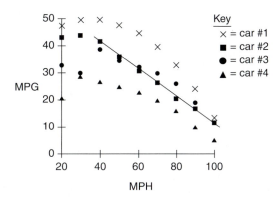

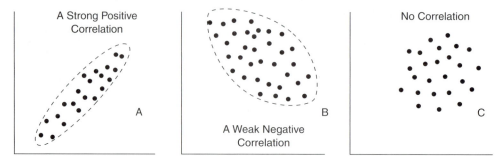

Figure 15–23
Scatter Diagrams of Various Correlations

the correlation would have suggested that the faster you travel, the better the fuel mileage. Suppose, however, that the data points did not form any recognizable linear or elliptical pattern but were simply in a disorganized configuration. This would suggest that there is no correlation between speed and fuel consumption.

Figure 15–23 is a collection of scatter diagrams illustrating strong *positive correlation* (Diagram A), weak *negative correlation* (Diagram B), and *no correlation* (Diagram C). To be classified as a strong correlation, the data points must be tightly grouped in a linear pattern. The more loosely grouped, the less correlation, and therefore the term *weak correlation*. When a pattern has no discernible linear component, it is said to show no correlation.

Scatter diagrams are useful in testing the correlation between process factors and characteristics of product flowing out of the process. Suppose you want to know whether conveyor speed has an effect on solder quality in a machine soldering process. You could set up a scatter diagram with conveyor speed on the x-axis and solder rejects or nonconformities on the y-axis. By plotting sample data as the conveyor speed is adjusted, you can construct a scatter diagram to tell whether a correlation exists.

In this case, Figure 15–24 suggests that the correlation is a curve, with rejects dropping off as speed is initially raised but then increasing again as the conveyor speed continues to increase. This is not atypical of process factors that have optimum operating

Figure 15–24
Scatter Diagram: Conveyor
Speed versus Rejects

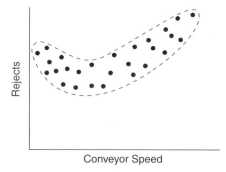

Figure 15–25
Scatter Diagram: Cleaning
Solution versus Cleaning Time

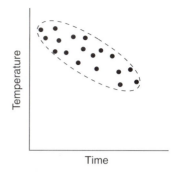

points. In the case of the conveyor, moving too slowly allows excess heat to build up, causing defects. So increasing speed naturally produces better results, until the speed increases to the point where insufficient preheating increases the number of defects. Figure 15–24, then, not only reveals a correlation but also suggests that there is an optimum conveyor speed, operation above or below which will result in increased product defects.

It is also possible to determine a correlation between two process factors. If your manufacturing process includes the washing of parts in a cleaning agent, and you are interested in reducing the time the parts are in the cleaning tank, you might want to know whether the temperature of the solution is correlated with the time it takes to get the parts thoroughly clean. The scatter diagram could have temperature of the cleaning agent on one axis and time to clean on the other. By adjusting the temperature of the solution and plotting the cleaning time, a scatter diagram will reveal any existing correlation.

Assume that the scatter diagram shows a discernible slope downward to the right, as in Figure 15–25. This shows that over the temperature range tested, there *is* a correlation between cleaning solution temperature and cleaning time. With this information, you might be able to reduce the cycle time of the product. *Cycle time* in manufacturing is basically elapsed time from the start of your build process until the product is finished. Cycle time is becoming more important as manufacturers adopt world-class techniques to compete in the global marketplace. If you can find a safe, cost-effective way to raise the cleaning agent temperature to some more efficient level, and in the process shorten the cycle (or perhaps maintain the cycle and do a better job of cleaning), doing so might provide a competitive advantage.

Not all scatter diagrams require that special tests be run to acquire raw data. The data are frequently readily available in a computer. Few companies would have to record new data to determine whether a correlation exists between the day of the week and employee performance. Such data are often available from the day-to-day inspection reports. In fact, where people are involved, it is advisable to use existing data rather than collecting new data to be sure that the data were not influenced by the test itself. Imagine people being told they were to be part of a test to determine whether their performance was as good on Friday or Monday as the rest of the week. This knowledge would undoubtedly affect their performance.

RUN CHARTS AND CONTROL CHARTS

The run chart is straightforward, and the control chart is a much more sophisticated outgrowth of it. Therefore, the two are usually thought of together as a single tool. Both can be very powerful and effective for the tracking and control of processes, and they are fundamental to the improvement of processes.

Run Charts

The *run chart* records the output results of a process over time. The concept is strikingly simple, and indeed it has been used throughout modern times to track performance of everything from AAA membership to zwieback production. Because one axis (usually the *x*-axis) represents time, the run chart can provide an easily understood picture of what is happening in a process as time goes by. That is, it will cause trends to "jump" out at you. For this reason the run chart is also referred to as a *trend chart*.

Consider as an example a run chart set up to track the percentage or proportion of product that is defective for a process that makes ballpoint pens. These are inexpensive pens, so production costs must be held to a minimum. On the other hand, many competitors would like to capture our share of the market, so we must deliver pens that meet the expectations of our customers—as a minimum. A sampling system is set up that requires a percentage of the process output to be inspected. From each lot of 1,000 pens, 50 will be inspected. If more than 1 pen from each sample of 50 is found defective, the whole lot of 1,000 will be inspected. In addition to scrapping the defective pens, we will attempt to discover why the defects were there in the first place and to eliminate the cause. Data from the sample will be plotted on a run chart. Because we anticipate improvements to the process as a result of this effort, the run chart will be ideal to show whether we are succeeding.

The run chart of Figure 15–26 is the result of sample data for 21 working days. The graph clearly shows that significant improvement in pen quality was made during the 21 working days of the month. The trend across the month was toward better quality (fewer defects). The most significant improvements came at the twelfth day and the seventeenth day, as causes for defects were found and corrected.

The chart can be continued indefinitely to keep us aware of performance. Is it improving, staying the same, or losing ground? Scales may have to change for clarity.

Figure 15–26
Run Chart: Pen Defect Rate for
21 Working Days

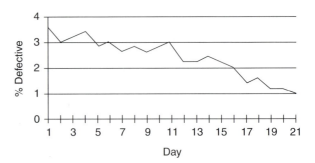

For example, if we consistently found all samples with defects below 2%, it would make sense to change the y-axis scale to 0%–2%. Longer-term charts would require changing from daily to weekly or even monthly plots.

Performance was improved during the first month of the pen manufacturing process. The chart shows positive results. What cannot be determined from the run chart, however, is what *should* be achieved. Assuming we can hold at 2 defective pens of 100, we still have 20,000 defective pens out of a million. Because we are sampling only 5% of the pens produced, we can assume that 19,000 of these find their way into the hands of customers—the very customers our competition wants to take away from us. So it is important to improve further. The run chart will help, but a more powerful tool is still needed.

Control Charts

The problem with the run chart and, in fact, many of the other tools, is that it does not help us understand whether the variation is the result of *special causes*—things such as changes in the materials used, machine problems, lack of employee training—or *common causes* that are purely random. Not until Dr. Walter Shewhart made that distinction in the 1920s was there a real chance of improving processes through the use of statistical techniques. Shewhart, then an employee of Bell Laboratories, developed the control chart to separate the *special causes* from the *common causes*.[5]

In evaluating problems and finding solutions for them, it is important to distinguish between special causes and common causes. Figure 15–27 shows a typical control chart. Data are plotted over time, just as with a run chart; the difference is that the data stay between the upper control limit (UCL) and the lower control limit (LCL) while varying about the center line or average—*only so long as the variation is the result of common causes* (i.e., *statistical variation*). Whenever a special cause (nonstatistical cause) impacts the process, one of two things will happen: either a plot point will penetrate UCL or LCL, or there will be a "run" of several points in a row above or below the average line. When a penetration or a lengthy run appear, this is the control chart's signal that something is wrong that requires immediate attention.

As long as the plots stay between the limits and don't congregate on one side or the other of the process average line, the process is in statistical control. If either of these conditions is not met, then we can say that the process is not in statistical control or simply is "out of control"—hence the name of the chart.

Figure 15–27
Basic Control Chart

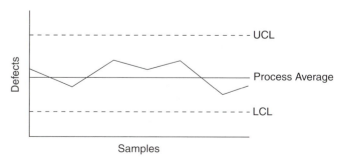

If you understand that it is the UCL, LCL, and process average lines added to the run chart that make the difference, you may wonder how those lines are set. The positioning of the lines cannot be arbitrary. Nor can they merely reflect what you want out of the process, for example, based on a specification. Such an approach won't help separate common causes from special causes, and it will only complicate attempts at process improvement. UCL, LCL, and process average must be determined by valid statistical means. Determination of UCL, LCL, and process average is fully covered in chapter 18, which is dedicated to the use of control charts in SPC.

All processes have built-in variability. A process that is in statistical control will still be affected by its natural random variability. Such a process will exhibit the normal distribution of the bell curve. The more finely tuned the process, the less deviation there will be from the process average, the narrower the bell curve. (Refer to Figure 15–19, Histogram A and Histogram B.) This is at the heart of the control chart and is what makes it possible to define the limits and process average.

Control charts are the appropriate tool to monitor processes. The properly used control chart will immediately alert the operator to any change in the process. The appropriate response to that alert is to stop the process at once, preventing the production of defective product. Only after the special cause of the problem has been identified and corrected should the process be restarted. Having eliminated a problem's root cause, that problem should never recur. (Anything less, however, and it is sure to return eventually.) Control charts also enable continuous improvement of processes. When a change is introduced to a process that is operated under statistical process control (SPC) charts, the effect of the change will be immediately seen. You know when you have made an improvement. You also know when the change is ineffective or even detrimental. This validates effective improvements, which you will retain. This is enormously difficult when the process is not in statistical control, because the process instability masks the results, good or bad, of any changes deliberately made.

To learn more about statistical process control and control charts, study Chapter 18, "Optimizing and Controlling Processes through Statistical Process Control (SPC)."

STRATIFICATION

Stratification is a simple tool in spite of its name. It involves investigating the cause of a problem by grouping data into categories. This grouping is called *stratification*. The groups might include data relative to the environment, the people involved, the machine(s) used in the process, materials, and so on. Grouping of data by common element or characteristic makes it easier to understand the data and to pull insights from it.

Consider an example from a factory floor. One of the factory's products requires six assemblers, all doing the same thing at the same rate. Their output flows together for inspection. Inspection has found an unacceptably high rate of defects in the products. Management forms a team to investigate the problem with the objective of finding the cause and correcting it. They plot the data taken over the last month (see Figure 15–28).

The chart in Figure 15–28 plots all operator-induced defects for the month. The team believes that for this product, zero defects can be approached. If you were going

Figure 15–28
Chart of Operator Defects for November

to react to this chart alone, how would you deal with the problem? You have five assemblers. Do they all contribute defects equally? This is hardly ever the case. The data can be stratified by the operator to determine each individual's defect performance. The charts in Figure 15–29 do this.

The five stratified charts in Figure 15–29 indicate that one operator, Assembler B, is responsible for more defects than the other four combined. Assembler A also makes more than twice as many errors as Assembler C or Assembler D and eight times as many as Assembler E, the best performer of the group.

The performance of Assembler A and Assembler B must be brought up to the level of the others. Possible causes of the operator-induced defects could be inherent skill, training, vision, attitude, attentiveness, and environmental factors such as noise, lighting, and temperature in the operator's work-station area. The charts provide an indication of the place to start making changes.

The Pareto charts of Figure 15–6 also represent *stratification*. Figure 15–6 started with a series of defect types that were the most costly (the first chart). Then it took the worst case, Miswires, and divided it into the *kinds* of miswires (the second chart). Then the worst kind, Hand-Wrap, was split into several categories (the third chart). The dominant Hand-Wrap defect category was operator induced. Finally, the Operator category was stratified by individual operator (the fourth chart).

The power of stratification lies in the fact that if you stratify far enough you will arrive at a *root cause* of the problem. Only when root causes are corrected will the problem be solved. Any other kind of solution is a *work-around* fix. Work-arounds are often used in the real world, but when they are, the underlying problem remains and will eventually cause disruption again.

In the present example, we probably did not go all the way to the root cause, unless Assembler B has serious mental, vision, or motor problems that could not be corrected. The most likely root cause is that Assembler B has not been adequately trained for the job, something readily ascertained when the focus is on that individual. One or two more charts looking at the time of day when the mistakes are being made might yield some information, but once the problem is isolated to a person, discussion will usually take you quickly to the root cause. If, on the other hand, Assembler B is a robot and not a human (which is entirely possible in today's automated environment), the stratification should go to at least one more level. We would have to determine the kinds of defects that Assembler B (the robot) is making. That may lead to an adjustment or repair of the machine.

Figure 15–29
Stratified Charts for Each
Operator

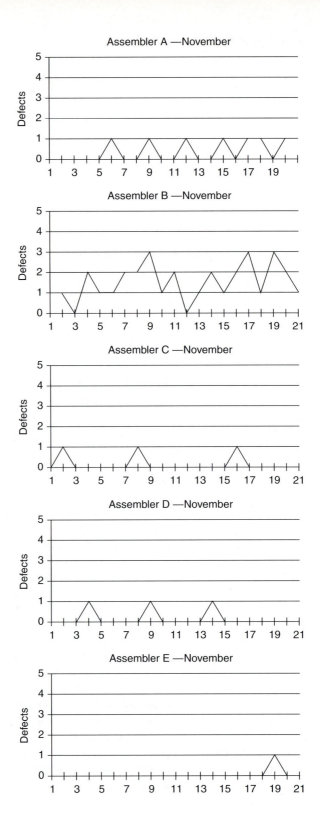

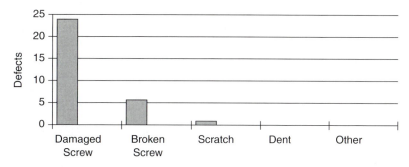

Figure 15–30
Robot B Defect Category for November

Figure 15–30 shows that the category of defects induced by this machine are almost all concerned with screws. The robot is either damaging the screws or breaking them off. Show this chart to the robot maintenance technician, and that person will immediately recognize that the robot needs an adjustment or replacement of its torque controller. The root cause of the problem is either misadjustment or a defective controller. The technician can confirm the diagnosis by running tests on the robot before certifying it for return to service.

Data collected for Pareto charts and run charts (Figure 15–28) can be stratified. Virtually any data can be subjected to stratification. This includes the data collected for control charts, check sheets, histograms, and scatter diagrams. Consider an example of a stratified scatter diagram.

Scatter diagrams, which show the relationship between the x- and y-axis, lend themselves well to stratification. In this example parts are being finished on two identical machines. A scatter diagram is plotted to correlate surface flatness and machine speed.

Figure 15–31 suggests that there is a correlation between machine speed (revolutions per minute, or rpm) and surface flatness between 500 and 1,000 rpm but no correlation at higher revolutions per minute. When the same data are *stratified* in the charts of Figure 15–32, the picture becomes clearer.

In Figure 15–32, the charts reveal that the two machines react similarly to speed increase, but Machine 1 is better than Machine 2 by about 0.0001" in its ability to produce a flat surface. The Machine 1 chart also suggests that increases beyond 1,000 rpm do not produce much improvement. A finish of 0.00095 is about as good as the machine will produce. On the other hand, the Machine 2 chart does show some improvement (two data points) past 1,300 rpm. Given the difference between the two machines, one message coming from the charts is that Machine 2 should be examined to determine the cause of its poorer performance. (More than likely it will be found that bearing wear is the factor in question and that can be corrected easily.) After the machine has been repaired, new data should be taken to verify that 1,000–1,100 rpm is the best practical machine speed.

The charts in Figure 15–32 indicate another message. Both machines had data points better than normal at 550 and 1,100 rpm. It appears that the machines have a

Figure 15–31
Scatter Diagram: Surface
Flatness versus Revolutions per
Minute

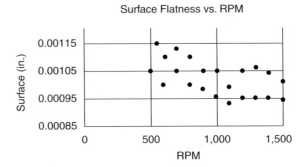

natural resonance that affects performance. The clue here is that both machines show it at 550 rpm and at double that speed (1,100 rpm). This should be checked out, because it could be adversely affecting performance across the range. If vibration and resonance could be "quieted" across the operating range as it apparently is at 550 and 1,100 RPM, the performance might be significantly improved in both machines. The data that gave us this signal are in the scatter diagram of Figure 15–31, but they don't jump out at you the way they do in the stratified charts of Figure 15–32.

In these examples, we have stratified assembly defects by operator, machine-induced defects by type of defect, and machine performance by machine. It was also pointed out that the earlier Pareto chart discussion involved stratification in which defects were stratified to types of defects, the worst of which was in turn stratified to the processes producing those defects. The process (Hand-Wrap) producing the most defects was stratified to process factors, and finally the factor revealed as the most significant (Operator) was stratified to individual operators.

There is virtually no limit on the directions stratification can take. For example, the operators could have been stratified by age, training, gender, marital status, teams, experience, or other factors. The machines could have been stratified by age, date of maintenance, tools, and location (and in the case of similar but not identical machines, by make and model number). In similar fashion, operating procedures, environment, inspection, time, materials, and so forth, can be introduced.

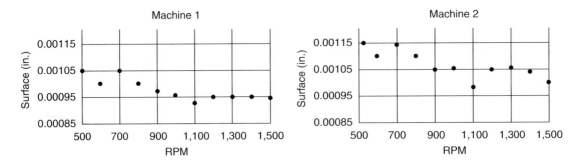

Figure 15–32
Stratified Scatter Diagrams: Surface Flatness versus Revolutions per Minute

SOME OTHER TOOLS INTRODUCED

The preceding sections have discussed the statistical tools that have come to be known as the Seven Tools. One should not conclude, however, that these seven are the only tools needed for pursuing world-class performance. These seven are the ones that have been found most useful for the broadest spectrum of users. Ishikawa refers to them as the "seven indispensable tools for quality control."[6] He goes on to say that they are being used by everyone from company presidents to line workers and across all kinds of work—not just manufacturing. These seven probably represent the seven basic methods most useful to all the people in the workplace. We recommend three more as necessary to complete the tool kit of any business enterprise, if not each of the players within the business:

- The flow diagram
- The survey
- Design of experiments (DOX or DOE)

Both W. Edwards Deming[7] and Joseph Juran[8] promote the use of flow diagrams. Ishikawa includes surveys and design of experiments in his Intermediate Statistical Method and Advanced Statistical Method,[9] respectively.

Flowcharts

A *flowchart* is a graphic representation of a process. A necessary step in improving a process is to flowchart it. In this way, all parties involved can begin with the same understanding of the process. It may be revealing to start the flowcharting process by asking several different team members who know the process to flowchart it independently. If their charts are not the same, one problem is revealed at the outset. Another strategy is to ask team members to chart how the process *actually* works and then chart how they think it *should* work. Comparing the two versions can be an effective way to identify causes of problems and to suggest improvement possibilities. The most commonly used flowcharting method is to have the team, which is made up of the people who work within the process and those who provide input to or take output from the process, work together to develop the chart. It is important to note that to be effective, the completed flowchart must accurately reflect the *way the process actually works*, not how it should work. After a process has been flowcharted, it can be studied to determine what aspects of it are problematic and where improvements can be made.

You may already be familiar with the flowchart, at least to the point of recognizing one when you see it. It has been in use for many years and in many ways. The application we have in mind here is for flowcharting the inputs, steps, functions, and outflows of a process to more fully understand how the process works, who or what has input to and influence on the process, its inputs and outputs, and even its timing.

A set of standard flowcharting symbols for communicating various actions, inputs, outflows, and so forth, are used internationally. They may be universally applied to any process. The most commonly used symbols are shown in Table 15–1. To illustrate their

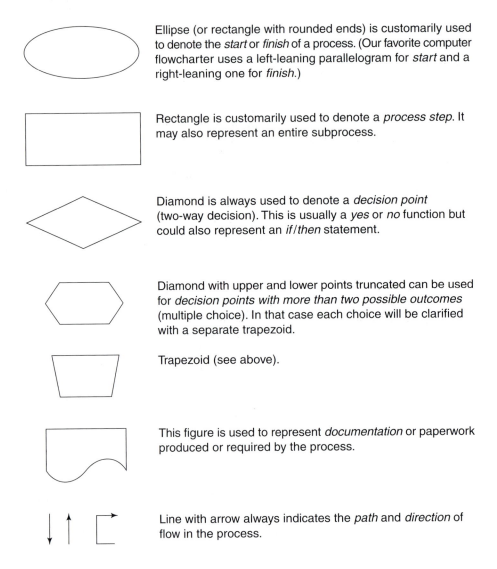

Ellipse (or rectangle with rounded ends) is customarily used to denote the *start* or *finish* of a process. (Our favorite computer flowcharter uses a left-leaning parallelogram for *start* and a right-leaning one for *finish*.)

Rectangle is customarily used to denote a *process step*. It may also represent an entire subprocess.

Diamond is always used to denote a *decision point* (two-way decision). This is usually a *yes* or *no* function but could also represent an *if/then* statement.

Diamond with upper and lower points truncated can be used for *decision points with more than two possible outcomes* (multiple choice). In that case each choice will be clarified with a separate trapezoid.

Trapezoid (see above).

This figure is used to represent *documentation* or paperwork produced or required by the process.

Line with arrow always indicates the *path* and *direction* of flow in the process.

These are the symbols you will use most often, and they will fit virtually any situation.

Table 15–1
Standard Symbology for Flowcharts

use, a simple flow diagram using the most common symbol elements is given in Figure 15–33. Flow diagrams may be as simple or as complex as you may need. For example, in Figure 15–33 the rectangle labeled "Troubleshoot" represents an entire subprocess that itself can be expanded into a complex flowchart. If an intent of the flowchart had been to provide information on the troubleshooting process, then each troubleshooting step would have to be included. Our purpose for Figure 15–33 was merely to chart the *major*

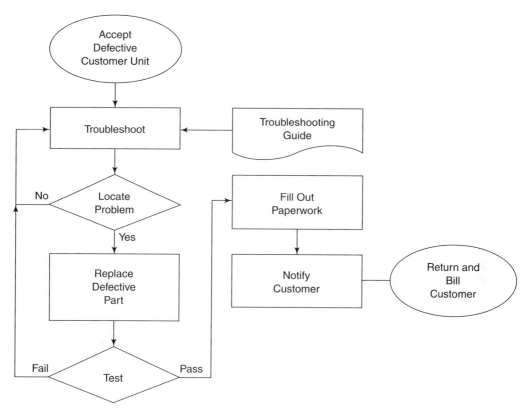

Figure 15–33
Typical Processes Flow Diagram

process steps for receiving and repairing a defective unit from a customer, so we did not require subprocess detail. This is a common starting point. From this high-level flow-chart, it may be observed that the customer's defective unit is (a) received, (b) the problem is located and corrected, (c) and the repaired unit is tested. (d) If the unit fails the test, it is recycled through the repair process until it does pass. (e) Upon passing the test, paperwork is completed. (f) Following that, the customer is notified, and (g) the unit is returned to the customer along with a bill for services. With this high-level flow-chart as a guide, your next step will be to develop detailed flowcharts of the sub-processes you want to improve. Only then can you understand what is really happening inside the process, see which steps add value and which do not, find out where the time is being consumed, identify redundancies, and so on. Once you have a process flow-charted, it is almost always easy to see potential for improvement and streamlining. Without the flowchart it may be impossible.

More often than not, people who work directly with a process are amazed to find out how little understanding of their process they had before it had been flow-charted. Working with any process day in and day out tends to breed a false sense of familiarity.

We once took over a large manufacturing operation that was having major problems with on-time delivery of systems worth $500,000 to $2,000,000 apiece. Several reasons accounted for the difficulty, but a fundamental problem was that we were not getting the input materials on time—even with a 24-month lead time for delivery. One of the first things we did was flowchart the entire material system. We started the chart at the signing of our customer's order and completed it at the point where the material was delivered to the stockroom. The chart showed dozens of people involved, endless loops for approval and checking, and flawed subprocesses that consumed time in unbelievable dimensions.

When the flowchart was finished, it was clear that the best case from the start of the order cycle until material could be expected in our hands required 55 weeks. The worst case could easily double that. With this knowledge, we attacked the material process and quickly whittled it down to 16 weeks and from there to 12 weeks. The point is this: here was a process that had grown over the years to the point that it was no longer tolerable, much less efficient. But the individual players in the process didn't see the problem. They were all working very hard, doing what the process demanded, and fighting the fires that constantly erupted when needed material was not available. The flow diagram illuminated the process problems and showed what needed to be done.

If you set out to control or improve any process, it is essential that you fully understand the process and why it is what it is. Don't make the assumption that you already know or that the people working in the process know, because chances are good that you don't, and they don't. Work with the people who are directly involved, and flowchart the process as a first step in the journey to world-class performance. Not only will you better understand how the processes work, but you will spot unnecessary functions or weaknesses and be able to establish logical points in the process for control chart application. Use of the other tools will be suggested by the flowchart as well.

Surveys

At first glance, the survey may not seem to be useful. When you think about it, though, all of the tools are designed to present information—information that is pertinent, easily understood by all, and valuable for anyone attempting to improve a process or enhance the performance of some work function. The purpose of a survey is to obtain relevant information from sources that otherwise would not be heard from—at least not in the context of providing helpful data. Because you design your own survey, you can tailor it to your needs. We believe that the survey meets the test of being a total quality tool. Experience has shown that the survey can be very useful.

Surveys can be conducted internally, as a kind of employee feedback on problem areas, or as *internal customer* feedback on products or services. They can also be conducted with *external customers*, your business customers, to gain information about how your products or services rate in the customers' eyes. The customer (internal or external) orientation of the survey is important, because the customer, after all is said and done, is the only authority on the quality of your goods and services. Some companies conduct annual customer satisfaction surveys. These firms use the input from customers to focus their improvement efforts.

Surveys are increasingly being used with suppliers as well. We are finally coming to the realization that having a huge supplier base is not the good thing we thought it was. The tendency today is to cut back drastically on the number of suppliers utilized, retaining those that offer the best *value* (not best price, which is meaningless) and that are willing to enter into partnership arrangements. If a company goes this route, it had better know how satisfied the suppliers are with the past and present working relationship and what they think of future prospects. The survey is one tool for determining this. It is possibly the best initial method for starting a supplier reduction/supplier partnership program.

Even if you are not planning to eliminate suppliers, it is vital to know what your suppliers are doing. It would make little sense for you to go to the trouble of implementing total quality if your suppliers continue to do business as usual. As you improve your processes and services and products, you cannot afford to be hamstrung by poor quality from your suppliers. Surveys are the least expensive way of determining where suppliers stand on total quality and what their plans are for the future. The survey can also be a not-too-subtle message to the suppliers that they had better "get on the bandwagon."

A typical department in any organization has both internal suppliers and internal customers. Using the same customer-oriented point of view in a survey has proven to be a powerful tool for opening communications among departments and getting them to work together for the common goal, rather than for department glory—usually at the expense of the overall company.

The downside of surveys is that the right questions have to be asked, and asked in ways that are unambiguous and designed for short answers. A survey questionnaire should be thoroughly thought out and tested before it is put into use. Remember that you will be imposing on the respondents' time, so make it easy and keep it simple.

Design of Experiments

Design of experiments (DOX/DOE) is a very sophisticated method for experimenting with processes with the objective of optimizing them. If you deal with complicated processes that have multiple factors affecting them, DOX/DOE may be the only practical way of bringing about improvement. Such a process might be found in a wave soldering machine, for example. Wave solder process factors include these:

Solder type	Conveyor speed	Flux specific gravity
Solder temperature	Conveyor angle	Wave height
Preheat temperature	PC board layer count	Flux type
PC board groundplane mass		

These 10 factors influence the process, often interacting with one another. The traditional way to determine the proper selection/setting was to vary one factor while holding all others fixed. That kind of experimentation led to making hundreds of individual runs for even the simplest processes. With that approach it is unusual to arrive at the optimum setup, because a change in one factor frequently requires adjustment of one or more of the other factors for best results.

Design of experiments reduces the number of runs from hundreds to tens as a rule, or by an order of magnitude. This means of process experimentation allows multiple factor adjustment simultaneously, shortening the total process, but equally as important, revealing complex interaction among the factors. A well-designed experiment can be concluded on a process such as wave soldering in 30–40 runs and will establish the optimum setting for each of the adjustable parameters for each of the selected factors. For example, optimal settings for conveyor speed, conveyor angle, wave height, preheat temperature, solder temperature, and flux specific gravity will be established for each PC board type, solder alloy, and so on.

DOX/DOE will also show which factors are critical and which are not. This information will enable you to set up control charts for those factors that matter, while saving the effort that might have been expended on the ones that don't. While design of experiments is beyond the scope and intent of this book, the DOX/DOE work of Deming, Taguchi, and others may be of help to you. Remember that DOX/DOE is available as a tool when you start trying to improve a complex process.

MANAGEMENT'S ROLE IN TOOL DEPLOYMENT

Management's role is changing from one of directing to one of facilitating. Since the Industrial Revolution, management has supplied the place of work, the machinery and tools, and the work instructions. The concept has been that management knows what the job is and needs only to hire the muscle power to get it accomplished. The workers were there only because management could not get the job done without their labor. Workers were not expected to think about doing things differently but simply to follow the boss's orders. Work was typically divided into small tasks that required minimal training, with little or no understanding on the part of laborers as to how their contribution fit into the mosaic of the whole.

During much of the 20th century, and certainly since World War II, changes have been creeping into the management–labor relationship. Some people think that the labor unions were responsible for these changes, and they did help obtain better pay, shorter hours, workplace improvements, and other benefits for workers. However, the relationship changes between management and labor have happened largely in spite of the unions. Unions have had at least as difficult a time as management has had in dealing with employee involvement. Nor has management at large been responsible for the changes sweeping across the industrial world today. Certainly there are champions representing management, but the changes are coming about for one reason and one reason only: they are necessary in order for businesses to survive in an increasingly competitive marketplace.

After World War II, when Deming went to Japan to teach industrialists about quality and the use of statistics for achieving it, Japan had just lost the war. Its industrial base was a shambles. The Japanese needed to resurrect their factories and put people to work quickly. That meant they had to be able to sell their products abroad—to the same people who had defeated them. To do that, it was essential that their products be of high quality. Their survival depended on it. You know the rest of the story.

Not only did the Japanese listen to Deming and Juran, but they embraced them and their philosophy (whereas in the United States we were abandoning their teaching amid a seemingly insatiable market for manufactured goods). Japan developed its own quality gurus (Ishikawa, Taguchi, Shingo, and others) who expanded the work of Deming and Juran. For 30 years, into the 1980s, Japanese manufacturers perfected their quality and production methods. The 1980s found Japan ahead of the rest of the world, not just the United States, in product quality and value. During that decade, companies in the United States began to wake up to the fact that Japan's products were the best in the world and that they were running roughshod over U.S. companies not only in the world markets but right here at home. Whole markets were conceded to Japan as U.S. companies found they could not compete.

The survival mentality finally surfaced. We woke up to the fact that not only our industrial survival but perhaps even our national survival was at stake. Either we became competitive in the global marketplace, or we lost the first war fought without bullets since the invention of gunpowder.

Now that the wake-up call has been received, many people have come to realize that we have been managing poorly for a very long time—say, since 1945. We (those of us who have heard the alarm) have come to understand that management's proper role is to facilitate, not to direct. Management provides the place of work and the machines and tools as before, but in addition, we do everything we can to *help* our employees do the job. That means training. It means listening to their thoughts and ideas—more than that—*seeking* their thoughts and ideas. It means acting on them. It means giving them the power to do their jobs without management interference. It means giving them time to think and discuss and suggest and experiment. It means communicating—fully and honestly. No secrets, no smoke screens. It means accepting every employee as a valued member of the corporate team.

This approach does not mean that management abdicates its responsibility to set the direction for the enterprise, to establish the corporate vision, to steer the course. But with the enlistment of all the brain power that had formerly gone untapped, even this job becomes easier than it was before.

It is management's responsibility to train employees to use not only physical tools (and that is very important) but also intellectual tools. The seven tools discussed in this chapter should eventually be used by every employee—*eventually* because it is a mistake to schedule all employees for training on the tools if they will not be using them very soon. You would not train a person on a new machine a year before the machine arrives, because without putting the training to practice its effect will be lost. So it is with the total quality tools. When a group of people are ready to put some of the tools to practice, that is when they should be trained. As the total quality concept takes root, it will be only a matter of time until everyone has the need. Train them as required.

Management must also provide the internal experts, often called *facilitators*, to help the new teams get started and to develop their expertise. Facilitation is probably a never-ending function, because the total quality envelope is constantly being expanded, and there will always be the need for a few to be on the leading edge, to bring the others along.

It is management's responsibility to ensure that the people who are solving the problems have the proper training and facilitation. It is also management's responsibility to make sure the problems being attacked are of interest to the enterprise and not trivial.

Management must populate the problem-solving team with the cross-functional expertise the problem requires. The team must be given the power and support necessary to see the effort brought to its conclusion.

Management must be vigilant that data used in problem solving are valid, which is a function that usually falls to the facilitator. Especially when teams are immature in total quality, they have a tendency to grab at the first set of data that comes along. Management must ensure that the data and the statistical techniques employed are appropriate for the problem at hand.

Finally, management must ensure that there are results. Too many problem-solving, process improvement, and related efforts take on a life of their own, and go on forever. This cannot be allowed. People are watching. Especially in the early stages, some people will hold the view that "This too shall pass." If results do not come rather quickly, the detractors will be given the ammunition they need to subvert the whole total quality effort. For this reason, it is important that the first projects attempted have a high probability of success, and management must monitor them closely, even to the point of being involved in the activity. As the process matures and successes are tallied, an occasional failure will not be an issue. In fact, people must be given the chance to fail, and failure must be free of repercussions for the team or its members.

Precautions

Implementing the use of statistical tools and the whole concept of process improvement, problem solving by the rank and file, empowerment—in short, the total quality culture—represents a profound change from the way things have been done in the past. People generally resist change until they see that it will benefit them. For that reason, management must become the champions of change and convince everyone that the effort will benefit all. Those who would undermine the effort must rapidly be converted or removed from the operation. People will be looking to management for evidence that management really believes in total quality. If for no other reason than that, it must be obvious to all that management is using the same techniques the other employees are being taught. Above all, management must support and facilitate the employees as they use the techniques of total quality to solve problems and improve processes.

Although results should be evident quickly, do not expect the necessary cultural change to occur overnight. This is a long process, requiring several years to get to the point where total quality is considered "just the way we do things" and not some special "project." Even so, during all that time, problems are being solved, improvements made, and efficiency, productivity and competitiveness increased.

- ■ *Communicate.*   Let everyone know what is going on and what the results are. Help them understand why it is good for them, the whole enterprise, and, yes, even for the nation.
- ■ *Never assume that you know it all.*   The people who live with the processes day in and day out know far more about what is wrong with them and how to improve them than any manager. Never delude yourself that you have learned all you need to

know about total quality. It will never happen because total quality is a dynamic and ever-expanding concept.

■ *Start slowly.* Don't try to organize an entire factory or office complex into improvement teams and train everyone in sight on day 1. Take it one or two steps at a time, training as you go. Be careful to pick early projects that have high prospects for success.

■ *But start.* The worst choice a manager could make today is to decide that total quality is not for his or her business. It is for every conceivable kind of business, large or small, whether public, private, military, civilian, mass production, job shop, classroom, or office. It would be a tragedy to decide not to start this journey when so much is at stake.

SUMMARY

1. Pareto charts are useful for separating the important from the trivial. They are named after Italian economist and sociologist Vilfredo Pareto, who developed the theory that a majority of problems are caused by a minority of causes. Pareto charts are important because they can help an organization decide where to focus limited resources. On a Pareto chart, data are arrayed along an *x*-axis and a *y*-axis.

2. The cause-and-effect diagram was developed by the late Dr. Kaoru Ishikawa, a noted Japanese quality expert; others have thus called it the Ishikawa diagram. Its purpose is to help identify and isolate the causes of problems. It is the only one of the seven basic quality tools that is not based on statistics.

3. The check sheet is a tool that helps separate unimportant data from important information. Check sheets make it easy to collect data for specific purposes and to present it in a way that facilitates its conversion into useful information.

4. Histograms have to do with variability. Two kinds of data are commonly associated with processes: attributes data and variables data. An *attribute* is something that the output product of the process either has or does not have. *Variables data* are data that result when something is measured. A *histogram* is a measurement scale across one axis and a frequency of like measurements on the other.

5. The scatter diagram is the simplest of the seven basic quality tools. It is used to determine the correlation between two variables. It can show a positive correlation, a negative correlation, or no correlation.

6. Run charts and control charts are typically thought of as being one tool together. The control chart is a more sophisticated version of the run chart. The *run chart* records the output results of a process over time. For this reason, the run chart is sometimes called a *trend chart*. The weakness of the run chart is that it does not tell whether the variation is the result of special causes or common causes. This weakness gave rise to the control chart. On such a chart, data are plotted just as they are on a run chart, but a lower control limit, an upper control limit, and a process average are added. The plotted data stays between the upper control limit and lower control limit while varying about the center line or average only so long as the variation is the result of common causes such as statistical variation.

7. Stratification is a tool used to investigate the cause of a problem by grouping data into categories. Grouping of data by common elements or characteristics makes it easier to understand the data and to draw insights from it.

8. Other useful quality tools are the flow diagram, survey, and design of experiments (DOX/DOE). Flowcharts are used in a total quality setting for charting the inputs, steps, functions, and outflows of a process to understand more fully how the function works and who or what has input into and influence on the process, its inputs and outputs, and even its timing. The survey is used to obtain relevant information from sources that otherwise would not be heard from in the context of providing helpful data. Design of experiments is a sophisticated method for experimenting with processes for the purpose of optimizing them.

 KEY TERMS AND CONCEPTS

Attributes data	Process average
Cause-and-effect diagram	Process variability
Check sheet	Run chart
Common causes	Scatter diagram
Control chart	Six Sigma
Control limits (UCL, LCL)	Special causes
Design of experiments	Specification limits
Flowchart	Standard deviation
Frequency distribution	Stratification
Histogram	Survey
In statistical control	Total quality tools
Lower control limit (LCL)	Trend chart
npmo	Upper control limit (UCL)
Pareto chart	Variables data
Pareto principle	Variation

 FACTUAL REVIEW QUESTIONS

1. Explain the purpose of a Pareto chart. Give an example of when one would be used.
2. Describe the origin and use of cause-and-effect diagrams.
3. How would a check sheet be used in a modern production facility?
4. What is a histogram, and how is one used?
5. What accounts for the disparity in npmo between the statistical 6σ and the popular Six Sigma approach?
6. Describe two methods for improving the yield of a process (i.e., taking it to a higher sigma value).
7. Explain the purpose of the scatter diagram. Give an example of how one would be used.
8. Contrast and compare run charts and control charts.

9. What is the most common use of stratification?
10. What purpose is served by flowcharts?
11. Give an example of how a survey might be used in a modern production setting.
12. What is the purpose of design of experiments?

 CRITICAL THINKING ACTIVITIES

Which Tool to Use

As the manager of a department that is beginning to use the total quality tools, Marion thinks a wall chart in the team meeting room would be helpful when the teams decide on which tools to apply to their problem-solving or decision-making efforts. She has listed the most commonly used tools and wants a simple one-line "purpose statement" for each tool, similar to the one she developed for the flow diagram. Your task: Develop the rest of the purpose statements for the wall chart here.

Total Quality Tools: What They Are Used For	
Tool	**Purpose Statement**
Flow diagram	Helps us understand our processes; facilitates thinking about improving them.
Pareto chart	
Cause-and-effect dia.	
Histogram	
Stratification	
Run chart	
Control chart	
Scatter diagram	
Check sheet	
Survey	

Constructing a Flow Diagram: Scheduling a Meeting

Marion read that even the most routine tasks can be flow diagrammed and that when they are, the diagram may reveal complexity that is both surprising and unnecessary. She thought she would test this by flow diagramming a process that seemed to be the most used in her organization, scheduling a meeting. She convened a meeting of her direct subordinates, and together they have listed the steps involved in scheduling a meeting. Now they are ready to develop the diagram. You are the facilitator. Lead them through the diagramming process. (Draw the diagram.)

Steps in Scheduling a Meeting	The Flow Diagram

• Select a topic.
• Select participants.
• Select date.
• Select time.
• Select place
  (accommodate any conflicts in above).
• Prepare draft agenda.
• Distribute draft for comments.
• Review comments
  (accommodate as appropriate).
• Prepare meeting room.
• Hold meeting.	

Selecting Improvement Subjects
Using Pareto Charts and Stratification

Starlight Homes, Inc., is a building contractor specializing in upscale homes in the Southwest. Before each new home is sold, Starlight conducts a final inspection of the home and repairs any defects. In addition, Starlight receives a "punch list" of defects (to be corrected) compiled by the buyers following the sale. Ricardo Alvarez, Starlight's general foreman, has concluded that even if it should cost more to do the work right in the first place, it will be a lot cheaper than going back later to fix the defects. In an effort to reduce costs, improve the quality of Starlight's homes, and reduce the number of complaints after the sale, Ricardo has assembled data from his final inspections and the punch lists for the last 20 homes sold. The data are as listed in the following chart:

Defect Type	Occurrences	Defect Type	Occurrences
Damaged walls	13	Doors	14
Exterior paint	5	HVAC	11
Plumbing	33	Roof	3
Caulking	28	Masonry	2
Electrical	25	Interior paint	61
Cabinetry	12	Landscaping	16
Woodwork	46	Fixtures	7

Construct a Pareto chart to illustrate the defect types by number of defects. Which two defect types appear to be the most significant? Should Ricardo focus his attention on these two categories of defects? Ricardo decided to stratify the Pareto chart by cost before making any decision. His bookkeeper developed the average cost per repair event per category. The data are shown here:

Average Cost per Repair Event			
Defect Type	**Average Cost**	**Defect Type**	**Average Cost**
Damaged walls	$126	Doors	$11
Exterior paint	25	HVAC	110
Plumbing	78	Roof	72
Caulking	7	Masonry	290
Electrical	74	Interior paint	4
Cabinetry	88	Landscaping	34
Woodwork	5	Fixtures	31

The second-level chart developed from these numbers should show Ricardo which two or three defect types are the most significant in terms of cost, enabling him to put his efforts where they will do the most good. What are they? Was Ricardo justified in going to the second level (of stratification) before making a decision on where to focus his efforts? Why?

Constructing a Cause-and-Effect Diagram

Your team has been given the charter to make recommendations for improving the cleanliness of the company's rest rooms. The team has finished compiling a list that it considers to be the possible contributors to less than desirable rest room cleanliness:

Janitor service	Use of paper towels	Lack of paper towels
Slobs	User carelessness	User attitude
Janitor attitude	Paper dispensers	Poor lighting
Too small	Unreliable plumbing	Type of floor material
Janitor pay	Management inattention	Janitor supervision
Air conditioning	Insufficient exhaust	In-house plumbers
Cleaning materials	Cleaning equipment	Paper receptacles
Maintenance	Cleaning schedule	Leaking faucets
Janitor procedures		

Construct a cause-and-effect diagram incorporating all these ideas.

Translating Your Histogram

Your process has a normal histogram with μ located at 10 millimeters, 1σ points at 9.9 and 10.1 millimeters, 2σ points at 9.8 and 10.2 millimeters, and 3σ points at 9.7 and 10.3 millimeters.

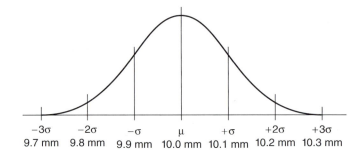

If your customer will accept parts measuring between 9.7 and 10.3 millimeters, how many parts of every 1,000 produced would you expect to scrap? If your customer notified you that henceforth it would only accept parts between 9.9 and 10.1 millimeters, what would you anticipate your scrap rate to be? What would have to be done to bring the scrap rate at this new customer requirement back down to what it had been at the former customer specification? Compared with the bell curve above, what would the new histogram have to look like?

DISCUSSION ASSIGNMENT 15–1

Reacting to a Process Gone Wrong

Cignet Plastics Corporation is a contract plastics die-casting house serving a wide range of clients. Over the years Cignet has been a favored supplier of precision die castings for a major producer of model airplane kits. In recent days the defect rate of these parts has increased. (Acceptance is based on a visual inspection of the parts for appearance.) After a thorough audit of the process, Quality Assurance has concluded that there has been no change to the process. It claims that the increase in defects must be variation that is related to some assignable cause. The president of Cignet Plastics does not have a clue as to what that means, and he has called you in for an explanation.

Discussion Questions
Discuss the following questions in class or outside of class with your fellow students.
   1. What will you tell the president?
   2. He wants you to change the process to get the defects back down, but you know that is the wrong approach. How do you talk him out of it?
   3. What approach would you use to get the operation back to normal?

ENDNOTES

1. Kaoru Ishikawa, *Guide to Quality Control* (Tokyo: Asian Productivity Organization, 1976).
2. The 80–20 rule is an approximation, and one should not expect the numbers to land exactly at 80% or 20%.
3. Ishikawa, *Guide to Quality Control*, 24–26.
4. Joseph M. Juran, *Juran on Planning for Quality* (New York: Free Press, 1988), 180.
5. W. W. Scherkenbach, *The Deming Route to Quality and Productivity* (Rockville, MD: Mercury, 1991), 100.
6. Ishikawa, *What Is Total Quality Control?* (Upper Saddle River, NJ:Prentice Hall, 1985), 198.
7. Scherkenbach, 104.
8. Juran, 18.
9. Ishikawa, *What Is Total Quality Control?* 199.

CHAPTER SIXTEEN

Problem Solving and Decision Making

"Some people approach every problem with an open mouth."
Adlai Stevenson

MAJOR TOPICS

- Problem Solving for Total Quality
- Solving and Preventing Problems
- Problem-Solving and Decision-Making Tools
- Decision Making for Total Quality
- The Decision-Making Process
- Objective versus Subjective Decision Making
- Scientific Decision Making and Problem Solving
- Employee Involvement in Decision Making
- Role of Information in Decision Making
- Using Management Information Systems (MIS)
- Creativity in Decision Making

Problem solving and decision making are fundamental to total quality. On the one hand, good decisions will decrease the number of problems that occur. On the other hand, the workplace will never be completely problem free. The purpose of this chapter is threefold:

- To help readers learn how to solve problems effectively, positively, and in ways that don't create additional problems
- To help readers become better decision makers
- To help readers learn to make decisions and handle problems in ways that promote quality

PROBLEM SOLVING FOR TOTAL QUALITY

If you ask the typical manager to describe his or her biggest problem in today's workplace, the response will probably include one or more of the following:

- We spend all our time in meetings trying to resolve problems.
- We are constantly fighting problems, and that doesn't leave us time to do our real jobs, such as planning, leading, and so forth.
- As soon as we "put out one fire," another pops up.
- We've got more problems than we can handle, and it bogs us down.

The actual words may vary, but the message is the same. The workplace can be so burdened with problems that managers and others spend so much time trying to fix them that nothing gets done right. Leadership suffers—there is just no time to lead. Performance suffers, from the standpoint of both the individual and the organization. Quality of product or service deteriorates. Competitiveness is negatively impacted. Failure of the organization becomes a real probability, especially if its competitors have turned to total quality and its philosophy for solving problems—once and for all. Why is it that with all the effort we put into it, consuming so much time in the process, we cannot solve our problems and get on with the jobs we are paid to do? The answer is simply that most of our problem solving is anything but that.

Consider this. We were once driving along in our car and the engine quit. Turn the key and it fired up; release the key and it quit again. Somehow we found that there was a resistor in the electrical system that was only used after the engine had been started, and it had failed; hence, no electrical energy to the spark plugs. We found that by replacing the resistor with a simple piece of wire, the engine would run fine. We put the wire in place and drove home. Had we corrected the problem? We suspected that we had not, because if a penny's worth of copper wire could have been used in place of a part that cost a couple of dollars, then surely Chrysler would have opted for the wire. We suspected that our fix was just a Band-Aid. Sure enough, the next day the ignition coil failed. That was a $10 part, and it failed because by substituting the wire for the defective resistor we had put too much voltage on the coil, causing it to self-destruct. We had not understood the reason for the resistor being there. We did not have all the data we should have had before using that piece of wire. The lessons here: (a) Get data before you try to solve any problem, or you may make things even worse; (b) Band-Aid fixes do not solve problems and may cause new, unforeseen problems.

At this point we spent $12 for a replacement resistor and an ignition coil (note that this was six times what we would have spent if we had properly replaced the resistor in the first place). The car ran fine. Problem solved? Most people would say so.

This is the level of problem solving in most organizations. When something breaks, fix it or replace it. Job done, problem solved. The most that we should claim for this type of problem solving is that we are back where we started (i.e., before the problem came up). But remember, if it happened once it can happen again.

Getting back to the Chrysler, the resistor failed twice more while we owned the car. And over the next 25 years there were two more Chrysler products in our family, both of which had multiple failures of the same resistor. Replacing the resistor did not solve the problem. For the problem to be truly solved so that the failure rate of that part would be sufficiently reduced as to be satisfactory would have required Chrysler to gather all the electrical, physical, and reliability data relating to the resistor, and the circuit it operated in, and then redesign the circuit or make use of a more robust resistor or some other change, as the data required. Had Chrysler done that, we could justifiably call it a solved problem. We could also call such a solution a product *improvement*, because the probability of failure would be greatly reduced. (We have had no firsthand experience with Chrysler cars for several years, so they may have indeed solved it. We hope so.)

The point we want to make here is that in total quality jargon, a problem is only solved when its recurrence has become impossible or significantly less probable. That will always be the objective of total quality problem solving. Any problem that is merely fixed by restoring the situation to what it was before the problem was manifested will return again. That is why our managers spend so much time with problem issues. The problems are not being solved, just put into a recycle loop. In those organizations that have adopted total quality, problems are solved once and for all. The same problems do not return time and time again. That means that there will be fewer problems tomorrow than there were today, fewer next month than this month, fewer next year than this year. Managers will have more time to manage, leaders to lead. With problem solutions leading to process or products/service improvement,

- product or service quality improves,
- costs decrease (through less waste and warranty action),
- customer satisfaction improves,
- competitiveness improves, and
- the probability for success improves.

Clearly all of these outcomes are desirable. And they are all achievable by applying the total quality principles to problem solving.

SOLVING AND PREVENTING PROBLEMS

Even the best-managed organizations have problems. A problem is any situation in which what exists does not match what is desired or, put another way, the discrepancy between the current and the desired state of affairs. The greater the disparity between the two, the greater the problem. Problem solving in a total quality setting is not just "putting out fires" as they occur. Rather, it is one more way to make continual improvements in the workplace and its products or services. This section contains two models for solving problems in ways that simultaneously lead to workplace improvements: the PDCA cycle and the Perry Johnson method.

The Plan-Do-Check-Adjust (PDCA) Cycle

This continual improvement model goes by several names. The Japanese call it the *Deming Cycle* after Dr. W. Edwards Deming, who introduced it to them. Deming himself referred to it as the *Shewhart Cycle* after its originator, Dr. Walter Shewhart. In the West it is commonly called simply the *PDCA cycle,* standing for plan-do-check-act. In this book we have taken the liberty to suggest that the letter *A* more correctly means *adjust.* Whatever we call it, the PDCA cycle consists of four major components, each of which can be subdivided into step-by-step activities. Deming disciple William W. Scherkenbach explains the model as follows:[1]

1. *Plan: develop a plan to improve.* Even before problems occur, create a plan for improving your area of responsibility, particularly the processes in that area. Then, when problems occur, they can be handled within the context of Deming's model for continual improvement. Developing such a plan involves completing the following steps:
   - Identify opportunities for improvement.
   - Document the current process.
   - Create a vision of the improved process.
   - Define the scope of the improvement effort.

2. *Do: carry out the plan.* Implement the plan for improvement. The recommended approach is to first implement on a small scale over a specified period of time. This is the equivalent of developing and testing a prototype of a design before moving to full production.

3. *Study [Check]: examine the results.* Examine and record the results achieved by implementing the plan. The recorded results form the basis for carrying out the steps in the next component.

QUALITY TIP

Don't Just Solve Problems, Make Improvements

"Over the years, various experts have proposed numbers of approaches to problem solving. You may recognize names such as Kepner-Trego, Alamo, Quality Improvement Process, Team-Oriented Problem Solving, Creative Problem Solving, Analytical Problem Solving, Breakthrough Process, QC Story, and the like. While each has particular strengths, each also has particular weaknesses. All of them have the shortcoming of needing a 'problem' to solve. Fortunately, for the problem-solving industry, there is no shortage of problems! Unfortunately for the rest of us, just solving problems, or reducing waste, or eliminating defects will not make us competitive in this new economic age. We need to go beyond problems and look for opportunities for continual improvement."

Source: William W. Scherkenbach, *Deming's Road to Continual Improvement* (Knoxville, TN: SPC, 1991), 61.

4. *Act: adjust as necessary.* Make adjustments as necessary based on what was learned in the previous component. Then repeat the cycle for the next planned improvement by returning to the first component of the model.

The PDCA cycle has evolved from that which Deming presented to his Japanese audience in the summer of 1950. The cycle started with *design the product*, followed in order by *production, sales,* and *market research.* His emphasis was on developing products that would be accepted in the world's markets, and that was precisely the requirement of the moment in Japan. In the design phase, he stressed finding out what is needed by potential customers, designing a product to meet the need, and planning sufficient production to validate the product's viability. That has become the Plan part of the cycle. The production plan was to be executed in the second quadrant of the cycle. That has become the Do phase. After producing the product, the firms were to sell it. Whether it sold well or poorly provided information on whether they had correctly chosen a product type. This has become the Check phase. Having sold the product, they were admonished to find out from their customers whether the product met their expectations and how it could be changed to better serve the customer. That has become the Act or Adjust phase. The concept was that a second cycle would commence immediately, taking into consideration everything that was learned from the first cycle; then a third, fourth, and so on, continually applying information learned to redesign the product and find ways to make production more efficient, always with customer requirements as a very important input to the process.

Kaoru Ishikawa's version is essentially that which we call the PDCA cycle. He specifies six steps:[2]

1. Determine goals and targets.
2. Determine methods of reaching goals. (Steps 1 and 2 constitute the Plan.)
3. Engage in education and training.
4. Implement work. (Steps 3 and 4 constitute the Doing.)
5. Check the effects of implementation (Check phase).
6. Take appropriate action (Action phase).

Where Deming's initial emphasis was on the product, Ishikawa's version seems to lend itself more clearly to processes as well as products. We know that Deming came to the same place, perhaps having been there all along, and there is no significant difference in philosophy, only in the choice of words.

The Perry Johnson Method

Perry Johnson, Inc., of Southfield, Michigan, recommends an approach to problem solving that works well in a total quality setting because of its three main characteristics:[3]

■ It promotes teamwork in problem solving.
■ It leads to continual improvement rather than just "putting out fires."
■ It approaches problems as normal by-products of change.

The Perry Johnson Method for problem solving is as follows:

QUALITY TIP

Problems Are Neither Good nor Bad

"Problems are not bad or good. They just are. They are normal, natural by-products of change. What we commonly think of as a problem is simply performance which varies from what is expected in an undesirable way."

Source: Perry L. Johnson, Rob Kantner, and Marcia A. Kikora, *TQM Team-Building and Problem-Solving* (Southfield, MI: Perry Johnson, Inc., 1990), 1-2.

1. *Establish a problem-solving team.* The reason for using a team in solving problems is the same as that for using a team in any undertaking: no single person knows as much as a team. Team members have their own individual experiences, unique abilities, and particular ways of looking at things. Consequently, the collective efforts of a team are typically more effective than the sole efforts of one person.

A problem-solving team can be a subset of one department or unit, or it can have members from two or more different departments. It can be convened solely for problem solving, or it can have other duties. Decisions about how to configure the team should be based on the needs, size, and circumstances of the organization.

2. *Brainstorm the problem list.* It is important to get out in front of problems and deal with them systematically. For example, the military doesn't just sit back and wait for the next trouble spot on the world scene to boil over. Rather, potential trouble spots are identified and entered onto a problem list. The potential problems are then prioritized and plans are developed for handling them. The same approach can be used in any organization. The problem-solving team should brainstorm about problems that might occur and create a master list.

3. *Narrow the problem list.* The first draft of the problem list should be narrowed down to the entries that are really problems. To accomplish this, evaluate each entry on the list by means of three criteria:
   • There is a standard to which the entry can be compared.
   • Actual performance varies from the standard in an undesirable way.
   • The variance is supported by facts.
Any entry that does not meet all three criteria should be dropped from the list.

4. *Create problem definitions.* All problems remaining on the list should be clearly defined. A problem definition has two parts: a description of the circumstance and a description of the variance. Figure 16–1 contains sample problem definitions that have the desirable characteristics of being thorough, brief, and precise.

A teamwork tool that can be helpful in clarifying the definition of a problem is *Why-Why.* This method involves asking why until the team runs out of answers. Figure 16–2 illustrates how Why-Why works. Going through this process can be an effective way to

- Job 21A is over budget by 22 percent.
  Circumstance:
  Job 21A is over budget.
  Variance:
  By 22 percent
- Reject rate of the machining department is too high by 12 percent.
  Circumstance:
  Reject rate of the machining department is too high.
  Variance:
  By 12 percent

Figure 16–1
Sample Problem Definitions

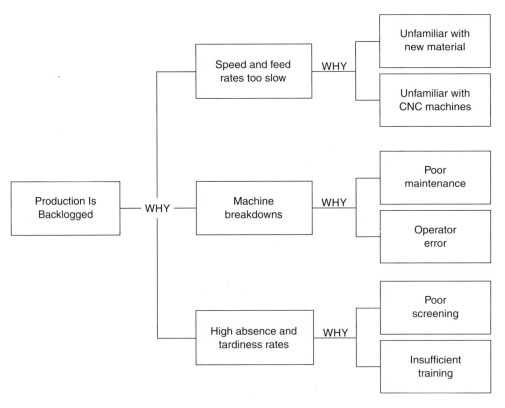

Figure 16–2
The Why-Why Method of Problem Definition

clearly define the real problem so that time and resources are not wasted working on symptoms.

5. *Prioritize and select problems.* With all problems on the list defined, the team can prioritize them and decide which one to pursue first, second, and so on. Perry Johnson recommends using a Problem Priority Matrix. The matrix is created as follows:

- Divide the problem-solving team into two groups and put them in separate rooms.
- Have Group A rank the problem list in terms of benefit to the organization.
- Have Group B rank the problems in terms of how much effort will be required to solve them.
- Set up the Problem Priority Matrix as shown in Figure 16–3. The extended number is arrived at by multiplying the benefit ranking by the effort ranking for each problem. The lowest extended number is the problem that should be solved first. Solving it will yield the most benefit to the organization with the least expenditure of effort. Other problems will yield more benefit or will require less effort. But when both benefit and effort required are taken into account, the lowest extended number would be the first to be solved.

6. *Gather information about the problem.* When the problems have been prioritized, the temptation will be to jump right in and begin solving them. This can be a mistake. The better approach is to collect all available information about a problem before pursuing solutions. Two kinds of information can be collected: objective and subjective. *Objective information* is factual. *Subjective information* is open to interpretation.

Rarely will the information collected be only objective in nature. Nothing is wrong with collecting subjective information, as long as the following rules of thumb are adhered to for both objective and subjective information:

- Collect only information that pertains to the problem in question.
- Be thorough (it's better to have too much information than too little).
- Don't waste time re-collecting information that is already on file.
- Allow sufficient time for thorough information collection, but set a definite time limit.
- Use systematic tools such as those explained in Chapter 15.

Specifying the Problem

Specifying the problem means breaking it down into its component parts. All problems can be broken into five basic components as follows:

- Who is the problem affecting?
- What is the problem? This is a restatement of the final results of the Why-Why process.
- Where did the problem occur first, when did it occur first, and when did it occur last? How often is it occurring?
- Where does the problem occur? The answer to this question should be specific. If a machine in Department A is the problem, where should be specified not just as

Problem Number	Ranking by Benefit	Ranking by Effort	Extended Number (Multiplied)	* Final Ranking
Problem A	2	5	10	2
Problem B	6	3	18	4
Problem C	1	9	9	1
Problem D	8	2	16	3
Problem E	10	1	10	2
Problem F	3	8	24	6
Problem G	5	4	20	5
Problem H	4	10	40	7
Problem I	7	7	49	8
Problem J	9	6	54	9

Figure 16–3
Problem Priority Matrix
Lowest number = highest priority

Department A, but also where in the machine the problem occurs (i.e., in the electrical system, in the hydraulic system, etc.).

■ How much? What is the extent of the problem? Are there defective parts in every 50 produced? Are there two breakdowns per shift? This question should be answered in quantifiable terms whenever possible.

Identifying Causes

Identifying causes is a critical step in the process. It involves the pairing off of causes and effects. *Effects* are the problems that have already been identified. Say that one such problem has been targeted for solving. An effective tool for isolating the causes of this problem is the cause-and-effect, or fishbone, diagram introduced in Chapter 15.

A fishbone diagram is illustrated in Figure 16–4. The six spines on this particular diagram represent the six major groupings of causes: manpower (personnel), method,

Figure 16–4
Sample Cause-and-Effect
Diagram

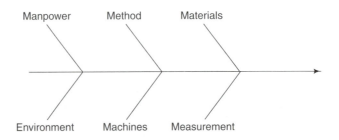

QUALITY TIP

Shortcoming of Perry Johnson Method

A shortcoming of the Perry Johnson method is that it does not focus on eliminating root causes. The method is useful for many types of problems, but the user should always ensure that the root cause of the problem is addressed, or the problem will not be solved—only changed.

Source: Goetsch and Davis

materials, machines (equipment), measurement, and environment. All causes of workplace problems fall into one of these major groupings. Using the diagram, team members brainstorm causes under each grouping. For example, under the machine grouping, a cause might be insufficient maintenance. Under the manpower grouping, a cause might be insufficient training.

Isolating the Most Probable Cause

The causes identified may or may not be the specific causes of the problem in question. To isolate the most probable cause or causes, each cause identified on the fishbone diagram is compared against the problem specifications developed earlier (i.e., who, what, when, where, how much).

When comparing potential causes and problem specifications for each cause, you will have three possibilities: the cause will fully explain the specification, the cause will partially explain the specification, or the cause will not explain the specification. A cause that can fully explain the specifications is a likely candidate to be the most probable cause. If more than one cause fully explains the specifications, there may be more than one cause to the problem.

QUALITY TIP

Separating Problems and Symptoms

"To narrow our problem list, the first thing we do is determine which items on our list are really problems and which are not. Sometimes, what we think at first glance to be problems are actually 'symptoms' of problems."

Source: Perry L. Johnson, Rob Kantner, and Marcia A. Kikora, *TQM Team-Building and Problem-Solving* (Southfield, MI: Perry Johnson, Inc., 1990), 3-3.

Finding the Optimum Solution

With the problem and its most probable cause identified, the next step is to find the optimum solution. Note: What Perry Johnson refers to as a "problem solution" is really the desired result of the solution. The first task in this step is to develop a solution definition that clearly explains the effect the solution is to have. A solution definition should be the opposite of the problem definition. Figure 16–5 contains examples of problem definitions and their corresponding solution definitions.

With the definition in place, the team brainstorms possible solutions and creates a list. Perry Johnson, Inc., has developed a tool known as SCAMPER that can improve the team's effectiveness in building a solutions list. It is explained as follows:[4]

- *Substituting.* Can the problem be solved by substitution? Can a new process be substituted for the old? Can one employee be substituted for another? Can a new material be substituted for the old?

- *Combining.* Can the problem be solved by combining two or more tasks, processes, activities, operations, or other elements?

- *Adapting.* Can the problem be solved by adapting an employee, a process, a product, or some other element to another purpose?

- *Modifying.* Can the problem be solved by modifying a process, job description, design, or something else?

- *Putting to other uses.* Can the problem be solved by putting a resource to other uses?

- *Eliminating.* Can the problem be solved by eliminating a position, part, process, machine, product, or something else?

- *Replacing.* Can the problem be solved by replacing an individual, part, process, machine, product, or something else?

With the solutions list developed, the next step is to identify the optimum entry on the list. One way to do this is by group consensus. A more objective approach is to undertake a cost-to-benefit analysis. This involves setting up a cost–benefit matrix. The matrix should contain cost categories and show the actual dollar costs in each category for each potential solution. The matrix is repeated from the perspective of benefits. Then, the total cost of each solution is compared against the total benefit to derive a cost-to-benefit ratio for each potential solution.

Problem Definition	**Problem Solution (Desired Result)**
• Job 21 A is over budget by 22%.	• Job 21A must be brought under budget by 2%.
• Reject rate of the machining department is too high by 12%.	• Reject rate of the machining department must be brought to 0%.

Figure 16–5
Problem Definitions and Problem Solutions

Figure 16–6 shows a cost–benefit matrix based on three proposed solutions to the following problem: the plastic coating unit is backlogged by 2 months. The three potential solutions are as follows:

■ Purchase two additional plastic coating machines.

■ Retrofit existing plastic coating machine with computer controls and automate the process.

■ Add a second shift and run the plastic coating unit an additional 8-hour shift each day.

The cost–benefit matrix in Figure 16–6 shows the total cost for each potential solution. The costs range from a low of $60,200 to a high of $196,300. The dollar value in benefits reveals that adding a second shift will produce costs that outweigh the benefits. The other two options are both feasible. However, the retrofit and automate option has a better cost-to-benefit ratio. With this option, every dollar spent will produce $2.23 in benefits. Based on the cost-to-benefit analysis, the team would recommend retrofitting and automating the existing process.

Implementing the Optimum Solution

The implementation phase of the process is critical. If handled properly, the problem will be solved in a way that results in an improvement to the process in question. However, if implementation is not handled properly, new and even more serious problems can be created.

The key to effective implementation of a solution is to take a systematic approach. Johnson recommends developing an action plan with the following components:

■ Actions to be taken

■ Methods for taking each action

Categories of Cost/Benefits	Purchase New Machines	Retrofit/ Automate	Add a Second Shift
COSTS			
Personnel Costs	$ -0-	$ -0-	$ 180,000
Equipment Costs	95,000	26,000	-0-
Downtime Costs	-0-	11,400	-0-
Installation Costs	15,200	7,600	-0-
Training Costs	-0-	15,200	16,300
	$ 110,200	$ 60,200	$ 196,300
BENEFITS			
Reduced Overhead	$ 27,500	$ 27,500	$ 27,500
Elimination of Late Fees	32,000	32,000	32,000
Elimination of Cancelled Orders	75,000	75,000	75,000
	$ 134,500	$ 134,500	$ 134,500
COST-TO-BENEFIT RATIO	1 to 2.2	1 to 2.23	1.46 to 1

Figure 16–6
Cost–Benefit Matrix

Action to be Taken	Method	Resources Required	Special Needs	Person Responsible	Deadline
Order replacement parts	Go to purchasing office and work with L. Smith	Approximately $1,500	None	Bao Du Vo	Tuesday the 8th
Prepare machine for repair	Remove old part and clean up	Basic toolbox	None	Louise Crockett	Wednesday the 9th
Schedule repair personnel	Call maintenance supervisor, M. Washington	None	None	Jose Ortega	Tuesday the 8th

Figure 16–7
Action Plan Matrix

- Resources needed for each action
- Special needs for each action
- Person responsible for each action
- Deadline for each action

Ease of implementation can be enhanced by building an action plan matrix that commits the plan to one single sheet of paper whenever possible (see Figure 16–7). Before implementing the action plan, it is important to gain the support of any individual or group that might be affected by implementation. For example, the action plan in Figure 16–7 requires the expenditure of $1,500 for replacement parts and the time of maintenance personnel.

Unless the team leader has the authority to spend $1,500, he or she will need to gain the support of someone who does. In addition, the maintenance manager should be consulted to ensure that his or her crew is available when needed. Any other individual or department

QUALITY TIP

Solve Problems without Creating New Ones

"Finding the solution is terrific. But if we implement it haphazardly, we could get into bigger trouble than we were in before. The solution may only partially solve the problem. Even worse, if poorly implemented, it may trigger all kinds of new problems."

Source: Perry L. Johnson, Rob Kantner, and Marcia A. Kikora, *TQM Team-Building and Problem-Solving* (Southfield, MI: Perry Johnson, Inc., 1990), 10-1.

that will be affected in any way by action taken to implement the solution should be brought "into the loop" before the plan is implemented. Finally, after the solution has been implemented, the results should be monitored and adjustments made as necessary.

PROBLEM-SOLVING AND DECISION-MAKING TOOLS

The models presented in the previous section can help problem-solving teams make better decisions provided that the decisions are based on facts. Decisions based on information that is inaccurate or tainted by personal opinions, exaggeration, or personal agendas are not likely to be good decisions regardless of the problem-solving model used. The information collection step of the Perry Johnson model can be made more effective through the use of the quality tools introduced in Chapter 15.

Organizational decisions can no longer be made the way we have been making them for the last 100 years. Today's business decisions cannot be made without sufficient knowledge of all the relevant factors, which often means that the collective knowledge of the organization must be tapped. At the least, we must be smart in our decision making, or we may find ourselves on the path to ruin.

DECISION MAKING FOR TOTAL QUALITY

All people make decisions. Some are minor. (What should I wear to work today? What should I have for breakfast?) Some are major. (Should I accept a job offer in another city? Should I buy a new house?) Regardless of the nature of the decision, decision making can be defined as follows:

> *Decision making* is the process of selecting one course of action from among two or more alternatives.

Decision making is a critical task in a total quality setting. Decisions play the same role in an organization that fuel and oil play in an automobile engine: they keep it running. The work of an organization cannot proceed until decisions are made.

Consider the following example. Because a machine is down, the production department at DataTech, Inc., has fallen behind schedule. With this machine down, DataTech cannot complete an important contract on time without scheduling at least 75 hours of overtime. The production manager faces a dilemma. On the one hand, no overtime was budgeted for the project. On the other hand, there is substantial pressure to complete this contract on time because future contracts with this client may depend on it. The manager must make a decision.

In this case, as in all such situations, it is important to make the right decision. But how do managers know when they have made the right decision? In most cases, there is no single right choice. If there were, decision making would be easy. Typically several alternatives exist, each with its own advantages and disadvantages.

For example, in the case of DataTech, Inc., the manager had two alternatives: authorize 75 hours of unbudgeted overtime or risk losing future contracts. If the man-

QUALITY TIP

Decision Making in Today's Business Environment

"Today's business environment:

- Demands more large-scale change via new strategies, reengineering, restructuring, mergers, acquisitions, downsizing, new product or market development, and so forth.

Decisions made inside the firm:

- Are based on bigger, more complex, more emotionally charged issues made more quickly.
- Are made in a less certain environment.
- Require more sacrifice from those implementing the decisions.

A new decision-making process:

- Is required because no one individual has the information needed to make all major decisions or the time and credibility needed to convince lots of people to implement the decisions.
- Must be guided by a powerful coalition that can act as a team."

Source: John P. Kotter, *Leading Change* (Boston: Harvard Business School Press, 1996), 56.

ager authorizes the overtime, his or her company's profit for the project in question will suffer, but its relationship with a client may be protected. If the manager refuses to authorize the overtime, the company's profit on this project will be protected, but the relationship with this client may be damaged. These and other types of decisions must be made all the time in the modern workplace.

Managers should be prepared to have their decisions evaluated and even criticized after the fact. Although it may seem unfair to conduct a retrospective critique of decisions that were made during the "heat of battle," having one's decisions evaluated is part of accountability, and it can be an effective way to improve a manager's decision-making skills.

Evaluating Decisions

There are two ways to evaluate decisions. The first is to examine the results. In every case when a decision must be made, there is a corresponding result. That result should advance an organization toward the accomplishment of its goals. To the extent that it does, the decision is usually considered a good decision. Managers have traditionally had their decisions evaluated based on results. However, this is not the only way that decisions should be evaluated. Regardless of results, it is wise also to evaluate the process used in making a decision. Positive results can cause a manager to overlook the fact that a faulty process was used, and, in the long run, a faulty process will lead to negative results more frequently than to positive ones.

For example, suppose a manager must choose from among five alternatives. Rather than collect as much information as possible about each, weigh the advantages and disadvantages of each, and solicit informed input, the manager chooses randomly. He or she has one chance in five of choosing the best alternative. Such odds occasionally produce a positive result, but typically they don't. This is why it is important to examine the process as well as the result, not just when the result is negative but also when it is positive.

THE DECISION-MAKING PROCESS

Decision making is a process. For the purpose of this textbook, the decision-making process is defined as follows:

> The decision-making process is a logically sequenced series of activities through which decisions are made.

Numerous decision-making models exist. Although they appear to have major differences, all involve the various steps shown in Figure 16–8 and discussed next.

Identify or Anticipate the Problem

If managers can anticipate problems, they may be able to prevent them. Anticipating problems is like driving defensively: never assume anything. Look, listen, ask, and sense. For example, if you hear through the grapevine that a team member's child has been severely injured and hospitalized, you can anticipate the problems that may occur. She is likely to be absent, or if she does come to work, her pace may be slowed. The

QUALITY TIP

Decisions Must Be Made Continually

"During conduct of operations certain cardinal decisions must be made, over and over again:

- The process—Should it run or stop?
- The resulting product—Does it conform to goals?
- Nonconforming product—What disposition should be made?"

Source: Joseph M. Juran, *Juran on Leadership for Quality: An Executive Handbook* (New York: Free Press, 1989), 163.

Figure 16–8
Decision-Making Model

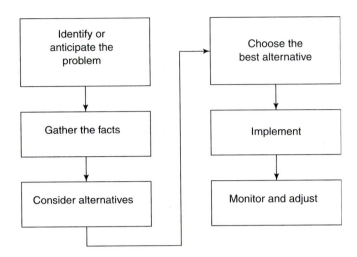

better managers know their employees, technological systems, products, and processes, the better able they will be to anticipate problems.

Gather the Facts

Even the most perceptive managers will be unable to anticipate all problems or to understand intuitively what is behind them. For example, suppose a manager notices a "who cares?" attitude among team members. This manager might identify the problem as poor morale and begin trying to improve it. However, he or she would do well to gather the facts first to be certain of what is behind the negative attitudes. The underlying cause(s) could come from a wide range of possibilities: an unpopular management policy, dissatisfaction with the team leader, a process that is ineffective, problems at home, and so forth. Using the methods and tools described earlier in this chapter and in Chapter 15, the manager should separate causes from symptoms and determine the root cause of the problem. Only by doing so will the problem be permanently resolved. The inclusion of this step makes possible *management by facts*—a cornerstone of the total quality philosophy.

It should be noted that the factors that might be at the heart of a problem include not only those for which a manager is responsible (policies, processes, tools, training, personnel assignment, etc.) but possibly also ones beyond the manager's control (personal matters, regulatory requirements, market and economic influences, etc.). For those falling within the manager's domain of authority, he or she must make sound, informed decisions based on fact. For the others, the organization has to adapt.

Consider Alternatives

Considering the alternatives involves two steps: (a) list all of the various alternatives available, and (b) evaluate each alternative in light of the facts. The number of alternatives identified in the first step will be limited by several factors. Practical considerations, the manager's range of authority, and the cause of the problem will all limit a manager's list of alternatives. After the list has been developed, each entry on it is evaluated. The main criteria against which alternatives are evaluated is the desired outcome. Will the alternative being considered solve the problem? If so, at what cost?

Cost is another criteria used in evaluating alternatives. Alternatives always come with costs, which might be expressed in financial terms, in terms of employee morale, in terms of the organization's image, or in terms of a client's goodwill. Such costs should be considered when evaluating alternatives. In addition to applying objective criteria and factual data, managers will also need to apply their judgment and experience when considering alternatives.

Choose the Best Alternative, Implement, Monitor, and Adjust

After all alternatives have been considered, one must be selected and implemented, and after an alternative has been implemented, managers must monitor progress and adjust appropriately. Is the alternative having the desired effect? If not, what adjustments should be made? Selecting the best alternative is never a completely objective process. It requires study, logic, reason, experience, and even intuition. Occasionally, the alternative chosen for implementation will not produce the desired results. When this happens and adjustments are not sufficient, it is important for managers to cut their losses and move on to another alternative.

Managers should avoid falling into the ownership trap. This happens when they invest so much ownership in a given alternative that they refuse to change even when it becomes clear the idea is not working. This can happen at any time but is more likely when a manager selects an alternative that runs counter to the advice he or she has received, is unconventional, or is unpopular. The manager's job is to solve the problem. Showing too much ownership in a given alternative can impede the ability to do so.

OBJECTIVE VERSUS SUBJECTIVE DECISION MAKING

All approaches to decision making fall into one of two categories: objective or subjective. Although the approach used by managers in a total quality setting may have characteristics of both, the goal is to minimize subjectivity and maximize objectivity. The approach most likely to result in a quality decision is the objective approach.

Objective Decision Making

The objective approach is logical and orderly. It proceeds in a step-by-step manner and assumes that managers have the time to systematically pursue all steps in the decision-making process (see Figure 16–9). It also assumes that complete and accurate information is available and that managers are free to select what they feel is the best alternative.

Measured against these assumptions, it can be difficult to be completely objective when making decisions. Managers don't always have the luxury of time and complete

Figure 16–9
Factors That Contribute to
Objective Decision Making

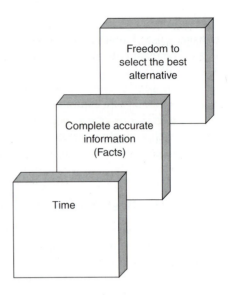

Freedom to
select the best
alternative

Complete accurate
information
(Facts)

Time

information. This does not mean that objectivity in decision making should be considered impossible. Managers should be as objective as possible. However, it is important to understand that the day-to-day realities of the workplace may limit the amount of time and information available. When this is the case, objectivity can be affected.

Subjective Decision Making

Whereas objective decision making is based on logic and complete, accurate information, subjective decision making is based on intuition, experience, and incomplete information. This approach assumes decision makers will be under pressure, short on time, and operating with only limited information. The goal of subjective decision making is to make the best decision possible under the circumstances. In using this approach, the danger always exists that managers might make quick, knee-jerk decisions based on no information and no input from other sources. The subjective approach does not give managers license to make sloppy decisions. If time is short, the little time available should be used to list and evaluate alternatives. If information is incomplete, use as much information as is available. Subjective decision making is an anathema in the total quality context, and it should be avoided whenever possible.

SCIENTIFIC DECISION MAKING AND PROBLEM SOLVING

As explained in the previous section, sometimes decisions must be made subjectively. However, through good management and leadership, such instances should and can be held to a minimum. One of the keys to success in a total quality setting is using a scientific approach in making decisions and solving problems.[5] One method is to use Joseph M.

Juran's 85/15 rule. Decision makers in a total quality setting should understand this rule. It is one of the fundamental premises underlying the need for scientific decision making.

Peter R. Scholtes explains the rationale for scientific decision making as follows:

> The core of quality improvement methods is summed up in two words: scientific approach. Though this may sound complicated, a scientific approach is really just a systematic way for individuals and teams to learn about processes. It means agreeing to make decisions based on data rather than hunches, to look for root causes of problems rather than react to superficial symptoms, to seek permanent solutions rather than rely on quick fixes. A scientific approach can, but does not always, involve using sophisticated statistics, formulas, and experiments. These tools enable us to go beyond Band-aid methods that merely cover up problems to find permanent, upstream improvements.[6]

Complexity and the Scientific Approach

In the language of scientific decision making, complexity means nonproductive, unnecessary work that results when organizations try to improve their processes without first developing a systematic plan.[7] Several different types of complexity exist, including the following: errors and defects, breakdowns and delays, inefficiencies, and variation. The Pareto Principle, explained in an earlier chapter, should be kept in mind when attempting to apply the scientific approach.

Errors and Defects

Errors cause defects and defects reduce competitiveness. When a defect occurs, one of two things must happen: the part or product must be scrapped altogether, or extra work must be done to correct the defect. Waste or extra work that results from errors and defects adds cost to the product without adding value.

QUALITY TIP

Juran's 85/15 Rule

"There is a widely held belief that an organization would have few, if any, problems if only workers would do their jobs correctly. As Dr. Joseph M. Juran pointed out years ago, this belief is incorrect. In fact, the potential to eliminate mistakes and errors lies mostly in improving the 'systems' through which work is done, not in changing the workers. This observation has evolved into the rule of thumb that at least 85% of problems can only be corrected by changing systems (which are largely determined by management) and less than 15% are under a worker's control—and the split may lean even more towards the system."

Source: Peter R. Scholtes, *The Team Handbook* (Madison, WI: Joiner Associates, 1992), 2-8.

Breakdowns and Delays

Equipment breakdowns delay work, causing production personnel either to work over-time or to work faster to catch up. Overtime adds cost to the product without adding value. When this happens, the organization's competitors gain an unearned competi-tive advantage. When attempts are made to run a process faster than its optimum rate, an increase in errors is inevitable.

Inefficiency

Inefficiency means using more resources (time, material, movement, or something else) than necessary to accomplish a task. Inefficiency often occurs because organizations fall into the habit of doing things the way they have always been done without ever asking why.

Variation

In a total quality setting, consistency and predictability are important. When a process runs consistently, efforts can begin to improve it by reducing process variations, of which there are two kinds:

- *Common-cause variation* is the result of the sum of numerous small sources of nat-ural variation that are always part of the process.
- *Special-cause variation* is the result of factors that are not part of the process and that occur only in special circumstances, such as a shipment of faulty raw material or the involvement of a new, untrained operator.

The performance of a process that operates consistently can be recorded and plot-ted on a control chart such as the one in Figure 16–10. The sources of the variation in this figure that fall within the control limits are likely to be common causes. The sources of variation in this figure that fall outside the control limits are likely to be special-causes sources. In making decisions about the process in question, it is impor-tant to separate common and special causes of variation.

Commenting on variation, Scholtes says:

QUALITY TIP

The Pareto Principle

"This principle is sometimes called the 80/20 rule: 80% of the trouble comes from 20% of the problems. Though named for turn-of-the-century economist Vilfredo Pareto, it was Dr. Juran who applied the idea to management. Dr. Juran advises us to concentrate on the 'vital few' sources of problems and not be distracted by those of lesser importance."

Source: Peter R. Scholtes, *The Team Handbook* (Madison, WI: Joiner Associates, 1992), 2-9.

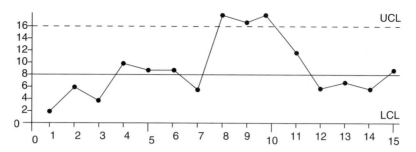

Figure 16–10
Control Chart

> If you react to common-cause variation as if it were due to special causes, you will only make matters worse and increase variation, defects, and mistakes. If you fail to notice the appearance of a special cause, you will miss an opportunity to search out and eliminate a source of problems.[8]

The concept of using control charts and statistical data in decision making is discussed in greater depth in Chapter 18.

EMPLOYEE INVOLVEMENT IN DECISION MAKING

Chapter 8 showed how employee involvement and empowerment can improve decision making. Employees are more likely to show ownership in a decision they had a part in making. Correspondingly, they are more likely to support a decision for which they feel ownership. There are many advantages to be gained from involving employees in decision making, as was shown in Chapter 8. There are also factors that, if not understood and properly handled, can lead to problems.

Advantages of Employee Involvement

Involving employees in decision making can have a number of advantages. It can result in a more accurate picture of what the problem really is and a more comprehensive list of potential solutions. It can help managers do a better job of evaluating alternatives and selecting the best one to implement.

Perhaps the most important advantages are gained after the decision is made. Employees who participate in the decision-making process are more likely to understand and accept the decision and have a personal stake in making sure the alternative selected succeeds.

Potential Problems with Employee Involvement

Involving employees in decision making can lead to problems. The major potential problem is that it takes time, and managers do not always have time. Other potential difficulties are that it takes employees away from their jobs and that it can result in conflict among team members. Next to time, the most significant potential problem is that employee involvement can lead to democratic compromises that do not necessarily represent the best decision. In addition, disharmony can result when a decision maker rejects the advice of the group.

Nevertheless, if care is taken, managers can gain all of the advantages while avoiding the potential disadvantages associated with employee involvement in decision making. Several techniques are available to help increase the effectiveness of group involvement. Prominent among these are brainstorming, the nominal group technique (NGT), and quality circles. Be particularly wary of the dangers of *groupthink* and *groupshift* in group decision making, as outlined in Chapter 8.

ROLE OF INFORMATION IN DECISION MAKING

Information is a critical element in decision making. Although having accurate, up-to-date, comprehensive information does not guarantee a good decision, lacking such information can guarantee a bad one. The old saying, that "knowledge is power" applies in decision making—particularly in a competitive situation. To make decisions that will help their organizations be competitive, managers need timely, accurate information.

Information can be defined as data that have been converted into a usable format that is relevant to the decision-making process.

Data that are relevant to decision making are those that might have an impact on the decision. Communication is a process that requires a sender, a medium, and a receiver. In this process, information is what is provided by the sender, transmitted by the medium, and received by the receiver. For the purpose of this chapter, decision makers are receivers of information who base decisions at least in part on what they receive.

Advances in technology have ensured that the modern manager can have instant access to information. Computers and telecommunications technology give decision makers a mechanism for collecting, storing, processing, and communicating information quickly and easily. The quality of the information depends on people (or machines) receiving accurate data, entering it into technological systems, and updating it continually. This dependence on accurate information gave rise to the expression "garbage in/garbage out" that is now associated with computer-based information systems. The saying means that information provided by a computer-based system can be no better than the data put into the system.

Data versus Information

Data for one person may be information for another. The difference is in the needs of the individual. Managers' needs are dictated by the types of decisions they make. For example, a computer printout listing speed and feed rates for a company's machine tools would contain valuable information for the production manager; the same printout would be just data to the warehouse manager. In deciding on the type of information they need, decision makers should ask themselves these questions:

- What are my responsibilities?
- What are my organizational goals?
- What types of decisions do I have to make relative to these responsibilities and goals?

Value of Information

Information is a useful commodity. As such it has value. Its value is determined by the needs of the people who will use it and the extent to which the information will help them meet their needs. Information also has a cost. Because it must be collected, stored, processed, continually updated, and presented in a usable format when needed, information can be expensive. This fact requires managers to weigh the value of information against its cost when deciding what information they need to make decisions. It makes no sense to spend $100 on information to help make a $10 decision.

Amount of Information

An old saying holds that a manager can't have too much information. This is no longer true. With advances in information technologies, not only can managers have too much information, they frequently do. This phenomenon has come to be known as *information overload*, the condition that exists when people receive more information than they can process in a timely manner. The phrase "in a timely manner" means in time to be useful in decision making (see Figure 16–11).

Figure 16–11
Information Overload

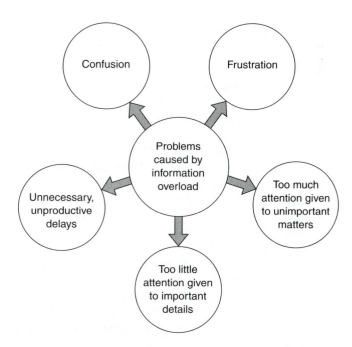

To avoid information overload, apply a few simple strategies. First, examine all regular reports received. Are they really necessary? Do you receive daily or weekly reports that would meet your needs just as well if provided on a monthly basis? Do you receive regular reports that would meet your needs better as exception reports? In other words, would you rather receive reports every day that say everything is all right or occasional reports when there is a problem? The latter approach *is reporting by exception* and can cut down significantly on the amount of information that managers must absorb.

Another strategy for avoiding information overload is formatting for efficiency. This involves working with personnel who provide information. If your organization has a management information systems (MIS) department, ensure that reports are formatted for your convenience rather than theirs. (MIS will be discussed in the next section) Decision makers should not have to wade through reams of computer printouts to locate the information they need. Nor should they have to become bleary-eyed reading rows and columns of tiny figures. Work with MIS personnel to develop an efficient report form that meets your needs. Also, have that information presented graphically whenever possible.

Finally, make use of on-line, on-demand information retrieval. In the modern workplace, most reports are computer generated. Rather than relying on periodic printed reports, learn to retrieve information from the MIS database when you need it (on demand) using a computer terminal or a networked personal computer (on-line).

USING MANAGEMENT INFORMATION SYSTEMS (MIS)

The previous section contained references to management information systems (MIS) and MIS personnel.

A management information system (MIS) is a system used to collect, store, process, and present information used by managers in decision making.

In the modern workplace, a management information system is typically a computer-based system. A management information system has three major components: hardware, software, and people. *Hardware* consists of the computer—be it a mainframe, mini-, or microcomputer—all of the peripheral devices for interaction with the computer, and output devices such as printers and plotters.

Software is the component that allows the computer to perform specific operations and process data. It consists primarily of computer programs but also includes the database, files, and manuals that explain operating procedures. *Systems software* controls the basic operation of the system. *Applications software* controls the processing of data for specific computer applications (word processing, databases, computer-assisted planning, spreadsheets, etc.).

A *database* is a broad collection of data from which specific information can be drawn. For example, a company might have a personnel database in which many different items of information about its employees are stored. From this database can be drawn a variety of different reports—such as printouts of all employees in order of employment date, by job classification, or by ZIP code. Data are kept on computer disks or tape on which they are stored under specific groupings or file names.

The most important component is the people component. It consists of the people who manage, operate, maintain, and use the system. Managers who depend on a management information system for part of the information needed to make decisions are *users*.

Managers should not view a management information system as the final word in information. Such systems can do an outstanding job of providing information about predictable matters that are routine in nature. However, many of the decisions managers have to make concern problems that are not predictable and for which data are not tracked. For this reason, it is important to have sources other than the management information system from which to draw information.

CREATIVITY IN DECISION MAKING

The increasing pressures of a competitive marketplace are making it increasingly important for organizations to be flexible, innovative, and creative in decision making. To survive in an unsure, rapidly changing marketplace, organizations must be able to adjust rapidly and change directions quickly. To do so requires creativity at all levels of the organization.

Creativity Defined

Like leadership, creativity has many definitions, and viewpoints vary about whether creative people are born or made. For the purposes of modern organizations, creativity can be viewed as an approach to problem solving and decision making that is imaginative, original, and innovative. Developing such perspectives requires that decision makers have knowledge and experience regarding the issue in question.

Creative Process

According to H. Von Oech, the creative process proceeds in four stages: preparation, incubation, insight, and verification.[9] What takes place in each of these stages is summarized as follows:

- *Preparation* involves learning, gaining experience, and collecting/storing information in a given area. Creative decision making requires that the people involved be prepared.
- *Incubation* involves giving ideas time to develop, change, grow, and solidify. Ideas incubate while decision makers get away from the issue in question and give the mind time to sort things out. Incubation is often a function of the subconscious mind.
- *Insight* follows incubation. It is the point in time when a potential solution falls in place and becomes clear to decision makers. This point is sometimes seen as a moment of inspiration. However, inspiration rarely occurs without having been preceded by perspiration, preparation, and incubation.
- *Verification* involves reviewing the decision to determine whether it will actually work. At this point, traditional processes such as feasibility studies and cost–benefit analyses are used.

Factors That Inhibit Creativity

A number of factors can inhibit creativity. Some of the more prominent of these are:[10]

- *Looking for just one right answer.*   Seldom is there just one right solution to a problem.
- *Focusing too intently on being logical.*   Creative solutions sometimes defy logic and conventional wisdom.
- *Avoiding ambiguity.*   Ambiguity is a normal part of the creative process. This is why the incubation step is so important.
- *Avoiding risk.*   When organizations don't seem to be able to find a solution to a problem, it often means decision makers are not willing to give an idea a chance.
- *Forgetting how to play.*   Adults sometimes become so serious they forget how to play. Playful activity can stimulate creative ideas.
- *Fear of rejection or looking foolish.*   Nobody likes to look foolish or feel rejection. This fear can cause people to hold back what might be creative solutions.
- *Saying "I'm not creative."*   People who decide they are not creative won't be. Any person can think creatively and can learn to be even more creative.

Helping People Think Creatively

In the age of high technology and global competition, creativity in decision making and problem solving is critical. Although it is true that some people are naturally more creative than others, it is also true that any person can learn to think creatively. In the modern workplace, the more people who think creatively, the better. Darrell W. Ray and Barbara L. Wiley recommend the following strategies for helping employees think creatively:[11]

- *Idea vending.*   This is a facilitation strategy. It involves reviewing literature in the field in question and compiling files of ideas contained in the literature. Periodically, circulate these ideas among employees as a way to get people thinking. This will facilitate the development of new ideas by the employees. Such an approach is sometimes called *stirring the pot.*
- *Listening.*   One of the factors that causes good ideas to fall by the wayside is poor listening. Managers who are perpetually too hurried to listen to employees' ideas do not promote creative thinking. On the contrary, such managers stifle creativity. In addition to listening to the ideas, good and bad, of employees, managers should listen to the problems employees discuss in the workplace. Each problem is grist for the creativity mill.
- *Idea attribution.*   A manager can promote creative thinking by subtly feeding pieces of ideas to employees and encouraging them to develop the idea fully. When an employee develops a creative idea, he or she gets full attribution and recognition for the idea. Time may be required before this strategy pays off, but with patience and persistence it can help employees become creative thinkers.

How does a football team that is no better than its opponent beat that opponent? Often, the key is more creative game planning, play calling, and defense. This phenomenon also occurs in the workplace every day. The organization that wins the competition in the marketplace is often the one that is the most creative in decision making and problem solving.

 SUMMARY

1. *Decision making* is the process of selecting one course of action from among two or more alternatives. Decisions should be evaluated not just by results but also by the process used to make them.

2. A *problem* is a situation in which what exists does not match what is desired or, put another way, the discrepancy between the current and the desired state of affairs. Problem solving in a total quality setting is not about putting out fires. It is about continual improvement. Two effective problem-solving models are the Deming Cycle and the Perry Johnson method.

3. Securing reliable information is an important part of problem solving and decision making. W. Edwards Deming recommends the use of the following tools: cause-and-effect diagrams, flowcharts, Pareto charts, run charts, histograms, control charts, and scatter diagrams.

4. The decision-making process is a logically sequenced series of activities through which decisions are made. These activities include identifying or anticipating the problem; gathering relevant facts; considering alternative solutions; choosing the best alternative; and implementing, monitoring, and adjusting. All approaches to decision making are objective, subjective, or a combination of the two.

5. Scientific decision making is a systematic way to learn about processes. It means making decisions based on data rather than on hunches. Complexity in scientific decision making means nonproductive, unnecessary work that results when organizations try to improve processes in a haphazard, nonscientific way. The different types of complexity include errors and defaults, breakdowns and delays, inefficiencies, and variation.

6. There are advantages and disadvantages to employee involvement. Techniques to enhance group decision making are brainstorming, NGT, and quality circles. Managers should be prepared to counteract groupshift and groupthink.

7. *Information* is data that have been converted into a usable format that is relevant to the decision-making process. Decision makers are receivers of information who base decisions in whole or in part on what they receive. Technological developments have introduced the potential for information overload, or the condition that exists when people receive more information than they can process in a timely manner.

8. A management information system is a system used to collect, store, process, and present information. Such a system has three components: hardware, software, and people. A management information system can do an outstanding job of providing information about predictable and routine matters. However, many decisions that managers have to make concern problems that are not predictable and for which no data are tracked.

9. *Creativity* is an approach to problem solving and decision making that is imaginative, original, and innovative. The creative process proceeds in four stages: preparation,

incubation, insight, and verification. Factors that inhibit creativity include looking for just one right answer, being too logical, avoiding ambiguity, avoiding risk, forgetting how to play, fearing rejection, and saying "I'm not creative." Three strategies for helping people think creatively are idea vending, listening, and idea attribution.

 KEY TERMS AND CONCEPTS

Brainstorming

Cause-and-effect (fishbone) diagram

Complexity

Control chart

Creativity

Decision making

Decision-making process

Deming Cycle

Flowchart

Groupshift

Groupthink

Histogram

Idea attribution

Idea vending

Incubation

Insight

Juran's 85/15 rule

Management by facts

Management Information Systems (MIS)

Nominal Group Technique (NGT)

Objective decision making

Pareto chart

PDCA cycle

Perry Johnson method

Preparation

Problem-solving team

Quality circles

Run chart

SCAMPER

Scatter diagram

Subjective decision making

Variation

Verification

Why-Why

 FACTUAL REVIEW QUESTIONS

1. Define *decision making* as it relates to total quality.
2. Explain how to evaluate decisions in a total quality environment.
3. Describe the PDCA cycle.
4. Describe the Perry Johnson method for problem solving.
5. Name and describe three problem-solving tools.
6. Define the decision-making process and explain each step in it.
7. Contrast and compare objective and subjective decision making.
8. What is Juran's 85/15 rule? State the rule.
9. Describe the scientific approach to decision making and problem solving.

10. Explain four types of complexity in the scientific approach.
11. What are the advantages and disadvantages of employee involvement in decision making?
12. Explain three techniques for increasing the effectiveness of group involvement.
13. What is the role of information in decision making?
14. Explain *creativity* as a concept and the role it can play in decision making.

 CRITICAL THINKING ACTIVITY

Using Data for Problem Solving

Rod Simmons is a civil service employee, managing a state Department of Motor Vehicles office. The DMV office issues annual auto registrations, renews driver's licenses, and conducts driver examinations. The office is organized into departments, each with a supervisor reporting to Rod. The Vehicle Registration Department has eight clerks. The Motor Vehicle Operator License Department is staffed by four clerks and four examiners.

Rod has become increasingly concerned because documented taxpayer complaints directed at his office have been increasing over the past year. This has recently come to the attention of his superiors at the state capitol, who have begun putting pressure on him to stop the complaints—somehow. Rod is not sure at this point how to go about it. He knows that it would be better to look at the available data to try to find the cause, or causes, rather than just giving a "pep talk" to his employees. He starts by assembling the following:

- The office handles about 35,000 license/registration transactions annually.
- Workforce history:

  Nine of the 12 clerks have been with the office 5 years or more. Three were added during the last year.

  Two of the examiners have over 5 years' tenure, while the other two have less than 1 year.

- Office work schedule and hours available:

  Office hours are 8:00 A.M. until 5:00 P.M. with one hour for lunch, Monday through Friday.

  Lunch hours are staggered for clerks and examiners, half from 11:00 A.M. until noon, and the other half from noon until 1:00 P.M. (By doing this the office is open for business 9 hours per day.)

  Clerk hours available/year: 23,000

  Examiner hours available/year: 7,800

  Supervisor hours available/year: 3,800

- Procedures:

  A driver's license must be renewed every 3 years during the driver's birth month.

  Vehicle registration must be renewed annually within the 12th month of the current registration.

■ Total complaints last 2 years:

Month	Complaints This Year	Complaints Last Year
January	14	6
February	10	7
March	14	5
April	20	10
May	19	9
June	21	10
July	27	13
August	31	16
September	36	19
October	24	13
November	28	14
December	30	12
Totals	274	134

■ Total taxpayer complaints by category:

Berating by examiners	21
Clerks ignoring customers (taxpayers)	35
Incompetence	33
Long waits in line	75
Rudeness	60
All others	50

If you were Rod, how would you use the problem-solving methods and tools covered in this chapter and Chapter 15 to highlight trends, determine how to proceed, and attempt to find root cause(s)?

DISCUSSION ASSIGNMENT 16–1

The Utility Company That Forgot to Ask Why

"Many years ago a utility company installed a vent in a power plant to comply with an environmental regulation. Later, the need for the vent, which allowed steam and heat to escape, disappeared when the company changed fuels and equipment. By then, no one bothered to question having a vent. Recently, a project team in one plant discovered that the vent had outlived its usefulness and could be closed. The company estimates the savings in heat and efficiency to be over $100,000 a year."

Discussion Question

Discuss the following question in class or outside of class with your fellow students:

1. How does this case illustrate the need to constantly review the status quo and ask "why"?

Source: Peter R. Scholtes, *The Team Handbook* (Madison, WI: Joiner Associates, 1992), 2–11.

DISCUSSION ASSIGNMENT 16–2

Soliciting Creative Ideas from Employees

"At an electric motor manufacturing plant, a crew ran out of wax in which to dip armatures. In the past, its supervisor, Brad, had shown an openness to new ideas from his subordinates so they frequently made suggestions about how to increase productivity. As a result, he was always ahead of schedule. This time Brad was faced with having his full second shift sit idle while falling far behind quota. Then one of his employees asked, 'Why couldn't we use the old wax? It doesn't get dirty, and all we do is throw it away.' It was a great idea and only required scraping out the vats and recycling the wax. The crew got through the shift with only an hour of downtime and actually exceeded its quota that evening. The net result was a savings of thousands of dollars in resources and productivity over the next few months."

Discussion Question

Discuss the following question in class or outside of class with your fellow students:

1. What type of things must Brad have done in the past to establish a work environment in which employees feel free to question and suggest innovations?

Source: Darell W. Ray and Barbara L. Wiley, "How to Generate New Ideas," *Management for the 90s: A Special Report from* Supervisory Management (Saranac Lake, NY: American Management Association, 1991), 4.

 ENDNOTES

1. William W. Scherkenbach, *Deming's Road to Continual Improvement* (Knoxville, TN: SPC, 1991), 63–66.
2. Kaoru Ishikawa, *What Is Total Quality Control? The Japanese Way* (Upper Saddle River, NJ: Prentice Hall, 1987), 59.

3. Perry L. Johnson, Rob Kantner, and Marcia A. Kikora, *TQM Team-Building and Problem-Solving* (Southfield, MI: Perry Johnson, Inc., 1990), 1-1–10-15.

4. Johnson et al., 11-2.

5. Joseph M. Juran, *Juran on Leadership for Quality: An Executive Handbook* (New York: Free Press, 1989), 163.

6. Peter R. Scholtes, *The Team Handbook* (Madison, WI: Joiner Associates, 1992), 2-8.

7. This section is based on Scholtes, 2-9–2-15.

8. Scholtes, 2-13.

9. H. Von Oech, *A Whack on the Side of the Head* (New York: Warner, 1983), 77.

10. Von Oech, 77.

11. Darell W. Ray and Barbara L. Wiley, "How to Generate New Ideas," *Management for the 90s: A Special Report from* Supervisory Management (Saranac Lake, NY: American Management Association, 1991), 6–7.

CHAPTER SEVENTEEN

Quality Function Deployment (QFD)

"The business process starts with the customer. In fact, if it is not started with the customer, it all too many times abruptly ends with the customer."
William W. Scherkenbach

MAJOR TOPICS

- What Is QFD?
- Benefits of QFD
- Customer Information: Feedback and Input
- QFD Tools
- Implementing QFD

A consistent theme throughout this book and a fundamental element of total quality is customer-defined quality. Chapter 7 dealt with how organizations can establish a customer focus. Chapter 19 explains several different approaches for achieving continuous improvement. A key element in each approach is customer involvement. Quality function deployment (QFD) is a specialized method for making customers part of the product development cycle. This chapter provides the information you need to understand and implement QFD.

WHAT IS QFD?

One of the keys to achieving continual improvement is getting customers involved as early in the product development process as possible. This is the main focus of QFD. Stephen Uselac defines QFD as follows:

> QFD is a practice for designing your processes in response to customer needs. QFD translates what the customer wants into what the organization produces. It enables an organization to prioritize customer needs, find innovative responses to those needs, and improve processes to maximum effectiveness. QFD is a practice that leads to process improvements that enable an organization to exceed the expectations of the customer.[1]

Historical Development of QFD

QFD was originally developed in Japan and used at the Kobe Shipyard in the 1960s. Its use spread throughout Japan, and it is still widely used there in both manufacturing and service settings. QFD was originally brought to the United States in the mid-1980s by Xerox. It has not yet achieved wide-scale adoption in the United States, but it is being used by both manufacturing organizations (such as Hewlett-Packard) and service organizations (such as St. Clair Hospital in Pittsburgh, Pennsylvania). The use of QFD is growing and will continue to increase as total quality inches closer to becoming the norm.

Structure of QFD

The most widely used analogy for explaining how QFD is structured is a house. Figure 17–1 shows how a basic QFD matrix is put together. The wall of the house on the left (Component 1) is customer input. This is the step in the process where customer requirements relating to the product are determined. Methods for collecting customer input are explained later in this chapter.

To meet customer requirements, the manufacturer works to certain performance specifications and requires suppliers to do the same. This is the ceiling of the house, or Component 2, and not the roof (which is Component 6). One of the questions the QFD process will answer is "Are our current manufacturing requirements sufficient to meet or exceed customer requirements?"

The wall of the house on the right (Component 3) is the planning matrix. This is the component most closely associated with QFD. The planning matrix is the component used to translate customer requirements into plans for meeting or exceeding those requirements. It involves plotting customer requirements on one matrix and manufacturing processes on the other, prioritizing customer requirements, and making decisions concerning improvements needed in manufacturing processes.

The middle of the house (Component 4) is where customer requirements are converted into manufacturing terms. If a customer wants the operating life of your product to

QUALITY TIP

QFD Identifies Customer Requirements

"The thing that makes QFD unique is that the primary focus is customer requirements. The process is driven by what the customer wants, not by innovations in technology. Consequently, more effort is involved in getting the information necessary for determining what the customer truly wants."

Source: James L. Bossert, *Quality Function Deployment: A Practitioner's Approach* (Milwaukee, WI: ASQC Quality Press, 1991), 1.

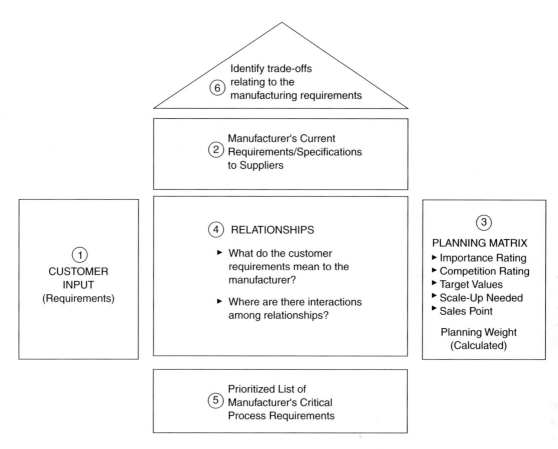

Figure 17–1
QFD Matrix Structure

be 12 months instead of 6, what does this mean in terms of the materials used? The design? Manufacturing processes? These types of questions are answered in this component.

The bottom of the house (Component 5) is where the manufacturer's critical process requirements are prioritized. Which process requirement is most important in terms of meeting or exceeding customer requirements? Which is next, and so on? Each prioritized process requirement is assigned a rating that depicts its level of difficulty or how hard it is to achieve.

The roof of the house (Component 6) is where trade-offs are identified. These are trade-offs relating to manufacturer's requirements. In view of its customer requirements and manufacturing capabilities, what is the best the organization can do? This type of question is answered here. This is the fundamental makeup of a QFD matrix.

QFD Process

Each matrix developed as part of the QFD process should be structured according to the house illustrated in Figure 17–1. There will be six such matrices developed in a complete cycle of the QFD process. Figure 17–2 shows the flow and focus of one complete cycle of the QFD process. The purpose of each matrix is as follows:

■ Matrix 1 is used to compare customer requirements with the related technical features of the product. All other matrices grow out of this first matrix.

■ Matrix 2 is used to compare the technical features in Matrix 1 with their related applied technologies. These two matrices produce the information needed to answer the following questions: (a) What does the customer want? (b) What are the technical requirements relating to features the customer wants? (c) What technologies are needed to meet or exceed customer requirements? (d) What are the trade-offs relating to technical requirements?

■ Matrix 3 is used to compare the applied technologies from Matrix 2 with their related manufacturing processes. This matrix helps identify critical variables in manufacturing processes.

■ Matrix 4 is used to compare the manufacturing processes from Matrix 3 with their related quality control processes (Little Q). This matrix produces the information

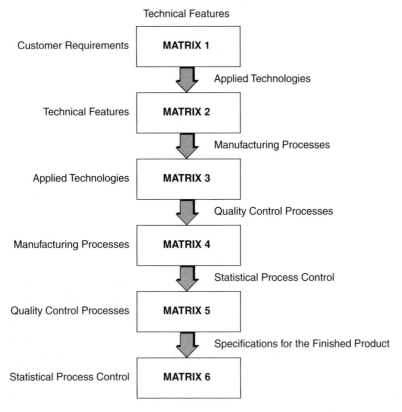

Figure 17–2
QFD Process: One Complete Cycle

needed to optimize processes. Through experimentation, the reliability and repeatability of processes are determined.

■ Matrix 5 is used to compare the quality control processes (Little Q) with their related statistical process control processes. This matrix helps ensure that the proper parameters and process variables are being used.

■ Matrix 6 is used to compare the statistical process control parameters with the specifications that have been developed for the finished product. At this point, adjustments are made to ensure that the product produced is the product the customer wants. The QFD process ensures that all resources are being optimally used in ways that maximize the organization's chances of meeting or exceeding customer requirements.

BENEFITS OF QFD

QFD brings a number of benefits to organizations trying to enhance their competitiveness by continually improving quality and productivity (see Figure 17–3). The process has the benefits of being customer focused, time efficient, teamwork oriented, and documentation oriented. These benefits are explained in the following paragraphs:

■ *Customer focused.* A total quality organization is a customer-focused organization. QFD requires the collection of customer input and feedback. This information is translated into a set of specific customer requirements. The organization's performance against these customer requirements as well as that of competitors is studied carefully. This allows the organization to know how it and the competition compare in meeting customer needs.

■ *Time efficient.* QFD can reduce development time because it focuses on specific and clearly identified customer requirements. Because of this, time is not wasted on developing features that have little or no value to customers.

■ *Teamwork oriented.* QFD is a teamwork-oriented approach. All decisions in the process are based on consensus and involve in-depth discussion and brainstorming. Because all

QUALITY TIP

QFD Identifies Customer Requirements

"This process aids the difficult transition of bringing a product from development to manufacturing. It also brings all the necessary information to manufacturing so that the line operator is capable of running the process as necessary to produce the highest-quality product."

Source: James L. Bossert, *Quality Function Deployment: A Practitioner's Approach* (Milwaukee, WI: ASQC Quality Press, 1991), 41.

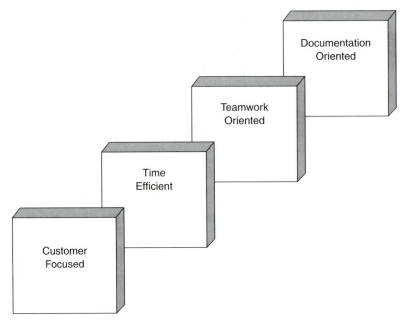

Figure 17–3
Benefits of QFD

actions that need to be undertaken are identified as part of the process, individuals see where they fit into the larger picture, thereby promoting teamwork even more.

■ *Documentation oriented.* QFD forces the issue of documentation. One of the products of the QFD process is a comprehensive document that pulls together all pertinent data about all processes and how they stack up against customer requirements. This document changes constantly as new information is learned and old is discarded. Having up-to-date information about customer requirements and internal processes is particularly helpful when turnover occurs.

QUALITY TIP

Customer Information: Different Types and Different Sources

"Customer information comes from a variety of sources; some are solicited and some are not, some are quantitative or measurable and some are qualitative, some are obtained in a structured manner and some are obtained in a random manner."

Source: James L. Bossert, *Quality Function Deployment: A Practitioner's Approach* (Milwaukee, WI: ASQC Quality Press, 1991), 15.

CUSTOMER INFORMATION: FEEDBACK AND INPUT

Customer information falls into two broad categories: feedback and input. *Feedback* is given after the fact. In a manufacturing setting, this means after a product has been developed, produced, and marketed. Feedback is valuable and should be collected. However, it comes too late in the process to help ensure that customer requirements are met. At best, it can help improve the product when the next batch is produced.

Input is obtained before the fact. In a manufacturing setting, this means during product development. Collecting customer input during product development allows changes to be made before producing, marketing, and distributing large quantities of a product. Collecting input is more valuable than collecting feedback. However, both input and feedback have value. Ideally, broad-based feedback should be used to verify input that is necessarily more narrow. Both types of information can be categorized further according to several characteristics (see Figure 17–4).

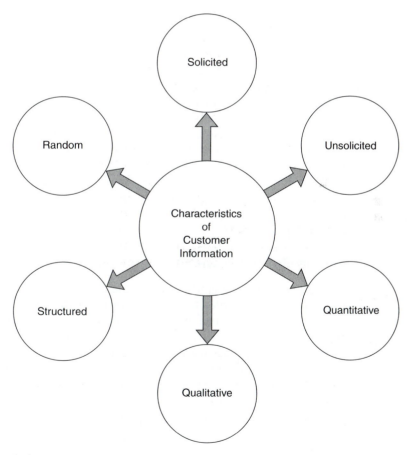

Figure 17–4
Characteristics of Customer Information

Solicited Information

Solicited information is information the organization asks for. Both input and feedback can be solicited. The most common approach for soliciting customer feedback is the customer satisfaction survey. Other methods include the publication of a toll-free hotline number that customers can use to express their satisfaction or air their complaints; focus groups composed of customers who try a new product and then give their feedback to a facilitator; and spot tests in which potential customers are selected at random, given a sample of a new product, and asked their opinion of it (e.g., asking people in a shopping mall to test a new soft drink).

Customer input can be solicited by forming focus groups, using surveys, and conducting spot tests, too. The difference is that with input, these things are done much earlier in the product development cycle. Focus groups deal with drawings, models, or prototypes instead of a finished product. In this way, the information they provide can be used to revise rather than correct the product.

Unsolicited Information

Unsolicited information is information the organization receives without asking for it. It often comes in the form of complaints, but not always. A customer complains to a member of the sales team. The product testing editor for a magazine or newspaper gives the product a try and writes an article pointing out weaknesses. At a trade show, a participant stops by the organization's booth to complain. Regardless of its origin, an unsolicited complaint should be brought into the system and treated just like solicited information. All customer information, whether solicited or unsolicited, should be used to improve the organization's product.

QUALITY TIP

Finding Out What the Customer Wants

"How do we find what is important to the customer? On the surface that seems to be the job of the sales force, and a well-trained team of representatives can help keep a company clued in and focused on customer objectives. But even the most perceptive representative can see things through rose-colored glasses or be the victim of his own defensiveness or biases.

"It is for this reason that formalized surveys before and during a quality implementation are integral parts of such programs. In reaching out in this way, the customers become partners."

Source: Jim Truesdell, "The Quality Survey: Assessing Customer Needs," *Quality Observer* (May 1993), 7.

Quantitative Information

Quantitative information is information that can be measured or counted. A particular type of automobile tire is supposed to last at least 40,000 miles under normal driving conditions. An automobile is supposed to get 20 miles to the gallon in city driving. A welding robot is supposed to put down a perfect seam weld at a rate of 1 linear foot per minute. These are criteria that can be measured. Input and feedback that are quantifiable are particularly helpful in improving a product.

When soliciting information from customers, it is important to structure the mechanism in such a way that quantifiable data are provided. For example, an automobile manufacturer might ask members of a focus group that have test-driven a prototype such questions as the following:

- How many miles to the gallon did you get in city driving conditions?
- How long did it take to go from a complete stop to 60 miles per hour?

These questions would provide quantifiable data the automobile manufacturer could then compare against customer expectations. If members of this same focus group had indicated that they wanted an automobile that could achieve 25 miles per gallon in city driving and the cars they tested had a mean performance level of 28 miles per gallon, the manufacturer would know that the production model would probably meet customer requirements. If, however, the mean performance level was 22 miles per gallon, the manufacturer could make the necessary improvements before putting the new model into production.

Qualitative Information

Qualitative information is subjective; it cannot be measured with the exactness of quantitative data. It can be solicited, or it can be received unsolicited. It comes in the form of opinions and preferences. What do you like? What don't you like? Which option do you prefer? Which option interests you least? These are the types of questions asked when soliciting qualitative information. The same type of information often comes from customers unsolicited. When soliciting information from customers, it is best to ask for both qualitative and quantitative information.

Structured Information

Structured information is information that comes from surveys, focus groups, and other mechanisms that pose specific questions in specific categories or measure customer opinions, satisfaction, or preferences against specific criteria. The structured approach ensures that input and feedback are given in the organization's particular areas of interest. This, in turn, makes identifying customer requirements easier. Figure 17–5 is an example of a document used by a textbook producer to collect structured feedback from a focus group.

Random Information

Unsolicited information is typically random in nature. Complaints to a sales representative, letters to the organization's chief executive officer, and comments from a person at

Focus Group Member	Criteria Numbers/Ratings					TOTAL Score
	1	2	3	4	5	
1						
2						
3						
4						
5						
6						
7						
8						
9						
10						
11						
12						
13						
14						
15						
16						
17						
18						

Criteria Key
1 = Reading Features (Level, Simplicity)
2 = Quantity of Illustrations
3 = Quality of Illustrations
4 = Chapter Format
5 = Pedagogical Features

Very Poor—0 1 2 3 4 5—Excellent
Rating Scale

Figure 17–5
Focus Group Survey

a social function are examples of unsolicited information. Don't ignore unsolicited feedback. Use it to help improve the mechanisms used for collecting structured information. Are we asking the right questions? Are we targeting the right product features? Are our customers who we think they are? Unsolicited random information can help answer these questions.

QFD TOOLS

Chapter 15 explained the most widely used quality tools. These tools may be used at various points in carrying out the QFD process. In addition to the tools explained in Chapter 15, the affinity diagram, interrelationship digraph, tree diagram, and matrix diagram are other tools frequently used with QFD (see Figure 17–6).[2]

Affinity Diagram

Affinity diagrams are used to promote creative thinking. They can be very helpful in breaking down barriers created by past failures and in getting people to give up ingrained paradigms that mitigate against finding new and different approaches. This is a critical element in achieving continual improvement. Affinity diagrams give structure

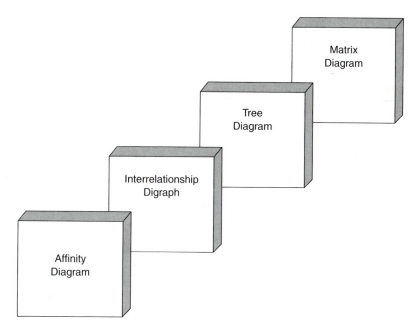

Figure 17–6
Basic QFD Tools

to the creative process by organizing ideas in a way that allows them to be discussed, improved, and interacted with by all participants. Affinity diagrams are used most appropriately when the following conditions exist:

■ When the issue in question is so complex and the known facts so disorganized that people can't quite "get their arms around" the situation

■ When it is necessary to shake up thought processes, get past ingrained paradigms, and get rid of mental baggage relating to past solutions that failed

■ When it is important to build a consensus for a proposed solution

Figure 17–7 is an affinity diagram developed by a textbook publisher. The publisher's goal was to collect creative information about why an engineering textbook is not selling. Such diagrams are developed using the following steps:

1. A team of employees familiar with the issue is formed. The team for Figure 17–7 included the following: sales/marketing personnel, production personnel, and editorial personnel.
2. The issue to be discussed is stated without detailed explanation. Too much detail can inhibit creative thinking and throw up barriers that prejudice participants. The issue was stated as follows: "Why does our engineering textbook not sell better?"

ARTWORK ISSUES	PEDAGOGICAL ISSUES	PRODUCTION ISSUES	WRITING ISSUES
▶ Drawings not produced on a CAD system	▶ Too limited in scope of coverage	▶ Poor registration on drawings	▶ Reading level too high
▶ Drawings do not comply with latest ANSI standards	▶ No chapter objectives	▶ Screens too dark	▶ Writing style complex
▶ Too simple; not enough complex examples	▶ No chapter summaries	▶ Page count too low	▶ Writing style among co-authors inconsistent
▶ Not enough variety (i.e., mechanical, civil, structural, etc.)	▶ Key terms	▶ Paperback cover	▶ Too few subheadings
▶ No color photos	▶ Superficial review questions	▶ Sloppy production	▶ Paragraphs too long
▶ Poor drawing package	▶ No design projects		▶ Poorly written
	▶ No glossary		
	▶ Poor pedagogy		

Figure 17–7
Affinity Diagram: Poor Sales Figures in an Engineering Textbook

3. Responses of participants are stated verbally and recorded on 3 × 5 cards. Participants should limit themselves to one idea per card. At this point, there should be no judgmental comments about the ideas proposed. The goal is to solicit as many ideas as possible. Judgmental comments will inhibit the process.
4. The cards are spread on a large table, and participants are asked to group them. Cards containing related ideas are grouped together. Cards that don't fit with any particular group can be put together as a miscellaneous group.
5. Participants examine the cards in each group and try to find a descriptive word that contains the essence of the various cards in that group. This word or brief phrase is written on a card that is placed at the top of the group and becomes the heading for that group of ideas.
6. The information on the cards is replicated on paper with boxes around each group of ideas. Copies of the draft affinity diagram are distributed to all participants for corrections, revisions, additions, or deletions. At this point, the diagram should resemble the one in Figure 17–7.

Interrelationship Digraph

The affinity diagram records the creative process. In doing so, it identifies issues and ideas relating to a specific goal or problem. The interrelationship digraph is used to bring logic to the process of identifying relationships among the various ideas recorded on the affinity diagram. To develop an interrelationship digraph, follow these steps:

1. Write the problem statement on a 3 × 5 card. Continuing the example in Figure 17–7, the problem is poor sales figures. Write this on a card and designate it as the problem statement by enclosing it in a double line (see Figure 17–8).
2. Place the problem statement card in the upper left-hand corner of a table. Then take out all of the cards used to develop Figure 17–7 and lay them on a table. Use the following order: the card most closely associated with the problem (most prominent cause) is placed closest to the problem card. The farther away a card is from the problem card, the less prominent it is as a cause. Discussion takes place as the cards are placed in their relative positions.
3. When the cards are all laid out in their relative positions, recreate them on paper. Distribute copies of the paper version showing each card as a small rectangle to all participants for final revisions. Do this in a group setting so that discussion can occur. Consensus as to the final arrangement of causes is important.
4. Distribute the final version to all participants and ask them to draw causal arrows showing what contributes to what. This is the step in which relationships between and among causes are established. Figure 17–9 is a partially completed interrelationship digraph built from the issues identified in Figure 17–7. By examining the completed digraph, you can learn a lot. For example, in Figure 17–9, the box containing the statement "poor drawing package" has four causal lines drawn to it. Based on this digraph, the poor drawing package is the result of the following factors: sloppy production, insufficient variety, failure to comply with ANSI standards, and failure to use a CAD system to produce the drawings.

Figure 17–8
Problem Statement Card

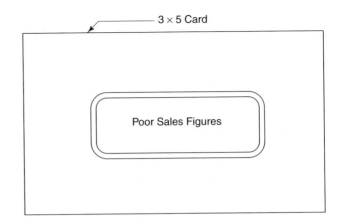

3 × 5 Card

Poor Sales Figures

Tree Diagram

The affinity diagram and the interrelationship digraph identify the issues or problems and how they interrelate. The tree diagram shows the tasks that must be accomplished to solve the problem in question. To develop a tree diagram, follow these steps:

1. Clearly identify the problem to be solved. It can be taken from the affinity diagram or from the interrelationship digraph. It can also be a problem that was identified without the use of either of these tools. Write it on a card and place the card on the left side of a large table.

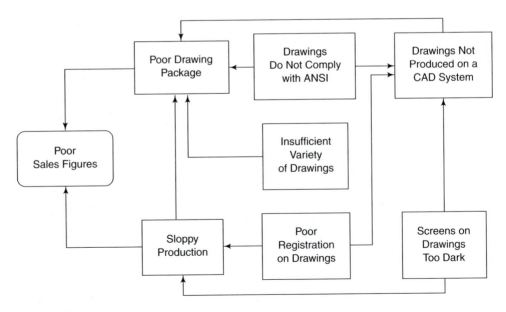

Figure 17–9
Interrelationship Diagram (Partial): Poor Sales Figures on an Engineering Textbook

2. Conduct a brainstorming session in which participants record on 3 × 5 cards all possible tasks, methods, and activities relating to the problem. Use the affinity diagram and interrelationship digraph as references, but don't allow participants to be limited or stymied by them. Continually repeat the following question: "For this to happen, what must happen first?" Continue this until all ideas have been exhausted.

3. Lay all cards on the table to the right of the problem card. Put them in order based on what must happen first, working from left to right. As this task progresses, it will probably be necessary to add task cards that were overlooked during the brainstorming session.

4. Duplicate the cards on the table on paper and distribute copies to all participants. Allow participants to revise and correct the document. Figure 17–10 is a partial tree diagram that was developed to address the problem identified in Figure 17–7.

Matrix Diagram

The matrix diagram is the most widely used of the QFD tools. It is a helpful tool for identifying and graphically displaying connections (seen as intersections on the diagram) among responsibilities, tasks, functions, and so forth. There are several different types of matrix diagrams. The format for the most widely used of these, the L-shaped matrix, is shown in Figure 17–11.

This type of matrix can be used in numerous ways. To continue the example of the publisher trying to produce an engineering book that will sell better than the current product, an L-shaped matrix such as the one in Figure 17–12 (on p. 624) could be developed.

Such a diagram is produced by listing one set of elements vertically and the other set horizontally. In Figure 17–12, the tasks to be accomplished are listed vertically, and the responsible departments/units are listed horizontally. Each intersection between the horizontal and vertical elements is coded using numbers, letters, or graphic symbols.

In Figure 17–12, numbers are used. Each number represents a level of responsibility. For example, the primary responsibility for the first task in the matrix—produce new drawings on a CAD system—falls to the art department. The production department has secondary responsibility. This means there is a relationship: in producing the new drawings, the art department will need to coordinate and communicate with the production department. However, it will have the primary responsibility for ensuring that the drawings are produced. Other such relationships can be identified by examining the matrix.

IMPLEMENTING QFD

QFD must be implemented in a systematic and orderly manner. Figure 17–13 (on p. 625) shows the six steps typically used when implementing QFD. These steps are explained in the remainder of this section.

Form the Project Team

The nature of the project will dictate the makeup of the project team. Is the team going to improve an existing product or develop a new one? If an existing product is to be improved, the team should consist of personnel from the marketing, engineering, quality,

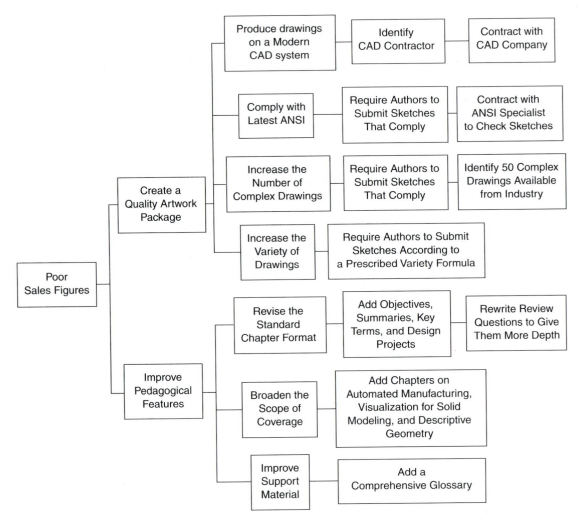

Figure 17–10
Tree Diagram

and manufacturing departments. If a new product is being developed, market research and development representatives should be added to the team. It is important to ensure that team members are able to commit the necessary time and that they have the support of their supervisors. It is also important to ensure that team members understand the purpose of the team and their individual roles on the team.

Establish Monitoring Procedures

Management will want to monitor the team's progress, and this is as it should be. However, micromanaging the team should be avoided. The proper balance between ignoring

Figure 17–11
L-Shaped Matrix

L-Shaped Matrix	Horizontal Entries				
	H1	H2	H3	H4	H5
V1					
V2					
V3					
V4					
V5					
V6					
V7					

(Vertical Entries)

and micromanaging can be achieved by carefully planning and establishing monitoring procedures. In doing this, the following three questions must be answered:

- What will be monitored?
- How will it be monitored?
- How often will it be monitored?

The team's mission will determine what should be monitored. For example, if the team has a product improvement mission, progress made in identifying improvements and developing plans for making them is what is monitored. Either a verbal or a written report format can be used. How frequently such reports should be made is a judgment call. However, experience has shown that once a week may be too frequent and once a month too infrequent. A report every 2 or 3 weeks may strike the proper balance.

Department/ Units / Tasks	Editorial	Art	Production
Produce New Drawings on a CAD System	–	1	2
Bring Drawings into ANSI Compliance	1	2	–
Add Color Photos	3	1	2
Add Chapter Objectives Summaries, Key Terms	1	–	2
Correct Registration on Drawings	–	2	1
Lighten Screens on Drawings	–	2	1
Lower the Reading Level	1	2	3
Add Design Projects	1	2	3

LEGEND

1 = Primary Responsibility
2 = Secondary Responsibility
3 = Tertiary Responsibility

Figure 17–12
L-Shaped Matrix: Improving Poor Sales Figures on an Engineering Textbook

Figure 17–13
Steps in Implementing QFD

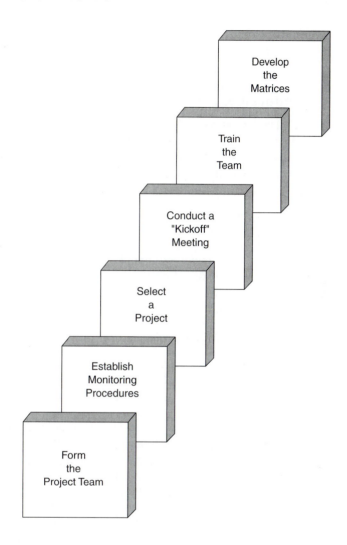

Decisions relating to frequency must take personalities, the nature of the team's mission, and other local factors into consideration. There are no hard-and-fast rules. What works well with one team may not with another. The key is to arrive at a frequency that keeps management properly informed without micromanaging the team.

Select a Project

It is a good idea to begin with an improvement project as opposed to a new product development project. Improvement projects have the advantage of an existing information/experience base. A new QFD team dealing with a new product may be too much newness at one time. With an improvement project, team members who may not be familiar with QFD are, at least, familiar with the product and customer information relating to it. This familiarity prevents a situation from developing in which team members are trying to learn about QFD and a new product simultaneously.

Conduct a Kickoff Meeting

The kickoff meeting is the first official meeting of the team. It is important to accomplish the following tasks during this meeting: (a) make sure all participants understand the mission of the project team, (b) make sure all participants understand their role on the team as well as the roles of other team members, and (c) establish logistical parameters (length, time, and frequency of meeting).

Train the Team

Before beginning the project, it is important to train all team members in the fundamentals of QFD. Team members should learn how to use the various quality tools as well as such specific tools as affinity diagrams, interrelationship digraphs, tree diagrams, and matrix diagrams. In addition, team members should learn how QFD works as a process (see Figure 17–2).

Develop the Matrices

Once team members understand QFD, QFD tools, and the format of a QFD matrix (Figure 17–1), the process of developing matrices can begin. A complete cycle of the process involves the development of six matrices, each structured according to the specifications in Figure 17–1.

The first matrix is the one most commonly associated with QFD. It compares customer requirements with technical features of the product. Outcomes that typically flow from the development of the first matrix include a summary of customer needs/requirements and a concept document that describes what features the product will need to have to meet customer expectations.

The second matrix compares technical features and applied technologies. At this point, decisions as to technical feasibility and trade-offs between what it will take to meet customer requirements and what capabilities currently exist are made.

The third matrix compares applied technologies and manufacturing processes. The fourth compares manufacturing processes and quality control processes. The fifth matrix compares quality control processes and statistical process control. The final matrix compares statistical process control and specifications for the finished product. In preparing all of these, such tools as affinity diagrams, tree diagrams, interrelationship digraphs, and matrix diagrams are used as needed.

SUMMARY

1. QFD is an approach to continual improvement that brings customers into the design of processes. It translates what the customer wants into what the organization produces. QFD was originally developed in Japan's Kobe Shipyard in the 1960s. A QFD matrix takes the shape of a house.

2. QFD yields the following benefits to organizations that are interested in continual improvement: customer focus, time efficiency, teamwork orientation, and documentation orientation.

3. Customer information falls into two broad categories: input and feedback. Feedback is given after the fact; input is given before the fact (early in the product development cycle). Both types of information can be further classified according to the following categories: solicited, unsolicited, quantitative, qualitative, structured, and/or random.

4. In addition to the traditional quality tools, QFD also makes use of several specialized tools including the affinity diagram, interrelationship digraph, tree diagram, and matrix diagram.

5. Affinity diagrams are used to promote creative thinking. The interrelationship digraph is used to bring logic to the process of identifying relationships among ideas. The tree diagram identifies all tasks that must be accomplished to solve a problem. Matrix diagrams are used to identify connections among responsibilities, tasks, functions, and so on.

6. The steps for implementing QFD are as follows: form the project team, establish monitoring procedures, select a project, conduct a kickoff meeting, train the team, and develop the matrices.

 KEY TERMS AND CONCEPTS

Affinity diagram	Product development process
Basic QFD matrix	Project team
Continual improvement	QFD
Customer focused	QFD process
Customer information	Qualitative information
Customer requirements	Quantitative information
Documentation oriented	Random information
Feedback	Solicited information
Input	Structured information
Interrelationship digraph	Teamwork oriented
Kickoff meeting	Time efficient
Matrix diagram	Tree diagram
Monitoring procedures	Unsolicited information

 FACTUAL REVIEW QUESTIONS

1. Define *quality function deployment*.
2. Describe the basic structure of a QFD matrix.
3. Explain QFD as a process.
4. List the benefits of QFD.
5. What are the two main categories of customer information? Differentiate between them.
6. List six other ways that information can be categorized.

7. What is an interrelationship digraph?
8. Explain the uses of a tree diagram.
9. List and briefly describe the steps in the implementation of QFD.

 CRITICAL THINKING ACTIVITY

Customer Demands versus Organizational Capabilities

"I don't know why we keep collecting all of this customer input and feedback. It's clear they want a better product than our processes can produce," said Derrick Kramer, CEO of Keltron, Inc.

"That's true," said Linda Carver, Keltron's director of quality. "But we are going to lose our customers if we don't improve our processes. We need to do more than collect customer feedback. We need to use it to keep our processes up to date."

Clearly Keltron needs to translate customer demands into process improvements. Explain how QFD could be used to help this company. How should Kramer and Carver proceed if they choose to apply QFD?

 # DISCUSSION ASSIGNMENT 17–1

Feedback versus Input

Jones Processing Company is the regional leader in the wholesale processing of beef and alternative meat products, including ostrich and emu. Its customers are grocery store chains throughout the southeastern United States. Jones Processing's CEO prides himself on collecting feedback from customers concerning cuts, packaging, delivery schedules, and all of the other factors that are important to them.

That is why he cannot understand the point his quality director is trying to make. Jones Processing's quality director wants to establish focus groups of customers to collect input *before* the fact, rather than feedback *after* the fact. The two executives have been debating the merits of feedback versus input for a week now.

Discussion Question
Discuss the following question in class or outside of class with your fellow students:

I.   Which do you think is more valuable, feedback or input? Why?

 ENDNOTES

1. Stephen Uselac, *Zen Leadership: The Human Side of Total Quality Team Management* (Loudonville, OH: Mohican, 1993), 52.
2. This section is based on James L. Bossert, *Quality Function Deployment: A Practitioner's Approach* (Milwaukee, WI: ASQC Quality Press, 1991), 52–118.

CHAPTER EIGHTEEN

Optimizing and Controlling Processes through Statistical Process Control (SPC)

"In the middle of difficulty lies opportunity."
   Albert Einstein

MAJOR TOPICS

- Statistical Process Control Defined
- Rationale for SPC
- Control Chart Development
- Management's Role in SPC
- Role of the Total Quality Tools
- Authority over Processes and Production
- Implementation and Deployment of SPC
- Inhibitors of SPC

The origin of what is now called statistical process control (SPC) goes back to 1931 and Dr. Walter Shewhart's book *The Economic Control of Quality of Manufactured Product*. Shewhart, a Bell Laboratories statistician, was the first to recognize that industrial processes themselves could yield data, which, through the use of statistical methods, could signal that the process was in control or was being affected by special causes (causes beyond the natural, predictable variation). The control charts used today are based on Shewhart's work. These control charts are the very heart of SPC. What may not be as obvious is that Shewhart's work became the catalyst for the quality revolution in Japan[1] and the entire movement now called total quality. We tend to look at SPC as one piece of the whole total quality picture, and it is, but it is also the genesis of total quality.

   In the 8 years since the first edition of this book was written, two very significant things have occurred in the SPC field. First, many organizations have adopted SPC as a preferred way of controlling manufacturing processes. Much of this has come about as a result of the quality quest by first-tier companies, making it necessary to require that their second-tier suppliers practice SPC. We have seen this ripple down to at least the fourth tier. Nowhere is this more evident than in the auto industry. But even beyond the mandate by corporate customers, more and more small companies are using SPC as part of their quality and competitiveness initiatives.

The second big change we have seen is that SPC users have backed away from the shotgun approach, where every process, no matter how trivial or foolproof, had to have SPC charts. Five years ago we visited a North American semiconductor plant and were overwhelmed by the sheer numbers of SPC charts. Everywhere you looked you saw control charts. The plant proudly admitted to having over 900 processes under control charts. When we visited the same plant a couple of years later the picture was very different. You could still find control charts, but only where they offered real benefit. The company had discovered that about 800 of its original charts had not been worthwhile. Control charts were being used with those processes that needed them, and no more. It is evident that this is the current thinking in industry. Don't waste time, energy, and money with more control charts than you need. In those process applications where you do need them, the control chart is invaluable. For all the rest, it is just window dressing. The important thing is to know the difference.

STATISTICAL PROCESS CONTROL DEFINED

Although SPC is normally thought of in industrial applications, it can be applied to virtually any process. Everything done in the workplace is a process. All processes are affected by multiple factors. For example, in the workplace a process can be affected by the environment and the machines employed, the materials used, the methods (work instructions) provided, the measurements taken, and the manpower (people) who operate the process—the Five M's. If these are the only factors that can affect the process output, and if all of these are perfect—meaning the work environment facilitates quality work; there are no misadjustments in the machines; there are no flaws in the materials; there are totally accurate and precisely followed work instructions, accurate and repeatable measurements, and people who work with extreme care, following the work instructions perfectly, concentrating fully on their work—if all of these factors come into congruence, then the process will be in statistical control. This means that there are no special causes adversely affecting the process's output. Special causes are (for the time being, anyway) eliminated. Does that mean that 100% of the output will be perfect? No, it does not. Natural variation is inherent in any process, and it will affect the output. Natural variation is expected to account for roughly 2,700 out-of-limits parts in every 1,000,000 produced by a 3-sigma process the ($\pm3\sigma$ variation), or 63 out-of-limits parts in every 1,000,000 produced by a 4 sigma process, and so on. Natural variation, if all else remains stable, will account for 2 out-of-limits parts per billion produced by a 6 sigma process.

SPC does not eliminate all variation in the processes, but it does something that is absolutely essential if the process is to be consistent, and if the process is to be improved. SPC allows workers to separate the special causes of variation (e.g., environment and the Five M's) from the natural variation found in all processes. After the special causes have been identified and eliminated, leaving only natural variation, the process is said to be in statistical control (or simply in control). When that state is achieved, the process is stable, and in a 3 sigma process, 99.73% of the output can be counted on to be within the statistical control limits. More important, improvement can begin. From this, we can develop a definition of statistical process control:

> Statistical process control (SPC) is a statistical method of separating variation resulting from special causes from variation resulting from natural causes, to eliminate the special causes and to establish and maintain consistency in the process, enabling process improvement.

Note: As explained in Chapter 1, the 6-sigma numbers given in this section differ from the Motorola Six Sigma numbers (2 parts per billion vs 3.4 parts per million).

RATIONALE FOR SPC

The rationale for SPC is much the same as that for total quality. It should not be surprising that the parallel exists, because it was Walter Shewhart's work that inspired the Japanese to invite W. Edwards Deming to help them get started in their quality program in 1949–1950. SPC was the seed from which the Japanese grew total quality.

The rationale for the Japanese to embrace SPC in 1950 was simple: a nation trying to recover from the loss of a costly war needed to export manufactured goods so that it could import food for its people. The Asian markets once enjoyed by Japan had also been rendered extinct by the war. The remaining markets, principally North America, were unreceptive to Japanese products because of poor quality. If the only viable markets rejected Japanese products on the basis of quality, then Japanese manufacturers had to do something about their quality problem. This is why Shewhart's work interested them. This also is why they called on Deming, and later Joseph Juran, to help them. That the effort was successful is well documented and manifestly evident all over the world. Deming told the Japanese industrialists in 1950 that if they would follow his teaching, they could become active players in the world's markets within 5 years. They actually made it in 4.

The Western world may not be in the same crisis Japan experienced following World War II, but the imperative for SPC is no less crucial. When one thinks of quality products today, Japan still comes to mind first. Many of the finest consumer products in the world come from Japan. That includes everything from electronics and optical equipment to automobiles, although U.S. car manufacturers, beginning with Ford, have narrowed the quality gap. Fine automobiles are also produced in Europe; however, there is a serious price disparity. The European equivalent of a top-of-the-line Lexus or Infiniti made in Japan is the high-end Mercedes-Benz or BMW. These European luxury cars cost much more than their Japanese counterparts. In the United States, on the other hand, although there is much ground to be made up, Cadillac and Lincoln are beginning to compete with Lexus and Infiniti on quality and price. Ford and General Motors are doing this by adopting such total quality strategies as SPC.

Most consumers are more interested in what Toyota, Honda, Ford, General Motors, Chrysler, and the others are doing at the middle and lower end of the automobile market. Until the early 1990s, the Japanese were the quality leaders in every strata of the automobile market. Cars made by Toyota, Nissan, Honda, Mazda, and Mitsubishi (including those produced in their North American factories) have been of consistently excellent quality. Now that manufacturers in the United States have adopted SPC and other total quality improvement strategies, the outcome of the race for quality leadership can no longer be predicted each new product year. Put simply, the rationale for Western manufacturers to embrace SPC has been not only to improve product quality and simultaneously reduce costs but also to improve their product image in order to compete successfully in the world's markets.

To comprehend how SPC can help accomplish this, it is necessary to examine five key points and understand how SPC comes into play in each one: control of variation, continual improvement, predictability of processes, elimination of waste, and product inspection.

Rationale: Control of Variation

The output of a process that is operating properly can be graphed as a bell-shaped curve, as in Figure 18–1. The x-axis represents some measurement, such as weight or dimension, and the y-axis represents the frequency count of the measurements. The desired measurement value is at the center of the curve, and any variation from the desired value results in displacement to the left or right of the center of the bell. With no special causes acting on the process, 99.73% of the process output will be between the $\pm 3\sigma$ limits. (This is not a specification limit, which may be tighter or looser.) This degree of variation about the center is the result of natural causes. The process will be consistent at this performance level as long as it is free of special causes of variation.

When a special cause is introduced, the curve will take a new shape, and variation can be expected to increase, lowering output quality. Figure 18–2 shows the result of a machine no longer capable of holding the required tolerance, or an improper work instruction. The bell is flatter, meaning that fewer parts produced by the process are at, or close to, the target, and more fall outside the original 3σ limits. The result is more scrap, higher cost, and inconsistency of product quality.

The curve of Figure 18–3 could be the result of input material from different vendors (or different batches) that is not at optimal specification. Again, a greater percentage of the process output will be displaced from the ideal, and more will be outside the original 3σ limits. The goal should be to eliminate the special causes to let the process operate in accordance with the curve shown in Figure 18–1, and then to improve the process, thereby narrowing the curve (see Figure 18–4).

Figure 18–1
Frequency Distribution Curve: Normal Curve

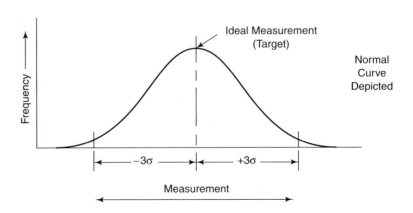

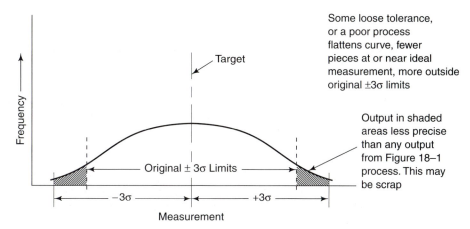

Figure 18–2
Frequency Distribution Curve: Process Not as Precise as Figure 18–1

When the curve is narrowed, more of the process output is in the ideal range, and less falls outside the original 3σ limits. Actually, each new curve will have its own 3σ limits. In the case of Figure 18–4, they will be much narrower than the original ones. If the original limits resulted in 2,700 pieces out of 1 million being scrapped, the improved process illustrated by Figure 18–4 might reduce that to 270 pieces, or even less, scrapped. Viewed from another perspective, the final product will be more consistently of high quality, and the chance for a defective product going to a customer is reduced by an order of magnitude.

Variation in any process is the enemy of quality. As we have already discovered, variation results from two kinds of causes: special causes and natural causes. Both kinds can be treated, but they must be separated so that the special causes—those associated with

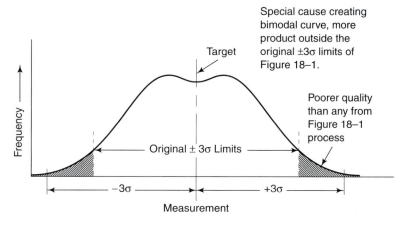

Figure 18–3
Bimodal Frequency Distribution Curve

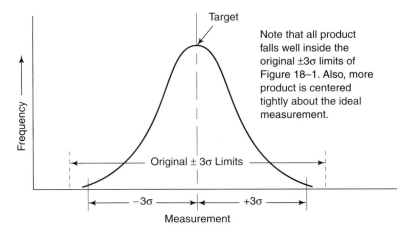

Figure 18–4
Frequency Distribution Curve: Narrowed (Less Variation) Relative to Figure 18–1

the Five M's and environment—can be identified and eliminated. After that is done, the processes can be improved, never eliminating the natural variation, but continuously narrowing its range, approaching perfection. It is important to understand why the special causes must first be eliminated. Until that happens the process will not be stable, and the output will include too much product that is unusable, therefore wasted. The process will not be dependable in terms of quantity or quality. In addition, it will be pointless to attempt improvement of the process, because one can never tell whether the improvement is successful—the results will be masked by the effect of any special causes that remain.

In this context, elimination of special causes is not considered to be process improvement, a point frequently lost on enthusiastic improvement teams. Elimination of special causes simply lets the process be whatever it will be in keeping with its natural variation. It may be good or bad or anything in between.

When thinking of SPC, most people think of control charts. If we wish to include the elimination of special causes as a part of SPC, as we should, then it is necessary to include more than the control chart in our set of SPC tools, because the control chart has limited value until the process is purged of special causes. If one takes a broad view of SPC, all of the statistical tools discussed in Chapter 15 should be included. Pareto charts, cause-and-effect diagrams, stratification, check sheets, histograms, scatter diagrams, and run charts are all SPC tools. Although the flowchart is not a statistical device, it is useful in SPC. The SPC uses of these techniques are highlighted later in this chapter. Suffice it to say for now that the flow diagram is used to understand the process better, the cause-and-effect diagram is used to examine special causes and how they impact the process, and the others are used to determine what special causes are at play and how important they are. The use of such tools and techniques makes possible the control of variation in any process to a degree unheard of before the introduction of SPC.

Rationale: Continual Improvement

Continual improvement is a key element of total quality. One talks about improvement of products, whatever they may be. In most cases, it would be more accurate to talk about continual improvement in terms of processes than in terms of products and services. It is usually the improvement of processes that yields improved products and services. Those processes can reside in the engineering department where the design process may be improved by adding concurrent engineering and design-for-manufacture techniques, or in the public sector where customer satisfaction becomes a primary consideration. All people use processes, and all people are customers of processes. A process that cannot be improved is rare. We have not paid sufficient attention to our processes. Most people have only a general idea of what processes are, how they work, what external forces affect them, and how capable they are of doing what is expected of them. Indeed, outside the manufacturing industry, many people don't realize that their work is made up of processes.

Before a process can be improved, it is necessary to understand it, identify the external factors that may generate special causes of variation, and eliminate any special causes that are in play. Then, and only then, can we observe the process in operation and determine its natural variation. Once a process is in this state of statistical control, the process can be tracked, using control charts, for any trends or newly introduced special causes. Process improvements can be implemented and monitored. Without SPC, process improvement takes on a hit-or-miss methodology, the results of which are often obscured by variation stemming from undetected factors (special causes). SPC lets improvements be applied in a controlled environment, measuring results scientifically, and with assurance.

Rationale: Predictability of Processes

A customer asks whether a manufacturer can produce 300 widgets within a month. If it can, the manufacturer will receive a contract to do so. The parts must meet a set of specifications supplied by the customer. The manufacturer examines the specifications and concludes that it can comply but without much margin for error. The manufacturer also notes that in a good month it has produced more than 300. So the order is accepted. Soon, however, the manufacturer begins having trouble with both the specifications and the production rate. By the end of the month only 200 acceptable parts have been produced. What happened? The same units with the same specification have been made before, and at a higher production rate. The problem is unpredictable processes. If the same customer had approached a firm that was versed in SPC, the results would have been different. The managers would have known with certainty their capability, and it would have been clear whether the customer's requirements could, or could not, be met. They would know because their processes are under control, repeatable, and predictable.

Few things in the world of manufacturing are worse than an undependable process. Manufacturing management spends half its time making commitments and the other half living up to them. If the commitments are made based on unpredictable processes, living up to them will be a problem. The only chance manufacturing managers have when their processes are not predictable is to be extra conservative when making commitments. Instead of keying on the best past performance, they look at the worst production month and base their commitments on that. This approach can relieve a lot of stress but can also lose a lot of business. In today's highly competitive marketplace (whether for a manufactured product or a service), organizations must have predictable, stable, consistent processes. This can be achieved and maintained through SPC.

Rationale: Elimination of Waste

Only in recent years have many manufacturers come to realize that production waste costs money. Scrap bins are still prominent in many factories. In the electronics industry, for example, it is not unusual to find that 25% of the total assembly labor cost in a product is expended correcting errors from preceding processes. This represents waste. Parts that are scrapped because they do not fit properly or are blemished represent waste. Parts that do not meet specifications are waste. To prevent defective products from going to customers, more is spent on inspection/reinspection. This, too, is waste. All of these situations are the result of some process not producing what was expected. In most cases, waste results from processes being out of control: processes are adversely influenced by special causes of variation. Occasionally, even processes that have no special causes acting on them are simply not capable of producing the expected result.

Two interesting things happen when waste is eliminated. The most obvious is that the cost of goods produced (or services rendered) is reduced—a distinct competitive advantage. At the same time, the quality of the product is enhanced.

Even when all parts are inspected, it is impossible to catch all the bad ones. When sampling is used, even more of the defective parts get through. When the final product contains defective parts, its quality has to be lower. By eliminating waste, a company reduces cost and increases quality. This suggests that Philip Crosby was too conservative: quality is not just free;[2] it pays dividends. This is the answer to the question of what happened to the Western industries that once led but then lost significant market share since the 1970s: total quality manufacturers simply built better products at competitive prices. These competitive prices are the result of the elimination of waste, not (as is often presumed) cheap labor. This was accomplished in Japan by applying techniques that were developed in the United States in the 1930s but ignored in the West after the crisis of World War II had passed. Specifically, through Japan's application of SPC and later the expansion of SPC to the broader concept of total quality, Japan went from a beaten nation to an economic superpower in just 30 years.

By concentrating on their production processes, eliminating the special causes as Shewhart and Deming taught, and bringing the processes into statistical control, Japanese manufacturers could see what the processes were doing and what had to be done to improve them. Once in control, a relentless process improvement movement was started, one that is still ongoing a half century later—indeed, it is never finished. Tightening the bell curves brought ever-increasing product quality and ever-diminishing waste (nonconforming parts). For example, while U.S. automakers were convinced that to manufacture a more perfect transmission would be prohibitively expensive, the Japanese not only did it but also reduced its cost. In the early 1980s, the demand for a particular Ford transmission was such that Ford second-sourced a percentage of them from Mazda. Ford soon found that the transmissions manufactured by Mazda (to the same Ford blueprints) were quieter, smoother, and more reliable than those produced in North America. Ford customers with Mazda transmissions were a lot happier than the others, as well. Ford examined the transmissions, and while they found that both versions were assembled properly with parts that met all specifications, the component parts of the Mazda units had significantly less variation piece to piece. Mazda employed SPC, and the domestic supplier did not. This demonstrated to Ford that the same design, when held to closer tolerances, resulted in a noticeably superior transmission that did not cost more. Shortly thereafter, Ford initiated an SPC program. To Ford's credit, its effort paid off. In 1993, the roles were reversed and Ford began producing transmissions for Mazda.

Statistical process control is the key to eliminating waste in production processes. It can do the same in virtually any kind of process. The inherent nature of process improvement is such that as waste is eliminated, the quality of the process output is correspondingly increased.

Rationale: Product Inspection

It is normal practice to inspect products as they are being manufactured (in-process inspection) and as finished goods (final inspection). Inspection requires the employment of highly skilled engineers and technicians, equipment that can be very expensive, factory space, and time. If it were possible to reduce the amount of inspection required, while maintaining or even improving the quality of products, money could be saved and competitiveness enhanced.

Inspection can be done on every piece (100% inspection) or on a sampling basis. The supposed advantage of 100% inspection is, of course, that any defective or nonconforming product will be detected before it gets into the hands of a customer (external or internal). The term *supposed advantage* is used because even with 100% inspection only 80% of the defects are found.[3] Part of the problem with 100% inspection is that human inspectors can become bored and, as a result, careless. Machine inspection systems do not suffer from boredom, but they are very expensive, and for many applications they are not a practical replacement for human eyes. It would be faster and less costly if it were possible to achieve the same level of confidence by inspecting only 1 piece out of 10 (10% sampling) or 5 out of 100 (5% sampling) or even less.

Such sampling schemes are not only possible but accepted by such critical customers as the U.S. government (see the applicable U.S. government military standards, MIL-STD-105 and MIL-STD-414 series) and the automobile industry, but there is a condition: for sampling to be accepted, processes must be under control. Only then will the processes have the consistency and predictability necessary to support sampling. This is a powerful argument for SPC.

After supplier processes are under control and being tracked with control charts, manufacturers can back off the customary incoming inspection of materials, resorting instead to the far less costly procedure of periodically auditing the supplier's processes. SPC must first be in place, and the supplier's processes must be shown to be capable of meeting the customer's specifications.

This also applies internally. When a company's processes are determined to be capable of producing acceptable products, and after they are in control using SPC, the internal quality assurance organization can reduce its inspection and process surveillance efforts, relying to a greater degree on a planned program of process audit. This reduces quality assurance costs and, with it, the cost of quality.

Control Chart Development

Just as there must be many different processes, so must there be many types of control charts. The table in Figure 18–19 (presented later) lists the seven most commonly used control chart types. You will note that the first three are associated with measured values, or variables data. The other four are used with counted values, or attributes data. It is important, as the first step in developing your control chart, to select the chart type that is appropriate for your data. The specific steps in developing control charts are different for variables data than for attributes data.

Control Chart Development for Variables Data (Measured Values)

Consider an example using $\bar{x}$ and R charts. $\bar{x}$ and R charts are individual, directly related graphs plotting the mean (average) of samples ($\bar{x}$) over time, and the variation in each sample (R) over time. The basic steps for developing a control chart for data with measured values are these:

1. Determine sampling procedure. Sample size may depend on the kind of product, the production rate, measurement expense, and likely ability to reveal changes in the process. Sample measurements are taken in subgroups of size (n), typically from 3 to 10. Sampling frequency should be often enough that changes in the process are not missed, but not so often as to mask slow drifts. If the object is to set up control charts for a new process, the number of subgroups for the initial calculations should be 25 or more. For existing processes that appear stable, that number can be reduced to 10 or so, and sample size (n) can be smaller, say, 3 to 5.
2. Collect initial data of 100 or so individual data points in k subgroups of n measurements.
    - Process must not be tinkered with during this time—let it run.
    - Don't use old data—it may not be relevant to the current process.
    - Take notes on anything that may have significance.
    - Log data on a data sheet designed for control chart use.
3. Calculate the mean (average) values of data in each subgroup ($\bar{x}$).
4. Calculate the data range for each subgroup (R).
5. Calculate the average of the subgroup averages ($\bar{\bar{x}}$). This is the *process average* and will be the centerline for the $\bar{x}$-chart.
6. Calculate the average of the subgroup ranges ($\bar{R}$). This will be the centerline for the R-chart.

7. Calculate UCL and LCL (using a table of factors, such as the one shown in Figure 18–6). UCL and LCL represent $\pm 3\sigma$ limits *of the process averages* and are drawn as dashed lines on the control charts.
8. Draw the control chart to fit the calculated values.
9. Plot the data on the chart.

What follows is a step-by-step example of how to construct a control chart. First we have to collect sufficient data from which to make statistically valid calculations. This means we will usually have to take at least 100 data measurements in at least 10 subgroups, depending on the process, rate of flow, and so forth. The measurements should be made on samples close together in the process to minimize variation between the data points within the subgroups. However, the subgroups should be spread out in time to make visible the variation that exists between the subgroups.

The process for this example makes precision spacers that are nominally 100 millimeters thick. The process operates on a two-shift basis and appears to be quite stable. Fifty spacers per hour are produced. To develop a control chart for the process, we will measure the first 10 spacers produced after 9 A.M., 1 P.M., 5 P.M., and 9 P.M. We will do this for 3 days, for a total of 120 data points in 12 subgroups.

At the end of the 3 days, the data chart is as shown in Figure 18–5. The raw data are recorded in columns x_1 through x_{10}.

Next we calculate the mean (average) values for each subgroup. This is done by dividing the sum of x_1 through x_{10} by the number of data points in the subgroup.

$$\bar{x} = \Sigma x \div n$$

where n = the number of data points in the subgroup. The $\bar{x}$ values are found in the Mean Value column.

The average ($\bar{\bar{x}}$) of the subgroup averages ($\bar{x}$) is calculated by summing the values of $\bar{x}$ and dividing by the number of subgroups (k):

$$\bar{\bar{x}} = \Sigma \bar{x} \div k.$$

In this case,

$$\bar{\bar{x}} = 1{,}200.8 \div 12$$
$$= 100.067$$

The range for each subgroup is calculated by subtracting the smallest value of x from the largest value of x in the subgroup.

$$R = (\text{maximum value of } x) - (\text{minimum value of } x)$$

Subgroup range values are found in the final column of Figure 18–5.

From the R values, calculate the average of the subgroup ranges.

$$\bar{R} = \Sigma R \div k$$

In this case,

$$\bar{R} = 68 \div 12$$
$$= 5.667$$

Date	Subgroup #	x_1	x_2	x_3	x_4	x_5	x_6	x_7	x_8	x_9	x_{10}	Sum Σx	Mean Value $\overline{X}$	Rng. R
						Measured Values								
7/6	1	101	98	102	101	99	100	98	101	100	102	1002	100.2	4
7/6	2	103	100	101	98	100	104	102	99	101	98	1006	100.6	6
7/6	3	103	101	99	102	100	99	102	98	103	100	1007	100.7	5
7/6	4	96	99	102	99	101	102	98	100	99	97	993	99.3	6
7/7	5	99	102	100	99	103	101	102	98	100	100	1004	100.4	5
7/7	6	101	103	99	100	99	98	101	100	99	100	999	99.9	5
7/7	7	100	103	101	98	99	100	99	102	100	98	1000	100.0	5
7/7	8	97	101	102	100	99	96	99	100	103	98	995	99.5	7
7/8	9	102	97	100	101	103	98	100	102	99	101	1003	100.3	6
7/8	10	100	105	99	100	98	102	97	97	99	101	998	99.8	8
7/8	11	101	99	98	101	104	100	98	100	102	98	1001	100.1	6
7/8	12	100	103	101	98	99	100	100	99	98	102	1000	100.0	5
											Total	1,200.8	68

$$k = 12, \qquad \overline{\overline{X}} = 100.067, \qquad \overline{R} = 5.667$$

Figure 18–5
Initial Data for Precision Spacer Process

Next we calculate the UCL and LCL values for the $\overline{x}$ chart.

$$\text{UCL}_{\overline{x}} = \overline{\overline{x}} + A_2\overline{R} \qquad \text{LCL}_{\overline{x}} = \overline{\overline{x}} - A_2\overline{R}$$

At this point, you know the origin of all the values in these formulas except A_2. A_2 (as well as D_3 and D_4, used later) is from a factors table that has been developed for control charts (see Figure 18–6). The larger the value of A_2, the farther apart the upper and lower control limits UCL_x and LCL_x will be. A_2 may be considered to be a confidence factor for the data. The table shows that the value of A_2 decreases as the number of observations (data points) in the subgroup increases. It simply means that more data points make the calculations more reliable, so we don't have to spread the control limits so much. This works to a point, but the concept of diminishing returns sets in around $n = 15$.

Applying our numbers to the UCL and LCL formulas, we have this:

$$\begin{aligned}
\text{UCL}_{\overline{x}} &= 100.067 + (0.31 \times 5.667) \\
&= 100.067 + 1.75677 \\
&= 101.82377 \\
\text{LCL}_{\overline{x}} &= 100.067 - 1.75677 \\
&= 98.31023
\end{aligned}$$

Now calculate the UCL and LCL values for the R chart.

$$\text{UCL}_R = D_4\overline{R} \qquad \text{LCL}_R = D_3\overline{R}$$

Number of data points in subgroup	Factors for $\bar{x}$ charts	Factors for R charts	
		LCL	UCL
(n)	A_2	D_3	D_4
2	1.88	0	3.27
3	1.02	0	2.57
4	0.73	0	2.28
5	0.58	0	2.11
6	0.48	0	2.00
7	0.42	0.08	1.92
8	0.37	0.14	1.86
9	0.34	0.18	1.82
10	0.31	0.22	1.78
11	0.29	0.26	1.74
12	0.27	0.28	1.72
13	0.25	0.31	1.69
14	0.24	0.33	1.67
15	0.22	0.35	1.65
16	0.21	0.36	1.64
17	0.20	0.38	1.62
18	0.19	0.39	1.61
19	0.19	0.40	1.60
20	0.18	0.41	1.59

Figure 18–6
Factors Table for $\bar{x}$ and R Charts

Like factor A_2 used in the $\bar{x}$ control limit calculation, factors D_3 and D_4 are found in Figure 18–6. Just as with A_2, these factors narrow the limits with subgroup size. With $n = 10$ in our example, $D_3 = 0.22$, and $D_4 = 1.78$. Applying the numbers to the LCL_R and UCL_R formulas, we have this:

$$UCL_R = 1.78 \times 5.667 \qquad LCL_R = 0.22 \times 5.667$$
$$= 10.08726 \qquad\qquad = 1.24674$$

At this point, we have everything we need to lay out the $\bar{x}$ and R charts (see Figure 18–7).

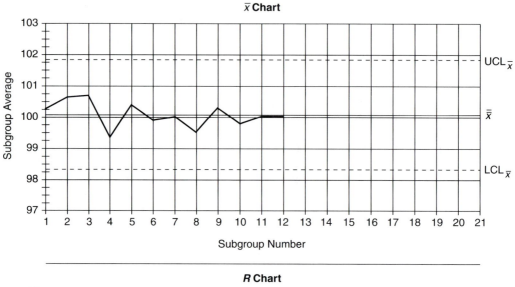

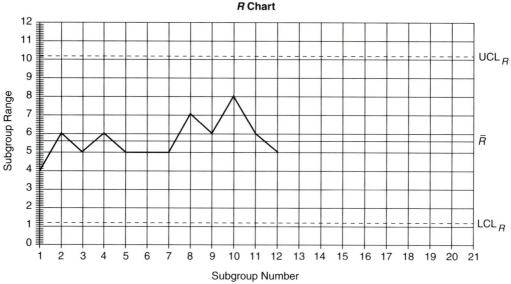

Figure 18–7
$\bar{x}$ and R Charts

The charts are laid out with y-axis scales set for maximum visibility consistent with the data that may come in the future. For new processes, it is usually wise to provide more y-axis room for variation and special causes. A rule of thumb is this:

■ (Largest individual value − smallest individual value) ÷ 2
■ Add that number to largest individual value to set the top of chart.

■ Subtract it from the smallest individual value to set the bottom of chart. (If this results in a negative number, set the bottom at zero.)

Upper and lower control limits are drawn on both charts as dashed lines, and $\bar{\bar{x}}$ and $\bar{R}$ centerlines are placed on the appropriate charts as solid lines. Then the data are plotted, subgroup average ($\bar{x}$) on the $\bar{x}$-chart, and subgroup range (R) on the R chart. We have arbitrarily established the time axis as 21 subgroups. It could be more or less, depending on the application. Our example requires space for 20 subgroups for a normal 5-day week.

Both charts in Figure 18–7 show the subgroup averages and ranges well within the control limits. The process seems to be in statistical control.

Suppose we had been setting up the charts for a new process (or one that was not as stable). We might have gotten a chart like the one in Figure 18–8.

Plotting the data shows that subgroup 7 was out of limits. This cannot be ignored, because the control limits have been calculated with data that included a nonrandom, special-cause event. We must determine and eliminate the cause. Suppose we were using an untrained operator that day. The operator has since been trained. Having established the special cause and eliminated it, we must purge the data of subgroup 7 and recalculate the process average ($\bar{\bar{x}}$) and the control limits. Upon recalculating, we may find that one or more of the remaining subgroup averages penetrates the new, narrower limits (as in Figure 18–9). If that happens, another iteration of the same calculation is needed to clear the data of any special-cause effects. We want to arrive at an initial set of charts that are based on valid data and in which the data points are all between the limits, indicating a process that is in statistical control (Figure 18–10). If after one or two iterations, all data points are not between the control limits, then we must stop. The process is too unstable for control chart application and must be cleared of special causes.

Control Chart Development for Attributes Data (Counted Data)

The p-chart

Attributes data are concerned not with measurement but with something that can be counted. For example, the number of defects is attributes data. Whereas the $\bar{x}$ and R charts are used for certain kinds of variables data, where measurement is involved, the

Figure 18–8
Chart for an Unstable Process

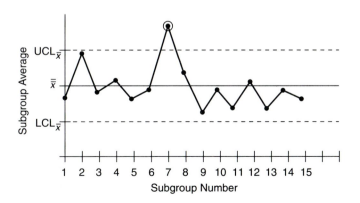

Figure 18–9

The New, Narrower Limits Are Penetrated

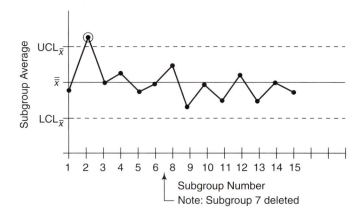

p-chart is used for certain attributes data. Actually, the *p*-chart is used when the data are the *fraction defective of some set of process output*. It may also be shown as *percentage defective*. The points plotted on a *p*-chart are the fraction (or percentage) of defective pieces found in the sample of *n* pieces. The "Run Charts and Control Charts" section of Chapter 15 began with the example of a pen manufacturer. Now let's take that example to the next logical step and make a *p*-chart.

When we left the pen makers, they seemed to have gotten their defective pens down to 2% or less. If we pick it up from there, we will need several subgroup samples of data to establish the limits and process average for our chart. The *p*-chart construction process is very similar to the $\bar{x}$ and R charts discussed in the preceding section. For attributes data, the subgroup sample size should be larger. We need to have a sample size (n) large enough that we are likely to include the defectives. Let's use $n = 100$. We want the interval between sample groups wide enough that if trends develop, we will see them. If the factory makes 2,000 pens of this type per hour and we sample the first 100 after the hour, in an 8-hour day we can obtain eight samples. Three days of sampling will give us sufficient data to construct our *p*-chart. After 3 days of collecting data, we have the data shown in Figure 18–11. To that data, we'll apply the *p*-chart formulas shown in Figure 18–12.

Constructing the *p*-chart, we have several things to calculate: the fraction defective by subgroup (p), the process average ($\bar{p}$), and the control limits (UCL_p and LCL_p).

Fraction Defective by Subgroup (p)

The p values given in Figure 18–11 were derived by the formula $p = np \div n$. For example, for subgroup one, $np = 1$ (one pen was found defective from the first sample of 100 pens). Because p is the fraction defective,

$$p = 1 \div 100$$
$$= 0.01$$

For the second subgroup:

$$p = 2 \div 100$$
$$= 0.02$$

and so on.

Figure 18–10
The Process Is in Statistical
Control

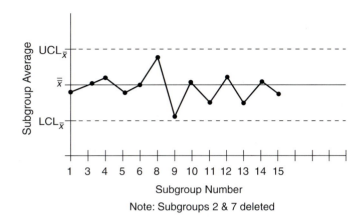

Note: Subgroups 2 & 7 deleted

Process Average ($\bar{p}$)

Calculate the process average by dividing the total number defective by the total number of pens in the subgroups:

$$\bar{p} = (np_1 + np_2 + \ldots + np_k) \div (n_1 + n_2 + \ldots n_k)$$
$$= 42 \div 2,400$$
$$= 0.0175$$

Date	Subgroup	np	p	Date	Subgroup	np	p
8/11	1	1	.01	8/12	13	4	.04
8/11	2	2	.02	8/12	14	0	0
8/11	3	0	0	8/12	15	1	.01
8/11	4	0	0	8/12	16	3	.03
8/11	5	2	.02	8/13	17	1	.01
8/11	6	0	0	8/13	18	3	.03
8/11	7	3	.03	8/13	19	0	0
8/11	8	2	.02	8/13	20	2	.02
8/12	9	1	.01	8/13	21	4	.04
8/12	10	5	.05	8/13	22	1	.01
8/12	11	0	0	8/13	23	3	.03
8/12	12	2	.02	8/13	24	2	.02

np = number defective in subgroup n = subgroup size = 100 k = # of subgroups

p = fraction defective $\bar{p}$ = process average

Figure 18–11
Collected Data for 3 Days

Figure 18–12
p-Chart Formulas

p = rejects in subgroup ÷ number inspected in subgroup
  = $np \div n$

$\bar{p}$ = total number of rejects ÷ total number inspected
  = $\Sigma np \div \Sigma n$

$$UCL_p = \bar{p} + \frac{3 \sqrt{\bar{p}(1-\bar{p})}}{\sqrt{n}}$$

$$LCL_p = \bar{p} - \frac{3 \sqrt{\bar{p}(1-\bar{p})}}{\sqrt{n}}$$

Control Limits (UCL$\bar{p}$ and LCL$\bar{p}$)

Because this is the first time control limits have been calculated for the process (as shown in Figure 18–13), they should be considered *trial limits*. If we find that there are data points outside the limits, we must identify the special causes and eliminate them. Then we can recalculate the limits without the special-cause data, similar to what we did in the series of Figures 18–8 through 18–10, but using the *p*-chart formulas.

In Figure 18–13, LCL_p is a negative number. In the real world, the fraction defective (p) cannot be negative, so we will set LCL_p at zero.

No further information is needed to construct the *p*-chart. The *y*-axis scale will have to be at least 0 to 0.06 or 0.07 because the $UCL_p = 0.0568$. The *p* values in Figure 18–11 do not exceed 0.05, although a larger fraction defective could occur in the future. Use the following steps to draw a control chart:

1. Label the *x*-axis and the *y*-axis.
2. Draw a dashed line representing UCL_p at 0.0568.
3. Draw a solid line representing the process average ($\bar{p}$) at 0.0175.
4. Plot the data points representing subgroup fraction defective (*p*).
5. Connect the points.

The *p*-chart (Figure 18–14) shows that there are no special causes affecting the process, so we can call it *in statistical control*.

Another Commonly Used Control Chart for Attributes Data

The *c*-Chart

The *c*-chart is used when the data is concerned with *the number of defects in a piece*—for example, the number of defects found in a tire or an appliance. In practice the data are collected by inspecting sample tires or toasters, whatever the product may be, on a scheduled basis, and each time logging the number of defects detected. Defects may also be logged by type (blemish, loose wire, and any other kind of defect noted), but the *c*-chart data are the simple sum of all the defects found in each sample piece. Remember, with the *c*-chart, a sample is one complete unit that may have multiple defect characteristics. The following example illustrates the development of a *c*-chart.

$$UCL_p = \bar{p} + \frac{3\sqrt{\bar{p}(1-\bar{p})}}{\sqrt{n}}$$

$$= .0175 + \frac{3\sqrt{.0175\,(1-.0175)}}{\sqrt{100}}$$

$$= .0175 + \frac{3\sqrt{.01719375}}{10}$$

$$= .0175 + \frac{3\times.1311}{10}$$

$$= .0175 + .0393$$

$$= .0568$$

$$LCL_p = \bar{p} - \frac{3\sqrt{\bar{p}(1-\bar{p})}}{\sqrt{n}}$$

$$= .0175 - \frac{3\sqrt{.0175\,(1-.0175)}}{\sqrt{100}}$$

$$= .0175 - \frac{3\sqrt{.01719375}}{10}$$

$$= .0175 - \frac{3\times.1311}{10}$$

$$= .0175 - .0393$$

$$= -.0218 \quad (\text{set at zero})$$

Figure 18–13
p-Chart Control Limit Calculations

A manufacturer makes power supplies for the computer industry. Rework to correct defects has been a significant expense. The power supply market is very competitive, and for the firm to remain viable, defects and rework must be reduced. As a first step the company decides to develop a c-chart to help monitor the manufacturing process. To compile the initial data, the first power supply completed after the hour was chosen as a sample and closely inspected. This was repeated each hour for 30 hours. Defects were recorded by type and totaled for each power supply sample. To develop the initial c-chart, the formulas of Figure 18–15 are applied to the power supply defect data recorded in Figure 18–16.

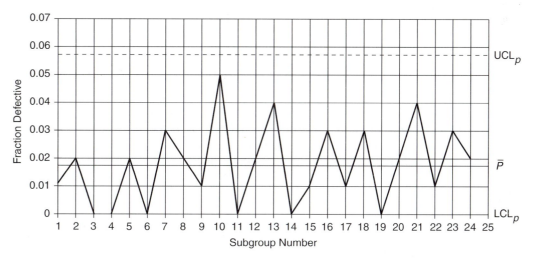

Figure 18–14
p-Chart

QUALITY TIP

Avoid This Control Chart Mistake

Upper control limit, lower control limit and *process average* are not arbitrary terms, nor are they the same as specifications and tolerances. They are statistically derived from the process's own running data. This cannot be emphasized too strongly. The problem is that if the control limits and process average are not statistically derived from the process, it is impossible to know whether the process is in control and hence makes it enormously difficult to institute or validate process improvements. Yet we find many public and private organizations, and especially the military, using arbitrary or specification limits rather than statistical limits on their "control charts." Using this approach requires less work in setting up the charts, and they may look impressive to the uninitiated, but they are not control charts and can perform none of the functions of a control chart.

Source: David Goetsch and Stanley Davis.

Calculating the *c*-chart parameters from the data:

Total defects = 47
Number of samples = 30

$$\bar{c} = 47/30 = 1.56667$$

Largest *c* = 3
Smallest *c* = 0

$$UCL_c = 1.56667 + 3\sqrt{1.56667}$$
$$= 4.32167$$
$$LCL_c = 1.56667 - 3\sqrt{1.56667}$$
$$= -2.18833 \text{ (Because this is negative, set to 0.)}$$

The *c*-chart of Figure 18–17 is constructed from these data. Notice that all data points fell within the control limits, and there were no protracted runs of data points above or below the process average line, $\bar{c}$. The process was "in control" and ready for SPC. Now, as the operators continue to inspect a sample power supply each hour, data will immedi-

Figure 18–15
c-Chart Formulas

$\bar{c}$ = total number of defects/number of samples

$$UCL_c = \bar{c} + 3\sqrt{\bar{c}}$$
$$LCL_c = \bar{c} - 3\sqrt{\bar{c}}$$

ately be plotted directly on the control chart, which of course will have to be lengthened horizontally to accept the new data. This is done with "pages" rather than physically lengthening the chart. Each new page represents a new control chart for the period chosen (week, month, etc.). However, the control limits and the average lines must remain in the same position until they are recalculated with new data. As process improvements are implemented and verified, recalculating the average and limits will be

Sample	Defects by Type						Number of Defects
	A	B	C	D	E	Other	
1			1	1			2
2		1	1		1		3
3	1						1
4							0
5			1				1
6		1		2			3
7			1			1	2
8			2	1			3
9							0
10	1		1			1	3
11		1					1
12				1	1		2
13	1	1					2
14			2				2
15				1		2	3
16							0
17		1					1
18			1	1			2
19							0
20			1				1
21							0
22						1	1
23		1			1		2
24			2		1		3
25	1						1
26			1				1
27			2			1	3
28							0
29				1		2	3
30			1				1
Sample: 1 Power Supply. Sample Rate: 1 Power Supply/hour							Total 47

Figure 18–16
c-Chart Data

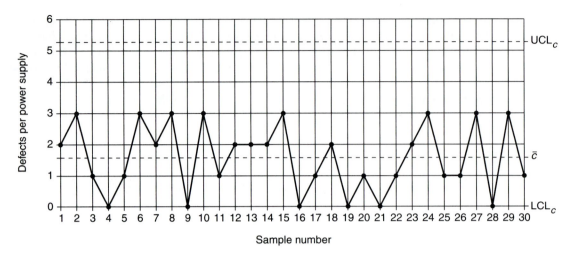

Figure 18–17
c-Chart: Power Supply Defects

necessary. That is because when a process is really improved, it will have less natural variation. The original average and control limits would no longer reflect the process, and hence their continued use would invalidate the control chart.

The Control Chart as a Tool for Continual Improvement

Control charts of all types are fundamental tools for continual improvement. They provide alerts when special causes are at work in the process, and they prompt investigation and correction. When the initial special causes have been removed and the data stay between the control limits (within $\pm 3\sigma$), work can begin on process improvement. As process improvements are implemented, the control charts will either ratify the improvement or reveal that the anticipated results were not achieved. Whether the anticipated results were achieved is virtually impossible to know unless the process is under control. This is because there are special causes affecting the process; hence, one never knows whether the change made to the process was responsible for any subsequent shift in the data or if it was caused by something else entirely. However, once the process is in statistical control, any change you put into it can be linked directly to any shift in the subsequent data. You find out quickly what works and what doesn't. Keep the favorable changes, and discard the others.

As the process is refined and improved, it will be necessary to update the chart parameters. UCL, LCL, and process average will all shift, so you cannot continue to plot data on the original set of limits and process average. The results can look like the succession of charts in Figure 18–18.

An important thing to remember about control charts is that once they are established and the process is in statistical control, the charting does not stop. In fact, only then can the chart live up to its name, *control chart*. Having done the initial work of

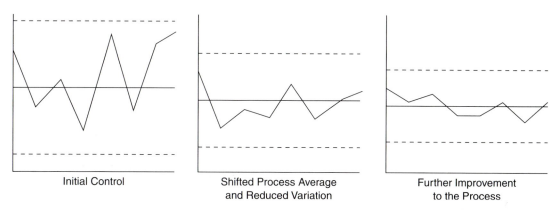

| Initial Control | Shifted Process Average and Reduced Variation | Further Improvement to the Process |

Figure 18–18
Succession of Control Charts

establishing limits and center lines and plotting initial data, and eliminating any special causes that were found, we have arrived at the starting point. Data will have to be continually collected from the process in the same way it was for the initial chart.

The plotting of these data must be done as it becomes available (in real time) so that the person managing the process will be alerted at the first sign of trouble in the process. Such trouble signals the need to stop the process and immediately investigate to determine what has changed. Whatever the problem, it must be eliminated before the process is restarted. This is the essence of statistical process control. The control chart is the statistical device that enables SPC on the shop floor or in the office.

This discussion of control charts has illustrated only the $\bar{x}$ and R charts, the p-chart, and the c-chart. Figure 18–19 lists common control charts and their applications. The methods used in constructing the other charts are essentially the same as for the four we discussed in detail. Each chart type is intended for special application. You must determine which best fits your need.

Statistical Control versus Capability

It is important to understand the distinction between a process that is *in statistical control* and a process that is *capable*. Asking the question, "Is our process in control?" is different from, "Is our process capable?" The first relates to the absence of special causes in the process. If the process is in control, you know that 99.7% of the output will be within the $\pm 3\sigma$ limits. Even so, the process may not be capable of producing a product that meets your customer's expectations.

Suppose you have a requirement for 500 shafts of 2-inch diameter with a tolerance of ± 0.02 inch. You already manufacture 2-inch diameter shafts in a stable process that is in control. The problem is that the process has control limits at 1.97 and 2.03 inches. The process is in control, but it is not capable of making the 500 shafts without a lot of scrap and the cost that goes with it. Sometimes it is possible to adjust the machines or

Data Category	Chart Type	Statistical Quantity	Application
Variables (measured values)	x bar-R ($\bar{x}$ & R)	Mean value and range	Charts dimensions and their precision, weight, time, strength, and other measurable quantities. Example: Anything physically measurable.
	x tilde-R ($\tilde{x}$ & r)	Median and range	Charts measurable quantities, similar to $\bar{x}$ & R, but requires fewer calculations to plot. Example: as above.
	x-Rs (also called x-chart)	Individual measured values	Used with long sample intervals: subgrouping not possible. Example: Products made in batches such as solutions, coatings, etc., or grouping too expensive (i.e., destructive testing). Histogram must be normal.
Attributes (counted values)	p-chart	Percentage defective (also *fraction* defective)	Charts the number of defects in samples of varying size as a percentage or fraction. Example: Anywhere defects can be counted.
	np-chart (also *pn*)	Number of defective pieces	Charts the number of defective pieces in samples of fixed size. Example: As above, but in fixed size samples.
	c-chart	Number of defects	Charts the number of defects in a product (single piece) of fixed size (i.e., like products). Example: Specific assemblies or products (i.e., PC boards, tires, etc.).
	u-chart	Number of defects per unit area, time, length, etc.	Charts the number of defects in a product of varying size (i.e., unlike products). Example: Carpet (area), extrusions (length).

Figure 18–19
Common Control Charts and Their Applications

procedures, but if that could have been accomplished to tighten the limits, it already should have been done. It is possible that a different machine is needed.

There are many variations on this theme. A process may be in control but not centered on the nominal specification of the product. With attributes data, you may want your in-control process to make 99.95% (1,999 out of 2,000) of its output acceptable, but it may be capable of making only 99.9% (1,998 of 2,000) acceptable. (Don't confuse that with $\pm 3\sigma$'s 99.73%: they are two different things.)

The series of charts in Figure 18–20 illustrates how *in statistical control* and *capable* are two different issues, but the control chart can clearly alert you to a capability problem. You must eliminate all special causes and the process must be in control before process capability can be established.

MANAGEMENT'S ROLE IN SPC

As in other aspects of total quality, management has a definite role to play in SPC. In the first place, as Deming has pointed out, only management can establish the production quality level.[4] Second, SPC and the continual improvement that results from it will transcend department lines, making it necessary for top management involvement. Third, budgets must be established and spent, something else that can be done only by management.

Commitment

As with every aspect of total quality, management commitment is an absolute necessity. SPC and continual improvement represent a new and different way of doing business, a new culture. No one in any organization, except its management, can edict such fundamental changes. One may ask why a production department cannot implement SPC on its own. The answer is that, providing management approves, it can. But the department will be prevented from reaping all the benefits that are possible if other departments are working to a different agenda. Suppose, for example, that through the use of SPC, a department has its processes under control and it is in the continual improvement mode. Someone discovers that if an engineering change is made the product will be easier to assemble, reducing the chance for mistakes. This finding is presented to the engineering department. However, engineering management has budgetary constraints and chooses not to use its resources on what it sees as a production department problem. Is this a realistic situation? Yes, it is not only realistic, but very common. Unless there is a clear signal from top management that the production department's SPC program, with its continual improvement initiative, is of vital interest, other departments will continue to address their own agendas. After all, each separate department knows what is important to the top management, and this is what they focus on because this is what affects their evaluations most. If SPC and continual improvement are not perceived as priorities of top management, the department that implements SPC alone will be just that, alone.

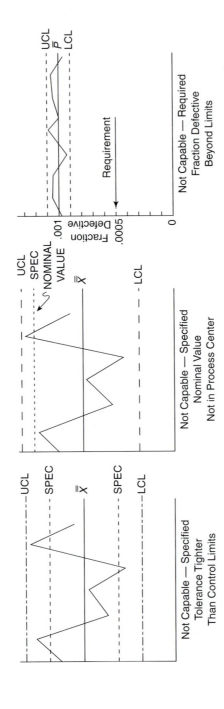

Figure 18–20
In Control and *Capable* Are Not the Same Thing

Training

It is management's duty to establish the policies and procedures under which all employees work and to provide the necessary training to enable them to carry out those policies and procedures. The minimum management involvement relative to SPC training involves providing sufficient funding. More often, though, management will actually conduct some of the training. This is a good idea. Not only will management be better educated in the subject as a result of preparing to teach it, but employees will be more likely to get the message that SPC is a priority to management.

Involvement

When employees see management involved in an activity, they get a powerful message that the activity is important. Employees tend to align their efforts with the things they perceive as being important to management. If managers want their employees to give SPC a chance, they must demonstrate their commitment to it. This does not mean that managers should be on the floor taking and logging data, but they should make frequent appearances, learn about the process, probe, and insist on being kept informed.

A major part of SPC is the continual improvement of processes. Deming pointed out that special causes of variation can be eliminated without management intervention. This is essentially true when it comes to correcting a problem. But when it comes to process improvement, management must be involved. Only management can spend money for new machines or authorize changes to the procedures and processes. Without management involvement, neither process nor product improvement will happen.

ROLE OF THE TOTAL QUALITY TOOLS

Some may consider it inappropriate to include tools other than control charts in a discussion of SPC. However, we take the broader view and include several tools:

- Pareto charts
- Cause-and-effect diagrams
- Stratification
- Check sheets
- Histograms
- Scatter diagrams
- Run charts and control charts
- Flowcharts
- Design of experiments

SPC does not start the moment a control chart is employed. Before SPC can be fully implemented, a lot of work must be done to eliminate the special causes of variation in the process concerned. Consequently, several quality tools will be used before it is time to develop and implement a control chart. When does SPC start? It starts when someone begins cleaning up the process. In the final analysis, this question is not that important, because the quality tools come into play either to support SPC or to be part of the SPC package, depending on the definition used.

With SPC, the total quality tools have a dual role. First, they help eliminate special causes from the process, so that the process can be brought under control. (Remember that a process that is in control has no special causes acting on it.) Only then can the control charts be developed for the process and the process monitored by the control charts. The second role comes into play when, from time to time, the control chart reveals a new special cause, or when the operator wants to improve a process that is in control. This is dealt with in a later section of this chapter, "Implementation and Deployment of SPC."

AUTHORITY OVER PROCESSES AND PRODUCTION

Operators who use SPC to keep track of their processes must have the authority to stop the production process when SPC tells them something is wrong. As long as the plots on the control chart vary about the process average but do not penetrate a control limit, the process is in control and is being influenced by the common causes of variation only. Once a penetration is made, or if the operator sees a run of several plots all on one side of the process average, he or she has good reason to believe that the process needs attention. The operator should be able to stop the process immediately.

A question that frequently arises in the early stages is, "Can a line stoppage be justified in terms of cost and, possibly, schedule?" Toyota found early on that the value in stopping the line for any problem was absolute. Not only does it prevent waste and defective products, but having production at a standstill is a very powerful incentive for finding the cause of the problem and eliminating it—quickly. The word *eliminate* implies that the problem is corrected for good. In a less enlightened factory, the fear of a line stop is so great that the standard procedure is to apply a quick fix to get the line moving again. The emphasis is on keeping production rolling. With this approach, the problem will be investigated and corrected after the line is back in motion. The difficulty is this: the data that might have been available at that moment may have disappeared by the time someone looks for it, and the trail to the root cause may be lost. The odds are good that the same event will recur later on with similar disruption and impact on quality. It will keep recurring until the cause is finally discovered and eliminated. Under SPC and total quality, the emphasis is not on maintaining production no matter what but on eliminating any cause of substandard quality the moment it comes up.

Attempting SPC without giving process-stopping authority to the operator is a serious error. Harry Truman once said that war was too important to leave in the hands of the generals. We believe that line stoppages are too important to leave in the hands of management. Give the authority to the operator, and the underlying problems will be eliminated.

QUALITY TIP

Two Attitudes toward Line Stoppages

Stopping production lines is seen differently by the traditional factory and the total quality factory.

Traditional Factory
- *Line stops because:* Broken machine, missing or incorrect parts, operator problem, and so forth.
- *Reaction:* Find a quick fix; get line moving again. Try to determine/correct the cause later.
- *Result:* Production of defective products and propensity for recurrence.
- *Attitude:* Line stoppages are to be avoided at nearly any cost.

Total Quality Factory
- *Line stops because:* Operator detected an indication of a process problem (e.g., SPC limit penetration or a run).
- *Reaction:* Determine cause and eliminate before restoring production.
- *Result:* Minimizes production of defective product; process becomes more robust.
- *Attitude:* Line stops represent opportunities for improvement.

Source: Stanley Davis.

IMPLEMENTATION AND DEPLOYMENT OF SPC

Going into SPC is not something to be taken lightly or approached in a halfhearted manner. It requires the time and commitment of key personnel. It involves training and the expenses that go with it. It may even involve hiring one or more new people with specialized skills. There may be expenses for consultants to help get the organization started and checked out in SPC. The organization may have to invest in some new tools/tooling, if what is already on hand turns out to be inadequate. But the single most important issue that must be faced when implementing SPC is the culture change that is implicit in using SPC.

Up to this point, the organization has relied on the quality department to assure the quality of products. With SPC, the process operator must be the one who assures product quality, and the quality department must step aside, taking on a significantly different role. Before, if operators could do the assembly steps necessary for their processes, that was adequate. Now their scope of activity must be expanded into new areas, and they must be helped to develop the skills needed to cope with the new requirements. Supervisors and middle managers must give operators the latitude and freedom required to perform the new functions effectively. This sounds easier than it actually is. Many people find it difficult to adjust and adapt to new procedures and attitudes.

In addition, when SPC is used, functions that were formerly carried out by individuals will increasingly be performed by collaborative teams. Employees learn that solving problems, using the quality tools, even defining their own processes are best done by teams of people who bring to the table an array of skills, knowledge, and viewpoints that would be impossible for the individual. Interpreting the control charts, finding root causes of any detected special-cause events, and developing ways to actually improve processes are examples of new tasks that come with SPC. All require team activity.

There is no single right way to implement SPC. What is presented here is a general road map for implementation, covering the major steps in the chronological order in which they should be introduced. The detail behind each of the steps must be worked out for each unique application. An SPC implementation is one area in which the retention of an expert consultant has merit. For SPC to provide any benefit, the program must be statistically valid, and it takes an expert to know whether it is or not.

Figure 18–21 summarizes the steps involved in implementing SPC. The implementation steps are divided into three phases: preparation, planning, and execution.

The Preparation Phase

The preparation phase for SPC includes three steps.

Step 1: Commit to SPC

Any endeavor that requires money, human resources, changing the organization's culture, hiring employees with new skills, or retaining consultants is something to which top management must be committed. The department that forges ahead without that commitment may find itself cut off in mideffort, a situation worse than not having started at all.

Step 2: Form an SPC Committee

SPC can take a lot of time, especially at first when employees are getting acquainted with it and are getting the processes on-line. Unlike total quality, however, SPC can be delegated to a cross-functional team that is tasked to oversee implementation and execution. The SPC team leader need not be the resident expert (at the beginning), but a statistics expert must be included on the team and that person must be heard. A typical team will be composed of representatives from manufacturing, quality assurance, engineering, finance, and statistics. In a manufacturing plant, the manufacturing member should be the team leader. The function of the team will be to plan and organize the implementation for its unique application, to provide training for the operators, and to monitor and guide the execution phase. Forming the SPC committee is top management's responsibility.

Phase	Responsibility	Action
Preparation	Top management	(1) Commitment to SPC
	Top management	(2) Organize SPC committee
	Consultant or in-house expert	(3) Train SPC committee
Planning	SPC committee assisted by consultant or expert	(4) Set SPC objectives
		(5) Identify target processes
	Consultant or in-house expert	(6) Train appropriate operators and support personnel
	QA	(7) Assure repeatability and reproduceability of instruments and methods
	Management	(8) Delegate responsibility for operators to play key role
Execution	SPC committee, operator suppliers, customers	(9) Flowchart the process
	Operator w/ expert assistance	(10) Eliminate the special causes of variation
	Consultant or in-house expert	(11) Develop control chart(s)
	Operator	(12) Collect and plot SPC data
	Operator w/ expert assistance	(13) Determine process capability*
	Operator	(14) Respond to trends and out-of-limits data
	SPC committee and management	(15) Track SPC data
	Operator w/ assistance as required	(16) Eliminate root causes of any special causes of variation
	All	(17) Continuously improve the process (narrow the limits)

Figure 18–21

The SPC Implementation Road Map

* If the process is not capable of meeting requirements, it must be changed or replaced; go back to step 9.

Step 3: Train the SPC Committee

The newly formed SPC committee must receive basic training before its work starts. In a typical situation, the committee members will have had little or no practical experience with statistics. The training must be done by an expert. It is possible to send employees to training courses or bring the expert to the company. At the conclusion of the training period, the members will not have become experts, but they will know enough to set objectives and to determine which processes should be targeted first. At this point, continued help from a statistics expert remains critical.

The Planning Phase

The planning phase includes the next five steps.

Step 4: Set SPC Objectives

The SPC committee should set objectives for the program. What do we hope to gain from SPC? How will we measure success (at the balance sheet's bottom line, customer feedback, reduction in scrap, lower cost of quality, or perhaps all of these)? If the team waits until the SPC machinery is in place and producing data to decide what gains are expected, consensus may never be reached on how well or how poorly it is working. Set the objectives. Measure against them. As with all objectives, they should be reviewed from time to time to make sure they are still valid and meaningful. Objectives may be added, eliminated, or changed, but they must be in place and understood by all.

Step 5: Identify Target Processes

It is not feasible to attempt to apply SPC to all processes at once. The people involved in designing the SPC application, collecting data and interpreting its meaning, getting the processes under control, and plotting and evaluating control chart data will be in a learning mode for the first several weeks. For that reason, it is important to select just a few pilot processes for the initial implementation. These should be processes that are well understood and that promise to be relatively easy to bring under control. They should also be important processes, ones that have meaning rather than something trivial. The key point to remember is this: Select initial processes from among those that stand the best chance of quick success. With some initial successes under its belt, the organization can go on with confidence to the processes that are the most critical.

Consideration should also be given to the flow of processes one to another. For example, if there is a production line with four processes, it makes sense to implement SPC in the order of production flow. Trying to introduce it at the end or in the middle of the four processes may prove difficult. If the first three processes feed their defects into the final process, it will be impossible to eliminate the special causes of variation of the fourth. On the other hand, by starting at the beginning of the flow, putting process 1 under control may eliminate one or more of the special causes affecting processes 2 through 4. The idea is to start implementation at the front of a series of processes, not at the back. Selection of the target processes should be done by the SPC committee, with comprehensive, open communication with the process operators.

Step 6: Train Appropriate Operators and Teams

The operators and teams who will be directly involved with the collection, plotting, and interpretation of SPC data, and those who will be involved in getting the targeted processes under control, will require training in the use of quality tools and in flowcharting. Some processes may require the use of design of experiments (DOE). If this is the case, the help of a specialist, both to provide training and to assist with the DOE process, may be needed. Training given at this point must make clear the significance and the objectives of the work to be undertaken. Participants will be the process operators and the engineers and quality specialists who support them. Only the employees who will be involved in the initial SPC projects should be included in the first class. As SPC is spread throughout the plant, it will be necessary to train other operators and teams and their support personnel. But by delaying training until it is time to expand beyond the initial processes, the advantage of just-in-time training will be gained. In addition, you will be able to capitalize on lessons learned from the initial projects. The training needed can typically be accommodated in a 1- or 2-day session.

Step 7: Assure Repeatability and Reproducibility (R&R)
of Gauges and Methods

All measuring instruments, from simple calipers and micrometers to coordinate measuring machines, must be calibrated and certified for acceptable R&R performance. For SPC to work, the measured data plotted on the control charts must be reliable. A gauge that cannot repeat the same measurement with the same operator consistently, or one that is so difficult or idiosyncratic in its use that no two operators can obtain the same data, will not work in an SPC environment. The particular application will determine the range of variability that is acceptable in measuring instruments. It must then be verified that each instrument to be used is capable—and that all the people who will be using the instrument are adequately trained in its use. This must be done before step 10.

Step 8: Delegate Responsibility for Operators to Play a Key Role

As the last step in the planning phase, just before SPC execution is to begin, management should delegate to the process operators responsibility for maintaining the SPC control charts, collecting and plotting the data, and taking appropriate action. Let the operators know that these functions are theirs, but make certain everyone else knows it too.

The Execution Phase

The execution phase includes nine steps.

Step 9: Flowchart the Process

The first step in the SPC execution phase, taking the broad perspective regarding SPC's boundaries, is flowcharting or characterizing the process to which SPC will be applied. Only when a graphic representation of the entire process exists—including its inputs, outputs, and all the steps between—can the process be fully understood. Invariably, flowcharting will reveal process features or factors that were not known to everyone.

After the flowchart has been completed and everyone agrees that it represents the way the process actually works, a large version should be produced on poster board and permanently placed in open view at the process location. It will provide invaluable information and may even suggest process improvements later on. Members of the SPC team should help, but the development of the process flowcharts should be the responsibility of special teams composed of the process operators, their internal suppliers and customers, and appropriate support members. (Support personnel may include engineers, materials specialists, financial specialists, etc., as needed.)

Step 10: Eliminate the Special Causes of Variation

Now that participants understand the process, it is time to identify and eliminate the special causes of variation. This is best begun through the use of the cause-and-effect diagram, which was discussed in Chapter 15 as one of the seven total quality tools. The cause-and-effect diagram will list all of the factors (causes) that might impact the output in a particular way (effect). Then by applying the other tools, such as Pareto charts, histograms, and stratification, the special causes can be identified and eliminated. Until the special causes that are working on the process are eliminated, the next steps will be difficult or impossible to complete. Elimination of the special causes should be a team effort among the process operators, internal process suppliers and customers, engineers, and quality assurance personnel, with additional help from other departments as required. For example, if materials are a factor, the purchasing department might become involved. Be sure to keep the operators at the center of the activity, as this will give them ownership as well as valuable experience.

Step 11: Develop Control Charts

With the absence of special causes, it is now possible to observe the process unencumbered by external factors. The statistics expert, or consultant, can now help develop the appropriate control charts and calculate valid upper and lower control limits and process averages. Selection of the control chart type will be determined by the kind of data to be used. (See Figure 18–19.)

Step 12: Collect and Plot SPC Data; Monitor

With the special causes removed, and with the process running without tweaking (frequent minor adjustments to one or more process factors), the process operator takes the sample data (as specified by the statistics expert) and plots it on the control chart at regular intervals. The operator carefully observes the location of the plots, knowing that they should be inside the control limits, with the pattern varying randomly about the process average if the process is in control.

Step 13: Determine Process Capability

Before going further, it is important to determine whether the process is capable of doing what is expected of it. For example, if the process output is to be metal parts with a specified length of between 5.999 and 6.001 inches, but the process turns out as many pieces outside those dimensions as it does within, the process is not capable. The process is capable if its frequency distribution is a bell-shaped curve centered on the specification average, in this case 6 inches, and with the $\pm 3\sigma$ spread coincident with, or narrower than, the specification limits. With the bell curve centered on the specification average, and the specification limits coincident with the $\pm 3\sigma$ spread of the bell curve, we could expect 3 nonconforming parts out of 1,000. If the specification limits are inside the $\pm 3\sigma$ spread, then the defect rate would be higher; if they are outside (the bell curve is narrower than the limits), the defect rate would be lower.

Two methods exist for determining process capability. The first assumes that the bell curve is centered on the specification average and is called C_p. The second does not assume alignment of the process average and the specification average, and it is called C_{pk}. Figure 18–22 explains the procedures for calculating these capability indices. As we have already learned, it is possible to have a process that is in control and still not capable of meeting the customer's specifications. When this is the case, it is up to management to replace or upgrade the process capability, which may require the purchase of new equipment.

Step 14: Respond to Trends and Out-of-Limits Data

As data are plotted, the operator must respond to any penetration of the control limits or to any run of data above or below the process average line. Either of these is an indication that something is wrong within the process or that some external factor (a special cause) has influenced the process. With experience, operators may be able to handle many of these situations on their own, but when they cannot, it is important that they summon help immediately. The process should be stopped until the cause is identified and removed. This is one of the most important functions of SPC—letting the operator know there is a problem early enough to prevent the production of defective products that must be scrapped or reworked. The only way to respond in such cases is by immediately eliminating the problem. This is another application for team (usually *ad hoc*) participation.

Step 15: Track SPC Data

The SPC committee and management should pay close attention to the SPC data that are generated on the production floor. Doing so will give them an accurate picture of their production capability, the quality of their processes, trends that may develop, and where they should concentrate resources for improvement. A secondary benefit of displaying this level of interest in SPC is that the operators and their support functions will know that management is truly interested in the program, and they will give it the attention and care appropriate to a high-visibility initiative.

Step 16: Eliminate the Root Causes of Any New Special Causes of Variation

From time to time, new special causes will come up, even in processes that have long been in control. When this happens, the operator will know it because the SPC data will go out of limits or skew to one side or the other of the control chart center line. It is important that the root causes of these special causes be eliminated to prevent their recurrence. For example, if the purchasing department placed an order for the next

Process Capability may be calculated in two ways. The first assumes that the process average is centered on the specification average, and is denoted as C_p, the process Capability Index.

$$C_p = \frac{USL-LSL}{6\hat{\sigma}}$$

Where: $USL = Upper\ Specification\ Limit$
         $LSL = Lower\ Specification\ Limit$

$\hat{\sigma} = Estimated\ Process\ Average$

$$\hat{\sigma} = \frac{\bar{R}}{d_2}$$

Where: d_2 **is a constant** (see table below)

$\bar{R} = the\ Process\ Average\ Range$

The second method is used when the process average is not assumed to be coincident with the specification average, and is denoted as C_{pk}.

$$C_{pk} = \frac{USL-\bar{\bar{X}}}{3\hat{\sigma}}$$

and

$$C_{pk} = \frac{\bar{\bar{X}}-LSL}{3\hat{\sigma}}$$

Where: $\bar{\bar{X}} = the\ Process\ Average$

C_{pk} is taken as the **smaller of the two values.**

For either case, a Capability Index of:

=1 means that the specification limits and average are coincident with the process ± 3σ limits and process average.

<1 means that the specification is tighter than the process spread. The defect rate will be greater than 3 parts per 1,000.

>1 means that the process spread is tighter than the specification limits. The defect rate will be less than 3 parts per 1,000.

NOTE: 1.33 IS THE PREFERRED MINIMUM CAPABILITY INDEX

Table for d_2 Values

# Observations in subgroup	2	3	4	5	6	7	8	9
d_2	1.128	1.693	2.059	2.326	2.534	2.704	2.847	2.970

Figure 18–22
Process Capability Calculation
664

shipment of raw material from a different vendor because its price was cheaper than the current supplier, it is possible that the material coming from the new source might react differently in the process, shifting the process average one way or the other. The root cause may not be the new material. If you scrap it, purchasing is more than likely going to order the replacement material from the same low bidder, and the problem will probably recur. It would seem that the root cause of the problem is purchasing's tendency to order from the cheapest source. Eliminating this root cause may require a management-approved procedure mandating the use of preferred suppliers. At the very least, there should be an ironclad agreement that purchasing would not order materials from a new supplier without having the material certified by quality assurance and manufacturing personnel. This is a case where the operator initiates the action, a team may identify the root cause, and management involvement may be required to eliminate the problem. This is the way the process is meant to proceed. Wherever the help must come from, it has to be readily available.

Step 17: Continual Improvement: Narrow the Limits

With the process under control and the special causes eliminated, continual improvement can be implemented. What this means is that the process average should be centered on the specification average, if that is not already the case, and more frequently, it means the narrowing of the $\pm 3\sigma$ limits (see Figure 18–23). Both of these improvements—centering the process on the specification average and narrowing the limits—will result in fewer parts failing to meet the specifications. Scrappage will be reduced, the process will become more robust, quality will improve, and costs will decrease. The key, of course, is finding ways to improve the processes, but with SPC, one has the understanding of the processes necessary to see and comprehend the problems. Only then can real improvement follow.

INHIBITORS OF SPC

As in the case of implementing just-in-time (JIT) and benchmarking, a number of factors can inhibit the implementation of SPC. With SPC there is not usually the kind of philosophical resistance that is common with JIT and benchmarking. However, it is true even with SPC that there must be a management commitment, because there will be start-up costs associated with implementation. The most common inhibitor of SPC is lack of resources.

Capability in Statistics

Many organizations do not have the in-house expertise in statistics that is necessary for SPC. As SPC is being introduced and decisions are made on where to sample, how much to sample, what kinds of control charts to use, and so forth, a good statistician is necessary to assure the validity of the program. If the organization does not employ such an expert, it should either hire one or retain the services of a consultant for the early phases of the SPC implementation.

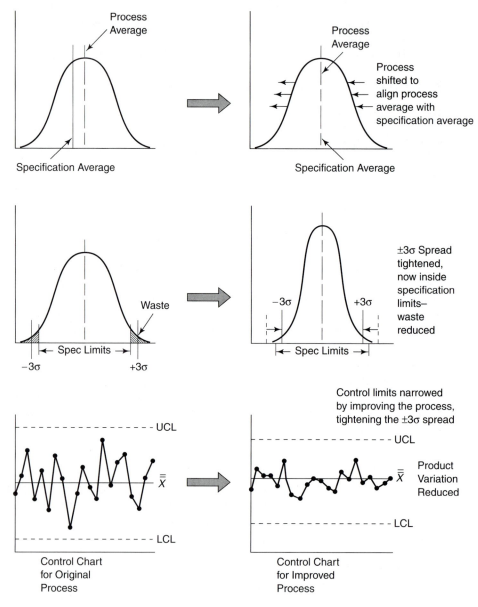

Figure 18–23
Process Improvement

The danger inherent in not having statistical expertise is developing an SPC program that is statistically invalid—a fact that can easily escape nonstatisticians. The organization will count on SPC to control processes when, in fact, it cannot. A flawed SPC implementation may send messages that make the process control situation worse than it was before. It is important that the initial design of the SPC program be valid. This requires someone with more than a passing knowledge of statistics. If there are any doubts, get help.

Misdirected Responsibility for SPC

Too many companies make the decision to use SPC but then turn it over to the statisticians or the quality assurance department. The value of those departments should not be minimized, but the owner of the process in question should be the person responsible for SPC. This person is the one who can make best use of SPC, and there will be no question about the validity of the data because he or she is the one collecting it. The process operators will require help from the statistician and others from time to time, but they are the appropriate owners of SPC for their processes.

When someone else is responsible for SPC (meaning the collection and logging of data and making corrections, stopping out-of-control processes, and getting them fixed), process operators see the entire SPC program as just another check on them, and they are very uncomfortable with it. Management tends to see it the same way, but from their particular perspective—a means of checking up on the operators. Nothing good will come from such a relationship.

Neither is ownership by the statistician the appropriate answer. If the statistician owns SPC, he or she is more apt to find fascination in the numbers themselves than in what they mean in terms of quality. Even if statisticians are tuned in to the objective, the operator will see them and their SPC charts as just another intrusion.

If operators have the responsibility for SPC, they will become familiar with the tasks involved and will see it as a means to help them get the most out of their processes. This is the payoff. All of the others need to observe and review, assisting when needed, but never usurping the operator's ownership.

Failure to Understand the Target Process

Unless a process has been flowcharted recently and characterized, the odds are good that the people designing the SPC system for it do not know how the process actually works. Most processes have evolved over many years, changing now and then to meet requirements of the market or the desires of management or operators. Few are adequately documented. People are usually astonished when flowcharts reveal the complexity of processes they thought to be straightforward. A good SPC system cannot be designed for a process that isn't fully understood.

Failure to Have Processes under Control

Before SPC can be effective, any special causes of variation must be removed. This was discussed earlier, but it is appropriate to mention it here again as an inhibitor of SPC. Remember, by definition, a process is not in control if any special cause is working on it.

The use of control charts assumes an in-control process. Their use will set off visual alarms whenever a new special cause is introduced. But the real work of process improvement can come about only when nothing but the common causes are active. This is why SPC is so powerful. It will show when common causes are the only causes of variation so that improvements to the fundamental process can be made. Will special causes still come up from time to time? Certainly, but this is different from trying to control a process with special causes constantly present, masking both the common causes and each other.

Inadequate Training and Discipline

Everyone who will be involved in the SPC program must be trained, not only in data acquisition, plotting, and interpretation for control charts but also in the use of the seven tools. Not everyone needs to be a statistics expert, but all need to know enough so that with a statistician's help, the program can be designed and operated.

Training should teach that SPC and tweaking do not make a good pair. If tweaking is permitted, SPC data will be meaningless. The process may appear to be more stable than it is if the person doing the tweaking is an expert, or it may show more variation than if left alone. It must be understood that operators and engineers alike are to let the process run essentially hands-off until an out-of-limits condition is detected. Variation between the limits is not to be tweaked out. The only acceptable means of reducing the variation is a real process improvement that will narrow the limits permanently.

Measurement Repeatability and Reproducibility

SPC data are the result of measurement or count. In the case of the variables data (the measurements), the data become meaningless when the measurements are not repeatable. For example, a worn instrument or a gauge with insufficient precision and resolution might yield measurements over an unacceptably wide range when measuring the same object repeatedly. This is not satisfactory. The data taken from all measurements must be accurate to the degree specified, and repeatable, or there is no point in recording it.

Nothing should be taken for granted. Before any gauge is used for SPC, it should be calibrated and its repeatability certified. It is also important that different operators obtain the same readings. This is known as *reproducibility*. Before getting them involved with SPC, certify all gauges and train all operators.

Low Production Rates

Although it is more convenient to implement SPC with processes that have continuous flow, or high rates of product output, it is by no means impossible to apply SPC to low-rate production of the type that is often found in a job shop setting. In a factory that produces several hundred printed circuit boards per day, sampling schemes are relatively easy. A job shop might produce only a few boards in a day, often with gaps between production days. Sampling there must be done differently. Low-rate production provides an opportunity for taking a 100% sample. It is possible to take a sample from every board. In such an application, a computer-generated random x–y table can designate a specific small area of a board for inspection of solder joints or other attributes. From that, a number representing the fraction defective may be developed. Control charts are easily constructed for fraction-defective data. Low production rates are not a good excuse for avoiding SPC.

 SUMMARY

1. The origin of SPC was in the work of Dr. Walter Shewhart at Bell Laboratories in 1931. Although SPC was ignored in the West after World War II, Japan adopted it, subsequently developing it into total quality. SPC is a statistical method of separating special-cause variation from natural variation to eliminate the special causes and to establish and maintain consistency in the process, enabling process improvement. SPC is essential in the West today to elevate the quality of products and services, while lowering costs, to compete successfully in the world's markets.

2. A total of 99.73% of the output of a process that is in statistical control will fall within the $\pm 3\sigma$ limits of the process. Do not confuse process average and limits with specification average and limits. It is usually desirable to make the process average coincident with the specification average and to make the process spread narrower than the specification limits.

3. Continual improvement of processes requires that special causes be eliminated first. Process improvement narrows the shape of the process's bell curve, resulting in less variation.

4. Continual improvement is a key element of SPC and total quality. SPC enhances the predictability of processes and whole plants. Elimination of waste is another key element of SPC. SPC can help improve product quality while reducing product cost.

5. SPC makes sampling inspection more reliable. SPC supports process auditing as a substitute for more expensive inspection. SPC requires a capability in statistics, either in-house or through a consultant. Process operators should be key players in any SPC program. Understanding the process is a prerequisite to SPC implementation. All employees involved in SPC must be trained for their involvement. Measurement repeatability and reproducibility is essential for SPC.

6. Management's role in SPC is similar to its role in total quality overall: commitment, providing training, and involvement. The seven tools, augmented by flowcharting and DOE, are required for SPC. SPC and the operator must have process-stop authority. SPC implementation must be carried out in an orderly, well thought out sequence.

7. SPC requires collaborative team activity.

 KEY TERMS AND CONCEPTS

Auditing

Authority over processes

Commitment

Continual improvement

Control charts

Control of variation

Elimination of waste

Five M's

Flowcharting

In-control process

Involvement

Narrow the limits

Natural causes

Out-of-limits data

Predictability of processes

Process capability

Repeatability

Reproducibility

Sampling

Seven tools

SPC

SPC committee

Special causes

Target processes

Training

Tweaking

 FACTUAL REVIEW QUESTIONS

1. Define the concept of statistical process control.
2. Explain briefly the rationale for SPC.
3. What is meant by variation in processes?
4. Define the following concepts:
   - Continual improvement
   - Predictability of processes
   - Elimination of waste
   - Sampling
   - Auditing
5. What is management's role in the implementation of SPC?
6. Describe how the seven tools are used when implementing SPC.
7. Why is it important to give operators authority over their processes?
8. List the various steps that should be followed when implementing SPC.
9. Why is management commitment so important when implementing SPC?
10. List and briefly explain the major inhibitors of SPC.

 CRITICAL THINKING ACTIVITY

SPC Review

1. Review the section "Statistical Process Control Defined." Explain how the Five M's and environment can affect processes used in the following:
   a. A hardware store
   b. A hospital
   c. An accounting firm
   d. A newspaper
   e. A factory
   f. A new-car dealership
2. Explain the relationship that exists between the histogram and the control chart.
3. Contrast the histogram's characteristic of representing a "snapshot" of a process with a control chart.
4. Defend the statement that the operator of the process should be the owner and data plotter of the control chart, as opposed to a person from quality assurance or engineering, for example.

5. Comment on the significance of the statement, "Control chart parameters must be statistically derived and cannot simply be specifications or some arbitrary values that are based on production expectations."

DISCUSSION ASSIGNMENT 18–1

Hi-Sport Manufacturing Co., Inc.

Hi-Sport is a small company that manufactures logo sporting jackets. A key goal of the company has always been excellent quality. This has been achieved largely through rigorous inspection, a process that has come to be known as "inspecting the quality in." As a result, the firm has always had a high reject rate at final inspection. This has necessarily resulted in too many jackets being scrapped or sold below cost as "seconds." It has also resulted in a bothersome percentage of imperfect jackets "slipping through" inspection and ending up in the hands of customers. The impact has been a so-so reputation with customers and prices too high to be competitive with the imports or major U.S. manufacturers.

Management tried and tried to get the production workers to do better, but it seemed that every effort to reduce defects came to nothing. Sometimes it appeared that good ideas and the best of intentions only made matters worse. A few weeks ago the managers retained a manufacturing consultant with statistics and process control credentials. He told them their first priority should be to get their processes under control.

With the consultant's help, they started their program by identifying and eliminating several special causes of variation. These special causes had included machines that needed maintenance and calibration, some employees with insufficient training, and the absence of written work instructions for certain procedures. By the 6-week mark, Hi-Sport's quality had noticeably improved. Management decided it was time to attempt the development of a control chart.

Because rejects were based on pass/fail criteria for various characteristics, the managers needed a control chart that could respond to nonmeasurable attributes. Three common charts meet that requirement: the p-chart, np-chart, and c-chart. The p-chart could help control the percentage of defective jackets. The np-chart could help control the number of defective jackets. The c-chart could help control the number of defects in a jacket. They decided the c-chart would give them what they wanted—namely, using one jacket as the sample, and tracking the defects found in the sample.

For the initial chart-development data, one jacket was inspected each hour for 30 consecutive working hours. The data are listed in the accompanying table. The data were recorded in the five most common defect categories, with a sixth column collecting all other types of defects encountered. At the end of the 30th hour, 46 total defects had been recorded from the 30 jacket samples.

Hi-Sport Manufacturing Co., Inc.							
	Defects by Type						Number of Defects
Sample	A	B	C	D	E	Other	
1		1	1		1		3
2		1					1
3							0
4	1		1				2
5					1		1
6		1					1
7	2		1		2		5
8				1			1
9							0
10	1					1	2
11		1		1			2
12							0
13							0
14				1			1
15		1					1
16			1			1	2
17			1				1
18	2					1	3
19							0
20				1			1
21	1						1
22							0
23			2				2
24	1						1
25	2	1		3			6
26	1			1			2
27	1	1		2			4
28			1				1
29							0
30			1	1			2
Defects per sample. Sample: 1 Jacket. Sample Rate: 1 Jacket/hour.							Total 46

Discussion Questions

Discuss the following questions in class or outside of class with your fellow students:

1. You are the consultant, and because this is Hi-Sport's introduction to control charts, you will have to help it develop the chart. From the data table, compute c, UCL_c, and LCL_c.
2. Next, construct a control chart. Be sure to include some "headroom" above the upper control limit to allow for any out-of-limit events that may be encountered. Also make the chart long enough horizontally to include all the data points in the table, plus a few more days of real-time data points that will need to be plotted as SPC tracking begins.
3. Now plot the appropriate data from the table on the chart. Is the process in control, or is one or more special causes still lingering? Where do you go from here? Complete your task by (a) providing Hi-Sport with a control chart that reflects an in-control process, or (b) abandoning the current data, starting over again by seeking out the special causes that prevented success the first time.

To see how the authors resolved this issue, see the next section.

Authors' Solution to Hi-Sport's Control Chart Exercise

In practice it is not uncommon to encounter a special cause while developing a control chart. This may happen if the special cause is so subtle or infrequently recurring it has gone unnoticed, or simply because there were others, now gone, that had masked this one. Suppose you have collected your data over 5 working days, made the necessary calculations, and plotted the new control chart. To your great consternation you find that one of your data points is beyond the upper control limit. Have you lost a week's work because there was still a special cause of variation lurking in your process? No. You can salvage it as we do here:

Total defects: 46 (This is the value for c.)

Number of samples: 30 (The number of jackets inspected.)

$\bar{c} = 46/30 = 1.533333$

Largest $c = 6$

Smallest $c = 0$

$$
\begin{aligned}
UCL_c &= \bar{c} + 3\sqrt{\bar{c}} \\
&= 1.533333 + 3\sqrt{1.533333} \\
&= 5.248168
\end{aligned}
$$

$$
\begin{aligned}
LCL_c &= \bar{c} - 3\sqrt{\bar{c}} \\
&= 1.533333 - 3\sqrt{1.533333} \\
&= -2.18150 \text{ (Since this is a negative number, use 0.)} \\
&= 0
\end{aligned}
$$

The control chart identified as Hi-Sport v1.0 was constructed using these $\bar{c}$ and UCL and LCL values. Note that sample 25 is above the upper control limit, which indicates that it is the result of special cause. Because the control chart must be developed from natural (common-cause) variation data only, sample 25 would seem to rule out the

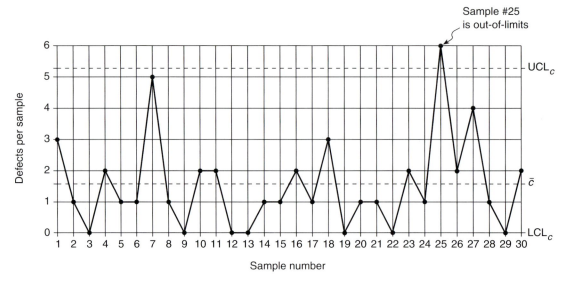

c-chart
Defects per jacket
Hi-Sport v1.0

$UCL_c = 5.248168$
$LCL_c = 0.0$
$\bar{c} = 1.533333$
OUT OF CONTROL

entire 30 hours of data. However, if sample 25 is eliminated from the data, and $\bar{c}$ and UCL and LCL are recalculated on the 29 remaining samples, the control chart will remain valid. Recalculating $\bar{c}$, UCL, and LCL without sample 25 yields the following:

$$\bar{c} = 40/29 = 1.37931 \qquad UCL_c = 4.902632 \qquad LCL_c = 0.0$$

The control chart constructed from these values is shown as Hi-Sport v1.1. Note that with the removal of sample 25 data, $\bar{c}$ (the average number of defects per sample) decreased from 1.533333 to 1.37931. At the same time UCL decreased from over 5 to under 5. This is a narrowing of limits, which is good. However, sample 7 is now beyond the upper control limit. The same procedure, deleting sample 7 data, can be repeated. Recalculating the remaining 28 samples yields the following:

$$\bar{c} = 35/28 = 1.25 \qquad UCL_c = 4.604102 \qquad LCL_c = 0.0$$

The corresponding control chart is shown as Hi-Sport v1.2. Note that all sample data points are now inside the control limits, and there are no protracted runs below or above the $\bar{c}$ line. With that Hi-Sport's consultant pronounced the chart ready for use. This was a close call. If you encounter an out-of-limits data point in your initial chart development data, it is OK to simply eliminate the offending sample data from the calculations and proceed. Doing the same again (as was done here) for a second out-of-limits sample is usually acceptable, but if a third one crops up in a set of 30 to 50

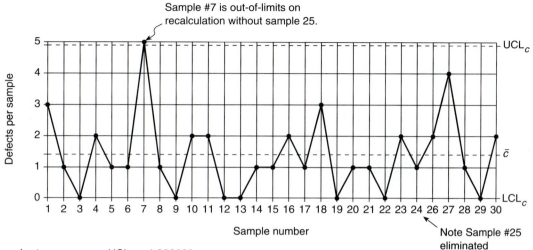

Sample #7 is out-of-limits on recalculation without sample 25.

Note Sample #25 eliminated

c-chart
Defects per jacket
Hi-Sport v1.1

$UCL_c = 4.902632$
$LCL_c = 0.0$
$\bar{c} = 1.37931$
OUT OF CONTROL

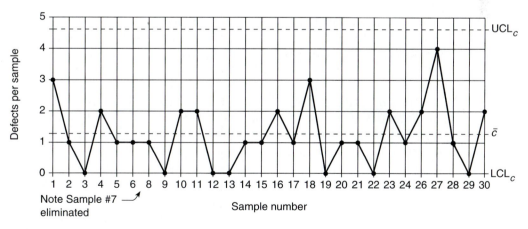

Note Sample #7 eliminated

c-chart
Defects per jacket
Hi-Sport v1.2

$UCL_c = 4.604102$
$LCL_c = 0.0$
$\bar{c} = 1.25$

data point samples, the process is too unstable for a control chart. Go back to eliminating special causes before collecting all-new data from which to construct your control chart.

You might ask, "If the control chart showed an out-of-limits condition for data sample 7, why not eliminate that special cause before you go on?" If the special cause was known, then certainly it should be eliminated forthwith. (We would still have to eliminate the data sample and recalculate on the remaining data.) The problem is that it occurred several hours (or maybe days) ago and may no longer be apparent. The trail may have become obscured. That is why, when a control chart is in place and an out-of-limits data point is taken, it is important to stop the process immediately to find the cause. The more time passes between the event and the search, the less likely the cause will be discovered.

DISCUSSION ASSIGNMENT 18–2

The Start of the Japanese Quality Movement

When the U.S. Forces occupied Japan at the end of World War II to set up the occupation government, they found the Japanese telephone system to be poor in quality and unreliable. General MacArthur's people knew this would be a major problem for them and for Japan in trying to get the country on its feet again. Bell Laboratories' people were brought in to assist the Japanese telecommunications industry, and starting in May 1946 they taught their Japanese counterparts the principles of modern quality control based on Dr. Shewhart's work.

While the Bell Labs people were in Japan, a copy of Shewhart's book *The Economic Control of Quality of Manufactured Product* was given to the Union of Japanese Scientists and Engineers (JUSE). One of its members (the organization had only 12 at the time) was so taken with Shewhart's ideas that he stenciled by hand a copy of the book onto mimeograph masters, so that it could be reproduced and circulated. These two events were the start of the quality movement in Japan.

Sources: Kaoru Ishikawa, *What Is Total Quality Control? The Japanese Way* (Upper Saddle River, NJ: Prentice Hall, 1985), 15. Walton, 12.

DISCUSSION ASSIGNMENT 18–3

SPC's Effect on Competitiveness

The government invites two companies to bid for a contract to produce 100 flight-line avionics maintenance systems. The design is owned by the Air Force, and the Air Force will provide all necessary documentation to the successful bidder. Both companies understand the requirements of the contract, and both are equipped and have the know-how to manufacture the devices.

Company ABC, with no SPC experience, develops a conservative proposal, accounting for 25% rework in its manufacturing labor costs, padding materials costs by 10% in anticipation of scrappage, and inspection sufficient to smoke out most of the defects—calculated at 20% of the basic manufacturing labor.

Company XYZ, which uses SPC in all its manufacturing processes, bids rework and scrap at much lower rates, and includes only enough inspection to audit processes and meet the customer's own minimum inspection criteria.

The following chart compares the bids from the two companies:

	Company ABC	Company XYZ
Assembly labor	$200,000	$200,000
Rework labor	50,000 (25%)	8,000 (4%)
Inspection labor	40,000 (20%)	4,000 (2%)
Materials	550,000	505,000
Totals	$840,000	$717,000

With a difference of $123,000, there can be no doubt that Company XYZ will win the contract. Not only is Company ABC's bid more than 17% higher, but one would be safe in predicting that its higher priced product would be inferior to XYZ's. SPC is the only difference here.

ENDNOTES

1. Mary Walton, *The Deming Management Method* (New York: Putnam, 1986), 12.
2. Philip B. Crosby, *Quality Is Free: The Art of Making Quality Certain* (New York: McGraw-Hill, 1979), 68.
3. Gregory B. Hutchins, *Introduction to Quality Control, Assurance, and Management* (New York: Macmillan, 1991), 162.
4. Nancy R. Mann, *The Keys to Excellence: The Story of the Deming Philosophy* (Los Angeles: Prestwick, 1989), 21.

CHAPTER NINETEEN

Continual Improvement

"People seldom improve when they have no other model but themselves to copy after."
   Oliver Goldsmith

MAJOR TOPICS

- Rationale for Continual Improvement
- Management's Role in Continual Improvement
- Essential Improvement Activities
- Structure for Quality Improvement
- The Scientific Approach
- Identification of Improvement Needs
- Development of Improvement Plans
- Common Improvement Strategies
- Additional Improvement Strategies
- The Kaizen Approach
- Goldratt's Theory of Constraints
- The CEDAC[1] Approach

One of the most fundamental elements of total quality is continual improvement. The concept applies to processes and the people who operate them, as well as to the products resulting from the process. A fundamental total quality philosophy is that all three, processes, people, and products, must be continually improved. This chapter provides the information needed to make continual improvements to the processes and products. (Improvement of employee skills and performance is the subject of Chapters 4,6, 8, 9, 10, 11, and 12.)

RATIONALE FOR CONTINUAL IMPROVEMENT

Continual improvement is fundamental to success in the global marketplace. A company that is just maintaining the status quo in such key areas as quality, new product development, the adoption of new technologies, and process performance is like a runner who is standing still in a race. Competing in the global marketplace is like competing in the Olympics. Last year's records are sure to be broken this year. Athletes who don't improve continually are not likely to remain long in the winner's circle. The same is true of companies that must compete globally.

Customer needs are not static; they change continually. A special product feature that is considered innovative today will be considered just routine tomorrow. A product cost that is considered a bargain today will be too high to compete tomorrow. A good case in point in this regard is the ever-falling price for each new feature introduced in the personal computer. The only way a company can hope to compete in the modern marketplace is to improve continually.

MANAGEMENT'S ROLE IN CONTINUAL IMPROVEMENT

In his book *Juran on Leadership for Quality*, Joseph Juran writes:

> The picture of a company reaping big rewards through quality improvement is incomplete unless it includes some realities that have been unwelcome to most upper managers. Chief among these realities is the fact that the upper managers must participate personally and extensively in the effort. It is not enough to establish policies, create awareness, and then leave all else to subordinates. That has been tried, over and over again, with disappointing results.[2]

Management can play the necessary leadership role—and that essentially is its role—in continual improvement by doing the following:

- Establishing an organization-wide quality council and serving on it
- Working with the quality council to establish specific quality improvement goals with timetables and target dates

QUALITY TIP

Customer Needs Change Continually

"Quality Improvement is needed for both kinds of quality: product features and freedom from deficiencies.

"To maintain and increase sales income, companies must continually evolve new product features and new processes to produce those features. Customer needs are a moving target.

"To keep costs competitive, companies must continually reduce the level of product and process deficiencies. Competitive costs are also a moving target."

Source: Joseph M. Juran, *Juran on Leadership for Quality: An Executive Handbook* (New York: Free Press, 1989), 31.

- Providing the necessary moral and physical support. Moral support manifests itself as commitment. Physical support comes in the form of the resources needed to accomplish the quality improvement objectives.
- Scheduling periodic progress reviews and giving recognition where it is deserved
- Building continual quality improvement into the regular reward system, including promotions and pay increases

ESSENTIAL IMPROVEMENT ACTIVITIES

Continual improvement is not about solving isolated problems as they occur. Such an approach is viewed as "putting out fires" by advocates of total quality. Solving a problem without correcting the fault that caused it—in other words, simply "putting out the fire"—just means the problem will occur again. Quality expert Peter R. Scholtes recommends the following five activities that he sees as crucial to continual improvement (see Figure 19–1):[3]

- *Maintain communication.* Communication is essential to continual improvement. This cannot be overemphasized. Communication within improvement teams and among teams is a must. It is important to share information before, during, and after attempting to make improvements. All people involved as well as any person or unit that might be impacted by a planned improvement should know what is being done, why, and how it might affect them.
- *Correct obvious problems.* Often process problems are not obvious, and a great deal of study is required to isolate them and find solutions. This is the typical case, and it is why the scientific approach is so important in a total quality setting. However, sometimes a process or product problem will be obvious. In such cases, the problem should be corrected immediately. Spending days studying a problem for which the solution is obvious just so that the scientific approach is used will result in $10 solutions to 10-cent problems.
- *Look upstream.* Look for causes, not symptoms. This is a difficult point to make with people who are used to taking a cursory glance at a situation and "putting out the fire" as quickly as possible without taking the time to determine what caused it.

QUALITY TIP

Improvement Must Be Continual

"Improve constantly and forever the system of production and service. Improvement is not a one-time effort. Management is obligated to continually look for ways to reduce waste and improve quality."

Source: W. Edwards Deming, cited in Mary Walton, *The Deming Management Method* (New York: Putnam, 1986), 35.

QUALITY TIP

Improvement Is Not Putting Out Fires

W. Edwards Deming says, "Putting out fires is not improvement. Finding a point out of control, finding the special cause and removing it is only putting the process back to where it was in the first place. It is not improvement of the process. [Deming attributed this conclusion to Joseph M. Juran, many years ago.]

"You are in a hotel. You hear someone yell fire. He runs for the fire extinguisher and pulls the alarm to call the fire department. We all get out. Extinguishing the fire does not improve the hotel.

"That is not improvement of quality. That is putting out fires."

Source: W. Edwards Deming, cited in Mary Walton, *The Deming Management Method* (New York: Putnam, 1986), 67.

■ *Document problems and progress.*   Take the time to write it down. It is not uncommon for an organization to continue solving the same problem over and over again because nobody took the time to document the problems that have been dealt with and how they were solved. A fundamental rule for any improvement project team is "document, document, document."

■ *Monitor changes.*   Regardless of how well studied a problem is, the solution eventually put in place may not solve it or may only partially solve it, or it may produce unintended consequences. For this reason, it is important to monitor the performance of a process after changes have been implemented. It is also important to ensure that pride of ownership on the part of those who recommended the changes does not interfere with objective monitoring of the changes. These activities are essential regardless of how the improvement effort is structured.

STRUCTURE FOR QUALITY IMPROVEMENT

Quality improvement doesn't just happen. It must be undertaken in a systematic, step-by-step manner. For an organization to make continual improvements, it must be structured appropriately. Quality pioneer Juran calls this "mobilizing for quality improvement."[4] It involves the following steps:[5]

■ *Establish a quality council.*   The quality council has overall responsibility for continual improvement. According to Juran, "The basic responsibility of this council is to launch, coordinate, and 'institutionalize' annual quality improvement."[6] It is essential that the membership include executive-level decision makers.

■ *Develop a statement of responsibilities.*   All members of the quality council, as well as employees who are not currently members, must understand the council's responsibilities. One of the first priorities of the council is to develop and distribute

Figure 19–1
Essential Improvement Activities

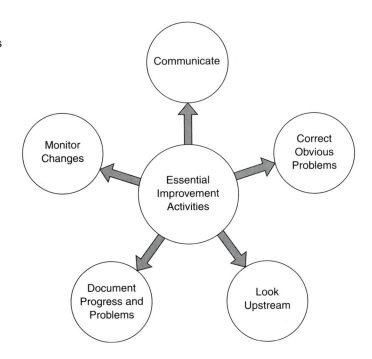

a statement of responsibilities bearing the signature of the organization's CEO. Responsibilities that should be stated include the following: (a) formulating policy as it relates to quality; (b) setting the benchmarks and dimensions (cost of poor quality, etc.); (c) establishing the team and project selection processes; (d) providing the necessary resources (training, time away from job duties to serve on a project team, and so on); (e) implementing the project; (f) establishing quality measures for monitoring progress and undertaking monitoring efforts; and (g) implementing an appropriate reward and recognition program.

■ *Establishing the necessary infrastructure.*   The quality council constitutes the foundation of an organization's quality effort. However, there is more to the quality infrastructure than just the council. The remainder of the quality infrastructure consists of subcommittees of the council that are assigned responsibility for specific duties, project improvement teams, quality improvement managers, a quality training program, and a structured improvement process.

THE SCIENTIFIC APPROACH

The scientific approach is one of the fundamental concepts that separates the total quality approach from other ways of doing business. Scholtes describes the scientific approach as "making decisions based on data, looking for root causes of problems, and seeking permanent solutions instead of relying on quick fixes."[7]

Scholtes developed the following four strategies for putting the scientific approach to work in a total quality setting:[8]

- Collect meaningful data.
- Identify root causes of problems.
- Develop appropriate solutions.
- Plan and make changes.

Collect Meaningful Data

Meaningful data are free from errors of measurement or procedure, and they have direct application to the issue in question.[9] It is not uncommon for an organization or a unit within it to collect meaningless data or to make a procedural error that results in the collection of erroneous data. In fact, in the age of computers, this is quite common. Decisions based on meaningless or erroneous data are bound to lead to failure. Before collecting data, decide exactly what is needed, how it can best be collected, where the data exists, how it will be measured, and how you will know the data are accurate.

Identify Root Causes of Problems

The strategy of identifying root causes is emphasized throughout this book.[10] Too many resources are wasted by organizations attempting to solve symptoms rather than problems. The total quality tools are helpful in separating problems from causes.

Develop Appropriate Solutions

With the scientific approach, solutions are not assumed.[11] Collect the relevant data, make sure it is accurate, identify root causes, and then develop a solution that is appropriate. Too many teams and too many people begin with, "I know what the problem is. All we have to do to solve it is . . . . " When the scientific approach is applied, the

QUALITY TIP

Importance of Statistical Thinking

"Statistical thinking is critical to improvement of a system. Only by use of properly interpreted data can intelligent decisions be made. But to depend on the use of statistics is a sure way to go out of business. A company must follow all of the Fourteen Points."

Source: W. Edwards Deming, cited in Mary Walton, *The Deming Management Method* (New York: Putnam, 1986), 57.

problem identified is often much different from what would have been suspected if acting on a hunch or intuition. Correspondingly, the solution is also different.

Plan and Make Changes

Too many decision makers use what is sometimes called the "Ready, fire, aim" approach rather than engaging in careful, deliberate planning.[12] Planning forces you to look ahead, anticipate needs and what resources will be available to satisfy them, and anticipate problems and consider how they should be handled.

Much of the scientific approach has to do with establishing reliable performance indicators and using them to measure actual performance. In his book *Total Manufacturing Management*, Giorgio Merli lists the following as examples of useful performance indicators:[13]

- Number of errors or defects
- Number of or level of need for repetitions of work tasks
- Efficiency indicators (units per hour, items per person)
- Number of delays
- Duration of a given procedure or activity
- Response time or cycle
- Usability/cost ratio
- Amount of overtime required
- Changes in workload
- Vulnerability of the system
- Level of criticalness
- Level of standardization
- Number of unfinished documents

QUALITY TIP

Improvement Can Be Measured

"Improvement can be measured and monitored by using 'performance indicators.' Performance indicators (PIs) are units of measurement pertaining specifically to improvements. Because it is not always possible to select a single PI that is valid for an entire improvement area, more than one indicator is often used. Moreover, it is not always possible to measure every aspect of improvement; in such cases the respective IA [improvement area] must remain 'open.'"

Source: Giorgio Merli, *Total Manufacturing Management* (Cambridge, MA: Productivity Press, 1990), 141.

This is not a complete list. Many other indicators could be added. Those actually used vary widely from organization to organization. However, such indicators, regardless of which ones are actually used, are an important aspect of the scientific approach.

IDENTIFICATION OF IMPROVEMENT NEEDS

Even the most competitive, most successful organizations have limited resources. Therefore, it is important to optimize those resources and use them in ways that will yield the most benefit. One of the ways to do this is to carefully select the areas of improvement to which time, energy, and other resources will be devoted. If there are 10 processes that might be improved on, which will yield the most benefit if improved? These are the processes that should be worked on first.

Scholtes recommends the following four strategies for identifying improvement needs:[14]

■ *Apply multivoting.* Multivoting involves using brainstorming to develop a list of potential improvement projects. Team members vote several times—hence the name—to decide which project or projects to work on first. Suppose the original list contains 15 potential projects. Team members vote and cut the list to 10. They vote again and cut it to five. The next vote cuts the list to three, and so on until only one or two projects remain. These are the first projects that will be undertaken.

■ *Identify customer needs.* An excellent way to identify an improvement project is to give the customer a voice in the process. Identify pressing customer needs and use them as projects for improvement.

■ *Study the use of time.* A good way to identify an improvement project is to study how employees spend their time. Is an excessive amount of time devoted to a given process, problem, or work situation? This could signal a trouble spot. If so, study it carefully to determine the root causes.

QUALITY TIP

Improvement Requires More Than Closer Control

"Continuous improvement of products requires 'continuous improvement of company processes.' Hoping to obtain qualitative improvement of products by means of closer control is a method that is not entrepreneurially valid, because it is in conflict with cost control. 'Higher quality = high costs' was actually a postulate of the earlier company model, in which quality was regarded as a factor that could be controlled by inspection procedures 'approval.'"

Source: Giorgio Merli, *Total Manufacturing Management* (Cambridge, MA: Productivity Press, 1990), 7.

■ *Localize problems.* Localizing a problem means pinpointing specifically where it happens, when it happens, and how often. It is important to localize a problem before trying to solve it. Problems tend to be like roof leaks in that they often show up at a location far removed from the source.

DEVELOPMENT OF IMPROVEMENT PLANS

After a project has been selected, a project improvement team is established. The team should consist of representatives from the units most closely associated with the problem in question including the process operator. It must include a representative from every unit that will have to be involved in carrying out improvement strategies. The project improvement team should begin by developing an improvement plan. This is to make sure the team does not take the "Ready, fire, aim" approach mentioned earlier.

The first step is to develop a mission statement for the team. This statement should clearly define the team's purpose and should be approved by the organization's governing board for quality (executive steering committee, quality council, or whatever the group is). After this has been accomplished, the plan can be developed. Scholtes recommends the following five stages for developing the plan:

1. *Understand the process.* Before attempting to improve a process, make sure every team member thoroughly understands it. How does it work? (This usually requires the development of a process flow diagram, see Chapter 15.) What is it supposed to do? What are the best practices known pertaining to the process? The team should ask these questions and pursue the answers together. This will give all team members a common understanding, eliminate ambiguity and inconsistencies, and point out any obvious problems that must be dealt with before proceeding to the next stage of planning.
2. *Eliminate errors.* In analyzing the process, the team may identify obvious errors that can be quickly eliminated. Such errors should be eradicated before proceeding to the next stage. This stage is sometimes referred to as "error-proofing" the process.
3. *Remove slack.* This stage involves analyzing all of the steps in the process to determine whether they serve any purpose and, if so, what purpose they serve. In any organization, processes exist that have grown over the years with people continuing to follow them without giving any thought to why things are done a certain way, whether they could be done better another way, or whether they need to be done at all. Few processes cannot be streamlined.
4. *Reduce variation.* Variation in a process results from either common causes or special causes. Common causes result in slight variations and are almost always present. Special causes typically result in greater variations in performance and are not always present. Strategies for identifying and eliminating sources of variation are discussed in the next section.
5. *Plan for continual improvement.* By the time this step has been reached, the process in question should be in good shape. The key now is to incorporate the

QUALITY TIP

Localizing a Problem

"A vehicle repair operation repeatedly found trucks with rusted fuel pumps. Before they could know what caused the rust, they had to localize the occurrence of the problem. They used stratification analysis of data they collected to localize the problem and found that:

- the problem occurred in vehicles using diesel fuel,
- and of those vehicles, only in ones operated in a certain geographical area
- and of those, only ones that got fuel from a specific pumping station.

"This information allowed them to zero in on the fuel pump site, and they discovered a small hole in a pipe feeding the underground tank. Rain water could seep into the tank through the hole. Repairing the leak and preventing its recurrence solved the problem of rusted fuel pumps."

Source: Joseph M. Juran, *Juran on Leadership for Quality: An Executive Handbook* (New York: Free Press, 1989), 28–29.

types of improvements made on a continuous basis so that continual improvement becomes a normal part of doing business. The Plan-Do-Check-Act cycle, disussed in Chapter 16, applies here. With this cycle, each time a problem or potential improvement is identified, an improvement plan is developed (Plan), implemented (Do), monitored (Check), and refined as needed (Act).

COMMON IMPROVEMENT STRATEGIES

Numerous different processes are used in business and industry; consequently, there is no single road map to follow when improving processes. However, a number of standard strategies can be used as a menu from which improvement strategies can be selected as appropriate. Figure 19–2 shows several standard strategies that can be used to improve processes on a continual basis. These strategies are explained in the following sections.[15]

Describe the Process

The strategy of describing the process is used to make sure that everyone involved in improving a process has a detailed knowledge of the process. Usually this requires some investigation and study. The steps involved are as follows:

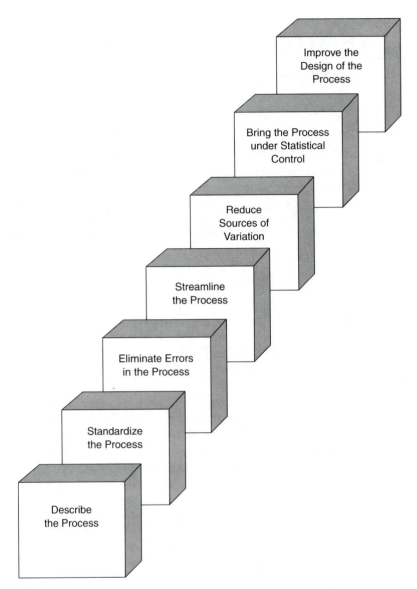

Figure 19–2
Standard Process Improvement Strategies

1. Establish boundaries for the process.
2. Flowchart the process. (As it is, not as it should be.)
3. Make a diagram of how the work flows.
4. Verify your work.
5. Correct immediately any obvious problems identified.

Standardize the Process

To continually improve a process, all people involved in its operation must be using the same procedures. Often this is not the case. Employee X may use different procedures than Employee Y. It is important to ensure that all employees are using the best, most effective, most efficient procedures known. The steps involved in standardizing a process are as follows:

1. Identify the currently known best practices and write them down.
2. Test the best practices to determine whether they are, in fact, the best, and improve them if there is room for improvement (these improved practices then become the final best practices that are recorded).
3. Make sure that everyone is using the newly standardized process.
4. Keep records of process performance, update them continually, and use them to identify ways to improve the process even further on a continual basis.

Eliminate Errors in the Process

The strategy of eliminating errors in the process involves identifying errors that are commonly made in the operation of the process and then getting rid of them. This strategy helps delete steps, procedures, and practices that are being done a certain way simply because that is the way they have always been addressed. Whatever measures can be taken to eliminate such errors are carried out as a part of this strategy.

Streamline the Process

The strategy of streamlining the process is used to take the slack out of a process. This can be done by reducing inventory, reducing cycle times, and eliminating unnecessary steps. After a process has been streamlined, every step in it has significance, contributes to the desired end, and adds value.

QUALITY TIP

Streamline by Reducing Inventory and Lot Sizes

"Large inventories of work-in-process cause many problems. They slow response time to customers, and are usually costly to maintain, and most deadly of all, hide other problems in the process. One cause of large inventories is slow change-overs from job to job, which also cause large lot sizes. This is true of both manufacturing and nonmanufacturing processes."

Source: Peter R. Scholtes, *The Team Handbook* (Madison, WI: Joiner Associates, 1992), 5-61.

Reduce Sources of Variation

The first step in the strategy of reducing sources of variation is identifying sources of variation. Such sources can often be traced to differences among people, machines, measurement instruments, material, sources of material, operating conditions, and times of day. Differences among people can be attributed to levels of capability, training, education, experience, and motivation. Differences among machines can be attributed to age, design, and maintenance. Regardless of the source of variation, after a source has been identified, this information should be used to reduce the amount of variation to the absolute minimum. For example, if the source of variation is a difference in the levels of training completed by various operators, those who need more training should receive it. If one set of measurement instruments is not as finely calibrated as another, they should be equally calibrated.

Bring the Process under Statistical Control

The strategy of bringing the process under statistical control is explained in detail in Chapter 18. For this discussion, it is necessary to know only that a control chart is planned, data are collected and charted, special causes are eliminated, and a plan for continual improvement is developed.

Improve the Design of the Process

There are many different ways to design and lay out a process. Most designs can be improved on. The best way to improve the design of a process is through an active program of experimentation. To produce the best results, an experiment must be properly designed, in the following steps:

1. Define the objectives of the experiment. (What factors do you want to improve? What specifically do you want to learn from the experiment?)
2. Decide which factors are going to be measured (cycle time, yield, finish, or something else).
3. Design an experiment that will measure the critical factors and answer the relevant questions.
4. Set up the experiment.
5. Conduct the experiment.
6. Analyze the results.
7. Act on the results.

ADDITIONAL IMPROVEMENT STRATEGIES

In his book *Total Manufacturing Management*, Merli lists 20 strategies for continual improvement that he calls "The Twenty Organizing Points of Total Manufacturing Management."[16] These strategies (see Figure 19–3) are explained in the following paragraphs:[17]

■ *Reduced lead time.* Raw materials sitting in a storeroom are not adding value to a product. Efficient management of the flow of materials is essential to competitiveness. Lead

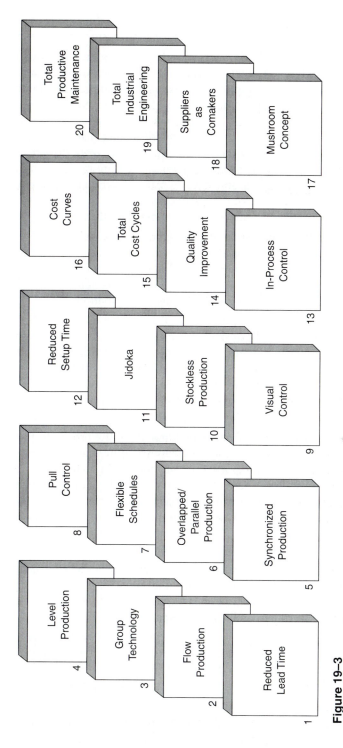

Figure 19–3
Strategies for Continual Improvement

Boxes (as labeled, numbered):

1. Reduced Lead Time
2. Flow Production
3. Group Technology
4. Level Production
5. Synchronized Production
6. Overlapped/Parallel Production
7. Flexible Schedules
8. Pull Control
9. Visual Control
10. Stockless Production
11. Jidoka
12. Reduced Setup Time
13. In-Process Control
14. Quality Improvement
15. Total Cost Cycles
16. Cost Curves
17. Mushroom Concept
18. Suppliers as Comakers
19. Total Industrial Engineering
20. Total Productive Maintenance

time can be reduced by evaluating the following factors: order processing time, waiting time prior to production, manufacturing lead time, storage time, and shipping time.

■ *Flow production.*   Traditionally, production has been a stop-and-go or hurry-up-and-wait enterprise. *Flow production* means production that runs smoothly and steadily without interruption. An example illustrates this point. A large manufacturer of metal containers had its shop floor arranged by type of machine (cutting, turning, milling, etc.). All cutting machines were grouped together, all turning machines were grouped together, and all milling machines were grouped together. However, this isn't how the flow of work went. Work flowed from cutting to turning, back to cutting, and on to milling. Arranging machines by type caused a great many interruptions and unnecessary material handling. To improve production efficiency, the machines were rearranged according to work flow. This smoothed out the rough spots and made work flow more smoothly.

■ *Group technology.*   Traditional production lines are straight. With group technology, processes are arranged so that work flows in a U-shaped configuration. This can yield the following benefits: shorter lead times, greater flexibility, less time in material handling, minimum work in progress, flexibility with regard to volume, less floor space used, and less need for direct coordination.

■ *Level production.*   This involves breaking large lots into smaller lots and producing them on a constant basis over a given period of time. For example, rather than producing 60 units per month in one large lot, production might be leveled to produce 3 units per day (based on 20 work days per month). This strategy can yield the added benefit of eliminating the need to store the materials needed for large lots. This, in turn, makes it easier to implement just-in-time manufacturing.

■ *Synchronized production.*   Synchronized production involves synchronizing the needs of the production line with suppliers of the materials needed on the line. For example, assume that a line produces computers in a variety of different internal configurations. The difference among the configurations is in the capacity of the hard drive installed. Such information as what type of hard drive is needed, in what quantities, at what time, and at what point on the line must be communicated to the hard-drive supplier. The supplier must, in turn, deliver the correct type of hard drive in the correct quantity at the correct time to the correct place on the line. When this happens, synchronized production results.

■ *Overlapped/parallel production.*   This strategy involves dismantling long production lines with large lot capacities and replacing them with production cells that turn out smaller lots. This allows production of different configurations of the same product to be overlapped and/or run parallel.

■ *Flexible schedules.*   Production cells and the ability to overlap production or run it parallel allow for a great deal of flexibility in scheduling. The more options available to production schedulers, the more flexible they can be in developing schedules.

■ *Pull control.*   Pull control is a concept applied to eliminate idle time between scheduling points in a production process, the need to maintain oversized inventories to offset operational imbalances, and the need to plan all target points within a process. With good pull control, work moves through a process uninterrupted by long waiting periods between steps.

■ *Visual control.* Visual control is an important aspect of just-in-time manufacturing. It is an information dissemination system that allows abnormalities in a process to be identified visually as they occur. This, in turn, allows problems to be solved as they occur rather than after the fact.

■ *Stockless production.* Stockless production is an approach to work handling, inventory, lead time planning, process balancing, capacity utilization, and schedule cycling that cuts down on work in progress. With stockless production, it is necessary to eliminate process bottlenecks, balance the process, and have an even work flow that eliminates or at least minimizes work in progress. Stockless production and just-in-time go hand-in-hand.

■ *Jidoka.* Jidoka means halting an entire process when a defect is discovered so that it won't cause additional problems further down the line. Jidoka can be accomplished manually, or the line can be programmed to stop automatically, or both.

■ *Reduced setup time.* This strategy consists of any activity that can reduce the amount of time required to break down a process and set it up again for another production run. Such things as quick changeovers of tools and dies are common with this strategy.

■ *In-process control.* Work-in-process (WIP) often means work that is sitting idle waiting to be processed. Controlling the amount of idle WIP involves organizing for a smoother flow, small lot sizes, process flexibility, and rapid breakdown and setup.

■ *Quality improvement.* In addition to improving productivity using the various strategies discussed in this chapter, it is important simultaneously to improve quality. This book is devoted to an approach for continually improving quality. The important point is that productivity and quality must be improved simultaneously.

■ *Total cost cycles.* This strategy involves basing decisions on the total cost cycle rather than isolated pieces of it. It is not uncommon for decisions to be based on reducing the costs associated with part of a process. True improvements have not been accomplished unless overall costs have been reduced.

■ *Cost curves.* A cost curve is a graphic representation of a time-based process wherein manufacturing costs accumulate relative to billing. Two types of costs are shown on a cost curve: materials and conversion costs. A cost curve shows graphically how much cost accumulates until the customer is billed for the product. It is a tool to help managers economize on the handling of orders.

■ *Mushroom concept.* This strategy is designed to broaden a company's customer base by creating a product that is open to diversification while, at the same time, being sufficiently standardized to minimize production costs. This is done by holding to standard processes as long as possible in the overall production cycle and adding different features only at the end of the process so that a variety of diversified products mushroom out at the end.

■ *Suppliers as comakers.* This strategy amounts to involving suppliers as partners in all phases of product development rather than keeping them in the dark and revealing your activities to them only through the low-bid process. If tested and trusted suppliers know what you are trying to do, they will be better able to maximize their resources in helping you do it.

■ *Total industrial engineering.*   Total industrial engineering integrates all elements—organizational, technical, and people related—in an effort to achieve continual improvement. Total industrial engineering focuses its efforts on the improvement of industrial systems rather than individual elements.

■ *Total productive maintenance.*   Total productive maintenance means maintaining all systems and equipment continually and promptly all of the time. In a rushed workplace, one of the most common occurrences is slacking off on machine and system maintenance. This is unfortunate because a poorly maintained system cannot achieve the quality and productivity needed to be competitive. Poor maintenance can result in the following problems: shutdowns from unexpected damage, increased setup and adjustment time, unused up time, speeds below the optimum, increased variations, increased waste from defects, and production losses during startup procedures.

THE KAIZEN APPROACH

Kaizen is the name given by the Japanese to the concept of continual incremental improvement. *Kai* means change, and *zen* means good. *Kaizen*, therefore, means making changes for the better on a continual, never-ending basis. The improvement aspect of kaizen refers to people, processes, and products.

If the kaizen philosophy is in place, all aspects of an organization should be improving all the time. People, processes, management practices, and products should improve continually; "good enough" is never good enough. In his landmark book, *KAIZEN: The Key to Japan's Competitive Success*, Masaaki Imai gives an overview of the concept that is summarized in the following paragraphs:[18]

■ *Kaizen value system.*   The underlying value system of kaizen can be summarized as continual improvement of all things, at all levels, all the time, forever. All of the

QUALITY TIP

Kaizen: Japan's Approach to Continuous Improvement

"KAIZEN strategy is the single most important concept in Japanese management—the key to Japanese competitive success. KAIZEN means 'on-going' improvement involving everyone—top management, managers, and workers."

Source: Masaaki Imai, *KAIZEN: The Key to Japan's Competitive Success* (New York: McGraw-Hill, 1986), xxix.

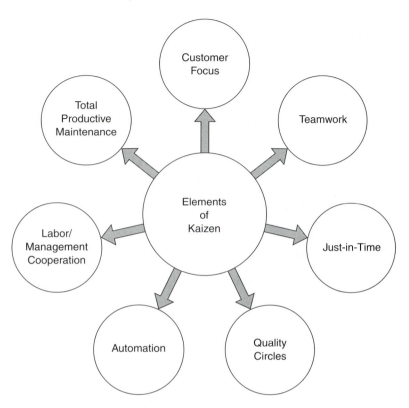

Figure 19–4
Elements of Kaizen

strategies for achieving this fall under the kaizen umbrella (see Figure 19–4). Executive managers, middle managers, supervisors, and line employees all play key roles in implementing kaizen (see Figure 19–5).

■ *Role of executive management.* Executive managers are responsible for establishing kaizen as the overriding corporate strategy and communicating this commitment to all levels of the organization; allocating the resources necessary for kaizen to work; establishing appropriate policies; ensuring full deployment of kaizen policies; and establishing systems, procedures, and structures that promote kaizen.

■ *Role of middle managers.* Middle managers are responsible for implementing the kaizen policies established by executive management; establishing, maintaining, and improving work standards; ensuring that employees receive the training necessary to understand and implement kaizen; and ensuring that employees learn how to use all applicable problem-solving tools.

■ *Role of supervisors.* Supervisors are responsible for applying the kaizen approach in their functional roles, developing plans for carrying out the kaizen approach at the functional level, improving communication in the workplace, maintaining

---

Directions

To promote continual improvement, consider the following factors every day and ask the following question: *"How can this be improved?"*

____ Personnel (at all levels)
____ Work techniques
____ Work methods
____ Work procedures
____ Time
____ Facilities
____ Equipment
____ Systems
____ Software
____ Tools
____ Material
____ Plant layout
____ Production levels
____ Inventory
____ Paradigms (mind-sets)

Comments

_____
_____
_____

Figure 19–5
Kaizen Checklist

morale, providing coaching for teamwork activities, soliciting kaizen suggestions from employees, and making kaizen suggestions.

■ *Role of employees.* Employees are responsible for participating in kaizen through teamwork activities, making kaizen suggestions, engaging in continual self-improvement activities, continually enhancing job skills through education and training, and continually broadening job skills through cross-functional training.

■ *Kaizen and quality.* In a total quality setting, quality is defined by customers. Regardless of how customers define quality, it can always be improved and it should be, continually. Kaizen is a broad concept that promotes quality from the all-encompassing Big Q perspective.

Kaizen Implementation Tools

All of the tools explained in Chapter 15 are used in kaizen, as are the tools explained elsewhere in this book. In addition, several are specifically thought of as kaizen implementation tools: kaizen checklists and the kaizen Five-Step Plan.

QUALITY TIP

Kaizen and Problem Solving

"The starting point for improvement is to recognize the need. This comes from recognition of a problem. If no problem is recognized, there is no recognition of the need for improvement. Complacency is the arch-enemy of KAIZEN. Therefore, KAIZEN emphasizes problem-awareness and provides clues for identifying problems."

Source: Masaaki Imai, *KAIZEN: The Key to Japan's Competitive Success* (New York: McGraw-Hill, 1986), 9.

Kaizen Checklists

Kaizen is about continual improvement of people, processes, procedures, and any other factors that can affect quality. One of the best ways to identify problems that represent opportunities for improvement is to use a checklist that focuses the attention of employees on those factors that are most likely in need of improvement. These factors include personnel, work techniques, work methods, work procedures, time, facilities, equipment, systems, software, tools, material, plant layout, production levels, inventory, and paradigms (see Figure 19–5).

Kaizen Five-Step Plan

The Five-Step Plan is the Japanese approach to implementing kaizen.[19] Posters bearing the words *seiri, seiton, seiso, seiketsu,* and *shitsuke* can often be found on the walls of Japanese plants. In English, these words mean "straighten up," "put things in order," "clean up," "personal cleanliness," and "discipline," respectively.

- *Step 1: Straighten up.* This step involves separating the necessary from the unnecessary and getting rid of the unnecessary in such areas as tools, work in process, machinery, products, papers, and documents.
- *Step 2: Put things in order.* This step involves putting such things as tools and material in their proper place and keeping things in order so that employees can always find what they need to do the job without wasting time looking.
- *Step 3: Clean up.* This step involves keeping the workplace clean so that work can proceed in an efficient manner, free of the problems that can result when the work site is messy.
- *Step 4: Personal cleanliness.* This step involves employees keeping themselves neat to present an appearance that promotes professionalism in performing work tasks.
- *Step 5: Discipline.* This step involves careful adherence to standardized work procedures. This requires personal discipline.

Five W's and One H

The Five W's and One H are not just kaizen tools. They are widely used as management tools in a variety of settings. The Five W's and one H (see Figure 19–6) are Who, What, Where, When, Why, and How. Using them encourages employees to look at a process and ask such questions as the following: Who is doing it? Who should be doing it? What is being done? What should be done? Where is it being done? Where should it be done? When is it being done? When should it be done? Why is it being done? Why do it that way? How is it being done? How should it be done?

Five-M Checklist

The Five-M Checklist is a tool that focuses attention on five key factors involved in any process.[20] The Five M's are man (operator), machine, material, methods, and measure-

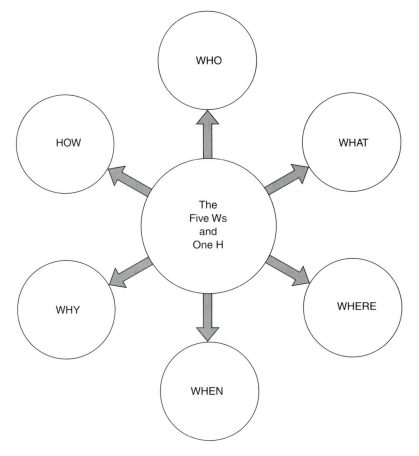

Figure 19–6
The Five W's and One H

Figure 19–7
The Five M's of Processes

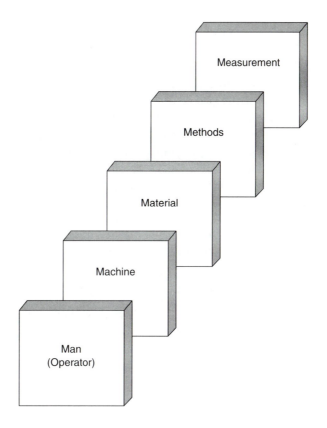

ment (see Figure 19–7). In any process, improvements can be made by examining these aspects of the process.

GOLDRATT'S THEORY OF CONSTRAINTS

Eliyahu M. Goldratt developed the Theory of Constraints as an approach to managing that helps organizations improve continually. Goldratt describes his theory as follows:

> The Theory of Constraints is an intuitive framework for managing an organization. Implicit in the framework is a desire to continually improve performance to have a process of on-going improvement. It starts, as it must, with clearly defining the goal of the organization and establishing measurements to determine the impact of any action on the goal.[21]

Goldratt does not question that the overriding goal of for-profit organizations is to make a profit. However, the Theory of Constraints applies a nontraditional system for measuring results. Rather than using the traditional net profit and return on investment measures, it uses throughput, inventory, and operating expense.

Goldratt's theory is based on the assumption that every organization faces constraints. However, the constraints that have the greatest negative impact on performance are policies as opposed to such physical entities as materials and resources. Goldratt defines a constraint as "anything that limits an organization from achieving higher performance vis-à-vis its goal."[22] The Theory of Constraints is applied in the following steps:[23]

- ■ *Identify.*   In this step, any factor that tends to constrain progress toward accomplishing the goal is identified.
- ■ *Exploit.*   This step involves determining how the constraints can be exploited (turned into positive factors, eliminated, or circumvented).
- ■ *Subordinate.*   Everything else is subordinated to making the decisions necessary to exploit the constraints. No other activity should have priority because the constraints are inhibiting progress toward accomplishing the organization's goal.
- ■ *Eliminate constraints/overcome inertia.*   In this step an examination is made to determine whether the constraints have been eliminated. If so, return to the first step and continue the improvement process. If not, inertia is the likely culprit. Attack this situation directly. An unwillingness to change should not be allowed to cause a constraint.

Because inertia is such an all-pervasive inhibitor of change, Goldratt recommends teaching people the following techniques for facilitating change:[24]

- ■ *Effect–cause–effect.*   This technique involves the use of brainstorming to verbalize an intuitive sense of problems and their causes.

QUALITY TIP

Theory of Constraints and Measuring Performance

"While the final judges of this goal are Net Profit and Return on Investment, the Theory of Constraints measurements for determining the impact of actions are Throughput, Inventory (Assets), and Operating Expense.

"One major difference between the Theory of Constraints approach to managing an industrial organization and our conventional approach is the relative priority given to these three measurements. While all three measurements are important, we have never regarded them as equal partners. Conventionally, we have tended to view Operating Expense (Cost) as the most important, with Throughput occupying a second position and Inventory only a remote third. The Theory of Constraints movement, along with the Just-In-Time and Total Quality movements, has adopted a different priority. These new management philosophies believe that Throughput should be at the top of the list, Inventory next, and Operating Expense in the last position. Moving from our conventional management philosophy to this new approach can be regarded as moving from the 'Cost World' to the 'Throughput World.'"

Source: Eliyahu M. Goldratt, "The Avraham Y. Goldratt Institute," undated paper explaining the Theory of Constraints and the services available from the Goldratt Institute, 2.

- *Evaporating clouds.* This technique involves stating a problem as a conflict and then isolating the related assumptions that can be challenged.

- *Prerequisite trees.* A prerequisite is something that must occur before something else can occur. In college, for example, Algebra I is a prerequisite for Algebra II. A prerequisite tree is a tool for identifying the path of least resistance from an old way of doing things to a new way.

- *Socratic method.* The Socratic method is one of the oldest and most time-honored ways of teaching. It consists of helping people find their own answers, draw their own conclusions, and form their own opinions. It involves a facilitator using devil's advocate–type questions rather than expressing his or her own views. The facilitator gives questions, not answers.

THE CEDAC APPROACH

CEDAC is an acronym for *Cause-and-Effect Diagram with the Addition of Cards.*[25] It was originally developed by Dr. Ryuji Fukuda of Sumitomo Electric, a Japanese manufacturing firm (see Discussion Assignment 19–1 at the end of this chapter). Its purpose is to facilitate continual improvement in the workplace.

CEDAC is based on the supposition that three conditions must exist in order for continual improvement to occur. Fukuda explains these conditions as follows:[26]

- *A reliable system.* For continual improvement to occur, there must be a standardized, reliable system. A system that is reliable will yield the same results regardless of who uses it, provided it is applied properly and according to standard procedures.

- *A favorable environment.* Continual improvement will not occur unless an environment favorable to it exists. The keys to creating and maintaining an environment favorable to continual improvement are leadership and education. Leadership manifests itself in the form of commitment, both to the concept of continual improvement and to the allocation of the necessary resources. Education is how employees become skilled in the use of the improvement system. The higher the density of employees who are skilled in the use of the improvement system, the better. Density is expressed as a percentage (see Figure 19–8).

- *Practicing as teams.* Like all endeavors requiring skills, continual improvement strategies must be practiced. Because in a total quality setting work is performed by teams of employees, it is important for team members to practice together.

As teams practice, it is important for individual members to understand that improvements may come only in small increments. This is not merely acceptable; it is desirable. In a competitive situation, the difference between winning and losing is often quite small.

In the United States, instant replay of sports events and big plays broadcast during the sports portion of the nightly news have accustomed people to last-minute game-winning home runs in baseball, desperation baskets just before the buzzer in basketball, and clock-beating 60-yard touchdown passes in football. What the camera does not show is all of the small, incremental improvements that put these teams in a position to

$$Density = \frac{Number\ of\ Employees\ with\ Mastery^*}{Total\ Number\ of\ Employees}$$

For Example:

Employees with Mastery   = 60
Total Number of Employees = 100

$$\frac{60}{100} = 60\%\ Density$$

*Employees who are skilled at using the improvement system.

Figure 19–8
Calculating the Density of Skilled Employees

win with one dramatic play in the final seconds of the game. This is unfortunate because a home-run mentality can mitigate against an organization's ability to make the small incremental improvements that can mean the difference between winning and losing.

The CEDAC system can be divided into two main parts (see Figure 19–9). Each part is subdivided into several related activities. In part 1, a problem is identified. Quality tools such as those explained in Chapter 15 are used. Additional quality tools are used to analyze the problem. Based on the analysis, solutions are implemented. Results are monitored and confirmed. If the results are positive, the procedures that solved the problem are standardized, and these new standard procedures are followed by all employees. If the desired results are not achieved, the team returns to part 1 to analyze the problem again.

QUALITY TIP

Leadership Means Committing Resources

"[I]t is the higher managers who have the decisive power over the allocation of the necessary resources such as manpower, material, money, and time. Therefore, one of the crucial ways that leaders demonstrate their commitment is to allocate those resources properly to the company's improvements."

Source: Ryuji Fukuda, *CEDAC: A Tool for Continuous Systematic Improvement* (Cambridge, MA: Productivity Press, 1989), 39.

Figure 19–9
CEDAC System

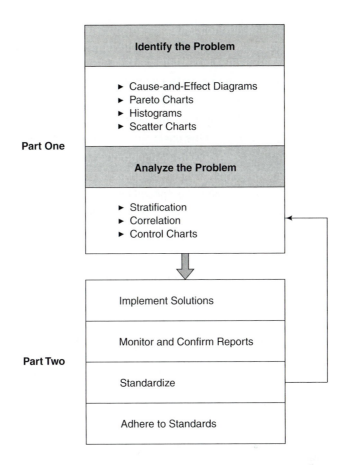

Identify the Problem

- ▶ Cause-and-Effect Diagrams
- ▶ Pareto Charts
- ▶ Histograms
- ▶ Scatter Charts

Part One

Analyze the Problem

- ▶ Stratification
- ▶ Correlation
- ▶ Control Charts

Implement Solutions

Monitor and Confirm Reports

Part Two

Standardize

Adhere to Standards

Developing a CEDAC Diagram

Figure 19–10 summarizes the main steps in developing a CEDAC diagram.[27] Instructions for these steps are contained in the following paragraphs:

1. *Draw the basic diagram.* The CEDAC diagram should be drawn with the cause side on the left and the effect side on the right. Using CEDAC is a team activity. Consequently, the basic diagram should be large enough to attach to a wall so that it can be reviewed by team members at any point in the process. The cause side is a fishbone diagram with the addition of cards to the left of each spine. The effect side can be any of the various tools explained in Chapter 15 (e.g., a control chart or a Pareto chart). Figure 19–11 is an example of a basic diagram laid out but not yet containing any information. Such a diagram might be as large as 4 × 6 feet.

2. *Select the focus of improvement efforts.* What is the focus of the CEDAC project? What problem is to be attacked? Quality tools such as those covered in Chapter 15 can be used to identify the improvement that has the most potential.

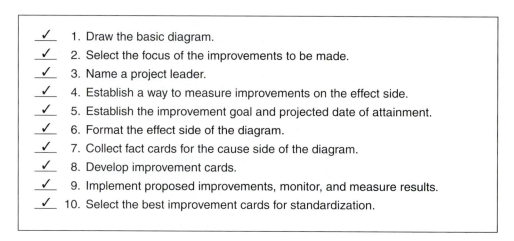

✓ 1. Draw the basic diagram.
✓ 2. Select the focus of the improvements to be made.
✓ 3. Name a project leader.
✓ 4. Establish a way to measure improvements on the effect side.
✓ 5. Establish the improvement goal and projected date of attainment.
✓ 6. Format the effect side of the diagram.
✓ 7. Collect fact cards for the cause side of the diagram.
✓ 8. Develop improvement cards.
✓ 9. Implement proposed improvements, monitor, and measure results.
✓ 10. Select the best improvement cards for standardization.

Figure 19–10
Ten-Step Checklist: Developing a CEDAC Program

3. *Name a project leader.* A CEDAC diagram relates to one specific improvement project and should have its own project leader. This is the person responsible for organizing, facilitating, monitoring, and completing the project.

4. *Establish a measurement method.* Improvements tried as part of the CEDAC process should result in improvements on the effect side of the diagram. These results must be measurable. Does the improvement decrease waste? Improve throughput?

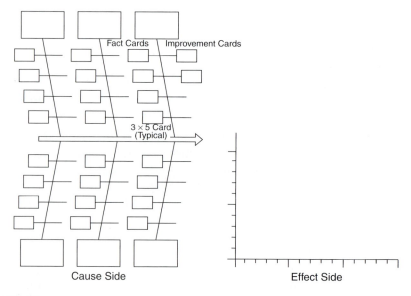

Figure 19–11
CEDAC Diagram

Reduce defects? These are improvements that can be measured. Regardless of the focus of the project, establish measures for documenting success or failure.

5. *Establish an improvement goal and date.* The improvement goal and projected target date are established by the project leader. It is important for this person to communicate why the goal and target date are important by relating them to the organization's ability to compete.

6. *Format the effect side of the diagram.* The actual format for the effect side of the CEDAC diagram must be decided on at this point. The preliminary chart prepared in the first step must be finalized. It can be a Pareto chart, a histogram, a control chart, or any other type of quality tool. The actual format selected should be the one that best communicates the necessary facts relating to the project. Figure 19–12 is a format that might be used if the project goal is to reduce throughput time by 50% over a 24-week period, or any similar goal.

7. *Collect fact cards for the cause side.* The goal is to reduce throughput time by 50% within 24 weeks. All members of the project team should examine the problem and commit their ideas as to why throughput time is as high as it is to cards (e.g., 3×5 cards). These cards are known as *fact cards*. The fact cards should be sorted into general categories (material, management, methods, etc.). After all cards have been sorted into categories, the spines on the cause side of the CEDAC diagram are labeled to corre-

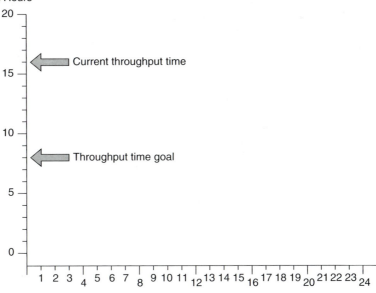

Figure 19–12
CEDAC Diagram: Effect Side

spond with these categories. Cards in each category are examined and combined where appropriate. Remaining cards are then attached to the left of the horizontal lines along the spines, as shown in Figure 19–11.

8. *Collect improvement cards.* Each fact card attached to the CEDAC diagram contains a description of some factor that contributes to increased throughput time. Each of these factors should be eliminated. To do this, improvement cards are collected from members of the project team. These cards correspond to the fact cards and contain a recommendation for eliminating the factor described on the corresponding fact card. Improvement cards are attached to the right of the horizontal lines along the spines, as shown in Figure 19–11.

9. *Implement and test improvement ideas.* Improvement ideas should be carefully evaluated for credibility before being attached to the CEDAC diagram. However, after an improvement card is attached, the idea it contains should be implemented and the results monitored and recorded on the effect side of the diagram.

10. *Select cards for standardization.* Improvement ideas that fail to reduce throughput time (see Figure 19–12) should be discarded. Those that work best should be written in as standard procedures in the process in question. After an idea is standardized, it should be adhered to strictly by all personnel.

 SUMMARY

1. The rationale for continual improvement is that it is necessary in order to compete in the global marketplace. Just maintaining the status quo, even if the status quo is high quality, is like standing still in a race.
2. Management's role in continual improvement is leadership. Executive-level managers must be involved personally and extensively. The responsibility for continual improvement cannot be delegated.
3. Essential improvement activities include the following: maintaining communication, correcting obvious problems, looking upstream, documenting problems and progress, and monitoring change.
4. Structuring for quality improvement involves the following: establishing a quality council, developing a statement of responsibilities, and establishing the necessary infrastructure.
5. Using the scientific approach means collecting meaningful data, identifying root causes of problems, developing appropriate solutions, and planning/making changes.
6. Ways of identifying improvement needs include the following: multivoting, seeking customer input, studying the use of time, and localizing problems.
7. Developing improvement plans involves the following steps: understanding the process, eliminating obvious errors, removing slack from processes, reducing variation in processes, and planning for continual improvement.
8. Commonly used improvement strategies include the following: describing the process, standardizing the process, eliminating errors in the process, streamlining

the process, reducing sources of variation, bringing the process under statistical control, and improving the design of the process.

9. Additional improvement strategies include the following: reducing lead time, flowing production, using group technology, leveling production, synchronizing production, overlapping production, using flexible scheduling, using pull control, using visual control, using stockless production, using jidoka, reducing setup time, applying in-process control, improving quality, applying total cost cycles, using cost curves, using the mushroom concept, making suppliers comakers, applying total industrial engineering, and applying total productive maintenance.

10. *Kaizen* is the name given by the Japanese to the concept of continual incremental improvement. It is a broad concept that encompasses all of the many strategies for achieving continual improvement and entails the following five elements: straighten up, put things in order, clean up, personal cleanliness, and discipline.

11. Goldratt's Theory of Constraints is another approach used to achieve continual improvement in the workplace. It involves the following steps: identify, exploit, subordinate, and eliminate restraints/overcome inertia. The following tools are used in applying Goldratt's Theory of Constraints: effect–cause–effect, evaporating clouds, prerequisite trees, and the Socratic method.

12. CEDAC is an acronym for *Cause-and-Effect Diagram with the Addition of Cards.* (This acronym is a registered trademark of Productivity, Inc.) With CEDAC, a cause-and-effect diagram is developed, but fact cards about problems and improvement cards containing ideas for solving the problems are used.

KEY TERMS AND CONCEPTS

CEDAC

Communication

Continual improvement

Cost curves

Customer needs

Five M's

Five W's and One H

Flexible schedules

Flow production

Group technology

Improvement plans

Improvement strategies

Improvement versus "putting out fires"

In-process control

Jidoka

Just-in-time

Kaizen

Lead time

Level production

Localize problems

Look upstream

Management's role in continual improvement

Multivoting

Mushroom concept

Necessary infrastructure

Obvious problems

Overlapped production

Plan-Do-Check-Act	Streamlining processes
Pull control	Suppliers as comakers
Quality council	Synchronized production
Reduced setup time	Theory of contraints
Root causes	Total cost cycles
Scientific approach to problem solving	Total industrial engineering
Slack	Total productive maintenance
Standardizing processes	Variation and sources of varation
Statement of responsibilities	Visual control
Stockless production	

 FACTUAL REVIEW QUESTIONS

1. Explain the rationale for continual improvement.
2. What is management's role in continual improvement?
3. Describe the five essential improvement activities.
4. If you were an executive manager in an organization, how would you structure the organization for quality improvement?
5. What is meant by using the scientific approach?
6. What is multivoting, and how is it used?
7. Describe the steps involved in developing an improvement plan.
8. List and explain three widely used improvement strategies.
9. Explain the following improvement strategies:
   a. Group technology
   b. Synchronized production
   c. Jidoka
   d. Cost curves
   e. Mushroom concept
10. Explain the concept of kaizen.
11. Describe Goldratt's Theory of Constraints.
12. What is CEDAC and how is it used?

 CRITICAL THINKING ACTIVITY

Which Approach Is Best?

Mark Berry, Sandra Griffith, and Juan Carlos are seniors at Florida Tech University, majoring in industrial technology with a quality emphasis. All three hope to be quality directors after graduation.

"I am going to use the kaizen approach when I graduate and land my first job," said Mark. "It's simple, easy to use, and effective."

"Not me," replied Sandra. "I like the CEDAC approach."

"I'm not going to use either of them," said Juan. "I'm going to make myself a master list of improvement strategies and use whatever is best for the individual situation."

Join their debate. What approach do you think is best for promoting continual improvement, and why?

DISCUSSION ASSIGNMENT 19–1

Origin of CEDAC

Dr. Ryuji Fukuda had worked at Sumitomo Electric for almost 20 years when in the mid-1970s he was asked to develop a method for effectively standardizing procedures and, then, continually improving them. Sumitomo executives had decided to expand into international markets and knew that competition would be intense. Fukuda began by forming a team of volunteers and asking them to identify a real problem against which his new method could be developed and tested.

They worked for a year before settling on the CEDAC approach. It was pilot-tested in all 40 of Sumitomo's plants with results that exceeded even Fukuda's expectations. A paper Fukuda published on his research into the results of applying CEDAC won Japan's coveted Nikai Award for 1978.

Discussion Questions
Discuss the following questions in class or outside of class with your fellow students:

1. Analyze Fukuda's approach to developing and testing the CEDAC approach to continual improvement. What did he do right?
2. Is there anything he could do or should have done differently?

Source: Based on Ryuji Fukuda, *(CEDAC: A Tool for Continuous Systemtic Improvement* (Cambridge, MA: Productivity Press, 1989), 10.

 ENDNOTES

1. CEDAC is a registered trademark of Productivity, Inc.
2. Joseph M. Juran, *Juran on Leadership for Quality: An Executive Handbook* (New York: Free Press, 1989), 72.
3. Peter R. Scholtes, *The Team Handbook* (Madison, WI: Joiner Associates, 1992), 5-6–5-9.
4. Juran, 35.
5. Juran, 42–46.
6. Juran, 43.
7. Scholtes, 5-9.
8. Scholtes, 5-9.

9. This section is based on Scholtes, 5-10.

10. This section is based on Scholtes, 5-10.

11. This section is based on Scholtes, 5-11.

12. This section is based on Scholtes, 5-11.

13. Giorgio Merli, *Total Manufacturing Management* (Cambridge, MA: Productivity Press, 1990), 143.

14. Scholtes, 5-17.

15. Scholtes, 5-54–5-67.

16. Merli, 163.

17. Merli, 153–165.

18. Masaaki Imai, *KAIZEN: The Key to Japan's Competitive Success* (New York: McGraw-Hill, 1986), 1-16.

19. This section is based on Imai, 233.

20. Imai, 237.

21. Eliyahu M. Goldratt, "The Avraham Y. Goldratt Institute," undated paper explaining the Theory of Constraints and the services available from the Goldratt Institute, 1.

22. Goldratt, 3.

23. Goldratt, 3.

24. Goldratt, 3.

25. Ryuji Fukuda, *CEDAC: A Tool for Continuous Systematic Improvement* (Cambridge, MA: Productivity Press, 1989), 3.

26. Fukuda, 3–10.

27. This section is based on Fukuda, 39–51.

CHAPTER TWENTY

Benchmarking

"[Benchmarking] is the difference between teaching yourself how to hit a golf ball and taking lessons from Jack Nicklaus."
Steven George[1]

MAJOR TOPICS

- Benchmarking Defined
- Rationale for Benchmarking
- Benchmarking Approach and Process
- Benchmarking versus Reengineering
- Role of Management in Benchmarking
- Prerequisites to Benchmarking
- Obstacles to Successful Benchmarking
- Benchmarking Resources
- Selection of Processes/Functions to Benchmark
- Acting on the Benchmark Data
- Perpetual Benchmarking

Benchmarking is becoming an increasingly popular tool among companies trying to become more competitive, striving for world-class performance. The vast majority of them are actively engaged in benchmarking. Benchmarking is a part of the total quality process, and anyone involved in total quality should have a solid understanding of this subject. This chapter is intended to help readers understand what benchmarking is all about, its benefits, and its pitfalls. The chapter includes sufficient information to enable any enterprise to make rational decisions concerning benchmarking, including whether or not to do it, and how to go about it.

Benchmarking was brought to our awareness through Robert C. Camp's 1989 landmark book.[2] Since then a number of variations have developed on the benchmarking theme. We have *benchmarking studies* in which there is no contact with an outside firm—information gained is strictly from the public domain. There is no question that this technique can be useful. It is something that the organization should be doing anyway. Sometimes third-party firms specializing in benchmarking studies are contracted for that work. There is considerable doubt that this is really benchmarking, however.

We also have *competitive benchmarking* in which a competitor's operation is studied from a distance *without* the cooperation of the target firm. The aim is to learn something that can help improve process or product quality. Competitive benchmarking uses publicly available data, and once again, it is possible to contract this work to specialist third-party firms. This approach, however, doesn't fit our definition of benchmarking.

Also in use are the unstructured plant visits in which the visitor firm has the intention of learning something that will help with its processes or products. This is often called benchmarking but has more aptly been named "industrial tourism." Such visits have some value, but they do not comprise benchmarking.

Many other variations exist, but the form of benchmarking addressed in this book is what has been called *cooperative benchmarking*, or *process benchmarking*, in which key processes are the focus for radical improvement. This involves a cooperative effort by two firms, the benchmarking firm wanting to bring a substandard process up to the world-class level of the partner firm's process.

BENCHMARKING DEFINED

Benchmarking has been around since the early 1980s, but it wasn't until the early 1990s that it became a widely accepted means of improving company performance. In 1985, almost no benchmarking activity existed among the Fortune 500 companies. By 1990, half of the Fortune 500 were using this technique. Today companies large and small are finding benchmarking to be an effective component in their total quality effort. If there is a single most likely reason for the slow rise in benchmarking popularity, it is a misunderstanding of the concept—misunderstanding of what benchmarking is, what it is not, and how to do it. It helps to begin with an examination of what benchmarking is not.

Benchmarking Is Not:

Cheating	Illegal
Immoral	Industrial espionage
Unethical	

All of these misconceptions about benchmarking assume that one party somehow takes advantage of an unsuspecting competitor by surreptitiously copying the competitor's product or processes. Nothing could be further from the truth. Benchmarking involves two organizations that have previously agreed to share information about processes or operations. The two organizations both anticipate some gain from the exchange of information. Either organization is free to withhold information that is considered proprietary. In addition, the two companies need not be competitors.

> Benchmarking is the process of comparing and measuring an organization's operations or its internal processes against those of a best-in-class performer from inside or outside its industry.

QUALITY TIP

A Psychological Barrier to Benchmarking

"We as Americans have a lot of psychological baggage to discard before we can successfully learn from others. Perhaps the greatest is the fear of being seen as 'copying.' From kindergarten onward, we have been punished for working together and rewarded for individual achievement."

Source: Carla O'Dell, foreword to Gregory H. Watson's book *The Benchmarking Workbook: Adapting Best Practices for Performance Improvement* (Cambridge, MA: Productivity Press, 1992), xvi.

Benchmarking is finding the secrets of success of any given function or process so that a company can learn from the information—and improve on it. It is a process to help a company close the gap with the best-in-class performer without having to "reinvent the wheel."

A distinction exists between benchmarking and competitive analysis. *Competitive analysis* involves comparing a competitor's product against yours. It compares the features and pricing of the product. Consumers perform competitive analysis when they compare competitors' products as they try to determine which brand of VCR, television, or automobile to purchase. Benchmarking goes beyond that to comparing how the product is engineered, manufactured, distributed, and supported. Benchmarking is interested not so much in what the product is and what it costs but rather in the underlying processes used to produce, distribute, and support it.

Finally, and most important, benchmarking is a tool to help establish where improvement resources should be allocated. If, for example, it is discovered that three of five processes are nearly as good as the best-in-class performers, but two are significantly off the best-in-class mark, the most resources should be allocated to these two. The most benefit for the dollars invested will come from changing those processes to conform more nearly to the best-in-class. Relatively little will be gained by drastically changing a process that is already close to the best there is. Key points to remember about benchmarking are as follows:

- Benchmarking is an increasingly popular improvement tool.
- Benchmarking concerns processes and practices.
- Benchmarking is a respected means of identifying processes that require major change.
- Benchmarking is done between consenting companies that may or may not be competitors.
- Benchmarking compares your process or practice with the target company's best-in-class process or practice.

QUALITY TIP

Today Change Is the Only Constant

"The future ain't what it used to be."

Source: Traditionally accredited to Yogi Berra.

■ The goal of benchmarking is to find "secrets of success" and then adapt and improve them for your own application.

■ Benchmarking is equally beneficial for both large and small businesses.

RATIONALE FOR BENCHMARKING

Never was Yogi Berra's comment, "The future ain't what it used to be," as relevant as it is to modern industry. The future for companies today seems far different from what it has been since World War II. The first real questions regarding the future and the ability of the United States to sustain its industrial leadership seem to have resulted from the oil crisis of 1974. By then, the United States had lost much of the commercial electronics business to Sony, Hitachi, and Panasonic, but the most important industry in the United States, the automobile, seemed secure. However, when the oil embargo struck, Americans quickly traded their big domestic cars for small, fuel-efficient Japanese models. When the embargo ended, Americans continued buying Japanese cars because at that time they were better than their American counterparts. The Japanese quickly claimed about 30% of the U.S. automobile market (and possibly could have gained much more except for voluntary restraints adopted out of fear that severe trade restrictions would be imposed by Washington). Following these events, North America finally started to wake up to the fact that the world was changing. Whole industries were moving from one part of the world to another, and most of that movement was to Japan. There was good reason to look at the Japanese to see what they were doing differently that allowed them to accomplish this.

What was learned, of course, was that by following the teachings of W. Edwards Deming, Joseph Juran, Ishikawa, Taguchi, Ohno, and other quality pioneers, Japan had developed vastly superior practices and processes. These resulted in superior manufactured goods at competitive prices—everything from motorcycles, to cars, to cameras, to electronics of all kinds, even to shipbuilding. It took several years of looking at Japan to realize fully what had happened. For a long time, Western leaders rationalized Japan's success to low labor costs, the Japanese work ethic versus that of Detroit, lifetime employment, and other factors. Such rationalizations simply clouded the real issue: the superiority of the Japanese practices and processes. Now that industrial leaders worldwide are aware that better practices and processes can enhance competitiveness, it makes good business sense to determine where an organization stands relative to world-class standards and what must be done to perform at that level. That is what benchmarking is designed to do.

Twenty years ago, benchmarking was a case of comparing North American industry with the Japanese. Today, benchmarking is a case of comparing your company with the best in the world. The best in the world for a given comparison may be in Japan, or it may be next door. It may be your direct competition, or it may be in a completely different industry. In addition to companies all over the world emulating the Japanese, customers all over the world are demanding the highest quality in the products they buy. Business as usual is no longer sufficient. Organizations must be improving always and forever, or they will be out of business soon and forever.

The rationale for benchmarking is that it makes no sense to stay locked in an isolated laboratory trying to invent a new process that will improve the product, or reduce cost, when that process already exists. If one company has a process that is four times as efficient, the logical thing for other companies to do is to adopt that process. An organization can make incremental improvements to its process through continual improvement, but it might take years to make a $4\times$ improvement, and by then the competition would probably be at $6\times$ or better. Benchmarking is used to show which processes are candidates for continual (incremental) improvement and which require major (one-shot) changes. Benchmarking offers the fastest route to significant performance improvement. It can focus an entire organization on the issues that really count.

Some factors that drive companies to benchmark are commitment to total quality, customer focus, product-to-market time, manufacturing cycle time, and financial performance at the bottom line. Every company that has won the Malcolm Baldrige Award endorses benchmarking (see Discussion Assignment 20–1 later in this chapter). Key points to remember about benchmarking as it relates to continual improvement are as follows:

- Today's competitive world does not allow time for gradual improvement in areas in which a company lags far behind.

QUALITY TIP

Look Far and Wide

When you look for best-in-class process owners as possible benchmarking partners, don't restrict your search to your own industry. For example, when Southwest Airlines was looking for a faster way to offload passengers and cargo and get the planes ready for their next flights, it benchmarked Indianapolis 500 pit crews. When Xerox needed major improvements in its warehousing operations, it benchmarked L. L. Bean, one of the world's best catalog sales organizations. IBM studied Las Vegas casinos to find ways to reduce employee theft.

Source: David L. Goetsch and Stanley B. Davis.

- Benchmarking can tell a firm where it stands relative to best-in-class practices and processes, and which processes must be changed.
- Benchmarking provides a best-in-class model to be adopted or even improved on.
- Modern customers are better informed and demand the highest quality and lowest prices. Companies have a choice to either perform with the best or go out of business.
- Benchmarking supports total quality by providing the best means for rapid, significant process/practice improvement.

BENCHMARKING APPROACH AND PROCESS

The benchmarking process is relatively straightforward, but steps must flow in a sequence. A number of variations are possible, but the process should follow this general sequence:

1. Obtain management commitment.
2. Baseline your own processes.
3. Identify your strong and weak processes and document them.
4. Select processes to be benchmarked.
5. Form benchmarking teams.
6. Research the best-in-class.
7. Select candidate best-in-class benchmarking partners.
8. Form agreements with benchmarking partners.
9. Collect data.
10. Analyze data and establish the gap.
11. Plan action to close the gap/surpass.
12. Implement change.
13. Monitor.
14. Update benchmarks; continue the cycle.

These 14 implementation steps are explained in turn in the following sections. Figure 20–2 is provided to help maintain perspective and afford clarity.

Step 1: Obtain Management Commitment

Benchmarking is not something one approaches casually. It requires a great deal of time from key people, and money must be available for travel to the benchmarking partners' facilities. Both of those require management's approval. You expect to gain information from your benchmarking partner for which it will expect payment in kind; namely, information from you about your processes. This can be authorized only by management. Finally, the object of benchmarking is to discover processes to replace yours or at least to make major changes to them. Such changes cannot be made without management's approval. Without a mandate from top management, there is no point in attempting to benchmark. That is why the requirement for management commitment is at the top of the list. If you cannot secure that commitment, proceed no further.

Step 2: Baseline Your Own Processes

If your company is involved in total quality, chances are good that you have already done some baselining of processes, because before continual improvement can be used effectively, and certainly before statistical process control can be applied, the processes in question must be understood. That is, the processes must be characterized in terms of capability, their flow diagrams, and other aspects. If this has not been done before, it must be done now. It is critical that you understand your own processes thoroughly before attempting to compare them with someone else's. Most people think they know their processes well, but that is rarely the case if a deliberate process characterization has not recently been done. It is also important that an organization's processes be completely documented, not just for its own use but for the benefit of everyone associated with the process in any way. (See the discussion of flowcharting in Chapter 15.)

Step 3: Identify and Document Both Strong and Weak Processes

Strong processes will not be benchmarked initially; continual improvement techniques will be sufficient for them. Weak processes, however, become candidates for radical change through benchmarking, because incremental improvement would not be sufficient to bring them up to the necessary level in the time frame required.

It can be difficult to categorize an organization's processes as weak or strong. A process that creates 50% scrap is an obvious choice for the benchmarking list. On the other hand, a process may be doing what is expected of it and as a result be classified as strong. However, it could be that expectations for that process are not high enough. It is possible that someone else has a process that is much more efficient, but you just don't know about it. Never consider a process to be above benchmarking, no matter how highly it is rated. Concentrate on the weak ones, but keep an open mind about the rest. If research identifies a better process, add it to the list.

Above all, document all processes fully—even the strong ones. Keep in mind that as you are looking at one of your benchmarking partner's processes because it is superior to yours, your partner will be looking at your strong processes for the same reason. If the processes are not well documented, it will be very difficult to help your partner. It is impossible to compare two processes for benchmarking if both are not fully documented.

Step 4: Select Processes to Be Benchmarked

When you have a good understanding of your own processes and the expectations of them, decide which ones to benchmark. An important point to remember is this: never benchmark a process that you do not wish to change. There is no point in it. Benchmarking is not something you engage in simply to satisfy curiosity. The processes that are put on a benchmark list should be those known to be inferior and that you intend to change. Leave the others for incremental change through continual improvement—at least for the time being.

Step 5: Form Benchmarking Teams

The teams that will do the actual benchmarking should include people who operate the process, those who have input to the process, and those who take output from it. These people are in the best position to recognize the differences between your process and that of your benchmarking partner. The team must include someone with research capability because it will have to identify a benchmarking partner, and that will require research. Every team should have management representation, not only to keep management informed but also to build the support from management that is necessary for radical change.

Step 6: Research the Best-in-Class

It is important that a benchmarking partner be selected on the basis of being best-in-class for the process being benchmarked. In practical terms, it comes down to finding the best-in-class-you-can-find-who-is-willing. Because benchmarking is accomplished by process, best-in-class may be in a completely different industry. For example, say that an organization manufactures copy machines. It might consider potential benchmarking partners who are leaders in the copying industry. But if it is a warehousing process that is to be benchmarked, the company might get better results by looking at catalog companies that have world-class warehousing operations. If the process to be benchmarked is accounts receivable, then perhaps a credit-card company would be a good partner.

Processes are shared across many industries, so don't limit research to like industries or you might miss the best opportunities for benchmarking. Remember that best-in-class does not mean best-in-your-industry, but best regardless of industry for the process in question. If team members stay up to date with trade journals, they should be able to compile a good list of potential benchmarking partners. Research should cover trade literature, suppliers and customers, Baldrige Award winners, and professional associations. The Internet offers a seemingly endless stream of benchmarking information. Team members will find that the best-in-class processes become well known very fast.

Step 7: Select Candidate Best-in-Class Benchmarking Partners

When the best-in-class have been identified, the team must decide with which among them it would prefer to work. Consideration must be given to location and to whether the best-in-class is a competitor (remember, the team will have to share information with its partner). The best benchmarking partnerships provide some benefit for both parties. If the team can find a way to benefit its potential partner, the linkage between the two companies will be easier to achieve. Even without that, most companies with best-in-class processes are willing, often eager, to share their insights and experience with others, even if they gain nothing in return. Indeed, Baldrige Award winners are expected to share information with other U.S. organizations.

Step 8: Form Agreements with Benchmarking Partners

After the team has selected the candidates, it contacts the potential partner to form an agreement covering benchmarking activities. It can be useful to have an executive contact an executive of the target company, especially if there is an existing relationship or some other common ground. Often the most difficult part of the process is identifying the right person in the potential partnering company. Professional associations can sometimes provide leads to help the team contact someone in the right position with the necessary authority.

After such a contact has been made, the first order of business is to determine the company's willingness to participate. If it is not willing, the team must find another candidate. When a company is willing to participate, an agreement can usually be forged without difficulty. The terms will include visit arrangements to both companies, limits of disclosure, and points of contact. In most cases, these are informal. Even so, care must be exercised not to burden either benchmarking partner with excessive obligations. Make the partnership as unobtrusive as possible.

Step 9: Data Collection

The team has already agreed to discuss a specific process (or processes). Observe, collect, and document everything about the partner's process. In addition to that, try to determine the underlying factors and processes: what is it that makes the company successful in this area? For example, does it employ total productive maintenance, continual improvement, employee involvement, use of statistics, and various other approaches? Optimally, your process operators should talk directly with its operators. It is important to come away with a good understanding of what its process is (flow diagram) and its support requirements, timing, and control. The team should also try to gain some understanding of the preceding and succeeding processes, because if you change one, the others may require change as well. If the team knows enough when it leaves the partner's plant to implement its process back home, then it has learned most of what is needed. Anything less than this, and the team has more work to do.

While you are in a partner's plant, try to get a feel for how it operates. Be open-minded and receptive to new ideas that are not directly associated with the process in question. Observing a different plant culture can offer a wealth of ideas worth pursuing.

Step 10: Analyze the Data; Establish the Gap

With the data in hand, the team must analyze it thoroughly in comparison with the data taken from its own process. In most cases, the team will be able to establish the *gap* (the performance difference between the two processes) numerically—for example, 200 pieces per hour versus 110 pieces, 2% scrap versus 20%, or errors in parts per million rather than parts per thousand.

After the team concludes there is no doubt that the partner's process is superior, other questions arise: Can its process replace ours? What will it cost, and can we afford it? What impact will it have on adjacent processes? Can we support it? Only by answering these questions can the team conclude that implementation is possible.

Step 11: Plan Action to Close the Gap or Surpass

Assume the team concluded that the change to the new process is desirable, affordable, and supportable and that the team wants to adopt the process. In most cases, implementation will require some planning to minimize disruption while the change is being made and while the operators are getting used to the new process. It is very important to approach implementation deliberately and with great care. This is not the time for haste. Consider all conceivable contingencies and plan to avoid them, or at least be prepared for them. Physical implementation may be accompanied by training for the operators, suppliers, or customers. Only after thorough preparation and training should an organization implement the change to the new process.

A second aspect of benchmarking should be kept in mind. The objective is to put in place a process that is best-in-class. If the team merely transplants the partner company's process, it will not achieve the objective, although improvements may occur. To achieve best-in-class, an organization must surpass the performance of the benchmark process. It may not be possible to do this at the outset, but the team's initial planning should provide for the development work necessary to achieve it in a specified period of time (see Figure 20–1).

Step 12: Implement the Change

The easiest step of all may be the actual implementation, assuming that the team's planning has been thorough and that execution adheres to the plan. New equipment may or may not be involved, there may be new people, or more or fewer people—but there will certainly be new procedures that will take time to become routine. Therefore, it should not be a surprise if initial performance does not equal the benchmark. After people get used to the changes and initial problems get worked out, performance should be close to the benchmark. If it is not, an important factor was overlooked, and another visit to the benchmarking partner may be necessary to determine what it is.

Step 13: Monitor

After the process is installed and running, performance should come up to the benchmark quickly. Before long, continual improvement should enable the organization to surpass the benchmark. None of this is likely to happen without constant attention and monitoring. Never install a new process, get it on-line and performing to expectations, and then forget about it. All processes need constant attention in the form of monitoring. Statistical process control can be an invaluable tool for this purpose, as can other types of charting.

Step 14: Update Benchmarks; Continue the Cycle

As was explained in step 11, the intent of benchmarking is not only to catch up with the best-in-class but to surpass, thereby becoming best-in-class. This is a formidable undertaking, because those with best-in-class processes are probably not resting on their laurels. They too will continue to strive for continually better performance. However, you are now applying new eyes and brains to their processes, and fresh ideas may well yield a unique improvement, vaulting your organization ahead of the benchmark-

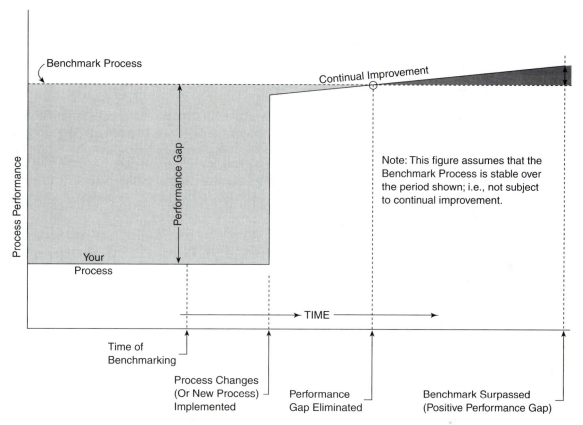

Figure 20–1
Effect of Benchmarking Process Change Followed by Continual Improvement

ing partner. Should that happen, your organization will be sought out as a best-in-class benchmarking partner by others who are trying to bootstrap their performance. Whether that happens or not, whether the benchmark is actually surpassed or not, the important thing is to maintain the goal of achieving best-in-class. Benchmarks must be updated periodically. Stay in touch with the best-in-class. Continue the process. Never be content with a given level of performance.

 ## QUALITY TIP

Continual Process Improvement Has No End

Any process that is "good enough" today will be inadequate tomorrow.

Source: David L. Goetsch and Stanley B. Davis

An important consideration, as you either achieve best-in-class or get close, is that limited resources have to be diverted to those processes that remain lowest in performance relative to their benchmarks. Let continual improvement take over for the best processes, and concentrate benchmarking on the ones that remain weak.

Three Phases of Benchmarking

This 14-step sequence represents the three phases of benchmarking: preparation, execution, and postexecution. Figure 20–2 illustrates the benchmarking process/sequence by phase and indicates action responsibility for each step. Figure 20–2 also makes it clear that the final step (14) causes the cycle to start over again at step 2, confirming the never-ending nature of the benchmarking process for companies that want to achieve or maintain leadership positions. Key points relating to the 14-step sequence of steps for implementing benchmarking are as follows:

- Benchmarking requires top management's commitment, participation, and backing.
- It is necessary that an organization thoroughly understand its own processes before attempting to benchmark.
- The processes that should be benchmarked are those that most need improvement.
- Benchmarking teams must include process operators.
- Benchmark best-in-class, not best-in-the-industry.
- Do not rush into new processes or major changes without thorough, thoughtful planning.
- Do not be satisfied with a zero gap—aim to surpass.
- Carefully monitor new processes or major process changes.
- Benchmarking is not a one-shot process: continue it forever.

BENCHMARKING VERSUS REENGINEERING

Benchmarking involves partnering with the owner of a best-in-class process so that you might adopt or adapt that process in your operation without having to spend the time and energy to try to design a duplicate of the superior process. Process reengineering requires you to do the latter, on your own. Therefore, in our view, process reengineering should only be considered when it is impossible to use benchmarking. That could happen for a number of reasons, including these:

- No known process available for benchmarking (rare)
- Best-in-class not willing to partner
- Best-in-class inaccessible due to geography or expense

If your subject process is unsatisfactory, and you cannot benchmark for any of these reasons, you may have to resort to reengineering. You should be careful to consider the reasons for the process being unsatisfactory. It may simply be the wrong process for the job, or it may be out of statistical control. Reengineering will not solve either of those problems. Be sure that the process is appropriate, and that it is in control

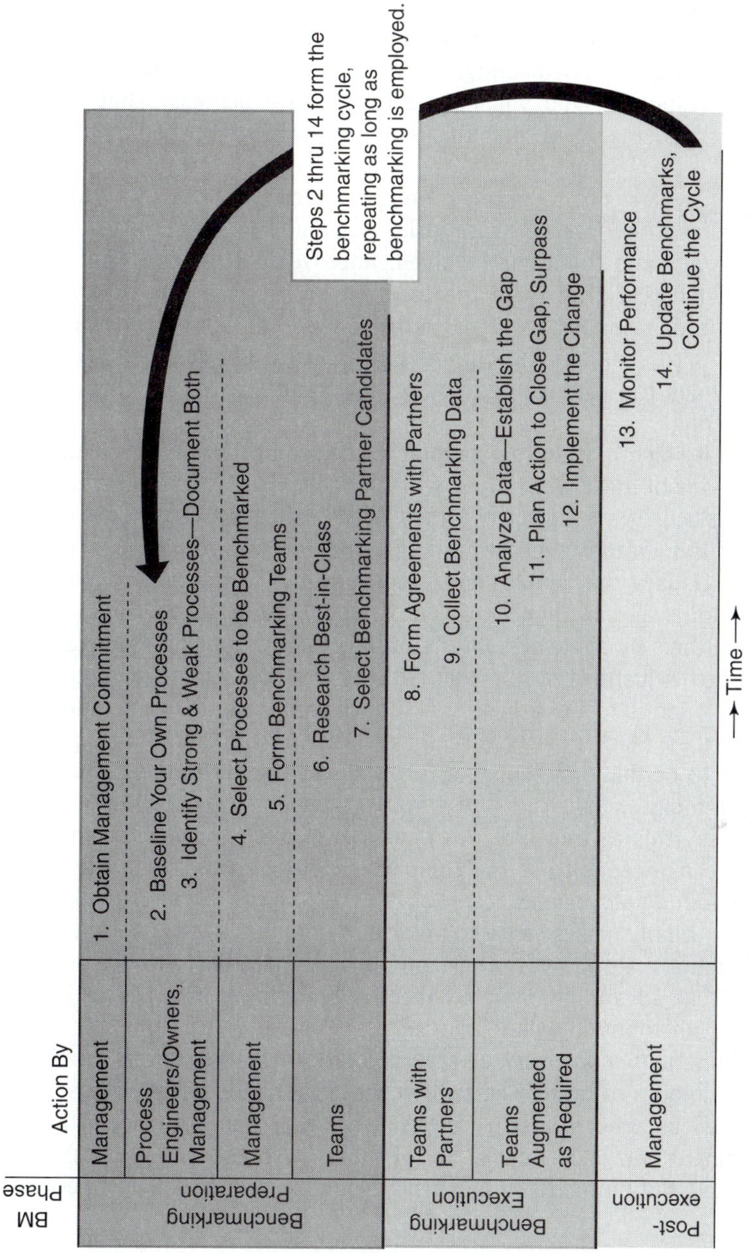

Figure 20–2
The Benchmarking Process/Sequence

first. If it is still not producing the desired results, suggesting that it is simply not capable, then redesigning it through reengineering is a good approach. One disadvantage with process reengineering is that there is no guarantee that after spending the time and resources, you will have a competitive process. That issue does not exist with benchmarking. With benchmarking you will have observed a competitive process in action.

When we set out to improve our processes, we normally flowchart them to help us understand how each process really works and to give us a visual impression of the steps, people, and functions involved. Improvement typically comes about by changing or eliminating activity in the process that does not add value or consumes too much time or resources, and so forth. There is an alternative way to go about this, and that is to abandon the current process and replace it with a brand-new process that provides the same functionality but better, faster, or cheaper. That is process reengineering.

Here is something to think about: If an organization could achieve the same results by either one of these two routes, which one would stand the best chance for success in the workplace? We believe the former—let's call it the continual improvement route—would be more readily accepted by the workforce and would be, therefore, more likely to succeed. Usually the people most closely related to the process have a major input to any continual improvement initiative, and it will not be perceived as something being forced on them by some person or group that doesn't really understand the process anyway. Whether justified or not, that is the way process reengineering has come across to workers. It tends to be radical and sudden, and seldom is consideration given to the human issues. Many times it is seen as a management tool for laying off workers. It does not have to be that way, but that is, we think, the way process reengineering is widely perceived today.

We say this to lead into our final thoughts on process reengineering. If you find process reengineering to be the approach for one of your processes, never let it be a surprise to your employees. In keeping with the philosophy we have promoted throughout this book, it only makes sense to involve the process owners and their internal suppliers and customers, along with other appropriate employees, in your process reengineering project. Take advantage of their collective brainpower and diverse perspectives, and in the doing, their buy-in will be assured.

In summary, if you have a very good process to begin with, use continual improvement techniques to make it better. On the other hand, if the process is clearly inferior to some used by other firms, try benchmarking. When you cannot achieve the kind of improvement you need from either of those methods, then process reengineering may be required. But no matter which way you go, be sure to get your people involved.

ROLE OF MANAGEMENT IN BENCHMARKING

Management plays a crucial role in the benchmarking process. In fact, without the approval and commitment of top management, benchmarking is not possible. Benchmarking is not something that can occur from the grassroots up without management's direct involvement. Several benchmarking considerations require management's approval before the process can start: commitment to change, funding, personnel, disclosure, and involvement.

Commitment to Change

Benchmarking is a serious undertaking for both benchmarking partners. Unless a firm commitment to change exists—unless the organization fully intends to radically improve its processes to come up to best-in-class standards—benchmarking should not be considered. Unfortunately, too many companies jump into benchmarking without that commitment, with the result that money and personnel are wasted by both parties. In addition, the hopes and expectations of employees are raised, only to be disappointed when nothing comes of it. To obtain any real benefit from benchmarking, an organization must resolve that when a best-in-class process is found, it will do what is necessary to incorporate it as a replacement (or radical improvement) model for its inferior process. That, after all, is what benchmarking is about.

Funding

Only management can authorize the expenditure of funds for benchmarking. These funds will support travel for teams visiting the organizations with best-in-class processes. Teams are usually composed of five to eight people. Visits may last from two days to two weeks. Travel destinations are inflexible, dictated by the location of the best-in-class firms. Clearly, travel expenses can be high. Management must make the funds available if benchmarking is to be carried out.

Human Resources

In similar fashion to funding, management must make the necessary human resources available for the benchmarking tasks. Although the costs for the human resources are usually far higher than for travel, the availability of personnel is seldom an issue except for the target company.

Disclosure

It may not be immediately obvious, but both companies—the benchmarker and the target—disclose information about their processes and practices. Management may be understandably hesitant to disclose such information to competitors, but what about the case of the noncompetitor benchmarking partner? Even there, management may be reluctant, because there can be no ironclad guarantee that information divulged to a noncompetitor will not find its way to the competition. The other side of the coin is that few processes or practices remain secret very long anyway. But if the organization has some unique process that gives it a competitive advantage, the process should be treated as proprietary and not be subjected to benchmarking. In any event, only management can make the decision to disclose information.

Involvement

Management must be actively and visibly involved in every aspect of the benchmarking process. Management should be involved in determining which processes are to be benchmarked and selecting benchmarking partner candidates. Management is in a unique position to establish the communication channels between the companies, because top managers tend to affiliate through professional organizations. Dialogue among top-level managers should be encouraged.

It is important for management to stay abreast of benchmarking events and to make certain that the effort supports the objectives and vision of the company. Management's ability to do this is greatly enhanced when it is directly involved. In addition, subordinates will recognize the importance placed on benchmarking by the degree to which management is visible in the process. With management active, all levels will be more productive in their benchmarking activities.

PREREQUISITES TO BENCHMARKING

Before getting involved in benchmarking, an organization should check the prerequisites—those philosophical and attitudinal mind-sets, skills, and necessary preliminary tasks that must precede any benchmarking efforts.

Will and Commitment

Without the will and commitment to benchmark, an organization cannot proceed. Don't waste time or the time of a benchmarking partner in the absence of a commitment and a will to benchmark on the part of the company's top management.

Vision/Strategic Objective Link

Benchmarking requires a strong focus, or it can go off in numerous different directions as benchmarkers get carried away in their enthusiasm. Before it is started, benchmarking objectives must be linked to the company's vision and strategic objectives, providing specific direction and focus for the effort. Failure to do this will almost certainly result in wasted resources and frustration.

Goal to Become the Best—Not Simply Improved

Nothing is wrong with incremental improvement—unless current performance is far below world class. However, if an organization is not at the world-class level, incremental improvement may only ensure that it remains inferior to the best-in-class forever. Benchmarking requires that the goal be to leap to the head of the field in one radical change, not just to be a few percentage points better than last year.

Openness to New Ideas

If a company is imbued with the "not-invented-here" syndrome, it will have a problem with benchmarking. The chief symptom of that affliction is a shortsighted mind-set that is characterized by a reluctance to consider other ways of doing things. Although few will admit it, many people are reluctant to consider ideas or approaches that are not their own. Organizations can be like individuals in this regard. Because the essence of benchmarking is capitalizing on the work and ideas of others, a company must be open to new ideas for benchmarking to provide any value. The benchmarking process may help bring about more receptivity to new ideas by demonstrating that they really work.

Understanding of Existing Processes, Products, Services, and Customer Needs

It is mandatory that an organization thoroughly understand its own processes, products, services, practices, and the requirements of its customers so that it can determine what needs to be benchmarked. In addition, it is necessary to have a solid understanding of your process in order to make meaningful measurements against that of the partner.

Processes Documented

It is not enough to understand the processes; they must be completely documented, for three reasons:

- All people associated with the process should have a common understanding of it, and that can come only from documentation.
- A documented starting point is needed against which to measure performance improvement after benchmarking changes have been implemented.
- The organization will be dealing with people (the partners) who are not familiar with its processes. Process documentation will help the partner understand the organization's processes. With an understanding of where the benchmarking organization is, the partner will be better able to help.

Process Analysis Skills

To achieve an understanding of your own processes, products, and services and to document those processes, you must have people with the skills to characterize and document processes. These same people will be needed to analyze the benchmarking partners' processes and to help adapt those processes to the organization's needs. Ideally, they should be employees, but it is possible to use consultants in this role.

Research, Communication, and Team-Building Skills

Additional skills required include research, communication, and team building. Research is required to identify the best-in-class process owners. Communication and team building are required to carry out the benchmarking both on an internal basis and with the partners.

OBSTACLES TO SUCCESSFUL BENCHMARKING

Like most human endeavors, benchmarking can fail. Failure in any activity usually means that the participant failed to prepare adequately for the venture—failed to learn enough about the requirements, the rules, and the pitfalls. So it can be with benchmarking. In this section, some of the common obstacles to successful benchmarking as drawn from the experiences of dozens of companies are explained.

Internal Focus

For benchmarking to produce the desired results, you have to know that someone out there has a far better process. If a company is internally focused (as many are), it may not even be aware that its process is 80% less efficient than the best-in-class. An internal focus limits vision. Is another firm better? Which is it? Such organizations don't even ask the question. This is complacency—and it can destroy a company.

Benchmarking Objective Too Broad

An overly broad benchmarking objective such as "Improve the bottom-line performance" can guarantee failure. This may well be the reason for benchmarking, but the team will need something more specific and oriented not to the *what* but to the *how*. A team could struggle with the bottom line forever without knowing with certainty that it achieved success or failure. The team needs a narrower target; for example, "Refine or replace the invoicing process to reduce errors by 50%." That gives team members something they can go after.

Unrealistic Timetables

Benchmarking is an involved process that cannot be compressed into a few weeks. Consider 4 to 6 months the shortest schedule for an experienced team, with 6 to 8 months the norm. Trying to do it in less time than that will force the team to cut corners, which can lead to failure. If you want to take advantage of benchmarking, be patient. On the other hand, any project that goes on for more than a year should be assessed; the team is probably floundering.

Poor Team Composition

When a process is benchmarked, those who own the process, the people who use it day in and day out, must be involved. These people may be production line operators or clerks. Management may be reluctant to take up valuable team slots with these personnel when the positions could otherwise be occupied by engineers or supervisors. Engineers and supervisors should certainly be involved, but not to the exclusion of process owners. The process owners are the ones who know the most about how the process really operates, and they will be the ones who can most readily detect the often subtle differences between your process and that of the benchmarking partner. Teams should usually be six to eight people, so be sure the first members assigned are the operators. There will still be room for engineers and supervisors.

Settling for "OK-in-Class"

Too often organizations choose benchmarking partners who are not best-in-class, for one of three reasons:

■ The best-in-class is not interested in participating.
■ Research identified the wrong partner.
■ The benchmarking company got lazy and picked a handy partner.

Organizations get involved in benchmarking when they decide that one or more of their processes is much inferior to the best-in-class. The intention is to examine that best-in-class process and adapt it to local needs, quickly bringing your organization up to world-class standards in that process area. It makes no sense to link with a partner the process of which is just good. It may be better than yours, but if adopted, it still leaves your organization far below best-in-class. For the same amount of effort, an organization could have made it to the top. Organizations should identify the best and go for it. Only if the absolute best will not participate can taking second-best be justified. Second-best should be used only if it is significantly superior to the process in question.

Improper Emphasis

A frequent cause of failure in benchmarking is that teams get bogged down in collecting endless data and put too much emphasis on the numbers. Both data collection and the actual numbers are important, of course, but the most important issue is the process itself. Take enough data to understand your partner's process on paper, and analyze the numbers sufficiently to be certain that your results can be significantly improved by implementing the new process. Unless the team has been deeply involved in the process, the practical knowledge to successfully adapt and implement it back home may be lacking. Keep the emphasis on the process, with data and numbers supporting that emphasis.

Insensitivity to Partners

Nothing will break up a benchmarking partnership quicker than insensitivity. Remember that a partner is doing your organization a favor by giving access to its process. You are taking valuable time from the partner's key people, and at best you are disrupting the routine of daily business. If you fail to observe protocol and common courtesy in all transactions, your organization runs the risk of being cut off.

Limited Top-Management Support

This issue keeps coming up because it is so critical to success at all stages of the benchmarking activity. Unwavering support from the top is required to get benchmarking started, to carry it through the preparation phase, and finally, to secure the promised gains.

BENCHMARKING RESOURCES

A number of sources of information can help organizations with their benchmarking efforts. These cover the spectrum from nonprofit associations to cooperative affiliations to for-profit organizations that sell information. In addition, of course, there are consulting firms with expertise and databases covering all aspects of benchmarking.

One of the most promising ventures is the American Productivity and Quality Center (APQC) Benchmarking Clearinghouse (123 N. Post Oak Lane, Houston, TX 77024; phone, [713] 681-4020; fax, [713] 681-5321). The APQC Benchmarking Clearinghouse has been set up to assist companies, nonprofit organizations, and government in the process of benchmarking. It works with an affiliation of organizations to collect and disseminate best practices through databases, case studies, publications, seminars, conferences, videos, and other media.

A wide range of benchmarking information is available on the Internet. Just ask your search engine to find "benchmarking" or "process benchmarking," and you will probably be rewarded with more information than you can use. This ranges from articles on the subject to promotions for books and consultants. Colleges list the contents of their libraries that are related to benchmarking. We would suggest a word of caution, however. Anyone can put anything on the Web without verification, so it is always a good idea to approach material from unfamiliar sources with a degree of skepticism. In spite of this, we consider the Web to be a valuable benchmarking resource center.

Excellent sources of information for benchmarking are trade and professional groups. They can often direct organizations to best-in-class practices, provide contacts, and offer valuable advice. Baldrige Award winners are committed to share information with other U.S. companies, and they hold periodic seminars for this purpose.

The trade literature publishes a wealth of relevant information, including lists of companies with best-in-class processes and practices. *Industry Week* is one example of an excellent source of benchmarking information. Companies such as Dun and Bradstreet and Lexis-Nexis maintain databases of potential benchmarking partners and share them for a fee.

Consultants and universities that are engaged in benchmarking can help organizations get started by providing initial training, offering advice and guidance, and directing organizations to benchmarking partner candidates.

A word of caution is in order at this point. Be sure that any information obtained is current. The very nature of benchmarking makes yesterday's data obsolete. To achieve maximum benefit, organizations must be sure that they are operating on current information.

SELECTION OF PROCESSES/FUNCTIONS TO BENCHMARK

Selection of processes or functions to benchmark would seem to be a straightforward decision but is in fact one that gives many would-be benchmarkers a great deal of trouble. If you keep in mind that the purpose of benchmarking is to make a radical improvement in the performance of a process—more improvement than could be made quickly through continual improvement techniques—it follows that most concern should be focused on the weakest processes and the functions that operate them.

The strongest processes are sometimes benchmarked as a means of obtaining a report card against the best-in-class. This is a waste of time and effort, to say nothing of money, on two counts. First, the organization is proud of this process and has no intention of replacing it or radically modifying it. What good does it do to determine that the process is within 10% of best-in-class? It may be intellectually gratifying, but the process will be no better for the effort. Second, the processes that are the weakest are the ones that are most detrimental to competitiveness, not those that are in the 90th percentile. Moreover, the weakest offer the most room for dramatic improvement, perhaps many times over. This is where the benchmarking effort should be focused. The reason companies get this wrong is that they are more inclined to talk about what they do right than what they do wrong. When attempting to benchmark, it is a good idea to leave vanity and pride out of the process.

ACTING ON THE BENCHMARK DATA

At the conclusion of the benchmarking project with your partner, data analysis will have produced both quantitative and qualitative information. The quantitative information is effectively the "stake driven into the ground" as the point from which future progress is measured. It is also used as the basis for improvement objectives. Qualitative information covers such matters as personnel policies, training, management styles and hierarchy, total quality maturity, and so on. This information provides insights on how the benchmarking partner got to be best-in-class.

The quantitative data are clearly the information sought and are always used. However, there may be more value in the qualitative information. It describes the atmosphere and environment in which best-in-class can be developed and sustained. Do not ignore it. Take it very seriously. Study it, discuss it in staff meetings, and explore the possibilities of introducing these changes into your culture.

In terms of the process that has been benchmarked, if the partner's process is significantly superior to your own—and we must assume that it is or it would not have been selected in the first place—you have to do something about implementing it. Perhaps you can modify your own process with some ideas picked up from benchmarking, or, more likely, you can adopt your partner's process, implementing it to replace yours. But whatever is indicated by the particular local situation, take decisive action and get it done.

PERPETUAL BENCHMARKING

If you have been through a series of benchmarking activities and have implemented changes that have significantly improved processes, your organization may develop a tendency to leave benchmarking. After all, there are other things that need attention and resources. But this can be a costly mistake. At this point, the organization not only has much-improved processes, but it has developed some valuable benchmarking experience. Keep in mind that best-in-class continues to be a dynamic and ever-changing mosaic. Processes are constantly being improved and altered. In a relatively short time, an organization can fall behind again. To prevent that from happening, the organization must take advantage of hard-won benchmarking experience and keep the effort moving. This means staying up to date with the best-in-class through all the means at your disposal, staying current with your own processes as they are continually improved, and benchmarking the weaker processes. This is a never-ending process.

 SUMMARY

1. Benchmarking is a process for comparing an organization's operations or processes with those of a best-in-class performer.
2. The objective of benchmarking is major performance improvement achieved quickly.
3. Benchmarking focuses on processes and practices, not products.
4. Benchmarking is done between consenting organizations.
5. Benchmarking partners are frequently from different industries.
6. Benchmarking is a component of total quality.
7. Benchmarking must be approached in an organized, planned manner, with the approval and participation of top management.
8. Benchmarking teams must include those who operate the processes.
9. Benchmarking is not restricted within industry boundaries, but only with best-in-class processes.
10. It is necessary for the benchmarker to understand its own process before comparing it with another.
11. Because best-in-class is dynamic, benchmarking should be seen as a never-ending process.
12. Management has a key role in the benchmarking process, including commitment to change, making funds available, authorizing human resources, being actively involved, and determining the appropriate level of disclosure.
13. The goal of benchmarking is to become the best-in-class, not simply improved.
14. The intent of benchmarking is to replace an inferior process with one rated best-in-class or to radically improve a process, bringing it up to best-in-class performance—and then to surpass best-in-class.
15. A number of obstacles to successful benchmarking exist, including internal focus, overly broad or undefined objectives, unrealistic timetables, improper team composition, failure to aim at best-in-class, diverted team emphasis, insensitivity toward the partner, and wavering support by top management.

KEY TERMS AND CONCEPTS

Benchmarking   Malcolm Baldrige Award
Benchmarking partner   Performance gap
Best-in-class   Process baselining
Continual improvement   Process capability
Customer focus   Process characterization
Internal focus   Process flow diagram
Key business process

 FACTUAL REVIEW QUESTIONS

1. Define *benchmarking*.
2. Explain the difference in objectives for continual improvement and benchmarking.
3. List five factors that lead organizations to benchmarking.
4. On which processes should an organization concentrate for benchmarking?
5. Why is it necessary that top management be committed as a prerequisite to benchmarking?
6. What are the reasons for characterizing and documenting an organization's processes before benchmarking?
7. Identify the critical members of the benchmarking team.
8. Explain why it is not enough to simply clone the benchmarking partner's process.
9. Explain the importance of linking the benchmarking objectives with the organization's strategic objectives.
10. Explain how the "not-invented-here" syndrome can be a hindrance to benchmarking effectiveness.
11. List and discuss the eight obstacles to successful benchmarking.

 CRITICAL THINKING ACTIVITY

A Competitive Crisis

(*Note:* This is a purely fictitious case study, both in terms of the company and the numbers.)

Empire Communications Products is a company of 420 employees engaged in designing and producing telephone equipment for the wireless telephone industry. It originally spun out of a large communications company 5 years ago as the wireless phone market began its rapid expansion. Today its major competition is from the North American, European, and Japanese telecom giants, all of which have now recognized the market potential and are competing furiously. Empire Communications has been a respected and successful niche market player, even selling some proprietary products to its big competitors. However, the pressure on profits has become extreme, and the company is searching for ways to cut its costs.

The CEO has had his management steering committee looking into benchmarking as a possible means of making processes more efficient. It has concluded that the company has five key business process areas, each with its own set of processes. The five key business process areas are as follows

Key Process Area	12-Month Expense	Primary Processes
Engineering	$1,000,000	Research & development
(45 employees)	6,005,000	Design
	300,000	Product improvement
	7,305,000	
Finance	500,000	Accounting
(12 employees)	100,000	Accounts payable
	130,000	Accounts receivable
	730,000	
Human Resources	300,000	Recruiting/hiring
(5 employees)	60,000	Compensation
	260,000	Employee development
	620,000	
Production	1,000,000	Procurement
(350 employees)	30,000,000	Materials
	3,000,000	Warehousing
	3,500,000	Material control
	550,000	Materials preparation
	3,800,000	Production control
	11,100,000	Assembly
	2,050,000	Integration and test
	55,000,000	
Quality Assurance	275,000	Incoming inspection
(8 employees)	500,000	In-process inspection
	150,000	Supplier auditing
	200,000	Internal auditing
	1,125,000	
Total annual expenses	$64,780,000	

The steering committee did some research to compare similar expenses with a typical firm in the industry. It found that the company's engineering expenses were close to the average. The same was true of the finance department. Human resources was on average except for expenses for employee development, where it was clear that Empire was spending less for training than the industry at large. What the committee found in the production and quality assurance areas was a major surprise. Empire spent far more than the industry average for warehousing, material control, materials preparation, production control, and even assembly. In quality assurance, Empire's costs were way above the average for inspection. On the other hand, Empire spent significantly less then the average for its materials, integration and test, and supplier audits and internal audits. The comparative data are given here:

Process	Industry Avg.	Empire	Difference
Employee development	450,000	260,000	−190,000
Materials	31,000,000	30,000,000	−1,000,000
Warehousing	975,000	3,000,000	+2,025,000
Material control	180,000	3,500,000	+3,320,000
Materials preparation	85,000	550,000	+465,000
Production control	725,000	3,800,000	+3,075,000
Assembly	10,000,000	11,100,000	+1,100,000
Integration & test	3,060,000	2,050,000	−1,010,000
Incoming inspection	75,000	275,000	+200,000
In-process inspection	200,000	500,000	+300,000
Supplier auditing	1,250,000	150,000	−1,100,000
Internal auditing	2,800,000	200,000	−2,600,000
	$50,800,000	$55,385,000	+$4,585,000

In total, Empire's costs were about $4.6 million above the annual average for its peers. If it could find a way to eliminate the excess cost, the $4.6 million would be added to before-tax profit, virtually doubling its current profit status. That would certainly be welcomed.

Exercise Questions

1. You are a benchmarking consultant. Empire staff has shared this information with you, and it wants to know why you think benchmarking is an appropriate course for it to follow. Is benchmarking likely to help Empire Communications? What will you tell it?
2. Does the pattern of spending more than the average in some areas and under the average in others suggest anything to you (relative to previous chapters)? How will this influence any benchmarking strategy?
3. How would you determine the best process areas for benchmarking?
4. Given the fact that the telecom industry is extremely competitive, is it likely that Empire will find willing benchmarking partners among its competitors? If not, what would you do?
5. Develop a recommendation for Empire Communications for how you propose to lead it through the benchmarking activity, and what it should expect to gain from it.

Author's Comments on the Empire Communications Exercise

Question 1. It seems clear that other companies are doing things much differently and, overall, much less expensively than Empire, so benchmarking would seem to be a good avenue for quick, major improvement.

Question 2. Spending less for training *may* mean that some of Empire's employees are not as well trained as they should be. That will have to be investigated in more depth. Beyond that there seems to be a pattern of over- and underspending that suggests Empire is not doing business like the average firms in production and quality assurance.

The gross difference in warehousing, materials control, and production control are probably the result of Empire's use of the old traditional manufacturing techniques, rather than using the just-in-time approach, which can be a huge cost saver. Lots of benchmarking possibilities here, if that is the case.

Empire spends 6.5 times the average for materials preparation. Perhaps it is doing it all in-house rather than having its supplier do it more efficiently. We need to find out exactly what Empire is doing, then look to see where there is a better process to be benchmarked.

Integration and test looks, on the surface, to be a winner. Find out why, and if it turns out to be a superior process, it can be used as an enticement for a prospective benchmarking partner.

Empire appears to have lower materials purchasing expense, but if that is coupled with overspending on incoming inspection and underspending in supplier audits, one gets the impression that Empire is still trying to sort defective materials out at its plant rather than having its suppliers do so before the materials ever get to Empire. That could also help explain why Empire's assembly costs are higher than average (rework due to defective materials).

Almost certainly benchmarking will show Empire new, more efficient ways to do things, but we need to fully understand how Empire works before we go benchmarking.

Question 3. Look for the production processes that indicate the biggest undesirable expense gap, then try to determine which will be easiest to benchmark and implement. Always take the easiest-to-benchmark, financially critical process first.

Question 4. The chances are slim that an outright competitor would become a benchmarking partner, although that does happen. Empire should look at the entire electronics industry (they only compete in telecom), and even outside it if appropriate for the process. Remember, it is best-in-class, not just best-among-competitors, in which we are interested.

Question 5. Your proposal should outline the work to be done internal to Empire; namely, checking out the issues raised by question 2, and determining where best to start regarding question 3. After comparing what Empire is doing against what the competition is apparently doing, do the research to find a best-in-class partner. Be sure to tell Empire that any process that is to be benchmarked will have to be thoroughly flow-charted and understood. (It would also be a good idea to point out that even those critical processes that are not on the benchmarking list at present should also be characterized as a first step in an internal continual improvement program.) You will need to talk about the benchmarking team composition, training, and the actual visit to the partner's site. Also, you will need to explain the process of adopting or adapting the new process to the Empire Communications environment and culture. It would be a good idea to go through each of the 14 steps with Empire's steering committee.

DISCUSSION ASSIGNMENT 20–1

Benchmarking at Motorola

Motorola, one of the original Malcolm Baldrige Award winners, has found that even after achieving best-in-class, there is much to be gained through continued benchmarking. Every new program, every new product, and every new improvement effort at Motorola is preceded by a search for the best-in-class. In this way, Motorola ensures that every new process introduced will be world class.

DISCUSSION ASSIGNMENT 20–2

Xerox

By the late 1970s, Xerox was losing significant market share to its Japanese competitors. Not only were the Japanese products excellent, but to Xerox's dismay, they were sold for less than Xerox could manufacture them. Xerox found that it had nine times as many suppliers as the Japanese companies and made seven times as many manufacturing defects. Lead times for new products were twice as long, and production setup times were five times as long as the competitors'.

Xerox introduced benchmarking in 1980. Its processes and practices were benchmarked against the best in and out of its industry. As a result of these efforts, Xerox saved itself. Today Xerox is a world-class competitor, capable of holding its own in terms of technology, price, service, and customer satisfaction against any competition. Benchmarking at Xerox has reached into every facet of the company, and remains a primary feature of the corporation.

 ENDNOTES

1. Steven George, *The Baldrige Quality System: The Do-It-Yourself Way to Transform Your Business* (New York: John Wiley & Sons, 1992).
2. Robert C. Camp, *Benchmarking: The Search for Industry Best Practices That Lead to Superior Performance* (Milwaukee: Quality Press/Quality Resources, 1989).

CHAPTER TWENTY-ONE

Just-in-Time Manufacturing (JIT)

"Time is money."
Greek proverb

MAJOR TOPICS

- Just-in-Time Defined
- Rationale for JIT
- Development of JIT
- Relationship of JIT to Total Quality and World-Class Manufacturing
- Benefits of JIT
- Requirements of JIT
- Automation and JIT

JUST-IN-TIME DEFINED

When people who should know are asked to define just-in-time (JIT), the typical response is that JIT "is getting your materials delivered just when you need them." Probing a little deeper may elicit a response that suggests JIT manufacturers let their suppliers keep their materials inventory until the manufacturers need it. The first statement demonstrates an inadequate understanding of JIT, and the second is simply wrong. Even so, many companies under the auspices of JIT have indeed pushed their warehousing back to the suppliers for a net gain of zero. If these are not the right answers to the question, "What is JIT?" then what is it? Although not exactly what was originally intended, just-in-time manufacturing has become a management philosophy that seeks to eliminate all forms of waste in manufacturing processes and their support activities. JIT permits the production of only what is needed, when it is needed, and only in the quantity needed. This must apply not only to the just-in-time manufacturer but also to its suppliers if the system is to eliminate all possible waste. Those companies that have required their suppliers to do their warehousing clearly have not gotten the point. The supplier should not produce the material until the JIT manufacturer needs it. In that mode there is no warehousing and therefore no wasted resources for buildings, maintenance, people to care for the material, spoilage, obsolescence, or other related problems.

JIT is not so much related to supplier activities, although they are important, but more to events on the manufacturing floor. For example, assume that a company manufactures motion sensors. There are five discrete processes involved, each carried out by one worker, as illustrated in Figure 21–1. The traditional production process places a big supply of input materials in the warehouse, doling them out to the production line at the rate of so many pieces per unit time. The electronic assembly and mechanical assembly processes convert their respective input materials into input materials for the electronic module assembly process. The electronic module assembly and the frame fabrication processes then convert their input materials into input materials for the final assembly process, which in turn converts them into completed motion sensors. Each of the five work areas produces at the rate necessary to meet a quota, or to consume all the input materials. The completed sensors are sent to the warehouse for storage until someone buys them.

Each preceding stage pushes its output into the succeeding stage. It is difficult to balance a line to the point that the succeeding stages need exactly what is produced by the preceding stages, so it is common to take the output of the preceding stages off the floor and store it in so-called staging areas. Staging areas are nothing more than miniwarehouses.

Just-in-time approaches the manufacturing process from the opposite end of the line. Rather than push materials into the processes, storing them whenever they cannot be accommodated, JIT controls the line from the output end. Indeed, it can be said that the customer controls the line, because nothing is built until there is an order for it. After an order is received for a product, the final assembly process is turned on to put together the required number. The assembler pulls the required input materials from the electronic module and frame fabrication processes—only enough to make the required number. Similarly, the electronic module assembly and frame fabrication processes pull input materials from their preceding processes, and so on back up the line. At the top of the line, input materials are pulled from suppliers in the exact quantity needed, and no more.

Following the JIT procedure, no step in the production process ever overproduces or produces before a demand is made. Therefore, there is no need for a staging area or the people required to move materials into it and out of it, account for it, and so on. No money is tied up in inventory of raw materials, partially built goods (known as work-in-process or WIP), or finished goods. If there are no stored materials, there is no spoilage or obsolescence. The elimination of these wastes alone makes JIT the most powerful manufacturing concept to come along since Henry Ford's moving assembly line of 1913. JIT contributes to the elimination of many more forms of waste, as discussed later in this chapter.

So the definition of JIT as used in this book is as follows:

Just-in-time (JIT) is producing only what is needed, when it is needed, and in the quantity that is needed.

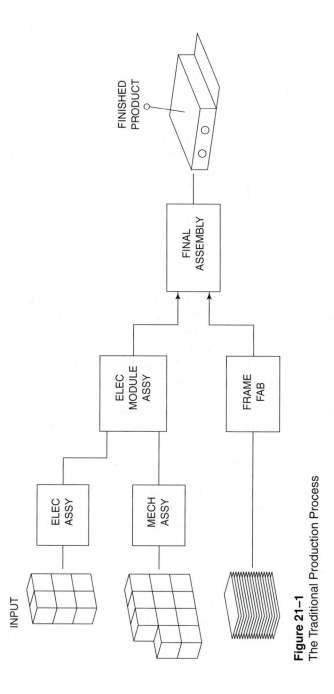

Figure 21–1
The Traditional Production Process

QUALITY TIP

What's in a Name?

Just-in-time is the name given the Toyota Production System developed by Taiichi Ohno. As is so often the case, we find that the same product is being repackaged under other names. JIT is also referred to as *lean production manufacturing*. The term *focused factory* is sometimes applied to JIT production cells. If you encounter a production system called *demand flow*, or *demand flow technology*, it is JIT with a new label. None of these are bad names, and in fact some may be more accurate. But in this book, and most others, the generic name for pull-system manufacturing, *just-in-time*, is preferred.

RATIONALE FOR JIT

Mass production manufacturers set their production schedules based on a forecast of future needs, which in turn is based on historical data and trend analysis (see Figure 21–2). The great weakness of this system is that no one can predict the future with sufficient certainty, even with a complete and perfect understanding of the past and a good sense of current trends in the marketplace. One does not have to search long to find examples of failed attempts to correctly project the marketability of products. The Edsel is one of many automobiles that were released with great fanfare to a disinterested public. A new formula for Coca-Cola introduced in the late 1980s is another example of market predictions gone awry. IBM has case after case of personal computers, such as the unlamented IBM PC Jr. (which failed in the marketplace in spite of the best market

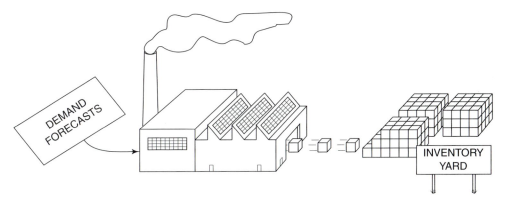

Figure 21–2
Factory Producing to Forecast Demand (Mass Production)

research IBM could muster). These failures demonstrate the difficulty of trying to determine beforehand what will sell and in what quantity.

Even products that are successful in the market have limits as to the quantities that buyers will absorb. When production is based on predictions of the future, risk of loss from overproduction is far greater than when production is based on actual demand. The previous section defined JIT as producing what is needed, when it is needed, and only in the quantity that is needed (see Figure 21–3). The result of JIT is that no goods are produced without demand. This, in turn, means no goods are produced that cannot be sold at a price that supports the viability of the company.

So far we have viewed JIT from the point of view of the manufacturer and the ultimate purchaser of the product—the producer and the customer. But if we look at the complete production process, we will find that it contains many producers and customers—internal customers (see Figure 21–4). Each preceding process in the overall system is a producer, or supplier, and each succeeding process is a customer (see Chapter 7). JIT fits here as well as or better than with the manufacturer-and-purchaser model. No process in the system produces its output product until it is signaled to do so by the succeeding process. This can eliminate waste on a grand scale. It is the elimination of waste that justifies JIT in any kind of manufacturing operation. Elimination of waste is translated to improved quality and lower costs. Improved quality and lower costs translate to becoming more competitive. Although improved competitiveness does not assure survival (the competition may still be ahead of you), being noncompetitive surely guarantees disaster.

Taiichi Ohno, the creator of the just-in-time system, saw that the mass production system produced waste at every step of the way. He identified seven wastes:[1]

1. Overproducing
2. Waiting (time)
3. Transporting
4. Processing itself
5. Having unnecessary stock on hand

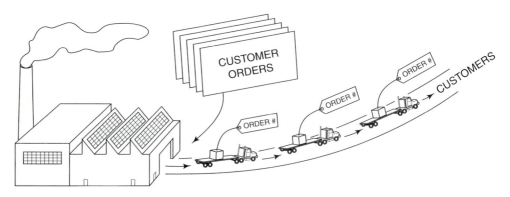

Figure 21–3
Factory Producing to Orders (JIT)

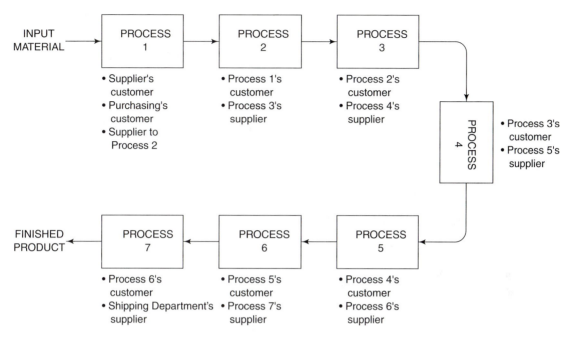

Figure 21–4
Internal Supplier–Customer Relationships

6. Using unnecessary motion
7. Producing defective goods

The elimination of these wastes is at the heart of the rationale for just-in-time: eliminate these wastes, and you will produce better products at lower cost. If the competition gets there first, your rationale for JIT is survival.

 ## QUALITY TIP

The Supermarket as a Just-in-Time System

Assume the supermarket is the preceding process in the production line. The subsequent process (the customer) goes to the supermarket to get exactly what it needs (in the case of the automobile assembly plant, auto parts) when it needs it. What should the preceding process then do? It must replenish that which has been withdrawn by the subsequent process.

Source: Taiichi Ohno.

DEVELOPMENT OF JIT

In the preceding section, Taiichi Ohno was identified as the creator of the just-in-time system, and it is true that he was responsible for developing the system as it is now known. However, other names should be remembered, at least to the extent to which they contributed by inspiration. The first is Henry Ford, creator of mass production. Because of Ford's great appreciation of the expense of waste, Ohno believes that if he were alive today he would have developed a system much like Toyota's. In his 1926 book *Today and Tomorrow*, Henry Ford talked about the waste of inventory in raw materials, work-in-process, and finished goods in the pipeline to market—and about the efforts taken to reduce the investment in this waste. Between 1921 and 1926 Ford output doubled, but investment in inventory of raw materials, semifinished goods, and finished goods actually declined. Based on 1921 performance, Ford should have had $170,000,000 tied up in this inventory but in fact had less than $50,000,000. Ford also recognized the wastes arising from transportation, waiting (time), and inefficiency on the factory floor. He believed in planning ahead to eliminate the waste before it happened. This is very contemporary thinking, and Ohno may be correct that Henry Ford, had he been living in the past 30 years, might well have developed a Toyota-like system. When Ohno wrote his book on the Toyota Production System, it was entitled *Just-in-Time for Today and Tomorrow*. It is not known if this was a tribute to Henry Ford's book, but it is at least an interesting coincidence.

Ford was a great influence on the Toyoda family—Sakichi, Kiichiro, and Eiji. Sakichi Toyoda, a designer of looms and founder of Toyota, is credited with the concept of *autonomation*, or automation with a human touch. His automatic loom could determine whether a thread was broken or missing, shutting itself down instead of making defective product.[2] Autonomation is one of the two pillars of the Toyota Production System, the other being just-in-time. Kiichiro Toyoda, Toyota's founding chairman, planted the seeds of the Toyota Production System prior to World War II with his planning for the introduction of the assembly line at Toyota's Kariya plant. He wrote a booklet about how production was to work, and it contains the words just-in-time. His original meaning in English was "just-on-time," intending that things be done exactly on schedule, with no surplus produced. World War II halted further work on the system, and after the war it was Taiichi Ohno who revived and developed it into the present-day Toyota Production System.[3]

Eiji Toyoda, Toyota's chairman and Taiichi Ohno's boss for 35 years, is credited with the JIT philosophy: "In broad industries, such as automobile manufacturing, it is best to have the various parts arrive alongside the assembly line just-in-time."[4] Eiji Toyoda's greatest contribution may have been his support for Ohno's trial-and-error approach, shielding him from the inevitable controversy of his endeavors. Ohno claims that Eiji never told him to back off or slow down. He absorbed the heat and let Ohno press on unimpeded.[5]

Taiichi Ohno's motivation, like that of the Toyodas, was to eliminate all forms of waste from the production process. He was well schooled in the Ford mass production system and observed that the system itself created waste in huge proportions. If one was determined to violate the seven wastes, a mass production line would do it. Mass production is prone to *overproducing*; having people or materials *waiting*; *moving* work in process back and forth across the plant; retaining inefficient *processes*; maintaining costly inventories of *stock on hand*; requiring non-value-added *movement* because lines were set up to accommodate product, not workers; and producing *defective* goods because the line must continue to move. The italicized words represent the seven wastes.

Ohno believed that a production system based on just-in-time could eliminate the wastes. To appreciate fully what is involved here, one must understand that the mass production system as defined by Henry Ford was not irrational. Ford's objective was to produce huge quantities of the same product using an assembly line technology that required little expertise of its workers. The result was a reliable, cheap car that millions of buyers could afford. In that, he and others who used his mass production technology were eminently successful. But mass production is inflexible and wasteful—inflexible because it is driven by the great stamping presses and other machines that do not easily accommodate a variety of products, and wasteful because the underlying philosophy of mass production is that the line must crank out products that spring from market forecasts in a never-ending high-volume stream. To support that high-volume stream, there must be stockpiles of the materials that go into the product, because the lack of a single part can shut down the mass production line. Machines must be capable of high output and are so costly they cannot sit idle without creating trauma in the accounting department. Therefore, even when fenders are not needed, the machines must continue to stamp them out. The overproduction will be warehoused until it is needed—perhaps when the press breaks down. So it is with all the parts and subassemblies that make up the complete product. They are stored in large quantities, just in case something goes wrong in their production or transportation cycle, when they might be needed to keep the final assembly line moving—fenders for a rainy day, so to speak.

This is the norm with mass production. The problem with this system is that the building space in which these parts and materials are warehoused is expensive. It requires a small army of people to care for the stored materials/parts, and these people add not a whit to the ultimate value of the product. Spoilage occurs by loss, damage, or obsolescence for stored parts—all waste. Part waste of inventory, part waste of overproduction.

Mass production advocates emphasize the need to keep the line moving and that the only way to do this is to have lots of parts available for any contingency that might arise. This is the fallacy of just-in-time according to mass production advocates. JIT, with no buffer stock of parts, is too precarious. One missing part or a single failure of a machine (because there are no stores of parts) causes the JIT line to stop. It was this very idea that represented the power of JIT to Ohno. It meant that there could be no work-arounds for problems that did develop, only solutions to the problems. It served to focus everyone concerned with the production process on anticipating problems before they happened, and developing and implementing solutions so that they would not cause mischief later on.[6] The fact is that as long as the factory has the security buffer of a warehouse full of parts that might be needed, problems that interrupt the flow of parts to the line do not get solved because they are hidden by the buffer stock. When that buffer is eliminated, the same problems become immediately visible, take on a new urgency, and solutions emerge—solutions that not only fix the problem for this time, but for the future as well. Ohno was absolutely correct. JIT's perceived weakness is one of its great strengths.

Mass production is a *push system* (see Figure 21–5). The marketing forecast tells the factory what to produce and in what quantity; raw materials and parts are purchased, stored, forced into the front end of the production process, and subsequently pushed through each succeeding step of the process, until finally the completed product arrives

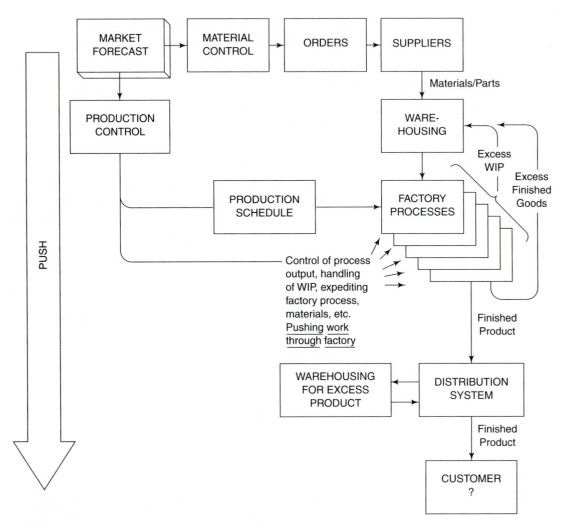

Figure 21–5
Mass Production Push System

at the shipping dock. It is hoped that by then there are orders for these goods, or they will have to be either stored or pushed (forced) into the dealers' hands, a widespread practice in the automobile business. The whole procedure, from imperfect forecast of marketability to the warehouse or the dealer, is one of pushing.

What if the market will only take half of the predicted amount, or wants none? What if the final assembly process can accommodate only two thirds of the preceding processes' output? These situations present big problems in terms of cost and waste, and they are common.

Just-in-time, on the other hand, is a *pull system* (see Figure 21–6; the term *kanban* in the figure will be clarified soon). The production schedule does not originate in a market forecast, although a great deal of market research is done to determine what customers want. The production demand comes from the customer. Moreover, the demand is made on the final assembly process by pulling finished products out of the factory. The operators of that process in turn place their pull demands on the preceding process, and that cycle is repeated until finally the pull demand reaches back to the material and parts suppliers. Each process and each supplier is allowed to furnish only the quantity of its output needed by the succeeding process.

Figures 21–5 and 21–6 also show a difference in the relationship between the customer and the factory. In the mass production system, no real relationship exists at all. The market forecasters take the place of the customers and place demands on the factory months in advance of production. In the JIT system, however, the customer's demand is felt throughout the system, all the way to the factory's suppliers, and even beyond that. The JIT system is simpler, eliminating entire functions such as material control, production control, warehousing/stocking, and so on.

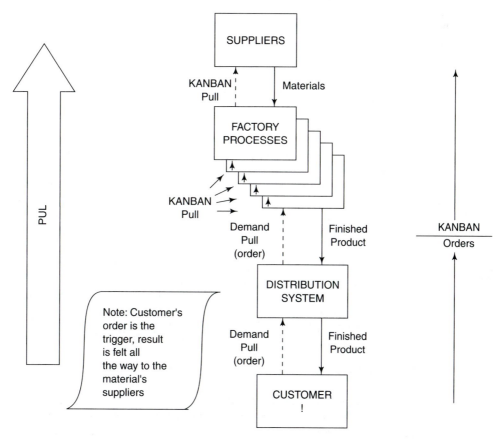

Figure 21–6
Just-in-Time Demand Pull System

The simplicity of JIT production is most evident on the factory floor. In a mass production plant, or even conventional job shops (low-volume, high-variety shops), it is almost never possible to tell from the factory floor how things are going relative to schedules. Parts of any product may be in any number of disparate locations in the plant at any given time—in the machine shop, welding shop, on the line, or in storage. Computers keep track of it all, but even then it is difficult to track a given product through the plant or to track its status at a given point in time. On the other hand, JIT, being a very visual process, makes tracking easy—even without computers. Parts have no place to hide in a JIT factory. The only work in process is that for which the process has a kanban (see the discussion of kanban in the section entitled "Process Problems").

The simplicity of today's JIT belies the difficulty Ohno encountered in developing the system. Because production must stop for a missing part, a process problem, or a broken machine, methods had to be developed to prevent these occurrences. These preventive strategies are explained in the following sections.

Machine Problems

There are two basic concerns about machines:

1. Is it running and turning out product?
2. If it is running, is the quality of its output product acceptable?

In a mass production environment, question 1 matters most. The tendency is to let the machine run as long as there is product, good or bad, coming out of it. Defective parts will cause problems farther down the line, but the consequences of shutting the machine down to fix it are seen as an even bigger problem. The JIT factory is more concerned about the second question, because allowing a machine to produce defective parts is to permit the production of waste, and that, above all, is forbidden.

Common sense dictates that machinery should always be maintained properly, but that can be very difficult in a mass production plant. Unfortunately, in many North American factories, machines tend to be ignored until they break down, in keeping with the grammatically incorrect but telling expression, "If it ain't broke, don't fix it." Toyota eliminated the machine problem through a systematic preventive maintenance process that keeps all machinery in top shape, modifying it for better reliability or performance, and even predicting when parts should be replaced or adjustments made to maintain the highest-quality output. This has come to be known as *total productive maintenance* or *total preventive maintenance* (TPM). It is finding widespread acceptance in forward-looking companies. TPM, by keeping the machines available for use when they are needed, eliminates a great many line stoppages. We will discuss TPM in more detail later in the chapter.

Process Problems

Process problems can be eliminated when people thoroughly understand the processes, optimize them, and use statistical methods (i.e., SPC) to keep them under control. In addition, the processes are continually improved, most often through the efforts of the same people who work with them every day. Time is allocated for these kinds of efforts in all JIT factories.

The most difficult conceptual problem with JIT is the precise control of production and material/parts flow through the complete production process. For that, Ohno developed the use of the *kanban* to signal the pulls through the system. Mass production demonstrated that one should not start the control at the beginning of the process. Too many things can go wrong at the bow wave of the flow. Ohno decided that the control had to start at the output end of the factory. From this concept, he introduced kanban. *Kanban* is a Japanese word meaning "card." Ohno used kanbans to trigger activity and the flow of materials/parts from one process to another. When a succeeding process uses the output of the preceding process, it issues a kanban to the preceding process to produce another.

Although Ohno describes the kanbans as slips of paper in a vinyl pouch—close enough for "card"—kanbans have evolved to a number of forms. A square painted or taped on a work station may be a very effective kanban. For example, a process produces a subassembly and places it on the marked area of the succeeding process work station. When the succeeding process uses the subassembly, the marked area, the kanban square, becomes empty and signals the preceding process to make another subassembly and fill the square. The same is done with totable bins. When the parts from a bin have been used, the empty bin is sent back to the preceding process as a signal for more production. Both of these kanban devices work when the part or subassembly in question is the only possible output of the preceding process. Should there be a variety of part or subassembly models, however, the kanban square alone will not provide sufficient information, and the bin with a descriptive card or the kanban card must be used. (More information about kanban is provided later in this chapter.)

Lot Size

A final issue for JIT production to overcome concerns lot size. Mass production is keyed to the largest possible lot sizes: set up the machines and parts streams to make as many as possible of the same item, like Henry Ford's identical black Model T's, before changing to another model or product. So-called economic lot size is still being taught in many universities. Just-in-time seeks to build in the smallest possible lots. The modern consumer demands variety. No auto company could survive today with a single car model, with each unit the same in all respects including equipment and color. JIT accommodates variety by being flexible. That is, the factory is set up so that changes can be rapidly implemented, and at little cost.

Traditionally, it has been a major problem to change models on a production line because breakdown and setup of the machines that have to be changed takes a lot of time. Hours and days and even longer for new setups are not uncommon. Ohno saw that the inherent inflexibility of the mass production line was in the setup time for the machines. Too much setup time meant that a manufacturer had to have a second line—or even a new factory—for the other model, or the customers' demand for the second model was simply ignored until the run on the current model was finished. By attacking the problem head-on, Toyota was able to reduce setup times to the point where they were no longer significant. Other companies, using the Toyota approach, found that they could quickly reduce setup times by 90% and even more with some effort.

Omark Industries was one of the first American companies to study the Toyota Production System. Using Toyota's techniques, it reduced the setup time for a large press from 8 hours to 1 minute and 4 seconds.[7] After setup time became irelevant, it was possible to manufacture in small lots—even lots of one—thereby permitting the intermixing of models on the same line. This meant that customer responsiveness was possible without huge inventories of prebuilt stock in all models. It also meant that one production line (or factory) could do the work of several. This ability is crucial if the factory is to respond to customer demand in a pull system.

The development of just-in-time production required more than the kanban, a point lost on many Westerners. JIT sprang from the understanding of the seven wastes and the need to eliminate them. The key elimination (almost) of material and parts inventories dictated the requirement for reliability and predictability of the plant's machinery and processes. This led to total productive maintenance and made necessary the use of statistical process control and continual improvement.

With the customer as the driver of production, the control technique for production changed from push to pull, and kanban was introduced as the controlling system. The requirement for small lot sizes, both for elimination of waste and for responsiveness and investment economy, led to the effort to reduce setup time. With all of these factors in place, JIT was born. Without doubt, JIT is the manufacturing system for today. It is adaptable to operations both large and small, high-volume/low-variety, and low-volume/ high-variety, as well as anything in between. In JIT, costs, lead time, and cycle time are reduced, quality is improved constantly, and both customers and the producers and their employees benefit.

RELATIONSHIP OF JIT TO TOTAL QUALITY AND WORLD-CLASS MANUFACTURING

The traditional production line pushes product from the front of the line to the final output, and even to the customers, whereas kanban is the controlling agent in a pull system. The two are incompatible. Similarly, implementing JIT in the absence of a comprehensive total quality system that includes the entire organization can be a problem. The traditional organization is incompatible with JIT, just as the traditional push production system is incompatible with kanban. In a typical manufacturing company, sepa-

QUALITY TIP

JIT as a Total Quality Concept

JIT was conceived as a total management system, not just for the manufacturing floor. Isolating JIT from the rest of the management system will not allow it to fully develop and mature. JIT needs to be a part of a total quality management system.

rate departments exist for engineering, manufacturing, purchasing, accounting, and so on, each with distinct boundaries and agendas. JIT is no respecter of boundaries. It requires all departments to respond to its needs. If the manufacturing department has embraced JIT, but the organization as a whole has not at least started a total quality effort, manufacturing personnel will soon encounter obstacles. More often than not there will be outright resistance because JIT's requirements represent change, and departments without a commitment to change will fight it at every step.

As an example, in the defense industry it is common to defray overhead expenses (buildings, utilities, indirect employees' salaries, all fringe benefits, and others) against direct labor dollars as a means of allocating the overhead burden across all contract programs. The more direct labor on a program, the more of the overhead cost accrues to that program. *Direct labor* is defined as the manufacturing, engineering, purchasing, and other labor charged to specific contract programs. The company may also have more than one pool for overhead defrayment, such as a manufacturing pool and an engineering pool. Virtually all of these companies, and the U.S. Department of Defense, pay a great deal of attention to what they call *overhead rate*. In a typical company in the defense industry, overhead rate is calculated by dividing overhead (indirect) expenses by direct labor cost.

Suppose that for an accounting period there were indirect expenses of $200,000. At the same time, the wages paid for direct labor amounted to $100,000. The overhead rate for the period is $200,000/$100,000 = 200%. Assume that we had been operating with that 200% rate for some time, and suddenly the manufacturing department discovered JIT. After the period of time necessary for the implementation to start showing results, manufacturing finds that it can eliminate direct labor positions for production control and material control and also use fewer assemblers on the production floor to get the same number of units out the door each period. A typical early change in the direct labor content of the work is 30%–35%. The next period's overhead expense is almost the same, decreasing slightly for removal of fringe benefits for the employees no longer needed, say, to $188,000. The direct labor is down by one-third to $67,000. This yields an overhead rate of $188,000/$67,000 = 281%. That kind of an increase in overhead rate, if sustained, will cause the head of manufacturing serious problems. The accounting department uses this overhead rate as proof that JIT doesn't work. All too often the accounting department blocks further progress in JIT. One might ask, "But isn't that valid if the overhead rate went out of control?" The answer is nobody should care about the overhead rate. It is simply the ratio of two numbers and carries no meaning without a thorough understanding of the two. What happened to the cost of goods sold in this example? Look at the numbers before and after JIT[8]:

	Before JIT	After JIT
Indirect Expense	$200,000	$188,000
Direct Labor	100,000	67,000
Materials	500,000	500,000
G&A	50,000	50,000
Cost of Goods Sold	$850,000	$805,000

In this example, it cost the company $45,000 less to produce the same goods after JIT implementation than it did before. Assuming the goods were sold for the same price, that $45,000 becomes pure profit. In the next competition for contracts, the lower cost becomes a competitive advantage (price to the customer can be lowered).

The solution to the overhead rate problem is to change from the obsolete accounting system and adopt an activity-based accounting system or some other more sensible method. In a total quality company, the accounting department is part of the team and would respond to the needs of a production system (JIT) that is actually improving company performance. But if the company as a whole is not involved in total quality, the accounting department, with its own walls and agendas, can be a formidable obstacle to progress. The same is true of other departments on whom manufacturing depends. This example could just as easily have been one involving the engineering department and a design philosophy called *concurrent engineering*. Concurrent engineering requires that from the beginning of a new product's design, manufacturing and other departments (and even suppliers) are directly involved with engineering to make sure, among other things, that the product can be manufactured efficiently when it finally goes into production. Traditional engineering departments do not like to have this kind of help from outsiders and will resist—but not in a total quality setting, where the departments all work for the common goal.

For JIT to bring about the benefits inherent in its philosophy, it must be part of a total quality system. To bring JIT into a company not otherwise engaged in total quality can be worthwhile (and may even enlighten the leadership), but it will be much more difficult, and its results severely restricted.

BENEFITS OF JIT

A discussion of the benefits of JIT must include four very important topics: inventory and work-in-process, cycle time, continual improvement, and elimination of waste. The discussion could be expanded to include such topics as reduced time-to-market, improvement of employee work life, flexibility, and employee ownership. All of these are definite benefits of JIT, but this discussion will be confined to the critical four mentioned. These are the usual targets of a JIT implementation.

Inventory and Work-in-Process

Just-in-time attempts to drive inventory to zero. But remember that this is a philosophical objective—an aiming point, if you will. In reality, zero inventory makes no sense. Without some inventory, you have nothing from which to produce your goods. The real objective is to minimize the inventory to the maximum possible extent without shutting down production. It is also important to recognize that there are at least three kinds of inventory. First there is the inventory of raw materials and parts needed to make the product. Traditionally these have filled warehouses, with enough on hand for several weeks of production, or longer. Second, there is the work-in-process (WIP) inventory of semifinished goods. WIP includes all materials and parts that have been put into the production system, including the various stages from the first process to the last within the factory. WIP may be at a work station undergoing one of the value-adding production processes, or it may be in storage between processes. In a mass production plant, the stored WIP can be substantial. Job shops—low-volume, high-variety shops not involved in mass production—are notorious for their WIP inventory. Third, there is the finished goods inventory. These finished goods are ready for customers, but the customers are not ready for them. Therefore, they are typically stored in warehouses, although some (most notably automobiles) must be stored in yards, unprotected from the elements. In the spring of 1992, a particularly vigorous hailstorm took its toll on automobile dealers in Orlando, Florida. People could buy brand-new cars for a fraction of their sticker price as a result.

One might ask, what is wrong with inventory? Having materials on hand allows you to produce without worrying about on-time material deliveries. Lots of WIP lets the assembly lines continue when a machine breakdown or some other problem occurs. Having an inventory of stored finished goods means that you can be responsive to customers. If those are positives (and we'll come back to that in a minute), there are also negatives. First, there are the costs of inventorying raw materials and parts, and finished goods. These are the costs of the materials and goods; the labor costs for storage, handling, and protection of the materials and goods; and the cost of warehouses, real estate, and capital equipment used in the inventorying of the materials and goods. Second, there is the cost of spoilage while in inventory. Spoilage can be due to damage, deterioration, corrosion, obsolescence, and so on. Third, there is the cost of taxes. While the product is in inventory, the manufacturer owns it, it has value, and the various governments want their share in the form of taxes.

Now go back to the suggestion made earlier that the three positives associated with inventory might not be so positive after all. The costs discussed earlier are all tangible costs. There are also intangible costs that, while difficult to measure precisely, are nevertheless significant. Foremost among the intangibles is the fact that as long as the manufacturer holds inventory of materials and WIP at high levels, it is not solving the problems and making the continual improvements that can bring efficiency. The very presence of these inventories masks the problems, so they go unnoticed and unresolved—being repeated over and over, consuming unnecessary labor, and preventing product quality improvement. Unmasking the production system's problems through the elimination of inventories is a major strength of JIT. Many North American and European companies still tend to see the elimination of inventories as a generator of problems. In reality, the problems are already there, and they are costing a great deal in terms of money and quality, but they are just not apparent with big inventories. Through inventories maintained, tons of money is spent, but no value is added, and needed improvements are not made in the production processes. The inevitable net result is loss of competitive position and market share as enlightened competitors use JIT and total quality to improve their positions.

If a plant could get its production processes under control to the point that they could be relied on to perform as intended, it would be logical to reduce WIP and material/part inventories. However, until the processes are well understood and in control, reducing inventories substantially will certainly result in production stoppages. One philosophy of reducing WIP and lot sizes is to do so in steps. By incrementally lowering WIP and lot sizes, the problems become apparent in a gradual, manageable stream rather than in a torrent, and they can be dealt with. Once through that process, the next logical step is to work with suppliers to deliver materials and parts in smaller, more frequent lots, until finally there is no need for warehousing at all. This clearly requires that the production processes be capable and reliable and that the suppliers be similarly capable and reliable.

This leaves only the finished goods inventory. As the processes and suppliers become more proficient, and the JIT line takes hold, production will be geared to customer demand rather than to sales forecasts. The ability of the JIT line to respond quickly to customer requirements means that it is no longer necessary to store finished goods. The only stored goods should be those in the distribution system, and that level will typically be far less than has been the case under mass production.

JIT strives for zero inventory of any kind. Achieving zero inventory is not a realistic intent, but by aiming at zero and continually reducing inventories, not only do manufacturers cut costs by significant numbers, but the whole continual improvement process comes to life, resulting in even more savings and improved product quality.

Cycle Time

Production cycle time is defined as the period bounded by the time materials are sent to the manufacturing floor for the making of a product and the time the finished goods are dispatched from the manufacturing floor to a customer or to finished goods storage. Generally speaking, the shorter the production cycle time, the lower the production cost. That may be reason enough to pay attention to cycle time, but there are other benefits. Short cycles improve a factory's ability to respond quickly to changing customer demands. The less time a product spends in the production cycle, the less chance there is for damage.

We are accustomed to thinking of a mass production line as having the shortest of cycle times, and there have been startling examples of this. Henry Ford's Model T lines (producing up to 2,000,000 cars per year, all the same, all black) achieved remarkable cycle times even by today's standards. For example, Ford's River Rouge facility took iron ore in the front door and shipped completed cars out the back door in four days.[9] When one considers that the Ford cycle included making the steel, in addition to stamping, casting, machining, and assembly, it is all the more amazing. One of his secrets was no variability in the product. Modern lines have the complication of different models and virtually unlimited options.

A modern auto assembly line cannot be compared with Ford's Model T line because the complexity and variability of the contemporary car is so much greater. However, the best lines beat Ford's cycle time for assembly. The differences in JIT lines and mass production lines is substantial. For example, comparisons between JIT plants and traditional mass production plants reveal that JIT plants can assemble automobiles in 52% of the time it takes traditional plants. Because there is very little waiting in a JIT line, one can assume the cycle time is one half of that for traditional lines. Interestingly, though not directly related to cycle time, traditional lines produce three times as many defects and require nearly twice the factory space. In addition, JIT plants can operate with a 2-hour parts inventory, while traditional plants typically need a 2-week supply.[10]

Consider the following example that helps bridge the issues of inventory and cycle time. The product was a line of very expensive military avionics test systems. The factories (two) were rather typical electronics job shops. Before being converted to JIT, they were struggling with a production schedule requiring the assembly of 75 large, complex printed circuit boards per day. They rarely met the goal, usually achieving about 50. The attempted solution involved pushing more parts into the front end of the assembly process, hoping that would force more out the other end as finished, tested boards. The computer system revealed that, at any point in time, about 3,500 boards were in the process. At the rate of 50 completed boards per day, and 3,500 boards in WIP, simple arithmetic showed that the cycle time for the average board was 13 weeks. Common sense said that 13 weeks was much too long for assembling these boards, but checking with others in the industry revealed that this was a typical cycle time. The company also found that it made absolutely no difference in final output rate to force more materials into the front of the process. This merely increased the number of boards in WIP.

With a production rate of 50 boards a day and 3,500 boards in process, one can imagine the difficulty in keeping track of where the boards were, scheduling them into and out of the various processes, and storing, retrieving, and safeguarding them. Such tasks were nearly impossible. More than 100 people were charged with handling and tracking the boards, adding no value whatever to the product. Further, because the assemblers were being pushed to their limits, quality suffered. The net result was that nearly half of total direct labor was spent repairing defects. That is not adding value either. Once again, however, checking with other manufacturers revealed that this was typical. A critical factor was that customer delivery schedules could not be met unless a solution was found. Initially, the company had to subcontract a great many boards, but that was a work-around, not a solution.

The eventual answer was to implement JIT techniques on the production floor. After a couple of quick pilot runs, in which it was discovered that the most difficult of the boards could be assembled and tested in 8 days (versus 13 weeks), management was convinced, and JIT was implemented at both plants, following the WIP reduction and lot -size scheme outlined in the previous section. In very short order, the board cycle time fell to about 5 days, and board quality improved dramatically. That enabled the company to eliminate the 100-plus positions that had handled the boards and eventually many other non-value-adding positions as well. The system delivery on-time rate went to 98% (unheard of for this kind of product), customer satisfaction improved, and a respectable profit was made.

The thing to remember about cycle time is this: any time above that which is directly required by the manufacturing process is not adding value and is costing money. For example, assume we use two processes to manufacture a product, and the total time consumed within the processes is 2 hours. It is determined that the actual cycle time is 3 hours. That means that 2 hours of the cycle is adding value and the other hour is not. Invariably this means a bottleneck is preventing the product from flowing from one process directly into the next without delay. The key is to detect the bottleneck and do something about it. It may be that a plant procedure requires inspection, logging, and a computer data entry. Are these tasks really necessary? Can they be eliminated? If they are necessary, can they be streamlined?

The extra hour may be the result of a problem in one of the processes. For example, it may be that the second process is no longer 1 hour in duration but 2. If the latter is the case, in a traditional production plant, the product flowing out of the first process will stack up at the input of the second process, because process 1 will continue to crank out its product at the rate of one unit per hour—whether process 2 is ready for it or not (see Figure 21–7). The surplus product at the input to process 2 will have to be stored for safety and housekeeping reasons, thus obscuring the fact that there is a problem.

As long as the problem persists, WIP will build, output will stay at one unit every 2 hours, but cycle time will increase as backlog builds up in front of process 2: the first unit went through the production system in 3 hours, and one unit per hour was expected after that, but the process is actually achieving one unit every 2 hours. Cycle time increases by 1 hour for each piece—for example, 8 hours later the sixth unit into process 1 will come out of process 2. Such an imbalance would not escape notice for long, and it would be corrected, but by then several pieces of WIP would be between the processes.

Figure 21–7
Cycle Time Example

Piece #	Process 1 In	Process 1 Out	Wait Time (In Hours)	Process 2 In	Process 2 Out	Cycle Time (In Hours)
1	7 A.M.	8 A.M.	0	8 A.M.	10 A.M.	3
2	8 A.M.	9 A.M.	1	10 A.M.	12 noon	4
3	9 A.M.	10 A.M.	2	12 noon	2 P.M.	5
4	10 A.M.	11 A.M.	3	2 P.M.	4 P.M.	6
5	11 A.M.	12 noon	4	4 P.M.	6 P.M.	7
6	12 noon	1 P.M.	5	6 P.M.	8 P.M.	8
7	1 P.M.	2 P.M.	6	8 P.M.	9 P.M.	8
8	2 P.M.	3 P.M.	6	9 P.M.	10 P.M.	8
9	3 P.M.	4 P.M.	6	10 P.M.	11 P.M.	8
10	4 P.M.	5 P.M.	6	11 P.M.	12 midn.	8
11	5 P.M.	6 P.M.	6	12 midn.	1 A.M.	8
12	6 P.M.	7 P.M.	6	1 A.M.	2 A.M.	8
13	7 P.M.	8 P.M.	6	2 A.M.	3 A.M.	8

Suppose that the problem in the second process was corrected as the sixth unit was completed. Everything is back to the original 2-hour process time, but by now there are seven more units through process 1, on which the cycle time clock has already started. If stable from this point forward, the cycle time will remain at 8 hours. We started with a process that had 2 hours of value-adding work and a 3-hour cycle. We now have a 2-hour value-adding process time and an 8-hour cycle. If some means is not taken to cause the second process to catch up, every time there is a glitch in process 2, the cycle time will grow. In a traditional plant, with literally dozens of processes, such conditions could go on forever. As observed earlier, some would hold that having the seven units from the first process sitting on the shelf means that process 1 could be down for a complete shift without causing a problem for the second process—it would merely draw from the seven.

In a JIT plant, the situation described here would never happen. Process 1 would not produce an additional piece until process 2 asked for it (kanban). At the start, process 1 produces one unit to enable process 2. When process 2 withdraws it, process 1 is signaled to produce another. If for any reason, when process 1 completes its second unit, process 2 is not ready to withdraw it, process 1 goes idle. It will stay idle until signaled to produce another—be it a few minutes or a week. No WIP inventory is produced. By process 1 going idle, alarms go off, quickly letting the appropriate people know that something has gone wrong. If there is a difficulty in the second process consuming too much time, it gets attention immediately. Similarly, if there is a delay getting the output of the first process to the second because of an administrative procedure, that too will be dealt with quickly, because it will cause problems throughout the overall process until it is solved.

Any contributor to cycle time is apparent in a JIT environment, and JIT philosophy calls for continual improvement and refinement. Wait time in storage is simply not a factor in JIT, because nothing is produced in advance of its need by the succeeding process. That single factor can easily remove 80%–90% of the cycle time in a traditional factory. In the earlier example of the printed circuit board factories, the initial reduction of cycle time from 13 weeks (65 working days) to 8 days was simply the elimination of storage time. That was a reduction of 88%. Further refinement, made possible because of the visibility afforded by JIT, brought the cycle to 4 days, or only 6% of the original cycle. Taking it further was restricted by procedural and governmental requirements. In a commercial setting, however, the same boards could probably have been produced in a 2-day cycle with no new capital equipment.

Before JIT, manufacturers tried to cut cycle time with automation. But that was not the answer. The solution was found in better control of production, and that was obtained with JIT. JIT is the most powerful concept available for reducing cycle time.

Continual Improvement

Continual improvement has been discussed in several other chapters and sections of this book. By now, you should have a good understanding of its meaning as applied in a total quality context. Continual improvement seeks to eliminate waste in all forms, improve quality of products and services, and improve customer responsiveness—and do all of this while at the same time reducing costs. A note of caution should be added in regard to interpretation of what constitutes improvement: Problem solving is not necessarily improvement. If a process that had previously been capable of producing 95 out of 100 good parts deteriorates to a level of 50 good parts, and the problem is found and corrected to bring the process back to where it had been, that is maintenance, not improvement. *Maintenance* is restoring a capability that previously existed. On the other hand, if a process was capable of 95 good parts out of 100 produced, and a team developed a way to change the process to produce 99 good parts, that would be improvement. It is important to differentiate between maintenance and improvement. Maintenance is important, and it must go on, but in the final analysis, you end up where you started. *Improvement* means becoming better than when you started. Continual improvement is to repeat that improvement cycle, in W. Edwards Deming's words, constantly and forever.[11]

The discussion of continual improvement in this chapter is to explain how JIT supports continual improvement. The traditional factory effectively hides its information through inventories of parts, WIP, and finished goods; people scurrying about, everybody busy, whether any value is being added or not. The JIT factory is visual: its information is there for everyone to see and use. Quality defects become immediately apparent, as do improper production rates—whether too slow or too fast. Either of these, for example, will result in people stopping work. While that is not acceptable behavior in a mass production factory, in a JIT plant it is encouraged and expected.

A true story from Toyota tells of two supervisors, one from the old school and unable to adapt to JIT, and the other ready to try JIT even if it did seem strange.[12] The first supervisor refused to allow his line to be stopped, whereas the second didn't hesitate to stop his. At first the line operated by the second supervisor was producing far fewer cars than the other line, because it was stopping for every little problem. These problems had been common knowledge among the workers but not to the supervisors. The problems were solved one by one as a result of stopping the line for each. After 3 weeks, the second supervisor's line took the lead for good. The first supervisor believed that stopping the line would decrease efficiency and cost the company money. As it turned out, the reverse was true. By stopping the line to eliminate problems, efficiency and economy were enhanced. The only reason for stopping a line is to improve it, eliminating the need for stopping again for the same reason.

In a mass production plant, the sight of idle workers will draw the ire of supervisors in no uncertain terms. But in a JIT situation, the rule is if there is a problem, stop. Suppose that a preceding process has responded to a kanban and provided a part to a succeeding process. The succeeding process finds that the part is not acceptable for some reason (fit, finish, improper model, or something else). The succeeding process worker immediately stops, reporting the problem to the preceding process and to supervision. Perhaps an *andon* (a Japanese word meaning "lamp") signal will be illuminated to call attention to the fact that his process is shut down. The problem is to be solved before any more work is done by the two processes, which means that downstream processes may soon stop as well, because their demands through kanban cannot be honored until the problem is fixed and the processes are once again running. This is high visibility, and it is guaranteed to get the proper attention not only to solve the immediate problem but to improve the process to make sure it does not happen again.

Consider the following example. A few weeks after JIT implementation was started in a New York electronics plant, there was a line shutdown. At the end of this line was a test station that was to do a comprehensive functional test of the product. There was an assembly all set up for test, but the technician had stopped. The line's andon light was illuminated. A small crowd gathered. The problem was that the test instructions were out of date. Over time the test instruction document had been red-lined with changes and had, up until that point, been used without apparent difficulty. But a company procedure required that any red-lined document be reissued to incorporate the approved changes within 1 year of the first red-line. The 1-year clock had expired months earlier, and the technician, with guidance from quality assurance, properly stopped testing. When management asked why the document had not long since been updated, it was found that the documents seldom were updated until the entire job was completed. In many cases, jobs lasted several years. Holding all formal revisions until a job was completed meant that documentation was revised just once, thereby saving considerable expense. Of course, in the meantime, manufacturing was using out-of-date or questionable information. The standard work-around seemed to be that when a system couldn't be completed for delivery, waivers were generated, allowing the tests to be conducted with the outdated red-lined procedures. This had been going on for years but never became apparent to the levels in manufacturing and engineering that could solve it. In this case, it took about 20 minutes to have the problem solved. Without JIT to highlight it, the problem would, in all probability, still exist.

What had happened because of JIT was a stop at the test station. That also shut off kanbans through the preceding processes. In short order the line stopped, getting the attention needed to eliminate the problem. If the plant had been operating in the traditional (non–JIT) way, the assemblies would have piled up at the test station for a while and then the production control people would have carted them off to a work-in-process storage area—out of sight. Eventually the inventory of previously tested assemblies would be consumed, and there would have been a "brushfire" from which a procedural waiver would emerge to enable the test technician to pull the untested assemblies from WIP stores and quickly get them tested so system deliveries could be made. And this would have been repeated time and again, just as had been happening surreptitiously in the past.

This is not an uncommon scenario. Fundamentally, it is the result of departments not communicating. Engineering is trying to save money by reducing the number of documentation revisions. Meanwhile, manufacturing may be producing obsolete and unusable product because their documentation is not up to date. At best, it results in the continual "firefighting" that saps the collective energy of the organization, leading to quick-fix, work-around "solutions" that let you get today's product out but only make each succeeding day that much more difficult. JIT, by highlighting problems, is quick to dispel the quick-fix mentality, demanding instead that problems be eliminated for today and tomorrow and forever.

A better illustration of JIT's ability to reveal real problems is the analogy of the lake (see Figure 21–8). You look out over a lake and see the calm, flat surface of the water and perhaps an island or two. From this observation, you conclude that the lake is navigable, so you put your boat in and cast off. You avoid running into the islands because they can be seen plainly and there is plenty of room to steer around them. However, a rock just below the surface is not evident until you crash into it. It turns out there are lots of rocks at various depths, but you can't see them until it is too late. This is like a traditional factory. The rocks represent problems that will wreak havoc on production (the boat). The water represents all the inventory maintained, raw materials and parts, WIP inventory, even finished goods. Now if you make the change to just-in-time, you start reducing those inventories. Every time you remove some, the level of the water in the lake is lowered, revealing problems that had been there all along but that were not eliminated because they couldn't be seen. You just kept running your boat into them, making repairs and sailing on to the next encounter. But with the lower water level, the problems

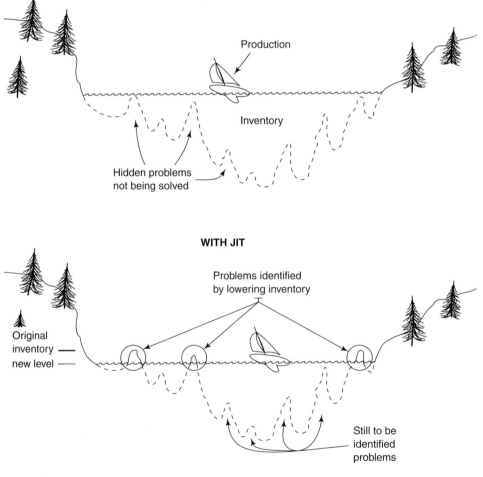

Figure 21–8
JIT Exposes Hidden Problems

become visible and can be eliminated. Clear sailing? Probably not. Other rocks are no doubt just below the new lower surface level, so you have to take some more water out of the lake (remove more inventory), enabling you to identify and eliminate them. Like most analogies, our lake doesn't hold all the way to the logical conclusion of zero inventory, because the lake would be dry by then. But remember, true zero inventory doesn't hold either. As was said before, it is a target to aim at but never to be fully reached.

JIT is by nature a visible process, making problems and opportunities for improvement obvious. Moreover, when problems do occur in a JIT setting, they must be solved and not merely patched up, or they will immediately reappear. Visibility to all levels, from the workers to the top executive, means that the power to make necessary changes to eliminate problems and improve processes is available.

Elimination of Waste

In the preceding three sections it was shown how just-in-time facilitates reduction of inventories and cycle time and promotes continual improvement. This section will show that JIT is also a powerful eliminator of waste. As discussed earlier in this chapter (in the "Rationale for JIT" section), Taiichi Ohno created the JIT system specifically to eliminate the seven wastes he saw in the Ford mass production system. Let's look at them one at a time and see how JIT helps in their elimination.

1. *Waste arising from overproducing.* Mass production pushes materials into the front of the factory in response to market forecasts. These raw materials are converted to finished goods and pushed through the distribution system. The first real customer input into the process is at the retail level. If customers don't want the goods, they will eventually be sold at prices much lower than anticipated, often below their actual cost. That is waste to the producer. In addition, producing goods for which there is not a matching demand is a waste to society by using resources to no purpose. In a JIT environment, the customers enter the system at the beginning, pulling goods from the distribution system and, in turn, from the manufacturer. The JIT factory produces nothing without a kanban, which, in effect, originates with a customer.

The same is true within the two kinds of factories. A fender stamping press in a mass production factory will continue to stamp out fenders even though the final assembly line, which uses the fenders, is stopped. The overproduction must then be handled by people who contribute nothing to the value of the product, stored in buildings that would otherwise be unnecessary, and tracked by people and systems that do nothing but cost money. In a JIT factory, the fender stamping press will shut down unless it receives kanbans requesting more fenders. No overproduction. Of all the wastes, overproduction is the most insidious because it gives rise to all the other types of waste.

2. *Waste arising from waiting (time).* Wait time can come from many causes: waiting for parts to be retrieved from a storage location, waiting for a tool to be replaced, waiting for a machine to be repaired or to be set up for a different product, waiting for the next unit to move down the line. JIT parts are typically located at the work station, not in some central staging area or warehouse. JIT sets time aside for tool and machine maintenance, so replacement or repair during a production period is rare.

Whereas setup times for machines in mass production plants tend to take hours (or even longer), JIT factories devote a great deal of attention to setup time, typically reducing it to a very few minutes. In a typical factory, an operator is assigned to each machine. While the machine is running under automatic control, the operator has nothing to do but wait. In a JIT factory, the same operator may run five machines, arranged so that he or she can easily see and control all five without much movement. As three machines are running automatically, the operator may set up a fourth and unload the fifth, for example. In this way, the day is no longer mostly wait time.

Perhaps the biggest waste associated with waiting involves not human waiting but inventory waiting. In the traditional setting, raw materials and parts can sit idle for weeks and months before they are needed. Work in process may wait weeks to have a few hours of value-adding work done. Finished goods may wait very long periods for customers. JIT does not allow any of these waits to occur, and the carrying expense is eliminated.

3. *Waste arising from transport.*   Mass production factories tend to buy their materials and parts in very large quantities from the lowest-price (as opposed to lowest-cost or best-value) source, regardless of the distance from the source to the factory. JIT factories of necessity must buy in small quantities (no warehousing) with frequent deliveries, often several times a day. That means that the suppliers should be relatively close to the factory, cutting transportation time and costs.

Transportation within plants can be a very high-cost item too. Moving things costs money and time and increases exposure to damage. Moving materials in and out of storage areas, to and from the floor, back and forth across the factory from process to process is waste. None of that happens with JIT. Production materials are delivered to the point of use in a JIT factory, so they are not shuttled in and out of storage or put in temporary storage to be moved again before use. Factories are arranged to minimize distances between adjacent processes, whereas the same product manufactured in the traditional factory could log thousands of feet, or even miles, of movement before completion.

4. *Waste arising from processing itself.*   Any process that does not operate smoothly as intended but instead requires extra work or attention by the operator is wasteful. An example is the necessity for the operator to override an automatic machine function to prevent defective products. Because one of the basic tenets of JIT is continual improvement of processes, wasteful processes are soon identified and improved to eliminate the waste. That is far more difficult in the traditional production environment because of its emphasis on output, not process improvement.

5. *Waste arising from unnecessary stock on hand.*   Any stock on hand has storage costs associated with it. When that stock is unnecessary, the costs are pure waste. Included in these costs are real estate, buildings, employees not otherwise needed, and the costs of tracking and administration. Because JIT attempts to eliminate stock, stock that is not necessary is just not tolerated.

6. *Waste arising from unnecessary motion.*   JIT plants are laid out to minimize motion of both workers and product. Motion takes time, adds no value, makes necessary additional workers, and hides waste. The contrast between a JIT plant laid out with product orientation and the traditional plant laid out with process orientation is profound (see Figure 21–9). In the traditional plant, there is much motion, with people and product

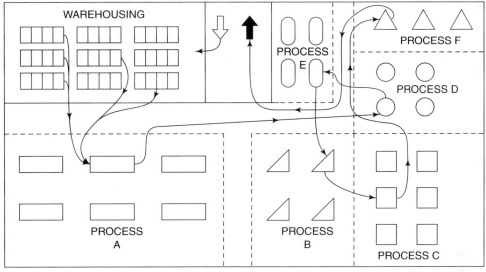

Traditional Factory Organized by Processes
(illustrating process control for one product)

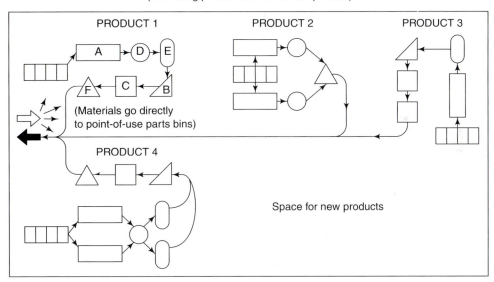

JIT Factory Organized by Product
(illustrating process flow for four products)

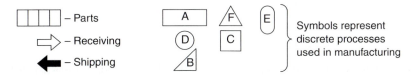

Figure 21–9
Comparison of Factory Floor Layouts: Traditional versus JIT

shuttling all over the place. In a JIT plant, motion is almost undetectable to a casual observer.

7. *Waste arising from producing defective goods.* Defective goods will surely cost money in one of three ways: (a) the product may be reworked to correct the deficiency, in which case the rework labor and material costs represent waste; (b) it may be scrapped, in which case the cost of the materials and the value added by labor has been wasted; or (c) it may be sold to customers who, on discovering that the product is defective, return it for repair under warranty and may be dissatisfied to the extent they will never buy this manufacturer's products again. Warranty costs represent a waste, and the potential for a lost customer is great, portending a future loss of sales.

In a traditional factory, it is possible to produce large quantities of products before defects are discovered and the line corrected. It is not uncommon in mass production for a company to keep the line running intentionally, producing defective products, while trying to figure out what has happened and devising a solution. It is considered less troublesome to fix the defective products later than to shut down the line. In JIT, however, because line stops are anticipated, and because the preferred lot size is one unit, it is improbable that more than one defective unit could be produced before shutting down the line.

Dr. M. Scott Myers, author of the landmark book *Every Employee a Manager*, made the case for an eighth waste: the waste arising from the underutilization of talent. Myers believed that the most damaging of the eight wastes is the waste of talent.[13] If all the talents of all employees were brought to bear on the problems and issues of production, the other wastes would probably disappear. This is the rationale for both employee involvement and teamwork. JIT is designed to make use of the ideas and talents of all employees through team activities and employee involvement, in an environment that fosters the open and free interchange of ideas, all of which are foreign to the traditional production systems. Elimination of waste is an integral focus of just-in-time by design. No other production system looks at waste except after the fact.

REQUIREMENTS OF JIT

For a factory to operate as a just-in-time production facility, a number of steps must be taken. It is very important that JIT implementation be a part of a larger total quality program; otherwise, many interdepartmental roadblocks will crop up as time passes. Like total quality, JIT requires an unwavering commitment from the top, because production is more than just the manufacturing department. If these two elements are in place (a total quality program and a commitment from the top), JIT implementation should be within reach. The following discussion touches on the issues that must be addressed as the implementation progresses.

Factory Organization

The JIT plant is laid out quite differently from that to which most people are accustomed. Most non–JIT factories are set up according to the processes that are used. For example, there may be a welding shop, a machine shop, a cable assembly area, a printed circuit board assembly area, a soldering area, and so on. Each of these discrete processes may be set up in separate parts of the factory (all machining operations done in the machine shop, all cable assembly done in the common cable and harness area), no matter which of many products it might be for (refer to Figure 21–9). The JIT plant attempts to set up the factory by product rather than process. All the necessary processes for a given product should be located together in a single area and laid out in as compact a manner as possible.

The chart at the top of Figure 21–9 represents the old process-oriented factory. Each of the processes has its own territory within the plant. Additionally, an area dedicated to warehousing is used for storage of production materials, work-in-process that is waiting for the next process, and perhaps finished goods awaiting orders. There is also an area set aside for shipping and receiving. Materials are received, inspected, processed, and sent to the warehouse area. Finished goods are taken from the warehouse or from final assembly, packed, and shipped. The upper illustration in Figure 21–9 maps the movement from the warehouse through the processes and finally to shipping in a traditional factory.

The lower illustration in Figure 21–9 represents a JIT factory that is set up to manufacture four different products. The warehousing area is gone. This cannot happen overnight, but an objective of JIT is to eliminate all inventories. The second thing to notice is that the factory is divided into discrete areas dedicated to the different products rather than to processes. Each product area is equipped with the processes required for that product. Parts bins are located right in the work area. These parts bins may have from a few hours' supply to a month or more, depending on the degree of maturity of the JIT implementation and the nature of the product and its anticipated production life.

Mapped out is a typical work-flow diagram for one product in the upper illustration of Figure 21–9. Parts and materials are pulled from several locations in the warehousing area, and moved to a Process A work station. These materials may be in kit form (all the parts needed to make one lot of a product). The work instructions call for process A first, followed by process D. If process D is busy when the lot is finished by process A, the lot, now WIP, may be stacked up in a queue at process D or taken back to the warehouse for safekeeping. Eventually process D will process the kit and it will be sent to Process E, perhaps waiting in queue or in the warehouse. This same sequence is repeated through process B, process C, and process F. From there it goes to shipping. The diagram does not show any trips back to the warehouse between processes, but that could very well happen after every step. The movement flow diagram represents a best-case scenario. (This was done purposely to ensure clarity.)

Now observe the flow in the JIT Factory of Figure 21–9. Product 1 is set up to follow exactly the same processing sequence (from parts bins to Process A, then process D, process E, process B, process C, process F, and to shipping). In this case, the parts come straight from the bins located in the work cell, not from the warehouse, and not in kit form, which is a waste of effort. The work cell is laid out in a U shape for compactness, to keep all the work cell members close to each other. The WIP flows directly from process to process without a lot of wasted movement. Moreover, because this is a JIT work cell, there will be small lot sizes, with work pulled through the process sequence by kanban. That means there will be no queue time on the floor or in the warehouse. Cycle time for this product in the JIT work cell can be expected to be less than half of that for the same product in the factory at the top of Figure 21–9. An 80%–90% reduction would not be unusual.

Before one can lay out a JIT factory, the processes required for the product must be known. This is usually not a problem. Typically, the greatest difficulty comes in determining how much of a process is needed. How many minutes of a process does the product use? One would think that if the product had been built before in the traditional way, one should know how much process time is required at each step. This may be a starting point, but typically it is not very accurate. With all the wasted motion and waiting time in queues and in the warehouse, the real processing time becomes obscured. However, you can use the best information available and refine it over time. Now that the processes are put right into the product work cell, having just the right amount is important.

In the case of product 4 in Figure 21–9's JIT factory, it was determined that the product required more capability in process A and process E than was available from single work stations, so they were doubled. Suppose that a product flow of 120 units per hour is anticipated. Each process has the following estimated capability for this product:

A: 75 units/hour	B: 150 units/hour	C: 130 units/hour
D: 120 units/hour	E: 70 units/hour	F: 135 units/hour

Because processes A and E are estimated to be capable of only about 60% of the anticipated demand, there is no point in trying to improve them. Rather, the process capability was doubled by putting in parallel equipment/work stations. This is a beginning. We now can watch for excess capacity that can be removed from the work cell, or for bottlenecks that require other adjustment. Work cells are coarsely tuned at first, with fine-tuning taking place during the initial runs. Excess capacity should be removed, just as required added capacity must be brought into the work cell. Bottlenecks will be quickly discovered and corrected. From there on, it is a matter of continual improvement to increase efficiency forever.

Training/Teams/Skills

Assuming an existing factory is converted to just-in-time, one would assume that the people who had been operating it would be capable of doing it under JIT. Naturally, many of the skills and much of the training necessary for the traditional factory are required under JIT, but JIT does require additional training. First, the transition from the traditional way of doing things in a factory to JIT involves profound changes. It will seem that everything has been turned upside down for a while. People should not be exposed to that kind of change without preparation. It is advisable to provide employees with training about why the change is being made, how JIT works, what to expect, and how JIT will affect them. Initial training should be aimed at orientation and familiarization. Detailed training on subjects such as kanban, process improvement, and statistical tools should be provided when they are needed—a sort of just-in-time approach to training.

Most factory workers are accustomed to working individually. That will change under JIT, which is designed around teams. A JIT work cell forms a natural team. The team is responsible for the total product, from the first production process to the shipping dock. Perhaps for the first time the workers will be able to identify with a product, something that they create, and the processes they own. This doesn't happen in a traditional factory. But with JIT, it is important to understand that workers must function as a team. Each will have his or her special tasks, but they work together, supporting each other, solving problems, checking work, helping out wherever they can. This may require some coaching and facilitating.

It was enough in the old way of production that workers had the skills for their individual processes. They did not need additional skills because they were locked into one process. This is not the case with JIT. Specialists are of far less value than generalists. Cross training is required to develop new skills. As a minimum, work cell members should develop skills in all the processes required by their product. Naturally, there are limits to this. We do not propose that all the members of a work cell become electronics technicians if their cell employs one for testing the product, but the cross training should broaden their skills as far as is reasonable. Even on the issue of technical skills, it is beneficial to move in that direction. For example, if an operator's task is to assemble an electronic assembly that will be part of an end-item device, there is no reason that operators couldn't test it when they complete the assembly. Go/no go testers can be built to facilitate testing any electronic assembly, and they can be simple enough to operate that the assembler can easily perform the test. This frees the technician for the more complicated tests downstream and ensures that the assembly is working before it is passed on to the next higher level. It also gives operators a sense of ownership and accomplishment. Over time they may even be able to troubleshoot an assembly that fails the test.

Requiring multiple skills in JIT teams is important for several reasons. First, when a team member is absent, the work cell can still function. Second, problem solving and continual improvement are enhanced by having more than one expert on whatever process is in question. New people will have fresh new ideas. Third, if one of the cell's processes starts falling behind, another member can augment the process until it is back on track.

Establish Flow/Simplify

Ideally, a new line could be set up as a test case to get the flow established, balance the flow, and generally work out initial problems. In the real world, this may not be feasible. Normally the new line is set up to produce deliverable goods. What typically happens is a line is set up, then operated with just a few pieces flowing through to verify the line's parameters. It is very important to maintain strict discipline on the line during pilot runs. Everyone must strictly adhere to procedures. Each operator must stay in his or her assigned work area, with no helping in another process. Only by pilot runs conducted this way will the information gained be meaningful and valid. This will allow process times to be checked, wait times to be assessed, bottlenecks to be identified, and workers to become synchronized. It is not necessary to have a pull system in place for these preliminary runs, because only a few pieces will be involved. In fact, until the flows have been established, kanban is not possible.

The second thing to look for in these pilot runs is how well the line accommodates the work. Are the work stations positioned for the least motion? Is there sufficient space, but not too much? Can the operators communicate easily with each other? Is the setup logical and simple? Can any changes be made to make it better, simpler? Don't overlook the processes themselves. Ultimately that is where most of the simplification will occur.

Kanban Pull System

Having established the flow and simplified it to the extent possible, the company can now introduce the kanban pull system. As the work cell is being designed, the kanban scheme should be developed. For example, will a single or double kanban card system be used, or kanban squares or bins? Or will some combination or a different variation be used? A kanban plan must be tailored to the application: there is no single, best, universally applicable kanban system.

Readers who are familiar with manufacturing may know that cards have been used in the manufacturing process as long as anyone can remember. They take the form of traveler tags, job orders, route sheets, and so on, but they are not at all the same as kanbans. These cards push materials and parts into a production process, such as PC-board stuffing. When the boards controlled by the card are all stuffed (the electronic components have been inserted into the boards), the entire batch is pushed to the next process—ready or not, here they come. The next process didn't ask for them and may not be ready for them—in which case, they will stack up in front of the process or be removed from the production floor and stored with other waiting WIP. By contrast, in a JIT line, the succeeding process signals the preceding process by kanban that it needs its output. Be sure to understand the distinction; with kanban, the succeeding process pulls from the preceding (supplying) process. The kanban always tells the supplying process exactly what it wants and how many. The supplying process is not authorized to make more product until the kanban tells it to do so—nothing waiting, no stored WIP.

The Toyota system uses two types of kanbans: the withdrawal kanban and the production kanban. The withdrawal kanban, also called the move kanban, is used to authorize the movement of WIP or materials from one process to another (see Figure 21-10). The kanban will contain information about the part it is authorizing for withdrawal, quantity, identification of containers used, and the two processes involved (supplying and receiving). The production kanban authorizes a process to produce another lot of one or more pieces as specified by the kanban (see Figure 21-11). The kanban also gives the description of the piece(s) authorized, identifies the materials to be used, designates the producing process work station, and tells the producing process what to do with it when it is completed.

Consider the operation of two processes in a manufacturing sequence to see how this works in practice. Figure 21-12 shows a preceding process that does grinding on metal parts. This is the supplier for the parts finishing work station, the succeeding process. Figure 21-12 shows five segments, described in the following paragraphs.

Segment 1 reveals that the finishing work station has containers at both its In and Out areas. The container at the In area carries a move kanban (MK), and has one part

PRODUCING PROCESS	LOCATION	PART NO.	WITHDRAWING PROCESS
HARNESS ASSY	BHA-15	3371-10130	PANEL INTEG
BHA-15	SQ. F 1		BPT-1
	CONTAINER TYPE	PART DESCR	
	N/A	BETA HARNESS	
		NO. WITHDRAWN	RECEIVING LOC
	CONTAINER CAPACITY	1	BPT-1 WS

Figure 21–10
Withdrawal Kanban

left to be used. The container at the Out area has five finished parts in it and is waiting for the sixth. Back at the grinding work station the Out container is filled with the six parts authorized by the production kanban (PK) attached. The container at the In area is empty, and work is stopped until another production kanban appears.

Segment 2 shows that the finishing work station has completed work on the six parts, emptying the container at its In area. The empty container with its attached MK for six parts is taken back to the grinding work station, which is ready to supply the parts.

Segment 3 shows that when the empty container is received at the grinding work station, the move kanban is removed from the empty container and attached to the full

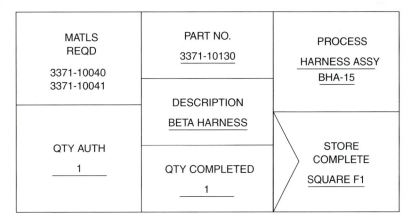

MATLS REQD	PART NO.	PROCESS
3371-10040	3371-10130	HARNESS ASSY
3371-10041		BHA-15
	DESCRIPTION	
	BETA HARNESS	
QTY AUTH		STORE COMPLETE
1	QTY COMPLETED	SQUARE F1
	1	

Figure 21–11
Production Kanban

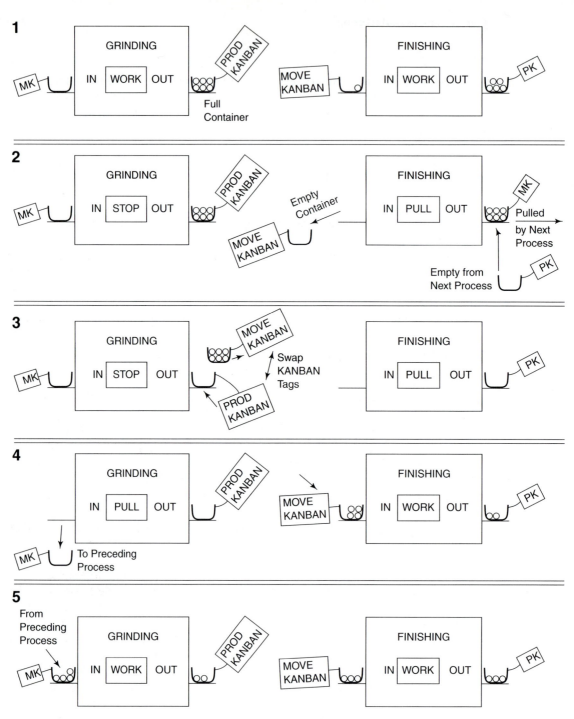

Figure 21–12
Dual Card Kanban System

770

container, which is sitting at the process's Out area. This authorizes movement of the six parts to the finishing work station. At the same time, the production kanban was removed from the full container and attached to the empty one, which is placed at the grinding work station's Out area. This authorizes the grinding process to grind six more pieces.

Segment 4 shows that the finishing process has now processed two parts. The empty container at the In area of the grinding process has been taken back to the preceding process in order to obtain the parts it needs to grind six new pieces.

Segment 5 shows the finishing work station halfway through its six pieces, with the grinding process started on its next six pieces. This cycle will repeat itself until there is no more demand pull from the right side (from the customers and the final processes).

The finishing work station had its Out parts pulled by the next process in Segment 2, triggering finishing's pull demand on grinding in Segment 3. That in turn resulted in grinding's pull from its previous process in Segment 4. The pulls flow from the right (customer side), all the way through the production processes to the left (supplier side). When demand stops at the customer side, pulling stops throughout the system, and production ceases. Similarly, increase or decrease in demand at the customer side is reflected by automatically adjusted pulls throughout the system.

As suggested earlier, it is not always necessary to use actual kanban cards. In many applications, it is necessary only to use kanban squares, kanban shelves, or kanban containers. In Figure 21-12, for example, the two processes could have used any of these devices. The Out side of the grinding work station could have the right side of its tabletop marked out in six kanban squares. One part ready for finishing would be placed on each square, like checkers on a checkerboard. The signal to grind six more parts would be the finishing work station's taking of the parts, leaving the kanban squares empty. In this case, the empty kanban square is the signal to produce more. Marked-off shelf areas, empty containers designated for so many parts, and various other devices can be used. Combinations are the rule.

Kanban is a shop floor control/management system. As such, it has some rules that must be observed:[14]

Rule 1: Do not send defective product to the subsequent process. Instead, stop the process, find out why it was made defective, and eliminate the cause. It will be much easier to find the cause immediately after it happened than it would be after time has elapsed and conditions have changed. Attention to the problem will escalate rapidly as subsequent processes come to a halt, forcing resolution. Only then should the subsequent process be supplied.

Rule 2: The subsequent process comes to withdraw only what is needed, when needed. There can be no withdrawal without a kanban (of some sort). The number of items withdrawn must match the number authorized by the kanban. A kanban must accompany each item.

Rule 3: Produce only the exact quantity withdrawn by the subsequent process. Never produce more than authorized by the kanban. Produce in the sequence the kanban are received (first in, first out).

Rule 4: Smooth the production load.   Production flow should be such that subsequent processes withdraw from preceding processes in regular intervals and quantities. If production has not been equalized (smoothed), the preceding process will have to have excess capacity (equipment and people) to satisfy the subsequent process. The earlier in the production process, the greater the need for excess capacity. Because excess capacity is waste, it is undesirable. The alternative would be for the processes to "build ahead" in anticipation of demand. This is not allowed by rule 3. Load smoothing will make or break the system, because it is the only way to avoid these two intolerable alternatives.

Rule 5: Adhere to kanban instructions while fine-tuning.   In the previous section, we said that for a kanban system to work, the flow must first be established. Kanban cannot respond to major change, but it is a valuable tool for the fine-tuning process. All the production and transportation instructions dealing with when, how many, where, and so on, are designated on the kanban. If the manufacturing process has not been smoothed, one cannot, for example, tell a preceding process to do something early to compensate. Instructions on the kanban must be observed. Adhering to the kanban's instructions while making small, fine-tuning adjustments will help bring about optimum load smoothing.

Rule 6: Stabilize and rationalize the process.   The processes need to be made capable and stable. Work instructions/methods must be simplified and standardized. All confusion and unreasonableness must be removed from the manufacturing system, or subsequent processes can never be assured of availability of defect-free material when needed, in the quantity needed.

Observing the six rules of kanban all the time is difficult, but it is necessary if the production flow in a JIT system is to mature and costs are to be reduced.

Kanban is often used by itself for shop floor control very effectively, but it can also be used in conjunction with automation, such as bar code and computer augmentation. Computer-based kanban systems exist that permit the fundamental kanban system in a paperless environment. As with automation in general, such a computerized system must be designed to suit the application. Applying technology simply for technology's sake is never a good idea. Whatever you do, it is best to have the system working in its basic form before automating; otherwise, you are likely to automate your problems.

The demand pull system has proven itself far more efficient than the traditional push system. If the advantages of just-in-time are wanted, there is no alternative but to use a pull system, and kanban, in one form or another, is what is needed.

Visibility/Visual Control

One of JIT's great strengths is that it's a visual system. It can be difficult to keep track of what is going on in a traditional factory, with people hustling to and fro storing excess WIP, stored WIP being brought back to the floor for the next stage of processing, caches of buffer WIP all over the place, and the many crisscrossing production routes. The JIT factory is set up in such a way that confusion is removed from the system. In a JIT factory, it is easy to tell whether a line is working normally or having a problem. A quick visual scan reveals the presence of bottlenecks or excess capacity. In addition to the obvious signals, such as an idle work station, JIT encourages the use of information boards to keep all the workers informed of status, problems, quality, and so on.

Each product work cell or team should have one or more boards, perhaps on easels, on which they post information. For example, if the schedule anticipates the production of 300 subassemblies for the day, the workers will check off the appropriate number each time a succeeding process pulls subassemblies from its output. This keeps the team apprised of how it is doing and presents the information to managers, who only have to glance at the chart to gauge the work cell activity and its kanbans to develop a very clear picture of how well the line is doing. Another board charts statistical process control data as the samples are taken in the work cell. Anyone can spot developing trends or confirm the well-being of the process with a quick look at the charts. Every time a problem beyond the control of the work cell or an issue with which the work cell needs help comes up, it is jotted down on a board. It stays there until resolved. If it repeats before it is resolved, annotations are made in the form of four marks and a slash for a count of five (see Figure 21–13). This keeps the concerns of the work cell in front of the managers and engineers who have the responsibility for resolution. The mark tally also establishes a priority for resolution. The longest mark "bar" gets the highest priority. Maintenance schedules for tools and machines are also posted in plain view, usually right at or on the machine, and normal maintenance activity, such as lubrication, cleaning, and cutter replacement are assigned to the work cell.

Consider what happens when these charts are used. Information is immediately available to the work cell. The team is empowered to perform maintenance and solve all problems for which it has the capability. With the information presented to the team in real time, the team solves the problems at once and performs maintenance at appropriate times. This approach minimizes waste, keeps the machines in top shape, and produces a flow of ideas for improvement. The shop floor control loop is as tight as it can get. The operator detects and posts the information. The operator reacts to the information to solve problems or take action.

If a problem is beyond the work cell team's capability, all the people who can bring skills or authority to bear are immediately brought in, presented with the data, and the problem gets solved—quickly. The control loop goes from information to action in one or two steps. In the traditional factory, the operator may not even be aware of a quality problem. It is usually detected by a quality assurance inspector hours or days after the defect was created. The inspector writes it up. The form goes to the management information system (MIS) department, where, after a period of time, the data are entered into the computer. Days or weeks later the computer prints a summary report including analysis of quality defects. The report is sent to management through the company mail or via an intranet system. The report may rest in an in-basket queue for a length of time before being examined. Managers in traditional plants are kept so busy with meetings and "firefighting" they hardly have time to read their mail, but eventually they will get around to looking at the report. They will see that the line is (or *was*) having a quality problem and pass the report to the floor supervisor for action. The floor supervisor will attempt to see whether the problem still exists. If it does not, case closed. It happened weeks ago, and the operator, who up until now was unaware of the defect(s), can't remember anything that would confirm the problem, let alone suggest a root cause. If

Figure 21–13
Work Cell Problem Status Board

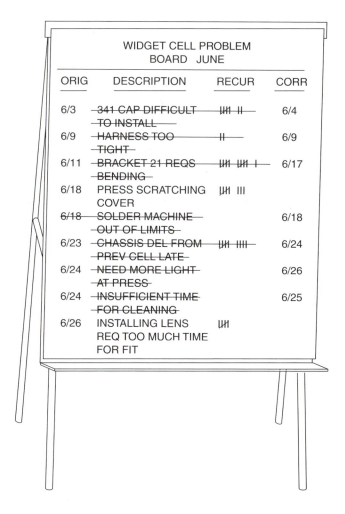

WIDGET CELL PROBLEM
BOARD   JUNE

ORIG	DESCRIPTION	RECUR	CORR
6/3	~~341 CAP DIFFICULT TO INSTALL~~	~~JHI II~~	6/4
6/9	~~HARNESS TOO TIGHT~~	~~II~~	6/9
6/11	~~BRACKET 21 REQS BENDING~~	~~JHI JHI I~~	6/17
6/18	PRESS SCRATCHING COVER	JHI III	
~~6/18~~	~~SOLDER MACHINE OUT OF LIMITS~~		6/18
6/23	~~CHASSIS DEL FROM PREV CELL LATE~~	~~JHI IIII~~	6/24
6/24	~~NEED MORE LIGHT AT PRESS~~		6/26
6/24	~~INSUFFICIENT TIME FOR CLEANING~~		6/25
6/26	INSTALLING LENS REQ TOO MUCH TIME FOR FIT	JHI	

the floor supervisor is lucky, the problem may still be there, and the cause may be found. But in the meantime, weeks of production may have been defective.

In this control loop, at least six functions are involved before the loop is closed. That is bad enough, but when the time delay factor is added, any chance of finding root causes of problems that come and go is eliminated. Process improvement is much more difficult in this kind of traditional production system. Having had personal experience with both, the authors can attest that the most expensive, most sophisticated computer-based defect analysis system, such as might be employed in the above example, is infinitely inferior to the simple one- or two-step, person-to-person, no-computers-involved control loop of JIT when it comes to presenting useful information on a timely basis for the purpose of problem solving and process improvement.

Before our plants changed over to JIT, a mainframe-based defect analysis system was used. The navy designated it as a best practice in the industry. Other companies came to see it, and many of them used it as a model for their own new systems. It could analyze data and present it in many different forms. But it had one flaw: time delay. From the time a process produced a defective part until the loop was closed with the operator of the process, several days (at best) had passed. We are not suggesting that the system was unable to make improvements, because it did. But the real revelation came with implementation of JIT and finding what could happen right inside the work cell when workers had the information they needed while it was fresh and vital, and they were empowered to do something with it. Immediately, defects dropped dramatically, and they continued to drop as continual improvement was established. Before JIT, these plants were never able to achieve results remotely comparable with their megadollar computer-based system.

Every JIT line develops its own versions of information display charts. But whatever the variation, everyone has valuable, useful information available at all times. That kind of information is extremely difficult to find in a traditional line and most often comes to light in the periodic (weekly or monthly) computer analysis reports. By then, the trail to the root cause may have become obliterated by the passage of time, other problems, or events. In the JIT factory, real-time visibility lets people know of the problem right then and there, while the cause is obvious. Coupled with the JIT philosophy that says that the problem must be solved before going any further, this visibility becomes a driver for elimination of problems, and for process improvement.

Elimination of Bottlenecks

Richard Schonberger makes the interesting point that only the bottlenecks in a traditional factory forward work to the next process just-in-time.[15] He explains that in a conventional manufacturing plant, the bottleneck process is one that goes as fast as it can all the time, barely keeping up with demand. If it breaks down, there is real trouble. To keep it running, and to attempt to find ways to increase its output, the bottleneck receives attention out of proportion to the rest of the plant, monopolizing the efforts of engineering and management.

In a JIT plant, all processes are potential bottlenecks in the sense just discussed, because there is little excess capacity, and there are no buffer stocks to fall back on when a process or machine shuts down. The upside of this is that all processes are constantly under scrutiny—none are ignored. As Schonberger also points out, the fact that all the processes must be watched carefully makes it imperative that the process operators play a

QUALITY TIP

Control Loop Effectiveness

The effectiveness of any control loop is inversely proportional to the number of functions in the loop and the time required to close the loop.

QUALITY TIP

Financial Data on the Factory Floor

Work cells develop many kinds of charts for display in their work areas. Recently we encountered financial charts in the production areas of a San Antonio plant where employee empowerment is in full bloom. The cells determine their contribution per unit produced to several financial factors, including waste, labor cost per unit, and so forth, and post a running record of the numbers on a chart in each work cell for all to see. The workers in the cells were obviously tuned in to this and were eager to explain both the data and the methodology used for collecting it. We found this interesting because workers in most factories do not have a clue as to their impact on cost or profit. Seldom do line workers take an interest, because management doesn't give them financial information. As a result, a connection between their work and the financial well-being of the company is seldom made. In this San Antonio company, however, with management's encouragement, the production workers themselves proudly keep track of how their cells contribute to the company's financial results. The benefits of this approach can be enormous. With the information constantly in front of the cells' workers, they always have an internal urge to improve. On the other hand, if something goes wrong, the cell is the first to know and react.

major role in the care and monitoring and improving of the processes. There cannot be enough engineers to go around when every process is a potential bottleneck.

For this discussion, though, the bottleneck is put into a slightly different frame of reference. We are talking primarily about the setup stage of a JIT operation when trying to establish a balanced, rational flow through the production system. In this early stage, it is not uncommon to have some real functional bottlenecks. For example, if the new JIT line is being established to produce as many as 1,000 parts per day, but the manual assembly process can turn out only 800, there is a bottleneck. One way or the other, the process must be brought up to 1,000 or more. If the process employs two people using hand tools, then the answer is simple: add a third person and the appropriate tools. Then the capacity for that process should be 1,200 per day. The extra capacity will have to be accepted until the process can be improved to bring the daily single-operator output up to 500 each, making it possible to go back to two operators.

Perhaps a machine can produce only 75% of the projected demand. Here the options are a little different. This may be a very expensive machine, too expensive to replicate. Is it possible to put that machine to work somewhere else and put two, lower capacity, less expensive machines in the line, or maybe a single new, higher capacity machine? Can the old machine be modified to increase its output? If setup time is a part of the machine's normal day, there is a potential for improvement. Another possibility may be adding a second, smaller machine to augment the existing machine's capacity, although two different machines on the same line making the same part/product is not a desirable solution.

Another kind of bottleneck can exist when a single physical process is shared by two or more JIT lines. It is preferable to make each JIT product line independent and self-sufficient, but this is not always possible. An example might be a single-wave solder machine servicing two or more JIT lines. Because of the cost, size, and maintenance requirements of such a machine, it may not be feasible to put one in each JIT product line. Rather, all the JIT lines take their PC boards to a single-wave solder service cell for soldering. The JIT lines operate independently of each other. Therefore, it is difficult to predict when conflicts might develop. If they all need servicing at the same time, there is a bottleneck. If soldering delays cannot be accommodated, then one or more of the lines must have its own soldering capability.

Technology can often provide solutions to such problems. For example, 20 high-quality drag soldering machines could be purchased for the price of one wave soldering machine. Production rates of drag solder machines are much lower than those of wave machines, but in many applications they are ideal for placement right in the JIT line, dedicated to the line's product, and controlled by the line. Such solutions are often feasible with other types of machines too.

Whether your bottlenecks appear during the setup phase or during production, the best approach is to assign a cross-functional team to solve the problem. The team should have representation from engineering, manufacturing, finance, and any other relevant function areas. Its job is to list all possibilities for eliminating the bottleneck. This can be done by brainstorming, setting aside those ideas that don't make sense, and finding the most satisfactory solution in terms of quality, expense, efficiency, and timing.

Frequently, the solution to a bottleneck results in some degree of excess capacity in the process, as occurred earlier when the third operator was added. This is not always bad. Although JIT always works to achieve more and more efficiency and, taken to the extreme, would have just exactly enough capacity to produce the demanded level and no more—in a practical sense, some excess capacity is desirable. If a line is running at top speed every day, the operators will have no time for problem solving or improvement activities. Some time should be set aside each week for those two items, as well as for maintenance and housekeeping. For most applications, 10%–15% excess capacity is acceptable.

Small Lot Sizes and Reduced Setup Times

For 80 years or more, industrial engineers have been taught that the larger the production lot size, the greater the benefit from economy of scale. If one wanted to hold down cost of production, bigger lot sizes were the answer. This was the conventional thinking until the JIT manufacturing bombshell landed on our shores from Japan in the early 1980s. Under the leadership of Toyota and Taiichi Ohno, Japanese manufacturers concluded that the ideal lot size is not the largest but the smallest. Is it possible that both the manufacturers and the universities could have been wrong all those years? (In fact, many Western manufacturers and universities are still bound to the big-lot philosophy.) Our conclusion is that the big lot was appropriate as long as mass production systems were used, although they certainly had major problems even then. But once the Toyota

Production System came into being, the big lot was not only out of step but impossible to justify.

It stands to reason that if a machine is used to produce different parts that are used in the subsequent processes of production, and that the time it takes to change the machine over from one part type to another is 6 to 8 hours, then once the machine is set up for a particular part type, one should make the most of it. It seems to make more sense to run the machine with the same setup for 4 days, setting up for the next part on the 5th day, than to run 1 day, spend the next on setup, and so forth. The 1-day runs result in about 50% utilization time for the machine, assuming a single shift for simplicity. The 4-day run yields about 80% utilization.

So what is the problem? If there are four different parts to make on the machine, simply make 20 days' supply in 4 days, then go to the next part. By the time production has used all the 20-day supply of the first part, the machine will have cycled back to make that part again. Perhaps a 30-day supply should be made, just in case the machine breaks down. Would a 40-day supply be better? Where does this stop? If we are willing to risk an occasional breakdown, the 20-day cycle is acceptable. A place to store a 20-day supply of not just one part but four parts will be needed. Then the capability to inventory, retrieve, and transport these parts will also be needed. That represents land, facilities, and labor that would not be otherwise needed. None of it adds value to the product, so it is pure waste. It is likely that these costs add up to more than the supposed inefficiency of running the machine with a 50% utilization factor, but these costs are more acceptable to accountants. Land, buildings, and people in motion are not as apparent as examples of waste as machines that are not making product. Traditional thinking says, "Because the machine is busy, people are busy, floor space is full, it can't be waste." But it is.

In addition, suppose that a production flaw is found in one of the parts, caused by the machine. Every part made in that lot is suspect. Samples will be tested, and maybe the whole lot will have to be scrapped. This could be 20 days' supply, representing significant cost. The line will be down until new parts can be made—a major disruption.

Suppose the engineering department corrects a design weakness in one of the parts. Is the entire inventory of parts already made scrapped, or do we use them up in production, knowing that they are not as good as the newly designed part? Either is a bad proposition.

Now assume that the 1-day 50% utilization cycle on the machine was employed. The greatest loss we could take would be 8 days' inventory for any of these cases. The 8-day supply can be stored easier than 20 days' supply. This would reduce the cost of warehousing, control, and transportation. Any design changes can be cut-in in 8 days. Everything seems positive except the 50% machine utilization.

Ideally, setup time might be reduced to 30 minutes, producing 1 day's supply of each part every day. Utilization will be 75% and need for any warehousing may be eliminated. This may seem to be out of reach, but manufacturers using JIT have done far better, often taking setups from many hours to a few minutes. For example, by 1973 Toyota had reduced the setup time for a 1,000-ton press from 4 hours to 3 minutes. Over a 5-year period, Yanmar Diesel reduced the setup time for a machining line from over 9 hours to just 9 minutes.[16] These are not isolated examples.

QUALITY TIP

Small Lot Sizes

Small lot sizes result in improved product quality, production flexibility, and customer responsiveness. Shortened setup times make small lots possible.

The general rule seems to be that organizing properly for the setup, making sure the tools and parts that will be needed are in place, and having the right people there at the appointed time, will yield an immediate 50% reduction. Then, by analyzing the setup process step by step, a company can usually streamline the process to cut time by half again. Ultimately, the machine itself may be modified to make setup faster and less difficult (by eliminating the need for adjustment). In any mature JIT factory, it would be a rare setup that took more than a few minutes, whereas the same setups were previously measured in hours.

The previously supposed advantage of manufacturing in big lots completely disappears when setup times are brought down to the kinds of times being discussed here. Machine utilization can be high to satisfy accounting criteria, and lots can be small to prevent waste and to enable kanban pulling straight from the machine to the next process. Short setup times coupled with kanban have the advantage of flexibility of production. For example, Harley-Davidson used to run its motorcycle line in long production runs of the same model. If a dealer placed an order for a model that had just finished its run, it might have been several weeks before that model could be run again, allowing the order to be filled. Harley was one of the first North American companies to adopt the total quality methods—as a means of survival.

For many years now, Harley has been able to mix models on the production line. It no longer has to produce its product in big lots because it was able to reduce setup times all along its line. Now when an order comes in, it is placed in the queue without regard for the model. Customers get their new bikes far sooner and are, therefore, less tempted by other brands in the meantime. Customer orientation is one major benefit of short setup times and small lot sizes, along with manufacturing flexibility, higher quality, and lower costs.

Total Productive Maintenance and Housekeeping

This is difficult to comprehend, but many manufacturers spend vast amounts on capital equipment and then ignore the machines until they self-destruct. By contrast, one can find relatively ancient machines in total quality Japanese factories that look like new and run even better. This must become the norm in the United States if U.S. companies are going to compete with the rest of the world. Because a JIT production line operates very close to capacity in every process, no tolerance exists for machine failure. When the machine is supposed to be running, it had better be, or the whole line will suffer. The

Japanese have virtually eliminated machine breakdowns by applying total productive maintenance (TPM) techniques to their machines. Machines are cleaned and lubricated frequently, most of that work being done by the operators who run the machines. More technical preventive maintenance routines are performed by experts at frequent intervals. The machines are continually upgraded and modified for closer tolerances, faster setup, and fewer adjustments. Not only do the machines last longer, but during their entire life span they perform as well or better than when new.

The difficulty with TPM is finding the time in which to perform the maintenance, especially in factories in which three shifts are the norm during times of prosperity. The third shift is rare in Japan and Europe, so companies there do not share this problem. Regardless of the workday schedule, it is imperative that maintenance time be provided. The operator-performed maintenance is done during the normal shift (one reason to have a bit more than just enough capacity—a half-hour to an hour a day of excess capacity should more than cover operator maintenance needs).

An added benefit of turning some of the maintenance responsibility over to the operators is that the operators develop a sense of ownership for the machines they use and care for. They pay keen attention to the looks, sounds, vibrations, and smells of the machines to spot problems before they develop. For the first time, the operators are in a position to call for maintenance before breakdown occurs. TPM is a must for JIT production systems.

Housekeeping is another area that is different under JIT. It is not unusual for the operators themselves to take on the responsibilities formerly associated with janitors. In the better JIT plants, one will see planned downtime being taken with cleaning chores—everything spotless, everything in its place. It follows that better performance will result from a clean, tidy, and well-organized work area than from one that is dirty and cluttered with tools scattered all over. People like a clean, bright, rational place in which to work. Again, time will have to be made available for this activity.

Process Capability, Statistical Process Control, and Continual Improvement

Process capability, statistical process control, and continual improvement have already been discussed in detail in this book, but it is important to understand the dependence

QUALITY TIP

A New Adage

"If it ain't broke, fix it anyway!"

Source: Alcoa CEO Paul O'Neill, cited in Tracy E. Benson, "Alcoa," *Industry Week* (April 19, 1993) 12.

on them by just-in-time. Is JIT a necessary prerequisite for process capability study and improvement, or for SPC, or continual improvement? The answer is no. At least one of the three is being done in the majority of traditional production plants. Still, there is a connection. The philosophy and discipline of just-in-time virtually demand that they be used in any JIT environment. While a traditional manufacturing operation *may* employ one or more of the three, the JIT manufacturing operation *must*, and it must be all three. The reason may be obvious to you by now. The JIT plant is fragile. Everything must work when it is supposed to, and it must work close to perfection. There are no warehouses of buffer stock to come to the aid of a broken-down process. There is never much excess capacity to help out in tight spots. All the processes with their machines and people must operate in top form all the time.

This is where process capability, SPC, and continual improvement come in. Even before the JIT line can be certified for full production, the line has to be balanced or rationalized, and a flow has to be established. Unless it is known what the processes are capable of doing in terms of quality and quantity, it will be difficult to achieve the even flow that is a necessary prerequisite of a kanban system. Without that, there is no JIT. In the traditional factory, there is not so much a problem: normally, gross overcapacity exists, so parts are stored for the day things go wrong, and the bad parts are sorted out because there will still be good ones to use. In JIT, no extra parts can be made, and they all have to be good. Workers must have a handle on the processes.

Because one cannot afford (from the time or cost standpoints) to make defective parts, the processes must be in control at all times. The only way to assure this is through statistical process control. This is not as necessary in a traditional plant, but it is absolutely essential in JIT. Perfection is difficult to achieve in any circumstances, so it follows that in a complex manufacturing situation, perfection is next to impossible. This is certainly true. We never quite get to the point where all the parts are perfect, but with solid, stable, in-control processes forming the basis of a relentless continual improvement program, we can come very, very close. (Some of the very best American plants target and achieve Six Sigma, 3.4 defects per million.) The best that can be achieved is the minimum that is acceptable for a JIT factory. In the process of continual improvement, ways are found to do things better, faster, cheaper, and with constantly improving quality. The process never ends, and the diminishing-return syndrome doesn't apply.

Suppliers

In the area of suppliers, JIT has different priorities from the traditional production system. The most obvious difference is the need for frequent, small-lot deliveries of parts, supplies, and materials, rather than the traditional infrequent, huge-volume deliveries. We are finding more and more JIT plants in which the suppliers deliver materials directly to the production cells, usually referred to as *point-of-use*. Several systems have been developed to cue the supplier that it is time to replenish materials. One is the *dual bin kanban* system. Two parts bins are used. Bin capacity may range from a few hours' to a couple of weeks' supply, depending on value, size, usage rate, and intended frequency of replenishment. When the cell has withdrawn all the parts from one bin, the empty bin itself is the signal that it is time to replenish. The supplier routinely checks the bins on the factory floor, and whenever he finds a bin empty, it is refilled with the exact number and kind of part designated on the bin label, usually in bar code. The supplier's bin checking must be scheduled frequently enough to assure that the second bin is never exhausted before the first is replenished. In a variation on the dual bin kanban scheme the cell's operators signal the supplier that a bin is empty, either by bar code transmission or automated electronic purchase order that is triggered by wanding the empty bin's bar code.

Clearly, for this kind of point-of-use materials delivery system to work, the supplier must be 100% reliable, the materials delivered must be of consistently high quality, and both the supplier and the manufacturing organization must be partners for the long haul. Consequently, choosing the suppliers for a JIT factory is a much more demanding job than it is for a traditional plant. Traditional factories are not so concerned with the delivery being on the dock at the precise date on the purchase order. It was going to be stored for a while anyway. And before that lot was used up, there would be another shipment in the warehouse. Neither do traditional factories concern themselves as much with quality from suppliers. The bad parts could always be sorted out, leaving enough good material to keep the line moving. The primary interest was price. Low price got the order. It quickly becomes apparent that this style of purchasing is incompatible with JIT.

The JIT plant must have its materials on the dock exactly on the day specified—in many cases at the hour and minute specified—or production may grind to a halt. Every part delivered must be a good part—there is no inventory cache from which to scrounge more parts to keep things moving. This means that the suppliers' quality must be consistently at or above specified requirements. Delivery and quality performance requirements of JIT effectively rule out buying for price. There is an oft-used phrase in JIT and TQM purchasing: "cost versus price." It suggests a holistic approach to the analysis of purchasing on the basis of total cost and value, not simply vendor price. How reliable is a particular vendor in terms of JIT deliveries? What kind of quality can be expected from the vendor? Does the vendor use JIT, SPC, and continual improvement? Are its processes stable and in control? A supplier that gives positive responses in these and other areas may not be the lowest price contender but may well be the lowest cost. Value is what the JIT purchasing manager must look for, not lowest price on a bid sheet, because in JIT that turns out not to be the whole story.

When a JIT factory finds a supplier that delivers excellent materials on time, every time, there is every reason to want to continue to do business with it. More and more companies are turning to supplier partnerships to cement these relationships. What this means is that the two companies agree to work together, not only as supplier and customer but as unstructured partners. The JIT manufacturer may, for example, provide training and technical assistance to the supplier to get it started in total quality, JIT, SPC, and other processes. The JIT firm may certify the supplier's quality system to the extent that incoming inspections are eliminated, relying on the partner supplier to provide acceptable quality in all its deliveries.

The supplier partner may assign one or more employees to take up residence in the JIT manufacturer's plant. Duties will include the continuing checks of the kanban bins mentioned above, having them replenished appropriately, and coordinating on-time deliveries of materials, parts, and other supplies provided under the JIT partnership agreement between the manufacturer and supplier. In addition, the resident supplier employee is empowered to do whatever is necessary to solve supply problems before they can cause disruption in the JIT factory. (While this practice has been around for more than a decade in the United States, and much longer in Japan, it is now sometimes referred to as *JIT II*. In the author's view, it is simply a logical variation of the materials element of just-in-time that can work very well in numerous situations.)

The supplier may also be called on to assist in the design phase of a new product, bringing its unique expertise into the design team. Such relationships usually carry a multiyear agreement, so the supplier can count on the business as long as its performance remains high.

There may be preferential bidding treatment, say, 10% or more advantage over non-partnership rivals. Effectively what happens is that the JIT manufacturer extends its factory right back into the supplier's premises. They operate to each other's requirements, and both are locked to each other. The results of this kind of arrangement have been excellent.

This kind of relationship is a far cry from the early ill-conceived attempts of some manufacturers to get into JIT before developing a full understanding of the concept. In those days, some companies would determine that by using JIT delivery of parts and materials, money could be saved. That part had some merit, but the execution was flawed. The companies simply told their suppliers to deliver a week's supply of materials once a week, rather than their customary 60 days' supply every 2 months. The suppliers' reaction is easy to imagine. They were being told, in effect, to store the materials in their own warehouses (which they didn't have) and to trickle the deliveries from the warehouses in small quantities weekly. This was simply a case of moving the storage

QUALITY TIP

Better, Faster, Cheaper

With the positive results of TQM being documented daily, some observers see themselves duty-bound to highlight every example of a problem with a TQM or JIT implementation. Horst Schulze is president and chief operating officer of the world-class Ritz-Carlton hotel chain, and an outspoken advocate of total quality. Whenever he is prodded by some detractor to concede that TQM is simply a fad or that it doesn't work, his response is that TQM results in three things: better products or services, faster delivery, and all this at a lower cost. He then asks, "Which is it you don't like: better, faster, or cheaper?" Because there is no logical answer, that invariably terminates that line of conversation.

Source: From a speech by Horst Schulze at a presentation of the Florida Sterling Awards for Quality, June 1995, Orlando, Florida.

facility from the manufacturer's plant to the suppliers'. A GM or a Ford has the power to do that to a supplier, but the suppliers, being smaller and with less influence, couldn't force the same back to their own suppliers, so they got caught in an intolerable situation. Only when the suppliers revolted and cried long and loud that this was not JIT—"and by the way, if you want me to store your goods for you, you're going to pay the tab anyway"—only then did the would-be JIT manufacturers see the error of their ways.

The new approach is working well because both parties benefit enormously. If a company wants JIT, then it must have the best possible suppliers, and both must want to work together for the long haul.

AUTOMATION AND JIT

Automation has not been discussed a great deal in this book. We have stuck to the fundamentals. One should not read into this, however, that JIT and automation are mutually exclusive. Rather, it is more meaningful to discuss the processes that use humans and manual machines, than the same processes powered by robots. If the fundamentals where humans apply are understood, the same fundamentals will be useful in an automated plant. All the same rules apply. We are not anti-automation.

We are, however, anti–"automation for the sake of automation." Many companies have made the costly mistake of thinking that automation will solve manufacturing problems. During the 1980s, manufacturers in the United States invested billions of dollars in automation. Cadillac built the most highly automated auto assembly plant in North America and probably the world. It turned into a nightmare of high-tech problems that took years to sort through. The plant that was to produce six cars per hour, after a year of operation, could do only half that and the quality of manufacture was, to put it charitably, questionable. Two years later, Toyota opened a new plant in Kentucky. Visitors to that plant, expecting to see a high-tech automated production line, were disappointed to find very little in the way of robotics.[17] The difference in the philosophies of the two companies becomes obvious. Executive managers at GM believed that by spending enough money, they could buy their way out of the trouble they were in. Toyota knew what it was capable of doing in one of its other low-tech plants that was operating successfully in Japan and simply cloned it down to the last detail in Kentucky. No razzle-dazzle, just good common sense.

Automation may be advantageous in many applications, but if you have not solved the problems in the human-operated versions of those same applications, you are not ready to automate them effectively. If you try, you will automate your problems and will find the robots far less adept at working around them than the humans they replaced.

It is frequently found that the need for automation is decreased or eliminated by converting to JIT. We certainly found that to be the case in two electronics plants. We were well into a program to build a factory of the future. The building was ready, much of the automation was on hand, and the rest—several million dollars' worth—was on order when we started the conversion to JIT. Within months it had become obvious to everyone, including the designers of the new factory, that we were getting more out of JIT for almost no investment than could be projected for the new automated plant. The outstanding orders were canceled, penalties paid, and we walked away from the whole idea. We had learned in those few weeks of exposure to JIT that world-class manufacturing equates to JIT in a total quality environment, not to a factory full of robots and automatic guided vehicles. JIT and automation are compatible, but one should look long and hard at the need, and the company's readiness for it, before automating processes.

Having said that, automation clearly has its place. There are many examples of very successful highly automated plants, especially for high-volume manufacturing. In such plants, JIT is at least as valuable as it is in the plants with less automation. JIT's pull system prevents overproduction of any manufacturing element and supplies needed materials at the front end of the processes when needed, and without the massive inventories of the pre–JIT period. Whether the processes are operated by humans or by robots makes no difference in this regard. Automation and JIT are completely compatible. Probably the best example of that is in today's auto industry. Although employing large numbers of workers, the industry is highly automated. All U.S. and Japanese auto plants use JIT successfully. Remember, JIT was originally designed for an auto producer, and as automation has been integrated as needed in the car-building processes, JIT has been there doing its job.

 ## SUMMARY

1. JIT is a management philosophy that seeks to eliminate all forms of waste. As a production system, JIT produces only what is needed, when it is needed, in the quantity needed.
2. Taiichi Ohno is credited with the development of the Toyota Production System and JIT.
3. The root justification for JIT is improved product quality with lower costs.
4. JIT began as a means of reducing the seven wastes. Over time, the JIT system came to be a pull system whose small lot production is supported by reduced setup times. Total productive maintenance and statistical process control were integrated to provide the necessary production reliability and predictability. Continual improvement provides the vehicle for the relentless attack on all wastes.
5. JIT is at its best as a part of a total quality system. Results can be severely restricted when JIT is operated without the total quality umbrella.
6. Inventory reduction, shortened cycle time, continual improvement, and elimination of waste are all inherent benefits of JIT.
7. JIT has a different set of requirements from traditional production systems: new skills training; rationalizing production flow for the pull system; empowering operators to take advantage of JIT's visibility features; guarding against bottleneck vulnerability through TPM, process capability study, SPC, and continual improvement; small lot sizes and short setup times; close working relationships with superior suppliers.
8. Although JIT is compatible with automation, some of the world's best plants use JIT with very little automation.
9. World-class manufacturing employs JIT as an integral part of a total quality system, producing the highest-quality products at competitive prices. It is not related to the presence or absence of automation.

KEY TERMS AND CONCEPTS

Automated electronic purchase order	Line stop
Automation	Lot size
Autonomation	Manufacturing process
Bar code	Marketability
Bottlenecks	Mass production
Buffer stock	Overhead/overhead rate
Continual improvement	Point-of-use
Cycle time	Production flow
Demand flow	Production smoothing
Dual bin kanban	Pull production
Focused factory	Push production
Internal/external customers	Setup time
JIT	Seven wastes
JIT II	Staging
Job shop	Statistical process control (SPC)
Kanban	Supplier partnerships
Kits/kitting	Total productive maintenance (TPM)
Lean production	Value-adding labor
Line balance	Work-in-process (WIP)

 FACTUAL REVIEW QUESTIONS

1. Define *JIT*.
2. Explain the difference between the traditional production system and JIT in terms of placement of production control.
3. Describe the bases for production scheduling for mass production systems and for JIT.
4. Explain how a JIT process knows when and how much to produce.
5. What two fundamental advantages are provided by JIT?
6. List the seven wastes.
7. Identify the two pillars of the Toyota Production System.
8. Explain how mass production contributes to the seven wastes.
9. Explain how JIT impacts each of the seven wastes.
10. Discuss JIT's vulnerability to parts shortages, breakdowns, and bottlenecks from the perspective of the mass production advocate and the JIT advocate.
11. Explain the push system and what triggers it to start.
12. Explain the pull system and how it is started.
13. Why is superior machine maintenance and improvement critical to JIT?

14. Describe how kanban supports the pull system.
15. What are the advantages of reduced setup time?
16. Why does this book recommend that JIT be a part of a total quality system?
17. Describe JIT's objectives relative to inventory and WIP.
18. Discuss the relative complexity of the two production systems.
19. Explain cycle time, and list its constituent elements.
20. Why is a JIT production line stop considered a positive phenomenon?
21. Explain how inventory can deter problem solving.
22. Describe the difference in plant organization/layout between the two production systems.
23. Explain how JIT can cause immediate cycle time reduction.
24. To whom is credit given for the development of the JIT system?

 ## CRITICAL THINKING ACTIVITIES

1. Study the operations of an electric utility company, and determine where the production of electricity stands vis-à-vis push or pull systems. Using charts and narrative, explain your finding.
2. Study the operation of a supermarket, and list the JIT features in use. Look for special JIT practices such as supplier partnerships, and describe how they operate in a supermarket environment.
3. Develop a chart contrasting the philosophies of mass production and just-in-time.
4. Mass production was the most successful production system from the time of Henry Ford's Model T until the 1960s. There is no question that this system permitted the manufacture of a wide range of goods at much lower prices than had been possible before then. Given this success, how do you explain that JIT is supplanting mass production?
5. You are the president of a new division of a major auto producer. Your organization has been chartered to design and build a line of cars to compete with intermediate-size Japanese and European imports. The division is to be located in a historically agricultural state that has availability of unskilled labor but no experience in auto manufacturing. Your key staff and midlevel managers will be hand-picked from among other divisions. The corporate headquarters staff has not been able to come to terms on how the new division should be set up. The options seem to be (a) set up as a total quality organization, using JIT in the factory, or, because the corporation has little experience with either, (b) set up like the other divisions in a traditional hierarchy and production system, converting to total quality and JIT little by little as time and experience permits. You believe that in the long run, total quality and JIT are necessary to compete, especially with the Japanese. But you and the people who will make up the division's management team are experts in mass production. Going the traditional route looks like an easier start-up path. Going directly to total quality/JIT has unknowns.

   The CEO is leaning toward letting you make the call, but he wants you to convince him and the senior staff. You have to prepare a briefing for the corporate staff for that purpose. Which course will you choose? Explain the advantages and possible pitfalls of both, and make the argument that supports your choice.

DISCUSSION ASSIGNMENT 21–1

German automobiles have, with some notable exceptions, always been held in high regard. German cars were, and in most cases are, still built under the concept of *Technik*, which combines art and technology as embodied in the craftsman worker. (This is essentially the same concept used by Ford and others prior to Henry Ford's 1913 assembly line.) Skilled workers are required to overcome the shortcomings of machines when parts do not fit, or finish is marred, or when paint needs to be touched up. The German philosophy was, and to a large extent continues to be, that if the machines do not do their jobs perfectly, the craftsmen will correct the mistakes. The Germans believed that technik would provide a continuing competitive edge, and Germany never evolved a production system that went beyond technik.

Through the 1960s and well into the 1970s, the Japanese auto industry mostly operated like the U.S. model; that is, using Ford's mass production concept. One lone auto producer, Toyota, had developed, and was using, the first new manufacturing system since Ford's moving assembly line of 1913. It was Tai-ichi Ohno's just-in-time system. When the 1973–1974 oil crisis hit, there was exactly one company operating JIT, and that was Toyota. Toyota was the only Japanese car company to remain profitable during the Japanese recession brought on by the oil crisis. Over several years after the oil crisis shock, the other Japanese companies adopted the Toyota Production System. Neither the American nor European companies followed suit, with the result that both American and European manufacturers became noncompetitive.

Ten years later, the American car companies, led by Ford, started converting to the Toyota System. Much of the JIT philosophy and many of the JIT techniques are in place in U.S. auto plants today. Clearly, General Motors's leadership has had a more difficult time with Ohno's and Deming's philosophy than has Ford's or Daimler-Chrysler's. The European plants lag farther behind, accounting for the much higher prices of their products.

1. Given that Chrysler's progress in JIT (and other areas such as cross-functional design teams and concept-to-market time) was made before the so-called "merger of equals" with the German company, discuss the possibility of having these advances reversed under Daimler's management.

2. Research the trade restrictions/agreements on imported autos during the 1970s and 1980s. How could they have contributed to the non-competitiveness of American automakers?

DISCUSSION ASSIGNMENT 21–2

By 1981–1982 the American auto industry was in crisis. Ford closed nine of its plants. GM would close more than twenty. Tens of thousands of long-term employees were affected. The industry was facing a full-scale invasion from Japan. Although imported cars had been around forever, they historically accounted for a very small share of the American market. But now car buyers had discovered that the cars from Japan were so superior to the American brands, that the auto industry got its first-ever taste of foreign competition; and a bitter taste it was. Suddenly the U.S. automakers found themselves in a market that demanded better quality and value for the dollar.

Ford realized sooner than the others that it would have to find out how the Japanese were able to produce such reliable, high-quality cars for the prices being charged. A few Ford managers were dispatched to Japan to study the methods of Toyota, Honda, Nissan and Mazda. They were followed by increasing numbers from the other automakers. Expecting to find a high degree of automation, they were surprised to find less than at Detroit. Nor was it that the Japanese worker was superior. Ford made a startling discovery. The difference was, it rightly concluded after long study, not automation or technology, nor the Japanese workers, but how the workers were managed.[18] All the Japanese automakers were practicing JIT on the Toyota model, and quality and productivity were improved incrementally every year. Ten or twenty years of this had created an enormous quality/productivity gap with the American industry.

The U.S. automakers had some gut-wrenching decisions to make, but it came down to the question, "Are we going to stay in the auto business?" In other words— survival. Twenty years later the U.S. auto industry has made considerable progress, but while the quality and value gap has been narrowed, it does not appear to have been erased. As evidence of this, the J. D. Power and Associates 2000 Initial Quality Study shows Japanese vehicles at the top of twelve of the fourteen study segments. J. D. Power also looks at auto assembly plants, and for the year 2000 the top award went to the Toyota plant at Kyushu, Japan. The three U.S. plants to win awards were Toyota Georgetown (Gold); NUMMI Fremont, which is jointly operated by Toyota and GM, using the Toyota system (Silver); and Toyota Cambridge (Bronze). The 2000 J. D. Power Vehicle Dependability survey ranks the top five as follows: Lexus, Porsche, Infiniti, Toyota, and Acura. All but Porsche are Japanese, and Porsche is much more expensive.

As Detroit has now largely embraced JIT, Japan has been a moving target, continuing year after year to improve all aspects of its production processes and products.

> *Discussion Assignment*
> 1. Discuss the J.D. Power and Associates ranking of U.S. auto plants for the year 2000.
>   a. Why is Toyota either the operator or co-operator of all three winners?
>   b. What is suggested by the fact that the Freemont facility was closed by GM because of poor quality and an unmanageable workforce, but under the Toyota/GM joint venture it is consistently at or near the top for quality and productivity?

ENDNOTES

1. Taiichi Ohno, *Just-in-Time for Today and Tomorrow* (Cambridge, MA: Productivity Press, 1990), 2.
2. Ohno, 31.
3. Ohno, 28–29.
4. Ohno, 9.
5. Ohno, 75.
6. James Womack, Daniel T. Jones, and Daniel Roos, *The Machine That Changed the World* (New York: HarperCollins, 1990), 62.
7. David Lu, *Kanban—Just-in-Time at Toyota* (Cambridge, MA: Productivity Press and Japanese Management Association, 1986), vi.
8. Materials and general and accounting expenses are held constant for this example to keep it simple, although both could be expected to decrease under JIT.
9. C. E. Sorensen, *My Forty Years with Ford* (New York: Norton, 1956), 174.
10. Womack, 83.
11. W. Edwards Deming, *Quality, Productivity and Competitive Position* (Cambridge, MA: Massachusette Institute of Technology Center for Advanced Engineering Study, 1982), 30.
12. Lu, 73.
13. M. Scott Myers, *Every Employee Is a Manager* (San Diego: Pfeiffer, 1991), 72.
14. Lu, 87–92.
15. Richard J. Schonberger, *World-Class Manufacturing* (New York: Free Press, 1986), 67.
16. Kiyoshi Suzaki, *The New Manufacturing Challenge: Techniques for Continuous Improvement* (New York: Free Press, 1987), 43.
17. Maryann Keller, *Rude Awakening: The Rise, Fall, and Struggle for Recovery of General Motors* (New York: Morrow, 1989), 206–209.
18. Robert L. Shook, *Turnaround: The New Ford Motor Company* (New York: Prentice-Hall, 1990), 81–83.

CHAPTER TWENTY-TWO

Implementing Total Quality Management

"All glory comes from daring to begin."
   Anonymous

MAJOR TOPICS

- Rationale for Change
- Requirements for Implementation
- Role of Top Management: Leadership
- Role of Middle Management
- Viewpoints of Those Involved
- Implementation Variation among Organizations
- Implementation Approaches to Be Avoided
- An Implementation Approach That Works
- Getting On with It
- What to Do in the Absence of Commitment from the Top
- Implementation Strategies: ISO 9000 and Baldrige

This chapter is intended to serve three purposes.

1. To collect in one chapter some of the salient points made elsewhere in the book concerning the key elements and philosophy of total quality—a kind of summary.
2. To develop a logical "road map", or structure for implementation in order for the student of total quality to better understand the scope and magnitude of total quality implementation.
3. To provide a practical, hands-on, how-to guide for implementing total quality in any organization, in the sincere hope that this book will have inspired some who are in positions of leadership to take this next step.

Total quality management has been accurately described as a journey, not a destination. It is the fervent hope of the authors that many who study the material presented in this book will embark on that journey. This chapter is offered as your guide.

Five decades ago Japan was in a state of crisis. Japan's industry had been decimated by World War II, and its economy was in shambles. It was struggling to rebuild its economy and put people to work. This involved more than just getting the factories running again. Even if they could manage to get production flowing, who would buy the goods that were produced? The vast majority of Japanese people had all they could do to put clothes on their backs and food in their bellies. Japan had to look beyond its own shores for markets. The most obvious market was the United States, the economy of which had burgeoned during the war. The U.S. market posed two problems for the Japanese, however: convincing Americans that they should buy goods from the nation that attacked Pearl Harbor, and overcoming the American association of Japanese goods with poor quality. Before World War II, Japan had been notably unsuccessful in American markets because of the perception of poor quality in its goods.

Enter Dr. W. Edwards Deming, an American statistician who had been in Japan in 1947 at the request of the Supreme Commander of Allied Powers to help prepare for a census to be taken in 1951. He had met some of the Japanese people who formed the Union of Japanese Scientists and Engineers (JUSE). As JUSE wrestled with the problems confronting Japanese industry and the economy, they were introduced to the 1931 McGraw-Hill book *The Economic Control of Manufactured Product* by Dr. Walter Shewhart of Bell Laboratories, the originator of the control chart. From their acquaintance with Deming, they thought he might help them apply Shewhart's techniques. JUSE wrote Deming in March 1950, asking him to give a series of lectures to plant managers, engineers, and research workers. Deming gave his first lecture on June 19. Some 500 people attended. Not willing to invest his time on a lost cause, Deming insisted that the top executives of Japanese industry get involved. JUSE arranged for that first high-level meeting on July 5, 1950. The top 21 Japanese company presidents attended. Deming told them that they could compete in the world's markets within 5 years if they followed his teachings. They did it in *4* years.

This chapter sets the stage for implementation of total quality in any organization. Had Japan not been in such dire straits after World War II—industry in shambles, people needing jobs, the nation with no money with which to import food—perhaps people there would not have listened to and acted on Deming's recommendations. They were at the point of *seeking a route to survival*. Your organization may or may not be in a similar fix. When an organization is truly facing the possibility of going out of business, there is a better chance that its management can be convinced to embrace the principles of total quality. On the other hand, when an organization is doing pretty well, then taking on the work that is involved in becoming a total quality organization is more difficult to sell—unless you are at the top of the organization chart.

Change is always difficult, and changing a culture that has been ingrained for many years is a monumental undertaking. When change is seen as the last hope for survival, it gets easier. People are more receptive to change when they realize that they will surely be out of jobs unless change is made. Is it worth the trouble? Unquestionably. Is survival assured with change? No. But the other side of the coin is that going out of business is virtually assured if you don't change. Every enterprise, no matter what the type, will be pressured more and more as total quality pervades industry, education, health care, government, merchandising, and services. Managers should consider whether they would prefer to be ahead of the quality groundswell or engulfed by it—out of control, fighting for survival with the odds against success much higher.

This chapter provides insights to help you implement total quality. No one best way fits the needs of all organizations. What you will find in this chapter are not prescriptions but suggestions and examples of what has worked, with the idea that you may find the inspiration that will lead you to success in your own organization.

RATIONALE FOR CHANGE

What's wrong with the traditional way we do business?

1. *We are bound to a short-term focus.* If the organization of which you are a part is similar to most in the West, it is driven by short-term objectives. This is true whether you are in industry, education, health care, services, or government. For more than 50 years, we have been the victims of Keynesian economics. Everything we do has to have a measurable payback in the next quarter or the next year, or it cannot be justified. Whether Keynes had that in mind or not, it has become a reality of Western management and business. It is the sentiment, "Don't tell me how good it will be in 5 years. What are you going to do for me today?"

2. *The traditional approach tends to be arrogant, rather than customer focused.* Most Western organizations are arrogant. They think they know more about what their customers need than their customers do. Or worse yet, they don't care about their customers' needs. To illustrate this point, go into a typical government office and try to get something done—get new license plates for your car or have some legal papers executed. Often you will find that the employees, whose salaries come from your taxes, are rude, inefficient, and totally disinterested in you or your needs. A case in point is the now publicly acknowledged misdeeds of the Internal Revenue Service, which is currently trying to reinvent itself. The same thing has happened in industry.

3. *We seriously underestimate the potential contribution of our employees, particularly those in hands-on functions.* The person who knows the most about a job—and the one who is most likely to know how to solve problems—is the person who is doing that job and facing the job's problems day in and day out. This truth is proven over and over, yet the typical traditional manager does not believe it. This factor alone is responsible for much of the poor job performance and ill will that exists between management and labor, the folks who have to do the work. People generally want to do a good job; but faced with processes that are not capable and management that will not listen, they soon determine the only way to get ahead, or stay employed, is to "live with it and don't make waves." The result is that the brainpower we employ is largely wasted. Think about it: If you are in a 100-person organization, and only two or three people can make changes to the procedures you work by and the processes you work with, 97% or 98% of the idea potential and creativity is silenced—but you still pay for it.

Let us bring this point home. Konosuke Matsushita, the head of Matsushita, the giant Japanese company that produces electronic equipment under the Panasonic brand name, has written:

> We are going to win and the industrial West is going to lose out; there's not much you can do about it because the reasons for failure are within yourselves.
>
> Your firms are built on the Taylor Model. Even worse, so are your heads. With your bosses doing the thinking while the workers wield the screwdrivers, you're convinced deep down that this is the right way to run a business. For you, the essence of management is getting the ideas out of the heads of the bosses and into the hands of labor.
>
> We [Japan] are beyond the Taylor Model. Business, we know, is now so complex and difficult, the survival of firms so hazardous in an environment increasingly unpredictable, competitive and fraught with danger, that their continued existence depends on the day-to-day mobilization of every ounce of intelligence.[1]

Considering what Matsushita, Sony, Hitachi, and other Japanese consumer electronic firms did to the American competition, his remarks, while chilling, seem reasonable. The Japanese certainly won that battle, but the war is not over yet. Many Western organizations have also concluded (if belatedly) that our traditional management system (the Taylor Model) wastes brainpower in unthinkable amounts, and is no longer appropriate; they have adopted the total quality model. If yours has not, now is the time.

4.   *The traditional approach equates better quality with higher cost.*   Phillip Crosby wrote a book entitled *Quality Is Free.*[2] The title was probably intended to catch the potential buyer's interest with its shock value. When the book was published in 1979, not many traditional managers would buy the idea that quality is free. In the ensuing years, however, that title has proven to be *understated.* Organizations that have successfully changed themselves into total quality enterprises have found not only that quality is free but that it also brings unforeseen benefits. Sadly though, many traditional managers still believe that if you want better quality, you have to pay more for it. But the marketplace has found that if you want better quality, you simply pick the supplier that has demonstrated superior quality at the same price. That is why the Japanese cars have been so successful. Unfortunately, it is also the reason so many industries have been lost—radios, televisions, VCRs, stereo equipment, just to name a few. Better quality was to be had from other suppliers—for the same cost—and that is where the buyers went.

This issue of better quality from foreign competitors for the same cost is not a matter of lower wage scales in those countries. Wage scales in Japan and Germany are not that different from those in the United States, and it has been a long time since there were meaningful differences. In fact, the U.S. Labor Department reported in August 1995 that Japan's average hourly wage surpassed the U.S. rate in 1993. The Japanese rate was $19.01 including benefits, such as health insurance paid by the employer. The U.S. average for the year: $16.73 an hour. In addition to that, labor costs represent roughly only 10%–15% of the cost of manufactured products and can be significantly less than that in factories with automation. When a Japanese company is able to produce a $15,000 car that is of higher quality than an equivalent domestic offering, it is simply because that company has embraced total quality methods and honed them over 30 to 40 years.

QUALITY TIP

Listen to Employees

From her research on team building, Purdue University professor Joan Chesterton says of factory workers, "I've found that many of them are highly creative individuals who like what they are doing and are the very heart and soul of their organizations." Chesterton has become convinced that, regardless of education, all employees can contribute to their organizations, and managers who think they are smarter than their blue-collar employees are acting counterproductively.

Source: Associated Press, September 13, 1995.

5. *The traditional approach is short on leadership and long on "bossmanship."* Far too many Western managers see their jobs as simply telling subordinates what to do and when to do it. It is their station in life to make sure that the procedures are followed, that quotas are met, and that no one makes waves. It is easy to be critical of this kind of "leadership," but for 90 years, it is what we have been taught. It is a product of mass production, springing out of Henry Ford's Highland Park assembly line in 1913 and being adopted in one form or another by just about every kind of production activity. What exactly did Ford do? Prior to 1908, all automobiles were manufactured in craft shops. In North America, Europe, wherever, craft production was how things were made. All the people engaged in the making of an automobile were skilled craftsmen. All parts had to be hand-fitted by filing, cutting, or shaping. No two of anything coming out of a craft shop were the same. Ford realized that if he could make parts interchangeable, thereby eliminating the filing and bending, he could produce his cars a lot cheaper—and achieve unit-to-unit consistency in the bargain.

For example, in 1908 a Ford assembler/fitter (notice the latter designation) spent 514 minutes to complete his task before repeating the same steps on the next car. His work included getting the parts, filing or shaping to fit, and bolting them on and adjusting and aligning as necessary. It also included maintaining his tools. These were multiple tasks—tasks requiring craftsman's skills. At about that time, Ford finally achieved perfect part interchangeability. Ford assemblers then went to a single task, with the cycle time dropping to 2.3 minutes—the assembler's assigned task took only 2.3 minutes to complete before the assembler was ready to repeat it again on the next car. Productivity went up in a dramatic fashion. Having to do only one simple task over and over meant that the assembler (he was no longer a fitter) got to be an expert at it very quickly. Ford took it to the next step in 1913 with his introduction of the moving assembly line. Now the assembler no longer had to move. The work came to him. Assembler cycle time dropped to 1.19 minutes.

We cannot give Ford all the credit for part interchangeability, because Cadillac apparently beat him to that goal by 2 years, achieving it in 1906. But Ford must be credited with the moving assembly line and what has been called the *interchangeable worker*.[3] Ford no longer needed the craftsmen. He could hire unskilled assemblers direct from the farm or immigrants who couldn't speak or read English. Within just a few minutes, they would be as expert as the assembly line demanded. This division of labor down to its simplest terms paid big dividends for Ford and for society in general. For the first time, the possibility of owning an automobile was not restricted to the wealthy. When the Model T was introduced in 1908 with its interchangeable parts, it cost far less than competing cars. In the early 1920s, Ford's interchangeable workers produced 2,000,000 identical cars each year, and the cost was reduced by another two-thirds. Ford's production techniques soon found their way into virtually all manufacturing activities in North America and Europe.

Mass production had arrived—and with it the elimination of skills. Soon, industry found ways to divide labor in other areas to minimize the need for worker skills and knowledge. We called it *specialization*. In this kind of environment, all you needed were simple work instructions, the right tool, and the requisite muscle to turn it. Follow the instructions—over and over and over again. Don't improvise; don't make waves; just follow orders. Supervisors and managers have been trained in this system for 90 years. It worked, at least for a while. For the last half of that 90-year period the Japanese have demonstrated a better approach.

REQUIREMENTS FOR IMPLEMENTATION

Some parts of your organization are concerned that the future holds little promise of prosperity unless fundamental changes are brought about. Perhaps your competition is taking market share. You know that your product quality is not good enough. There is strife within your firm, bickering among departments, endless "brushfires." The total quality approach is working for others. Maybe total quality is what is needed. What has to happen for total quality to take place? What are the requirements for its implementation?

Commitment by Top Management

First and foremost, for total quality to become *the way we do business*, an unwavering and unquestioned commitment is required at the top. The CEO, general manager, or whatever title the top person has, must commit not only resources but a considerable amount of his or her own time. Top executives should plan on a third to half their time being used in the total quality effort. Certainly less than a quarter of their time is not sufficient. Some say, "But the president is so busy, why can't he delegate the implementation?" Neither from personal experience nor from the recorded experiences of dozens of companies of which we are aware is there a single success story of a delegated total quality implementation. People expect the boss to put his or her efforts on the most important issues. If they don't see that effort as being total quality, the subliminal message is that total quality is not number one. Some departments will press on—for a while—until they get at cross purposes with other departments that are marching to the beat of a different drummer. Who has the authority to solve the impasse? The boss, and he or she is not involved in the process.

For an organization to completely embrace total quality from the mailroom to the executive office, a profound change is required in the corporate culture. Changing a culture is very difficult even when everyone is willing, and it is almost never the case that everyone will be willing. Some see danger in change, danger to their personal position, the threat of loss of power or prestige, perhaps even loss of employment. Some just like everything the way it is and see no reason to change. Some will be unwilling to put in the work required. Some cannot believe that total quality makes sense. If the message from the top is not crystal clear, and if the person at the top is not seen as being totally involved, that will be all the encouragement some will need to "toss wrenches into the gears."

But there is another reason the person at the top must be involved: the change to total quality is a learning experience. If the boss is not involved in it day to day, he or she will never know enough about what is happening to make rational decisions affecting the change. For example, suppose department heads have been meeting over the course of a month or two, wrestling with the issue of the organizational structure needing change to accommodate total quality. These people have aired the problem, perhaps developed suggestions for change, and generally understand the issue. However, a change as far-reaching as creating a new organizational structure is beyond the scope of the department heads. Only the person at the top can do that. How do the department heads get that person to understand everything that has happened in these meetings? They have been at it for weeks. A 1-hour briefing is not going to get the boss up to speed. In this setting, the naysayer's impassioned plea for the status quo takes on a credibility that would have been impossible if the boss had been involved in the meetings from the start. The boss hears from one side that the proposed change must be put in place if total quality is ever going to provide maximum benefit. From the other side, he or she hears that the proposed change would be disruptive at best, and possibly disastrous. The span of control will be too wide, allowing things to drop through the cracks. Perhaps the system currently in place is not perfect, but at least it is familiar—and it works. What would you do? If you were the boss and heard these arguments, would you risk the company and make the change? The easy thing to do is do nothing: Tell the department heads that you understand where they are coming from and that maybe sometime later it will be an appropriate thing to do, but in the meantime, they'll have to figure ways to work around the structure.

What happened in this example goes on all the time. The boss is given a briefing from which he or she is expected to know as much as the briefer. It cannot be. The briefer has been directly involved in weeks of discussion and has the benefit of long and thoughtful consideration and deliberation. The boss got a few minutes of encapsulated data and has had no opportunity to consider them. Should he or she decide against the change, the decision will make an immediate and lasting impact on the proponents. They won't make that mistake again. Wasted weeks of effort, and for what? Only to be told that the boss thought the organization had better stay the way it is and make the best of it. Total quality will probably come to a screeching halt then and there. Does this happen in the real world? Yes, it does!

Commitment of Resources

The other part of the commitment is resources. Total quality implementation need not be expensive, but everything has a cost. In this case, the cost will certainly include some training. It may also include some consultant expense. The dollars must be there when they are needed. The difficulty is that it will not be easy to project a payback; so many factors can affect a company's performance that it may be impossible to know with certainty that X dollars invested in training yielded Y dollars in performance gains. This area conforms to Deming's truth that some things are not measurable. Accountants don't like to hear that.

The test for commitment of money should be one of reasonableness. Does it make sense to do this? Is the timing right? Is the money available? Can we afford it? Is it the right thing to do? If the answer to these questions is yes, you should not worry unduly about trying to capture the payback. Chances are good that it would cost more to figure out what the payback should be than the project itself will cost, and you can never be certain of the data.

Organization-Wide Steering Committee

The third thing needed for company-wide implementation is a top-level steering committee. It may be called by a number of names, but it should be chaired by the person filling the top position in the structure, and its membership should be comprised of that person's direct subordinates. In a typical corporate setting, this would be the president as chair, with all of the vice presidents filling the membership. The function of this group is to establish how total quality is to be implemented and then to see that it happens. As the conversion process starts, it will be necessary to set up cross-functional teams, to establish the teams' objectives, and to monitor results. Ultimately this group will find itself operating as a team, rather than just as the staff. It will set the vision and goals for the organization, establish teams to pursue the goals, monitor the teams' progress, and reward them for their achievements. The important point, from the outset, is that implementation requires management. Otherwise, it can easily set off in too many directions at once, some of which may not even be in concert with the company's objectives. This cannot be allowed to happen. The energy that is going to be unleashed throughout the organization must be channeled. The steering committee does that.

Another important aspect of the steering committee is symbolic. If the employees observe the top management group functioning like a team and doing things differently from the way things used to be done, they will get a strong message that this time something really is happening. If, on the other hand, they see the staff operating just as they always have, they will know that failure is simply a matter of time. Why bother to get involved? Do not minimize the difficulty of doing this. The typical staff is made up of stars, not team players. They have insulated their respective departments with walls that can defy all efforts to penetrate them. Their interests usually lie in their own departments rather than in the long-term vision and objectives of the company. What is worse, they don't have a common language—having backgrounds as diverse as engineering, finance, management information systems, human resources, quality assurance, manufacturing, purchasing, and so on. In many cases, they do not trust each other. Is it any wonder that we have problems? The person with the biggest challenge is the one who has to forge this crowd into a cohesive, mutually supportive team. But it must be done. The upside is that almost invariably, once they really start to function as a team, staff members will never want to go back to the old ways again.

Planning and Publicizing

So far we have secured commitment from the top and established the steering committee. At this point, the real work has only begun. We've just said we're going to do it and determined who is going to manage it. Now we have to get down to the details. The steering committee must develop the vision statement and guiding principles, set the goals and objectives, put the TQ implementation plan in place, and then develop an awards and recognition program and other publicity efforts. All these matters will be discussed next.

Vision Statement and Guiding Principles

Where would the organization like to be 5 or even 10 years down the road, and what are the guiding principles for operating the business? The vision statement is a long-range strategic view. Total quality needs a long-range vision because total quality is only achieved over a relatively long period, although there will be visible improvements practically from the outset. We are really talking about fundamental changes in the way we do things, how people work together, involving customers and suppliers in ways never before considered, and putting values on matters that may never have been discussed. Not everything will come together overnight, so the vision must be of a distant target to provide a consistent course into the future. Without that, the company will find itself taking turns and detours with every new quarter or year. That will destroy the effort. Consistency is the watchword.

The vision statement need not be lengthy—in fact, the shorter the better. But it must represent the best collective thoughts of free and open discussion by the steering committee. If your organization is part of a larger entity (such as a division within a company) that has a vision, then you need only tailor yours to support that one. The total quality vision statement will usually include a recognition that only the customers make the final judgment of success or failure. If not stated in words, that idea must be implicit. Sample vision statements from a variety of businesses are found in Figure 22–1.

The guiding principles are the second element of the vision, and usually accompany the vision statement in a single document. The guiding principles establish the rules of conduct for the organization and its members. These principles may be concerned with honesty, ethics, respect, fairness, quality, suppliers, customers, community, environment, roles of management and employees, and so on. This can sound very lofty indeed, and that is not a bad thing. People want to be associated with organizations with lofty ideals. They want to be proud not only of their own contribution but also of the company. Sample guiding principles are listed in Figure 22–2.

Ford Motor Company:
To be the world's leading automotive company[a]

Maytag Corporation:
The Maytag Corporation's vision for the future is to be recognized as designing, building, marketing, and servicing the best appliances in the world.[b]

The Boeing Company:
People working together as one global company for aerospace leadership.[c]

Northwest Florida Daily News:
The ***Northwest Florida Daily*** news will be recognized by its customers as the region's news and information source of choice, and will promote the spirit of freedom in all communities it serves.[d]

Figure 22–1
Sample Vision Statements from Large and Small Companies
a. Source: Ted Derwa, "Reengineering Ford Motor Company," a presentation of Ford Motor Company at the 1995 Florida Governor's Sterling Awards Conference, Orlando.
b. Source: Maytag Corporation, First Quarter Report, 1995.
c. Source: The Boeing Company, 1997 Annual Report.
d. Source: *Northwest Florida Daily News*.

A well-written vision statement with its attending guiding principles has the following properties:

1. Easily understood by all stakeholders (employees, customers, suppliers, and others)
2. Briefly stated yet clear and comprehensive in meaning
3. Challenging yet possible to accomplish, lofty yet tangible
4. Capable of stirring excitement and unity of purpose among stakeholders
5. Sets the tone for how the organization and its employees conduct their business
6. Not concerned with numbers

The vision statement must be crafted in such a way that all employees can relate to it, and, in so doing, execute their work in a manner and direction that are consistent with its meaning and objectives.

Goals and Objectives

The broad strategic goals and objectives established by the steering committee must harmonize with the vision statement. These goals and objectives are for the total organization, rather than necessarily aimed at the individual operating departments. They flow from the vision statement and are frequently part of the organization's strategic plan. To achieve the vision, these are the objectives that must be achieved. From these goals and objectives, supporting specific tactical objectives will be developed for departments, teams, and even individuals. The vision points the company in the desired direction and girds employees with the principles they must use in pursuit of the vision. The broad goals and objectives represent the strategic targets along the way

---

IBM
- The marketplace is the driving force behind everything we do.
- At our core, we are a technology company with an overriding commitment to quality.
- Our primary measures of success are customer satisfaction and shareholder value.
- We operate as an entrepreneurial organization with a minimum of bureaucracy and a never-ending focus on productivity.
- We never lose sight of our strategic vision.
- We think and act with a sense of urgency.
- Outstanding, dedicated people make it happen, particularly when we work together as a team.
- We are sensitive to the needs of all employees and to the communities in which we operate.[a]

Manufacturing Technology, Inc.
- Customer satisfaction is MTI's primary objective. Meeting the expectations of our external and internal customers is the primary task of each MTI employee.
- Honesty and integrity are cornerstones of the MTI culture. MTI will always conduct its business with the highest standards of ethics.
- Management will provide the company's vision for the future, and the leadership to steer the course, while facilitating MTI employees by providing necessary training, tools, and a creative environment.
- Employee involvement will be a fundamental cultural feature of MTI's operation. All MTI people will be personally involved as team members and as individuals in establishing and achieving our goals.
- Continuous improvement is a primary business objective at MTI. This philosophy will be applied to all our products and services and to the processes and systems that produce them.
- Total Quality Management principles will be applied throughout MTI's operations.[b]

Figure 22–2
Sample Guiding Principles from a Large and a Small Business
a. Source: IBM, 1993 Annual Report.
b. Source: Manufacturing Technology, Inc.

to achieving the vision. Finally, a lower tier of specific tactical objectives describe what must be done as the company goes about achieving broad objectives and the vision. At both levels, objectives should be stated relating to total quality implementation. A word of caution: don't try to include every possibility and contingency. A few well-crafted goals are what you want. It may be that not all of your goals are measurable, but all should be defined such that you at least know when a goal has been achieved. See Figure 22–3 for the hierarchy and Figure 22–4 for sample objectives.

Total Quality Implementation Plan

The plan is driven by the vision, goals, and objectives. It spells out as precisely as possible the route the implementation will take. No two total quality implementations will be the

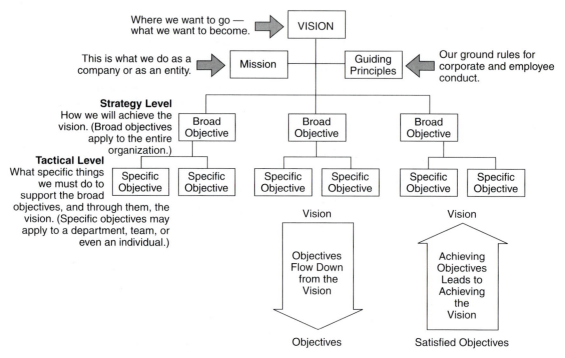

Figure 22–3
Hierarchy of Vision and Objectives

same. Your own organization, after considering your vision and objectives, studying the material, perhaps consulting with someone who has been there, and deliberating among the steering committee members, is best equipped to chart its own course. You may want to set some pilot projects in two or three departments. Proceed slowly and monitor closely. Another organization may establish the total quality initiative by setting teams to work at understanding functional processes. A manufacturing organization may start by introducing just-in-time concepts on the assembly floor—probably in pilot programs at first. At this point, everyone is learning. Don't fret about the mistakes, but do learn from them. Soon a couple of successes will be achieved, and then things will tend to fall into place more readily.

One thing the implementation plan must provide for is training. Before the top executive and the steering committee can function as a total quality team, they will require training. This can be obtained through a variety of sources: seminar courses are available, a consultant could be brought in, or self-teaching is possible (but is probably the least desirable approach). After the steering committee has been trained, it will be ready to start its work. Before the first pilot program is initiated, however, the people who are to be involved must be trained. That degree of training is usually minimal—perhaps as little as a half day. Don't let them venture forth without some training.

Figure 22–4
Sample Broad and Specific
Objectives
Source: Maytag Corporation, First
Quarter Report, 1995.

Maytag Corporation

- **Broad objective** (Strategy):
  Premium brand leadership

  Specific objective (Tactic):
  Extend the product offerings of our Maytag and Jenn-Air
  brands.

- **Broad objective** (Strategy):
  Continuous reinvestment in new product introductions and
  innovation.

  Specific objective (Tactic):
  A new Maytag dishwasher.
  A new Hoover deep carpet cleaner.
  A new Jenn-Air laundry line.

A common mistake in U.S. industry has been to go into an across-the-board training program before total quality is introduced. This is wrong on several counts. First, it is a very expensive approach. Second, not everyone needs the training at one time. Third, most of those trained will have forgotten what they learned before getting a chance to apply it. It is better to train only as needed, a kind of just-in-time approach to training.

Many companies have successfully used the approach of cascading training. First the steering committee is trained. They go off and do their planning to get ready for total quality. Then just prior to the first implementation, members of the steering committee train the total quality project leaders, who in turn train the team members.

Awards and Recognition Program

How will you recognize team achievement? An awards and recognition program should be in hand before the implementation starts. The typical award/recognition program found in the United States is out of step with the total quality concept. Our society is oriented toward individuals. Our award programs recognize individual achievement. Even in team sports that is true. Consider the Heisman trophy. There may be a *best college football player* out there, but unless he is a member of a team that supports his efforts in superlative fashion, he will never win that trophy. Virtually all team sports, amateur or professional, have most valuable player awards. Many manufacturing plants have similar programs to recognize exceptional performance by individuals who are nominated by their peers. For a time, this approach was well received, and perhaps it still has its place. But as total quality started to be a way of life, people began to view such systems as counterproductive. When trying to do things through teams, don't create superstars. We are not suggesting that exceptional individual accomplishment should not be recognized, but the focus of the reward system should be on the teams.

These award and recognition programs usually operate on two levels. At one level, the employee review establishes pay raises, frequently associated with an annual or semiannual performance review. The other is like the one mentioned earlier: a kind of spot award for having achieved something special. Both now need to be oriented toward the team rather than the individual. It is very difficult to do that in the former case. This is an area of continuing study in many U.S. firms. The obstacle is that as individuals, we do not like to think we might be penalized for actions that are not our own. Of course, we would not complain if the team carried us above our own limitations. The perceived problem is far smaller than it appears; it is not likely that any individual will be penalized because he or she happened to be assigned to a team of poor performers. Teams themselves tend to weed out the poor performers far more effectively than management ever could. Still, not many American or Canadian workers are paid on the basis of their teams' performance. Until that happens, we are sending the wrong message to them. This will be a good project for the human resources department.

Spot awards are easily set up to recognize team accomplishments. These need not be extravagant. Some companies provide a dinner for two to each of the winning team's members. Movie tickets are also popular. Some use cash awards. Some simply stress recognition, usually in a forum where all peers are present. Some companies have periodic awards banquets or similar affairs. The possibilities are endless, but employees must see the award process as being fair and equitable, and in tune with the organization's stated intentions

Publicity Approach

An approach for publicizing total quality activities and results is important. All the employees need to know what is going on—all the time. Every employee survey of which we are aware has placed communications at or very near the top of the problem list. It used to be that we tried to give employees *only as much information as they needed.* Now we know that if there is information, virtually all employees need it. They may not act on it directly in their daily job function, but not providing sufficient information to employees with which to make life's decisions is an unconscionable lapse in caring about people. In addition, only when employees are fully informed can they understand many management decisions and consequently support management when they otherwise might be hostile. When it comes to total quality implementation, every employee will be affected in dramatic ways. It is absolutely essential that they understand what is going on and why. There should be no sugarcoating; tell it straight. Let employees know ahead of time what is planned and how they will be affected. When results come in, good or bad, let them know about it. It will be far better to admit a failure in a project than to let the employees find out about it through the underground communications that *always* fill an information void. The failure can be used to demonstrate that there will be no recriminations.

Almost limitless possibilities exist for communicating: a company newspaper or newsletter, all-employee meetings, videos playing in the lunchroom, special total quality bulletins, the intercom system, and so on. Some companies make a very big deal of it and hold picnics, fairs, or even circuses for the express purpose of communicating about total quality. Pick your methods, and use them vigorously.

Infrastructure That Supports Deployment and Continual Improvement

With commitment from the top, a high-level steering committee, a set of plans for at least the early phases of implementation, and the means of providing the required training, the only thing we may lack is the infrastructure that will support the deployment of total quality throughout the organization and continual improvement on a never-ending basis. Actually, most of what we talked about in the previous section—the vision and its harmonized objectives, the awards and recognition program, and certainly communications—can be considered a part of the supporting infrastructure. Three other infrastructure features can offer support for your total quality implementation efforts: your operating procedures, organizational structure, and union situation—or they can get in your way. These considerations will be discussed next.

Procedures

Virtually all organizations should operate in accordance with published procedures. You will undoubtedly find that many of your procedures, having been developed in another culture, do not support total quality, and represent an opportunity for improvement. Be aware of this always, and never accept the old excuse, "We've got to do it this way because that's what the procedures say." Don't buy it. If there is a better way, change the procedures.

Organization

The typical Western organizational hierarchy does not fit with total quality (see Figure 22–5). Such organizations are arranged in departments that effectively raise all kinds of barriers to efficient operation. Communication among them is only the most notable of the problems. As total quality implementation progresses, you may find it necessary to alter your organization; it is an absolute certainty that you will run into walls that have been built up around the departments over the years.

There are a number of ways to eliminate the walls. Some firms have simply designed new organizational structures. Others develop minicompanies oriented toward specific products or customers by drawing cross-functional talent from the functional "home" departments—the engineering department or the accounting department, for example (see Figure 22–6.). Both approaches seem to work well. All members of the team (whether 6 or 600) must see their prospects tied to the project team, not to the home department. This is what "organizational reengineering" is about.

Union Considerations

In organizations with organized labor, the union is an inherent part of the infrastructure. The natural bent of the labor unions makes it difficult for them to accept the changes necessary for maximum benefit of total quality. Unions as we know them today are

Figure 22–5
Typical Traditional Organization

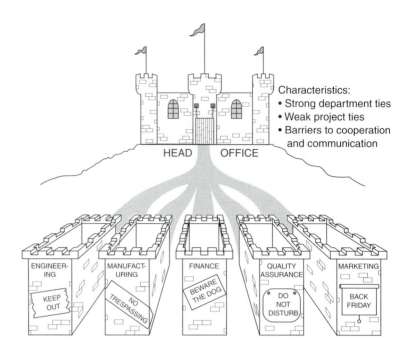

Characteristics:
• Strong department ties
• Weak project ties
• Barriers to cooperation and communication

HEAD OFFICE

ENGINEER-ING — KEEP OUT
MANUFACT-URING — NO TRESPASSING
FINANCE — BEWARE THE DOG
QUALITY ASSURANCE — DO NOT DISTURB
MARKETING — BACK FRIDAY

another product of the mass production era. Just as much of U.S. and European management is out of step with the realities of today's worldwide business environment, so are the unions. Fundamentally, the unions embrace the concept of division of labor to a fault. Henry Ford's assembler of 1913, tightening the same two bolts over and over again—never touching the third bolt—epitomizes the union's view of the worker. By

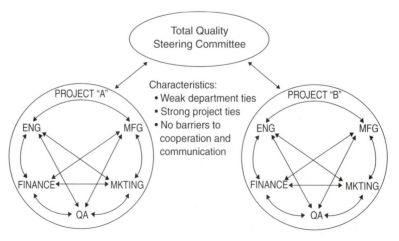

Total Quality Steering Committee

PROJECT "A"

ENG   MFG

FINANCE   MKTING

QA

Characteristics:
• Weak department ties
• Strong project ties
• No barriers to cooperation and communication

PROJECT "B"

ENG   MFG

FINANCE   MKTING

QA

Figure 22–6
Total Quality Organization

making sure management could not require him to service bolt 3, the union guaranteed a job for another assembler. And this is not confined to the assembly line. In an engineering laboratory of an electronics firm with organized labor, an engineer is forbidden to place a scope probe on a circuit test point. That is the job of the union's technician. So the engineer tells the technician where to probe and looks over the technician's shoulder to see the response on the oscilloscope screen. Does the engineer need the technician to do this probing? Not at all. Does the technician in this role add value to the process? No. Would they both prefer to work under different rules? Certainly the engineer would. But the union assures the technician of a job where one might argue none would exist without the union.

These kinds of work rules are incompatible with total quality. Division of labor becomes a thing of the past. People are asked to do multiple tasks and to consider matters that would never have been brought up before—such as, "Is this part perfect?" They are expected to stop the process when a problem occurs. Their ideas and suggestions are not only solicited but expected. Work is done in teams, and people are shifted from function to function to pick up peak demands or to fill in for someone who is absent. These are very difficult concepts for the unions.

But over the past few years unions have been coming around to the point of tolerance, at least. In shops where these techniques have been successfully applied, such as the Japanese automakers' transplants in the United States, their unions have become supporters of total quality. An illuminating example is the NUMMI plant in Fremont, California, which is jointly operated by Toyota and General Motors. The Fremont plant had been operated earlier by GM, and when GM closed the plant in 1982, it had the worst record in the GM family—an unmanageable workforce, rampant absenteeism, and a quality record that was an embarrassment. When the plant reopened in 1984 under joint management by Toyota and GM, the same equipment was in place, and the workforce was composed of the same people who had been laid off a few years earlier. But everything else was different. The United Auto Workers had signed a contract that permitted the plant to be run according to the Toyota model. Work was to be performed by teams. Employee participation was expected. Division of labor was a thing of the past. NUMMI had four job classifications in contrast to some GM plants with 183.[4] The UAW went along with the rules changes in return for a no-layoff policy. Everyone, it appears, turned out winners—the workers, the companies, the union, and, most important, the customers. In a very short time, the cars rolling off the NUMMI line, Toyota Corollas and Geo (now Chevy) Prizms, were receiving the top-quality ratings in GM—virtually defect free. See Figure 22–7A, "Initial Quality 1995 Cars," and note that the NUMMI product, the Prizm, had the best initial quality rating of all cars produced in the United States or Europe, and it was only marginally led by five very expensive Japanese models. Lest you attribute this to a fluke, the Prism, and sister car Corolla, have consistently been rated at or near the top of the J. D. Power and Associates studies. The 2000 study rated NUMMI's Corolla/Prism as the best compact car. (See Figure 22–7B.) J. D. Power and Associates also rates plants for product quality, and of all North American plants, NUMMI Fremont was second only to the Toyota Georgetown facility. (See Figure 22–7C.)

The absence of a union shop might make total quality implementation easier, but the presence of a union is certainly no excuse for living with the status quo. Unions

Figure 22–7A
Initial Quality, 1995 Cars
Source: J. D. Power and Associates.

Top 10 Car Models, 1995

Car	Problems per 100 Vehicles
1. Honda Prelude (Japan)	48
2. Infiniti J30 (Japan)	48
3. Lexus SC 300/SC 400 (Japan)	49
4. Acura Legend (Japan)	50
5. Lexus LS 400 (Japan)	51
6. Geo Prizm (U.S.)	56
7. Infiniti G 20 (Japan)	62
8. Volvo 940 (Sweden)	62
9. Honda Accord (U.S. and Japan)	63
10. Cadillac Deville/Concours (U.S.)	64
Industry average	103

Best overall car line in initial quality: Infiniti

Best in Their Price Class

Price Class	Car	Problems per 100 Vehicles
$13,000 and under	Mercury Tracer	95
$13,001–$17,000	Geo Prizm	56
$17,001–$22,000	Honda Accord	63
$22,001–$29,000	Honda Prelude	48
$29,001 and over	Infiniti J30	48

have found that the benefits outweigh the disadvantages and may even be able to help in the company's quest for total quality. The union must be a part of the team. In fact, many companies place the top union official on the total quality steering committee.

ROLE OF TOP MANAGEMENT: LEADERSHIP

Every organization must have a leader. That is what we pay our top managers to be. Yet our expectations are infrequently met. For some inexplicable reason, Westerners have difficulty defining leadership. Leaders are described as people that command or guide a group or activity—not a very illuminating definition. In management circles, the debate goes on about who should be called a leader, or if so-and-so has leadership ability. Somehow the point has been missed: If a person is in charge of any group, that person must lead.

From this perspective, every supervisor, manager, director, vice president, president, and CEO must be a leader. The problem is that most of them have never been told what their job is, and few figure it out for themselves. Their job is to *lead.* You may be think-

Figure 22–7B
Initial Quality, 2000 Vehicles
Source: J. D. Power and Associates.

Top Three Vehicles Per Segment

Car Segments	Truck Segments
Compact Car	**Compact Pickup**
Best: Toyota Corolla	Best: Mazda B-Series
Chevrolet Prizm	GMC Sonoma
Mitsubishi Mirage	Chevrolet S-10 Pickup
Entry Midsize Car	**Full-Size Pickup**
Best: Plymouth Breeze	Best: Toyota Tundra
Chevrolet Malibu	Ford F-150 Light Duty
Dodge Stratus	Ford F-250 Super Duty
Premium Midsize Car	**Mini Sport Utility Vehicle**
Best: Toyota Avalon (tie)	Best: Honda CR-V
Best: Toyota Camry (tie)	Toyota RAV4
Nissan Maxima	Subaru Forester
Sporty Car	**Compact Sport Utility Vehicle**
Best: Acura Integra	Best: Nissan Pathfinder
Honda Prelude	Mercury Mountaineer
Dodge Avenger	Toyota 4Runner
Entry Luxury Car	**Full-Size Sport Utility Vehicle**
Best: Lexus ES 300 (tie)	Best: Toyota Land Cruise
Best: Acura TL (tie)	Ford Expedition
Infiniti I 30	Chevrolet Suburban GMT800
Premium Luxury Car	**Luxury Sport Utility Vehicle**
Best: Lexus LS 400	Best: Lexus LX 470
Acura RL	Infiniti QX4
Lexus GS Sedan	Lexus RX 300
Sports Car	**Compact Van**
Best: Porsche 911	Best: Toyota Sienna
Honda S2000	Plymouth Voyager
Chevrolet Corvette	Dodge Caravan

ing that this is rather obvious. It is not. The way our society has traditionally organized its enterprises, the farther up the organizational structure you climb, the less leadership is demanded. When new floor supervisors in a factory have just been promoted from a position on the line, they intuitively know that they have to help the less experienced people, and usually they do so. That is a major role of leadership. On the other hand, when a company president becomes the CEO, he or she is likely to spend less time in a leadership role in the company, making it more competitive, than in lobbying Washington to do something about "unfair foreign competition."

Any person in charge of any group is a leader. That person may or may not be an *effective* leader. Too many times he or she is not. But someone put the person in a

Figure 22–7C
2000 Vehicle Assembly Plant
Awards
Source: J. D. Power and Associates
2000 Initial Quality Study (SMI).

Based on Vehicles Produced for U.S. Market	
Platinum Award	Toyota—Kyushu, Japan (Car)
North America	Gold: Toyota Georgetown (Car)
	Silver: NUMMI Fremont (Car)
	Bronze: Toyota Cambridge (Car)
Europe	Gold (tie): Jaguar—Browns Lane (Car)
	Gold (tie): Porsche—Stuttgart (Car)
	Bronze: BMW—Dingolfing (Car)
Asia	Silver: Toyota—Tahara (Car)
	Bronze: Honda—Sayama (Car)

leadership position—it did not happen by self-proclamation. Who did it? Management, of course. Which brings us back to the question, What is a leader? What skills, what natural abilities, what aura are prerequisites of leadership? While some individuals seem to possess qualities that are useful in leadership, we have all known excellent leaders who did not look the part. It helps to be attractive and to have charisma, but these traits are not necessary, nor are they a guarantee of leadership ability. So what is it that separates the good leaders from the ineffective?

1. *Leaders pull rather than push.* They are out in front leading the effort, not back in the office (foxhole) yelling, "Charge!" Wess Roberts says it well: "A Chieftain can never be in charge if he rides in the rear."[5] What this really means is that the leader is *visibly involved in the effort he or she is leading.* The antithesis of this is the general manager who gathers the people together and says, "We are going to be a total quality organization, and I'm putting Jim in charge of getting it done." That is abdication of leadership. A corollary to this definition of leadership is: If you're not personally involved in an effort, you cannot be leading it.

2. *Leaders know where they want to go.* They set the vision for their organizations and chart the course to achieve the vision. Moreover, they make the choices of how to achieve the vision and *stick with it.* They provide the constancy of direction and purpose necessary for success in the long run. They keep their eyes on the prize and are not buffeted by the tempests that are often confronted. A second corollary is: If you don't know where you're going, you cannot lead the expedition.

3. *Leaders must be courageous and trustworthy.* There are snares and obstacles along any new pathway. Leaders cannot turn back every time they encounter one. If the goal is worth going after, they must stay the course—even at the risk of enduring hardship along the way. Short-term objectives must be sacrificed if they become obstacles to achieving the vision—and in our society, that takes courage. In similar fashion, leaders can be trusted to come to your aid when one of those obstacles springs up in your area. You have to know beyond any doubt that if, when following their marching orders, something goes wrong, they will be there to protect you and

QUALITY TIP

The Aim of Leadership

"The aim of leadership should be to improve the performance of man and machine, to improve quality, to increase output, and simultaneously to bring pride of workmanship to people. Put in a negative way, the aim of leadership is not merely to find and record failures of men, but to remove the causes of failure: to help people to do a better job with less effort."

Source: Dr. W. Edwards Deming, cited in Mary Walton, *Deming Management at Work* (New York: Perigee, 1991), 237.

won't "hang you out to dry." When people see courage flagging or subordinates taking the heat when things get tough, they will vote the leader out of the leadership office regardless of "rank." The third corollary is: If you don't have faithful followers, you cannot be a leader.

4. *A leader's most important role after forming the vision and setting the course is helping people to do their jobs with pride.* This is about training and nurturing. It is about giving employees the necessary tools, both physical and intellectual. It is about encouraging when something is difficult and praising when something is accomplished. It is about, as the army commercial says, helping your people to be all that they can be. The role of the leader is not to dictate but to facilitate. It is not to know it all, for you cannot, but to find out what others know and put it to use. It is not to "keep people in line" but to involve them to the fullest extent of their capabilities. The fourth corollary is: A group that is not trained and equipped for a task cannot be led to accomplish it.

ROLE OF MIDDLE MANAGEMENT

The middle manager is not in a position to initiate the kind of cultural change required by total quality. The middle manager must deal with the facilities, equipment, and processes put in place by higher management. He or she must operate within budget constraints for training self and subordinates. The middle manager is to a greater or lesser extent stuck with the infrastructure established by higher management. The manager in the middle cannot commit company resources, cannot establish the corporate vision, and cannot set up recognition and publicity programs. Nor can the middle manager arbitrate interdepartmental friction. These are the very reasons that no successful total quality program can exist without the full backing and involvement of the top levels. This is not to say that the middle manager (and let's include all levels between the hands-on workers and the department heads) has no role to play in total quality; far from it. These are the people who will carry the brunt of the work as the path to total quality unfolds.

From personal experience, and from the experience of others, it is clear that the middle management levels present the greatest obstacles to success in total quality. It always seems easier to sell total quality to the top managers and the hands-on people than to the middle managers. There are several reasons for this:

■   A good many people in these positions have been there a long time, recognize that they will progress no further, and see total quality's sweeping changes as threats to maintaining their status quo. Many times their insecurity is well founded, because Western hierarchies typically have far too many layers, and total quality makes that obvious.

■   Many middle managers moved into those positions after long apprenticeships in the hands-on level. They feel that they know more about their subordinates' jobs than the subordinates themselves. One of the basics of total quality is that the expert in any job is the person doing it day in and day out—not the one who did it 10 years ago.

■   Most of the middle managers came up doing only what they were told to do, making no waves, playing the company game. That they were successful at it is demonstrated by their eventual promotions to their current stations. They may really believe that is the way it should be.

■   Middle managers as a group tend to study less than managers at the top. Often the events that are reshaping the world's industry pass them by, they simply do not know what total quality is about. People seldom support any concept they do not understand.

There are also many bright, forward-thinking people in the middle manager levels. These are the ones who will become total quality leaders, while the others fade away. The people in this category will take on a role similar to the top managers—the role centers on leadership. We can take the previous section, "Role of Top Management"; strip away functions that apply only to top managers, such as creating the vision, broad objectives, committing resources; and apply the rest directly to the middle manager. The middle manager must facilitate his or her people to do their jobs better, easier, and with increased satisfaction. He or she must help, teach, encourage, praise, and, most important of all, *listen to* these people. He or she must build trust and work for the success of the team.

Middle managers will often function as project team leaders, seeking to define and characterize processes, and finding ways to improve them, or to take on a wide variety of special total quality projects. With their teams, they will find new ways to do things and new things to do. They will find themselves on the firing line, for it is at this level that products are produced and information is collected and analyzed. This is, in other words, where the action is, the very *raison d'être* for the enterprise. How effective middle managers are in adapting to total quality, and how successful their leadership, will have the greatest possible impact on the company's ultimate success.

VIEWPOINTS OF THOSE INVOLVED

The journey into total quality will be accompanied by fundamental changes in the culture of the organization. That being so, it is helpful to examine the perspectives of people at various levels, and at various milestones along the way. This is presented not so much as a warning, but more to raise the sensitivity level of the manager of a total quality effort. Naturally, no two cultures will show identical reactions, but we believe these to be typical.

Factory or Office Worker

Initial reaction: Here we go again. Another company buzzword.

After some experience: Hey, maybe there's something to this total quality.

Six months later: I'd never go back. We're proud of what we do. We're a team.

Middle Management (Unenlightened)

Initial reaction: We've been through "zero defects" and "do it right the first time." This too shall pass.

After some experience: They are having lots of problems. They must be nuts to think involving lower levels will do anything but cost money.

Six months later (the still unenlightened): They're just trying to do away with our jobs.

Six months later (the newly enlightened): It's hard to deny success. I'm beginning to understand.

Middle Management (Enlightened)

Initial reaction: Time will tell whether top management is really behind this.

After some experience: We're finding problems we never knew we had. Our team is excited.

Six months later: We're operating better than ever, but we've just scratched the surface. Total quality is the way to go.

Top Management

Initial reaction: How much will this cost, how difficult will the cultural transition be, and will the employees buy in?

After some experience: Most of the staff (now steering committee) are on board. Lower levels seem to be developing enthusiasm, but there is some resistance in the middle levels. Overall, we're making progress, but it is sure taking a lot of attention.

Six months later: Definite progress in several areas. Almost everyone involved and excited. A few holdouts in the middle. Fewer crises to deal with. More time to put into total quality.

The Customer

After the first year: Far better delivery performance. Quality improved. If they can maintain this kind of improvement, they've got my business.

It is our experience that once a total quality effort gets off to a good start and the successes begin to add up, a kind of critical mass develops that causes the whole effort to gain momentum and enthusiasm. From that point on, it is as difficult to slow the effort as it was initially to get it started. Nearly everyone becomes a proponent of total quality.

IMPLEMENTATION VARIATION AMONG ORGANIZATIONS

People who are about to undertake the leadership of total quality implementation in their organizations invariably look to the published literature or the experiences of others for the recipe that will result in success for them. Unfortunately, that magic, succeed-every-time formula does not exist. Our organizations and their cultures are all different, they are staffed with people who are all different from each other, and their business situations are always unique. Therefore, the implementation plan that worked well for XYZ company will never fit exactly with the needs of ABC company. However, in the literature and the experiences of other organizations you will certainly find the ideas and techniques that can be tailored to your own situation. You will find that the approaches to implementation that have been successfully used cover the spectrum of possibilities. The point is this: There is no *one right way*. For a given organization with its special strengths and weaknesses, its peculiar business situation, its unique culture, there may well be some *wrong ways*. There will also be more than one *right way*.

We have already discussed some of the starting tasks: making a commitment at the top, forming a steering committee of the top management staff, and defining the organization's vision and broad objectives. These are musts. There are some other necessary steps.

1. *Train the steering committee.*   The basics of this training should include these things:
   - Deming's Fourteen Points, the Seven Deadly Diseases (see Figures 22–8 and 22–9)
   - The seven tools and the add-ons (see Chapter 15)
   - Team building
2. *Identify organizational strengths and weaknesses.*
   - Statistical capability
   - Data collection and analysis capability
3. *Identify the probable advocates of total quality.*
   - Which departments are most likely to be advocates of total quality?
   - Who will resist total quality?
4. *Identify customers, both external and internal.*
   - Who are the organization's real, ultimate customers?
   - Who are the internal customers of the various departments or processes?
   - Who are the customers of the individual employees?
5. *Develop a means for determining customer satisfaction (external/internal).*
   - Establish the current baseline against which you will measure improvement.

By completing these tasks, the steering committee will be able to make rational judgments about how the journey should be started. For example, if you conclude that one of your weaknesses is in data collection and analysis, it would probably not be

1. **Create constancy of purpose for improvement of product and service.** Dr. Deming suggests that the role of any company should be to stay in business and provide jobs. Research, innovation, and continual improvement are mandatory in order to do that.

2. **Adopt the new philosophy.** No longer put up with poor quality and bad attitudes. These must be unacceptable.

3. **Cease dependence on mass inspection.** Quality cannot be inspected into a product. All inspection can do is cull out most of the defective ones, which will be reworked or thrown out. That is too expensive, and not satisfactory. Quality comes from relentlessly improving the processes which make the product.

4. **End the practice of awarding business on price tag alone.** Buying materials for your products on the basis of lowest price is fraught with problems. Instead you should seek quality and value, and establish long-term relationships with your good suppliers.

5. **Improve constantly and forever the system of production and service.** It is management's responsibility to constantly improve processes, products, and services while reducing waste.

6. **Institute training.** We have neglected this extremely important function to the extent that most Western workers do not know how to do their jobs.

7. **Institute leadership.** Find out what leadership is, and do it. It is not giving orders or threatening. It is leading, helping, facilitating.

8. **Drive out fear.** Too many workers continue to do their jobs poorly because they are afraid to ask or suggest. Fear has a huge economic and quality impact.

9. **Break down barriers between staff areas.** We must get people working as a team for the goals of the enterprise, not working to protect and maximize department objectives. Structures must support the whole, not isolate into fiefdoms.

10. **Eliminate slogans, exhortations, and targets for the workforce.** They do no good and are often seen as putting down the workforce, treating members like children. If the teams want to create their own slogans, let them.

11. **Eliminate numerical quotas.** Quotas send the signal to the people that volume is what counts, not quality or processes. They force people to achieve the quotas no matter what the cost in terms of waste or company reputation.

12. **Remove barriers to pride of artisanship.** Everyone wants to do a good job, but too often we do not provide people with the leadership, training, tools, or processes necessary. These barriers to pride of artisanship must be eliminated.

13. **Institute a vigorous program of education and retraining.** Every employee from top to bottom will have to be educated in total quality, the statistical tools, and teamwork.

14. **Take action to accomplish the transformation.** Make this everyone's job, from the top executive to the hands-on labor. Everyone must be involved, and the top levels must be committed to support and facilitate the effort.

Figure 22–8
Deming's Fourteen Points

1. **Lack of constancy of purpose.** You must have a long-range plan and stick to it. Otherwise, you are buffeted by every new influence, and no one knows where he or she is supposed to be headed.
2. **Emphasis on short-term profits.** The course must be on the long-range vision, not this quarter's numbers, which often lead to counterproductive decisions.
3. **Evaluation by performance, merit rating, or annual review of performance.** We need to promote teamwork, not individual contribution. Everything about most evaluation systems in use today is counterproductive and demoralizing to employees.
4. **Mobility of management.** Managers need to stay in jobs long enough to learn them and then offer stability over a longer period. Total quality is a long-term project, and there must be continuity to see it through.
5. **Running a company on visible figures alone.** Dr. Deming rightly claims that the most important figures are unknown and unknowable. What numbers do you assign to a delighted customer, or to a team of employees who are fired up to solve process problems? Running by the numbers, especially short-term numbers, can take you in the wrong direction.
6. **Excessive medical costs.** (U.S. only) When the average American automobile's price includes $1,500 for medical insurance costs for the workers who built it, we are fighting an uphill battle to be competitive with the rest of the world.
7. **Excessive costs of warranty, fueled by lawyers who work on contingency fees.** (U.S. only) No comment needed. It is a competitiveness issue. The entire United States light aircraft industry has been wiped out by this disease. And this could be only a hint of what comes next.

Figure 22–9
Deming's Seven Deadly Diseases

advisable to start into total quality with a wholesale leap into process improvement. If you cannot establish the baseline data and collect and analyze data as changes are implemented, how will you know whether you are doing the right things? If that is a weakness, you will have to overcome it before you can do much with your complex processes. On the other hand, if this is one of your strengths, you might make your first venture the establishment of one or more process teams, the task of which will be to flowchart a key process, understand how it works, characterize it to baseline its present capability, and then set about refining the process. In one high-tech company that used this approach, the steering committee identified the key processes and established cross-functional teams for each one. The committee capitalized on their data and analysis strengths to get the movement started. Each of the key process teams established lower level process teams as required. Because everything we do is associated with some process, this company soon found that total quality had permeated its operations. Its Baldrige-based score, a score derived using the criteria for the Malcolm Baldrige Quality Award, more than doubled in a year and a half, placing it at the world-class level.

If you have a manufacturing group that you believe will advocate total quality, you might start by introducing just-in-time production techniques on the factory floor. Manufacturing has led the way to total quality in a great many companies. On the other hand, you may have some particular problem that has not been solved by the traditional methods. The approach used by many organizations is to enter total quality with the establishment of cross-functional teams (perhaps including the steering committee itself) applying total quality techniques to analyze and solve problems. This problem-solving team approach is probably the most common introduction to total quality. It is important that the problems to be attacked and the team structures be controlled by the steering committee—especially at first. In the beginning, everyone will lack experience. At the same time, enthusiasm can build very quickly. The combination of inexperience and overenthusiasm can produce chaos and, if unchecked, will surely fail to produce the desired results. During the first 6 months or so, nothing should be done that is not directed by the steering committee. In fact, the steering committee needs to stay on top of all total quality activities forever, at least to the point of receiving regular input from all the teams. Nevertheless, the steering committee must remain receptive to suggestions for problems to be solved. Some of those *ad hoc* problems probably should be on the list of early candidates.

Another valid approach, if you are strong in the statistical area, is starting total quality by implementing the use of the seven tools. You know by now that the application of these tools is not restricted to the manufacturing floor. They may be applied wherever processes are at work, and that is everywhere. Before this can happen, the people affected will require a few hours of instruction on the use of the tools.

QUALITY TIP

The Eighth Deadly Disease

We would like to be presumptuous enough to add an eighth Deadly Disease to Deming's list of seven: *Executive incentive programs that involve stock that may be sold in less than 5 years after retirement.* Stock-option programs are the vehicle through which many executives acquire wealth, presumably by being wise managers, thereby enhancing the stock's value. However, there can be enormous pressure to make decisions that drive up the stock's price in the short term, sacrificing the long term, when executives are able to dispose of their stock holdings for personal gain while still employed by the company, or within 1 or 2 years after retirement. The result can be that the corporate vision is abandoned, and with it the company's best prospects for the future. Executives may engage in schemes such as massive downsizing, selling off units, even selling out to another company, and many other stratagems that can quickly, if temporarily, raise the stock price. If they could not sell the stock until the fifth anniversary of their retirement, the incentive would be to make decisions compatible with the company's long-term interests (i.e., its vision).

The customer satisfaction approach is another valid entry. For this, the employees will have to identify their internal customers and determine their basis for customer satisfaction. Then teams can be deployed to find ways to improve their processes in light of their internal customers' expectations. Results can be seen almost at once with this approach, because little time elapses between the completing of a process and getting the output of that process to the internal customer. This approach also works with external customers and is very important, but the cycle time in the external customer loop is much longer, so the information feedback is slower. Many times it does not exist at all except in customer satisfaction's impact to orders and the bottom line.

Although there is no one right way to implement total quality, the fundamentals apply in all cases. You have to approach it in a structured manner that takes advantage of your strengths, culture, business situation, and the personalities involved. Your first steps should be careful, deliberate, and well monitored. Study the data and listen to the people. Use the feedback you gain to help make the midcourse corrections that will assuredly be required. Build on your successes and learn from your failures. Keep your eye on the vision, and never let Deming's Fourteen Points be out of mind. Communicate, communicate, and communicate some more.

IMPLEMENTATION APPROACHES TO BE AVOIDED

Before we immerse ourselves in the right way to approach a total quality implementation, let's dispense with some wrong approaches. There are surely more examples of inappropriate total quality implementation than there are of superlative ones. Not surprisingly, however, we do not learn about most of the failures without being directly involved. Still, there is a lot of information around from which we can identify approaches to be avoided.

The following discussion gives you some ideas of implementation approaches from which you should stay away.

- *Don't train all your employees at once.*   It became popular in the mid-1980s to play numbers games with the number of employees who had received training in total quality, the seven tools, SPC, and so on. Some organizations (including government) spent megadollars training thousands of employees from the top of the organization to the bottom as a first step. Then they found that the vast majority of those employees would have no use for the training for months or years. By that time, not having applied their new skills, the employees had forgotten most of the training. People find it discouraging to be trained in a new subject but unable to do anything with it. The right way to do it is to train small groups of your people just-in-time—just as they need it.

- *Don't rush into total quality by putting too many people in teams.*   Another total quality numbers game was keeping score by the number of teams that were deployed or the percentage of employees on teams. Management wanted numbers, and because it was often difficult to develop meaningful numbers in terms of increased profitability, customer satisfaction, reduction of waste—all attributable to total quality—the number of teams was something—but no one knew what. The one thing that was certain in these cases was that aside from the teams themselves, no one knew what they were doing, except spending money. This was a throwback to the early days of American quality circles. The idea then was to get everyone into quality circles and let the circles pick projects on which to work. This is not at all like Japanese quality circles. Taiichi Ohno started quality circles at Toyota as production process teams.[6] He postulated that rather than have assembly line operators work as individuals, as they do in a mass production factory, small teams possessing the skills demanded by their broader process(es) could be more efficient. In addition to the normal assembly process, Ohno also assigned quality-checking, routine machine maintenance, and housekeeping duties to the teams. He also allocated time for them to discuss their work and develop ways to improve their processes. This was the genesis of the Japanese kaizen, continual incremental improvement.

The people in early quality circles in the United States didn't necessarily work together, they had none of the additional duties, and they were not focused on kaizen. They did take an hour or so a week to meet and discuss problems, but few of the problems they discussed had much relevance to quality improvement. Teams should be formed deliberately as needed to take on specific issues or problems as directed by the steering committee. Don't worry about how many teams you have, only about results.

■ *Total quality implementation must not be delegated.* One approach has been for top management to commit the organization to total quality and then to delegate the implementation to the quality assurance department. Beyond demonstrating colossal ignorance of the meaning of total quality, this also reveals something about the leadership at the top. In this case, top management sees it as another program to be endured, and since it is about quality, who better than the QA director to run it? One very competent, nationally known, quality assurance professional and advocate of total quality got caught in this trap. He tried very hard for 2 or 3 years to make a successful total quality implementation, but everyone in the company knew the top management was not involved, so the roadblocks held and the program was a failure. A successful total quality implementation requires both complete commitment and active, personal, day-in, day-out involvement by top management and staff.

■ *Don't start an implementation before you are prepared.* Sometimes higher level managers find it difficult to acknowledge that they don't know something. It should be obvious that plunging into anything as technically and sociologically complex as total quality without having a grasp of the subject will guarantee failure. A lot of plants can be found throughout the United States with walls plastered with control charts, where weekly meetings are held to review the data, and where phrases such as "employee involvement" abound. Yet in these same plants, there is no real continual improvement, no real involvement. What has happened is that the traditional department heads, without any total quality leadership, have given the boss what he or she wanted—charts and buzzwords. Be sure the first step is becoming educated on the subject of total quality before attempting to implement. This must include the top manager and his or her immediate subordinates.

AN IMPLEMENTATION APPROACH THAT WORKS

We have just discussed implementation approaches that have demonstrated through practice that they should be avoided. Experience has also provided a wealth of information on successful implementations and the techniques that made them that way. Much can be learned from successful implementations, but we must always be mindful that we will not be able to duplicate another's success by following exactly the same path; there are simply too many variables. Our 20-step total quality implementation process will work for any organization with a little tailoring here and there to accommodate the specific organization.

We have said that no two total quality implementations will be the same. However, every implementation will require certain steps, and these steps must be taken in a logical order. Refer to Figure 22–10. Our implementation model has three phases:

- Preparation
- Planning
- Execution

The preparation phase is necessary for any organization, and the steps are listed in the appropriate order. With the skills honed, and critical information developed in the preparation phase, we can enter the planning phase. It is here that the unique strengths and weaknesses and other characteristics of the organization are accommodated by tailoring. Then we go into the execution phase, carrying out the planning just completed. At this point we find ourselves in a continuous loop between planning and execution as we go further along the total quality pathway. Even when we can say that we have fully implemented total quality, this loop will continue to operate, making continual improvement and superior competitiveness a reality. As we go step-by-step through this 20-step implementation process, continue your reference to Figure 22–10 so that you develop a clear picture of how the steps interrelate, and how they should be sequenced in time.

We have discussed things that must be done and other things you need to consider for your journey into total quality. Now these things must be put into a time-phased sequence: a schedule. You need to have a clear picture of what comes first and what follows, for some steps are prerequisites for others. The following statement is not meant to be a disclaimer, but the schedule cannot have a well-defined time scale. Only you can determine the time you require to do these things, how long certain steps will take. For that reason, the time scale must be undefined; however, when you apply it to your own implementation, you should apply a more precise scale to target milestone events. Be prepared to allow some flexibility, because even within a specific organization, it may be difficult at first to project time requirements for some of the changes. Refer to Figure 22–10 for clarification of the phases of entry and execution of total quality.

The Preparation Phase (Steps 1 through 11)

This phase includes steps labeled 1 through 11 in Figure 22–10, and they are carried out in sequence. Even before step 2 (forming the total quality steering committee) can take place, one all-important event must occur: the top executive must make the commitment

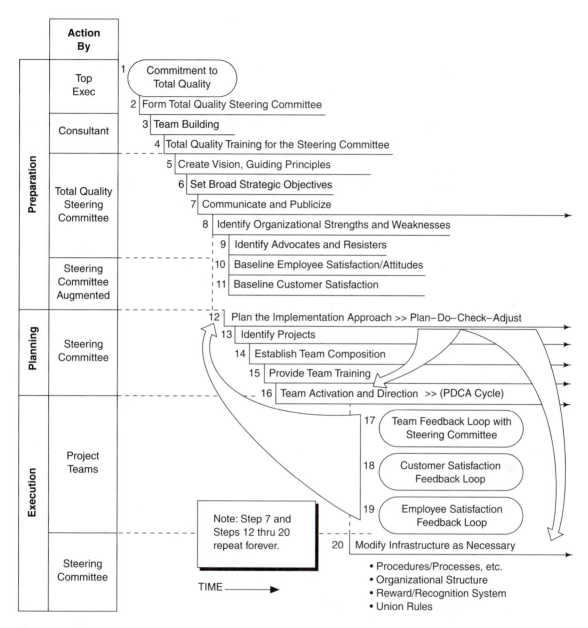

Action By

Top Exec	1 Commitment to Total Quality
	2 Form Total Quality Steering Committee
Consultant	3 Team Building
	4 Total Quality Training for the Steering Committee
Total Quality Steering Committee	5 Create Vision, Guiding Principles
	6 Set Broad Strategic Objectives
	7 Communicate and Publicize
	8 Identify Organizational Strengths and Weaknesses
Steering Committee Augmented	9 Identify Advocates and Resisters
	10 Baseline Employee Satisfaction/Attitudes
	11 Baseline Customer Satisfaction
Steering Committee	12 Plan the Implementation Approach >> Plan–Do–Check–Adjust
	13 Identify Projects
	14 Establish Team Composition
	15 Provide Team Training
	16 Team Activation and Direction >> (PDCA Cycle)
Project Teams	17 Team Feedback Loop with Steering Committee
	18 Customer Satisfaction Feedback Loop
	19 Employee Satisfaction Feedback Loop
Steering Committee	20 Modify Infrastructure as Necessary

Preparation / Planning / Execution

Note: Step 7 and Steps 12 thru 20 repeat forever.

TIME ⟶

• Procedures/Processes, etc.
• Organizational Structure
• Reward/Recognition System
• Union Rules

Figure 22–10
The Goetsch-Davis 20-Step Total Quality Implementation Process

of time and resources. Without that commitment you should go no further. Assuming that commitment, proceed as follows:

Step 2: Formation of the Total Quality Steering Committee

Action: Top executive designates immediate staff (direct reports) to be the total quality steering committee, with himself or herself as chair. If a union is involved, the senior union official should also be a member of the steering committee.

Note: Several names have been used for this committee. Whatever name you choose, the function will be the same.

Duration: The steering committee will be a permanent entity and will replace the former staff organization.

Step 3: Steering Committee Team Building

Action: The steering committee needs to go through a team-building session before it starts any total quality work. This will usually require an outside consultant.

Duration: This usually requires 1 to 3 days, preferably done away from the work environment.

Step 4: Steering Committee Total Quality Training

Action: The steering committee will require training in total quality philosophy, techniques, and tools before it starts any total quality work. Usually requires an outside consultant.

Duration: Two or 3 days of intensive training. This should be followed through in the long run with self-study and appropriate seminars.

Step 5: Creation of the Vision Statement and Guiding Principles

Action: The first real total quality work effort is creating the organization's vision statement and putting on paper the guiding principles under which the company is to operate. Typically the top executive initiates discussion by using "strawman" vision and principles. The objective is getting the steering committee's thoughts, refining the language, and concluding with short, meaningful documents that embody the hopes and aspirations of the company.

Duration: Plan on at least 1 full day.

Step 6: Establishment of Broad (Strategic) Objectives

Action: The steering committee flows the vision statement into a set of broad company objectives. These are by nature on a grand scale; for example, "Become the dominant player in our market in 5 years." These are strategic objectives. From these flow a set of supporting tactical objectives that go into specifics— for example, " . . . by introducing new products on a 9-month cycle over the next 3 years."

Duration: This will take at least a full week but probably will be spread over several weeks. Take the time to do this step with consideration and deliberation, but on the other hand, set your schedule and stick to it.

Step 7: Communication and Publicity

Action: The top executive and the steering committee should communicate information about steps 2 to 4 as they occur. At this point, however, there should be a communication blitz. Make sure that everyone in the organization knows about the vision, the guiding principles, the objectives, and total quality. It is very important that they know why total quality is being implemented. If you don't tell them, the rumor mill will fill the void. Employees should see the top executive as the champion, with the support of the steering committee. This is very important.

Duration: Starts now and goes on forever.

Step 8: Identification of Organizational Strengths and Weaknesses

Action: The steering committee must objectively identify the strengths and weaknesses of the organization. This information will help guide it to the best total quality implementation approach and may also highlight deficiencies that must be corrected.

Duration: Plan on a full day.

Step 9: Identification of Advocates and Resisters

Action: (May be parallel to or after step 8.) The steering committee should try to identify those in key slots who are likely to be total quality advocates and those who are likely to resist total quality. This will help in selecting the early projects and team members.

Duration: Should require no more than an hour or two, if the members independently prepare their assessments prior to the meeting.

Step 10: Baseline Employee Satisfaction/Attitudes

Action: (May be parallel to or after step 8.) With the help of the human resources department or an outside consultant, the steering committee should attempt to gauge the current state of employee satisfaction and attitudes. Although there are sophisticated devices for determining this information, it is probably only necessary to make an objective judgment. Having that established, you will later be able to determine whether your total quality changes are working effectively, as shown by improving satisfaction and attitudes.

Duration: Allow a week to do it in-house, at least a month if you have an outside firm do it. Should be repeated annually.

Step 11: Baseline Customer Satisfaction

Action: (May be parallel to or after step 8.) The steering committee, perhaps augmented by the department that works closest with customers, should attempt to obtain objective feedback from customers to determine their level of satisfaction. Depending on the size of the customer base, the selection of customers to be surveyed may be random. Be certain that someone doesn't pick only those known to be favorably disposed. Having this information will allow you to judge the effectiveness of your total quality efforts as seen by the ones who make the final determination—your customers.

Duration: Allow 2 months if you send out survey forms, 2 weeks if you do it by phone. Should be repeated annually.

The Planning Phase (Steps 12 through 15)

Step 12: Plan the Implementation Approach— Then Use Plan–Do–Check–Adjust (PDCA Cycle)

Action: Now is the time for the steering committee to start planning the implementation of total quality. This step becomes continuous, because after initial projects are under way, information will be fed back to this step to accommodate course correction, adjustment, and so on. In addition, this step will continue to spin out new projects and teams. Also, once total quality is in motion, the step effectively shifts to operate within the PDCA cycle, as originated by Walter Shewhart.[7]

Duration: Never ends. This is the step from which the total quality process is managed, not only at the implementation stage but for as long as the process exists.

Step 13: Identification of Projects

Action: The steering committee is responsible for selecting the initial total quality projects, based on the strengths and weaknesses of the company, the personalities involved, the vision and objectives, and the probability of success. The early projects must be selected to assure success in order to set a foundation of positive experience from which to move to the more difficult challenges later on. The steering committee should be open to suggestions for projects from all sources.

Duration: Initial projects selected over a few days. Process continues forever.

Step 14: Establish Team Composition

Action: After the projects have been selected, the steering committee establishes the composition of the teams that will execute them. Most teams will be cross functional, having representation from multiple departments or disciplines, as appropriate for the project at hand. This is one step where it is handy to know who the advocates are.

Duration: This task goes on forever.

Step 15: Provide Team Training

Action: Before a new team can go to work, it must be trained. Training should cover basics of total quality and tools appropriate to the project. Training may be done by a member of the steering committee.

Duration: At least one-half day, followed by facilitation. As new teams are formed, the need for training will continue until eventually all employees are trained and experienced.

The Execution Phase (Steps 16 through 20)

Step 16: Team Activation and Direction (Use PDCA Cycle)

Action: The steering committee gives each team its direction and activates it. Teams work on their assigned projects using the total quality techniques they have learned. They use the plan–do–check–adjust cycle as their total quality process model.

Duration: Project teams, depending on the project, may have life spans of weeks, months, or longer. Measurable results must continue to flow, however, to ensure that the team is being effective.

Step 17: Team Feedback Loop to the Steering Committee

Action: Through this step, the project team closes the loop with the steering committee by providing feedback information on progress and results. This feedback is usually in the form of presentations to the steering committee. Early on, this loop should be tight, with feedback coming in frequent packets—perhaps weekly. As the project gets on track and stabilizes, monthly feedback is appropriate. Never let it go longer than that. The steering committee uses this feedback to determine if adjustments or changes in direction are required. Any changes desired are fed back to the project team, which carries out the new instructions. Both the team and the steering committee use the PDCA cycle.

Duration: Specific projects may have finite lives of weeks, months, or longer, but this process across all the projects goes on forever.

Step 18: Customer Satisfaction Feedback Loop

Action: Special project teams are deployed to obtain customer feedback information, covering both external and internal customers. Formal external customer surveys should be conducted annually, and other customer satisfaction data (sales results, warranty data, customer service input, data from customer visits, etc.) are collected and processed on a continual basis. Internal customer satisfaction is baselined for key processes and monitored continually. (This latter may be done by the project team assigned to the process in question.) All of this information is fed back to the steering committee on a regular basis, certainly no less frequently than quarterly. This information is digested in the steering committee's PDCA cycle and influences direction changes issued to the project teams, and also the formation of new project teams.

Duration: Forever.

Step 19: Employee Satisfaction Feedback Loop

Action: Another special project team periodically takes the pulse of employee attitude and satisfaction. Formal surveys may be done annually, and in between it is only necessary for the steering committee and other managers to stay close to the employees to develop rather accurate information on attitude and satisfaction. This information is fed to the steering committee as another leg of data on which to evaluate progress and determine any necessary course corrections.

Duration: Forever.

Step 20: Modify Infrastructure as Necessary

Action: Feedback to the steering committee from steps 17, 18, and 19 (from project teams, customers, and employees) will guide the steering committee to address necessary changes in the corporate infrastructure—procedures and processes, organization structure, awards and recognition programs, union rules, and so on. Many times the changes will have to be made by the steering committee itself (e.g., organization structure). On other occasions, it is appropriate to authorize the project teams to make the changes *that are needed* (e.g., to the processes on which they are working).

Duration: Because we are talking about continual improvement, this goes on forever.

We have now walked through the various implementation phases. In this exercise, we have followed the process from the time the top executive decided to embrace total quality through working teams. Starting with step 12 and going through step 20, we have a closed-loop system that will continue for as long as the total quality process is in use. The feedback to the steering committee provides the information it needs to manage the process. The steering committee in turn issues new instructions, establishes new teams, and so on, to keep efforts going in the direction that will achieve the objectives and remain in harmony with the vision.

Step 7 (communication and publicity) also goes on forever. We cannot overemphasize the importance of keeping employees informed. Many companies have found that through innovative communication, it is possible to stimulate the kind of enthusiasm throughout the ranks that really produces advocates.

The material covered above is necessarily concise. For a complete in-depth treatment of the Goetsch-Davis 20-step total quality implementation process, you are invited to read our companion book, *Implementing Total Quality* (Prentice Hall, 1995).

GETTING ON WITH IT

Whether an organization can go through this kind of an implementation without the help of a consultant will depend on the internal availability of one or more experts. Students who study total quality in college can become these internal experts. Rather than trying to muddle through, it would be far better to enlist the aid of a consultant to get you through the implementation and into execution. It is certainly possible to pay a lot of money for such services, but it is by no means necessary. Most consultants would prefer to work with you on a part-time basis (simultaneously servicing other clients). A typical 6-month implementation would probably require 1 person-month or less of consulting services. This could be a wise investment for your organization, because the "muddle through" method is almost sure to fail, after which it will be doubly difficult to recover, to try again.

Virtually every type of enterprise can benefit from total quality. Starting on the journey now may put you ahead of your competition. Failure to start now may leave you behind and doomed to failure as the whole world embraces the principles of total quality.

What to Do in the Absence of Commitment from the Top

We have said repeatedly throughout this book that no entity can truly become a total quality organization without complete, unwavering, participative commitment from the very top. For those of you who have that commitment, your path is clear—you know what to do and how to proceed. On the other hand, many of you will not be so fortunate, and you face a dilemma. Should you try to push total quality from where you are, or should you forget it and wait for top management to come around? There can be no hard and fast answer to this. So much depends on where you are in the organization and whether the top managers are outwardly hostile to total quality or merely unknowledgeable. Even the geographical structure of the organization is a factor.

If your top management is hostile to total quality, it would not seem prudent for you to push the subject much beyond trying to enlighten them. Even that could be hazardous depending on your relationship with the management group. Certainly any overt total quality initiatives on your part in defiance to management's wishes would not be in your best interest. If enlightenment does not work, it may be time to consider moving on to different employment. That is not always a reasonable option, but long-term prospects for your current employment are not bright either, given top management's attitude toward total quality.

If your top managers are simply ambivalent toward total quality, perhaps because they do not know much about it, you have a whole different ball game. Assuming your position gives you some latitude of operation, you may be able to construct a mini-implementation within your department that can improve performance in many areas. To do this you must be the head of a department or must be able to influence a department head who will work with you. Because departments can range in size from just a handful of people to a complete stand-alone facility far removed from the home office, the magnitude of accomplishment of a departmental total quality implementation can vary greatly. The head of a stand-alone facility can look at the implementation almost as if he or she headed the company. The more common situation, where the department is one of many, proximally located, will not offer the same freedom of action, but it is still worth pursuing. Generally speaking, all processes contained within the department, even though they have extradepartmental suppliers and customers, are good candidates for a total quality effort. You should concentrate your efforts there first. After you have gained some successes that have been noticed by top management and other departments, you can begin to push for expansion into the cross-functional processes, requiring other departments to join you in total quality initiatives.

Recognizing that *departmental total quality* is a contradiction in terms (i.e., it cannot be *total quality* until every aspect of the organization is involved and committed to it), implementing total quality in a single department is better than not doing it at all. If the implementation is well done, the gains will be significant enough to attract the attention of other departments and of top management. Your department may provide the seeds for the larger organization to become involved.

Getting started in a single department is not much different from the process we have just studied for a company-wide implementation. The numbers of people involved in every step may be smaller, but most of the steps are required. The following list briefly addresses each of the 20 steps for applicability in a departmental implementation:

- *Step 2.*   Unless it is a very large department, such as a stand-alone plant or something similar, you may not require a steering committee. For a large department we would recommend one.

- *Steps 3 and 4.*   If you have a steering committee, you will need to give the members some training. Because you are doing this on your own without top management's involvement, don't expect a lot of support money at this point.

- *Steps 5, 6, and 7.*   Setting the vision, mission, and guiding principles for your department is a good idea. If these exist at the company level, yours should flow directly from them and support them. If the company has no vision/mission/guiding principles, be sure to have yours approved by top management before proceeding. Once that is done, communicate them to every department employee.

- *Steps 8 and 9.*   This is just as important at the department level as it is for the larger organization. Departmental strengths and weaknesses, and the identification of those likely to help you will be of invaluable assistance in selecting your initial projects.

- *Steps 10 and 11.*   We believe that it will be important for you to know your employee attitude index within the department. In terms of baselining customer satisfaction, you will probably be looking at internal customers in other departments, but the same considerations apply. In this case it should be easy to survey for the information.

- *Step 12.*   Must be done. No matter what the scale, the implementation approach should always be thoroughly planned.

- *Step 13.*   You should identify your initial projects based on your department's strengths and weaknesses, where your support is and, very important, the processes and activities that are contained completely within your department. Try to pick initial projects that can yield measurable gains without significant expense.

- *Steps 14, 15, and 16.*   Follow the same procedure as for the larger organization, but recognize that you may have to improvise in the training step in the absence of dollar support.

- *Steps 17, 18, and 19.*   These steps follow the same procedures as for the larger organization except the customer feedback will probably be from your internal customers in other departments. If you elect not to use a steering committee, the feedback path will be directly to you.

- *Step 20.*   Although you may find it advantageous to make some intradepartmental infrastructure changes, you will probably have to secure approval of top management before the fact.

As in the earlier discussion, steps 12 through 20 become a repeating cycle even within a departmental implementation. Eventually it will become clear whether your implementation has piqued the interest of top managers, encouraging them to spread the implementation to other departments, or whether total quality will go no further in the organization. Another possibility is that your peers may join you department by department. This is far from the ideal way to implement total quality in any organization, but it is preferable to discarding total quality out of hand.

IMPLEMENTATION STRATEGIES: ISO 9000 AND BALDRIGE

Organizations that are not ready to undertake a full total quality implementation but want to move in that direction can use the criteria of several different award and certification programs as a starting point. The two programs that we recommend are ISO 9000 and the Malcolm Baldrige Award.

ISO 9000 as a Starting Point

If an organization has not yet committed to total quality but is in a business that could benefit from ISO 9000 registration, going through the preparation steps will automatically provide a start into total quality. Although ISO 9000 and total quality are not equivalent, as we discussed in Chapter 14, any total quality organization should apply the kinds of procedures, checks, and management involvement required by ISO 9000. ISO 9000 is a subset of total quality, but it has come much closer with its 2000 version.

ISO 9000 registration requires, for most organizations, a lot of work. If starting without any written procedures, if processes are not well documented, if there is no quality system or quality manual, an organization faces an uphill battle. But the work required is work that must be done for total quality anyway. ISO 9000 preparation can get an organization started on a total quality implementation.

Self-Assessment Based on Baldrige or Similar Criteria as a Start

The Malcolm Baldrige National Quality Award program, which operates under the auspices of the National Institute of Standards and Technology (NIST), was established in 1987 by legislation (P.L. 100–107) and is intended to do the following:

- Promote awareness of the importance of quality improvement to the national economy.
- Recognize organizations that have made substantial improvements in products, services, and overall competitive performance.
- Foster sharing of best-practices information among U.S. organizations.

From 1988 through 2000, 43 organizations have received the coveted Baldrige Award. Competitors for the award are evaluated according to a seven-category criteria. The categories are weighted in value, with a maximum score being 1,000. The list of evaluation criteria, which was updated in 2001, is shown in Figure 22–11. As you can see from the figure, each of the seven categories is broken down into two or more subcategories, so that the evaluation covers the 18 areas of the organization's operations considered most meaningful to an evaluation of world-class ranking. Each subcategory is further broken down two more levels to enable the evaluation into the actual processes used.

Virtually all of the Baldrige competitors, both the winners and those who have not won, say that the real value of the Baldrige Award is not in the award itself but in the preparation. Consequently, more and more organizations are going through the prepara-

---

ITEM LISTING

2001 Malcolm Baldridge National Quality Award Criteria

1.0	**Leadership**		**120 Points**
	1.1 Organizational Leadership	80	
	1.2 Public Responsibility and Citizenship	40	
2.0	**Strategic Planning**		**85 Points**
	2.1 Strategy Development	40	
	2.2 Strategy Deployment	45	
3.0	**Customer and Market Focus**		**85 Points**
	3.1 Customer and Market Knowledge	40	
	3.2 Customer Relationships and Satisfaction	45	
4.0	**Information and Analysis**		**90 Points**
	4.1 Measurement and Analysis of Organizational Performance	50	
	4.2 Information Management	40	
5.0	**Human Resource Focus**		**85 Points**
	5.1 Work Systems	35	
	5.2 Employee Education, Training, and Development	25	
	5.3 Employee Well-Being and Satisfaction	25	
6.0	**Process Management**		**85 Points**
	6.1 Product and Service Processes	45	
	6.2 Business Processes	25	
	6.3 Support Processes	15	
7.0	**Business Results**		**450 Points**
	7.1 Customer-Focused Results	125	
	7.2 Financial and Market Results	125	
	7.3 Human Resource Results	80	
	7.4 Organizational Effectiveness Results	120	
		Total Points	**1,000**

Revised 2001

Figure 22–11
Malcolm Baldrige National Quality Award Criteria Item Listing

Quality Tip

Seeing It as It Is

As a word of caution, a self-assessment score would probably not hold up against an assessment by qualified outside assessors. Much of our consulting effort has been involved in these assessments. Whenever we have conducted a Baldrige-type assessment, we have always required the executive team to conduct a self-assessment first. Without exception, the self-assessment score is higher than our assessment team's—usually by 200 to 300 points. Many executives think their organizations' performance is better than it really is. The facts emerge from the middle and lower levels of the organization; hence the difference in scores.

Source: David L. Goetsch and Stanley Davis.

tion phase without any intention of competing for the prize. What does this do for them? It gives them a scorecard, ranking them against a world-class standard, and provides them with a list of needed improvements. The scores even let them prioritize their efforts. From this starting point, organizations can develop projects to address their deficient areas and assign them to project teams (Step 13 of the 20-step implementation process).

Competing for the Baldrige Award can be quite expensive, but at the time of this writing, 46 states (lacking Montana, Nebraska, Ohio, and West Virginia) sponsor quality programs based on the Baldrige. The state competitions offer most of the benefits at a much lower cost. (The most significant difference is in the prestige of the national award.) Most states also offer self-assessment criteria derived from the Baldrige. These, or the Baldrige criteria itself, can be used by organizations wishing to assess themselves against world-class standards without going through the actual competition.

If an organization has not yet committed to total quality, conducting one of these self-assessments will provide a scorecard for comparison against the best in the world. More than that, it will clearly identify the areas that most need to be improved, thus providing the impetus for implementing total quality and putting all your people to work on the areas of need. Do that in the context of the 20-step total quality implementation process.

Do not get the idea from this discussion that an organization can implement total quality without the order, discipline, and planning suggested by this book. Ultimately organizations will need everything presented here from cover to cover. By using ISO 9000 or quality award assessment criteria, an organization can get started. But a full-fledged implementation as we have described will be needed if an organization is to realize the full benefit of total quality.

Addresses for ISO 9000 and Quality Awards Information

For ISO 9000, ANSI/ISO/ASQ Q9000:

> American Society for Quality
> 611 E. Wisconsin Avenue
> PO Box 3005
> Milwaukee, WI 53201-3005

For the Malcolm Baldrige National Quality Award:
> United States Department of Commerce
> Technology Administration
> National Institute of Standards and Technology
> Route 270 and Quince Orchard Road
> Gaithersburg, MD 20899
> Phone: National Quality Program (NQP) at NIST, (301) 975-2036
> Web: NQP@nist.gov

You may also contact the ASQ, as listed earlier.

 ## SUMMARY

1. The traditional way of doing business presents the following problems:
   - We are bound to a short-term focus.
   - The traditional approach tends to be arrogant rather than customer focused.
   - We seriously underestimate the potential contribution of our employees, particularly those in hands-on functions.
   - The traditional approach equates better quality with higher cost.
   - The traditional approach is short on leadership and long on bossmanship.
2. The requirements for implementation are as follows: commitment by top management, creation of an organization-wide steering committee, planning and publicizing, and establishing an infrastructure that supports deployment and continual improvement.
3. The role of top management can be summarized as providing leadership and resources. The role of middle management is facilitation.
4. Although implementation must vary with each organization, the 20 fundamental steps offered in this chapter must be followed, generally in the order given. Tailoring to the organization's specific culture, values, strengths, and weaknesses is done in the *planning* phase, steps 12 through 16.
5. Implementation approaches that should be avoided are as follows: don't train all employees at once, don't rush into total quality by putting too many people in too many teams too soon, don't delegate implementation, and don't start an implementation before you are prepared.
6. Implementation phases are as follows: preparation phase, planning phase, and execution phase.
7. Going through the ISO 9000 registration steps will give an organization a good start on implementing total quality. ISO 9000 is an international standard for providers of

goods and services that sets broad requirements for the assurance of quality and for management's involvement.

8. The Malcolm Baldrige National Quality Award evaluates candidates for the award according to criteria in seven categories as follows: leadership, strategic planning, customer and market focus, information and analysis, human resource focus, process management, and business results.

 KEY TERMS AND CONCEPTS

Advocates

Baseline customer satisfaction

Baseline employee satisfaction/attitudes

Commitment by top management

Communicate and publicize

Customer feedback

Customer focused

Employee feedback

Evaluation criteria

Execution phase

Infrastructure

Interchangeable worker

ISO 9000

Leadership

Malcolm Baldrige National Quality Award

National Institute for Standards and Technology (NIST)

Organization

Organization-wide steering committee

Planning and publicizing

Planning phase team composition

Preparation phase

Probable advocates

Resisters

Short-term focus

Strategic (broad) objectives

Strategic planning

Tactical (specific) objectives

Team activation–PDCA cycle

Team training

Union considerations

Vision statement and guiding principles

World-class standard

 FACTUAL REVIEW QUESTIONS

1. What is meant by the statement, "We are bound to a short-term focus"?
2. How does the traditional approach to doing business equate quality with higher cost?
3. Differentiate between leadership and bossmanship.
4. List and explain the requirements for total quality implementation.
5. Describe the necessary components of an infrastructure that supports goal deployment and continual improvement.
6. What is the role of top management in the implementation of total quality?
7. What is the role of middle management in the implementation of total quality?
8. List the implementation steps that follow after the vision statement and broad objectives have been developed.
9. List and briefly explain implementation approaches that should be avoided.

10. Implementation of total quality happens in phases. Explain each phase in the order it occurs.
11. What is ISO 9000 registration, and how does it relate to total quality?
12. List the various categories of criteria for the Malcolm Baldrige National Quality Award.

CRITICAL THINKING ACTIVITY

1. The chart on the next page was developed from the J. D. Power and Associates 1998 Initial Quality Study 2.[8]

    Notice that this 1998 Vehicle Initial Quality chart shows eight North American, eight Japanese, and six German makes. Now compare this with the 2000 chart in Figure 22–7B. Do the 2000 results suggest that the West is gaining on Japan's manufacturing quality? What other conclusions can you draw from the data?

2. E-Z Open Manufacturing Company is a leading maker of manual can openers. In the year just closed, E-Z Open controlled 17.2% of the manual can opener market in North America. That placed the company in the number two sales position for manual can openers. The company in the number one position, Saf-T Products Co., had a 22.3% share of the North American market. E-Z Open was eager to expand its market with a new rechargeable battery-operated can opener. Saf-T also has an electric can opener, but it is a countertop model that must be plugged in to operate. E-Z Open thinks it has a more desirable product with a battery-operated model, because it will not tie up kitchen counter space, nor will it be encumbered by an electric cord. In addition, both the manual and electric models now have removable cutter heads that can easily be cleaned. (Difficulty in keeping can openers clean has been a problem since the first rotary cutter models appeared 70 years ago.)

    Already well into the early work of a TQM implementation, E-Z Open is gearing up for a big year. It is at the point where it needs a vision statement, mission, guiding principles, and a set of broad objectives. You have been retained as a consultant to help develop all of these. The company will hold an off-site meeting of the steering committee in 2 weeks, and you are to attend. It has asked you to provide the "straw-man" documents to kick off the meeting. Your task is to develop the initial versions of each of the documents except the objectives.

3. E-Z Open Manufacturing's organization structure is straight out of the 1950s. The president is the senior executive, and he has a secretary and five department heads reporting to him. The departments are product development, manufacturing, finance, marketing, and human resources. Each is headed by a vice president. Quality assurance is headed by a manager who reports to the manufacturing VP. The VP of product development has 35 people working on designs for a new family of small kitchen appliances, which the company hopes will render the firm immune to the dreaded can opener demand cycle.

    Manufacturing is are aware of the new-product development effort and is concerned that it might face problems getting these products into production. Finance

Car Segments	Rank	Make/Model
Compact Car	1	Ford Escort
	2	Mercury Tracer
	3 (tie)	Honda Civic
	3 (tie)	Saturn SL Sedan
Entry Midsize Car	1	Chrysler Cirrus
	2	Nissan Altima
	3	Ford Contour
Premium Midsize Car	1	Chrysler Concorde
	2	Chevrolet Lumina
	3	Mercury Sable
Sporty Car	1	Honda Prelude
	2	Acura Integra
	3	BMW 318 Ti
Entry Luxury Car	1	Lexus ES300
	2	BMW 3-Series
	3	Acura TL
Premium Luxury Car	1	Lexus LS400
	2	Acura RL
	3	BMW 5-Series
Sports Car	1	Mercedes-Benz SLK
	2 (Tie)	Porsche Boxter
	2 (Tie)	Porsche 911

figures that the company cannot build mixers with the tools used for making can openers but has no clue as to the investment size. Meanwhile, marketing, gleeful that it will soon have something new to sell, keeps sending ideas for still more new products into product development. QA is totally occupied inspecting can openers and is out of the loop for the new products. You have gotten a sense for this, and you think the infrastructure is an impediment to the growth the firm is anticipating. What would you have the company do?

4. Using your place of work as the model, how does (or could) TQM benefit it in general, and in particular, what improvements would you expect from following the 20-step implementation process? (If you are a full-time student, and not employed, skip this question and go on to number 5.)

5. Using your college as the model, explain how implementing a total quality program might benefit both the institution and its students. Include specific areas for improvement.

DISCUSSION ASSIGNMENT 22–1

McDonnell Douglas Corporation

McDonnell Douglas is now part of the Boeing Company, but that does not diminish in any way the turnaround that took place at McDonnell Douglas between 1992 and 1997. Times were not good for the company in the early 1990s. With the military buildup a thing of the past, the huge military division watched as sales plummeted. The commercial aircraft division struggled to be competitive with Boeing and Airbus. Waste and inefficiency were rampant. McDonnell Douglas stock sat at $9.00 a share in 1992—the lowest in anyone's memory. The company, once the nation's largest defense contractor, and the world's number two supplier of commercial aircraft, was in serious trouble.

The senior management staff under the leadership of John McDonnell, then chairman of the board, decided to try total quality management. They got off to a very rocky start but learned from their mistakes and by 1992 were making good progress. Starting that year, executives were measured on three items: cash flow, return on net assets, and TQM, with the latter being tied to improvement on a Baldrige self-assessment score. In 1992 the self-assessment score was 200 (on the 0 to 1,000 scale). Over the next 3 years, the score increased by 100 points per year. This took the company from being a so-so performer in the traditional ranks to a high midlevel in the TQM realm.

McDonnell Douglas found that as its Baldrige self-assessment score improved each year, its key business performance indicators tracked in parallel fashion. By 1995 their stock valuation reached $70 per share, profits were several times greater than in 1992, cash was up, debt was down—all in a vastly smaller market. TQM literally turned this giant company around to the extent that the Boeing Company considered it imperative to merge in order to save Boeing. (The merger became final August 1, 1997.)

Discussion Assignment
1. Explain how the use of TQM could contribute to the improvement noted in McDonnell Douglas Corporation's key business performance indicators.
2. What was John McDonnell's motive in making improvement in the company's Baldrige self-assessment score a part of executive performance evaluations?

From a paper delivered by John McDonnell, chairman of the board, McDonnell Douglas Corporation, at the Florida Sterling Awards for Quality, June 1995, Orlando, Florida.

ENDNOTES

1. Helio Gomes, *Quality Quotes* (Milwaukee, WI: ASQ Quality Press, 1996), 34.
2. Phillip B. Crosby, *Quality Is Free: The Art of Making Quality Certain* (New York: McGraw-Hill, 1979).
3. James Womack, Daniel T. Jones, and Daniel Roos, *The Machine That Changed the World* (New York: HarperCollins, 1990), 30.
4. Maryann Keller, *Rude Awakening: The Rise, Fall, and Struggle for Recovery of General Motors* (New York: Morrow, 1989), 131.
5. W. Roberts, *Leadership Secrets of Attila the Hun* (New York: Warner, 1991), 107.
6. Womack, 56.
7. Mary Walton, *The Deming Management Method* (New York: Perigee, 1986), 87.
8. *J. D. Power and Associates 1998 Initial Quality Study 2* (Los Angeles: J. D. Power and Associates, 1998), 3.

APPENDIX

Manufacturing Networks in the United States

Accessible Housing Components
The Appalachian Center for Economic Networks (ACENET)
94 N. Columbus Road
Athens, OH 45701
614-592-3854

Center for Design and Manufacturing in Greenpoint
North Brooklyn Development Corporation
6th Floor
1155 Manhattan Avenue
Brooklyn, NY 11222
718-383-3935

Erie Bolt (EBC Industries)
1325 Liberty Street
Erie, PA 16502
814-456-4287

The FlexCell Group
1537 Hutchins Avenue
Columbus, IN 47201
812-376-0200

Flexible Manufacturing Network Project
Trio-State Manufacturers Association
Scaffold Building/Central Avenue, N.
PO Box 150
Elbow Lake, MN 56531
218-685-5356

Garment Industry Development Corp. (GIDC)
275 Seventh Avenue, Eighth Floor
New York, NY 10001
212-366-6160

Greater Syracuse Metalworking Industry
1053 Erie Street
Macedon, NY 14502
315-986-9096

Heat Treating Network, Inc.
Managing and Marketing Office
17200 Pearl Road, Suite 202
Cleveland, OH 44136
216-572-7995

Kentucky Wood Manufacturers Network
Cowart & Co.
Cabinet for Economic Development
Office of Business and Technology
Capital Plaza Tower, 24th Floor
Frankfort, KY 40601
502-564-7670

Knitwear Industry Center
57-14 Myrtle Avenue
Ridgewood, NY 11385
718-381-7080

Lehigh Valley Apparel and Textile Network
Participative Systems, Inc.
PO Box 181
Princeton, NJ 08542
609-921-1770

Louisiana Furniture Industry Association
PO Box 2972
Hammond, LA 70404
504-542-9618

Mahoning Valley Aluminum Extruders Network
Youngstown State University Technology Development Corp.
Youngstown State University
Youngstown, OH 44455
234-742-1472

Manufacturing Innovation Network Plastics Initiative
Office of Research and Economic Development
Penn State Erie
Station Road
Erie, PA 16563-0101
814-898-6270

MechTech, Inc.
PO Box 295
Greenville, RI 02828
401-949-2562

The Metalworking Connection, Inc.
Economic Development Center
Southern Arkansas University
PO Box 1239
Magnolia, AR 71753
501-235-4375

The Metalworking Consortium
Targeted Development Project
1800 W. Cuyler Avenue
Chicago, IL 60613
312-871-1151

Needle Trades Action Project
PO Box 627
Fall River, MA 02722-0627
508-678-1991

Ohio Forging Network
Advanced Manufacturing Center
1751 E. 23rd Street
Cleveland, OH 44004
216-687-5273

Pacific Wood Products Co-op
PO Box 1422
Chehalis, WA 98532
206-748-1597

The Philadelphia Collection
Delaware Valley Industrial Resource Center
12265 Townsend Road, Suite 500
Philadelphia, PA 19143
215-464-8550

The Philadelphia Guild
7348 Malvern Avenue
Philadelphia, PA 19151
215-473-5393

Technology Coast Manufacturing and Engineering Network (TeCMEN)
Okaloosa Economic Development Council
1170 Martin Luther King Road
Fort Walton Beach, FL 32547
850-651-7374

Western Pennsylvania Regional Metalworking Network
Jordan-Krauss Associates
995 S. Cameron Street
Harrisburg, PA 17101
717-257-5451

Wood Products Manufacturing Network
PO Box 906
Bemidji, MN 56601
218-751-3108

Index